SYMBOLS AND UNITS OF PHYSICAL QUANTITIES

QUANTITY	SYMBOL*	UNITS	QUANTITY	SYMBOL*	UNITS
Acceleration	$\mathbf{a}$	m/s^2	Entropy	S	J/K
Angle	θ, ϕ	rad	Force	$\mathbf{F}$	N
Angular acceleration	$\boldsymbol{\alpha}$	rad/s^2, s^{-2}	Frequency	f	Hz
Angular frequency, angular speed, angular velocity	$\omega, \boldsymbol{\omega}$	rad/s, s^{-1}	Heat	Q	J
			Heat flow	H	J/s
Angular momentum	$\mathbf{L}$	$kg \cdot m^2/s$	Inductance	L	H
Atomic number	Z		Intensity	$I, S, \mathbf{S}$	W/m^2
Capacitance	C	F	Magnetic field	$\mathbf{B}$	T
Charge	q, Q, e	C	Magnetic flux	ϕ_B	$T \cdot m^2$
Charge density			Mass	m, M	kg
Line	λ	C/m	Mass number	A	
Surface	σ	C/m^2	Molar specific heat	C_V, C_P	$J/mol \cdot K$
Volume	ρ	C/m^3	Momentum	$\mathbf{p}$	$kg \cdot m/s$
Conductivity	σ	$\Omega^{-1} \cdot m^{-1}$	Period	T	s
Current	I	A	Power	P	W
Current density	$\mathbf{J}$	A/m^2	Pressure	P	Pa
Density	ρ	kg/m^3	Resistance	R	Ω
Dielectric constant	κ		Resistivity	ρ	$\Omega \cdot m$
Dipole moment, electric	$\mathbf{p}$	$C \cdot m$	Rotational inertia	I	$kg \cdot m^2$
Dipole moment, magnetic	$\boldsymbol{\mu}$	$A \cdot m^2$	Temperature	T	K
Distance, displacement, length, position	$x, y, z, s, d,$ $\ell, w, h, \mathbf{r}$	m	Time	t	s
			Torque	$\boldsymbol{\tau}$	$N \cdot m$
Electric field	$\mathbf{E}$	N/C, V/m	Specific heat	c	$J/kg \cdot k$
Electric flux	ϕ, ϕ_E	$N \cdot m^2/C$	Speed, velocity	$v, \mathbf{v}$	m/s
Electric potential	V	V	Volume	V	m^3
Electromotive force	$\mathscr{E}$	V	Wavelength	λ	m
Energy	E, U, K	J	Wavenumber	k	m^{-1}
			Work	W	J

*Boldface indicates vector quantities.

TRIGONOMETRY

Definition of angle (in radians): $\theta = \dfrac{s}{r}$

2π radians in complete circle
1 radian $\simeq 57.3° = 2.06$ arc sec

TRIGONOMETRIC FUNCTIONS

$\sin\theta = \dfrac{y}{r}$

$\cos\theta = \dfrac{x}{r}$

$\tan\theta = \dfrac{\sin\theta}{\cos\theta} = \dfrac{y}{x}$

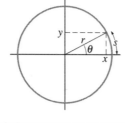

THE GREEK ALPHABET

	UPPERCASE	LOWERCASE
Alpha	A	α
Beta	B	β
Gamma	Γ	γ
Delta	Δ	δ
Epsilon	E	ε
Zeta	Z	ζ
Eta	H	η
Theta	Θ	θ
Iota	I	ι
Kappa	K	κ
Lambda	Λ	λ
Mu	M	μ
Nu	N	ν
Xi	Ξ	ξ
Omicron	O	o
Pi	Π	π
Rho	P	ρ
Sigma	Σ	σ
Tau	T	τ
Upsilon	Υ	υ
Phi	Φ	ϕ
Chi	X	χ
Psi	Ψ	ψ
Omega	Ω	ω

PHYSICS

WITH MODERN PHYSICS
FOR SCIENTISTS AND ENGINEERS

Second Edition

Volume II

Richard Wolfson
Middlebury College

Jay M. Pasachoff
Williams College

HarperCollins*College*Publishers

To the User

Physics with Modern Physics for Scientists and Engineers, Volume II, second edition, is composed of Chapters 23 through 45 (pp. 555–1243) and the complete appendices from *Physics with Modern Physics for Scientists and Engineers,* second edition, by Richard Wolfson and Jay M. Pasachoff. Cross references, Answers to Odd-Numbered Problems, Credits, and the Index are comprehensive and apply to both volumes of the text.

Executive Editor: Doug Humphrey

Senior Development Editor: Kathy Richmond

Project Editor: Carol Zombo

Design Administrator: Jess Schaal

Text and Cover Design: Ellen Pettengell

Cover Photos: Hang gliding: Bill Ross/All Stock;
    Cloud chamber: Photo Researchers, Inc.

Photo Researchers: Lynn Mooney, Roberta Knauf

Production Administrator: Randee Wire

Project Coordination: Elm Street Publishing Services, Inc.

Compositor: Interactive Composition Corporation

Art Studio: Precision Graphics

Printer and Binder: R. R. Donnelley & Sons Company

For permission to use copyrighted material, grateful acknowledgment is made to the copyright holders on pp. A-38 to A-43, which are hereby made part of this copyright page.

Physics with Modern Physics for Scientists and Engineers, Volume II, Second Edition
Copyright © 1995 by HarperCollins College Publishers

Library of Congress Cataloging-in-Publication Data

Wolfson, Richard.
    Physics with modern physics : for scientists and engineers /
Richard Wolfson, Jay M. Pasachoff. — 2nd ed.
        p. cm.
    Includes index.
    ISBN 0–06–502490–7
    1. Physics. I. Pasachoff, Jay M. II. Title.
QC21.2.W655 1995
530—dc20   93–31085

94 95 96 97 9 8 7 6 5 4 3 2 1

BRIEF CONTENTS

CONTENTS

PART 5 895

OPTICS

EXAMPLES AND APPLICATIONS

PHYSICS: CHALLENGE AND SIMPLICITY

Physics is fundamental. To understand physics is to understand how the world works, both at the everyday level and on scales of time and space so small and so large they defy intuition.

To the student, physics can be at once fascinating, challenging, subtle, and yet simple. Physics fascinates with its applications throughout science and engineering and in its revelation of unexpected phenomena like superconductivity, black holes, and chaos. It challenges with its need for precise thinking and skillful application of mathematics. It can be subtle in its interpretation, especially in describing phenomena at odds with everyday intuition. Most importantly, physics is simple. Its few fundamental laws are stated in the simplest of terms, yet they encompass a universe of natural phenomena and technological devices. Students who recognize the simplicity of physics develop confidence that stems from understanding the physical universe at a fundamental level.

This text is for science and engineering students. The standard version covers a full sequence of calculus-based university physics, and the extended version adds seven chapters on modern physics. The extended version is also available as a two-volume set.

Coverage The book is organized into six parts. Part 1 (Chapters 2 through 14) covers the basics of mechanics. Part 2 (Chapters 15 through 18) studies motion in terms of oscillations, waves, and fluids. Part 3 (Chapters 19 through 22) is on thermodynamics. Part 4 (Chapters 23 through 34) deals with electricity and magnetism. Part 5 (Chapters 35 through 37) treats optics. Part 6 (Chapters 38 and 39) briefly introduces modern physics. Each part ends with Cumulative Problems that help students synthesize concepts from several chapters. Part 6 of the extended version (Chapters 38 through 45) begins with relativity and continues with quantum mechanics and its applications to atoms, molecules, and the solid state; nuclear physics and its applications; and particle physics and cosmology.

DISTINGUISHING FEATURES OF THE TEXT

In the second edition, like the first, we emphasize careful and thorough explanations. We have pared down wordiness without sacrificing clarity of explanation. And we've added many new features to help you learn.

Contents This revision includes many substantial changes and improvements, most of which were suggested by instructors who used the text or reviewed the manuscript. Here are the most important changes that were implemented in this edition:

- We added a separate chapter (Chapter 3) on vectors.
- We reorganized Chapter 9 on gravitation so that it is more focused than in the first edition.
- We added a second chapter on wave motion (Chapter 17) that includes extensive new material on sound.
- We reorganized the material on electricity and magnetism (Part 4) for a shorter, clearer presentation.
- The optics chapters in Part 5 are completely rewritten, and we added a chapter to this edition for more thorough coverage.

- Relativity (Chapter 38) is now treated after optics, for an introduction to modern physics at the end of the text.
- The number of problems in the second edition is double that of the previous edition.
- A complete package of supplements is offered for this edition.

Applications We include a rich array of practical applications—from the workings of the compact disc through skyscraper engineering, biomedical technology, antilock brakes, space exploration, global warming, microelectronics, lasers, and much more. We chose to integrate many shorter applications into the text, where they are more likely to be read, rather than presenting a few applications as guest essays. An index of examples and applications appears at the beginning of the text. High-quality color photographs and figures enhance the applications.

Questions These follow the chapter synopsis and can be used for class discussion or to get students thinking about concepts in the chapter before they start on the problem sets.

Problems Science and engineering students learn physics best by working physics problems. The second edition contains nearly *3,000* end-of-chapter problems—double the number from the first edition. Problems range from simple confidence builders, to more complex and realistic problems involving the application of multiple concepts, to difficult **Supplemental Problems** that will challenge the best students. A section of **Paired Problems** for each chapter lets students practice problem-solving techniques on a pair of problems whose solutions involve closely related physical concepts or mathematical approaches. **Cumulative Problems** at the end of each part of the text integrate the material from several chapters.

In-Text Exercises Reinforce the Examples Worked examples in the text are generally followed by an exercise to reinforce concepts or processes. A **Similar Problems** line after each exercise points out problems at the end of the chapter that relate to the in-text exercises.

Tip Boxes Tip boxes point out useful problem-solving techniques and warn against common pitfalls. Frequent text references to specific problems link text and problems in the common purpose of enhancing the student's understanding of physics.

Pedagogical Use of Color The book's design enhances its readability; from the carefully planned use of color in figures and to highlight important equations and definitions to the selection of photographs, design is an essential pedagogical feature. Physical elements are coded in color to make them more apparent. A list of the elements and the colors used for them follow this preface.

Chapter Synopses Chapter summaries emphasize key concepts and remind students of new terms and mathematical symbols. Even in a field as fundamental as physics, many theories and their equations have limited applicability, and each chapter concludes with a reminder of limitations students should keep in mind.

Appendices and Endpapers The book's appendices and endpapers contain a mathematical review and a wealth of up-to-date physical data, conversion factors, and information on measurement systems.

PHYSICS AND MATHEMATICS

For many students, the university physics course is their first contact with practical applications of calculus. We recognize that our readers bring a range of mathematical abilities, from those taking their first calculus course concurrently with physics to those fluent in both differential and integral calculus. The former will find tip boxes and figures to build understanding of and confidence in their new mathematical skills; the latter will find a selection of challenging calculus-based problems.

For all students, we refuse to let mathematical derivations dangle. We ask frequently after the derivation of an equation or an example solution: "Does this result make sense?" We show that it does by examining easily understood special cases, thus building both physical intuition and confidence in the application of mathematics. We explore the meaning of equations verbally and through figures, ensuring that concepts are clear as students begin to use the material qualitatively.

THE SUPPLEMENT PACKAGE TO ACCOMPANY THIS TEXT

Professors and students alike often find it useful to have supplements designed to complement their text. For the second edition, we provide a new expanded package.

For the Instructor

The *Solution Manual*, prepared by Edward S. Ginsberg, University of Massachusetts-Boston, includes worked solutions to all problems in the text.

TestMaster software, available in IBM and Macintosh formats, enables instructors to select problems for any chapter, scramble them as desired, or create problems. A print version of the Test Generator is also available.

Overhead transparencies are available to instructors who adopt the text. This set of 150 color acetates of figures from the text will be useful in classroom discussions.

For the Student

The *Student Study Guide*, by Jeffrey J. Braun, University of Evansville, briefly summarizes the text discussion and important equations in each chapter. It also gives some common pitfalls for students to avoid and includes plenty of practice problems (with solutions). To order, use ISBN 0-673-52369-1.

The *Student Solution Manual*, by Edward S. Ginsberg, University of Massachusetts-Boston, includes complete, worked-out solutions to some of the odd-numbered problems in the text. To order, use ISBN 0-06-501873-7.

PhysiCad Explorer, by Tara C. Woods, consists of approximately 200 examples from the text that have been adapted for use with Mathcad® for Windows software (version 4.0 or higher). The student can call up representa: tive examples from the text and substitute new variables to see how the results differ. Mathcad® will perform all the calculations, generating numerical solutions and graphics where appropriate. Besides offering additional problem-solving drills, *PhysiCad Explorer* gives the student hands-on practice using one of the most powerful numerical tools available, Mathcad®. *PhysiCad Explorer* is self-contained and includes the Mathcad Reader, so students need not own the full Mathcad® software package.

A Calculus Laboratory Workshop with Applications to Physics, by Joan R. Hundhausen and F. Richard Yeatts, both from the Colorado School of Mines, contains 30 self-contained projects that can be used to strengthen the calculus skills of introductory physics students. To order, use ISBN 0-06-501719-6.

ACKNOWLEDGMENTS

A project of this magnitude is not the work of its authors alone. Here we acknowledge the many people whose contributions made this book possible. Colleagues using the first edition at universities and colleges throughout the world have volunteered suggestions, and most will find those incorporated here. We are especially grateful to Bob Gould at Middlebury College and Al Bartlett at the University of Colorado for their many thoughtful comments. Students in Middlebury's Physics 109–110 courses have made significant contributions to the book's accuracy and readability. Middlebury's laboratory supervisor, Cris Butler, skillfully prepared lab demonstrations illustrated in the book, and photographer Erik Borg captured them on film. Linda Eisenhart, Ann Broughton, and Tucky Ceballos at Middlebury and Susan Kaufman at Williams College provided invaluable assistance with many details. Willie S. M. Yong of Singapore, a physics educator and publisher with extensive involvement in the International Physics Olympiad, suggested new problems and helped keep the authors aware of our international readership.

It is frustrating for students and professors to find numerical errors in a textbook, especially in answers to problems. We have gone to considerable lengths to make this book as free from error as possible, and we credit the people who helped achieve this goal. Edward Ginsberg, University of Massachusetts-Boston, meticulously checked all numerical results in examples, exercises, and end-of-chapter problems. Alan Goldman of Iowa State University, Sven Rudin of Ohio State University, Kent Scheller of the University of Evansville, and Thomas Suleski of the Georgia Institute of Technology rechecked these numerical results. Claire D. Dewberry, Florida Community College at Jacksonville, checked the art proofs and made numerous suggestions. Naomi Pasachoff provided expert proof-reading.

Every chapter of the second edition was reviewed by physics professors who examined part or all of the manuscript. Their comments were incorporated into the final product. We are very grateful to them all:

Edward Adelson, *Ohio State University*
Vijendra Agarwal, *Moorhead State University*
William Anderson, *Phoenix College*
Gordon J. Aubrecht, *Ohio State University-Marion*
Paul Avery, *University of Florida*
John Bartelt, *Vanderbilt University*
Marvin Blecher, *Virginia Polytechnic Institute and State University*
John Brient, *University of Texas at El Paso*
James H. Burgess, *Washington University*
Bernard Chasan, *Boston University*
Roger Clapp, *University of South Florida*
Claire D. Dewberry, *Florida Community College at Jacksonville*

Robert J. Endorf, *University of Cincinnati*
Heidi Fearn, *California State University-Fullerton*
Shechao Feng, *University of California, Los Angeles*
Albert L. Ford, *Texas A & M University*
Ian Gatland, *Georgia Institute of Technology*
James Goff, *Pima Community College*
Alan I. Goldman, *Iowa State University*
Philip Goode, *New Jersey Institute of Technology*
Denise S. Graves, *Clark Atlanta University*
Donald Greenberg, *University of Alaska*
Phillip Gutierrez, *University of Oklahoma*
Stephen Hanzely, *Youngstown State University*
Randy Harris, *University of California, Davis*
Warren Hein, *South Dakota State University*
Gerald Harmon, *University of Maine at Orono*
Roger Herman, *The Pennsylvania State University*
Francis L. Howell, *University of North Dakota*
J. N. Huffaker, *University of Oklahoma*
Wayne James, *University of South Dakota*
Karen Johnston, *North Carolina State University*
Evan Jones, *Sierra College*
Dean Langely, *St. John's University*
Chew-Lean Lee, *Florida Community College*
Brian Logan, *University of Ottawa*
Peter Loly, *University of Manitoba*
Hilliard K. Macomber, *University of Northern Iowa*
David Markowitz, *University of Connecticut*
Daniel Marlow, *Princeton University*
Nolan Massey, *University of Texas at Arlington*
Ralph McGrew, *Broome Community College*
Victor Michalk, *Southwest Texas State University*
P. James Moser, *Bloomsburg University*
Vinod Nangia, *University of Minnesota*
Robert Osborne, *Yakima Valley Community College*
Michael O'Shea, *Kansas State University*
George Parker, *North Carolina State University*
R. J. Peterson, *University of Colorado*
Joseph F. Polen, *Shasta College*
Talat Rahman, *Kansas State University*
Dennis Roark, *The King's College*
Roger F. Sipson, *Moorhead State University*
John Sperry, *Sierra College*

Lon D. Spight, *University of Nevada at Las Vegas*
Robert Sprague, *Foothill College*
J. C. Sprott, *University of Wisconsin-Madison*
Konrad M. Stein, *Golden West College*
Bryan H. Suits, *Michigan Technological University*
Leo Takahashi, *The Pennsylvania State University-Beaver Campus*
Frank Tanherlini, *College of the Holy Cross*
Karl Trappe, *University of Texas*
Michael Trinkala, *Hudson Valley Community College*
Loren Vian, *Centralia College*
Clarence Wagener, *Creighton University*
Robert Weidman, *Michigan Technological University*
George Williams, *University of Utah*
Robert J. Wilson, *Colorado State University*
Arthur Winston, *Gordon Institute*
David Yee, *City College of San Francisco*
John Yelton, *University of Florida*
Dale Yoder-Short, *Michigan Technological University*

We also thank editors Kathy Richmond, Jane Piro, and Doug Humphrey at HarperCollins for their vigorous support of this project, and Sue Nodine of Elm Street Publishing Services for her skillful efforts in bringing it to fruition. Finally, we thank our families for their patience during the intense process of revising this book.

We invite suggestions from readers—students, professors, and others—for improvements to our text. We promise a speedy reply to each correspondent and an effort to incorporate appropriate suggestions in subsequent printings. Please write us at our respective institutions or contact us by electronic mail.

Richard Wolfson
Middlebury College
Middlebury, Vermont 05753
wolfson@middlebury.edu

Jay M. Pasachoff
Williams College
Williamstown, Massachusetts 01267
pasachoff@williams.edu

Richard Wolfson is Professor of Physics and George Adams Ellis Professor of the Liberal Arts at Middlebury College, where he has taught since 1976. He did undergraduate work at the Massachusetts Institute of Technology and Swarthmore College and holds the M.S. degree from the University of Michigan and Ph.D. from Dartmouth. He has published widely in scientific journals, including works ranging from medical physics research to experimental plasma physics, electronic circuit design, solar energy engineering, and theoretical astrophysics. He is also an interpreter of science for the nonspecialist, a contributor to *Scientific American*, and author of the book *Nuclear Choices: A Citizen's Guide to Nuclear Technology*. Wolfson has spent sabbatical years as Visiting Scientist at the National Center for Atmospheric Research in Boulder, Colorado, and in 1993 was Visiting Scientist at St. Andrews University in Scotland.

Jay M. Pasachoff is Field Memorial Professor of Astronomy and Director of the Hopkins Observatory at Williams College. He was born and brought up in New York City. After attending the Bronx High School of Science, he received his A.B. degree from Harvard College and his A.M. and Ph.D. from Harvard University. He then held postdoctoral fellowships at Harvard and at the California Institute of Technology before going to Williams in 1972. His research has dealt mainly with solar physics and nuclear astrophysics, namely, the solar atmosphere, and with the abundances of the light elements and their formation in the first minutes of the universe. Pasachoff has spent sabbatical leaves at the University of Hawaii, at l'Institut d'Astrophysique in Paris, at the Institute for Advanced Study in Princeton, and at the Harvard-Smithsonian Center for Astrophysics. He is also author or co-author of major texts in physics, calculus, physical science, and astronomy.

The artwork for this second edition is carefully designed to make effective use of color printing as a learning aid. In particular, we have assigned colors to the vector quantities that are so important in physics, and we have used those colors consistently throughout the book. The table below lists some important physical quantities, along with their text and graphic symbols and color assignments. We also include electric circuit symbols, which, to be consistent with usage in engineering, are printed in black.

Vector	Text Symbol	Graphic Symbol
Displacement	$\mathbf{r}, \boldsymbol{\ell}$	
Velocity	$\mathbf{v}$	
Acceleration	$\mathbf{a}$	
Force	$\mathbf{F}$	
Linear momentum	$\mathbf{p}$	
Angular velocity	$\boldsymbol{\omega}$	
Torque	$\boldsymbol{\tau}$	
Angular momentum	$\mathbf{L}$	
Electric field	$\mathbf{E}$	
Magnetic field	$\mathbf{B}$	
Electric dipole moment	$\mathbf{p}$	
Magnetic dipole moment	$\boldsymbol{\mu}$	
Electric charge		
Positive charge	q, Q	$\oplus$
Negative charge	q, Q	$\ominus$
Circuit symbols		
Battery, emf	$\mathcal{E}$	
Resistor	R	
Capacitor	C	
Inductor	L	
Switch	S	

PART 4

ELECTROMAGNETISM

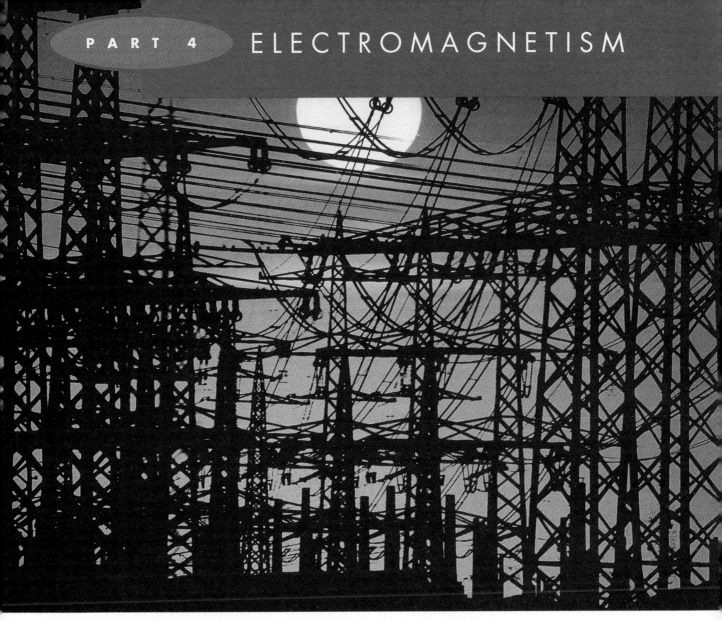

Electricity and magnetism mediate the generation, transmission, and use of energy. They also govern the structure of matter at the atomic and molecular level. Even sunlight, shown here as the Sun sets behind an electric power substation, is an electromagnetic phenomenon.

CHAPTER 23

ELECTRIC CHARGE, FORCE, AND FIELD

Lightning strikes at Kitt Peak National Observatory in Arizona relieve the buildup of electric charge in the atmosphere.

What force keeps the molecules in your body together? What force keeps a skyscraper standing or prevents a mountain from spreading into a flat blob? What force holds your car on the road as you round a turn? What force accelerates the electrons that paint the picture on your TV screen? What force underlies the awesome beauty of a thunderstorm?

Remarkably, these and all other forces except gravity that we encounter in our everyday lives—and in nearly all scientific work—are manifestations of a single force: the **electromagnetic force.** The friction, tension, and normal forces of mechanics are ultimately electromagnetic. So are interactions as diverse as the focusing of light in the lens of your eye, the extraction of information from a computer disk, or the formation of a water molecule from separate atoms. Electromagnetism is so important in science and engineering that we devote the next 12 chapters to it.

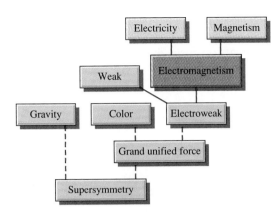

FIGURE 23-1 The place of electromagnetism among the fundamental forces. Electromagnetism comprises electricity and magnetism, once thought to be separate. Similarly, electromagnetism and the weak force are both aspects of the electroweak force. Grand unification and supersymmetry are two theories being considered for further unification of the fundamental forces.

23-1 ELECTROMAGNETISM

Electromagnetism is among the fundamental forces of nature that we introduced in Chapter 5 (see Fig. 23-1) and is the dominant force in a vast range of natural and technological phenomena. Other than gravity, electromagnetism is the only force most of us will ever deal with.

Three distinct themes motivate our study of electromagnetism:

1. The electromagnetic force is solely responsible for the structure of matter from atoms to objects of roughly human size. Much of physics, all of chemistry, and most of biology deal in this realm. Only at much smaller scales do the color and weak forces become important; only at larger scales is gravity significant.

   The wonderful diversity of chemical compounds is testimony to the rich possibilities contained in the electromagnetic interaction. Even life itself, and the DNA replicating mechanism at its heart, are manifestations of electromagnetism (Fig. 23-2). For students of physical and biological sciences, understanding electromagnetism means understanding the most fundamental basis of these disciplines.

2. We live in a technological world increasingly dominated by devices that operate on electromagnetic principles. Electric lights, motors, batteries, and generators have been essential throughout the twentieth century. More recently, electronic technology has led to the proliferation of devices for storing and processing information, for sensing and measuring, and for sophisticated control of industrial, scientific, medical, and even household systems.

3. Studying electromagnetism leads to an understanding of the nature of light and from there to the theory of relativity. Relativity profoundly alters our ideas of space and time—the very basis of physical reality.

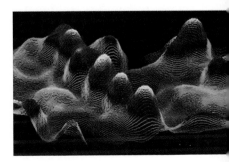

FIGURE 23-2 Electric forces govern the structure and replication of the DNA molecule, shown here in a scanning tunneling microscope image.

We've been speaking of electromagnetism, yet you're probably more familiar with electricity and magnetism separately. Although we begin with separate studies of these seemingly distinct phenomena, the relation between the two will become increasingly central. Eventually you will understand electricity and magnetism as two aspects of a single phenomenon that is basic to the workings of the universe.

23-2 ELECTRIC CHARGE

Electric charge is a fundamental property of nature. Of the three building blocks of ordinary matter—the electron, the proton, and the neutron—two carry electric charge. What is electric charge? At the most fundamental level, we don't know. We don't know what mass "really" is either, but we're familiar with it because we've spent our lives pushing objects around. By studying electrical interactions we gain a similar familiarity with charge that is as close as we can get to understanding what it "really" is.

Two Kinds of Charge

Electric charge comes in two varieties, which Benjamin Franklin designated **positive** and **negative.** These are convenient labels, but they have no physical significance. There's nothing "missing" about negative charge. Positive and negative charge are complementary properties, not the presence and absence of something. The utility in the names is mathematical since it's the algebraic sum of charges—described with positive and negative numbers—that has physical significance.

Quantities of Charge

All electrons carry the same charge, and all protons carry the same charge. The proton's charge has *exactly* the same magnitude as the electron's, but with opposite sign. Given that electrons and protons differ substantially in other properties—like mass—this electrical relation is remarkable. Problem 1 shows how dramatically different our world would be if there were even a slight difference between the magnitudes of the electron and proton charges.

The magnitude of the electron or proton charge is the **elementary charge, e.** Electric charge is **quantized**—that is, it comes only in discrete amounts. In a famous experiment in 1909, the American physicist R. A. Millikan used electric forces to suspend small oil drops. From the electric force he computed the charge on each drop and found it was always a multiple of a basic value we now know as the elementary charge (Fig. 23-3).

Modern elementary particle theories suggest that the most basic unit of charge is actually $\frac{1}{3}e$. Such "fractional charges" reside on quarks, the basic building blocks of protons, neutrons, and many other particles. Quarks always join to produce particles with integer multiples of the full elementary charge, and it seems impossible to isolate individual quarks.

The SI unit of charge is the **coulomb** (C), named for the French physicist Charles Augustin de Coulomb (1736–1806). Although the coulomb's formal definition is in terms of electric current, it's convenient to describe one coulomb as being about 6.25×10^{18} elementary charges, making the elementary charge approximately 1.60×10^{-19} C.

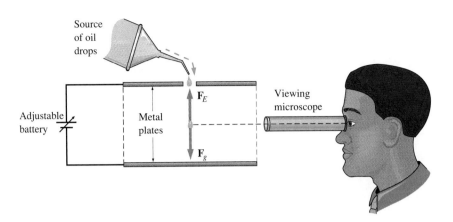

FIGURE 23-3 Millikan's oil-drop experiment. By balancing the electric force $\mathbf{F}_E$ and gravitational force $\mathbf{F}_g$ on oil drops, Millikan showed that charge is quantized.

Charge Conservation

Electric charge is a conserved quantity, meaning that the algebraic sum of the electric charges—i.e., the **net charge**—in a closed region remains constant. Charged particles may be created or annihilated, but always in pairs of equal and opposite charge (Fig. 23-4). The net charge always remains the same.

23-3 COULOMB'S LAW

You can transfer charge to a balloon by rubbing it on your clothing; you'll then find that the balloon sticks to you. Charge another balloon and the two balloons will repel each other (Fig. 23-5). You're seeing a manifestation of the fundamental fact that unlike charges attract and like charges repel. Socks clinging to other clothes when they come out of a dryer, dust attracted to the front of a TV screen or computer monitor, and the shocks you get when you cross a carpet and touch a doorknob are other common examples where you're directly aware of this electrical interaction.

But electricity would be rather unimportant if the only significant electrical interactions were these obvious ones. In fact, all interactions of everyday matter—from the motion of a car to the movement of a muscle to the growth of a tree—are dominated by the electric force. It's just that matter on the large scale is nearly perfectly neutral, so electrical effects in bulk matter are not obvious. At the molecular or even cellular level the appearance of individual charged particles makes the electrical nature of matter immediately obvious (Fig. 23-6).

The attraction and repulsion of electric charges implies that a force acts between them. The physicist Coulomb first investigated this force in the late 1700s. He found that the force between two charges acts along the line joining them, with a magnitude proportional to the product of the charges and inversely proportional to the square of the distance between them. These results are summarized in **Coulomb's law:**

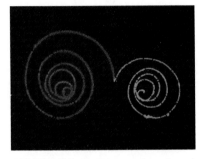

FIGURE 23-4 The creation of an electron and its antiparticle, a positron. The two particles' paths are the oppositely directed spirals that originate in a common point where the pair was created. The net charge remains zero before and after the particle creation.

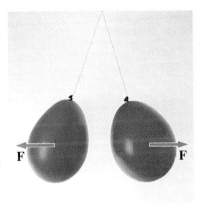

FIGURE 23-5 Two balloons carrying similar electric charge repel each other.

$$\mathbf{F}_{12} = \frac{kq_1q_2}{r^2}\hat{\mathbf{r}} \qquad \text{Coulomb's law.} \qquad (23\text{-}1)$$

FIGURE 23-6 (*a*) The salt shaker and even a single salt grain are electrically neutral, so the role of electric forces in these structures is not obvious. (*b*) At the atomic scale, electric forces bind individual sodium and chlorine ions together and are, in fact, responsible for the cubical shape of the salt crystal.

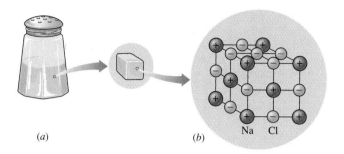

(*a*)   (*b*)   Na   Cl

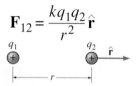

$$\mathbf{F}_{12} = \frac{kq_1q_2}{r^2}\hat{\mathbf{r}}$$

FIGURE 23-7 Quantities in Coulomb's law used in calculating the electric force of charge q_1 on q_2. The unit vector $\hat{\mathbf{r}}$ always points in the direction from q_1 toward q_2, regardless of the signs of the charges. The force $\mathbf{F}_{12}$ points in the same direction as $\hat{\mathbf{r}}$ if the two charges have the same sign, and otherwise in the opposite direction.

Here $\mathbf{F}_{12}$ is the force exerted *by* the charge q_1 *on* the charge q_2 and r is the distance between the two. The quantity k is a proportionality constant whose value in SI units is approximately 9.0×10^9 N·m²/C². The force $\mathbf{F}_{12}$ is a vector, and $\hat{\mathbf{r}}$ is a unit vector giving its direction. Figure 23-7 shows that $\hat{\mathbf{r}}$ lies on the line passing through the two charges and points in the direction *from q_1 toward q_2*.

Coulomb's law is a vector equation that covers all possible combinations of charges and the associated attractive and repulsive forces. If q_1 and q_2 have the same sign—either positive or negative—then the product q_1q_2 is positive, and the force points in the same direction as $\hat{\mathbf{r}}$. The force on q_2 is thus away from q_1, or repulsive, as it should be. But if q_1 and q_2 have opposite signs, then q_1q_2 is negative and the direction of the force is opposite that of $\hat{\mathbf{r}}$—that is, attractive.

What about the force $\mathbf{F}_{21}$ that q_2 exerts on q_1? Equation 23-1 shows it has the same magnitude as $\mathbf{F}_{12}$, but a drawing like Fig. 23-7 gives the opposite direction. Thus the electric force obeys Newton's third law.

● EXAMPLE 23-1 TWO CHARGES

A 1.0-μC charge is located at $x = 1.0$ cm, and a -1.5-μC charge at $x = 3.0$ cm. (a) What force does the positive charge exert on the negative one? (b) How would the force change if the distance between the charges were tripled?

Solution

(a) A vector from the positive charge toward the negative charge points in the $+x$ direction, so $\hat{\mathbf{r}}$ is simply the unit vector $\hat{\mathbf{i}}$. Then Equation 23-1 becomes

$$\mathbf{F} = \frac{kq_1q_2}{r^2}\hat{\mathbf{r}}$$

$$= \frac{(9.0 \times 10^9 \text{ N·m}^2/\text{C}^2)(1.0 \times 10^{-6} \text{ C})(-1.5 \times 10^{-6} \text{ C})}{(0.020 \text{ m})^2}\hat{\mathbf{i}}$$

$$= -34\hat{\mathbf{i}} \text{ N}.$$

The minus sign shows that the force is in the negative x direction—toward the positive charge, or attractive.

(b) If the distance were tripled the force would drop by a factor of $1/3^2$, to $-3.8\hat{\mathbf{i}}$ N.

TIP Let the Algebra Take Care of the Signs Equation 23-1 accounts for all aspects of the force calculation. If you correctly take $\hat{\mathbf{r}}$ to be a unit vector pointing *away* from the charge giving rise to the force—whatever its sign—and *toward* the charge being acted on—whatever its sign—then calculation of $\frac{kq_1q_2}{r^2}\hat{\mathbf{r}}$ gives correctly the magnitude and direction of the force.

EXERCISE A 2.5-μC charge is at the origin. Find the force it exerts on a 4.1-μC charge at the point $x = 2$ m, $y = 2$ m.

Answer: $8.15\hat{\mathbf{i}} + 8.15\hat{\mathbf{j}}$ mN

Some problems similar to Example 23-1: 7, 12, 13

●

Coulomb's law for the electric force is similar to Newton's law for the gravitational force. Both show the same inverse-square decrease with distance, and both are proportional to the product of the interacting charges or masses. But there is an important difference: There's only one kind of mass (even antimatter has "positive mass") and the gravitational force it produces is always attractive. That means large concentrations of mass give rise to large gravitational forces. But charge comes in two varieties, so large concentrations of charge tend to be electrically neutral and therefore give rise to weak electric forces. The electric force between individual particles is vastly stronger than the gravitational force between the same particles (see Problem 6), and it's only because of the nearly complete cancellation of positive and negative charge in bulk matter that gravity becomes important on the macroscopic scale.

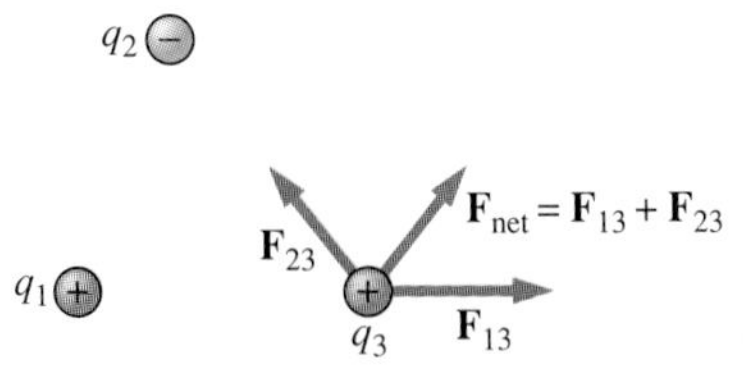

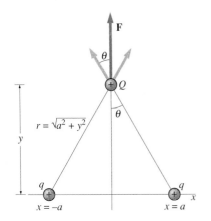

FIGURE 23-8 The superposition principle allows us to add vectorially the forces from two or more charges.

The Superposition Principle

Coulomb's law describes the force between *two* charges. But what if we have more than two charges? If we want the force on q_3 arising from two other charges q_1 and q_2, we simply calculate $\mathbf{F}_{13}$ and $\mathbf{F}_{23}$ from Equation 23-1 and add the resulting force vectors (Fig. 23-8). That is, the force that q_1 exerts on q_3 is unaffected by the presence of q_2, and the force that q_2 exerts on q_3 is unaffected by the presence of q_1. This may seem obvious, but nature need not have been that simple.

The fact that electric forces add vectorially is called the **superposition principle.** Our confidence in this principle is ultimately based on experiments that show electric and indeed electromagnetic phenomena behave according to the principle. With superposition we can solve relatively complicated problems by breaking them into simpler parts. If the superposition principle did not hold, the mathematical description of electromagnetism would be far more complicated than it is.

● **EXAMPLE 23-2** RAINDROPS

The charging of individual raindrops is ultimately responsible for the electrical activity of a thunderstorm. Suppose two drops with equal charge q are located on the x axis at $\pm a$, as shown in Fig. 23-9. Find the electric force on a third drop with charge Q located at an arbitrary point on the y axis.

Solution
Charge Q is the same distance $r = \sqrt{a^2 + y^2}$ from the two other drops, so the force from each has the same magnitude, given by Equation 23-1:

$$F = \frac{kqQ}{a^2 + y^2}.$$

But the directions of the two forces are different. It's evident from Fig. 23-9 that the x components cancel, while the y components add to give

$$F_y = 2\frac{kqQ}{a^2 + y^2}\cos\theta.$$

FIGURE 23-9 The force on Q is the vector sum of the forces from the individual charges.

But Fig. 23-9 shows that $\cos\theta = \dfrac{y}{r} = \dfrac{y}{\sqrt{a^2 + y^2}}$, so the force on Q becomes

$$\mathbf{F} = \frac{2kqQy}{(a^2 + y^2)^{3/2}}\hat{\mathbf{j}}.$$

Does this result make sense? Evaluating $\mathbf{F}$ at $y = 0$ gives zero force. Here, midway between the two charges, Q experiences equal but opposite forces from the two q's, so the net force must be zero. At very large distances such that $y \gg a$, on the other hand, we can neglect a^2 compared with y^2, and the force becomes

$$\mathbf{F} = \frac{k(2q)Q}{y^2}\hat{\mathbf{j}}. \qquad (y \gg a)$$

This is just the force we would expect from a single charge of magnitude $2q$ located a distance y from Q—showing that this system of two charges acts like a single charge $2q$ at distances large compared with the charge separation.

In drawing Fig. 23-9 we tacitly assumed that q and Q have the same signs. But our analysis holds even if they don't; in that case the product qQ is negative, and $\mathbf{F}$ therefore points in the opposite direction from $\hat{\mathbf{j}}$.

EXERCISE Find the net force on the charge located at the origin in Fig. 23-10.

Answer: $73\hat{\mathbf{i}} - 23\hat{\mathbf{j}}$ mN

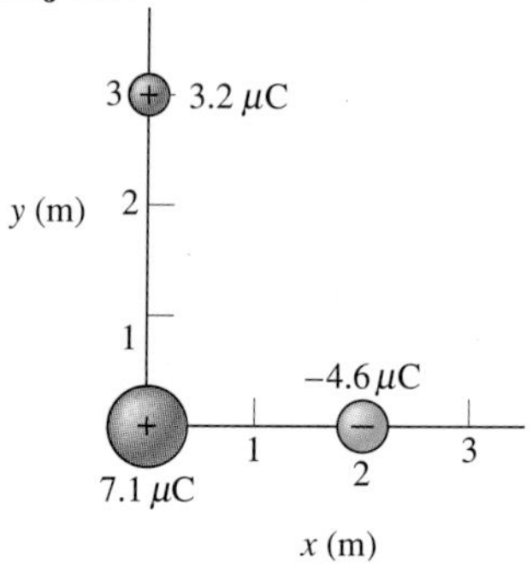

FIGURE 23-10 What is the force on the charge at the origin?

Some problems similar to Example 23-2: 19–22, 34

● EXAMPLE 23-3 BALANCING FORCES

A positive charge $+2q$ lies on the x axis at $x = -a$, and a charge $-q$ lies at $x = +a$, as shown in Fig. 23-11. Find a point where the electric force on a third charge Q would be zero.

Solution
The point must lie on the x axis since off axis the individual force vectors cannot point in opposite directions. But where on the axis? In between the two charges the repulsion of one and attraction of the other would add to give a nonzero net force. To the left of $+2q$, a third charge Q would always be closer to $+2q$ than to $-q$. The force from $+2q$ would always be greater, partly because Q would be closer to it and partly because of its greater charge. So the only place the forces from the two charges might cancel is to the right of $-q$. Even though any point in this region is closer to $-q$, the larger magnitude of $+2q$ might lead to the two forces being equal.

So suppose the third charge Q lies at some point x to the right of $-q$, so $x > a$. Then a unit vector from either $+2q$ or $-q$ toward Q points in the $+x$ direction, so is the vector $\hat{\mathbf{i}}$. Summing the two forces as given by Equation 23-1 and setting the result to zero then gives

$$\mathbf{F} = \mathbf{F}_{2q} + \mathbf{F}_{-q} = \frac{k(2q)Q}{(x + a)^2}\hat{\mathbf{i}} + \frac{k(-q)Q}{(x - a)^2}\hat{\mathbf{i}} = \mathbf{0},$$

where the denominators are the distances from any point $x > a$ to the charges located at $x = -a$ and $x = +a$, respectively. Note that Q cancels from the equation, showing that the

FIGURE 23-11 Where will a third charge experience no net force (Example 23-3)?

point we're finding will be a point of zero force for *any* charge, no matter what its sign or magnitude. The quantities k and q also cancel, so the x component of our equation for the net force becomes

$$\frac{2}{(x + a)^2} = \frac{1}{(x - a)^2}.$$

Inverting and taking square roots gives

$$\frac{x + a}{\sqrt{2}} = \pm(x - a).$$

We solve separately for the two possible signs. For the $+$ sign, we have

$$x = a\frac{\sqrt{2} + 1}{\sqrt{2} - 1} = a\frac{(\sqrt{2} + 1)^2}{(\sqrt{2} - 1)(\sqrt{2} + 1)}$$
$$= (3 + 2\sqrt{2})a = 5.83a.$$

Since this value of x is greater than a, this point does lie to the right and therefore is indeed a point of zero force. Physically, we can understand the location of this point by noting that very close to $-q$ its force dominates. Far to the right of $-q$, on the other hand, we're nearly equal distances from the two charges and therefore $+2q$ dominates by virtue of its greater charge. At some intermediate point the two forces must balance; this is the point we've found.

You can verify that the solution for the minus sign lies to the left of $-q$ and is therefore inconsistent with our choice of direction for the unit vector from $-q$. This second solution is therefore not a meaningful answer.

TIP Let the Algebra Take Care of Signs Once again, Equation 23-1 tells it all—including the direction of the force once you choose the unit vector cor-

rectly. You may be tempted in a problem like Example 23-3 simply to set the two forces equal and solve for x. That may work if you're careful, but it's safer to remember that you're really solving for a point where *the vector sum of two forces is zero*. You can't go wrong if you write those forces carefully with the correct vector directions and then sum them.

EXERCISE Repeat Example 23-3 with the charge $-q$ changed to $+q$.

Answer: $(3 - 2\sqrt{2})a = 0.17a$

Some problems similar to Example 23-3: 16–18, 32, 34 ●

Point Charges and Charge Distributions

Strictly speaking, Coulomb's law applies only to **point charges**—charged objects of arbitrarily small size. We're often interested in the forces arising from **charge distributions**—arrangements of charge spread over space. The two-charge systems of Examples 23-2 and 23-3 constitute simple charge distributions. The DNA molecule shown in Fig. 23-2 is a very complicated charge distribution. Other charge distributions include the electrodes in a TV tube, a memory cell in a computer memory chip, your heart, and a thundercloud. Ultimately these charge distributions consist of point-like electrons and protons to which Coulomb's law does apply, so we can in principle calculate the associated forces using the superposition principle.

Although the force of one point charge acting on another decreases with the inverse square of the distance, it's important to recognize that the same may or may not be true of the force arising from a charge distribution. For the distribution of two identical point charges in Example 23-2, for example, we found that the force on a third charge Q drops as the inverse square of the distance only at large distances; closer in, the dependence on distance is more complicated. This is generally true: When we're very far from a finite-sized charge distribution, it acts like a point charge with the total net charge of the distribution. But in closer the detailed arrangement of the individual charges becomes important, and the force associated with the distribution is generally no longer like that of a point charge.

23-4 THE ELECTRIC FIELD

We're often interested in the effect a charge distribution has, not on some particular charge, but on *any* charge we place in its vicinity. Accordingly, we define the **electric field:**

> *The electric field at any point is the force per unit charge experienced by a charge at that point. Mathematically,*

$$E = \frac{F}{q}. \qquad (23\text{-}2a)$$

The entire field is a set of vectors, one for each point in space, giving the magnitude and direction of the force per unit charge at that point. The units of electric field are newtons/coulomb. Fields of hundreds to thousands of N/C are commonplace, while fields of 3×10^6 N/C or more will tear electrons from air molecules. Electric fields within atoms may exceed 10^{12} N/C.

If we know the electric field $\mathbf{E}$ at a point we can compute the force on *any* charge q by rearranging Equation 23-2a:

$$\mathbf{F} = q\mathbf{E}. \tag{23-2b}$$

In Chapter 9 we gave a definition of the gravitational field as the gravitational force per unit mass. Conceptually, the gravitational field concept replaced the notion of action-at-a-distance—for example, Earth "reaching out" across empty space to pull on the moon—with the view that Earth creates a gravitational field in its vicinity and that the moon then responds to the gravitational field at its location. Defining the electric field implies the same conceptual shift; instead of thinking of one charge attracting or repelling another, we view a charge as creating an electric field throughout the space surrounding it. A second charge then responds to the field at its immediate location.

At this point you may regard the electric field as a mere mathematical construct we introduce for computational convenience or to satisfy some philosophical aversion to action-at-a-distance. But as you advance into the study of electromagnetism, you'll find that fields seem increasingly real. To a physicist, in fact, fields are every bit as real as matter itself.

There are two practical difficulties in using Equation 23-2a to measure electric fields. First, we must be sure the measured force is caused only by the electric field; if not, we must subtract the effect of other forces such as gravity. Second, the field we're trying to measure arises from one or more other charges. If the charge q is large, its own field may be strong enough to move the other charges, thereby altering the field we're trying to measure. For this reason we usually think of measuring the field with a very small "test charge."

● **EXAMPLE 23-4**   A THUNDERSTORM: FORCE AND FIELD

A charged raindrop carrying 10 μC experiences an electric force of 0.30 N in the $+x$ direction. What is the electric field at its location? What would be the force on a -5.0-μC drop at the same location?

Solution

Equation 23-2a defines the electric field; here the force is $0.30\hat{\imath}$ N, so

$$\mathbf{E} = \frac{\mathbf{F}}{q} = \frac{0.30\hat{\imath} \ \text{N}}{10 \times 10^{-6} \ \text{C}} = 30\hat{\imath} \ \text{kN/C}.$$

Acting on a -5.0-μC charge, this field would give rise to a force

$$\mathbf{F} = q\mathbf{E} = (-5.0 \ \mu\text{C})(30\hat{\imath} \ \text{kN/C}) = -0.15\hat{\imath} \ \text{N}.$$

TIP   The Field Is Independent of the Test Charge   You might wonder if the field should point in the $-x$ direction when we talk about putting a negative charge in the field. It doesn't—because the whole point of the field concept is to provide a description that's independent of the particular charge experiencing that force. The electric field in this example points in the $+x$ direction *no matter what charge* we may choose to put in the field. For a positive charge the force $q\mathbf{E}$ points in the *same* direction as the field; for a negative charge $q < 0$, and the force is *opposite* the field direction. As always, the algebra takes care of the signs.

EXERCISE An electric field $\mathbf{E} = -450\hat{\mathbf{i}}$ kN/C is used to accelerate electrons in a portion of a TV picture tube, where the x axis points from the back to the front of the tube. Find the force in this field experienced by (a) an electron and (b) an ion carrying $+2$ elementary charges. (c) Which particle will be accelerated toward the front of the tube?

Answers: (a) $7.2 \times 10^{-14}\hat{\mathbf{i}}$ N; (b) $-1.4 \times 10^{-13}\hat{\mathbf{i}}$ N; (c) the electron

Some problems similar to Example 23-4: 26–28

The Field of a Point Charge

Once we know the field of a charge distribution we can calculate its effect on other charges. The simplest charge distribution is a single point charge. Coulomb's law gives the force on a test charge q_1 located a distance r from a point charge q:

$$\mathbf{F} = \frac{kqq_1}{r^2}\hat{\mathbf{r}},$$

where $\hat{\mathbf{r}}$ is a unit vector pointing *away* from q. The electric field arising from q is the force per unit charge, or

$$\mathbf{E} = \frac{\mathbf{F}}{q_1} = \frac{kq}{r^2}\hat{\mathbf{r}}. \quad \text{(field of a point charge)} \quad (23\text{-}3)$$

Since it is so closely related to Coulomb's law for the electric force, we also refer to Equation 23-3 as Coulomb's law. Note that the equation contains no reference to the test charge q_1, since the field of q exists independently of any other charge. Since $\hat{\mathbf{r}}$ always points *away* from q, the direction of $\mathbf{E}$ is radially outward if q is positive and radially inward if q is negative. Figure 23-12 shows some field vectors for positive and negative point charges.

23-5 ELECTRIC FIELDS OF CHARGE DISTRIBUTIONS

The electric field is just the electric force per unit charge. Since the electric force obeys the superposition principle, so does the electric field. That means the field of a charge distribution is the vector sum of the fields of the individual point charges comprising the distribution:

$$\mathbf{E} = \mathbf{E}_1 + \mathbf{E}_2 + \mathbf{E}_3 + \cdots = \sum_i \mathbf{E}_i = \sum_i \frac{kq_i}{r_i^2}\hat{\mathbf{r}}_i, \quad (23\text{-}4)$$

where the $\mathbf{E}_i$'s are the fields of the point charges q_i located at distances r_i from the point where we're evaluating the field, and where the $\hat{\mathbf{r}}_i$'s are unit vectors pointing *from* each point charge *toward* where we're evaluating the field. In principle, Equation 23-4 gives the electric field of *any* charge distribution. In practice, the process of summing the individual field vectors is often complicated unless the charge distribution contains relatively few charges arranged in a symmetric way.

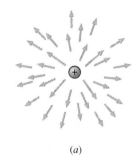

(a)

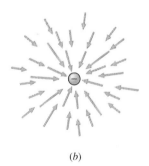

(b)

FIGURE 23-12 Field vectors for (a) positive and (b) negative point charges. Vector directions show that field vectors point radially outward or inward, and lengths show the magnitudes decreasing with the inverse square of the distance. In three dimensions the field vectors fill all space in a spherically symmetric fashion.

● EXAMPLE 23-5 TWO PROTONS

Two protons are 3.6 nm apart. (a) Find the electric field at the point P shown in Fig. 23-13. (b) Find the force on an electron at point P.

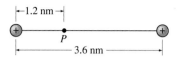

FIGURE 23-13 What is the electric field at P (Example 23-5)?

Solution

If we take the x axis along the line joining the protons, then $\hat{\mathbf{r}}_1$ is just $\hat{\mathbf{i}}$ and $\hat{\mathbf{r}}_2$ is $-\hat{\mathbf{i}}$, where the subscripts 1 and 2 refer to the left and right protons, respectively. Then Equation 23-4 gives

$$\mathbf{E} = \mathbf{E}_1 + \mathbf{E}_2 = \frac{ke}{r_1^2}\hat{\mathbf{i}} + \frac{ke}{r_2^2}(-\hat{\mathbf{i}}) = ke\left(\frac{1}{r_1^2} - \frac{1}{r_2^2}\right)\hat{\mathbf{i}}$$

$$= (9.0\times10^9 \text{ N·m}^2/\text{C}^2)(1.6\times10^{-19} \text{ C})$$

$$\times\left[\frac{1}{(1.2\times10^{-9} \text{ m})^2} - \frac{1}{(2.4\times10^{-9} \text{ m})^2}\right]\hat{\mathbf{i}}$$

$$= 750\hat{\mathbf{i}} \text{ MN/C}.$$

An electron in this field will experience a force

$$\mathbf{F} = q\mathbf{E} = (-1.6\times10^{-19} \text{ C})(750\times10^6 \hat{\mathbf{i}} \text{ N/C})$$

$$= -1.2\times10^{-10}\hat{\mathbf{i}} \text{ N}.$$

EXERCISE Find the magnitude of the electric field (a) 4.8 nm and (b) 200 nm to the right of the right-hand proton in Fig. 23-13. (c) Show that your answer to (b) is nearly equal to the field of a single charge $2e$ located midway between the two protons.

Answers: (a) 82.9 MN/C; (b) 70.7 kN/C

Some problems similar to Example 23-5: 31–34 ●

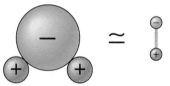

FIGURE 23-14 (left) A water molecule. Electrons spend more time in the vicinity of the single oxygen atom, giving rise to separate regions of negative and positive charge. The molecule therefore approximates the dipole shown at right, and at large distances gives rise to an electric field like that of Example 23-6.

The Electric Dipole

One of the most important charge distributions is the **electric dipole,** consisting of two point charges of equal magnitude but opposite sign held a fixed distance apart. Many molecules are essentially electric dipoles, so an understanding of the dipole helps explain molecular behavior (Fig. 23-14). During contraction the heart muscle becomes essentially a dipole, and physicians performing electrocardiography are measuring, among other things, the strength and orientation of that dipole. The dipole configuration is also used in a number of technological devices such as radio and TV antennas.

● EXAMPLE 23-6 MODELING A MOLECULE

A molecule consists of separated regions of positive and negative charge, modeled approximately as a positive charge q at $x = a$ and a negative charge $-q$ at $x = -a$, as shown in Fig. 23-15. Find a general expression for the electric field at any point on the y axis, and an approximate expression valid at large distances ($y \gg a$).

Solution

Figure 23-15 shows the individual field vectors $\mathbf{E}_-$ and $\mathbf{E}_+$, along with their sum. The y components cancel to give a net field parallel to the x axis. The x components of the two fields are clearly the same, so we have

$$E_x = E_{x-} + E_{x+} = -2\left(\frac{kq}{r^2}\sin\theta\right),$$

where the minus sign occurs because, as Fig. 23-15 shows, the net field points in the negative x direction. Fig. 23-15 also shows that the distance from both charges is $r = \sqrt{y^2 + a^2}$ and that $\sin\theta = a/r = a/\sqrt{y^2 + a^2}$. Then we have

$$\mathbf{E} = E_x\hat{\mathbf{i}} = -\frac{2kqa}{(y^2 + a^2)^{3/2}}\hat{\mathbf{i}}.$$

Does this result make sense? Midway between the charges the fields from each charge point in the same direction and have the same magnitude, so we expect a resultant field twice that of either charge alone. Setting $y = 0$ at the midpoint gives

$$\mathbf{E}(y=0) = -\frac{2kqa}{(a^2)^{3/2}}\hat{\mathbf{i}} = -\frac{k(2q)}{a^2}\hat{\mathbf{i}},$$

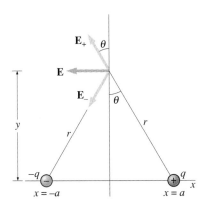

FIGURE 23-15 An electric dipole. The electric field on the dipole's perpendicular bisector (here the y axis) is parallel to the dipole axis.

which is indeed twice the field of either charge at the distance a.

We're frequently interested in the field far from a dipole, which is why this example asks for an approximate expression for $y \gg a$. (One needn't go very far from a molecule for its size to be insignificant.) Under this approximation we can neglect a^2 compared with y^2 in our expression for the field, giving

$$\mathbf{E} = -\frac{2kqa}{y^3}\hat{\mathbf{i}}, \qquad y \gg a.$$

We'll see later how forces associated with such molecular electric fields give rise to the van der Waals interaction that we considered in Chapter 20.

TIP Approximations Making approximations requires some care. Here we're basically asking for the field when y is so large that a is negligible *compared with y*. So we neglect a^2 compared with y^2 when the two are summed, but we *don't* neglect a when it appears in the numerator, where it's not summed with y.

EXERCISE Find (a) a general expression for the dipole field at points on the x axis to the right of $x = a$ in Fig. 23-15, and (b) an approximate expression valid for $x \gg a$.

Answers: (a) $\mathbf{E} = \dfrac{4kqax}{(x^2 - a^2)^2}\hat{\mathbf{i}}$; (b) $\mathbf{E} = \dfrac{4kqa}{x^3}\hat{\mathbf{i}}$

Some problems similar to Example 23-6: 31, 37, 38, 41 ●

The dipole fields in Example 23-6 and its exercise both decrease, at large distances, as the inverse *cube* of distance. Physically, this is because the dipole has zero *net* charge. Its field arises entirely from the slight separation of two opposite charges. Because of this separation the dipole field isn't exactly zero, but it is weaker and more localized than the field of a point charge. Many complicated charge distributions exhibit the essential characteristic of a dipole—namely, they're neutral but consist of separated regions of positive and negative charge—and at large distances such distributions all have essentially the same field configuration (see Problems 48 and 82).

At large distances the dipole's physical characteristics q and a enter the equations for the electric field only in the product qa. We call the product of the charge q and separation $2a$ the **dipole moment,** p. Its units are C·m. Using this definition, the fields given in Example 23-6 and its exercise for the dipole of Fig. 23-15 are

$$\mathbf{E} = -\frac{kp}{y^3}\hat{\mathbf{i}} \qquad \left(\begin{array}{c}\text{dipole field for } y \gg a, \\ \text{on perpendicular bisector}\end{array}\right) \qquad (23\text{-}5a)$$

and

$$\mathbf{E} = \frac{2kp}{x^3}\hat{\mathbf{i}} \qquad \left(\begin{array}{c}\text{dipole field} \\ \text{for } x \gg a, \text{ on axis}\end{array}\right). \qquad (23\text{-}5b)$$

Problem 41 generalizes these results to an arbitrary point not necessarily on the axis or bisector. Because the dipole isn't spherically symmetric, its field is a function of both distance and the angle between the position vector and the

FIGURE 23-16 The dipole moment vector has magnitude given by the product of the charge and separation, and it points from the negative toward the positive charge.

dipole axis. For this reason it's convenient to think of the dipole moment as a vector, **p**, whose magnitude is the product of the charge and separation and whose direction is from the negative to the positive charge (Fig. 23-16).

We emphasize that Equations 23-5 are approximations valid far from the dipole. It's useful to imagine a dipole whose separation $2a$ shrinks toward zero while the magnitude of its two charges grows to keep the product $p = 2qa$ constant. In the limit as $a \to 0$ we have a **point dipole.** Since the point dipole has zero size, Equations 23-5 become exact in this case. Although the point dipole is an idealization, it's a useful approximation when we're far from a real dipole.

Continuous Charge Distributions

Although any charge distribution ultimately consists of point-like electrons and protons, it would be impossible in practice to sum all the field vectors from the 10^{23} or so particles comprising a typical piece of matter. Instead, it's convenient to make the approximation that charge is spread continuously over the distribution. If the charge distribution extends throughout a volume, we describe it in terms of the **volume charge density,** ρ, with units of C/m³. For charge distributions spread over surfaces or lines the corresponding quantities are the **surface charge density,** σ, and **line charge density,** λ. Their units are C/m² and C/m, respectively.

To calculate the field of a continuous charge distribution, we break the charged region into very many small charge elements dq, each small enough that it is essentially a point charge. Each dq then produces an electric field $d\mathbf{E}$ given by Equation 23-3:

$$d\mathbf{E} = \frac{k\,dq}{r^2}\hat{\mathbf{r}}.$$

We then form the vector sum of all the $d\mathbf{E}$'s. In the limit of infinitely many infinitesimally small dq's and their corresponding $d\mathbf{E}$'s, that sum becomes an integral and we have

$$\mathbf{E} = \int d\mathbf{E} = \int \frac{k\,dq}{r^2}\hat{\mathbf{r}}. \tag{23-6}$$

The limits of this integral are chosen to include the entire charge distribution. Figure 23-17 shows the meaning of Equation 23-6; note in particular that both the distance r and the direction specified by $\hat{\mathbf{r}}$ generally vary with position.

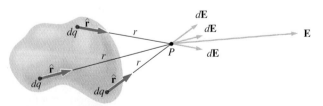

Charge distribution

FIGURE 23-17 The electric field at point P, given by Equation 23-6, is the sum of the vectors $d\mathbf{E}$ arising from the individual charge elements dq in the entire distribution, each calculated using the appropriate distance r and unit vector $\hat{\mathbf{r}}$.

● EXAMPLE 23-7 A CHARGED ROD

Wires, antennas, and similar elongated structures can often be considered as thin rods carrying electric charge. Suppose such a rod of length ℓ carries a positive charge Q distributed uniformly over its length. Find the electric field at point P in Fig. 23-18, a distance a from the end of the rod.

Solution

Let the y axis lie along the rod, with origin at P. Consider a small length dy of the rod, containing charge dq, and located a distance y from P. A unit vector from dq toward P is $-\hat{\mathbf{j}}$, so the field at P due to dq is

$$d\mathbf{E} = -\frac{k\,dq}{y^2}\hat{\mathbf{j}}.$$

The net field at P is the sum—that is, the integral—of all the fields $d\mathbf{E}$ arising from all the dq's along the rod:

$$\mathbf{E} = \int d\mathbf{E} = -\hat{\mathbf{j}}\int_{y=a}^{y=a+\ell} \frac{k\,dq}{y^2},$$

where we've chosen the limits to cover the entire rod. To evaluate the integral we must relate dq and y. How? The rod carries total charge Q, distributed uniformly over its length ℓ. The line charge density is therefore $\lambda = Q/\ell$. This is the charge per unit length; a length dy therefore carries charge $dq = \lambda\,dy$, or $dq = Q\,dy/\ell$. Then our integral becomes

$$\mathbf{E} = -\hat{\mathbf{j}}\int_a^{a+\ell} \frac{k(Q\,dy/\ell)}{y^2} = -\frac{kQ}{\ell}\hat{\mathbf{j}}\int_a^{a+\ell} \frac{dy}{y^2}$$

$$= -\frac{kQ}{\ell}\hat{\mathbf{j}}\left[-\frac{1}{y}\right]_a^{a+\ell} = -\frac{kQ}{\ell}\hat{\mathbf{j}}\left(\frac{1}{a} - \frac{1}{a+\ell}\right) = -\frac{kQ}{a(a+\ell)}\hat{\mathbf{j}},$$

where we took k, Q, ℓ, and $\hat{\mathbf{j}}$ outside the integral because they are constants.

Does this result make sense? First consider the direction: The negative sign shows that the field is downward for positive Q and upward for negative Q, as we should expect. Now suppose P is very far from the rod, so $a \gg \ell$. Then our result becomes approximately

$$\mathbf{E} = -\frac{kQ}{a^2}\hat{\mathbf{j}}, \qquad (a \gg \ell),$$

which is just what we expect for the field of a point charge Q. In this case we're so far from the rod that its length becomes negligible, and indeed it acts like a point charge. But as we move closer, the field becomes a more complicated superposition of the fields of all the dq's at different locations along the rod, and the field no longer exhibits the inverse-square dependence of the point-charge field.

Knowing the field of the charged rod, we can use superposition to find the fields of charge distributions involving more than one rod or a combination of rods and point charges (see Problems 44–46).

FIGURE 23-18 The field at P is the sum—or integral—of the fields arising from all the infinitesimal charge elements dq along the rod.

TIP Find a Single Integration Variable

Evaluating the integral in Equation 23-6 requires that we relate the charge element dq and the position variable r so we'll have the integral expressed in terms of a single variable. The charge density provides the link needed, since it allows us to write dq as the charge density multiplied by an appropriate element of volume, area, or length:

$$dq = \rho\,dV, \quad dq = \sigma\,dA, \quad \text{or} \quad dq = \lambda\,dx.$$

Which of these we use depends on whether charge is distributed over a volume, area, or length; for the thin rod of Example 23-7, the appropriate charge density was the line charge density λ and the position variable was y, so we had $dq = \lambda\,dy$. Sometimes you'll be given the charge density explicitly, and other times you can compute it from the charge and dimensions of the charge distribution.

EXERCISE A thin rod of length ℓ lies along the y axis with its bottom end at the origin. It carries a line charge density λ that varies with position, being given by $\lambda = Q_0\dfrac{y^3}{\ell^4}$, where Q_0 is a constant and y is the distance from the origin. Find the magnitude of the electric field at $y = 0$.

Answer: $\dfrac{kQ_0}{2\ell^2}$

Some problems similar to Example 23-7: 42–45, 73, 74, 82 ●

● **EXAMPLE 23-8** A CHARGED RING

A thin ring of radius a is centered on the origin and carries a total charge Q distributed uniformly around the ring, as shown in Fig. 23-19. Find the electric field at a point P located a distance x along the axis of the ring, and show that the result makes sense when $x \gg a$.

Solution

In Example 23-7 the magnitude but not direction of the individual field vectors $d\mathbf{E}$ from all the charge elements dq varied. Here we have the opposite situation: a point on the ring axis is equidistant from all points on the ring, so the field magnitudes dE are the same but their directions vary. Figure 23-19 shows, however, that components perpendicular to the x axis cancel for any pair of charge elements on opposite sides of the ring, leaving a net field in the x direction. Each charge element contributes an amount dE_x to the field:

$$dE_x = \frac{k\, dq}{r^2}\cos\theta = \frac{k\, dq}{x^2 + a^2}\frac{x}{\sqrt{x^2 + a^2}} = \frac{kx\, dq}{(x^2 + a^2)^{3/2}},$$

where the geometry of Fig. 23-19 gives $r = \sqrt{x^2 + a^2}$ and $\cos\theta = x/\sqrt{x^2 + a^2}$. We now need to integrate this expression over the entire ring. In this integration k, a, and x are all constants—they don't change as we move around the ring. So we have

$$E = E_x = \int_{\text{ring}} dE_x = \int_{\text{ring}} \frac{kx\, dq}{(x^2 + a^2)^{3/2}}$$

$$= \frac{kx}{(x^2 + a^2)^{3/2}} \int_{\text{ring}} dq = \frac{kx\, Q}{(x^2 + a^2)^{3/2}},$$

where the last step follows because $\int_{\text{ring}} dq$ simply means the total charge on the ring. For positive Q this field points away from the ring.

Does this result make sense? At large distances from the ring we can neglect its size a compared with the distance x, and our result reduces to $E = \dfrac{kQ}{x^2}$—just what we would expect for

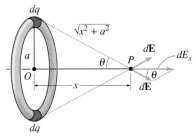

FIGURE 23-19 The electric field of a charged ring points along the ring axis since field components perpendicular to the axis cancel in pairs.

a point charge Q. As always, a finite-size charge distribution looks essentially like a point charge at distances large compared with its size.

> **TIP How Can x Be a Constant?** In this exercise x is constant because it's the distance from the center of the ring to the point where we're evaluating the field—that is, it's the distance from the origin to the *field point, P*. In the integration to find the field we're supposed to consider all *source points*—all points where charge is located that contributes to the field at P. Moving around the ring doesn't affect the value of x, so x is a constant for this integration. However, x is arbitrary, so our result holds for *any* value of x.

EXERCISE Find the point on the x axis where the electric field of the ring in Example 23-8 has its greatest magnitude.

Answer: $x = a/\sqrt{2}$

Some problems similar to Example 23-8: 48–51, 83 ●

● **EXAMPLE 23-9** A POWER LINE'S FIELD

A long, straight electric power line coincides with the x axis and carries a line charge density λ C/m. What is the electric field at a point P on the y axis? Use the approximation that the line is infinitely long.

Solution

Here *both* the direction and magnitude of the field element $d\mathbf{E}$ arising from a charge element on the line vary with the position x of the charge element. But Fig. 23-20 shows that charge elements on opposite sides of the y axis give rise to electric

fields whose x components cancel. Thus the net field points in the y direction—that is, away from the line if λ is positive. So each charge element contributes an amount dE_y to the net field:

$$dE_y = \frac{k\, dq}{r^2}\cos\theta = \frac{k\lambda\, dx}{x^2 + y^2}\frac{y}{\sqrt{x^2 + y^2}} = \frac{k\lambda y\, dx}{(x^2 + y^2)^{3/2}},$$

where we've written $dq = \lambda\, dx$ and used the geometry of Fig. 23-20 to write $r = \sqrt{x^2 + y^2}$ and $\cos\theta = y/r$. To find the net field, we integrate over the entire line, from $x = -\infty$ to

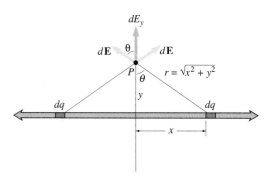

FIGURE 23-20 The field of a charged line is the vector sum of the fields $d\mathbf{E}$ from all the individual charge elements dq along the line. The x components from each pair of charge elements cancel, giving a net field that points directly away from the line.

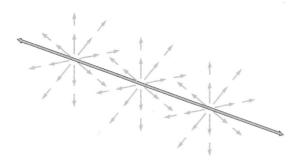

FIGURE 23-21 Field vectors for an infinite line of positive charge point radially outward from the line, with magnitude decreasing inversely with distance.

$x = +\infty$. The quantities k, λ, and y don't change as we move along the line, so we have

$$E = E_y = \int_{x=-\infty}^{x=\infty} \frac{k\lambda y\, dx}{(x^2 + y^2)^{3/2}} = k\lambda y \int_{-\infty}^{\infty} \frac{dx}{(x^2 + y^2)^{3/2}}$$

$$= k\lambda y \left[\frac{x}{y^2\sqrt{x^2 + y^2}} \right]_{-\infty}^{\infty} = k\lambda y \left[\frac{1}{y^2} - \left(-\frac{1}{y^2} \right) \right] = \frac{2k\lambda}{y}.$$

$$(23\text{-}7)$$

Here we evaluated the integral using the integral table in Appendix A and applied the limits $\pm\infty$ by noting that as $x \to \pm\infty$ the term y^2 becomes negligible compared with x^2, giving $x/\sqrt{x^2 + y^2} \to x/\sqrt{x^2} = \pm 1$. The integral could also be evaluated by rewriting it in terms of the angle θ (see Problem 87).

Since the line is infinite in both directions and has cylindrical symmetry, Equation 23-7 holds for *any* point a distance y from the line. Our result thus shows that the electric field of a long line of positive charge points radially away from the line, with magnitude that drops inversely with distance from the line (Fig. 23-21).

What about the field at large distances? Shouldn't it resemble that of a point charge, falling with the inverse square of the distance? No: The charged line is infinitely long, so no matter how far away we go it never resembles a point. Its slower dropoff reflects that fact.

Of course our infinite line is an impossibility. But many real charge distributions, including the power line of this example, have long, thin shapes that approximate an infinite line. Equation 23-7 is therefore a good approximation to the field of a *finite* line as long as we're much closer to it than its length, and not too near either end. Very far from a *finite* line, on the other hand, the field does approach that of a point charge.

EXERCISE A thin rod 2.0 m long carries 50 μC distributed uniformly over its length. Find the electric field strength (a) 1.0 cm from the rod axis, not near either end and (b) 500 m from the rod. Make suitable approximations in both cases.

Answers: (a) 45 MN/C; (b) 1.8 N/C

Some problems similar to Example 23-9: 52–55 ●

23-6 MATTER IN ELECTRIC FIELDS

We're ultimately interested in electric fields since they give rise to forces on charged particles. Because matter consists of such particles, much of the behavior of matter is fundamentally determined by electric fields.

Point Charges in Electric Fields

The motion of a single charge in an electric field is governed by the definition of the electric field (Equation 23-2):

$$\mathbf{F} = q\mathbf{E}$$

and Newton's law:

$$\mathbf{F} = m\mathbf{a}.$$

Combining these equations gives the acceleration of a particle with charge q and mass m in an electric field $\mathbf{E}$:

$$\mathbf{a} = \frac{q}{m}\mathbf{E}. \qquad (23\text{-}8)$$

This equation shows that it's the charge-to-mass ratio that determines a particle's response to an electric field. Electrons, nearly 2000 times less massive than protons but carrying the same charge, are readily accelerated by electric fields. Many practical devices, from x-ray machines to TV tubes, make use of electrons accelerated in electric fields.

When the electric field is uniform, problems involving the motion of charged particles reduce to the constant-acceleration problems we considered in Chapter 2. We'll see in the next chapter that uniform fields are easily produced by flat, uniformly charged plates.

● EXAMPLE 23-10   INSIDE A HEART MONITOR

A heart monitor used in a hospital's intensive care unit includes a cathode-ray tube that gives continuous visual display of a patient's heartbeat (Fig. 23-22). An electron beam "paints" the display on a phosphor screen at the front of the tube, being "steered" to different parts of the screen by electric forces. Suppose the beam is initially heading horizontally to the right at 4.0 Mm/s when it enters the "steering" electric field of 1.0 kN/C pointing downward. The field region extends horizontally for 2.0 cm, as shown in Fig. 23-23. In what direction is the electron moving when it leaves the field region?

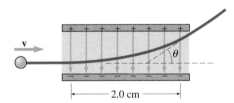

FIGURE 23-23   A uniform electric field deflects an electron from its straight-line path (Example 23-10).

Solution
As usual in two-dimensional motion, the vertical force due to the electric field does not affect the horizontal component of the electron's velocity. Therefore the electron spends a time $t = \Delta x/v_x$ in the field region. During this time it experiences a vertical acceleration $q\mathbf{E}/m$, and therefore gains a vertical velocity component given by

$$v_y = a_y t = \frac{qE_y}{m}\frac{\Delta x}{v_x}$$

$$= \frac{(-1.6\times10^{-19}\text{ C})(-1.0\times10^3\text{ N/C})}{9.11\times10^{-31}\text{ kg}}\left(\frac{2.0\times10^{-2}\text{ m}}{4.0\times10^6\text{ m/s}}\right)$$

$$= 8.78\times10^5\text{ m/s}.$$

Note that the field points downward, as reflected by the minus sign, and combined with the electron's negative charge thus results in an upward velocity. The electron thus leaves the field region moving at an angle of

$$\theta = \tan^{-1}\!\left(\frac{v_y}{v_x}\right) = \tan^{-1}\!\left(\frac{8.78\times10^5\text{ m/s}}{4.0\times10^6\text{ m/s}}\right) = 12.4°$$

to the horizontal.

As it traverses the field region the electron describes the parabolic trajectory of an object undergoing constant acceleration in two dimensions; once outside the field it continues in a straight line with its new velocity, as indicated in Fig. 23-23.

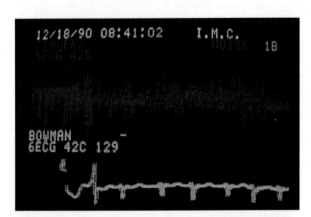

FIGURE 23-22   This heart monitor uses an electron beam deflected by electric fields to display patients' heartbeats.

Field configurations like this one are used to "steer" electron beams not only in heart monitors but also in oscilloscopes and other electronic instrumentation.

EXERCISE What electric field strength would be needed to cause the electron of Example 23-10 to leave the field region at a point 4.6 mm above where it entered? The electron's initial velocity and the length of the field region are the same as in Example 23-10.

Answer: 2.1 kN/C

Some problems similar to Example 23-10: 59, 60, 77, 78
●

When the field is not uniform it's generally difficult to calculate particle trajectories. An important exception is the case of a particle moving at right angles to a field that points radially. In that case—the subject of the following example—the electric force changes the particle's direction but not its speed, so the motion is uniform circular.

● **EXAMPLE 23-11** AN ELECTROSTATIC ANALYZER

Two curved metal plates are used to establish an electric field given by

$$E = E_0 \frac{b}{r},$$

where $E_0 = 24$ kN/C and $b = 5.0$ cm. The field points toward the center of curvature, as shown in Fig. 23-24, and r is the distance from that center. A beam of protons with a mix of speeds is incident on the device. Find the speed v for which an incident proton will leave the analyzer moving horizontally in Fig. 23-24.

Solution

To exit with its velocity horizontal, a proton must describe a circular arc while inside the analyzer. The field provides the v^2/r acceleration required for that circular motion:

$$a = \frac{v^2}{r} = \frac{eE}{m} = \frac{e}{m} E_0 \frac{b}{r}.$$

Solving for v gives

$$v = \sqrt{\frac{eE_0 b}{m}} = \sqrt{\frac{(1.6 \times 10^{-19} \text{ C})(2.4 \times 10^4 \text{ N/C})(0.050 \text{ m})}{1.67 \times 10^{-27} \text{ kg}}}$$

$$= 3.4 \times 10^5 \text{ m/s}.$$

Note that it doesn't matter where the protons enter the analyzer since the $1/r$ decrease in field strength matches the $1/r$ depen-

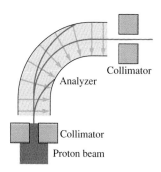

FIGURE 23-24 An electrostatic analyzer, showing the trajectories of protons in the radial electric field. Only those entering with the right speed will emerge at the top moving horizontally. Collimators block protons not moving at right angles to the field.

dence of the acceleration. Devices of this sort have been used on spacecraft to analyze charged particles in interplanetary space.

EXERCISE A proton is in circular motion centered on a long charged wire carrying uniform negative line charge density $-\lambda$. Find its speed. *Hint:* Consult Example 23-9 for the field of the charged wire.

Answer: $v = \sqrt{2k\lambda e/m}$, with m the proton mass

Some problems similar to Example 23-11: 61–65
●

Dipoles in Electric Fields

Earlier in this chapter we calculated the field of an electric dipole, which consists of two opposite charges of equal magnitude. Here we study a dipole's response to electric fields. Since the dipole configuration provides a simple model for molecules, our results help explain molecular behavior.

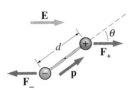

FIGURE 23-25 A dipole in a uniform electric field experiences no net force, but it does experience a torque.

Figure 23-25 shows a dipole with charges $\pm q$ separated by a distance d, located in a uniform electric field. The dipole moment vector **p** has magnitude qd and points from the negative to the positive charge (recall Fig. 23-16). Since the field is uniform it's the same at both ends of the dipole. Since the dipole charges are equal in magnitude but opposite in sign, they experience equal but opposite forces $\pm q\mathbf{E}$—and therefore there's no net force on the dipole.

However, Fig. 23-25 shows that the dipole does experience a torque that tends to align it with the field. In Chapter 13 we described torque as the cross product of the position vector with the force: $\boldsymbol{\tau} = \mathbf{r} \times \mathbf{F}$, where the magnitude of the torque vector is $rF\sin\theta$ and its direction is given by the right-hand rule. Figure 23-25 thus shows that the torque about the center of the dipole due to the force on the positive charge has magnitude

$$\tau_+ = rF\sin\theta = \frac{1}{2}d\,qE\sin\theta.$$

The torque associated with the negative charge has the same magnitude and both torques are in the same direction since both tend to rotate the dipole in Fig. 23-25 clockwise. Thus the net torque has magnitude $\tau = qdE\sin\theta$; applying the right-hand rule shows that this torque is into the page. But qd is the magnitude of the dipole moment **p**, and Fig. 23-25 shows that θ is the angle between the dipole moment vector and the electric field **E**; therefore we can write the torque vectorially as

$$\boldsymbol{\tau} = \mathbf{p} \times \mathbf{E}. \qquad \text{(torque on a dipole)} \qquad (23\text{-}9)$$

Because of this torque, it takes work to rotate a dipole in an electric field. If we start with the dipole oriented at right angles to the field ($\theta = \pi/2$), then Equation 12-28b gives the work required to rotate it to a new angle θ:

$$W = \int_{\pi/2}^{\theta} \tau\,d\theta = \int_{\pi/2}^{\theta} pE\sin\theta\,d\theta = pE\,[-\cos\theta]_{\pi/2}^{\theta} = -pE\cos\theta,$$

where the last step follows because $\cos(\pi/2) = 0$. This work ends up as stored potential energy U. Since the product of two vector magnitudes with the cosine of the angle between the vectors defines the dot product, we can write the potential energy in compact form as

$$U = -\mathbf{p} \cdot \mathbf{E}, \qquad (23\text{-}10)$$

where the zero of potential energy corresponds to the dipole at right angles to the field.

■ **APPLICATION** MICROWAVE COOKING AND LIQUID CRYSTALS

The torque on dipoles in electric fields forms the basis of two widespread contemporary technologies: the microwave oven and the liquid crystal display (LCD) (Fig. 23-26).

A microwave oven works by generating an electric field whose direction changes several billion times per second. Water molecules, whose dipole moment is much greater than that

(a)   (b)

FIGURE 23-26 (a) Microwave ovens and (b) liquid crystal displays make use of the torque that dipolar molecules experience in electric fields.

of most other molecules, respond by attempting to align with the field. But the field is constantly changing, so the molecules swing rapidly back and forth. As they jostle against each other, the energy they gain from the field is dissipated as heat that cooks the food.

Calculators, laptop computers, gas pumps, and many other devices display numerical and alphabetic information with liquid crystals. This unique state of matter combines the fluidity of a liquid with the order of a solid. The liquid crystal consists of long molecules whose chemical structure gives rise to a

dipole-like charge separation. In response to each others' electric fields, the molecules tend to align with each other (Fig. 23-27). But an external electric field can rotate the liquid crystal dipoles, altering the optical properties of the material. Figure 23-28 shows how this effect is exploited to make practical displays. Liquid crystal displays have the advantage that they consume very little power. On the other hand they generate no light of their own, and therefore an external light source is needed to illuminate the display when it's dark.

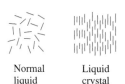

FIGURE 23-27 Alignment of dipole-like molecules in a liquid crystal. The dipole's direction can be rotated by applying an electric field.

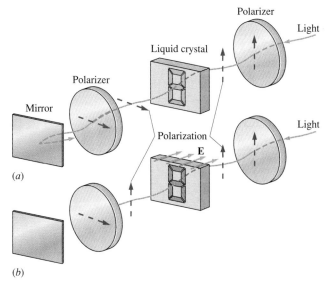

FIGURE 23-28 A typical liquid-crystal display consists of seven small bands of liquid crystal sandwiched between optical polarizers. Applying an electric field to any segment rotates its liquid crystal molecules, changing the polarization and making the segment appear dark. (We deal further with polarization in Chapter 34). (a) Electric field off. (b) Electric field on.

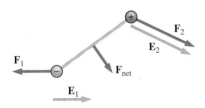

FIGURE 23-29 When the electric field differs in magnitude or direction at the two ends of the dipole, then the dipole experiences a nonzero net force as well as a torque.

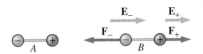

FIGURE 23-30 Dipole B is aligned with the electric field of dipole A. Since the field of A is stronger at the negative end of B, F_- is greater in magnitude than F_+, so dipole B experiences a net force toward dipole A. This is the origin of the van der Waals force between gas molecules.

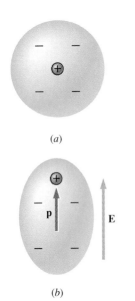

FIGURE 23-31 (*a*) A molecule with no dipole moment has negative charge concentric with positive charge. (*b*) In an applied electric field, the molecule stretches and acquires a dipole moment.

When the electric field is not uniform, the charges at opposite ends of the dipole experience forces that differ in magnitude and/or are not exactly opposite in direction. Then the dipole experiences a net force as well as a torque (Fig. 23-29). An important instance of this effect is the force on a dipole in the field of another dipole (Fig. 23-30). Because the dipole field falls off rapidly with distance and because the dipole responding to the field has closely spaced charges of equal magnitude but opposite sign, the dipole-dipole force is quite weak and falls extremely rapidly with distance. This weak force, which Fig. 23-30 shows to be attractive, is the basis of the van der Waals interaction between gas molecules that we considered in Chapter 20. Problems 70 and 85 deal with forces on dipoles in nonuniform fields.

Conductors, Insulators, and Dielectrics

Bulk matter consists ultimately of vast numbers of point charges, namely electrons and protons. In some matter—notably metals, ionic solutions, and ionized gases—individual charges are free to move throughout the material. In such materials—called **conductors**—the application of an electric field results in the ordered motion of electric charge that we call **electric current.** We'll consider the behavior of conductors and related materials called semiconductors in subsequent chapters.

Materials in which charge is not free to move are called **insulators,** since they do not support electric current. Insulators, however, still contain charges—it's just that their charges are bound into neutral molecules. Some molecules, like water, have intrinsic dipole moments and therefore rotate in response to an applied electric field. Even if they don't have dipole moments, molecules may respond to an electric field by stretching and acquiring **induced dipole moments** (Fig. 23-31). In either case, the application of an electric field results in the alignment of molecular dipoles with the field (Fig. 23-32). The fields of the dipoles, pointing from their positive to their negative charges, then reduce the applied electric field within the material. We'll explore the consequences of this effect further in Chapter 26. Materials in which molecules have either intrinsic dipole moments or acquire induced moments are called **dielectrics.**

If the electric field applied to a dielectric becomes too great, individual charges are ripped free, and the material then acts like a conductor. Such **dielectric breakdown** can cause severe damage in electrical equipment (Fig. 23-33). On a larger scale, lightning results from dielectric breakdown in air.

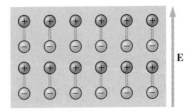

FIGURE 23-32 The alignment of molecular dipoles in a dielectric results in a reduction of the electric field within the dielectric.

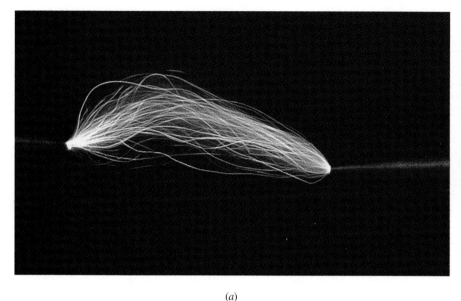

(a)

(b)

FIGURE 23-33 (a) Dielectric breakdown in air results in sparks jumping between two highly charged wires. (b) Dielectric breakdown in a solid produced this tree-like pattern.

CHAPTER SYNOPSIS

Summary

1. **Electromagnetism** is among the fundamental forces of nature. Electromagnetism comprises electricity and magnetism, seemingly distinct phenomena that are actually intimately related.

2. **Electric charge** is a fundamental property of matter. Charge comes in two types, arbitrarily called positive and negative. Charge is **quantized,** with one elementary charge—the magnitude of the electron or proton charge—equal to 1.60×10^{-19} C, where the coulomb (C) is the SI unit of charge. Charge is also **conserved,** in that the algebraic sum of the charges in a closed system never changes.

3. Charges interact via the **electric force.** This force is proportional to the product of the charges and inversely proportional to the square of the distance between them. **Coulomb's law** provides a mathematical description of the electric force between point charges:

$$\mathbf{F}_{12} = \frac{kq_1q_2}{r^2}\hat{\mathbf{r}}.$$

4. The **superposition principle** states that the electric force on a charge arising from two or more other charges is the vector sum of the force arising from each according to Coulomb's law. The superposition principle greatly simplifies the calculation of electrical effects of various **charge distributions.**

5. The electric field at a point is a vector giving the electric force per unit charge that would be experienced by a charge at that point:

$$\mathbf{E} = \frac{\mathbf{F}}{q}.$$

The electric field of a point charge q is therefore

$$\mathbf{E} = \frac{kq}{r^2}\hat{\mathbf{r}}.$$

6. The superposition principle shows that the electric field of a charge distribution is the sum of the fields of its individual point charges:

$$\mathbf{E} = \sum_i \mathbf{E}_i = \sum_i \frac{kq_i}{r_i^2}\hat{\mathbf{r}}_i.$$

A particularly important charge distribution is the **electric dipole,** consisting of two point charges with equal magnitude but opposite sign, separated by a fixed distance. At large distances from a dipole the field decreases as the inverse cube of the distance.

7. With continuous distributions of charge, the sum over all point charges becomes an integral, giving

$$\mathbf{E} = \int d\mathbf{E} = \int \frac{k\,dq}{r^2}\hat{\mathbf{r}}.$$

For any finite charge distribution with nonzero net charge, the field approaches that of a point charge at large distances.

8. A point charge in an electric field experiences a force $\mathbf{F} = q\mathbf{E}$; if this is the only force acting, the charge undergoes an acceleration $\mathbf{a} = (q/m)\mathbf{E}$ in accordance with Newton's law.

9. Because it consists of two equal but opposite point charges, an electric dipole experiences no net force in a uniform electric field. It does, however, experience a torque given by $\tau = \mathbf{p} \times \mathbf{E}$, where $\mathbf{p}$ is the dipole moment vector. A dipole in an electric field has potential energy given by $U = -\mathbf{p} \cdot \mathbf{E}$, where the zero of potential energy corresponds to the dipole oriented perpendicular to the field. In a nonuniform field, a dipole experiences both a torque and a net force.

Terms You Should Understand

(Pairs are closely related terms whose distinction is important; number in parentheses is chapter section where term first appears.)

electromagnetism (23-1)
electric charge (23-2)
coulomb (23-2)
Coulomb's law (23-3)
superposition principle (23-3)
point charge, charge distribution (23-3)
electric field (23-4)
electric dipole (23-5)
dipole moment (23-5)
volume, surface, and line charge density (23-5)
conductor, insulator, dielectric (23-6)
dielectric breakdown (23-6)

Symbols You Should Recognize

C (23-2)
e (23-2)
k (23-3)
q (23-3)
$\hat{\mathbf{r}}$ (23-3)
$\mathbf{E}$ (23-4)
$\mathbf{p}$ (23-5)
ρ, σ, λ (23-5)

Problems You Should Be Able to Solve

calculating electric forces arising from one or more charges acting on another (23-3)
calculating electric fields of distributions of discrete charges (23-5)
calculating electric fields of continuous charge distributions (23-5)
approximating electric fields at large distances from charge distributions (23-5)
evaluating forces on charges in electric fields (23-4, 23-6)
describing the motion of charged particles in uniform and radial electric fields (23-6)
evaluating torques on dipoles in electric fields (23-6)

Limitations to Keep in Mind

Coulomb's law applies strictly only to one point charge acting on another. The forces and electric fields arising from more than one point charge or from a continuous distribution must be calculated using the superposition principle.
At large distances, the field of a charge distribution approaches that of a point charge only if the distribution is both finite in size and has nonzero net charge.
The dipole field decreases with the inverse cube of the distance only for distances large compared with the dipole's charge separation.

QUESTIONS

1. How might a universe with only one kind of electric charge differ from our universe?
2. Discuss this statement: It is precisely because the electric force is so strong that the electrical nature of most everyday interactions is not obvious.
3. The gravitational force between an electron and proton is about 10^{-40} times weaker than the electrical force between the two. Since matter consists largely of electrons and protons, why is the gravitational force ever important?
4. You are given two electric charges. Could you determine whether they had the same or opposite signs? Could you determine the signs of each?
5. In Example 23-3 we found a point where the electric force on a third charge would be zero. Would a charge placed at that point be in stable equilibrium? Why or why not?
6. The gravitational force between an electron and a proton is about 10^{-40} times the electrical force between them. Does this ratio depend on how far apart they are? Explain.
7. In which of the following phenomena does electromagnetism play a dominant role?
   (a) Gasoline burns in a car engine.
   (b) The moon orbits Earth.
   (c) A nerve impulse travels from your brain to a muscle.
   (d) Protons and neutrons join to form an atomic nucleus.
   (e) A chemist synthesizes a new polymer.
   (f) You sit in a chair and the chair doesn't collapse.
8. A free neutron is unstable, and soon decays into other particles. One of the decay products is a proton. Must there be others? If so, what electrical properties must they have?
9. Where in Fig. 23-9 could you put a third charge so it would experience no net force? Would it be in stable or unstable equilibrium?
10. Why should the test charge used to measure an electric field be small?
11. Equation 23-3 gives the electric field of a point charge. Does the direction of $\hat{\mathbf{r}}$ depend on whether the charge is

positive or negative? Does the direction of **E** depend on the sign of the charge?

12. Is the electric force on a charged particle always in the direction of the field? Explain.

13. Why does a dipole produce an electric field at all? After all, the dipole has no net charge.

14. The rod in Example 23-7 carries total charge Q, and the point P is a distance a from the rod. So why isn't the electric field of the rod just kQ/a^2?

15. The field of a dipole decreases with the inverse cube of the distance. Why doesn't this violate our assertion that the field of a finite size charge distribution with nonzero net charge approaches that of a point charge at large distances?

16. A spherical balloon is initially uncharged. If you spread positive charge uniformly over the balloon's surface, would it expand or contract? What would happen if you spread negative charge instead?

17. Suppose someone argued that the force we call gravity is really an electric force, arising from a net electric charge on Earth. How could you disprove this?

18. Two cubical blocks of wood are each 10 cm on a side and carry electric charge spread over their surfaces. If they're 5 cm apart, would you be justified in writing kq_1q_2/r^2 for the force between them? How about if they're 5 m apart? Explain the difference.

19. Two charged particles are suspended in the same electric field, the electric force on each balancing the gravitational force. What quantity must they have in common?

20. A deuteron (heavy hydrogen nucleus) has twice the mass but the same charge as a normal hydrogen nucleus (a proton). Both are released from rest in the same uniform electric field. Compare the distances each goes in the same time.

21. Under what circumstances is the path of a charged particle a parabola? A circle?

22. Explain why a nonuniform field is required for a net force on a dipole.

23. Why should there be a force between two dipoles? After all, each has zero net charge.

24. Dipoles A and B are both located in the field of a point charge Q, as shown in Fig. 23-34. Does either experience a net torque? A net force? If each dipole is released from rest, describe qualitatively its subsequent motion.

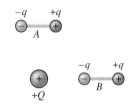

FIGURE 23-34 Question 24.

PROBLEMS

Section 23-3 Electric Charge

1. Suppose the electron and proton charges differed by one part in one billion. Estimate the net charge you would carry.

2. A typical lightning flash delivers about 25 C of negative charge from cloud to ground. How many electrons are involved?

3. Protons and neutrons are made from combinations of the two most common quarks, the u quark and the d quark. The u quark's charge is $+\frac{2}{3}e$ while the d quark carries $-\frac{1}{3}e$. How could three of these quarks combine to make (a) a proton and (b) a neutron?

4. A 2-g ping-pong ball rubbed against a wool jacket acquires a net positive charge of 1 μC. Estimate the fraction of the ball's electrons that have been removed.

Section 23-3 Coulomb's Law

5. If the charge imbalance of Problem 1 existed, what would be the approximate force between you and another person 10 m away? Treat the people as point charges, and compare the answer with your weight.

6. Compare the gravitational force between an electron and a proton with the electrical force between the two. At what distance(s) is your answer correct?

7. The electron and proton in a hydrogen atom are 52.9 pm apart. What is the magnitude of the electric force between them?

8. How far apart should an electron and proton be so the force of Earth's gravity on the electron is equal to the electric force arising from the proton? Your answer shows why gravity is unimportant on the molecular scale!

9. Two charges, one twice as large as the other, are located 15 cm apart and experience a repulsive force of 95 N. What is the magnitude of the larger charge?

10. Earth carries a net charge of -4.3×10^5 C. The force due to this charge is the same as if it were concentrated at Earth's center. How much charge would you have to place on a 1.0-g mass in order for the electrical and gravitational forces on it to balance?

11. A 6.5-μC charge is held at rest, while a small charged sphere of mass 2.3 g is released 50 cm away. Immediately after its release, the sphere accelerates toward the charge at 340 m/s^2. What is the sphere's charge?

12. A proton is at the origin and an electron is at the point $x = 0.41$ nm, $y = 0.36$ nm. Find the electric force on the proton.

13. A 9.5-μC charge is at $x = 16$ cm, $y = 5.0$ cm, and a -3.2-μC charge is at $x = 4.4$ cm, $y = 11$ cm. Find the force on the negative charge.

14. A spring of spring constant 100 N/m is stretched 10 cm beyond its 90-cm equilibrium length. If you want to keep it stretched by attaching equal electric charges to the opposite ends, what magnitude of charge should you use?

15. Two small spheres with the same mass m and charge q are suspended from massless strings of length ℓ, as shown in Fig. 23-35. Each string makes an angle θ with the vertical. Show that the charge on each sphere is $q = \pm 2\ell \sin\theta \sqrt{mg \tan\theta/k}$.

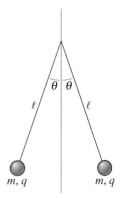

FIGURE 23-35 Problem 15.

16. A charge $3q$ is at the origin, and a charge $-2q$ is on the positive x axis at $x = a$. Where would you place a third charge so it would experience no electric force?

17. A 60-μC charge is at the origin, and a second charge is on the positive x axis at $x = 75$ cm. If a third charge placed at $x = 50$ cm experiences no net force, what is the second charge?

18. You have two charges $+4q$ and one charge $-q$. (a) How would you place them along a line so there's no net force on any of the three? (b) Is this equilibrium stable or unstable?

19. In Fig. 23-36 take $q_1 = 68 \mu$C, $q_2 = -34 \mu$C, and $q_3 = 15 \mu$C. Find the electric force on q_3.

20. In Fig. 23-36 take $q_1 = 25 \mu$C and $q_2 = 20 \mu$C. If the force on q_1 points in the $-x$ direction, (a) what is q_3 and (b) what is the magnitude of the force on q_1?

21. Four identical charges q form a square of side a. Find the magnitude of the electric force on any of the charges.

22. Three identical charges $+q$ and a fourth charge $-q$ form a square of side a. (a) Find the magnitude of the electric force on a charge Q placed at the center of the square. (b) Describe the direction of this force.

23. Three charges lie in the x-y plane: $q_1 = 55 \mu$C at $x = 0$, $y = 2.0$ m; q_2 at $x = 3.0$ m, $y = 0$; and q_3 at $x = 4.0$ m, $y = 3.0$ m. If the force on q_3 is $8.0\hat{\imath} + 15\hat{\jmath}$ N, find q_2 and q_3.

24. Two identical small metal spheres initially carry charges q_1 and q_2, respectively. When they're 1.0 m apart they experience a 2.5-N attractive force. Then they're brought together so charge moves from one to the other until they have the same net charge. They're again placed 1.0 m apart, and now they repel with a 2.5-N force. What were the original values of q_1 and q_2?

Section 23-4 The Electric Field

25. An electron placed in an electric field experiences a 6.1×10^{-10} N electric force. What is the field strength?

26. What is the magnitude of the force on a 2.0-μC charge in a 100 N/C electric field?

27. A 68-nC charge experiences a 150-mN force in a certain electric field. Find (a) the field strength and (b) the force that a 35-μC charge would experience in the same field.

28. A -1.0-μC charge experiences a $10\hat{\imath}$-N electric force in a certain electric field. What force would a proton experience in the same field?

29. The electron in a hydrogen atom is 0.0529 nm from the proton. What is the proton's electric field strength at this distance?

30. A 65-μC point charge is at the origin. Find the electric field at the points (a) $x = 50$ cm, $y = 0$; (b) $x = 50$ cm, $y = 50$ cm; (c) $x = -25$ cm, $y = 75$ cm.

Section 23-5 Electric Fields of Charge Distributions

31. In Fig. 23-37, point P is midway between the two charges. Find the electric field in the plane of the page (a) 10 cm directly above P, (b) 10 cm directly to the right of P, and (c) at P.

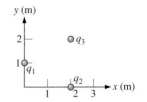

FIGURE 23-36 Problems 19, 20.

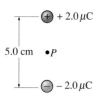

FIGURE 23-37 Problem 31.

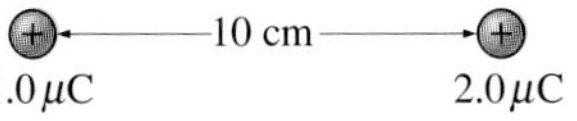

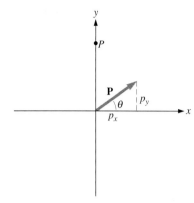

FIGURE 23-38 Problem 32.

32. A 1.0-μC charge and a 2.0-μC charge are 10 cm apart, as shown in Fig. 23-38. Find a point where the electric field is zero.

33. A proton is at the origin and an ion is at $x = 5.0$ nm. If the electric field is zero at $x = -5$ nm, what is the charge on the ion?

34. For the situation of Example 23-3, (a) write an expression for the electric field as a function of x for points to the right of the charge $-q$ shown in Fig. 23-11. (b) Taking $q = 1.0$ μC and $a = 1.0$ m, plot the field as a function of position for $x = 5$ m to $x = 25$ m.

35. (a) Find an expression for the electric field on the y axis due to the two charges q in Fig. 23-9. (b) At what point does the field on the y axis have its maximum strength?

36. Write an expression for the dipole moment vector of the dipole shown in Fig. 23-15.

37. A dipole lies on the y axis, and consists of an electron at $y = 0.60$ nm and a proton at $y = -0.60$ nm. Find the electric field (a) midway between the two charges, (b) at the point $x = 2.0$ nm, $y = 0$, and (c) at the point $x = -20$ nm, $y = 0$.

38. What is the electric field strength 10 cm from a point dipole with dipole moment 3.8 μC·m (a) on the dipole's perpendicular bisector and (b) on its axis?

39. The dipole moment of the water molecule is 6.2×10^{-30} C·m. What would be the separation distance if the molecule consisted of charges $\pm e$? (The effective charge is actually less because electrons are shared by the oxygen and hydrogen atoms.)

40. You're 1.5 m from a charge distribution whose size is much less than 1 m. You measure an electric field strength of 282 N/C. You move to a distance of 2.0 m and the field strength becomes 119 N/C. What is the net charge of the distribution? *Hint:* Don't try to calculate the charge. Determine instead how the field decreases with distance, and from that infer the charge.

41. A point dipole lies at the origin, with its dipole moment vector **p** making an angle θ with the x axis, as shown in Fig. 23-39. By resolving **p** into components and applying Equations 23-5a and 23-5b to the x and y components, respectively, show that the electric field at an arbitrary point P on the y axis is given by

$$\mathbf{E} = \frac{kp}{y^3}(-\hat{\mathbf{i}}\cos\theta + 2\hat{\mathbf{j}}\sin\theta).$$

42. Three identical charges q form an equilateral triangle of side a, with two charges on the x axis and one on the positive y axis. (a) Find an expression for the electric field at points on the y axis above the uppermost charge.

FIGURE 23-39 Problem 41. (The dipole is a true point dipole, and lies exactly at the origin.)

(b) Show that your result reduces to the field of a point charge $3q$ for $y \gg a$.

43. A 30-cm-long rod carries a charge of 80 μC spread uniformly over its length. Find the electric field strength on the rod axis, 45 cm from the end of the rod.

44. A thin rod of length ℓ carries charge Q distributed evenly over its length. A point charge with the same charge Q lies a distance b from the end of the rod, as shown in Fig. 23-40. Find a point where the electric field is zero.

FIGURE 23-40 Problem 44.

45. The rods shown in Fig. 23-41 are both 15 cm long and both carry 1.2 μC of charge. Find the magnitude and direction of the electric field at point P.

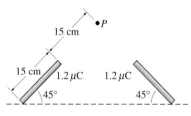

FIGURE 23-41 Problem 45.

46. Repeat the preceding problem for the case where the right-hand rod carries -1.2 μC.

47. A thin rod of length ℓ has its left end at the origin and its right end at the $x = \ell$. It carries a line charge density given by $\lambda = \lambda_0 \frac{x^2}{\ell^2}\sin(\pi x/\ell)$, where λ_0 is a constant. Find the electric field strength at the origin.

48. Two identical rods of length ℓ lie on the x axis and carry uniform charges $\pm Q$, as shown in Fig. 23-42. (a) Find an

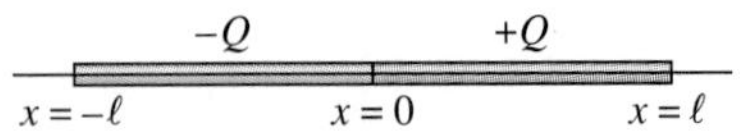

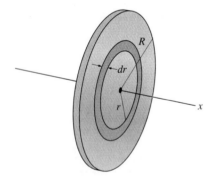

FIGURE 23-42   Problem 48.

expression for the electric field strength as a function of position x for points to the right of the right-hand rod. (b) Show that your result has the $1/x^3$ dependence of a dipole field for $x \gg \ell$. (c) What is the dipole moment of this configuration? *Hint:* See Equation 23-5b.

49. A uniformly charged ring is 1.0 cm in radius. The electric field on the axis 2.0 cm from the center of the ring has magnitude 2.2 MN/C and points toward the ring center. Find the charge on the ring.

50. Figure 23-43 shows a thin, uniformly charged disk of radius R. Imagine the disk divided into rings of varying radii r, as suggested in the figure. (a) Show that the area of such a ring is very nearly $2\pi r \, dr$. (b) If the surface charge density on the disk is σ C/m², use the result of (a) to write an expression for the charge dq on an infinitesimal ring. (c) Use the result of (b) along with the result of Example 23-8 to write the infinitesimal electric field dE of this ring at a point on the disk axis, taken to be the positive x axis. (d) Integrate over all such rings (that is, from $r = 0$ to $r = R$), to show that the net electric field on the disk axis is

$$ E = 2\pi k \sigma \left(1 - \frac{x}{\sqrt{x^2 + R^2}} \right). $$

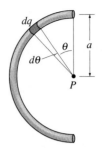

FIGURE 23-44   Problem 52.

53. The electric field 22 cm from a long wire carrying a uniform line charge density is 1.9 kN/C. What will be the field strength 38 cm from the wire?

54. What is the line charge density on a long wire if the electric field 45 cm from the wire has magnitude 260 kN/C and points toward the wire?

55. A straight wire 10 m long carries 25 μC distributed uniformly over its length. (a) What is the line charge density on the wire? Find the electric field strength (b) 15 cm from the wire axis, not near either end and (c) 350 m from the wire. Make suitable approximations in both cases.

56. Figure 23-45 shows a thin rod of length ℓ carrying charge Q distributed uniformly over its length. (a) What is the line charge density on the rod? (b) What must be the electric field direction on the rod's perpendicular bisector (taken to be the y axis)? (c) Modify the calculation of Example 23-9 to find an expression for the electric field at a point P a distance y along the perpendicular bisector. (d) Show that your result for (c) reduces to the field of a point charge Q for $y \gg \ell$.

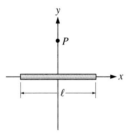

FIGURE 23-45   Problem 56.

FIGURE 23-43   Problem 50.

51. Use the result of the preceding problem to show that the field of an *infinite*, uniformly charged flat sheet is $2\pi k \sigma$, where σ is the surface charge density. Note that this result is independent of distance from the sheet.

52. A semicircular loop of radius a carries positive charge Q distributed uniformly over its length. Find the electric field at the center of the loop (point P in Fig. 23-44). *Hint:* Divide the loop into charge elements dq as shown in Fig. 23-44, and write dq in terms of the angle $d\theta$. Then integrate over θ to get the net field at P.

Section 23-6 Matter in Electric Fields

57. In his famous 1909 experiment that demonstrated quantization of electric charge, R. A. Millikan suspended small oil drops in an electric field. With a field strength of 20 MN/C, what mass drop can be suspended when the drop carries a net charge of 10 elementary charges?

58. How strong an electric field is needed to accelerate electrons in a TV tube from rest to one-tenth the speed of light in a distance of 5.0 cm?

59. A proton moving to the right at 3.8×10^5 m/s enters a region where a 56 kN/C electric field points to the left. (a) How far will the proton get before its speed reaches zero? (b) Describe its subsequent motion.

60. An oscilloscope display requires that a beam of electrons moving at 8.2 Mm/s be deflected through an angle of 22° by a uniform electric field that occupies a region 5.0 cm long. What should be the field strength?

61. A uniform electric field **E** is set up between two metal plates of length ℓ and spacing d, as shown in Fig. 23-46. An electron enters the region midway between the plates moving horizontally with speed v, as shown. Find an expression for the minimum speed the electron needs to get through the region without hitting either plate. Neglect gravity.

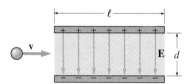

FIGURE 23-46 Problem 61.

62. An electrostatic analyzer like that of Example 23-11 has $b = 7.5$ cm. What should be the value of E_0 if the device is to select protons moving at 84 km/s?

63. An electron is moving in a circular path around a long, uniformly charged wire carrying 2.5 nC/m. What is the electron's speed?

64. Figure 23-47 shows a device its inventor claims will separate isotopes of a particular element. (Isotopes of the same element have nuclei with the same charge but different masses). Atoms of the element are first stripped completely of their electrons, then accelerated from rest through an electric field chosen to give the desired isotope exactly the right speed to pass through the electrostatic analyzer (see Example 23-11). Prove that the device won't work—that is, that it won't separate different isotopes.

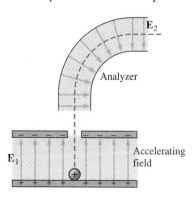

FIGURE 23-47 An isotope separator that won't work (Problem 64).

65. What is the line charge density on a long wire if a 6.8-μg particle carrying 2.1 nC describes a circular orbit about the wire with speed 280 m/s?

66. The electron in a hydrogen atom has kinetic energy 2.18×10^{-18} J. Assuming the electron is in a circular orbit around the central proton, estimate the size of the atom. (Although this problem gives a reasonable answer, the simple model of an electron orbiting a proton is not consonant with the quantum mechanical description of the atom.)

67. A dipole with dipole moment 1.5 nC·m is oriented at 30° to a 4.0-MN/C electric field. (a) What is the magnitude of the torque on the dipole? (b) How much work is required to rotate the dipole until it's antiparallel to the field?

68. A molecule has its dipole moment aligned with a 1.2-kN/C electric field. If it takes 3.1×10^{-27} J to reverse the molecule's orientation, what is its dipole moment?

69. Two identical dipoles, each of charge q and separation a, are a distance x apart as shown in Fig. 23-48. By considering forces between pairs of charges in the different dipoles, calculate the net force between the dipoles. (a) Show that, in the limit $a \ll x$, the force has magnitude $6kp^2/x^4$, where $p = qa$ is the dipole moment. (b) Is the force attractive or repulsive?

FIGURE 23-48 Problem 69.

70. A dipole with charges $\pm q$ and separation $2a$ is located a distance x from a point charge $+Q$, with its dipole moment vector perpendicular to the x axis, as shown in Fig. 23-49. Find expressions for the magnitude of (a) the net torque and (b) the net force on the dipole, both in the limit $x \gg a$. (c) What is the direction of the net force?

FIGURE 23-49 Problem 70.

Paired Problems

(Both problems in a pair involve the same principles and techniques. If you can get the first problem, you should be able to solve the second one.)

71. An electron is at the origin and an ion with charge $+5e$ is at $x = 10$ nm. Find a point where the electric field is zero.

72. A proton is at the origin and an ion is at $x = 5.0$ nm. If the electric field is zero at $x = -6.83$ nm, what is the charge on the ion?

73. A thin rod of length ℓ has its left end at $x = -\ell$ and its right end at the origin. It carries a line charge density given by $\lambda = \lambda_0 \dfrac{x^2}{\ell^2}$, where λ_0 is a constant. Find the electric field at the origin.

74. Repeat the preceding problem for the case when $\lambda = \lambda_0 \dfrac{x^4}{\ell^4}$.

75. A thin, flexible rod carrying charge Q spread uniformly over its length is bent into a quarter circle of radius a, as shown in Fig. 23-50a. Find the electric field strength at the point P, which is the center of the circle. *Hint:* Consult Problem 52.

76. A thin, flexible rod carrying charge Q spread uniformly over its length is bent into a circular arc of radius a, as shown in Fig. 23- 50b. Find the electric field strength at the point P, which is the center of the circular arc.

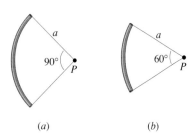

(a)   (b)

FIGURE 23-50 Problems 75, 76.

77. Ink-jet printers work by deflecting moving ink droplets with an electric field so they hit the right place on the paper. Droplets in a particular printer have mass 1.1×10^{-10} kg, charge 2.1 pC, speed 12 m/s, and pass through a uniform 97-kN/C electric field in order to be deflected through a $10°$ angle. What is the length of the field region?

78. If the drop speed in the printer of Problem 77 is doubled, what should be done to the electric field to have the drops hit the same point on the paper?

Supplementary Problems

79. A charge $-q$ and a charge $\tfrac{4}{9}q$ are located a distance a apart, as shown in Fig. 23-51. Where would you place a third charge so that all three are in static equilibrium? What should be the sign and magnitude of the third charge?

80. Two 34-μC charges are attached to the opposite ends of a spring of spring constant 150 N/m and equilibrium length 50 cm. By how much does the spring stretch?

FIGURE 23-51 Problem 79.

81. A 3.8-g particle with a 4.0-μC charge experiences a downward force of 0.24 N in a uniform electric field. Find the electric field, assuming that the gravitational force is *not* negligible.

82. A rod of length 2ℓ lies on the x axis, centered on the origin. It carries a line charge density given by $\lambda = \lambda_0 \dfrac{x}{\ell}$, where λ_0 is a constant. (a) What is the net charge on the rod? (b) Find an expression for electric field strength at all points $x > \ell$. (c) Show that your result has the $1/x^3$ dependence of a dipole field when $x \gg \ell$. *Hint:* For $\ell \ll x$, $\ln\left(\dfrac{x - \ell}{x + \ell}\right)$ becomes approximately $-\dfrac{2\ell}{x} - \dfrac{2\ell^3}{3x^3}$. (d) By comparing with Equation 23-5b, determine the dipole moment of the rod.

83. The electric field on the axis of a uniformly charged ring has magnitude 380 kN/C at a point 5.0 cm from the ring center. The magnitude 15 cm from the center is 160 kN/C; in both cases the field points away from the ring. Find the radius and charge of the ring.

84. Use the binomial theorem to show that, for $x \gg R$, the result of Problem 50 reduces to the field of a point charge whose total charge is the charge density σ times the disk area.

85. The dipole moment of a water molecule is 6.2×10^{-30} C·m. A water molecule is located 1.5 nm from a proton, with its dipole moment vector aligned as shown in Fig. 23-52. (a) Use Equation 23-5b to find the force the molecule exerts on the proton. (b) Now find the net force on the dipole in the proton's nonuniform electric field by considering that the dipole consists of two opposite charges q separated by a distance d, such that $qd = 6.2 \times 10^{-30}$ C·m. Take the limit as d becomes very small, and show that the force has the same magnitude as that of part (a), as required by Newton's third law.

FIGURE 23-52 Problem 85.

86. An *electric quadrupole* consists of two oppositely directed dipoles in close proximity. (a) Calculate the field of the quadrupole shown in Fig. 23-53 for points to the right of $x = a$, and (b) show that for $x \gg a$ the quadrupole field falls off as $1/x^4$.

87. Derive Equation 23-7 in Example 23-9 by making θ the integration variable, then evaluating the resulting integral.

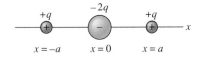

FIGURE 23-53 Problem 86.

CHAPTER 24

GAUSS'S LAW

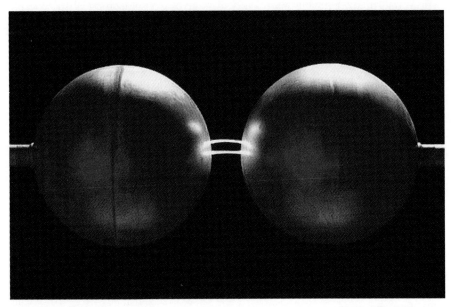

Sparks jump between two highly charged conducting spheres. Gauss's law determines the distribution of charge on these and other conductors.

We've seen how it's possible, in principle, to calculate the electric field of any charge distribution by summing the contributions of the many individual charges comprising the distribution. But in practice that process involves a vector integration that becomes difficult for all but the simplest charge distributions. How can we hope to calculate the field of a solid ball of charge, for example, when the individual charge elements are varying distances from the field point and their field vectors point in different directions (Fig. 24-1)?

In this chapter we introduce an elegant way of describing electric fields that makes almost trivial the calculation of fields from certain charge distributions. In the process we will formulate one of the four fundamental laws of electromagnetism—a statement that is equivalent to Coulomb's law but that gives deeper insights into the behavior of the electric field.

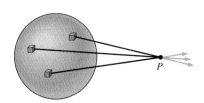

FIGURE 24-1 It would be difficult to find the field of a charged ball by summing vectorially the contributions of all the individual charge elements, three of which are shown here.

FIGURE 24-2 Electric field lines of an isolated positive point charge. The lines spread farther apart with increasing distance from the charge.

24-1 ELECTRIC FIELD LINES

The electric field is a set of vectors defined at all points in space, and we've therefore been representing fields by drawing a number of field vectors. A simpler way to visualize electric fields is with **electric field lines,** continuous lines whose direction is everywhere that of the electric field. To draw a field line, start at some point, and determine the field direction there. Move a small distance in the direction of the field, and evaluate the field direction at the new point. Extending this process in both directions from the starting point traces out an electric field line. The resulting line is a path whose direction at any point is that of the electric field at that point. Drawing many such lines gives a visualization of the overall field structure.

Tracing the field lines of a point charge is particularly simple. Starting at any point near a positive point charge, we find field vectors pointing radially outward from the charge. Move a little way outward, and the field still points in the same direction. So the field lines are straight lines, starting at the point charge extending radially outward indefinitely (Fig. 24-2).

Field lines show the direction of the field, but what about its magnitude? In Fig. 24-2 the field lines spread apart as they extend farther from the point charge. Coulomb's law tells us that the field weakens farther from the charge. So in Fig. 24-2 the field is stronger where the lines are closer and weaker where they're farther apart. This qualitative statement is always true of electric field lines, and allows us to infer relative field strength as well as field direction from field line pictures.

The relation between field strength and number of field lines is in fact quantitative. Fig. 24-3 shows a point charge field and two concentric spheres surrounding the point charge. The same number of field lines crosses the surface of each sphere, and the lines are perpendicular to the spherical surfaces. The larger sphere has twice the radius and therefore four times the surface area as the smaller one. Therefore the number of field lines *per unit area* crossing the outer sphere's surface is one-fourth that of the inner sphere. This is just the decrease in field strength given by the $1/r^2$ dependence in Coulomb's law. The number of field lines per unit area crossing a surface in the electric field is therefore proportional to the field strength. Remember in looking at two-dimensional pictures that it's the number of lines per unit *area* that counts, and that you need to consider that field lines generally spread in all three dimensions.

In tracing the field lines of charge distributions we must add vectorially the contributions from all the charges comprising the distribution. Usually the direction of the field varies as we move along a field line, so the line itself is curved (Fig. 24-4). Nevertheless, the number of field lines per unit area at right angles to the field remains proportional to the strength of the field—a fact that is ultimately grounded in Coulomb's law and the superposition principle.

You might argue that "number of field lines" is vague because we can always draw as many field lines as we want. To make the field-line picture more precise, we associate a fixed number of field lines with a charge of given magnitude. In Fig. 24-5, for example, eight field lines correspond to a charge of magnitude q. Then eight lines *begin* on the *positive* charge $+q$ (Fig. 24-5a), and 16 on $+2q$ (Fig. 24-5b). Eight lines *end* on the *negative* charge $-q$ in Fig. 24-5c. Figures 24-5d–f show the fields of some two-charge distributions, drawn with the same eight-line convention. Note that field lines always *begin* on *positive* charges and *end* on *negative* charges.

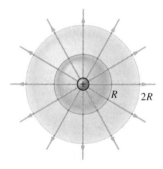

FIGURE 24-3 The outer sphere has twice the radius of the inner sphere. Since the area of a sphere is $4\pi R^2$, the outer sphere's area is four times that of the inner sphere. The same number of field lines crosses each, so the number of field lines per unit area crossing the outer sphere is one-fourth that of the inner sphere. The field strength decreases in the same way, so the number of field lines crossing a unit area is proportional to the field strength.

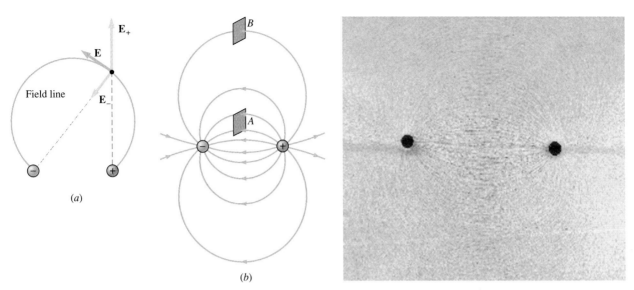

FIGURE 24-4 (*a*) Tracing a field line for an electric dipole. At each point the direction of the field line is that of the *net* electric field $\mathbf{E} = \mathbf{E}_+ + \mathbf{E}_-$. (*b*) Tracing several field lines gives an overall sense of the dipole field. Near each charge the field has the radial structure of a point-charge field, but farther away the influence of both charges becomes important and the field lines curve. Field strengh is proportional to the number of field lines per unit area crossing perpendicular to the field, so the field at *A* is stronger than at *B*. (In three dimensions the dropoff in field strength would be more dramatic.) (*c*) Field line pattern made visible by floating small fibers on a liquid in which two opposite charges are immersed.

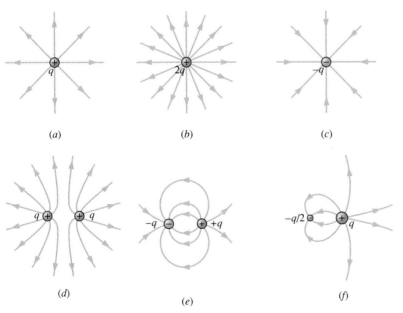

FIGURE 24-5 Field lines for (*a*) a positive charge *q*, (*b*) 2*q*; (*c*) −*q*, (*d*) two identical charges *q*, (*e*) a dipole, consisting of two equal but opposite charges ±*q*, and (*f*) opposite and unequal charges *q* and −*q*/2. In each drawing eight lines are used to represent a charge of magnitude *q*. Note in (*d*) and (*f*) that the field at large distances begins to resemble that of a single point charge.

FIGURE 24-6 In all cases, the number of field lines emerging from a closed surface is proportional to the charge enclosed.

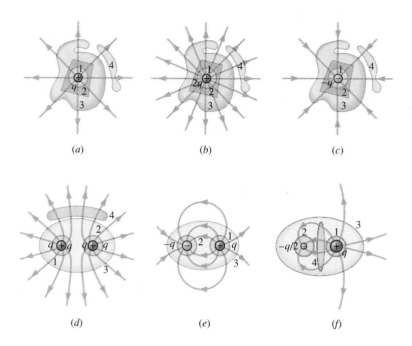

(a)   (b)   (c)

(d)   (e)   (f)

24-2 ELECTRIC FLUX

Counting Field Lines

Figure 24-6 shows the charge distributions of Fig. 24-5, each surrounded by several surfaces. (The figure shows only the two-dimensional cross section of each surface.) Each surface is closed, meaning it's impossible to get from inside to outside without crossing the surface. We now ask a simple question: How many field lines emerge from inside each surface?

In Fig. 24-6a the answer for surfaces 1 and 2 is obvious: eight. With surface 3 one field line crosses three times, twice going out and once going in. If we count a field line going inward as negative, then the algebraic sum of field lines is again eight. In fact, any *closed* surface you can draw around $+q$ will have eight field lines emerging from within. That's because eight lines begin on the charge and extend indefinitely outward, so they cross *any* closed surface surrounding the charge.

What about surface 4? Two lines cross going inward and two going outward, so the net number of field lines emerging from this surface is zero. What's different about surface 4 is that it doesn't enclose the charge. By drawing other surfaces you can convince yourself that any surface not enclosing the charge will have as many lines going in as out, and will therefore have zero net field lines emerging.

Figure 24-6b is identical except that now those surfaces enclosing the charge have 16 field lines emerging, reflecting the greater magnitude of the charge. Surfaces that don't enclose the charge still have zero net field lines emerging. Figure 24-6c is also like Fig. 24-6a, except that now the charge is negative so all field lines point inward. According to our sign convention, -8 field lines emerge from any surface enclosing the charge $-q$.

In Fig. 24-6d, surfaces 1 and 2 each enclose one of the charges q, and each has eight field lines emerging. Surface 3 encloses *both* charges, for a total enclosed charge $+2q$, and has 16 field lines emerging. Finally, surface 4 encloses no charge and has zero net field lines emerging.

On to Fig. 24-6e, the dipole. Surface 1 encloses charge q and has eight field lines emerging. Surface 2 encloses $-q$ and has -8 field lines emerging. Surface 3 encloses both $+q$ and $-q$, giving zero net charge enclosed. And as many field lines enter surface 3 as leave it, giving zero net field lines emerging.

Finally, in Fig. 24-6f, eight field lines emerge from surface 1—and that surface encloses $+q$. Surface 2 encloses $-q/2$, and has -4 field lines emerging. Surface 3 encloses both charges, for a net enclosed charge $+q/2$—and four field lines emerge from this surface. Surface 4 encloses no charge and has zero net field lines emerging.

Counting the field lines in Fig. 24-6 leads to a simple statement about how electric fields must behave:

> **The number of electric field lines emerging from any closed surface is proportional to the charge enclosed.**

This statement is very general: It doesn't matter what shape the surface is or whether the enclosed charge is a single point charge or a lot of charges adding to the same net charge. Nor does it matter how the charges are arranged, as long as they're *enclosed* by the surface in question. And the presence of charges *outside* the surface doesn't alter the conclusion about the number of field lines emerging—even though it may alter the shape of the individual lines.

We'll now rephrase our statement in a more mathematically rigorous way, obtaining one of the four fundamental laws of electromagnetism.

TIP **Remember Fig. 24-6** As we define new terms and write equations involving integrals, remember that the mathematics just reflects in a concise way the truth that's so obvious in Fig. 24-6—that the number of field lines emerging from a closed surface depends only on the net charge enclosed. Go back to that figure any time you begin to lose the physical significance of the mathematics.

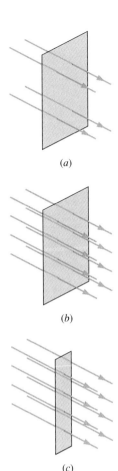

(a)

(b)

(c)

FIGURE 24-7 (a) Four field lines cross the surface shown. (b) Here the field strength has doubled, and eight lines cross the surface. (c) With the same field strength as in (b), but with half the area, only four lines cross the surface. In general, the number of field lines crossing a surface is proportional to the surface area and to the field strength.

Electric Flux

We can make the "number of field lines" more rigorous with Fig. 24-7, which shows several flat surfaces in uniform electric fields. Study the figure and its caption, and you'll see that the number of field lines crossing each surface is proportional to the surface area A and the field strength E. Figure 24-8 shows that it also depends on the orientation of the surface, specified by a vector normal to the surface. As the figure suggests, the number of field lines crossing the surface is proportional to the cosine of the angle between that normal vector **A** and the field **E**. Putting this all together, we have

$$\text{Number of field lines} \propto E\, A \cos\theta.$$

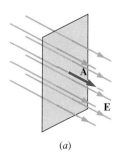

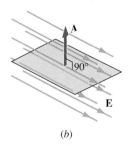

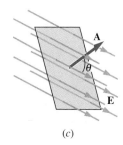

(a)   (b)   (c)

FIGURE 24-8 The number of field lines crossing a surface also depends on the surface orientation relative to the field **E**. The orientation is specified by a vector **A** normal to the surface. In (a) the surface is perpendicular to the field, so **A** and **E** are parallel; then $\theta = 0$, $\cos\theta = 1$, and the number of field lines crossing the surface is a maximum. In (b) the surface is parallel to the field, so **A** and **E** are perpendicular; here $\theta = 90°$, $\cos\theta = 0$, and no lines cross the surface. (c) In general, the number of field lines varies as $\cos\theta$, where θ is the angle between **E** and **A**.

The quantity on the right-hand side of this equation has a definite value that captures the spirit of the more vague expression "number of field lines crossing a surface." We call this quantity the **electric flux,** ϕ, through the surface. If we make the magnitude of the surface normal vector **A** equal to the surface area A, then we can define the flux compactly using the vector dot product:

$$\phi = \mathbf{E} \cdot \mathbf{A}, \tag{24-1}$$

where the dot product, defined in Chapter 7, is the product of the two vector magnitudes with the cosine of the angle between them. Since the units of **E** are N/C, flux is measured in N·m²/C.

For the open surfaces of Fig. 24-7 and 24-8 there's an ambiguity in the sign of ϕ, since we could have taken **A** in either of the two directions along the perpendicular to the surface. But for *closed* surfaces, we unambiguously define the direction of **A** as that of the outward-pointing normal to the surface.

> **TIP** **The Flux Is Not the Field** The flux ϕ and field **E** are related but distinct quantities. The field is a vector defined at each point in space. Flux is a global property of the field, depending not on a single point but on how the field behaves over an extended surface. Unlike field, flux is a scalar quantity; it's simply a quantification of the "number of field lines crossing a surface."

What if a surface is curved and/or the field varies with position? Then we divide the surface into many small patches, each small enough that it's essentially flat and that the field is essentially uniform over each. If a patch has area dA, then Equation 24-1 gives the flux through it:

$$d\phi = \mathbf{E} \cdot d\mathbf{A},$$

where the vector $d\mathbf{A}$ is normal to the patch (Fig. 24-9). The total flux through the surface is then the sum over all the patches. If we make the patches arbitrarily small that sum becomes an integral, and the flux is

$$\phi = \int_{\text{surface}} \mathbf{E} \cdot d\mathbf{A}. \tag{24-2}$$

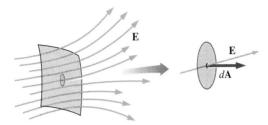

FIGURE 24-9 Even though the surface is curved and the field varies, a small enough patch of surface is essentially flat and the field is uniform over it, so the flux through the patch is $d\phi = \mathbf{E} \cdot d\mathbf{A}$.

The limits of the integral range over the entire surface, picking up contributions from all the patches $d\mathbf{A}$. Although the integral can be difficult to evaluate, we'll find it most useful in cases where its evaluation is almost trivial. Again, remember what Equation 24-2 means: The flux ϕ simply serves as a more precise measure of the "number of field lines crossing a surface."

24-3 GAUSS'S LAW

We showed in the preceding section that the number of field lines emerging from a closed surface is proportional to the charge enclosed. Now that we've developed electric flux to express more rigorously the notion "number of field lines," we can state the following:

The electric flux through any closed surface is proportional to the charge enclosed by that surface.

Writing the same thing mathematically gives

$$\phi = \oint \mathbf{E} \cdot d\mathbf{A} \propto q_{\text{enclosed}},$$

where the circle on the integral sign indicates that the integral is over a *closed* surface.

To evaluate the proportionality between flux and enclosed charge, consider a positive point charge q and a spherical surface of radius r centered on the charge (Fig. 24-10a). The flux through this surface is given by Equation 24-2:

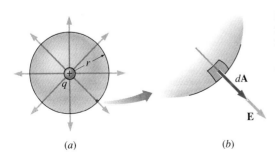

(a)

(b)

FIGURE 24-10 (a) The electric field of a point charge has the same magnitude everywhere on a spherical surface centered on the charge. (b) At each point on the surface the field and the surface normal $d\mathbf{A}$ are parallel.

$$\phi = \oint \mathbf{E} \cdot d\mathbf{A} = \oint E \, dA \cos \theta.$$

But Fig. 24-10b shows that the surface normal $d\mathbf{A}$ and the electric field $\mathbf{E}$ are parallel at any point on the sphere, so $\cos \theta = 1$. Since the electric field varies as $1/r^2$ its magnitude is the same everywhere at the fixed radius r of our sphere. Thus, we can take E outside the integral, giving

$$\phi = \oint_{\text{sphere}} E \, dA = E \oint_{\text{sphere}} dA = E(4\pi r^2),$$

where the last step follows because $\oint dA$ just means the surface area of the sphere. Now, the electric field of a point charge is given by Equation 23-3: $E = kq/r^2$. Before using this expression, we introduce the so-called permittivity constant, ε_0, defined by the relation

$$k = \frac{1}{4\pi \varepsilon_0}. \tag{24-3}$$

There's no new physics here; it's just that ε_0 will prove more convenient mathematically than k. The constant ε_0 has the value 8.85×10^{-12} C^2/N·m^2. With the definition 24-3, the electric field of the point charge q becomes $E = q/4\pi \varepsilon_0 r^2$, and our equation for flux reads

$$\phi = E(4\pi r^2) = \left(\frac{q}{4\pi \varepsilon_0 r^2}\right)(4\pi r^2) = \frac{q}{\varepsilon_0}.$$

So the proportionality constant between the flux ϕ through a closed surface and the enclosed charge is just $1/\varepsilon_0$, and our statement relating flux and charge becomes

$$\oint \mathbf{E} \cdot d\mathbf{A} = \frac{q}{\varepsilon_0}, \qquad \text{(Gauss's law)} \tag{24-4}$$

where the integral is taken over *any closed surface* that *encloses* the charge q.

Equation 24-4 is **Gauss's law,** and is one of four fundamental relations that govern the behavior of electromagnetic fields throughout the universe. Whether you journey into a star in some remote galaxy, down among the strands of a DNA molecule, or into the microprocessor chip at the heart of your computer, you will find that the flux of the electric field through any closed surface depends only on the enclosed charge. In over a century of experiments, no electric field has ever been observed to violate Gauss's law.

We stress that Gauss's law, although clothed in the mathematical finery of a surface integral, is just a more rigorous way of saying what's obvious in Fig. 24-6: that the number of field lines emerging from a closed surface is proportional to the enclosed charge. This, in turn, is true because electric field lines don't begin or end in empty space, but only on point charges—a fact that reflects the inverse-square nature of the electric force.

24-4 USING GAUSS'S LAW

Gauss's law is true for *any* surface enclosing *any* charge distribution. When the charge distribution has sufficient symmetry we can choose a surface—called a **gaussian surface**—over which evaluation of the flux integral becomes simple. Then Gauss's law allows us to calculate the field far more easily than we could using Coulomb's law and superposition. We now illustrate the use of Gauss's law for three important symmetries.

Spherical Symmetry

A charge distribution is spherically symmetric if the charge density depends only on the distance from a central point. A point charge, a uniformly charged solid sphere, and a spherical surface carrying uniform surface charge density are all spherically symmetric charge distributions. Spherical symmetry implies that the magnitude of the electric field depends only on the distance r from the center of symmetry, and that the field direction is radial (Fig. 24-11).

Gauss's law applies to *any* surface we might draw around the spherically symmetric charge distribution. But the most useful surface is a sphere of arbitrary radius r centered on the center of symmetry. Then the magnitude E is the same at all points on this gaussian surface. Furthermore, **E** is everywhere in the same direction as the perpendicular to the surface, so $\cos\theta = 1$. Thus, although the direction of **E** and of the surface vary, the product $E\cos\theta$ remains constant and is simply the field magnitude E. Then the flux through our gaussian sphere becomes

$$\phi = \oint \mathbf{E} \cdot d\mathbf{A} = \oint E\cos\theta \, dA = E \oint dA = 4\pi r^2 E, \qquad (24\text{-}5)$$

where the last step follows because $\oint dA$ is just the surface area of the sphere, $4\pi r^2$. This expression for the flux does not depend on the details of the charge distribution, so long as it is spherically symmetric.

Gauss's law says that the flux through the sphere is given by q/ε_0, where q is the net charge *enclosed* by the sphere. Suppose our spherically symmetric charge distribution carries total charge Q and has radius R. That is, whatever the particular distribution of charge for $r \leq R$, there is no charge at $r > R$. For any gaussian sphere with $r > R$, like the surface 1 in Fig. 24-12, the enclosed charge is the total charge Q. Equating the flux in Equation 24-5 to Q/ε_0 gives

$$4\pi r^2 E = \frac{Q}{\varepsilon_0},$$

or

$$E = \frac{1}{4\pi\varepsilon_0}\frac{Q}{r^2}. \quad (r > R) \qquad (24\text{-}6)$$

This is just the field of a point charge! (Recall that $1/4\pi\varepsilon_0$ is the coulomb constant, k.) Equation 24-6 says that *outside* any spherically symmetric distribution of charge, the field is identical to that of a point charge located at the center of symmetry (Fig. 24-13). This is *not* an approximation—it is exactly true right up to the surface $r = R$! Imagine how hard it would have been to calculate this

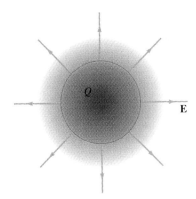

FIGURE 24-11 For a spherically symmetric charge distribution, the field vectors at a given radius all have the same magnitude and point in the radial direction—outward for a positive charge, as shown, or inward for a negative charge.

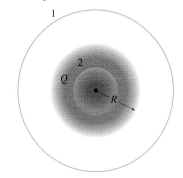

FIGURE 24-12 Two gaussian surfaces surrounding a spherical charge distribution. Surface 1 lies outside the distribution, and encloses all the charge Q. Surface 2 lies inside the distribution, and encloses only part of the charge.

FIGURE 24-13 Hairs on a highly charged person's head trace out the electric field. The essentially spherical head produces a field like that of a point charge.

field using the superposition principle! Yet somehow all the charge elements throughout the spherically symmetric distribution produce $d\mathbf{E}$'s that add vectorially to give the same field as a single point charge. Like Gauss's law itself, this result is a manifestation of the inverse-square force law.

The field *inside* the charge distribution depends on how charge is distributed. This is because a gaussian sphere with $r < R$, such as surface 2 of Fig. 24-12, does not enclose the entire charge Q. How much it encloses depends on the charge distribution. In the examples below, we consider two important special cases.

● **EXAMPLE 24-1**   A UNIFORMLY CHARGED SPHERE

A total charge Q is spread uniformly throughout a sphere of radius R. What is the electric field at all points in space?

Solution
This charge distribution is spherically symmetric, so the field for $r > R$ is like that of a point charge, given by Equation 24-6.

Inside the sphere, Equation 24-5 for the flux still holds, but now the charge enclosed is some fraction of Q. What fraction? The volume of the sphere is $\frac{4}{3}\pi R^3$, and it contains a total charge Q. Since charge is spread uniformly throughout the sphere, the volume charge density is given by

$$\rho = \frac{Q}{V} = \frac{Q}{\frac{4}{3}\pi R^3}.$$

The charge enclosed by a sphere of radius r is just the volume of that sphere multiplied by the volume charge density:

$$q_{\text{enclosed}} = V\rho = \frac{4}{3}\pi r^3 \frac{Q}{\frac{4}{3}\pi R^3} = Q\frac{r^3}{R^3}.$$

Equating the flux from Equation 24-5 to $q_{\text{enclosed}}/\varepsilon_0$, we have

$$4\pi r^2 E = \frac{Qr^3}{\varepsilon_0 R^3},$$

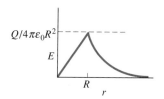

FIGURE 24-14 Field strength versus radial distance for a uniformly charged sphere of radius R. For $r > R$ the field has the inverse-square dependence of a point-charge field.

or

$$E = \frac{1}{4\pi\varepsilon_0}\frac{Qr}{R^3} = \frac{\rho r}{3\varepsilon_0}, \qquad (r < R) \qquad (24\text{-}7)$$

where we've written the field in terms of both the total charge Q and the charge density $\rho = Q/\frac{4}{3}\pi R^3$. Equation 24-7 shows that the field *inside* the charge distribution increases linearly with distance from the center. This result is entirely consistent with the inverse-square law for point charges. Although the field of each charge element decreases as $1/r^2$, in this case the amount of charge enclosed increases more rapidly—as r^3—resulting in a field that increases linearly with r. Figure 24-14 shows the combined results for the fields both inside and outside the sphere. The field direction is, of course, radial, pointing outward if Q is positive and inward if Q is negative.   ●

● **EXAMPLE 24-2**   A THIN SPHERICAL SHELL

A thin spherical shell of radius R carries a total charge Q distributed uniformly over its surface. What is the electric field inside and outside the shell?

Solution
Since this distribution is spherically symmetric, we already know that the field outside is just the point-charge field of Equation 24-6.

For any gaussian sphere inside the shell, the enclosed charge

is zero (Fig. 24-15). Equating the flux from Equation 24-5 to this zero enclosed charge gives

$$4\pi r^2 E = 0$$

so the field is zero everywhere inside the shell! How can this be? Again, it's a manifestation of the inverse-square law. At any point inside the shell, the larger fields of nearby portions of the shell are exactly canceled by the weaker fields of more distant, but more extensive, parts of the shell (Fig. 24-16).

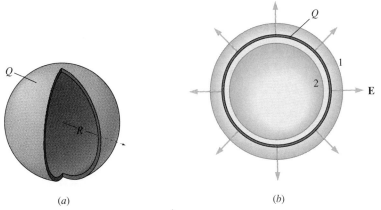

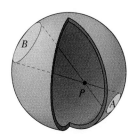

FIGURE 24-16 At any point P inside a charged shell, the field arising from the relatively few but nearby charges in region A is exactly canceled by the field arising from the more numerous but more distant charges in region B.

FIGURE 24-15 (a) A thin spherical shell carries charge Q distributed uniformly over its surface of radius R. (b) Cross-sectional view, showing two spherical gaussian surfaces. Surface 1 encloses the entire charge Q, while surface 2 encloses zero charge.

● **EXAMPLE 24-3**   A POINT CHARGE INSIDE A SHELL

A point charge $+q$ is at the center of a spherical shell of radius R carrying total charge $-2q$, distributed uniformly over its surface. (a) Draw the electric field lines for this configuration, using eight lines to represent a charge of magnitude q. Find expressions for the field strength for (b) $r < R$ and (c) $r > R$.

Solution
(a) The situation has spherical symmetry and so must the field. Gauss's law tells us that eight field lines must emerge from any surface surrounding $+q$ alone, and that eight field lines must go into any surface surrounding the entire distribution with its net charge $-q$. Figure 24-17 shows the only way to draw the field that's compatible with Gauss's law. Notice that a total of 16 lines end on the charged shell, consistent with its charge of $-2q$.

(b) For $r < R$ we're inside the shell, so the enclosed charge is just q. Solving Equation 24-4 for E then gives

$$E = \frac{q}{4\pi\varepsilon_0 r^2}.$$

What about the shell? Didn't we forget to take it into account? No! The shell and its charge are *irrelevant* as long as they preserve the spherical symmetry. Example 24-2 showed that the field inside a charged shell *due to the shell itself* is zero. Here we're inside the shell, so the only field we see is that of the point charge.

(c) Outside the shell, a spherical gaussian surface encloses net charge $-q$, so the field we see is that of a point charge $-q$; it has magnitude $E = q/4\pi\varepsilon_0 r^2$, and points radially inward.

What about the field at $r = R$? That's ambiguous; just inside the shell the field points outward, while just outside it points inward. In fact, the field undergoes a discontinuous jump across the infinitesimally thin surface charge layer on the shell.

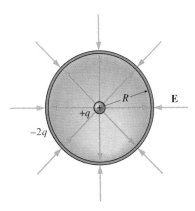

FIGURE 24-17 Field lines for a charge q surrounded by shell carrying $-2q$ (Example 24-3).

TIP   Trust Gauss Gauss's law says that the flux—or number of field lines—emerging from any closed surface depends *only* on the *enclosed* charge—not on any other charge that may happen to be outside the surface. (Consult Fig. 24-6 to convince yourself of this.) *When there is enough symmetry* that often means that external charges are totally irrelevant in a field calculation, as is the shell in Example 24-3 for points with $r < R$. But *symmetry matters*: Without enough symmetry, zero net charge inside a gaussian surface is *not* sufficient to ensure that the field on the surface is zero (Fig. 24-18).

EXERCISE A solid sphere 10 cm in radius carries a uniform 40-μC charge distributed throughout its volume. It is surrounded by a concentric shell 20 cm in radius, also uniformly charged with 40 μC. Find the electric field (a) 5.0 cm, (b) 15 cm, and (c) 30 cm from the center.

Answers: (a) 18 MN/C; (b) 16 MN/C; (c) 8.0 MN/C

Some problems similar to Examples 24-1 through 24-3: 17–26, 66

FIGURE 24-18 The net charge enclosed by the gaussian sphere (gray) is zero, but the field on the sphere is not zero. Here the charge distribution—a dipole—is not spherically symmetric, so Equation 24-5 is not a valid expression for the flux.

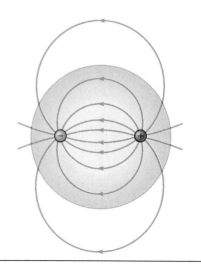

> **TIP** **Using Gauss's Law** This section illustrated the steps needed to calculate the electric field using Gauss's law:
>
> 1. Study the symmetry to see if you can construct a gaussian surface on which the field magnitude and its direction relative to the surface are constant. (With spherical symmetry that surface was spherical.) If you can't find such a surface then Gauss's law, while still true, won't help in calculating the field.
> 2. Evaluate the flux. This should be easy because your choice of gaussian surface makes $E \cos\theta$ constant over the surface. This term then comes outside the integral, leaving the integral equal to the surface area.
> 3. Evaluate the *enclosed* charge. This is not the same as the total charge if the gaussian surface lies *within* the charge distribution.
> 4. Equate the flux to $q_{\text{enclosed}}/\varepsilon_0$, and solve for E. The direction of **E** should be evident from the symmetry.
>
> Once steps 1 and 2 are done for a particular symmetry, you'll have an equation like Equation 24-5 for the flux, and you can jump right to step 3 to calculate the field in a specific case—just as we did in Examples 24-1 and 24-2.

Line Symmetry

A charge distribution has line symmetry when its charge density depends only on the perpendicular distance r from a line, called the symmetry axis (Fig. 24-19). Symmetry then requires that the field point radially and that the field magnitude depend only on distance from the axis. An appropriate gaussian surface is the cylinder of length ℓ and radius r shown in Fig. 24-19. Being radial, field lines don't cross the circular ends of the cylinder, so there's no flux through these ends. The field is everywhere perpendicular to the curved part of the surface, so

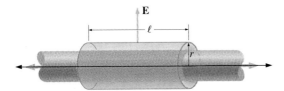

FIGURE 24-19   A positive charge distribution with line symmetry (red) extends infinitely in both directions, and its charge density depends only on the perpendicular distance from the symmetry axis. A cylindrical gaussian surface (gray) of length ℓ and radius r surrounds the charge distribution. Symmetry requires that the electric field on the surface point radially outward, and that the field magnitude E be constant over the surface. With a negative charge the field would point radially inward.

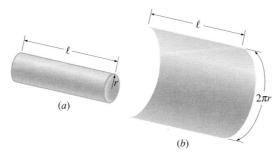

FIGURE 24-20   (a) A cylindrical gaussian surface. (b) Unrolling the cylinder gives a flat sheet of area $2\pi r \ell$.

$\cos\theta = 1$ in the expression for flux through this part. Since the field magnitude is constant at the fixed radius r of the curved surface, the flux becomes

$$\phi = \int \mathbf{E} \cdot d\mathbf{A} = \int E \, dA \cos\theta = E \int dA = 2\pi r \ell E, \qquad (24\text{-}8)$$

where the last step follows because the cylinder "unrolls" into a rectangular sheet of length ℓ and width $2\pi r$ (Fig. 24-20). Since there is no flux through the cylinder ends, Equation 24-8 gives the total flux through our entire gaussian surface. Solving for the electric field in any situation with line symmetry then amounts simply to equating the flux given in Equation 24-8 to the enclosed charge divided by ε_0, then solving for E—exactly as we did for spherical symmetry using the analogous flux equation, Equation 24-5.

● **EXAMPLE 24-4**   AN INFINITE LINE OF CHARGE

Use Gauss's law to calculate the electric field of an infinite line carrying line charge density λ.

Solution

This is the same problem we solved in Example 23-9 through a tedious Coulomb's law calculation. With line charge density λ, the charge enclosed by a gaussian cylinder of length ℓ is $q_{\text{enclosed}} = \lambda \ell$. Setting the flux given in Equation 24-8 to $q_{\text{enclosed}}/\varepsilon_0$ and solving for E then gives

$$E = \frac{\lambda \ell}{2\pi \varepsilon_0 r \ell} = \frac{\lambda}{2\pi \varepsilon_0 r}.$$

Since $1/2\pi \varepsilon_0 = 2k$, this is the same result we found in Example 23-9. The Gauss's law calculation is far simpler; symmetry and an intelligent choice of gaussian surface helped us bypass the entire integration of Example 23-9.

Although this problem dealt with an infinitesimally thin charged line, you can easily convince yourself that the same result must hold *outside* any charge distribution with line symmetry. And, as we argued in Example 23-9, the result is a good approximation for long, thin structures of finite length provided we're not too far away nor too close to the ends.

● **EXAMPLE 24-5** A HOLLOW PIPE

A thin-walled pipe 3.0 m long and 2.0 cm in radius carries a net charge $q = 5.7 \ \mu C$, distributed uniformly over its surface. Find the electric field (a) 8.0 mm and (b) 8.0 cm from the pipe axis, not near either end.

Solution

Since the pipe is much longer than its diameter, we can approximate its field as that of an infinitely long charge distribution with line symmetry.

(a) A point 8.0 mm from the axis lies inside the pipe. An 8.0-mm-radius gaussian cylinder therefore encloses zero net charge; equating the flux from Equation 24-8 to zero then shows that the field at this radius—and indeed anywhere deep within the hollow pipe—is zero.

(b) For a point outside the pipe, a gaussian cylinder of length ℓ encloses charge $\lambda \ell$, where the line charge density λ is $5.7 \ \mu C / 3.0$ m $= 1.9 \ \mu C/m$. Setting the flux from Equation 24-8 to this $q_{enclosed}$ and solving for E then gives

$$E = \frac{q_{enclosed}}{2\pi \varepsilon_0 r \ell} = \frac{\lambda}{2\pi \varepsilon_0 r}$$

$$= \frac{1.9 \times 10^{-6} \ \text{C/m}}{(2\pi)(8.85 \times 10^{-12} \ \text{C}^2/\text{N·m}^2)(8.0 \times 10^{-2} \ \text{m})}$$

$$= 4.3 \times 10^5 \ \text{N/C}.$$

In this region the field points radially outward and falls inversely with distance from the axis.

TIP That's Distance from the Symmetry Axis The distance r that arises in applying Gauss's law to situations with spherical or line symmetry is always the distance *from the point or line of symmetry*—as you can see from the derivations of Equations 24-4 and 24-8. Equations for the field are therefore most simply expressed in terms of that distance—*not* in terms of distance from the edge of the charge distribution.

EXERCISE Suppose the pipe in Example 24-5 were surrounded concentrically by a second pipe of the same length and 5.0 cm in diameter. What should be (a) the total charge and (b) the surface charge density (assumed uniform) on this outer pipe in order that there be no electric field outside the entire structure?

Answers: (a) $-5.7 \ \mu C$; (b) $-12 \ \mu C/m^2$

Some problems similar to Examples 24-4 and 24-5: 28, 29, 31, 32

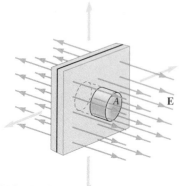

FIGURE 24-21 A charge distribution with plane symmetry. The charge density depends only on the distance from the plane of symmetry (black), and extends infinitely in both directions parallel to that plane. Also shown are the electric field and a gaussian surface.

Plane Symmetry

A charge distribution has plane symmetry when its charge density depends only on the distance from a plane. The only electric-field direction consistent with this symmetry is perpendicular to the symmetry plane (Fig. 24-21). We can evaluate the flux integral in Gauss's law using a gaussian surface whose sides are perpendicular to the symmetry plane and whose ends are parallel to it, as shown in Fig. 24-21. Our surface straddles the symmetry plane, extending equal distances on either side of it. Since no field lines cross the sides, the flux through them is zero. Symmetry of the situation implies that the field magnitude E cannot depend on position parallel to the symmetry plane. Therefore E is uniform over each end of the gaussian cylinder so, with the field perpendicular to the ends, the flux through each end is just EA, where A is the end area. Since the ends are the same distance from the symmetry plane, they must have the same field strength E. The total flux through the gaussian surface is therefore

$$\phi = 2EA. \tag{24-9}$$

This equation holds for any charge distribution with plane symmetry; to find the field we must evaluate the charge enclosed by the gaussian surface, then apply Gauss's law.

● EXAMPLE 24-6 A SHEET OF CHARGE

An infinite sheet of charge carries a uniform surface charge density σ. What is the electric field arising from this sheet?

Solution
Since the surface charge density is uniform, this charge distribution has plane symmetry. Figure 24-22 shows the sheet and an appropriate gaussian surface. The sheet area enclosed by the gaussian surface is clearly equal to the end area A. The surface charge density—charge per unit area—is σ, so the enclosed charge is $q_{enclosed} = \sigma A$. Setting $q_{enclosed}/\varepsilon_0$ to the flux $\phi = 2EA$ given by Equation 24-9 and solving for E then gives

$$E = \frac{\sigma}{2\varepsilon_0}. \tag{24-10}$$

This simple result says that the field strength does not depend on distance from the sheet. How can this be? By symmetry, the field must point perpendicular to the sheet. There is no charge anywhere but on the sheet, so that's the only place where field lines can begin or end. Therefore the density of field lines—the measure of field strength—is the same everywhere. Figure 24-23 shows how this result is fully consistent with Coulomb's law.

Although this example treated only an infinitesimally thin sheet of charge, we would find a uniform electric field *outside* any charge distribution with plane symmetry. The field *inside*

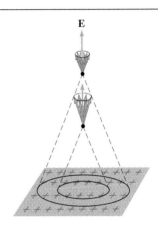

FIGURE 24-23 As we rise above an infinite sheet, the amount of charge within a given angular region increases just enough to compensate for the inverse-square decrease in field strengths of the individual charges. The result is a field that does not depend on distance from the sheet.

such a distribution would depend on how the charge density varies in the direction perpendicular to the symmetry plane (see Problem 36).

Perfect plane symmetry requires a charge distribution that is infinite in extent. But close to any large, flat, uniformly charged surface and not near an edge, the assumption of plane symmetry becomes a good approximation, and Equation 24-10 becomes reasonably accurate. We make the same approximation when we treat the acceleration of Earth's gravity as a constant; we're neglecting Earth's curvature and therefore its finite size, and approximating its gravitational field as the uniform field of an infinite sheet of mass.

EXERCISE An electron close to a large, flat sheet of charge is repelled from the sheet with a 1.8×10^{-12} N force. Find the surface charge density on the sheet.

Answer: $-200 \ \mu C/m^2$

Some problems similar to Example 24-6: 34–37

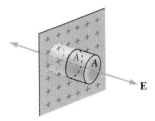

FIGURE 24-22 The area of the charged sheet enclosed by the gaussian surface is the same as the area A of its ends; the enclosed charge is therefore σA.

24-5 FIELDS OF ARBITRARY CHARGE DISTRIBUTIONS

The examples of Section 24-4 show how easy Gauss's law can sometimes make problems that would be difficult to solve using Coulomb's law. In each case the symmetry allowed us to construct a gaussian surface on which $E\cos\theta$ was constant. Only then could we take E outside the integral and solve for it. But many situations do not possess the symmetry needed to apply Gauss's law in calculating the field. Try, for example, to calculate the field of a dipole using Gauss's law. The attempt fails because it is impossible to draw an appropriate surface.

We can understand the fields associated with more complicated charge distributions by considering the fields of simpler distributions that we have already calculated using either Coulomb's law or Gauss's law. Figure 24-24 summarizes the fields of a dipole, a point charge, a uniformly charged line, and a uniformly charged plane. For the last three, note the simple relation between the number of

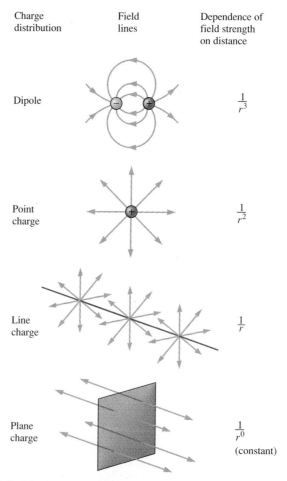

Charge distribution	Field lines	Dependence of field strength on distance
Dipole		$\frac{1}{r^3}$
Point charge		$\frac{1}{r^2}$
Line charge		$\frac{1}{r}$
Plane charge		$\frac{1}{r^0}$ (constant)

FIGURE 24-24 The fields of a dipole, a point charge, a charged line, and a charged plane.

dimensions in the charge distribution and the way the field strength depends on distance. The plane has two dimensions and its field strength is independent of distance. The line has one dimension and its field falls as $1/r$. The point has no dimensions and its field falls as $1/r^2$. In a sense, the dipole continues this progression, for it consists of two opposite point charges whose effects very nearly cancel. No wonder its field falls even faster, as $1/r^3$. In fact, one can construct a hierarchy of charge distributions whose fields fall off ever faster as dipoles nearly cancel dipoles, and so on. Such distributions are useful in the mathematical analysis of complicated charge distributions such as complex molecules or radio antennas.

Frequently we have a charge distribution that lacks the symmetry required to make Gauss's law useful and for which a Coulomb's law calculation would be impossibly difficult. Good thinking coupled with knowledge of simpler charge distributions can go a long way toward providing a reasonable approximation to the field. Consider, for example, the uniformly charged disk shown in Fig. 24-25. For points much closer to the disk surface than to the edge, the disk looks almost like an infinite flat plane of charge. For these points the field is approximately that of an infinite plane—a field that points directly away from the plane and does not fall off with distance. Far from the disk, meanwhile, its exact size and shape are unimportant. Its field closely resembles that of a single charged point: far from the disk the field points radially outward in all directions and falls off as the inverse square of the distance from the disk. The field at intermediate distances is harder to determine. But somehow the infinite-plane field lines close to the disk must connect smoothly to the point-charge field lines far away. If we sketch these in, as in Fig. 24-25, we have a rough picture of the field everywhere. Don't underestimate the value of a simple approximation like this one! It can often tell all we need to know about a situation and may provide a much clearer understanding than would a detailed calculation.

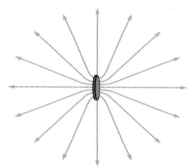

FIGURE 24-25 The field of a uniformly charged disk.

24-6 GAUSS'S LAW AND CONDUCTORS

Electrostatic Equilibrium

In the preceding chapter we defined electrical conductors as materials containing free charges—like the free electrons in metals. Figure 24-26 shows what happens when an electric field is applied to a piece of conducting material. Free charges respond to the electric force $q\mathbf{E}$ by moving—in the direction of the field if they are positive, opposite the field if negative. The resulting charge separation gives rise to an electric field within the material that is opposite to the applied field. As more charge moves this internal field becomes stronger until its magnitude eventually equals that of the applied field. At that point free charges within the conductor experience zero net force, and the conductor is in **electrostatic equilibrium.** Although individual charges continue to move about in random thermal motion, there is no longer any net motion of charge. Once equilibrium is reached the internal and applied fields are equal but opposite, and therefore,

> **The electric field is zero inside a conductor in electrostatic equilibrium.**

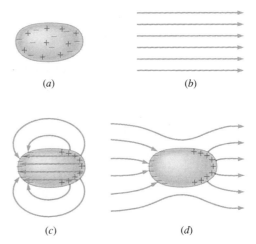

(a)

(b)

(c)

(d)

FIGURE 24-26 (a) A piece of conducting material contains positive and negative charge. (b) A uniform electric field, arising from charges outside the region shown. (c) Charges in the conductor separate in response to the field, resulting in an internal field that cancels the applied field. (d) The net field is the vector sum of the applied field and the field resulting from the redistribution of charge within the conductor. Note that field lines, as always, begin and end on charges.

It could not be otherwise: Since a conductor contains free charges, the presence of any internal electric field would result in bulk charge motion, and we would not have equilibrium. This result does not depend on the size or shape of the conductor, the magnitude or direction of the applied field, or even the nature of the material as long as it's a conductor. This ability of a conductor to cancel applied fields is the basis of shielding—the use of conductive enclosures to keep out unwanted electric fields (Fig. 24-27).

This discussion of equilibrium is a macroscopic one; it considers only overall average fields within the material. At the atomic and molecular level there are still strong electric fields near individual electrons and protons. But the *average* field,

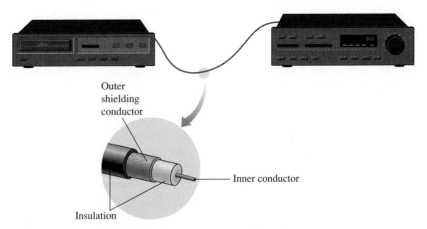

Outer shielding conductor

Inner conductor

Insulation

FIGURE 24-27 The shielded cable connecting a CD player to a stereo receiver keeps out stray electric fields that would otherwise introduce noise. Although the situation is not strictly static, the fields change slowly enough that charges in the shield can move to cancel them.

taken over distances many times the separation between individual charges, is zero in electrostatic equilibrium.

Charged Conductors

Although they contain free charges, conductors are normally electrically neutral since they include equal numbers of electrons and protons. But suppose we give a conductor a nonzero net charge, for example, by injecting excess electrons into its interior. There will be a mutual repulsion among the electrons and, because these are *excess* electrons, there is no compensating attraction from protons. We might expect, therefore, that the electrons will move as far apart as possible—namely to the surface of the conductor (Fig. 24-28). (Electrons might even leave the material—but that only occurs with very high charge densities.)

We now use Gauss's law to prove rigorously that excess charge *must* be at the surface of a conductor in electrostatic equilibrium. Figure 24-29 shows a piece of conducting material with a gaussian surface drawn just below the material surface. In equilibrium there is no electric field within the conductor, and thus the field is zero everywhere on the gaussian surface. The flux, $\oint \mathbf{E} \cdot d\mathbf{A}$, through the gaussian surface is therefore also zero. But Gauss's law says that the flux through a closed surface is proportional to the net charge enclosed, and therefore the net charge within our gaussian surface must be zero. This is true no matter where the gaussian surface is as long as it is *inside* the conductor. We can move the gaussian surface arbitrarily close to the conductor surface and it still encloses no net charge. If there is a net charge on the conductor it lies outside the gaussian surface, and therefore we conclude that

> **If a conductor in electrostatic equilibrium carries a net charge, all excess charge resides on the conductor surface.**

FIGURE 24-28 Excess charge accumulates at the surface of a charged conductor. In this elongated conductor, mutual repulsion of the excess electrons results in the greatest charge accummulation at the opposite ends.

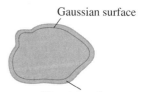

FIGURE 24-29 Since the electric field inside the conductor is zero, a gaussian surface within the conductor encloses zero net charge. Any excess charge therefore resides on the conductor surface.

● EXAMPLE 24-7 A HOLLOW CONDUCTOR

An irregularly shaped conductor has a hollow cavity, as shown in Fig. 24-30. The conductor carries a net charge of 1.0 μC. A 2.0-μC point charge is inside the cavity, not touching the conductor. Find the net charge on the cavity wall and on the outer surface of the conductor, assuming electrostatic equilibrium.

Solution

The electric field is zero everywhere within the conductor, in particular on the gaussian surface shown in Fig. 24-30. The flux through this surface is thus zero, and Gauss's law tells us that the surface therefore encloses zero net charge. But there is a 2.0-μC charge within the cavity. In order for the gaussian surface to enclose zero net charge, the cavity wall must therefore carry -2.0 μC.

Since the conductor's net charge is 1.0 μC and there is -2.0 μC on its inner wall, the outer surface of the conductor surface must carry 3.0 μC.

EXERCISE A point charge $+q$ is placed inside a hollow conducting shell carrying a net charge $-3q$. What is the total charge on the outside surface of the shell?

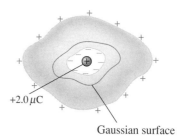

+2.0 μC

Gaussian surface

FIGURE 24-30 A conductor with a hollow cavity containing a charge. The gaussian surface shown encloses no net charge, so the charge on the inner wall must be equal but opposite that of the charge within the cavity.

Answer: $-2q$

Some problems similar to Example 24-7: 46–49

●

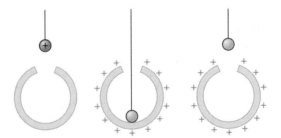

FIGURE 24-31 Experimental test of Gauss's law. When the small charged conductor contacts the interior of the hollow conductor, all its charge moves to the outside of the hollow conductor.

Experimental Tests of Gauss's Law

The fact that excess charge resides only on a conductor surface provides a very sensitive test of Gauss's law, and thus of the inverse-square law for the electric field. Figure 24-31 shows a charged conducting ball being placed inside a hollow, initially neutral conductor. When the two conductors touch, all the excess charge flows to the outer surface of the hollow conductor, leaving no net charge on the ball. In practice the experiment is often done in reverse, with an uncharged conducting ball placed within a hollow conductor. The outer conductor is then charged, and sensitive instruments used detect any charge moving to the ball. Absence of such charge motion confirms the inverse-square law. Recent experiments of this type show that the exponent 2 appearing in the inverse-square law is indeed 2 to within 3×10^{-16}. Such tests are far more sensitive than direct measurements of how the electric force varies with distance.

The Field at a Conductor Surface

There can be no electric field *within* a conductor in electrostatic equilibrium, but there *may* be a field right *at* the conductor surface. Such a field, though, cannot have a component parallel to the surface; if it did, charge would move along the surface and we would not have equilibrium. So the field at a conductor surface must be perpendicular to the surface (Fig. 24-32*a*).

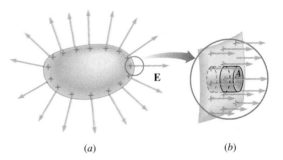

(*a*)   (*b*)

FIGURE 24-32 (*a*) The electric field at the surface of a charged conductor is perpendicular to the conductor surface. (*b*) A small gaussian surface straddles the conductor surface.

We can compute the field strength by considering a small gaussian surface that straddles the conductor surface, as shown in Fig. 24-32b. We make the gaussian surface so small that curvature of the conductor becomes negligible, and we orient the gaussian surface with its sides perpendicular and its top parallel to the conductor surface. Since the field is perpendicular to the conductor, there is then no flux through the sides of the gaussian surface. Since the field is zero inside the conductor, there is also no flux through the inner end of the gaussian surface. The only flux is through the outer end, whose area is A. Since the field is essentially uniform and perpendicular to this end, the flux is just EA. The only charge enclosed is right at the conductor surface, where it occupies the same area A. If the surface charge density is σ, then the enclosed charge is σA. Gauss's law equates the flux to $q_{enclosed}/\varepsilon_0$, so we have

$$EA = \frac{\sigma A}{\varepsilon_0}$$

or

$$E = \frac{\sigma}{\varepsilon_0}. \qquad \text{(field at conductor surface)} \qquad (24\text{-}11)$$

This result applies to any conductor in electrostatic equilibrium and shows that large electric fields develop where the charge density is high. Engineers designing electrical devices must avoid high charge densities whose associated fields could lead to sparks, arcing, and breakdown of electrical insulation.

● **EXAMPLE 24-8** EARTH'S FIELD

The Earth, which is an electrical conductor, carries a net charge of -4.3×10^5 C distributed approximately uniformly over its surface. Find the surface charge density, and use Equation 24-11 to calculate the electric field at Earth's surface.

Solution

Let Q be Earth's charge and R_E its radius (which is given inside the front cover and in Appendix E). Then the surface charge density is

$$\sigma = \frac{Q}{A} = \frac{Q}{4\pi R_E^2} = \frac{-4.3 \times 10^5 \text{ C}}{(4\pi)(6.37 \times 10^6 \text{ m})^2}$$

$$= -8.43 \times 10^{-10} \text{ C/m}^2.$$

Equation 24-11 then gives the electric field at Earth's surface:

$$E = \frac{\sigma}{\varepsilon_0} = \frac{-8.43 \times 10^{-10} \text{ C/m}^2}{8.85 \times 10^{-12} \text{ C}^2/\text{N} \cdot \text{m}^2} = -95 \text{ N/C},$$

where the minus sign indicates that the field direction is downward. This modest field is present near Earth's surface in fair weather; in thunderstorms the local field exceeds this value by many orders of magnitude.

Does our result make sense? We could also treat Earth as a spherical charge distribution, whence its surface field strength would be $E = Q/4\pi\varepsilon_0 R_E^2$. Using the symbolic form of our result for σ in Equation 24-11 gives precisely this expression, showing that these two approaches to the field are indeed consistent.

EXERCISE Dielectric breakdown of air occurs with fields of about 3×10^6 N/C, and results in sparks jumping through the air. What is the maximum surface charge density permissible on a conductor if dielectric breakdown of the surrounding air is to be avoided?

Answer: 27 μC/m^2

Some problems similar to Example 24-8: 43, 44, 51

●

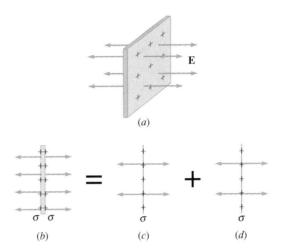

FIGURE 24-33 (*a*) An isolated, charged conducting plate. Its field points outward from both faces. (*b*) Edge-on view of the plate. If the plate is isolated, symmetry requires that there be equal charge densities on both faces. (*c*) The field anywhere is the sum of the fields of the two faces, each treated as a single charged sheet. Within the plate the fields cancel, while outside they sum to give a net field σ/ϵ_0.

Equation 24-11 gives a field that depends only on the local charge density. Does that mean the field at any point on a conductor surface arises only from the charge right at that point? No! As always, the field at any point is the vector sum of contributions from all charge elements making up the charge distribution. Remarkably, Gauss's law requires that charges on a conductor arrange themselves in such a way that the field at any point on the conductor surface depends only on the surface charge density right at that point—even though that field arises from *all* the charges on the surface (as well as from charges elsewhere if there are such)!

Consider a thin, flat, isolated, conducting sheet that has charge density σ on one of its two faces (Fig. 24-33*a*). Equation 24-11 shows immediately that the field at the surface of this plate is σ/ε_0. But if the plate is large and flat we can approximate it as an infinite sheet of charge—for which we found earlier (Equation 24-10 in Example 24-6) that the field should be $\sigma/2\varepsilon_0$. Is there a contradiction here? No! If the plate is isolated from other conductors or charges, then symmetry requires that charge spread itself evenly over *both* faces. If one face has charge density σ, so must the other—so we really have *two* charge sheets, each with density σ (Fig. 24-33*b*). Each gives rise to a field of magnitude $\sigma/2\varepsilon_0$, and *outside* the conductor those fields superpose to give the net field σ/ε_0 (Fig. 24-33*c*). *Inside* the conductor they also superpose, but here their directions are opposite and the result is that there is no field inside the conductor. Application of Equation 24-11 skips all these details. But because Equation 24-11 was derived on the assumption that the field inside the conductor is zero, it "knows" about charges everywhere on the conductor—and in this case that means on the second face of the conductor.

Equation 24-11 also applies to the pair of oppositely charged conducting plates shown in Fig. 24-34; the result, for the field between the plates, is σ/ε_0, where σ is the surface charge density on either plate. Why not $2\sigma/\varepsilon_0$? Again, Equation 24-11 always gives the field at a conductor surface—and it takes into account other charges that may be present. In this case the symmetry is broken, and charge

FIGURE 24-34 Edge view of two parallel conducting plates carrying opposite charges. Electrical attraction brings the excess charges to the inner faces. Each face constitutes a charge layer whose surface charge density has magnitude σ; between the plates their fields add to give field strength σ/ε_0.

builds up only on the inner faces of the two plates. Now each plate is a single charge layer, giving rise to a field $\sigma/2\varepsilon_0$, and between the plates the fields sum to Equation 24-11's result, σ/ε_0. Beyond the plates the fields sum to zero—a result that also follows from Equation 24-11 because now there is zero surface charge on the outer faces.

CHAPTER SYNOPSIS

Summary

1. **Electric field lines** provide a visual representation of the electric field. The direction of the field line passing through a given point is the direction of the electric field vector at that point. The number of field lines per unit area crossing an area perpendicular to the field is a measure of the field strength. Electric field lines begin and end only on charges.

2. **Electric flux** quantifies the notion "number of field lines crossing a surface." For a flat surface in a uniform field, the flux ϕ is given by $\phi = \mathbf{E} \cdot \mathbf{A}$, where $\mathbf{A}$ is a vector whose magnitude is the surface area A and whose direction is perpendicular to the surface. When the surface is curved and/or the field varies over the surface, the flux must be calculated by integration:

$$\phi = \int \mathbf{E} \cdot d\mathbf{A},$$

   where $d\mathbf{A}$ is an infinitesimal vector perpendicular to the surface at each point.

3. **Gauss's law** is a fundamental relation governing the behavior of electric fields throughout the universe. Loosely, Gauss's law states that the number of field lines emerging from a closed surface depends only on the charge enclosed—itself a reflection of the inverse-square dependence of the electric force. More rigorously, Gauss's law states that the electric flux emerging from any closed surface is proportional to the charge enclosed:

$$\oint \mathbf{E} \cdot d\mathbf{A} = \frac{q_{\text{enclosed}}}{\varepsilon_0},$$

   where $\varepsilon_0 = 1/4\pi k$.

4. Gauss's law is true for any surface and any distribution of charge, but it proves useful in calculating the electric field only in cases with sufficient symmetry—in particular, spherical symmetry, line symmetry, and plane symmetry. *Outside* charge distributions with these three symmetries, respectively, the field varies as $1/r^2$, as $1/r$, and exhibits no variation. The fields of more realistic charge distributions are often approximated by the fields associated with these symmetries.

5. There is no net charge motion and no electric field inside a conductor in **electrostatic equilibrium.** Gauss's law shows that any excess charge on the conductor resides on the conductor surface, and that the field at the conductor surface is perpendicular to the surface and has magnitude σ/ε_0, with σ the surface charge density.

Terms You Should Understand

(Pairs are closely related terms whose distinction is important; number in parentheses is chapter section where term first appears.)

electric field lines (24-1)
electric flux (24-2)
Gauss's law (24-3)
electrostatic equilibrium (24-6)

Symbols You Should Recognize

ϕ (24-2)
$\mathbf{A}$, $d\mathbf{A}$ (24-2)
$\int$ (24-2)
$\oint$ (24-2)
$\oint \mathbf{E} \cdot d\mathbf{A}$ (24-2)
σ (24-6)

Problems You Should Be Able to Solve

drawing and interpreting field line patterns for simple charge distributions (24-1)
calculating electric flux (24-2)
using Gauss's law to calculate electric fields with spherical, line, or plane symmetry (24-4)
describing charge distributions in electrostatic equilibrium (24-6)
calculating electric fields at conductor surfaces (24-6)

Limitations to Keep in Mind

Gauss's law is universally true, but it can be used to calculate electric fields only with sufficient symmetry.

QUESTIONS

1. Can electric field lines ever cross? Why or why not?
2. If identical charged particles are placed at points A and B in Fig. 24-35, which will experience the greater force?

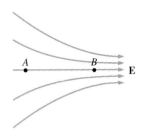

FIGURE 24-35 Question 2.

3. The electric flux through a closed surface is zero. Must the electric field be zero on that surface? If not, give an example.
4. If the flux of the gravitational field through a closed surface is zero, what can you conclude about the region interior to the surface?
5. Under what conditions can the electric flux through a surface be written as EA, where A is the surface area?
6. Eight field lines emerge from a closed surface surrounding an isolated point charge. Would this fact change if a second identical charge were brought to a point just *outside* the surface? If not, would anything change? Explain.
7. In what sense is Gauss's law equivalent to the inverse-square law?
8. If a charged particle were released from rest on a curved field line, would its subsequent motion follow the field line? Explain.
9. Gauss' law describes the flux of the electric field through a surface that may enclose charge. Must the field in Gauss's law arise only from charges within the closed surface?
10. In a certain region the electric field points to the right and its magnitude increases as you move to the right, as shown in Fig. 24-36. Does the region contain net positive charge, net negative charge, or zero net charge?

FIGURE 24-36 Question 10.

11. A spherical shell carries a nonuniform charge density. Why can't you conclude that the field inside the shell is zero?

12. A point charge is located a fixed distance from a uniformly charged sphere, outside the sphere. If the sphere shrinks in size without losing any charge, what happens to the force on the point charge?
13. In applying Equation 24-6 for the field outside a spherically symmetric charge distribution, is r the distance from the center or from the edge of the distribution?
14. The field of an infinite line of charge falls as $1/r$. How is this not a violation of the inverse-square law?
15. Why can't you use Gauss's law to determine the field of a uniformly charged cube? Why wouldn't it work to draw a cubical gaussian surface?
16. You're sitting inside an uncharged hollow spherical shell. Suddenly someone dumps a billion coulombs of charge on the shell, distributed uniformly. What happens to the electric field at your location?
17. No matter how far you get from an infinite sheet of charge, its field never changes. Why doesn't this violate our statement that far from a charge distribution its field resembles that of a point charge?
18. Why is it that the field inside a uniformly charged sphere actually increases with distance from the center? How is this consistent with Coulomb's law?
19. There is a nonzero flux through each of the surfaces in Fig. 24-7, yet there is no charge in the region shown. Why is this not a violation of Gauss's law?
20. Does Gauss's law apply to a spherical surface not centered on a point charge, as shown in Fig. 24-37? Would this be a useful surface to use in calculating the electric field?

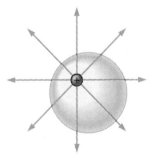

FIGURE 24-37 Question 20.

21. An insulating sphere carries charge spread uniformly throughout its volume. A conducting sphere has the same radius and net charge, but of course the charge is spread over its surface only. How do the electric fields outside these two charge distributions compare?
22. Why must the electric field be zero inside a conductor in electrostatic equilibrium?
23. Why must the electric field at the surface of a conductor in electrostatic equilibrium be perpendicular to the surface?

24. In electrostatic equilibrium, the electric field at the surface of an insulator need not be perpendicular to the insulator surface. Why not?
25. Where in Fig. 24-28 would you find the strongest electric field?
26. The electric field of a flat sheet of charge is $\sigma/2\varepsilon_0$. Yet the field of a flat conducting sheet—even a thin one, like a piece of aluminum foil—is σ/ε_0. Explain this apparent discrepancy.
27. A metal contains free electrons not bound to individual atoms. Does Gauss's law require that all these free electrons be on the metal surface?

PROBLEMS

Section 24-1 Electric Field Lines

1. What is the net charge shown in Fig. 24-38? The magnitude of the middle charge is 3 μC.

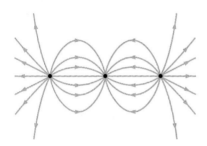

FIGURE 24-38 Problem 1.

2. A charge $+2q$ and a charge $-q$ are near each other. Sketch some field lines for this charge distribution, using the convention of eight lines for a charge of magnitude q.
3. Two charges $+q$ and a charge $-q$ are at the vertices of an equilateral triangle. Sketch some field lines for this charge distribution.
4. The net charge shown in Fig. 24-39 is $+Q$. Identify each of the charges A, B, C shown.

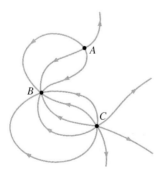

FIGURE 24-39 Problem 4.

Section 24-2 Electric Flux

5. A flat surface with area 2.0 m² is in a uniform electric field of 850 N/C. What is the electric flux through the surface when it is (a) at right angles to the field, (b) at 45° to the field, and (c) parallel to the field?
6. What is the electric field strength in a region where the flux through 1.0 cm $\times$ 1.0 cm flat surface is 65 N·m²/C, if the field is uniform and the surface is at right angles to the field?
7. A flat surface with area 0.14 m² lies in the x-y plane, in a uniform electric field given by $\mathbf{E} = 5.1\hat{\mathbf{i}} + 2.1\hat{\mathbf{j}} + 3.5\hat{\mathbf{k}}$ kN/C. Find the flux through this surface.
8. The electric field on the surface of a 10-cm-diameter sphere is perpendicular to the sphere and has magnitude 47 kN/C. What is the electric flux through the sphere?
9. What is the flux through the hemispherical open surface of radius R shown in Fig. 24-40? The uniform field has magnitude E. *Hint:* Don't do a messy integral! Think about the flux through the open end of the hemisphere.

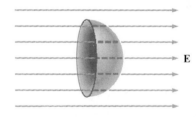

FIGURE 24-40 Problem 9.

10. The electric field shown in Fig. 24-41 is given by $\mathbf{E} = E_0\dfrac{y}{a}\hat{\mathbf{k}}$, where E_0 and a are constants. Find the flux through the square of side a shown.

Section 24-3 Gauss's Law

11. What is the electric flux through each closed surface shown in Fig. 24-42?
12. A 6.8-μC charge and a -4.7 μC charge are inside an uncharged sphere. What is the electric flux through the sphere?
13. A 2.6-μC charge is at the center of a cube 7.5 cm on each side. What is the electric flux through one face of the cube? *Hint:* Think about symmetry, and don't do an integral.

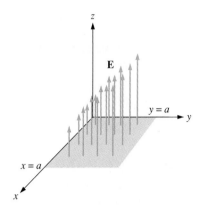

FIGURE 24-41 Problem 10.

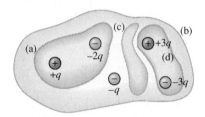

FIGURE 24-42 Problem 11.

14. If the charge in the preceding problem is still inside the cube but not at the center, (a) what is the flux through the *entire* cube? (b) Could you still calculate the flux through one face without doing an integral?

15. A dipole consists of two charges ± 6.1 μC located 1.2 cm apart. What is the electric flux through each surface shown in Fig. 24-43?

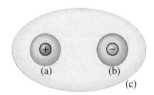

FIGURE 24-43 Problem 15.

16. The electric field in a certain region is given by $\mathbf{E} = 40x\hat{\imath}$ N/C, with x in meters. What is the volume charge density in the region? *Hint:* Apply Gauss's law to a cube 1 meter on a side.

Section 24-4 Using Gauss's Law

17. The electric field at the surface of a uniformly charged sphere of radius 5.0 cm is 90 kN/C. What would be the field strength 10 cm from the surface?

18. A solid sphere 25 cm in radius carries 14 μC, distributed uniformly throughout its volume. Find the electric field strength (a) 15 cm, (b) 25 cm, and (c) 50 cm from the sphere's center.

19. A crude model for the hydrogen atom treats it as a point charge $+e$ (the proton) surrounded by a uniform cloud of negative charge with total charge $-e$ and radius 0.0529 nm. What would be the electric field strength inside such an atom, halfway from the proton to the edge of the charge cloud?

20. Positive charge is spread uniformly over the surface of a spherical balloon 70 cm in radius, resulting in an electric field of 26 kN/C at the balloon's surface. Find the field strength (a) 50 cm from the balloon's center and (b) 190 cm from the center. (c) What is the net charge on the balloon?

21. A 10-nC point charge is located at the center of a thin spherical shell of radius 8.0 cm carrying -20 nC distributed uniformly over its surface. What are the magnitude and direction of the electric field (a) 2.0 cm, (b) 6.0 cm, and (c) 15 cm from the point charge?

22. A solid sphere 2.0 cm in radius carries a uniform volume charge density. The electric field 1.0 cm from the sphere's center has magnitude 39 kN/C. (a) At what other distance does the field have this magnitude? (b) What is the net charge on the sphere?

23. A point charge $-2Q$ is at the center of a spherical shell of radius R carrying charge Q spread uniformly over its surface. What is the electric field at (a) $r = \frac{1}{2}R$ and (b) $r = 2R$? (c) How would your answers change if the charge on the shell were doubled?

24. A spherical shell of radius 15 cm carries 4.8 μC, distributed uniformly over its surface. At the center of the shell is a point charge. (a) If the electric field at the surface of the sphere is 750 kN/C and points outward, what is the charge of the point charge? (b) What is the field just inside the shell?

25. A spherical shell 30 cm in diameter carries a total charge 85 μC distributed uniformly over its surface. A 1.0-μC point charge is located at the center of the shell. What is the electric field strength (a) 5.0 cm from the center and (b) 45 cm from the center? (c) How would your answers change if the charge on the shell were doubled?

26. The thick, spherical shell of inner radius a and outer radius b shown in Fig. 24-44 carries a uniform volume charge density ρ. Find an expression for the electric field strength

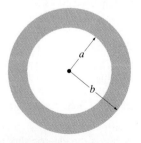

FIGURE 24-44 Problems 26, 32.

in the region $a < r < b$, and show that your result is consistent with Equation 24-7 when $a = 0$.

27. How should the charge density within a solid sphere vary with distance from the center in order that the magnitude of the electric field in the sphere be constant?

28. A long, thin wire carrying 5.6 nC/m runs down the center of a long, thin-walled, hollow pipe with radius 1.0 cm carrying -4.2 nC/m spread uniformly over its surface. Find the electric field (a) 0.50 cm from the wire and (b) 1.5 cm from the wire.

29. A long solid rod 4.5 cm in radius carries a uniform volume charge density. If the electric field strength at the surface of the rod (not near either end) is 16 kN/C, what is the volume charge density?

30. The electric field strength outside a charge distribution and 18 cm from its center has magnitude 55 kN/C. At 23 cm the field strength is 43 kN/C. Does the distribution have spherical or line symmetry?

31. An infinitely long rod of radius R carries a uniform volume charge density ρ. Show that the electric field strengths outside and inside the rod are given, respectively, by $E = \rho R^2/2\varepsilon_0 r$ and $E = \rho r/2\varepsilon_0$, where r is the distance from the rod axis.

32. Repeat Problem 26, assuming that Fig. 24-44 represents the cross section of a long, thick-walled pipe. Now the case $a = 0$ should be consistent with the result of Problem 31 for the interior of the rod.

33. A long, thin wire carries a uniform line charge density $\lambda = -6.8 \ \mu\text{C/m}$. It is surrounded by a thick concentric cylindrical shell of inner radius 2.5 cm and outer radius 3.5 cm. What uniform volume charge density in the shell will result in zero electric field outside the shell?

34. A square nonconducting plate measures 4.5 m on a side and carries charge spread uniformly over its surface. The electric field 10 cm from the plate and not near an edge has magnitude 430 N/C and points toward the plate. Find (a) the surface charge density on the plate and (b) the total charge on the plate. (c) What is the electric field strength 20 cm from the plate?

35. If you "painted" positive charge on the floor, what surface charge density would be necessary in order to suspend a 15-μC, 5.0-g particle above the floor?

36. A slab of charge extends infinitely in two dimensions and has thickness d in the third dimension, as shown in Fig. 24-45. The slab carries a uniform volume charge density ρ. Find expressions for the electric field strength (a) inside and (b) outside the slab, as functions of the distance x from the center plane.

37. Figure 24-46 shows sections of three infinite flat sheets of charge, each carrying surface charge density with the same magnitude σ. Find the magnitude and direction of the electric field in each of the four regions shown.

FIGURE 24-45 Section of an infinite slab of charge (Problem 36).

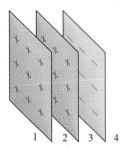

FIGURE 24-46 Problem 37.

Section 24-5 Fields of Arbitrary Charge Distributions

38. A rod 50 cm long and 1.0 cm in radius carries a 2.0-μC charge distributed uniformly over its length. What is the approximate magnitude of the electric field (a) 4.0 mm from the rod surface, not near either end, and (b) 23 m from the rod?

39. A nonconducting square plate 75 cm on a side carries a uniform surface charge density. The electric field strength 1 cm from the plate, not near an edge, is 45 kN/C. What is the approximate field strength 15 m from the plate?

40. Two circular plates 10 cm in diameter and 2.0 mm apart carry equal but opposite charges $\pm 0.50 \ \mu$C distributed uniformly over their facing surfaces. What is the electric field strength (a) between the plates but not near either edge? (b) 2.5 m from the plates on a plane passing midway between them? *Hint* for (b): See Example 23-6.

41. The electric field strength on the axis of a uniformly charged disk is given by $E = 2\pi k\sigma(1 - x/\sqrt{x^2 + a^2})$, with σ the surface charge density, a the disk radius, and x the distance from the disk center. If $a = 20$ cm, (a) for

what range of x values does treating the disk as an infinite sheet give an approximation to the field that is good to within 10%? (b) For what range of x values is the point-charge approximation good to 10%?

42. A nonconducting square 2.0 cm on a side carries a 45-nC charge spread uniformly over its surface. The x axis runs through the plate center, perpendicular to the plate, with $x = 0$ at the plate center. A -45-nC point charge is at $x = 5.0$ cm. Find approximate values for the electric field strength on the x axis at (a) $x = 1.0$ mm; (b) $x = 4.8$ cm; (c) $x = 2.5$ m. *Hint* for (c): Consult Example 23-6.

Section 24-6 Gauss's Law and Conductors

43. What is the electric field strength just outside the surface of a conducting sphere carrying surface charge density 1.4 $\mu C/m^2$?

44. Calculate the acceleration of a proton at the surface of a conductor carrying surface charge density 0.60 C/m^2.

45. A net charge of 5.0 μC is applied on one side of a solid metal sphere 2.0 cm in diameter. After electrostatic equilibrium is reached, what are (a) the volume charge density inside the sphere and (b) the surface charge density on the sphere? Assume there are no other charges or conductors nearby. (c) Which of your answers depends on this assumption, and why?

46. A point charge $+q$ lies at the center of a spherical conducting shell carrying a net charge $\frac{3}{2}q$. Sketch the field lines both inside and outside the shell, using 8 field lines to represent a charge of magnitude q.

47. A 250-nC point charge is placed at the center of an uncharged spherical conducting shell 20 cm in radius. (a) What is the surface charge density on the outer surface of the shell? (b) What is the electric field strength at the shell's outer surface?

48. A point charge is placed at the center of an uncharged spherical conducting shell of inner radius 2.5 cm and outer radius 4.0 cm (Fig. 24-47). As a result, the outer surface of the shell acquires a surface charge density $\sigma = 71$ nC/cm^2. Find (a) the value of the point charge and (b) the surface charge density on the inner wall of the shell.

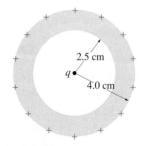

FIGURE 24-47   Problem 48.

49. An irregular conductor containing an irregular, empty cavity carries a net charge Q. (a) Show that the electric field inside the cavity must be zero. (b) If you put a point charge inside the cavity, what value must it have in order to make the surface charge density on the outer surface of the conductor everywhere zero?

50. A neutral dime is placed in a uniform electric field of 6.2×10^5 N/C, with its faces perpendicular to the field. (a) What is the approximate charge density on the faces of the dime? (b) What is the total charge on each face? (Measure a dime!)

51. A total charge of 18 μC is applied to a thin, square metal plate 75 cm on a side. Find the electric field strength near the plate's surface.

52. Two closely spaced parallel metal plates carry surface charge densities ± 95 nC/m^2 on their facing surfaces, with no charge on their outer surfaces. Find the electric field strength (a) between the plates and (b) outside the plates. Treat the plates as infinite in extent.

53. A conducting sphere 2.0 cm in radius is concentric with a spherical conducting shell with inner radius 8.0 cm and outer radius 10 cm. The small sphere carries 50 nC charge and the shell has no net charge. Find the electric field strength (a) 1.0 cm, (b) 5.0 cm, (c) 9.0 cm, and (d) 15 cm from the center.

54. A coaxial cable consists of an inner wire and a concentric cylindrical outer conductor (Fig. 24-48). If the conductors carry equal but opposite charges, show that there is no surface charge density on the *outside* of the outer conductor.

FIGURE 24-48   Problem 54.

Paired Problems

(Both problems in a pair involve the same principles and techniques. If you can get the first problem, you should be able to solve the second one.)

55. A point charge $-q$ is at the center of a spherical shell carrying charge $+2q$. That shell, in turn, is concentric with a larger shell carrying charge $-\frac{3}{2}q$. Draw a cross section of this structure, and sketch the electric field lines using the convention that 8 lines correspond to a charge of magnitude q.

56. A point charge $-q$ is at the center of a spherical shell carrying charge $-\frac{3}{2}q$. That shell, in turn, is concentric with a larger shell carrying charge $+2q$. Draw a cross

section of this structure, and sketch the electric field lines using the convention that 8 lines correspond to a charge of magnitude q.

57. A point charge q is at the center of a spherical shell of radius R carrying charge $2q$ spread uniformly over its surface. Write expressions for the electric field strength at (a) $\frac{1}{2}R$ and (b) $2R$.

58. A point charge q is at the center of a spherical shell of radius R carrying charge $5q$. At what other distance does the electric field have the same value it does at a point halfway from the center to the shell?

59. A long, thin hollow pipe 4.0 cm in diameter carries charge at a density of -2.6 μC/m, uniformly distributed over the pipe. It is concentric with 10-cm diameter pipe carrying $+2.6$ μC/m, also uniformly distributed. Find the magnitude of the electric field at (a) 0.50 cm, (b) 3.5 cm, and (c) 12 cm from the axis of the pipes.

60. Two concentric hollow pipes are 5.0 cm and 12 cm in diameter, respectively. Both carry uniformly distributed electric charges. The electric field 4.0 cm from their common axis is 630 kN/C, radially outward. The field 10 cm from their common axis is 126 kN/C, radially outward. (a) Find the linear charge densities on the two pipes. (b) How would the electric field strengths at 4.0 cm and 10 cm change if the charge density on the outer pipe were doubled?

61. An early (and incorrect) model for the atom pictured its positive charge as spread uniformly throughout the spherical atomic volume. For a hydrogen atom of radius 0.0529 nm, what would be the electric field due to such a distribution of positive charge (a) 0.020 nm from the center and (b) 0.20 nm from the center?

62. A solid sphere of radius R carries a charge spread uniformly throughout its volume. At what point outside the sphere is the electric field strength equal to that at a point halfway from the center to the edge? Express your answer as a distance from the center.

63. A sphere of radius $2a$ has a hole of radius a, as shown in Fig. 24-49. The solid portion carries a uniform volume charge density ρ. Find an expression for the electric field strength within the solid portion, as a function of the distance r from the center.

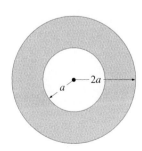

FIGURE 24-49   Problems 63, 64.

64. Repeat the previous problem, now considering that the figure represents the cross section of a thick cylindrical pipe.

Supplementary Problems

65. Repeat Problem 10 for the case $\mathbf{E} = E_0\left(\dfrac{y}{a}\right)^2\hat{\mathbf{k}}$.

66. The volume charge density inside a solid sphere of radius a is given by $\rho = \rho_0 r/a$, where ρ_0 is a constant. Find (a) the total charge and (b) the electric field strength within the sphere, as a function of distance r from the center.

67. A proton is released from rest 1.0 cm from a large sheet carrying a surface charge density of -24 nC/m^2. How much later does it strike the sheet?

68. Fig. 24-50 shows a rectangular box with sides $2a$ and length ℓ surrounding a line of charge with uniform line charge density λ. The line passes directly through the center of the box faces. Using an expression for the field of a line charge, integrate over strips of width dx as shown to find the electric flux through one face of the box. Multiply by 4 to get the total flux through the box, and show that your result is consistent with Gauss's law.

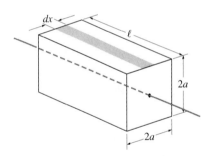

FIGURE 24-50   Problem 68.

69. Repeat Problem 36 for the case when the charge density in the slab is given by $\rho = \rho_0|x/d|$, where ρ_0 is a constant.

70. The charge density within a uniformly charged sphere of radius R is given by $\rho = \rho_0 - ar^2$, where ρ_0 and a are constants, and r is the distance from the center. Find an expression for a such that the electric field outside the sphere is zero.

71. A small object of mass m and charge q is attached by a thread of length ℓ to a large, flat, nonconducting plate carrying a uniform surface charge density σ with the same sign as q (Fig. 24-51). If the object is displaced slightly sideways from its equilibrium, show that it undergoes simple harmonic motion with period $T = 2\pi\sqrt{2\epsilon_0 m\ell/q\sigma}$. Assume the gravitational force is negligible.

72. An infinitely long nonconducting rod of radius R carries a volume charge density given by $\rho = \rho_0(r/R)$, where ρ_0 is a

FIGURE 24-51   Problem 71.

constant. Find the electric field strength (a) inside and (b) outside the rod, as functions of the distance r from the rod axis.

73. A thick spherical shell of inner radius a and outer radius b carries a charge density given by $\rho = \dfrac{ce^{-r/a}}{r^2}$, where a and c are constants. Find expressions for the electric field strength for (a) $r < a$, (b) $a < r < b$, and (c) $r > b$.

74. A solid sphere of radius R carries a nonuniform volume charge density given by $\rho = \pi^2 \rho_0 \sin(\pi r/R)$, where r is the distance from the center and ρ_0 is a positive constant. Find the magnitude and direction of the electric field at the sphere's surface.

75. A solid sphere of radius R carries a uniform volume charge density ρ. A hole of radius $R/2$ occupies a region from the center to the edge of the sphere, as shown in Fig. 24-52. Show that the electric field everywhere in the hole points horizontally and has magnitude $\rho R/6\epsilon_0$. *Hint:* Treat the hole as a superposition of two charged spheres of opposite charge.

76. You're 5.0 m from a charge distribution and you measure an electric field strength of 850 N/C. At 2.5 m the field

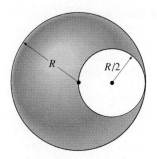

FIGURE 24-52   Problem 75.

strength has increased to about 3.4 kN/C. When you're 5.0 mm from the center of the distribution the field strength is 42.5 MN/C, and it increases to 85 MN/C at 2.5 mm. Describe the distribution as fully as you can, including its shape, any dimensions you can find, its total charge, and any appropriate charge density.

77. Two flat, parallel, closely spaced metal plates of area 0.080 m^2 carry total charges of -2.1 μC and $+3.8$ μC. Find the surface charge densities on the inner and outer faces of each plate.

78. Since the gravitational force of a point mass goes as $1/r^2$, the gravitational field **g** also obeys a form of Gauss's law. (a) Formulate this law, and (b) use it to find an expression for the gravitational field strength *within* the Earth, as a function of distance r from the center. Treat Earth as a sphere of uniform density.

CHAPTER 25

ELECTRIC POTENTIAL

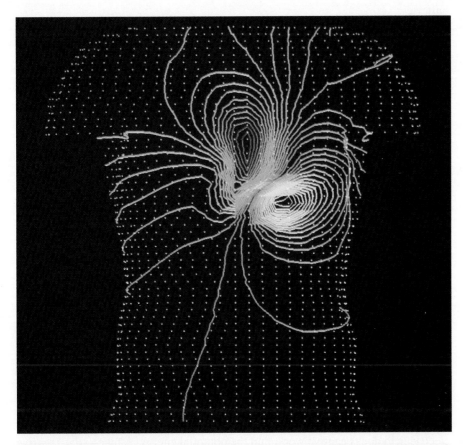

Lines of constant electric potential on the surface of the human body reflect the electric dipole structure of the heart.

You have to do work to lift your book against Earth's gravitational field. The work you do gets stored as potential energy, which is released if you lower the book. In Chapter 8 we introduced the term **conservative** to describe a force or field that, like gravity, "gives back" all the stored energy. An important property of conservative fields is path independence: The work required to move an object from one point to another does not depend on the path taken, but only on the endpoints of the path (Fig. 25-1).

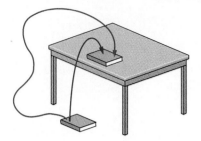

FIGURE 25-1 The gravitational field is conservative, and therefore the net work required to move between two points is independent of the path taken.

The electric field of a static charge distribution is also conservative. You can see this because Coulomb's law has essentially the same form as Newton's law of universal gravitation, and we'll soon prove path independence for the work done in moving one point charge about in the field of another. Because the electric field is conservative, it is meaningful to talk about potential energy. If you do work moving a charge against the electric force, for example, that work ends up stored as potential energy.

In this chapter we deal with electric potential, a useful and easily measured quantity that derives from the concept of potential energy. We'll see how the use of electric potential can provide a simpler approach to calculating electric fields, and how electric potential relates to the properties of everyday devices like batteries.

25-1 POTENTIAL ENERGY, WORK, AND THE ELECTRIC FIELD

In Chapter 8 we defined potential energy difference as the negative of the work $W_{A \to B}$ done by a conservative force $\mathbf{F}$ on an object moved from point A to point B:

$$\Delta U_{A \to B} = U_B - U_A = -W_{A \to B} = -\int_A^B \mathbf{F} \cdot d\boldsymbol{\ell}, \qquad (25\text{-}1)$$

where $d\boldsymbol{\ell}$ is a small element of the path from A to B.

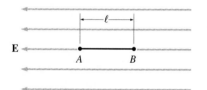

FIGURE 25-2 The work required to move a charge q from A to B in a *uniform* electric field is $qE\ell$.

Consider a positive charge q being moved between two points A and B a distance ℓ apart in a uniform electric field $\mathbf{E}$, as shown in Fig. 25-2. A conservative electric force $q\mathbf{E}$ acts on the charge, and since the field is uniform that force doesn't vary as we move from A to B. Therefore, we can dispense with the integral and write

$$\Delta U_{A \to B} = -W_{A \to B} = -\mathbf{F} \cdot \boldsymbol{\ell} = -q\mathbf{E} \cdot \boldsymbol{\ell} = -qE\ell \cos(180°) = qE\ell,$$

where $\boldsymbol{\ell}$ is a vector from A to B and where the factor $\cos 180° = -1$ appears because the vectors $\mathbf{E}$ and $\boldsymbol{\ell}$ are in opposite directions. Does this result make sense? Pushing a positive charge against the electric field is like pushing a car up a hill: We do positive work, gravity does negative work, and the potential energy increases. Here we must do positive work to move the charge from A to B, the electric field does negative work, and the potential energy change—the *negative* of the work done by the conservative electric force—is positive. Let go of the charge and the field will accelerate it back toward B, changing potential to kinetic energy.

25-2 POTENTIAL DIFFERENCE

We've just found the potential energy change in moving a charge q from point A to point B in Fig. 25-2. If we had moved a charge $2q$ the potential energy change would have been twice as great, and it would have required twice as much work; $\frac{1}{4}q$ would have one-fourth the potential energy change and would require one-fourth the work. Since the potential energy change is directly proportional to the charge, it is convenient to consider the *potential energy change per unit charge* involved in moving between two points. We call this quantity the **electric potential difference.** Mathematically, we express the potential difference by writing $\mathbf{F} = q\mathbf{E}$ in Equation 25-1 and dividing by q:

$$\Delta V_{A \to B} = \frac{\Delta U_{A \to B}}{q} = -\frac{1}{q}\int_{A}^{B} q\mathbf{E} \cdot d\boldsymbol{\ell} = -\int_{A}^{B} \mathbf{E} \cdot d\boldsymbol{\ell}.$$

Then our definition of potential difference becomes:

> **The electric potential difference from point A to point B is the potential energy change per unit charge in moving from A to B:**

$$\Delta V_{A \to B} = -\int_{A}^{B} \mathbf{E} \cdot d\boldsymbol{\ell}. \qquad \text{(electric potential difference)} \qquad (25\text{-}2a)$$

In other books you may see our $\Delta V_{A \to B}$ written as V_{AB} or V_{BA} or $V_B - V_A$. We use the Δ here to show explicitly that we're talking about a *change* or *difference* from one point to another, and we use the subscript $A \to B$ to make it clear that this is the potential difference going *from A to B*. In the next section we'll show how our notation is equivalent to the commonly used $V_B - V_A$, and in subsequent chapters we'll sometimes use just the symbol V for potential difference.

An external agent moving charge at constant speed between A and B would do work equal to the change in the charge's potential energy; therefore we can think of potential difference as the work per unit charge done by an external agent moving charge at constant speed between two points. We'll often say, loosely, that potential difference is the energy per unit charge involved in moving between two points; by "energy" we can mean either the work done by an external agent or the potential energy change in the electric field.

In the special case of a uniform field, Equation 25-2a reduces to

$$\Delta V_{A \to B} = -\mathbf{E} \cdot \boldsymbol{\ell}, \quad \text{(uniform field)} \qquad (25\text{-}2b)$$

where $\boldsymbol{\ell}$ is a vector from A to B. Figure 25-2 shows the special case when the field $\mathbf{E}$ and path $\boldsymbol{\ell}$ are in opposite directions.

Equations 25-2 show that potential difference can be positive or negative, depending on whether the path goes against or with the field. Moving a positive charge through a positive potential difference is like going uphill: We must do work on the charge, and its potential energy increases. Moving a positive charge through a negative potential difference is like going downhill: We do negative work or, equivalently, the field does work on the charge, and its potential energy decreases. In both cases the opposite is true for a negative charge; even though the potential difference remains the same, the work and potential energy reverse because of the negative sign on the charge.

The Volt and the Electron Volt

The definition of potential difference shows that its units are joules/coulomb. Potential difference is important enough that 1 J/C has a special name—the **volt** (V). To say that a car has a 12-V battery, for example, means that the

FIGURE 25-3 Potential difference depends on *two* points. This parasailer landed on a 138,000-V power line, but he's not being electrocuted because his body is not contacting *two* points with a potential difference between them.

battery does 12 J of work on every coulomb that moves between its two terminals.

We often use the term **voltage** to speak of potential difference, especially in describing electric circuits. Strictly speaking the two terms are not synonymous, since voltage is used even in nonconservative situations that arise when fields change with time. But in common usage this subtle distinction is usually not bothersome.

TIP **Potential Difference Depends on Two Points**
Specifically, it is the energy per unit charge involved in moving *between those points*. Always think of potential difference in terms of two points. This is ultimately a very practical matter; if you forget it you won't be able to hook up a voltmeter properly, or connect jumper cables safely to your car battery! Figure 25-3 provides a dramatic illustration of this point.

Sometimes we speak of "the potential (or the voltage) at point *P*." This is *always* a shorthand way of talking, and we *must* have in mind some other point. What we mean is the potential difference going from that other point to point *P*.

In molecular, atomic, and nuclear systems it's often convenient to measure energy in **electron volts** (eV), defined as follows:

> One electron volt is the energy gained by a particle carrying one elementary charge when it moves through a potential difference of one volt.

Since one elementary charge is 1.6×10^{-19} C, 1 eV is 1.6×10^{-19} J. Energy in eV is particularly easy to calculate when charge is given in elementary charges. However, the eV is *not* an SI unit and should be converted to joules before calculating other quantities.

● **EXAMPLE 25-1** A TV PICTURE TUBE: POTENTIAL DIFFERENCE, WORK, AND ENERGY

At the back end of a TV picture tube, a uniform electric field of 600 kN/C extends over a distance of 5.0 cm and points toward the back of the tube (Fig. 25-4). (a) Find the potential difference between the back and the front end of this field region. (b) How much work would it take to move an ion with charge $+2e$ from the back to the front of the field region? (c) What would happen to an electron released at the back of the field region?

Solution
(a) With a uniform field the potential difference is given by Equation 25-2b:

$$V_{A \to B} = -\mathbf{E} \cdot \boldsymbol{\ell} = E\ell = (600 \times 10^3 \text{ N/C})(0.050 \text{ m})$$
$$= 30 \text{ kV},$$

FIGURE 25-4 A potential difference on the order of 30 kV is used to accelerate the electrons that "paint" the picture on the screen of a TV tube, shown here in a cutaway view.

where the second equality follows because a path from the back toward the front is opposite the field direction, as shown in Fig. 25-2, so $\cos\theta = -1$.

(b) Potential difference is the potential energy change per unit charge, or, equivalently, the work per unit charge an external agent must do to move charge between two points. Thus, the work needed to move the ion of charge $q = 2e$ against the 30-kV potential difference is

$$W_{ion} = q\Delta V = (2)(1.6\times10^{-19}\,\text{C})(30\,\text{kV}) = 9.6\times10^{-15}\,\text{J},$$

where the units work out because $1\,\text{V} = 1\,\text{J/C}$. Since the ion carries two elementary charges, we could also express this energy as $(2e)(30\,\text{kV}) = 60\,\text{keV}$.

(c) With its negative charge, the electron would *gain* energy given by $e\Delta V$, or 4.8×10^{-15} J (more simply calculated as 30 keV).

EXERCISE A 1.2-μC charge is accelerated through a 3400-V potential difference. How much energy does it gain?

Answer: 4.1 mJ ●

● **EXAMPLE 25-2** POTENTIAL OF A CHARGED SHEET

An isolated, infinite charged sheet carries a uniform surface charge density σ. (a) Find an expression for the potential difference from the sheet to a point a perpendicular distance x from the sheet. (b) What is the potential difference between two points the same distance from the sheet?

Solution

Equation 24-10 gives $E = \dfrac{\sigma}{2\varepsilon_0}$ for the field of a single, isolated sheet of charge. Since the field is uniform we can apply Equation 25-2b. Moving away from the sheet in either direction is going *with* the field (assuming positive σ), so the dot product in Equation 25-2b is positive and therefore the potential difference is negative. Taking $\ell = x$ in Equation 25-2b then gives

$$\Delta V_{0\to x} = -Ex = -\frac{\sigma x}{2\varepsilon_0},$$

where the notation $0 \to x$ means we're taking the potential difference from the sheet ($x = 0$) to the point x. Thus the potential decreases linearly with distance for positive σ; for

negative σ it would increase. The important point is that the potential in a *uniform* field varies *linearly* with distance along the field direction.

If, on the other hand, we consider moving charge between two points equidistant from the sheet, then the charge moves perpendicular to the field and there is no change in its potential energy; thus, the potential difference between such points is zero.

EXERCISE Two nonconducting charged sheets carry equal but opposite surface charge densities ± 53 nC/m^2. The negative sheet is located at $x = 0$, the positive sheet at $x = 10$ cm. Find expressions for the potential difference from the negative sheet to the points (a) $x = 2.0$ cm, (b) $x = 5.0$ cm; (c) $x = 25$ cm, and (d) $x = -10$ cm. *Hint:* Think about the different fields *between* and *beyond* the plates.

Answers: (a) 120 V; (b) 300 V; (c) 600 V; (d) 0 V

Some problems similar to Examples 25-1 and 25-2: 5, 7–9, 18 ●

Curved Paths and Nonuniform Fields

Equations 25-2 contain a dot product that accounts for the orientation of the path relative to the field. Figure 25-5, for example, shows several straight paths of the same length ℓ in a uniform electric field. Path AB is the same path we considered in Fig. 25-2; the potential difference between its ends is just $\Delta V_{A\to B} = E\ell$ because the angle between **E** and ℓ is 180°. Path AC is at 135° to the field direction, giving a potential difference $\Delta V_{A\to C} = -E\ell \cos 135° = E\ell/\sqrt{2}$. Finally, path AD is perpendicular to the field, giving $\Delta V_{A\to D} = 0$. Quite generally, as Fig. 25-5 suggests, the potential difference depends only on the component of the path *along* the field direction. This is analogous to the situation with gravity, where the work mgh needed to lift a mass depends only on the *vertical* distance h and not on any horizontal component of the motion.

If the field is not uniform or the path is not straight, then we must use the integral form of Equation 25-2a to calculate the potential because the magnitude

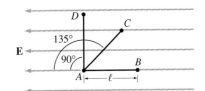

FIGURE 25-5 The potential difference depends only on the component of the path *along* the field. Mathematically, $\Delta V = -\mathbf{E} \cdot \boldsymbol{\ell} = -E\ell \cos\theta$, where the quantity $-\ell \cos\theta$ can be interpreted as the path component along the field.

FIGURE 25-6 The line integral in Equation 25-2a is the sum of infinitely many infinitesimally small potential differences dV.

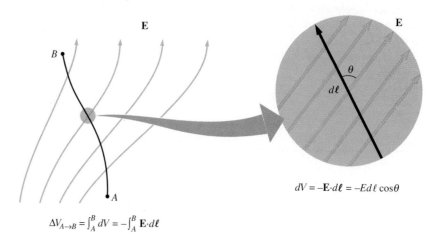

FIGURE 25-6 The line integral in Equation 25-2a is the sum of infinitely many infinitesimally small potential differences dV.

$$dV = -\mathbf{E} \cdot d\boldsymbol{\ell} = -E \, d\ell \cos\theta$$

$$\Delta V_{A \to B} = \int_A^B dV = -\int_A^B \mathbf{E} \cdot d\boldsymbol{\ell}$$

of $\mathbf{E}$ and/or the angle between $\mathbf{E}$ and the path is changing. Figure 25-6 shows the meaning of Equation 25-2a. If we look at a sufficiently small part of the curve in Fig. 25-6, so small that the field is essentially uniform and the path essentially straight, then Equation 25-2b for a straight path in a uniform field should apply. Describing the short path segment by a small vector $d\boldsymbol{\ell}$, we can write Equation 25-2b as

$$dV = -\mathbf{E} \cdot d\boldsymbol{\ell},$$

where dV is the potential difference from tail to head of $d\boldsymbol{\ell}$. The integral in Equation 25-2a is the sum of all the dV's over the path:

$$\Delta V_{A \to B} = \int_A^B dV = -\int_A^B \mathbf{E} \cdot d\boldsymbol{\ell}.$$

Like the work integral we introduced in Chapter 7, this **line integral** is simply a sum of scalar quantities—dot products of the vectors $\mathbf{E}$ and $d\boldsymbol{\ell}$—over some path. The limits of the integral are the endpoints of the path. Because the electrostatic field is conservative, it is not necessary to specify which of the many paths between these endpoints is taken; all such paths give the same result. Several examples in the next section illustrate the use of Equation 25-2a.

25-3 CALCULATING POTENTIAL DIFFERENCE

The Potential of a Point Charge

The electric field of a point charge q is given by Equation 23-3:

$$\mathbf{E} = \frac{kq}{r^2}\hat{\mathbf{r}},$$

where $\hat{\mathbf{r}}$ is a unit vector from the charge toward the point where the field is being evaluated. Consider two points A and B at distances r_A and r_B from a positive point charge, as shown in Fig. 25-7. What is the potential difference between these points? The distance between them is $r_B - r_A$, but we cannot just multiply this distance by the electric field because the field varies with position. Instead we integrate, following Equation 25-2a:

$$\Delta V_{A \to B} = -\int_{r_A}^{r_B} \mathbf{E} \cdot d\boldsymbol{\ell} = -\int_{r_A}^{r_B} \frac{kq}{r^2} \hat{\mathbf{r}} \cdot d\boldsymbol{\ell}.$$

As we move from r_A toward r_B, the path element vectors $d\boldsymbol{\ell}$ correspond to small increments dr in the radial direction, and therefore, we can write $d\boldsymbol{\ell} = \hat{\mathbf{r}} dr$. Then the potential becomes

$$\Delta V_{A \to B} = -\int_{r_A}^{r_B} \frac{kq}{r^2} \hat{\mathbf{r}} \cdot \hat{\mathbf{r}} dr = -kq \int_{r_A}^{r_B} r^{-2} dr,$$

since the dot product of the unit vector $\hat{\mathbf{r}}$ with itself is simply 1. Evaluating the integral gives

$$\Delta V_{A \to B} = -kq \left[-\frac{1}{r} \right]_{r_A}^{r_B} = kq \left(\frac{1}{r_B} - \frac{1}{r_A} \right). \qquad (25\text{-}3)$$

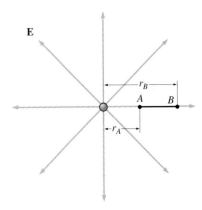

FIGURE 25-7 The potential difference between points A and B is found by integrating between r_A and r_B.

Does this result make sense? For $r_B > r_A$ the potential difference is negative, showing that a positive test charge at r_A would "fall down" the potential "hill" toward r_B. Going the other way would require that positive work be done on a positive charge, as it's pushed "up" the potential "hill" against the repulsive force of the charge q. Although we considered q to be positive, our result holds as well for $q < 0$, in which case the sign of the potential difference changes.

Although we derived Equation 25-3 for two points on the same radial line, Fig. 25-8 shows that the result holds for *any* two points in the field of a charge q. It doesn't matter which point is at the greater distance either; if $r_B < r_A$ Equation 25-3 still gives the correct potential difference, which then becomes positive to indicate that work must be done moving a positive test charge *toward* a positive q.

The Zero of Potential

So far we've only talked about potential differences, symbolized by the expression $\Delta V_{A \to B}$. That's because only differences in potential energy—and thus in electric potential, which is potential energy per unit charge—have physical significance. But as we did with potential energy, it's often convenient to define the **electric potential** at some point as zero and then to measure potential differences relative to that point. We then speak of "the potential at point P," designated $V(P)$ or V_P, and meaning the potential difference $V_{0 \to P}$ *from* our reference point *to* point P. Once we've defined a zero of potential, we can then write potential differences as differences in potential between two points; thus, our terminology $\Delta V_{A \to B}$ can equally well be written $V(B) - V(A)$ or $V_B - V_A$. The choice for the zero of potential is arbitrary and is usually made on the basis of mathematical or physical convenience. In electric power systems, the Earth, called "ground," is usually taken as the zero. In automobile electric systems, the

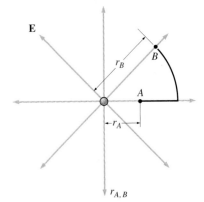

FIGURE 25-8 Here points A and B do not lie on the same radial line. But potential difference is independent of path, and one path between A and B consists of a radial segment and a circular arc. Since $\mathbf{E}$ is perpendicular to the arc, it takes no work to move a charge along the arc. The potential difference $\Delta V_{A \to B}$ therefore arises only from the radial segment, and is therefore given by Equation 25-3.

car's metal structure makes a convenient zero; this is usually connected electrically to the negative battery terminal. (As we'll soon see, every point on a conductor is at the same potential, so it's appropriate to consider an entire conductor like Earth or a car's metal structure as the "point" of zero potential.)

In dealing with isolated point charges, Equation 25-3 shows that it is convenient to choose the zero of potential at infinity. If we let r_A become arbitrarily large, and drop the subscript on r_B because it can be at any radial distance r, Equation 25-3 becomes

$$V_{\infty \to r} = V(r) = \frac{kq}{r}. \qquad \text{(point-charge potential)} \qquad (25\text{-}4)$$

When we call this expression $V(r)$ "the potential of a point charge," we really mean that $V(r)$ is the potential difference going from a point very far from a charge q to a point a distance r from the charge—an interpretation that is consistent with our definition of potential difference as depending on *two* points. Because the field outside any spherically symmetric charge distribution is that of a point charge, Equation 25-4 also gives the potential outside a spherically symmetric charge distribution.

Does it bother you that potential difference can be finite over an infinite distance? The reason lies in the inverse-square dependence of the field, which drops so rapidly that the work done in moving a charge from infinity to the vicinity of a point charge remains finite. We found an analogous result in Chapter 9, where it took only a finite amount of energy—and therefore a finite "escape speed"—to escape completely from a planet's gravitational attraction. As long as a charge distribution is finite in size—so its field at large distances falls at least as fast as $1/r^2$, then it makes sense to take the zero of potential at infinity.

● **EXAMPLE 25-3**   THE POTENTIAL OF A SPHERICAL CHARGE DISTRIBUTION

A charge Q is distributed in a spherically symmetric way over a sphere of radius R. (a) What is the potential at the sphere's surface, with the zero of potential taken at infinity? (b) How much work would it take to move a proton from infinity to the sphere's surface? (c) What is the potential difference from the sphere's surface to a point located $2R$ from the center?

Solution

(a) The field outside the sphere is exactly that of a point charge Q located at the sphere's center, so the potential difference between infinity and any point *outside* the sphere is given by Equation 25-4. Therefore the potential at the sphere's surface is

$$V(R) = \frac{kQ}{R}.$$

(b) This quantity is the work per unit charge needed to move a charge from infinity to the sphere's surface; since a proton carries charge e, the work involved in bringing the proton to the sphere's surface is just $W = kQe/R$.

(c) To get the potential difference from the surface ($r = R$) to $2R$, we could use Equation 25-3 with $r_B = 2R$ and $r_A = R$.

But it's perhaps easiest to keep in mind just Equation 25-4 and then find the potential difference by subtracting the potentials at the two points:

$$\Delta V_{R \to 2R} = V(2R) - V(R) = \frac{kQ}{2R} - \frac{kQ}{R} = -\frac{kQ}{2R}.$$

Does this result make sense? Yes. The negative result (for positive Q) shows that the potential energy of a positive charge would decrease as it moved away from the sphere, going in the direction of the sphere's electric field.

EXERCISE   The potential at the surface of a 10-cm-radius sphere is 4.8 kV. (a) What is the charge on the sphere, assuming it's distributed in a spherically symmetric fashion? (b) What is the electric field at its surface? Assume here—and anytime it's not specified—that potential differences are taken from infinity.

Answers: (a) 53 nC; (b) 48 kN/C

Some problems similar to Example 25-3:   19–23, 25, 26, 65, 71   ●

Potentials of Arbitrary Charge Distributions

If we already know the field of a charge distribution, we can calculate potential differences by applying Equation 25-2a, as we did for the point-charge field. Example 25-4 illustrates this approach.

● **EXAMPLE 25-4** POTENTIAL DIFFERENCE IN THE FIELD OF A LINE CHARGE

An infinite line of charge carries line charge density λ. What is the potential difference between two points at distances r_A and r_B from the line?

Solution
In the preceding chapter we used Gauss's law to obtain the result

$$\mathbf{E} = \frac{\lambda}{2\pi\varepsilon_0 r}\hat{\mathbf{r}}$$

for the field of a line charge. As we move from r_A to r_B in Fig. 25-9, we can again write $d\boldsymbol{\ell} = \hat{\mathbf{r}}\,dr$ just as we did in evaluating the point-charge potential. Then Equation 25-2a becomes

$$\Delta V_{A\to B} = -\int_{r_A}^{r_B} \mathbf{E} \cdot d\boldsymbol{\ell} = -\int_{r_A}^{r_B} \frac{\lambda}{2\pi\varepsilon_0 r}\hat{\mathbf{r}} \cdot \hat{\mathbf{r}}\,dr$$

$$= -\frac{\lambda}{2\pi\varepsilon_0}\int_{r_A}^{r_B} \frac{dr}{r} = -\frac{\lambda}{2\pi\varepsilon_0}\ln r \Big|_{r_A}^{r_B} \qquad (25\text{-}5)$$

$$= \frac{\lambda}{2\pi\varepsilon_0}\ln\!\left(\frac{r_A}{r_B}\right),$$

where the last step follows because $\ln x - \ln y = \ln(x/y)$. If λ is positive and $r_A < r_B$, then $\Delta V_{A\to B}$ is negative—indicating, as expected, that the electric field does work on a positive charge moving from A to B (recall that $\ln x < 0$ for $x < 1$). Conversely, moving a positive charge from B to A requires that the agent moving the charge do work since the potential difference $\Delta V_{B\to A}$ is positive.

Note that we cannot let r_A go to infinity in this case, for this would give an infinite potential difference. Physically, this reflects the fact that our charge distribution is itself of infinite

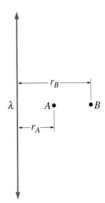

FIGURE 25-9 Two points in the field of a line charge.

extent. Mathematically, it reflects the slow $1/r$ decrease in field strength.

Although we derived Equation 25-5 for an infinitesimally thin line of charge, considerations of Section 24-4 show that this result holds outside *any* charge distribution with line symmetry.

EXERCISE An infinitely long rod of radius R carries a uniform volume charge density ρ; Problem 24-31 shows that the electric field *inside* this rod points radially outward and has magnitude $E = \rho r/2\varepsilon_0$. Use Equation 25-2a to find the potential difference from the rod's surface to its axis.

Answer: $\rho R^2/4\varepsilon_0$

Some problems similar to Example 25-4: 28, 29, 66, 75   ●

Finding Potential Differences Using Superposition

When we don't know the field of a charge distribution, or the field is too complicated to integrate easily, we can find the potential using superposition. As we will see in the next section, this often provides an easier approach to the field as well.

Consider a charge q being brought from infinity to a point P in the vicinity of some other charges. We want to know the potential at P—by which we mean the work per unit charge required to move from infinity to P. The superposition principle states that the electric field of a charge distribution is the sum of the fields of the individual charges comprising the distribution. Therefore the work per unit charge—that is, the potential difference—between infinity and P is just the sum of the potential differences associated with the individual point charges. Mathematically, we find $V(P)$ by summing Equation 25-4 over the individual point charges q_i:

$$V(P) = \sum_i \frac{kq_i}{r_i}, \tag{25-6}$$

where the r_i's are the distances from each of the charges to the point P. Equation 25-6 has one enormous advantage over its counterpart for the electric field, Equation 23-4. Electric potential is a *scalar*, so the sum in Equation 25-6 is a scalar sum, and there's no need to consider angles or vector components.

● **EXAMPLE 25-5** THE DIPOLE POTENTIAL

The dipole of Fig. 25-10 consists of two point charges $\pm q$ separated by a distance $2a$. Find the potential at an arbitrary point P, taking the zero of potential at infinity.

Solution
We sum the potentials of the individual point charges, as Equation 25-6 suggests:

$$V(P) = \sum_i \frac{kq_i}{r_i} = \frac{kq}{r_1} + \frac{k(-q)}{r_2} = kq\left(\frac{1}{r_1} - \frac{1}{r_2}\right) = \frac{kq(r_2 - r_1)}{r_1 r_2},$$

where r_1 and r_2 are the distances from P to the positive and negative charges, respectively.

We have seen that in practical situations we're often interested in electrical effects a great distance from the dipole. If r is the distance from the dipole's center to P, then in the limit $r \gg a$ the quantity $r_1 r_2$ becomes approximately r^2, and, as Fig. 25-10 shows, $r_2 - r_1 \simeq 2a \cos\theta$. Then the dipole potential for $r \gg a$ becomes

$$V(r, \theta) = \frac{k(2aq)\cos\theta}{r^2} = \frac{kp\cos\theta}{r^2}, \tag{25-7}$$

with $p = 2aq$ the dipole moment.

Note that the dipole potential drops more rapidly with distance than the point-charge potential, just as the dipole field drops more rapidly with distance than the point-charge field. Note also that Equation 25-7 gives $V = 0$ along the perpendicular bisector of the dipole ($\theta = 90°$). That makes sense because a charge approaching the dipole along its bisector moves at right angles to the dipole field, so no work need be done. Since potential difference is path independent, that means it takes no

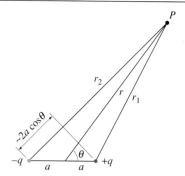

FIGURE 25-10 A dipole and a point P where its potential is to be evaluated. When P is far from the dipole, then r_2 is longer than r_1 by approximately $2a\cos\theta$. (The angle whose vertex is $-q$ is very nearly θ.)

net work to move on *any* path from infinity to a point on the dipole's bisector (see Question 20).

EXERCISE The dipole moment of a water molecule is 6.2×10^{-30} C·m. (a) Find the potential difference $V_B - V_A$ between two points on the axis of the molecular dipole, where points A and B are 8.2 nm and 5.1 nm, respectively, from the center. Both points are closer to the positive end. (b) How much work would it take to move a proton from A to B?

Answers: (a) 1.32 mV; (b) 1.32 meV $= 2.1 \times 10^{-22}$ J

Some problems similar to Example 25-5:
31–35, 76   ●

Continuous Charge Distributions

We can calculate the potential of a continuous charge distribution by considering it to be made up of infinitely many infinitesimal charge elements dq. Each acts like a point charge and therefore contributes to the potential at some point P an amount dV given by

$$dV = \frac{k\,dq}{r},$$

where the zero of potential is at infinity. The potential at P is the sum—in this case an integral—of the contributions dV from all the charge elements:

$$V = \int dV = \int \frac{k\,dq}{r}, \tag{25-8}$$

where the integration is over the entire charge distribution.

● **EXAMPLE 25-6**   A CHARGED RING

A total charge Q is distributed uniformly around a thin ring of radius a, as shown in Fig. 25-11. What is the potential on the axis of this charged ring?

Solution
Let x be the distance from the center of the ring to some arbitrary point on the axis. The distance from each point on the ring to a point on the axis is the same, and is given by $r = \sqrt{x^2 + a^2}$. The potential on the axis is the sum of the potentials dV of all the charge elements dq around the ring, as described by Equation 25-8:

$$V = k \int_{ring} \frac{dq}{r} = \frac{k}{\sqrt{x^2 + a^2}} \int_{ring} dq,$$

where we have taken $r = \sqrt{x^2 + a^2}$ outside the integral because it's the same for all charge elements. The remaining integral is simply the total charge Q, so we have

$$V = \frac{kQ}{\sqrt{x^2 + a^2}}. \tag{25-9}$$

Does this result make sense? At great distances from the ring $(x \gg a)$, a^2 in the denominator becomes negligible, and our result becomes

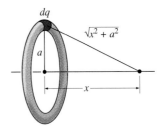

FIGURE 25-11   A charged ring (Example 25-6).

$$V = \frac{kQ}{x},$$

which is just the potential of a point charge Q—as we would expect when we're so far from the ring that its size is no longer significant. At the center of the ring, on the other hand, Equation 25-9 gives

$$V = \frac{kQ}{a}.$$

Here we're a distance a from all parts of the ring, and, since potential is a *scalar* the directions to those parts don't matter. The result is therefore the same as being a distance a from a point charge Q.   ●

● **EXAMPLE 25-7**   A CHARGED DISK

A charged disk of radius a carries a total charge Q distributed uniformly over its surface. What is the potential at a point P on the disk axis, a distance x from the disk?

Solution
To use Equation 25-8, we must divide the disk into charge elements dq. In the preceding example we found the potential of a charged ring, so we can take the charge elements of our disk to be thin rings (see Fig. 25-12), and integrate over all the rings comprising the disk. If a ring-shaped charge element has charge dq and radius r, then Equation 25-9 gives its potential dV a distance x from the disk center:

$$dV = \frac{k\,dq}{\sqrt{x^2 + r^2}}.$$

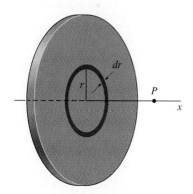

FIGURE 25-12 A charged disk, showing a ring-shaped charge element dq of radius r and width dr.

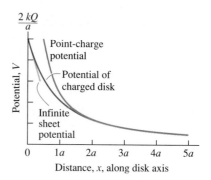

FIGURE 25-14 Charged-disk potential of Equation 25-10 approaches the potential of an infinite sheet for points close to the disk, and that of a point charge far from the disk.

Then the potential of the entire disk is

$$V = \int_{ring} dV = \int_{r=0}^{r=a} \frac{k\,dq}{\sqrt{x^2 + r^2}}.$$

To evaluate this integral, we must relate r and dq. "Unwinding" the ring gives a strip of area $2\pi r\,dr$ (Fig. 25-13). The surface charge density σ is the total charge divided by the disk area: $\sigma = Q/\pi a^2$. Then the charge dq on our infinitesimal ring of area $2\pi r\,dr$ is

$$dq = \sigma 2\pi r\,dr = \frac{Q}{\pi a^2} 2\pi r\,dr = \frac{2Q}{a^2} r\,dr.$$

Using this result in the integral for the potential gives

$$V = \int_0^a \frac{2kQ}{a^2} \frac{r\,dr}{\sqrt{x^2 + r^2}} = \frac{kQ}{a^2} \int_0^a \frac{2r\,dr}{\sqrt{x^2 + r^2}}.$$

Note that $2r\,dr = d(r^2) = d(x^2 + r^2)$ since x is a constant with respect to the integration. The integral therefore has the form $u^{-1/2}\,du$, where $u = x^2 + r^2$, and the result is $2u^{1/2}$ or

$$V = \frac{2kQ}{a^2}\sqrt{x^2 + r^2}\,\Big|_{r=0}^{r=a} = \frac{2kQ}{a^2}\left(\sqrt{x^2 + a^2} - |x|\right).$$

Figure 25-14 shows that this complicated-looking result makes sense: Close to the sheet, the potential resembles that of an infinite sheet, while far from the disk it approaches the potential of a point charge. Example 25-9 and Problem 72 explore these limiting cases further.

EXERCISE Point P in Fig. 25-15 lies a perpendicular distance y from the end of a uniformly charged rod of length ℓ and total charge Q. Find an expression for the potential at P, taking the zero of potential at infinity.

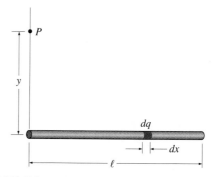

FIGURE 25-15 What is the potential at P? Figure shows a charge element dq, of length dx, to use in the integration for the potential.

Answer: $V = \dfrac{kQ}{\ell}\ln\left(\dfrac{\ell + \sqrt{\ell^2 + y^2}}{y}\right)$

Some problems similar to Examples 25-6 and 25-7: 36–39, 73, 80

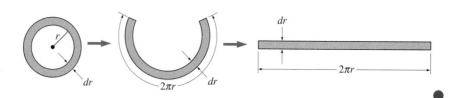

FIGURE 25-13 Unwinding the thin ring gives a strip of width dr and length $2\pi r$.

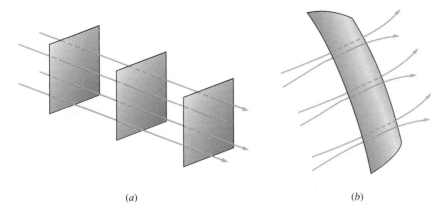

FIGURE 25-16 (*a*) Equipotential surfaces in a uniform electric field are planes perpendicular to the field. (*b*) In a nonuniform field the equipotential surfaces are curved, but are still perpendicular to the field.

(*a*)　　　　　　　　　　(*b*)

25-4 POTENTIAL DIFFERENCE AND THE ELECTRIC FIELD

Equipotentials

It takes no work to move a charge at right angles to an electric field. Therefore there can be no potential difference between two points on a surface that is everywhere perpendicular to the electric field. Such surfaces are called **equipotential surfaces,** or simply **equipotentials.** Figure 25-16 shows some equipotential surfaces for both uniform and nonuniform electric fields.

Equipotentials are like contour lines used on a map to show land elevation (Fig. 25-17). A contour line is a line of constant elevation, and therefore, it takes no work to move along a contour line. Contour lines are usually spaced at even increments of elevation. Where lines are closely spaced, the elevation changes quickly. Similarly, closely spaced equipotentials indicate large potential differences between nearby points. That, in turn, means it takes a lot of work to move charge between those points—and therefore there must be a large electric field present. Figure 25-17 might just as well represent electric potential, in which case regions with closely spaced equipotentials—steep slopes on the "potential hill"—indicate large electric fields. Similarly, the equipotentials for a dipole describe the steep "hill" of the positive charge and a correspondingly deep "hole" of the negative charge (Fig. 25-18).

Calculating the Field from the Potential

Given electric field lines, we can construct equipotentials. Conversely, given equipotentials we can reconstruct the field by sketching field lines at right angles to the equipotentials. Specifying the potential at each point thus conveys all the information needed to determine the field.

We can quantify the relation between potential and field by writing the potential difference dV between two points separated by an infinitesimal displacement $d\boldsymbol{\ell}$:

$$dV = -\mathbf{E} \cdot d\boldsymbol{\ell} = -E_\ell \, d\ell,$$

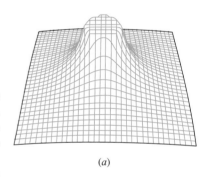

(*a*)

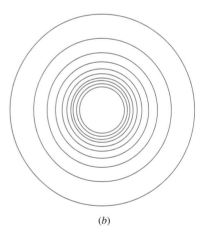

(*b*)

FIGURE 25-17 (*a*) A flat-topped hill and (*b*) its representation as a contour map. Closely spaced contours indicate steep slopes. Figure can also represent the potential of a uniformly charged spherical shell, whose potential is constant inside (since the electric field is zero) and falls as $1/r$ outside. Closely spaced contours then represent regions of strong electric field.

FIGURE 25-18 (*a*) Equipotentials and field lines for a dipole. The two sets of curves are everywhere perpendicular. (*b*) A plot of the potential as a function of position in the *x-y* plane shows steep "hill" for the positive charge and deep "hole" for the negative charge. The lines shown in (*b*) are *not* equipotentials.

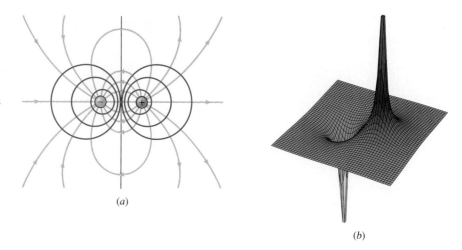

(*a*)

(*b*)

where E_ℓ designates the field component in the direction of $d\ell$. Rearranging this equation gives

$$E_\ell = -\frac{dV}{d\ell}. \qquad (25\text{-}10)$$

This equation confirms our statement that the electric field is strong where potential changes rapidly with distance. The minus sign in Equation 25-10 is the same one that appears in Equation 25-2; here it tells us that if we move in the direction of *increasing* potential, then we must be moving *against* the electric field. If we want to find the field components in a chosen coordinate system, we simply choose $d\ell$ along one of the coordinate axes and apply Equation 25-10; the *x* component of the field, for example, is given by $E_x = -dV/dx$. (If *V* is a function of all three coordinates we should, strictly speaking, write E_x in terms of the *partial* derivative, $\partial V/\partial x$, that we introduced in Chapter 16.) Equation 25-11, incidentally, shows that the units of electric field can be written as V/m.

● **EXAMPLE 25-8** THE FIELD OF A POINT CHARGE

Use the point-charge potential of Equation 25-4 to derive the electric field of a point charge.

Solution

The point-charge potential, $V(r) = kq/r$, depends only on r. Therefore the electric field points in the radial direction, and has the form $\mathbf{E} = E_r\hat{\mathbf{r}}$. Choosing $d\ell = dr$ in Equation 25-10 gives the field component:

$$E_r = -\frac{dV}{dr} = -\frac{d}{dr}\left(\frac{kq}{r}\right) = -kq\frac{d(r^{-1})}{dr} = kqr^{-2} = \frac{kq}{r^2}.$$

Thus $\mathbf{E} = \frac{kq}{r^2}\hat{\mathbf{r}}$, as expected.

●

● EXAMPLE 25-9 THE FIELD OF A CHARGED DISK

Use the result of Example 25-7 to find the electric field on the axis of a charged disk.

Solution

Symmetry shows that the field must point along the disk axis, which is the x axis in Fig. 25-12. So the field has only an x component, given by applying Equation 25-10 to the disk potential:

$$E_x = -\frac{dV}{dx} = -\frac{d}{dx}\left(\frac{2kQ}{a^2}(\sqrt{x^2 + a^2} - |x|)\right)$$

$$= \frac{2kQ}{a^2}\left(1 - \frac{|x|}{\sqrt{x^2 + a^2}}\right).$$

To see that this makes sense, consider the case $x \ll a$, for which the field becomes approximately $2kQ/a^2$. Writing Q as the surface density σ times the area πa^2 gives $E_x = 2\pi k\sigma = \sigma/2\varepsilon_0$, which is the field of an infinite charged sheet. Of course—very close to the disk it looks effectively infinite, and its field should be well approximated by that of an infinite sheet. Problem 72 shows that the field far from the disk approaches that of a point charge Q.

EXERCISE Use Equations 25-7 and 25-10 to calculate the electric field on the axis of a point dipole, and show that your result is equivalent to Equation 23-5b.

Some problems similar to Examples 25-8 and 25-9: 49–51, 80

Examples 25-8 and 25-9 show that it is often much easier to calculate the electric field by first finding the potential and then differentiating, rather than doing a vector integration to get the field of a complicated charge distribution.

It's important to recognize that the *values* of the field and potential are not directly related; rather, as Equation 25-11 indicates, the field measures the *rate of change* of the potential. Field and potential are like acceleration and velocity; the *values* of the two are quite independent, with the former depending on the *rate of change* of the latter. Figure 25-19 and Example 25-10 illustrate the relation between potential and field.

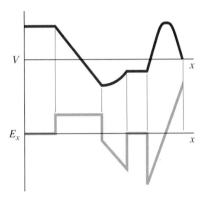

FIGURE 25-19 The x component of the electric field is the negative rate of change of the potential with respect to x.

● EXAMPLE 25-10 POTENTIAL AND FIELD

A positive charge $+2q$ lies at $x = -a$, and a charge $-q$ lies at $x = +a$. (a) Derive an expression for the potential on the x axis, and find a point on the axis in the region $x > a$ where the potential (with respect to infinity) is zero. (b) Use your expression for potential to find the electric field for $x > a$. Is the electric field zero where the potential is zero? If not, where is the field zero?

Solution

(a) The distance from any point x to the positive charge at $-a$ is $r_+ = |x - (-a)| = |x + a|$; similarly, the distance to the negative charge is $r_- = |x - a|$. Thus, the potential on the axis is

$$V(x) = \frac{k(2q)}{r_+} + \frac{k(-q)}{r_-} = kq\left(\frac{2}{|x + a|} - \frac{1}{|x - a|}\right).$$

We can find where $V = 0$ by setting the quantity in parentheses to zero. For $x > a$ both denominators are positive and we can remove the absolute value signs. For $V = 0$ we then have

$$\frac{2}{x + a} = \frac{1}{x - a}.$$

Solving for x gives $x = 3a$.

(b) The field clearly has only an x component, and Equation 25-10 then gives

$$E_x = -\frac{dV}{dx} = -kq\frac{d}{dx}\left(\frac{2}{x + a} - \frac{1}{x - a}\right)$$

$$= \frac{2kQ}{(x + a)^2} - \frac{kQ}{(x - a)^2}.$$

This result is hardly surprising: It's just the sum of two point-charge fields. We actually solved for the zero-field point in this configuration in Example 23-3; there we found that the electric force would be zero at $x = 5.83a$, which is *not* the same place where the potential is zero.

Figure 25-20 shows the potential of this charge distribution in the region $x > a$, showing clearly the point $x = 3a$ where $V = 0$. Getting a charge to this point from infinity would take

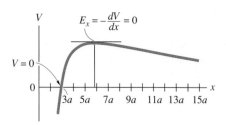

FIGURE 25-20   Potential on the x axis for a charge distribution consisting of $+2q$ at $x = -a$ and $-q$ at $x = +a$. Shown here is the region $x > a$, including the point $x = 3a$ where the potential is 0 and the point $x = 5.83a$ where the electric field—proportional to the *slope* of the potential curve—is zero. Note the deep "hole" associated with the negative charge.

no *net* work, although it would require going up and then down a potential "hill." Also clear in Fig. 25-20 is the point $x = 5.83a$ where the electric field is zero. Note that this is the point where the *slope* of the potential curve is zero, *not* where the potential itself is zero.

There are actually two points on the x axis where $V = 0$; the second lies between the charges. Both lie on an equipotential surface of zero potential that surrounds the negative charge; Fig. 25-21 shows some equipotentials and field lines in the x-y plane.

> **TIP   Field and Potential Are Not Proportional**   In particular, where one is zero, the other need not be zero. You can see that in Fig. 25-20, where it clearly takes work to get from infinity to the point where $E = 0$. Just because a mountaintop is flat doesn't mean it didn't take work to climb it! Potential depends not on the field at a point but on the field over an entire path from infinity to that point. Similarly, the potential can be zero at points where the field is not, as evidenced by the steep slope of the potential curve in Fig. 25-20 at the point where it crosses zero.

EXERCISE   Find the second point on the x axis in Example 25-10 where the potential is zero.

Answer: $x = a/3$

Some problems similar to Example 25-10:   48, 52, 79

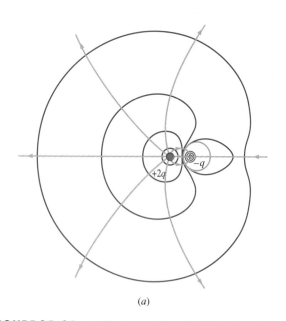

(a)

(b)

FIGURE 25-21   (a) Equipotentials and field lines for the two opposite but unequal charges of Example 25-10. Blue circle is the $V = 0$ equipotential whose x-axis intersections were calculated in Example 25-10. Values of potential inside this circle are negative; all others are positive. Note that at large distances positive equipotentials surround the entire charge distribution. (b) Three-dimensional plot of the potential over the same region shown in (a). "Hill" of the positive charge $2q$ is larger than the "hole" of the negative charge q because of the difference in charge magnitudes. Note that one would have to go "uphill" coming in from infinity before dropping into the region of negative potential near the negative charge. That means there is an equipotential with $V = 0$ surrounding the negative charge, as shown in (a).

25-5 POTENTIALS OF CHARGED CONDUCTORS

That there is no electric field inside a conductor in electrostatic equilibrium means it takes no work to move a test charge around inside the conductor; that the field at the conductor surface is perpendicular to the surface means it takes no work to move a test charge along the surface, either. Therefore the potential difference between two points in or on a conductor must be zero, and thus,

▌ **A conductor in electrostatic equilibrium is an equipotential.**

Consider an isolated, spherical conductor of radius R carrying charge Q. Since the conductor is isolated, charge is distributed uniformly over its surface, and for $r > R$ it therefore acts like a point charge. The potential at its surface is then

$$V(R) = \frac{kQ}{R}, \tag{25-11}$$

as we found in Example 25-3. The sphere itself is an equipotential, so the potential *difference* between two points *on the sphere* is zero. But that doesn't mean the potential difference between *infinity* and the sphere is zero; since the sphere is charged, it takes work to move charge to its surface from infinity, and that's what Equation 25-11 says.

Now consider two widely separated spheres of different sizes. If we connect them by a thin conducting wire, as shown in Fig. 25-22, then the system constitutes a single conductor, and charge will move through the wire until both spheres are at the same potential. But since the spheres are widely separated, each still has an essentially spherical charge distribution, so Equation 25-11 applies to each. Since the spheres have the same potential, Equation 25-11 implies that

$$\frac{kQ_1}{R_1} = \frac{kQ_2}{R_2},$$

where the subscripts label the two spheres. We can write each charge as the surface area of the sphere multiplied by the surface charge density: $Q = 4\pi R^2 \sigma$. Substituting for the Q's in the above equation and solving for the ratio of surface charge densities then gives

$$\frac{\sigma_1}{\sigma_2} = \frac{R_2}{R_1}.$$

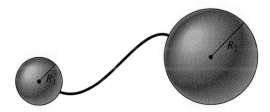

FIGURE 25-22 Two conducting spheres held at the same potential by a conducting wire. The surface charge density is greater on the smaller sphere, in inverse proportion to its radius.

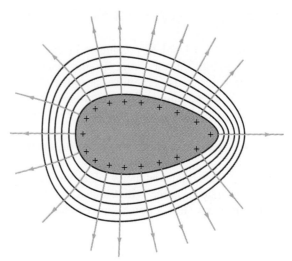

FIGURE 25-23 An irregular charged conductor. Equipotentials near the conductor have approximately its shape, and the field is strongest—and therefore the equipotentials closest—where the conductor curves most sharply.

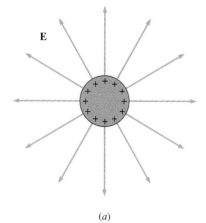

(a)

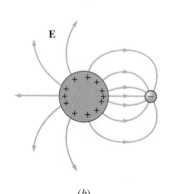

(b)

FIGURE 25-24 (a) An isolated conducting sphere carries a uniform surface charge density and has a spherically symmetric electric field. (b) The presence of a nearby charge distorts the surface charge distribution and therefore the electric field.

Thus the *smaller* sphere has the *larger* surface charge density. Since the electric field at a conductor surface has magnitude $E = \sigma/\varepsilon_0$, the field must be stronger at the smaller sphere.

This discussion of spherical conductors provides a qualitative description of nonspherical conductors as well. All parts of an irregularly shaped conductor must be at the same potential. Where the conductor surface curves sharply, it is like a small sphere and therefore has a higher surface charge density and a stronger electric field. In general, the field is strongest where the surface curves most sharply.*

Because a conductor surface is an equipotential and the electric field is perpendicular to the conductor surface, equipotentials just above the surface must have approximately the same shape as the surface. Because the electric field is stronger where the conductor surface curves sharply, there must be more field lines emerging from such regions. Far from a charged conductor, on the other hand, its field must resemble that of a point charge, with radial field lines and circular equipotentials. With these limiting cases in mind, we can sketch the approximate form for the field of an arbitrarily shaped conductor (Fig. 25-23).

We stress that our conclusion about surface charge density and curvature applies only to *isolated* conductors—those far from any other charges. The field of a nearby charge will modify the charge distribution of a conductor, altering the surface charge distribution (Fig. 25-24).

* This association of strong field and sharp curvature is only approximate and, in some unusual configurations, may not hold at all. See "The Lightning Rod Fallacy," R. H. Price and R. J. Crowley, *American Journal of Physics,* vol. 53, September 1985, p. 843.

■ APPLICATION CORONA DISCHARGE, POLLUTION CONTROL, AND XEROGRAPHY

The large electric fields that develop where a charged conductor is sharply curved can cause serious problems in electrical equipment; in other applications those fields are put to good use. Fields above 3 MN/C are strong enough to strip electrons from air molecules, making the air a conductor. Breakdown of air is often evidenced by a blue glow around sharply-pointed conductors (Fig. 25-25). Called **corona discharge,** this glow results from the recombination of electrons with atoms. Corona discharge causes loss of power from high-voltage transmission lines, and engineers try to avoid it by eliminating sharp edges on wires and other conducting structures.

Corona discharge is put to good use in the **electrostatic precipitator,** a pollution-control device used especially on coalburning power plants. A typical precipitator consists of parallel metal plates with thin wires running between them. Application of a high voltage between plates and wires sets up large electric fields near the wires. Exhaust gases flow between the plates and the field ionizes some gas molecules. These charged molecules, in turn, attach themselves to pollutant particles. The charged particles are driven to the collecting plates by the electric field. Every few minutes a mechanical vibrator taps the plates and the particles fall into a hopper, where they can be trucked away to use for fill or in making products like cinder blocks. A typical power plant produces some 30 pounds of particulates every second, and precipitators keep most of this out of the air. In the process, the precipitators may consume several per cent of the power plant's electrical output. Fig. 25-26 shows electrostatic precipitators at a large power plant.

Xerography (literally, "dry writing") used in copiers, laser printers, and similar devices is another widespread application of corona discharge. Copying starts as corona discharge between a charged wire and a special light-sensitive plate spreads a layer of positive charge on the plate (Fig. 25-27a). Light, imaged from the document being copied or scanned by a computer-driven laser mechanism, strikes the plate and causes charge to flow from the illuminated regions (Fig. 25-27b). Negatively charged particles called toner are then spread on the plate, where they adhere to the charged regions that correspond to dark areas in the image (Fig. 25-27c). Finally, the toner is transferred to paper and heated to fuse it into the paper, making a permanent copy.

(a)

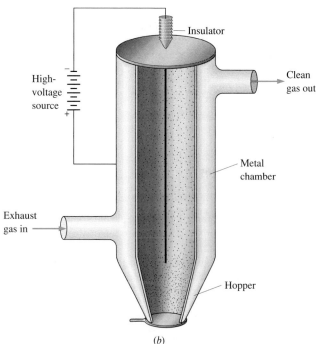

(b)

FIGURE 25-26 (a) Electrostatic precipitators at a 1300-MW coal-burning power plant. (b) In the electrostatic precipitator, high voltage applied between the metal chamber and the thin central wire results in a strong, nonuniform electric field near the wire. This field ionizes air molecules, which attach to soot particles with the result that the particles are accelerated to the chamber walls. Mechanical vibration then dislodges the particles into the hopper for eventual collection. Parallel metal plates are often used instead of the cylindrical chamber shown here.

FIGURE 25-25 Corona discharge on a power-line insulator.

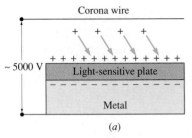 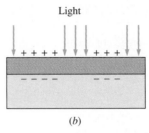 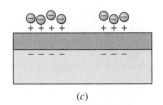

(a)   (b)   (c)

FIGURE 25-27 The xerography process, widely used in copiers and laser printers. (a) Corona discharge from a thin wire lays positive charge on a light-sensitive surface. (b) Where light hits, charge is driven into the metal substrate, leaving a charge pattern corresponding to the image being reproduced. (c) Negatively charged toner particles stick to the plate, and are later transferred and fused to paper.

CHAPTER SYNOPSIS

Summary

1. The **electric potential difference** between two points is the change in potential energy per unit charge for a charge moved between those points. Because the electric field is conservative, the potential difference between two points is independent of the path taken, and is also equal to the work per unit charge that an external agent must do in moving charge at constant speed between the two points. The potential difference between two points A and B is calculated by evaluating the **line integral** of the electric field over any path between those points:

$$\Delta V_{A \rightarrow B} = -\int_A^B \mathbf{E} \cdot d\boldsymbol{\ell}.$$

When the field is uniform, this expression reduces to

$$\Delta V_{A \rightarrow B} = -\mathbf{E} \cdot \boldsymbol{\ell}.$$

2. Defining the potential to be zero at some point allows us to speak of "the potential at a point," meaning the potential difference from the reference point to the point in question. For isolated point charges, a convenient zero is infinitely far from the charge; then the potential at an arbitrary point a distance r from the point charge q is

$$V(r) = \frac{kq}{r}. \qquad \text{(point-charge potential)}$$

The potentials of charge distributions may be found by taking the line integral of the field, if the latter is known, or by summing the potentials of the point charges making up the distribution:

$$V = \sum_i \frac{kq_i}{r_i}, \quad \begin{pmatrix} \text{discrete} \\ \text{charges} \end{pmatrix} \quad \text{or}$$

$$V = \int \frac{k\,dq}{r}. \quad \begin{pmatrix} \text{continuous charge} \\ \text{distribution} \end{pmatrix}$$

3. **Equipotentials** are surfaces over which the potential has a constant value. Equipotentials are everywhere perpendicular to the electric field. The field is strong where equipotentials are closely spaced and vice versa. Mathematically, the field component in a given direction is related to the rate of change of potential with position in that direction:

$$E_\ell = -\frac{dV}{d\ell}.$$

4. A conductor in electrostatic equilibrium is an equipotential. The surface charge density and therefore the electric field at the conductor surface are usually greatest where the conductor curves most sharply. Very strong electric fields occur at sharp bends; if strong enough, these fields can result in **corona discharge,** in which the surrounding air becomes a conductor and charge leaks off the charged conductor.

Terms You Should Understand

(Pairs are closely related terms whose distinction is important; number in parentheses is chapter section where term first appears.)
conservative field (introduction)
potential difference (25-2)
volt, electron volt (25-2)
line integral (25-2)
equipotential (25-4)
corona discharge (25-5)

Symbols You Should Recognize

$\Delta V_{A \rightarrow B}$ (25-2)
V, eV (25-2)
$V(P)$ (25-3)
$d\ell$ (25-2)

Problems You Should Be Able to Solve

calculating the work needed to move a given charge through a given potential difference (25-2)

calculating potential differences in uniform electric fields (25-2)

calculating potential differences using the line integral of a known electric field (25-3)

evaluating potentials by summing or integrating over point charges (25-3)

finding electric field components given potential as a function of position (25-4)

sketching equipotentials of simple charge distributions (25-4)

sketching equipotentials and fields around conductors (25-5)

Limitations to Keep in Mind

Electric potential difference depends on *two points*. Phrases like "the potential at point P" or "the potential of a point charge" are always shorthand ways of talking about the potential difference between two points. When the second point is not specified, it is often taken to be at infinity.

QUESTIONS

1. Why can a bird perch on a high-voltage power line without getting electrocuted?

2. One proton is accelerated from rest by a uniform electric field, the other by a nonuniform electric field. If they move through the same potential difference, how do their final speeds compare?

3. Would a free electron move toward higher or lower potential?

4. The potential difference from A to B in Fig. 25-28 is zero since the two points are equidistant from the charge Q. How can this be, when a charge moving along the path shown clearly experiences an electric force not perpendicular to the path?

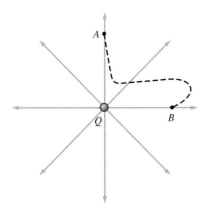

FIGURE 25-28   Question 4.

5. A proton and a positron (a particle with the electron's mass carrying charge $+e$) are accelerated through the same potential difference. How do their final energies compare? Their final speeds?

6. The electric field at the center of a uniformly charged ring is obviously zero, yet Example 25-6 shows that the potential at the center is not zero. How is this possible?

7. Must the potential be zero at any point where the electric field is zero? Explain.

8. Must the electric field be zero at any point where the potential is zero? Explain.

9. The potential is constant throughout an entire volume. What must be true of the electric field within that volume?

10. In considering the potential of an infinite flat sheet, why is it not useful to take the zero of potential at infinity?

11. The potential of a point charge is given by kq/r. Is r the distance between the two points for which this is the potential difference? Explain.

12. "Cherry picker" trucks for working in trees or power lines often carry the warning sign shown in Fig. 25-29. Explain how this hazard arises and why it might be more of a danger to someone on the ground than to a worker on the truck.

FIGURE 25-29   Question 12.

13. Two positive point charges are located a small distance apart. Are there any points, other than at infinity, where the potential is zero?

14. Is it possible for equipotential surfaces to intersect? Explain.

15. Is the potential at the center of a hollow, uniformly charged spherical shell higher, lower, or the same as at the surface?

16. A solid sphere contains charge uniformly distributed throughout its volume. Is the potential at its center higher, lower, or the same as at the surface?

17. Why do the spheres in Fig. 25-21 need to be far apart for the conclusion that surface charge density is inversely proportional to radius to hold accurately?

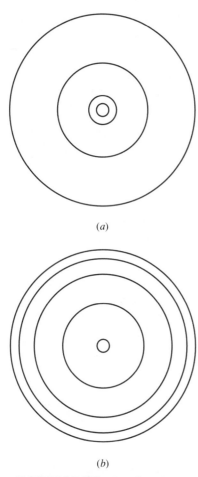

(a)

(b)

FIGURE 25-30 Question 18.

18. Figure 25-30 shows cross sections of two sets of spherical equipotentials, spaced in even increments of potential difference. Describe qualitatively how the charge distributions in the two regions differ.
19. Two equal but opposite charges form a dipole. Describe the equipotential surface on which $V = 0$.
20. Figure 25-31 shows three paths leading from infinity to a point P on the perpendicular bisector of a dipole. For each path, how much work is needed to bring a charge q from infinity to P?

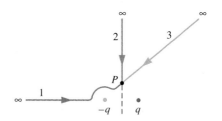

FIGURE 25-31 Question 20.

21. The electric potential in a region increases linearly with distance. What can you conclude about the electric field in this region?
22. In Fig. 25-23b the charge density on the conducting sphere is not uniform, yet the field inside must still be zero since the system is in electrostatic equilibrium. How is this possible?
23. Why is lightning likely to strike an isolated tree?
24. What is the difference between a volt and an electron volt?

PROBLEMS

Section 25-2 Potential Difference

1. How much work does it take to move a 50-μC charge against a 12-V potential difference?
2. The potential difference between the two sides of an ordinary electrical outlet is 120 V. How much energy does an electron gain when it moves from one side to the other?
3. It takes 45 J to move a 15-mC charge from point A to point B. What is the potential difference $\Delta V_{A \to B}$?
4. Show that 1 V/m is the same as 1 N/C.
5. Find the magnitude of the potential difference between two points located 1.4 m apart in a uniform 650 N/C electric field, if a line between the points is parallel to the field.
6. A charge of 3.1 C moves from the positive to the negative terminal of a 9.0-V battery. How much energy does the battery impart to the charge?

7. Two points A and B lie 15 cm apart in a uniform electric field, with the path AB parallel to the field. If the potential difference $\Delta V_{A \to B}$ is 840 V, what is the field strength?
8. Figure 25-32 shows a uniform electric field of magnitude E. Find expressions for (a) the potential difference $\Delta V_{A \to B}$ and (b) $\Delta V_{B \to C}$. (c) Use your result to determine $\Delta V_{A \to C}$.

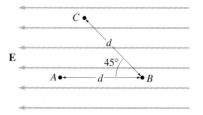

FIGURE 25-32 Problem 8.

9. A proton, an alpha particle (a bare helium nucleus), and a singly ionized helium atom are accelerated through a potential difference of 100 V. Find the energy each gains.

10. Two points A and B lie 77 cm apart in a uniform 540 V/m electric field. If $\Delta V_{A \to B} = 390$ V, what angle does a line from A to B make with the field?

11. What is the potential difference between the terminals of a battery that can impart 7.2×10^{-19} J to each electron that moves between the terminals?

12. Electrons in a TV tube are accelerated from rest through a 25-kV potential difference. With what speed do they hit the TV screen?

13. A 12-V car battery stores 2.8 MJ of energy. How much charge can move between the battery terminals before it is totally discharged? Assume the potential difference remains at 12 V, an assumption that is not realistic.

14. What is the charge on an ion that gains 1.6×10^{-15} J when it moves through a potential difference of 2500 V?

15. Two large, flat metal plates are a distance d apart, where d is small compared with the plate size. If the plates carry surface charge densities $\pm \sigma$, show that the potential difference between them is $V = \sigma d / \varepsilon_0$.

16. An electron passes point A moving at 6.5 Mm/s. At point B the electron has come to a complete stop. Find the potential difference $\Delta V_{A \to B}$.

17. A 5.0-g object carries a net charge of 3.8 μC. It acquires a speed v when accelerated from rest through a potential difference V. A 2.0-g object acquires twice the speed under the same circumstances. What is its charge?

Section 25-3 Calculating Potential Difference

Note: In these problems, the zero of potential is taken at infinity unless noted otherwise.

18. An electric field is given by $\mathbf{E} = E_0 \hat{\mathbf{j}}$, where E_0 is a constant. Find the potential as a function of position, taking $V = 0$ at $y = 0$.

19. The classical picture of the hydrogen atom has a single electron in orbit a distance 0.0529 nm from the proton. Calculate the electric potential associated with the proton's electric field at this distance.

20. Earth carries an electric charge of -4.3×10^5 C, distributed essentially uniformly over its surface. What is the potential difference between Earth's surface and the base of the ionosphere, about 80 km above the surface?

21. Points A and B lie 20 cm apart on a line extending radially from a point charge Q, and the potentials at these points are $V_A = 280$ V, $V_B = 130$ V. Find Q and the distance r between A and the charge.

22. What is the maximum potential allowable on a 5.0-cm-diameter metal sphere if the electric field at the sphere's surface is not to exceed the 3 MV/m breakdown field in air?

23. A 3.5-cm-diameter isolated metal sphere carries a net charge of 0.86 μC. (a) What is the potential at the sphere's

surface? (b) If a proton were released from rest at the sphere's surface, what would be its speed far from the sphere?

24. A sphere of radius R carries a negative charge of magnitude Q, distributed in a spherically symmetric way. Find the "escape speed" for a proton at the sphere's surface—that is, the speed that would enable the proton to escape to arbitrarily large distances.

25. A thin spherical shell of charge has radius R and total charge Q distributed uniformly over its surface. What is the potential at its center?

26. A solid sphere of radius R carries a net charge Q distributed uniformly throughout its volume. Find the potential difference from the sphere's surface to its center. *Hint:* Consult Example 24-1.

27. Find the potential as a function of position in an electric field given by $\mathbf{E} = ax\hat{\mathbf{i}}$, where a is a constant and where $V = 0$ at $x = 0$.

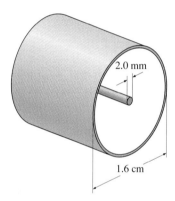

FIGURE 25-33 Problem 28.

28. A coaxial cable consists of a 2.0-mm-diameter inner conductor and an outer conductor of diameter 1.6 cm and negligible thickness (Fig. 25-33). If the conductors carry line charge densities ± 0.56 nC/m, what is the magnitude of the potential difference between them?

29. The potential difference between the surface of a 3.0-cm-diameter power line and a point 1.0 m distant is 3.9 kV. What is the line charge density on the power line?

30. Three equal charges q form an equilateral triangle of side a. Find the potential at the center of the triangle.

31. A charge $+Q$ lies at the origin, and $-3Q$ at $x = a$. Find two points on the x axis where $V = 0$.

32. Two identical charges q lie on the x axis at $\pm a$. (a) Find an expression for the potential at all points in the x-y plane. (b) Show that your result reduces to the potential of a point charge for distances large compared with a.

33. Find the potential 10 cm from a dipole of moment $p = 2.9$ nC$\cdot$m (a) on the dipole axis, (b) at 45° to the axis, and (c) on the perpendicular bisector. The dipole separation is much less than 10 cm.

34. Two points A and B lie 55 cm from a dipole of moment $p = 6.4$ nC·m, whose charge separation is much less than 55 cm. A line from the dipole to A makes a 20° angle with the dipole axis, and a line to B makes a 50° angle. Find the potential difference $V_B - V_A$.

35. The potential at point P in Fig. 25-34 is 37 kV. Find the charge separation a, assuming it is much smaller than the 1.6-cm distance to P.

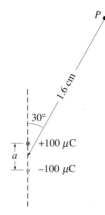

FIGURE 25-34 Problem 35.

36. A thin plastic rod 20 cm long carries 3.2 nC distributed uniformly over its length. (a) If the rod is bent into a circular ring, find the potential at its center. (b) If the rod is bent into a semicircle, find the potential at the center (i.e., at the center of the circle of which the semicircle is part).

37. A thin ring of radius R carries a charge $3Q$ distributed uniformly over three-fourths of its circumference, and $-Q$ over the rest. What is the potential at the center of the ring?

38. The potential at the center of a uniformly charged ring is 45 kV, and 15 cm along the ring axis the potential is 33 kV. Find the ring's radius and its total charge.

39. The annulus shown in Fig. 25-35 carries a uniform surface charge density σ. Find an expression for the potential at an arbitrary point P on its axis.

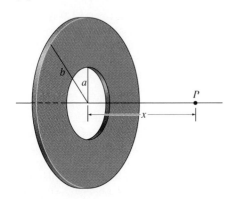

FIGURE 25-35 Problem 39.

40. A thin rod of length ℓ carries a charge Q distributed uniformly over its length. (a) Show that the potential in the plane that perpendicularly bisects the rod is given by

$$V(r) = \frac{2kQ}{\ell} \ln\left[\frac{\ell}{2r} + \sqrt{1 + \frac{\ell^2}{4r^2}}\,\right],$$

where r is the distance from the rod center. (b) Show that this expression reduces to an expected result when $r \gg \ell$. *Hint:* See Appendix A for a series expansion of the logarithm.

41. (a) Find the potential as a function of position in the electric field $\mathbf{E} = E_0(\hat{\mathbf{i}} + \hat{\mathbf{j}})$, where $E_0 = 150$ V/m. Take the zero of potential at the origin. (b) Find the potential difference from the point $x = 2.0$ m, $y = 1.0$ m to the point $x = 3.5$ m, $y = -1.5$ m.

Section 25-4 Potential Difference and the Electric Field

42. In a uniform electric field, equipotential planes that differ by 1.0 V are 2.5 cm apart. What is the field strength?

43. Figure 25-36 shows a plot of potential versus position along the x axis. Make a plot of the x component of the electric field for this situation.

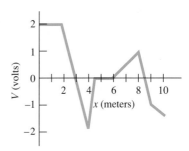

FIGURE 25-36 Problem 43.

44. Figure 25-37 shows some equipotentials in the x-y plane. (a) In what region is the electric field strongest? What are (b) the direction and (c) the magnitude of the field in this region?

45. The potential in a certain region is given by $V = axy$, where a is a constant. (a) Determine the electric field in the region. (b) Sketch some equipotentials and field lines.

46. Sketch some equipotentials and field lines for a distribution consisting of two equal point charges.

47. Figure 25-38 shows some equipotentials in the x-y plane. The equipotentials shown are 10 V apart, as indicated. Find an expression for the electric field in the region.

48. Sketch some equipotentials and field lines for a distribution consisting of two point charges $+3Q$ and $-Q$.

49. The electric potential in a region of space is given by $V = 2xy - 3zx + 5y^2$, with V in volts and the coordinates in

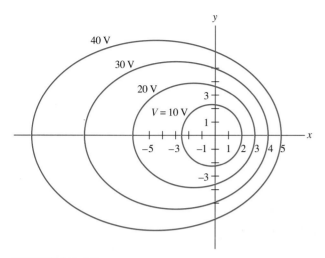

FIGURE 25-37 Problem 44.

meters. If point P is at $x = 1$ m, $y = 1$ m, $z = 1$ m, find (a) the potential at P and (b) the x, y, and z components of the electric field at P.

50. Use Equation 25-7 to calculate the electric field on the perpendicular bisector of a point dipole, and show that your result is equivalent to Equation 23-5a.

51. Use the result of Example 25-6 to determine the on-axis field of a charged ring, and verify that your answer agrees with the result of Example 23-8.

52. A charge $+4q$ is located at the origin and a charge $-q$ is on the x axis at $x = a$. (a) Write an expression for the potential on the x axis for $x > a$. (b) Find a point in this region where $V = 0$. (c) Use the result of (a) to find the electric field on the x axis for $x > a$ and (d) find a point where $\mathbf{E} = \mathbf{0}$.

53. The electric potential in a region is given by $V = -V_0(r/R)$, where V_0 and R are constants, r is the radial distance from the origin, and where the zero of potential is taken at $r = 0$. Find the magnitude and direction of the electric field in this region.

Section 25-5 Potentials of Charged Conductors

54. (a) How much charge can be placed on a metal sphere 1.0 cm in diameter before corona discharge occurs to the surrounding air? (b) What is the sphere's potential at this maximum charge?

55. The spark plug in an automobile engine has a center electrode made from wire 2.0 mm in diameter. The electrode is worn to a hemispherical shape, so it behaves approximately like a charged sphere. What is the minimum potential on this electrode that will ensure the plug sparks in air? Neglect the presence of the second electrode.

56. A large metal sphere has three times the diameter of a smaller sphere and carries three times as much charge. Both spheres are isolated, so their surface charge densities are uniform. Compare (a) the potentials and (b) the electric field strengths at their surfaces.

57. Two metal spheres each 1.0 cm in radius are far apart. One sphere carries 38 nC of charge, the other -10 nC. (a) What is the potential on each? (b) If the spheres are connected by a thin wire, what will be the potential on each once equilibrium is reached? (c) How much charge must move between the spheres in order to achieve equilibrium?

58. Sketch some equipotentials and field lines for the isolated, charged conductor shown in Fig. 25-39.

59. Two conducting spheres are each 5.0 cm in diameter and each carries 0.12 μC. They are 8.0 m apart. Determine (a) the potential on each sphere; (b) the field strength at the surface of each sphere; (c) the potential midway between the spheres; (d) the potential difference between the spheres.

60. Two small metal spheres are located 2.0 m apart. One has radius 0.50 cm and carries 0.20 μC. The other has radius 1.0 cm and carries 0.080 μC. (a) What is the potential

FIGURE 25-38 Problem 47.

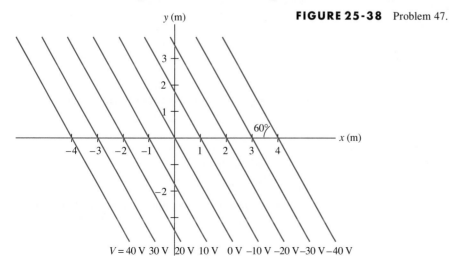

FIGURE 25-39 Problem 58.

difference between the spheres? (b) If they were connected by a thin wire, how much charge would move along it, and in which direction?

Paired Problems

(Both problems in a pair involve the same principles and techniques. If you can get the first problem, you should be able to solve the second one.)

61. Three 50-pC charges sit at the vertices of an equilateral triangle 1.5 mm on a side. How much work would it take to bring a proton from very far away to the midpoint of one of the triangle's sides?

62. Repeat the preceding problem for the case when one of the charges is -50 pC and the proton is brought to the midpoint of the side between the two positive charges.

63. A pair of equal charges q lies on the x axis at $x = \pm a$. (a) Find expressions for the potential at points on the x axis for which $x > a$ and (b) show that your result reduces to a point-charge potential for $x \gg a$.

64. (a) For the charge distribution of the preceding problem, find an expression for the potential at *all* points on the y axis. (b) Show that your result reduces to a point-charge potential for $y \gg a$.

65. A 2.0-cm-radius metal sphere carries 75 nC and is surrounded by a concentric spherical conducting shell of radius 10 cm carrying -75 nC. (a) Find the potential difference between the shell and the sphere. (b) How would your answer change if the shell charge were changed to $+150$ nC?

66. A coaxial cable consists of a 2.0-mm-radius central wire carrying 75 nC/m, and a concentric outer conductor of radius 10 mm carrying -75 nC/m. (a) Find the potential difference between the outer and inner conductor. (b) How would your answer change if the outer conductor were charged to $+150$ nC/m?

67. On the x axis, the electric field of a certain charge distribution is given by $\mathbf{E} = a/x^4\hat{\mathbf{i}}$, where $a = 55$ V·m³. Find the potential difference from the point $x = 1.3$ m to the point $x = 2.8$ m.

68. A sphere of radius R carries a nonuniform but spherically symmetric volume charge density that results in an electric field in the sphere given by $\mathbf{E} = E_0(r/R)^2\hat{\mathbf{r}}$, where E_0 is a constant. Find the potential difference from the sphere's surface to its center.

69. The potential as a function of position in a certain region is given by $V(x) = 3x - 2x^2 - x^3$, with x in meters and V in volts. Find (a) all points on the x axis where $V = 0$, (b) an expression for the electric field, and (c) all points on the x axis where $\mathbf{E} = \mathbf{0}$.

70. The potential in a certain region is given by $V(x) = -[2x^2 + (y - 1)^2 - 1]$. Find (a) a point where $V = 0$, (b) an expression for the electric field, and (c) a point in the x-y plane where $\mathbf{E} = \mathbf{0}$.

Supplementary Problems

71. A conducting sphere 5.0 cm in radius carries 60 nC. It is surrounded by a concentric spherical conducting shell of radius 15 cm carrying -60 nC. (a) Find the potential at the sphere's surface, taking the zero of potential at infinity. (b) Repeat for the case when the shell also carries $+60$ nC.

72. Show that the result of Example 25-9 approaches the field of a point charge for $x \gg a$. *Hint:* You will need to apply the binomial theorem to the quantity $1/\sqrt{x^2 + a^2}$.

73. The potential on the axis of a uniformly charged disk at 5.0 cm from the disk center is 150 V; the potential 10 cm from disk center is 110 V. Find the disk radius and its total charge.

74. A uranium nucleus (mass 238 u, charge 92e) decays, emitting an alpha particle (mass 4 u, charge 2e) and leaving a thorium nucleus (mass 234 u, charge 90e). At the instant the alpha particle leaves the nucleus, the centers of the two are 7.4 fm apart and are essentially at rest. Find their speeds when they are a great distance apart. Treat each particle as a spherical charge distribution.

75. A power line consists of two parallel wires 3.0 cm in diameter spaced 2.0 m apart. If the potential difference between the wires is 4.0 kV, what is the charge per unit length on each wire? The wires carry equal but opposite charges. *Hint:* The wires are far enough apart that they don't greatly affect each other's fields.

76. For the dipole of Example 25-5, show that the electric field at an arbitrary point far from the dipole can be written
$$\mathbf{E} = \frac{kp}{r^3}[(3\cos^2\theta - 1)\hat{\mathbf{i}} + 3\sin\theta\cos\theta\hat{\mathbf{j}}].$$

77. A thin rod of length ℓ lies on the x axis with its center at the origin. It carries a line charge density given by $\lambda = \lambda_0(x/\ell)^2$, where λ_0 is a constant. (a) Find an expression for the potential on the x axis for $x > \ell/2$. (b) Integrate the charge density to find the total charge on the rod. (c) Show that your answer for (a) reduces to the potential of a point charge whose charge is the answer to (b), for $x \gg \ell$.

78. Repeat the preceding problem for the case $\lambda = \lambda_0(x/\ell)$. Why is your answer for $x \gg \ell$ different? *Hint:* What does this charge distribution resemble at large distances?

79. For the situation of Example 25-10, find an equation for the equipotential with $V = 0$ in the *x-y* plane. Plot the equipotential, and show that it passes through the points described in Example 25-10 and its exercise.

80. A disk of radius a carries a nonuniform surface charge density given by $\sigma = \sigma_0(r/a)$, where σ_0 is a constant. (a) Find the potential at an arbitrary point on the disk axis, a distance x from the disk center. (b) Use the result of (a) to find the electric field on the disk axis, and (c) show that the field reduces to an expected form for $x \gg a$.

81. An open-ended cylinder of radius a and length $2a$ carries charge q spread uniformly over its surface. Find the potential on the cylinder axis at its center. *Hint:* Treat the cylinder as a stack of charged rings, and integrate.

CHAPTER 26

ELECTROSTATIC ENERGY AND CAPACITORS

A test firing of the Particle Beam Fusion Accelerator at Sandia National Laboratories involves the sudden release of energy stored in electric fields.

Suppose you hold two positive charges in your outstretched arms (Fig. 26-1). Bringing them closer takes work, as you move each charge against the other's electric field. That work is stored as potential energy associated with the new distribution of charge you create by moving the charges closer together. Because the static electric field is conservative, you could recover the stored energy by releasing the charges and letting them accelerate.

The example of Fig. 26-1 is trivial, but its implications are not. Energy storage in configurations of electric charge is a vital aspect of the natural and technological worlds. The energy produced in chemical reactions—including the metabolizing of food and the burning of coal, oil, and other fuels—is electrical energy released in the rearrangement of molecular charge distributions. Energy storage in systems of charged conductors is essential to the workings of electronic equipment and is important in devices that require large amounts of energy delivered in a short time. In this chapter we explore the energy of charge distributions and their electric fields, and we introduce a practical device—the capacitor—whose function is electrical energy storage.

26-1 ENERGY OF A CHARGE DISTRIBUTION

In the preceding chapter we defined the electric potential difference between two points as the change in potential energy per unit charge associated with moving charge between those points. To move charge between the two points takes work equal to the change in potential energy—equal, that is, to the electric potential difference multiplied by the charge being moved. In the simple case of two point charges, suppose charge q_1 is initially an infinite distance from a fixed charge q_2. You can assemble a new charge distribution by moving q_1 to a distance r from q_2. Equation 25-4 gives the electric potential difference from infinity to any point a distance r from q_2:

$$V_{\infty \to r} = \frac{kq_2}{r}.$$

Multiplying by q_1 gives the work you must do to bring q_1 in from infinity—that is, to assemble the new charge distribution. Since that work, W, gets stored as potential energy, U, we can write

$$W = U = \frac{kq_1q_2}{r}, \tag{26-1}$$

where we've taken the zero of potential energy when the charges are infinitely far apart. Note that W here is the work that *you* or some other agent has to do to assemble the charge distribution; it's not the work done by the electric field, which, as Chapter 8's definition of potential energy shows, is $-W$.

Equation 26-1 gives the **electrostatic potential energy** of two point charges. The equation shows that the potential energy is positive if the charges have the same sign and negative if they have opposite signs. In the latter case, it would take positive work to separate the charges. We would have obtained the same potential energy had we moved q_2 in the field of q_1, showing that the potential energy of a charge distribution is independent of how it is assembled.

These considerations hold for any charge distribution. In general, it takes work to assemble a charge distribution and that work is stored as potential energy. The potential energy can be positive, zero, or negative. Because the electric force obeys the superposition principle, the potential energy is independent of how the charge distribution is assembled: the total energy is simply the sum of the potential energies of every charge pair making up the distribution.

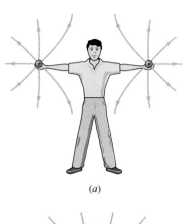

(a)

(b)

FIGURE 26-1 (a) Widely separated charges exert little force on each other. (b) Moving them together takes work, which is stored as potential energy. Note how the bending of the field lines suggests the repulsive force.

● **EXAMPLE 26-1**   THE ENERGY OF A CHARGE DISTRIBUTION

Three point charges each carrying $+q$ and a fourth carrying $-q/2$ are initially infinitely far apart. They are brought together to form the square charge distribution shown in Fig. 26-2. What is the electrostatic potential energy of this charge distribution?

Solution

We can assemble the charge distribution in any order. Assume that the positive charge at the upper left is brought in first. This takes no work because no other charge is in place. Next the

positive charge q at the upper right is brought to a distance a from the first positive charge. Equation 26-1 tells us that the work required is

$$W_2 = k\frac{q^2}{a}.$$

Now the third positive charge is brought to its place at the lower left. This point is a distance a from the first charge q and $\sqrt{2}a$ from the second, so the work required is

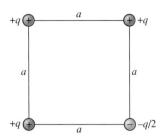

FIGURE 26-2 How much energy is stored in this square charge distribution? (Example 26-1)

$$W_3 = k\left(\frac{q^2}{a} + \frac{q^2}{\sqrt{2}a}\right).$$

Finally, the negative charge $-q/2$ is brought to a point a distance a from the second and third charges and $\sqrt{2}a$ from the first charge. The work required is again the sum of the potential differences multiplied by the charge $-q/2$:

$$W_4 = k\left(-\frac{q^2}{2a} - \frac{q^2}{2a} - \frac{q^2}{2\sqrt{2}a}\right).$$

This work is negative, indicating that it would take positive work to remove the negative charge. Adding the work required to bring in the second, third, and fourth charges gives the electrostatic potential energy of the charge distribution:

$$W = W_2 + W_3 + W_4$$

$$= k\left(\frac{q^2}{a} + \frac{q^2}{a} + \frac{q^2}{\sqrt{2}a} - \frac{q^2}{2a} - \frac{q^2}{2a} - \frac{q^2}{2\sqrt{2}a}\right)$$

$$= \frac{kq^2(2\sqrt{2}+1)}{2\sqrt{2}a}.$$

That this is a positive quantity indicates that the work needed to assemble the three positive charges is greater than the energy gained bringing in the negative charge.

E X E R C I S E Repeat Example 26-1 for the case when the charge in the upper left corner is changed to $-q$.

Answer: $-3kq^2\left[1 - 1/(2\sqrt{2})\right]/a$

Some problems similar to Example 26-1:
1–4, 7, 79

26-2 TWO ISOLATED CONDUCTORS

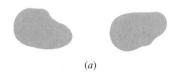

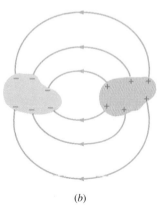

FIGURE 26-3 A pair of isolated conductors. (*a*) Initially they are uncharged, and there is no potential difference between them. (*b*) When they carry opposite charges, there is an electric field and consequently a potential difference between them.

An important charge distribution consists of two isolated conductors carrying equal but opposite charges. Figure 26-3 shows two such conductors, each initially uncharged. Imagine moving a small quantity of charge from one conductor to the other, giving rise to a net positive charge on one and a net negative charge on the other. This results in an electric field and therefore in a potential difference between the conductors. If we try to transfer more charge between the conductors, we must do work traversing this potential difference. The more charge we move, the harder it gets to transfer additional charge. The work it takes to transfer the charge is stored as potential energy of the charge distribution.

It is generally difficult to calculate the stored potential energy for a pair of irregularly shaped conductors like those of Fig. 26-3. An important practical case for which the potential energy may be calculated is a pair of identical, flat, parallel conducting plates whose separation is small compared with their width (Fig. 26-4a). We start with the plates uncharged, and then transfer charge Q from one plate to the other. (In practice, we would accomplish this by connecting the plates to the terminals of a battery.) Charging the plates results in an electric field between them. For closely spaced plates, this field is essentially uniform except very near the edges (Fig. 26-4b), and we may neglect this nonuniform "fringing field."

What is the electric field strength between the plates? In Chapter 24 we used Gauss's law to show that the field near the surface of a conductor carrying surface charge density σ is given by

$$E = \frac{\sigma}{\varepsilon_0}.$$

As we discussed in Section 24.6, charge gathers entirely on the facing surfaces of the two plates, giving rise to a charge density of magnitude $\sigma = q/A$, where A is the plate area. So the electric field between the plates is

$$E = \frac{\sigma}{\varepsilon_0} = \frac{q}{\varepsilon_0 A},$$

where q is the magnitude of the charge on either plate. Shouldn't this result be doubled because there are two plates? No! Review the discussion accompanying Figs. 24-32 and 24-33 to convince yourself of this point.

The presence of the electric field means there is a potential difference between the plates. Since the field is uniform, this potential difference is a simple product of the field strength with the distance d between the plates:

$$V = Ed = \frac{qd}{\varepsilon_0 A},$$

where we're now using V rather than $\Delta V_{A \to B}$ for the potential difference.

Now imagine moving an additional very small positive charge dq from the negative to the positive plate. How much work does this take? That depends on the potential difference between the plates, which, as our expression for V shows, depends on how much charge has already been transferred. Because potential difference is work per unit charge, the work dW required to move the charge dq between the plates is

$$dW = V \, dq = \frac{qd}{\varepsilon_0 A} \, dq.$$

Suppose we start with zero net charge on either plate and gradually transfer a total charge Q from one plate to the other. Each dq that we move requires work dW as given above, so the total work is the sum of all the dW's associated with all the small quantities of charge dq that make up Q. In the limit of infinitely many infinitesimal charges dq, this sum becomes an integral, and we have

$$W = \int_0^Q dW = \int_0^Q \frac{qd}{\varepsilon_0 A} \, dq = \frac{d}{\varepsilon_0 A} \int_0^Q q \, dq.$$

That the variable q remains under the integral sign reflects the physical fact that the continually increasing charge on the plates results in an increasing potential difference and therefore makes it harder to move each additional charge dq. Continuing the integration gives

$$W = \frac{d}{\varepsilon_0 A} \frac{q^2}{2} \Big|_0^Q = \frac{d}{2\varepsilon_0 A} Q^2.$$

Thus the work required to charge the plates increases as the square of the charge Q. The work done in charging the plates ends up as stored potential energy of the final charge distribution, so the stored energy is

$$U = W = \frac{d}{2\varepsilon_0 A} Q^2. \tag{26-2}$$

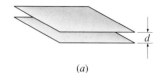

(a)

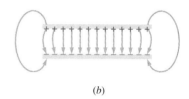

(b)

FIGURE 26-4 (a) A pair of closely spaced conducting plates. (b) Edge-on view of the plates when oppositely charged. For closely spaced plates the electric field is essentially uniform, except very near the edges, and the field outside is very small. Neglecting these small "fringing fields" is a good approximation.

The quadratic dependence of the stored energy on charge suggests that a pair of parallel plates is an excellent device for storing electrostatic energy.

26-3 ENERGY AND THE ELECTRIC FIELD

We have seen that the work required to assemble a distribution of electric charge ends up as electrostatic potential energy. Just where is the energy stored? In Example 26-1, we considered the assembly of four point charges to form a square. Surely the point charges themselves did not change; we only moved them closer together. Similarly, when we charged our pair of metal plates we did not alter the individual charges; we only moved them from one plate to another. What has changed in both these cases? The electric field has changed. In the first case we started with four isolated point charges and an electric field that looked like four isolated point-charge fields. We ended with a new charge distribution whose field did not look at all like a point-charge field.

In the case of the parallel plates, we started with uncharged plates and no electric field. As soon as we began transferring charge from one plate to the other, an electric field appeared between the plates, and this field grew in strength as more charge was transferred.

So where is the energy stored? It is stored in the electric field. As we create or alter a charge distribution, we do work and an altered electric field configuration develops. The work we do in moving the charges ultimately goes into the alteration of the field. If the work done by the applied force is positive, we have added energy to the field. If the work is negative, we have removed energy from the field.

Every electric field represents stored energy. If the field is altered, energy is either accumulated or released, depending on whether the work done is positive or negative. If the field disappears entirely, all its energy is released in some other form. Because electric forces are primarily responsible for the behavior of everyday matter, many seemingly different forms of energy storage really involve electric field energy. When you burn gasoline or metabolize food, for example, you are rearranging the charge distributions we call molecules into new configurations whose electric fields contain less energy (Fig. 26-5).

If electric fields store energy, then the amount of stored energy should depend on the field strength. Since the field strength may vary with position, we describe the stored energy in terms of **energy density,** or energy stored per unit volume. We can readily determine the energy density for our parallel plates. There we found that the field strength is given by $E = Q/\varepsilon_0 A$; solving for Q and using the result in Equation 26-2 for the stored energy gives

FIGURE 26-5 Combustion involves the rearrangement of atoms into new molecular structures. Energy released in the process comes ultimately from the electric fields associated with the charge distribution in the molecules.

$$U = \frac{d}{2\varepsilon_0 A} Q^2 = \frac{d}{2\varepsilon_0 A} (\varepsilon_0 A E)^2 = \tfrac{1}{2} \varepsilon_0 E^2 A d.$$

Our assumption that the plates are very close together allowed us to conclude that the field is very nearly uniform between the plates and essentially zero outside the plates. Therefore, the energy U is stored in the region between the plates, and is distributed uniformly because the field is uniform. The volume between the plates is just the plate area times the separation, or Ad, and therefore the energy density u_E is given by $u_E = U/Ad$, or

$$u_E = \tfrac{1}{2}\varepsilon_0 E^2. \qquad \text{(electric energy density)} \qquad (26\text{-}3)$$

Although we derived this expression for the uniform field between two parallel plates, it is in fact a universal expression that holds for *any* electric field. At any point where an electric field exists, there is stored energy whose density, in J/m³, is given by Equation 26-3.

The deepest significance of Equation 26-3 lies in its statement that every electric field represents stored energy. As we observe a variety of physical phenomena, from everyday happenings on Earth to events in distant galaxies, we can understand that the driving energy for many of these phenomena comes from the release of energy stored in electric fields.

EXAMPLE 26-2 ELECTRICAL ENERGY OF A THUNDERSTORM

Electric fields inside a thunderstorm have typical values of 10^5 V/m and get even higher just before electrical energy is unleashed as lightning (Fig. 26-6). The origin of these fields and hence of the energy stored in them is believed to be associated with charge transfer to rising and falling water droplets or ice crystals in the intense updrafts and downdrafts of the thunderstorm. Consider a typical thundercloud that rises to an altitude of 10 km and has a diameter of 20 km. Assuming an average field strength of 10^5 V/m, estimate the total electrostatic energy stored in the cloud. How many gallons of gasoline would you have to burn to release the same amount of energy?

Solution
The energy density is given by Equation 26-3:

$$u_E = \tfrac{1}{2}\varepsilon_0 E^2 = \tfrac{1}{2}(8.85\times10^{-12}\ \text{C}^2/\text{N}\cdot\text{m}^2)(10^5\ \text{V/m})^2$$

$$= 4.4\times10^{-2}\ \text{J/m}^3.$$

(You should verify that the units work out!) We are assuming that this energy density is the same throughout the storm, so we find the total energy by multiplying the energy density by the volume. The storm is roughly cylindrical in shape, so its volume is

$$V = \pi r^2 h = \pi(10\ \text{km})^2(10\ \text{km}) = 3100\ \text{km}^3$$

$$= 3.1\times10^{12}\ \text{m}^3.$$

Then the total stored energy is

$$U = u_E V = (4.4\times10^{-2}\ \text{J/m}^3)(3.1\times10^{12}\ \text{m}^3) = 1.4\times10^{11}\ \text{J}.$$

A gallon of gasoline contains about 10^8 J (see Appendix C), so the electrical energy stored in a thunderstorm at any given

FIGURE 26-6 Lightning is the sudden release of energy stored in atmospheric electric fields.

instant is equivalent to about 1000 gallons or 4000 L of gasoline. This comparison is not quite fair to the thunderstorm, though, because its electrical energy is continually dissipated in lightning strikes and at the same time renewed by the violent motion of the air. Problem 77 explores thunderstorm energetics in more detail.

EXERCISE In fair weather, Earth's atmospheric electric field is about 100 V/m. Find the energy stored in each km³ of the fair-weather atmosphere.

Answer: 44 J

Some problems similar to Example 26-2: 16–18 ●

When the electric field is uniform, as in our thunderstorm example, we can find the stored energy simply by multiplying the energy density by the volume. But when the field changes with position we must resort to calculus. Consider a small volume element dV, so small that the electric field is essentially uniform over this volume. The stored energy dU in the volume element is just the energy density times the volume, or

$$dU = u_E \, dV = \tfrac{1}{2} \varepsilon_0 E^2 \, dV.$$

The total energy U is then the sum of all the dUs. In the limit of infinitesimally small volumes dV and energies dU, this sum becomes an integral:

$$U = \tfrac{1}{2} \varepsilon_0 \int E^2 \, dV, \qquad (26\text{-}4)$$

where the limits on the integral are chosen to cover the entire region in which the electric field of interest exists.

We derived Equation 26-4 for the electric field energy using our previously determined expression for the work needed to assemble a simple charge distribution. We can also reverse that process, using the electric field of a charge distribution to calculate the energy density and from it the stored energy and therefore the work needed to assemble the distribution. Example 26-3 illustrates this procedure for a case when the energy density varies with position.

● **EXAMPLE 26-3**   A SHRINKING SPHERE

A sphere of radius R_1 carries a total charge Q distributed evenly over its surface (Fig. 26-7a). How much work does it take to shrink the sphere to a smaller radius R_2? Practical applications in which this question might prove important include the behavior of cell membranes, charged bubbles, and raindrops in thunderclouds.

Solution

Shrinking the sphere moves all the charge elements on its surface closer together, and therefore requires positive work. That work is equal to the change in stored electric field energy given by Equation 26-4. With all charge distributed evenly over the sphere's surface, Gauss's law tells us that there is no electric field within the sphere. Because of the spherical symmetry, Gauss's law also tells us that the field outside the sphere is identical to that of a point charge Q at the sphere's center. This means that the field at and beyond the original radius R_1 does not change as we shrink the sphere. What does change is the field between R_1 and R_2 (Fig. 26-7b). Originally this field was zero. After the sphere has shrunk, this region, too, is filled with a point-charge field. This newly created field is the site of the additional energy stored in shrinking the sphere.

Because the point-charge field between R_1 and R_2 changes with position, we must use the integral form 26-4 to calculate

the stored energy. The electric field outside the sphere is that of a point charge:

$$E = \frac{kQ}{r^2},$$

so, from Equation 26-3, the energy density as a function of r is

$$u_E(r) = \tfrac{1}{2} \varepsilon_0 E^2 = \tfrac{1}{2} \varepsilon_0 \left(\frac{kQ}{r^2} \right)^2 = \frac{kQ^2}{8\pi r^4},$$

where we used $k = 1/4\pi\varepsilon_0$.

To determine the total stored energy we integrate this energy density over the volume between R_2 and R_1. Because of the spherical symmetry we consider volume elements made of thin spherical shells of thickness dr. Figure 26-8 shows that the volume of the shell is very nearly $dV = 4\pi r^2 dr$, so Equation 26-4 becomes

$$U = \int_{R_2}^{R_1} u_E \, dV = \int_{R_2}^{R_1} \frac{kQ^2}{8\pi r^4} \, 4\pi r^2 \, dr$$

$$= \frac{kQ^2}{2} \int_{R_2}^{R_1} r^{-2} \, dr = \frac{kQ^2}{2} \left(-\frac{1}{r} \right) \Bigg|_{R_2}^{R_1}$$

$$= \frac{kQ^2}{2} \left(\frac{1}{R_2} - \frac{1}{R_1} \right).$$

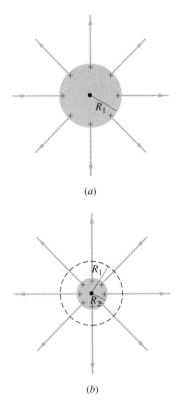

(a)

(b)

FIGURE 26-7 (a) A charged sphere and its electric field. (b) Shrinking the sphere creates new field in the region $R_2 < r < R_1$.

FIGURE 26-8 A thin spherical shell of thickness dr and radius r. Because the shell is very thin, its inner and outer surfaces have essentially the same area, namely $4\pi r^2$. Its volume is therefore $dV = 4\pi r^2\,dr$.

EXERCISE A long coaxial cable consists of an inner cylindrical conductor of radius a and an outer cylindrical conducting shell of radius b (Fig. 26-9). The conductors carry equal but opposite line charge densities $\pm\lambda$. Find the electric energy stored in a length ℓ of this cable.

This is the total energy stored in the new electric field between R_2 and R_1, and is therefore also the work done in shrinking the sphere from R_1 to R_2. If we let R_1 go to infinity, our result becomes the work required to assemble a sphere of radius R_2 carrying surface charge Q, or equivalently, the energy stored in the field of the sphere. Because the stored energy becomes infinite as R_2 approaches zero, our result suggests that the notion of a point charge is an impossible idealization. Problem 90 explores some implications of this result in the theory of elementary particles.

FIGURE 26-9 How much energy is stored in a length ℓ of the cable? (The cable is much longer than the outer conductor radius b.)

Answer: $k\lambda^2\ell\,\ln(b/a)$

Some problems similar to Example 26-3: 21–23, 26, 27

26-4 CAPACITORS

In electrical and electronic equipment, electrical energy is often stored using a pair of charged conductors separated by an insulator. Such a device is called a **capacitor.** Capacitors are typically used for short-term energy storage in situations where it is necessary to store or release electrical energy quickly. Most practical electronic devices, including radio, TV, computers, and audio equipment, would be virtually impossible to construct without capacitors. When you tune a radio, you are adjusting a capacitor. Failure of a capacitor in your car's ignition system could leave you stranded on the highway. And many high-energy experiments in physics and engineering use so much power that, were it not for capacitors, they could not be done without disrupting the supply of electric power to the rest of the world!

In an uncharged capacitor both conductors are neutral. Therefore there is no electric field in the capacitor, and no stored energy. A capacitor is charged by transferring charge (usually electrons) from one conductor to the other. The work required is generally supplied by some other source of electrical energy connected to the capacitor through wires leading to its two conductors. An electric field develops as a result of the charge separation, and the stored energy resides in this field. Although we speak of a capacitor as being charged, the capacitor remains overall electrically neutral, in that the net charge on the whole capacitor is zero. But the two individual conductors making up a charged capacitor are not neutral: one carries a charge $+Q$, the other $-Q$. When we say that the charge on a capacitor is Q, we really mean that Q is the magnitude of the charge on either conductor.

Once a capacitor is charged there is an electric field and therefore a potential difference between its two conductors. As the charge is increased the potential difference increases proportionately. Conversely, imposing a potential difference on a capacitor (by connecting it to a battery, for example) causes the capacitor to become charged in proportion to the potential difference imposed. The ratio of charge to potential difference is characteristic of a given capacitor and is called its **capacitance:**

$$C = \frac{Q}{V}.$$

(26-5)

Here Q is the magnitude of the charge on either conductor and V the potential difference (or voltage) between the conductors. Clearly the units of capacitance are coulombs/volt. One coulomb/volt is given the name **farad** (F), in honor of the nineteenth century scientist Michael Faraday. One farad is so huge a capacitance that the smaller units microfarad (10^{-6} F; abbreviated μF) and picofarad (10^{-12} F; abbreviated pF and often pronounced "puff") are widely used.

Capacitance depends on the physical construction of a capacitor—the shapes of its two conductors, their separation, and the choice of insulating material between them. Although Q and V enter the defining relation 26-5, capacitance is a constant. If V is increased, Q increases proportionately, maintaining the constant ratio C that characterizes the capacitor.

Any arrangement of two insulated conductors constitutes a capacitor. Practical capacitors are manufactured in a variety of configurations. Often they are made from two long strips of aluminum foil separated by thin layers of plastic or paper. This foil "sandwich" is then rolled into a compact cylinder, wires attached, and the whole assembly covered with a protective coating. Another common arrangement is the variable capacitor, whose configuration can be altered to change its capacitance. This change can be accomplished mechanically or, as in many modern electronic devices, electrically. Very large capacitances are achieved with so-called electrolytic capacitors, in which a thin insulating layer develops chemically under the influence of the applied voltage. Figure 26-10 shows some typical capacitors.

Calculating Capacitance

Capacitance is defined through Equation 26-5 as the ratio of charge to potential difference. To calculate the capacitance of a particular configuration of two conductors, we assume there is a charge Q on the capacitor and calculate the corresponding potential difference. Because the capacitance depends only on the physical configuration of the capacitor, it doesn't matter what value we assume for the charge—that will cancel when we take the ratio of charge to potential difference. For two irregularly shaped conductors we are back to the problem of determining the potential difference between the conductors from the distribution of charge on their surfaces. When the capacitor design includes sufficient symmetry, this calculation becomes straightforward.

By far the most important capacitor design is the parallel-plate configuration. In Section 26-2 we examined such a capacitor in some detail, although at that time we did not call the configuration a capacitor. There we found that the potential difference between plates of area A separated by a distance d is

$$V = \frac{Qd}{\varepsilon_0 A}.$$

Solving for the ratio $C = Q/V$ gives

$$C = \frac{Q}{V} = \frac{\varepsilon_0 A}{d}. \qquad \text{(parallel-plate capacitor)} \qquad (26\text{-}6)$$

Equation 26-6 gives the capacitance of a parallel-plate capacitor in terms of the universal constant ε_0 and factors that describe the physical configuration of the capacitor. (Strictly speaking, this expression holds only for capacitors insulated by vacuum. Later we will modify Equation 26-6 to account for other insulating materials.) Note that neither charge nor potential difference enters the final expression for capacitance, showing that the capacitance is indeed a constant. Equation 26-6 suggests that the way to make a capacitor with large capacitance is to use two plates of large area but small separation. Incidentally, Equation 26-6 shows that the units of ε_0 may be expressed as farads/meter (F/m); see Problem 32.

FIGURE 26-10 Typical capacitors. The large blue unit is an 18-mF electrolytic capacitor. At top right is an air-insulated variable capacitor in which a set of metal plates rotates with respect to fixed plates in order to change the capacitance. The remaining smaller capacitors range from 43 pF to 10 μF.

● **EXAMPLE 26-4** A PARALLEL-PLATE CAPACITOR

A capacitor consists of two circular metal plates of radius 10 cm separated by an air gap of 5.0 mm (Fig. 26-11). What is its capacitance? When a 12-volt battery is connected to the capacitor, how much charge appears on the plates?

Solution

Since the plate spacing is much smaller than the plate size, Equation 26-6 holds and the capacitance is

$$C = \frac{\varepsilon_0 A}{d} = \frac{\varepsilon_0 \pi r^2}{d} = \frac{(8.85 \times 10^{-12} \text{ F/m})(\pi)(0.10 \text{ m})^2}{5.0 \times 10^{-3} \text{ m}}$$

$$= 5.6 \times 10^{-11} \text{ F} = 56 \text{ pF}.$$

Equation 26-5 defines capacitance as the ratio of charge to potential difference. We can rewrite this defining relation to solve for the charge:

$$Q = CV = (56 \text{ pF})(12 \text{ V}) = 670 \text{ pC}.$$

What this really means, of course, is that the positive plate carries 670 pC and the negative plate −670 pC. Overall, the capacitor remains neutral. Note that by working with the capacitance in pF, the charge automatically comes out in pC.

EXERCISE A parallel-plate capacitor is to be made from two square pieces of aluminum foil each 8.0 cm on a side. (a) What should be the spacing between them if the capacitance is to be 47 pF? (b) What applied voltage will put ±95 nC on the plates?

Answers: (a) 1.2 mm; (b) 2.0 kV

FIGURE 26-11 A parallel-plate capacitor connected to a battery. Drawing is not to scale (Example 26-4).

Some problems similar to Example 26-4: 33, 34

● **EXAMPLE 26-5** A CYLINDRICAL CAPACITOR

A capacitor consists of two long concentric metal cylinders of length L, as shown in Fig. 26-12. The inner and outer cylinders have radii a and b, respectively. What is the capacitance?

Solution
Equation 26-6 does not apply to this configuration because the field between the cylinders is not uniform. To find the capacitance, we need a relation between charge and potential difference for the cylindrical configuration. In Example 25-4, we found that the potential difference between two points outside a charge distribution with line symmetry can be written

$$V(a) - V(b) = \frac{\lambda}{2\pi\varepsilon_0} \ln\left(\frac{b}{a}\right),$$

where λ is the line charge density. Because our capacitor is long compared with its diameter, this expression is a good approximation to the potential difference due to the field of the inner conductor. What about the outer conductor? Recall (Example 24-5) that the electric field inside an empty, hollow pipe is zero; therefore, the outer conductor contributes nothing to the electric field or the potential difference between the conductors. If the magnitude of the charge on either conductor is Q, then the line charge density is $\lambda = Q/L$, and our expression for potential difference becomes

$$V = V(a) - V(b) = \frac{Q}{2\pi\varepsilon_0 L} \ln\left(\frac{b}{a}\right).$$

Capacitance is the ratio of charge to potential difference, so we have

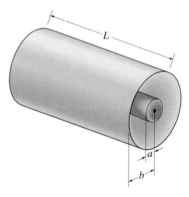

FIGURE 26-12 A cylindrical capacitor (Example 26-5).

$$C = \frac{Q}{V} = \frac{Q}{(q/2\pi\varepsilon_0 L)\ln(b/a)} = \frac{2\pi\varepsilon_0 L}{\ln(b/a)}. \quad (26\text{-}7)$$

Does this result make sense? We already found that the capacitance of a parallel-plate capacitor increases with increasing plate area or with decreasing plate separation. With the cylindrical capacitor we can increase the area of both conductors by increasing the length L of the capacitor, and indeed Equation 26-7 shows the capacitance increasing proportionately. We can decrease the spacing of the conductors by making the radii a and b more nearly equal. This makes b/a closer to one, and $\ln(b/a)$ closer to zero, again increasing the capacitance. Although the geometries of the cylindrical and parallel-plate capacitors are quite different, the same physical considerations apply to both: a large capacitance is achieved with large conductor areas and small separation. When the separation is very small, the curvature of the cylindrical capacitor cannot matter, and Equation 26-7 should reduce to Equation 26-6 for the parallel-plate capacitor (see Problem 84).

EXERCISE A conducting sphere of radius R is enclosed in a concentric spherical conducting shell of radius $\frac{3}{2}R$. What is the capacitance of this configuration?

Answer: $3R/k$

Some problems similar to Example 26-5: 35, 36, 67 ●

26-5 ENERGY STORAGE IN CAPACITORS

In Section 26-3 we found that any electric field represents stored energy. The example that guided us to that conclusion was a parallel-plate capacitor. For that configuration, the stored energy U is given by Equation 26-2:

$$U = \frac{d}{2\varepsilon_0 A}Q^2.$$

Since $\varepsilon_0 A/d$ is the capacitance of the parallel-plate capacitor, this stored energy may be written

$$U = \frac{Q^2}{2C}. \qquad \text{(energy in a capacitor)} \qquad (26\text{-}8a)$$

It is usually easier to measure voltage than charge. To express the stored energy in terms of voltage, we can solve the equation defining capacitance, $C = Q/V$, for Q and use the result in Equation 26-8a:

$$U = \frac{Q^2}{2C} = \frac{(CV)^2}{2C} = \tfrac{1}{2}CV^2. \qquad \text{(energy in capacitor)} \qquad (26\text{-}8b)$$

FIGURE 26-13 The large black cylinders are 4700-μF electrolytic capacitors used in the power supply of a stereo amplifier.

Although Equations 26-8a and b were derived for a parallel-plate capacitor, they hold for any capacitor regardless of its configuration (see Problem 49).

That the stored energy depends on the *square* of the potential difference implies that more energy can be stored in a small capacitor at high voltage than in a larger one at low voltage. Practically, the difficulties of handling high voltages mitigate this conclusion somewhat, but the fact remains that the stored energy in a capacitor increases rapidly with increasing voltage.

Large capacitors can store energy for a long time. In VCRs, for example, capacitors of several farads are used to maintain the program memory in the event of power failures lasting as much as an hour or so. Energy storage in capacitors has important safety implications since large capacitors may retain dangerous voltages even after the equipment containing them is turned off (Fig. 26-13). TVs, stereos, computers, and other electronic devices use large capacitors to produce steady direct current, and before beginning work on such equipment technicians often discharge these capacitors by touching a screwdriver across their terminals. The resulting spark and loud noise are testimony to the amount of stored energy (Fig. 26-14).

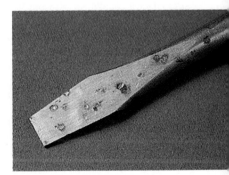

FIGURE 26-14 This screwdriver has been used repeatedly to discharge a large capacitor. Steel vaporized in each of the pitted areas.

● **EXAMPLE 26-6** WHICH CAPACITOR?

A 100-μF capacitor can tolerate a maximum potential difference of 20 V, while a 1.0-μF capacitor can tolerate 300 V. Which can store the most energy? The most charge?

Solution
Applying Equation 26-8b with V set to the maximum tolerable voltage gives

$$U_{100\,\mu F} = \tfrac{1}{2}CV^2 = \tfrac{1}{2}(100\ \mu F)(20\ V)^2 = 20\times10^3\ \mu J = 20\ mJ$$

and

$$U_{1.0\,\mu F} = \tfrac{1}{2}CV^2 = \tfrac{1}{2}(1.0\ \mu F)(300\ V)^2$$
$$= 45\times10^3\ \mu J = 45\ mJ\,.$$

Because of its higher voltage rating, the smaller capacitor can store more energy. On the other hand, the larger capacitor stores more charge, as shown by solving the defining relation $C = Q/V$:

$$Q_{100\,\mu F} = CV = (100\ \mu F)(20\ V) = 2000\ \mu C = 2.0\ mC$$

and

$$Q_{1.0\,\mu F} = CV = (1.0\ \mu F)(300\ V) = 300\ \mu C = 0.30\ mC\,.$$

Again, these numbers refer to the magnitude of the charge on each plate; overall, each capacitor remains neutral.

TIP Capacitance Isn't Capacity A capacitor is not like a bucket that can hold a fixed amount of water. Instead the charge and energy that a capacitor can hold depend on both the capacitance and on the applied voltage. In comparing the "capacities" of two capacitors one needs to know *both* the capacitance and the voltage that will be applied.

EXERCISE The "memory" capacitor in a VCR stores 25 J of energy with a potential difference of 3.5 V. (a) What is its capacitance? (b) What is the magnitude of the charge on each plate?

Answers: (a) 4.1 F; (b) 14 C

Some problems similar to Example 26-6: 38–41, 43, 44 ●

■ **APPLICATION** CAMERA FLASHES, TOILET FLUSHES, AND LASER FUSION

You've probably used a camera equipped with an electronic flash. The flash unit contains a special tube filled with xenon gas. When a large potential difference is applied across the tube, dielectric breakdown occurs and the xenon suddenly ionizes. Recombination of electrons with xenon ions then results in a bright pulse of white light. After a flash picture has been taken the photographer must wait a while—typically around 10 seconds—before the flash is ready again. Why is this? Although the total energy used by the flash is small, during the short interval that the xenon is being ionized the rate of energy use far exceeds the maximum power output of the camera's battery (see Problem 45). So the battery is used to charge a capacitor at a slow rate. Once the capacitor is charged its stored energy is dumped suddenly into the flashtube, providing the short burst of high power needed to give the intense light. The flashtube cannot be used again until the capacitor is recharged.

This situation is exactly analogous to one you encounter every day in household plumbing. Flushing a toilet requires a large amount of water in a short time—far more than typical household plumbing could supply in that time. So the water is accumulated gradually in the toilet tank, then suddenly dumped when needed for flushing. In this analogy the household plumbing, with its relatively narrow pipes, corresponds to the small battery of the camera flash. The toilet tank, which gradually accumulates water, corresponds to the capacitor,

which gradually accumulates electrical energy. Of course you need to wait between flushes for the toilet tank to fill, just as you need to wait between flash pictures for the capacitor to charge.

Professional photographers needing to take flash pictures in rapid succession often carry around a large, heavy battery pack capable of supplying the flash power directly. Similarly, institutional buildings with large, high-pressure water pipes often have toilets without tanks that can be flushed in rapid succession.

The simple example of the camera flash, scaled up many times in size, shows how very large amounts of power may be obtained briefly for industrial or scientific applications. Indeed, some experiments involving high-power pulsed lasers for nuclear fusion and ballistic missile defense research require more power than all the world's electric generating stations produce (Fig. 26-15). The required energy is accumulated in huge banks of capacitors, which are suddenly discharged to provide energy to the laser. Think here about the difference between energy and power! The pulsed laser is only on for about 10^{-9} seconds, so although it consumes energy at an enormous rate while on, it does not use all that much total energy (see Problem 46). The laser is not fired very often (at least in today's test configurations), so that there is plenty of time to charge the capacitors. The *average* power consumption of the experiment is modest.

(a)

(b)

FIGURE 26-15   (a) This huge capacitor bank stores 60 MJ of energy to drive the Nova laser fusion experiment at Lawrence Livermore National Laboratory. A fraction of the stored energy can be delivered to the lasers in less than 1 ns. (b) A 2.3-F capacitor bank for a laser fusion experiment at Los Alamos National Laboratory.

26-6 CONNECTING CAPACITORS TOGETHER

The most important consideration in using a capacitor is whether its capacitance is right for the particular application. But it's also important to respect a capacitor's **working voltage,** or the maximum potential difference that should be applied across the capacitor if dielectric breakdown is to be avoided.

Large capacitances are most easily achieved using small plate separations, as Equation 26-6 suggests. But a small plate separation implies a large electric field for a given voltage. Thus in practical capacitors there is a trade-off between capacitance and working voltage. High working voltage and high capacitance together require large plate separation to keep the electric field small and avoid dielectric breakdown, while at the same time requiring large plate area to keep the capacitance up. Thus large, high-voltage capacitors are physically bulky and expensive to build. Often economics as well as physics dictates the final design of a circuit involving capacitors.

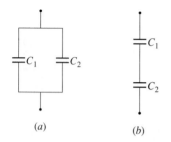

(a)   *(b)*

FIGURE 26-16 Connecting capacitors togther. (*a*) Parallel; (*b*) series. $\perp$ is the standard circuit symbol for a capacitor.

When a single capacitor with a desired combination of capacitance and working voltage is not available, we can often obtain the desired combination by connecting two or more capacitors together. There are only two ways to connect two capacitors (and indeed any electronic components with only two wires coming from them). Called **parallel** and **series,** these two possible connections are shown in Fig. 26-16. We would like to know the equivalent capacitance of each combination.

Capacitors in Parallel

Consider first the parallel combination of Fig. 26-16a. If we impose a potential difference V across the two wires coming from the combination, what will be the potential difference across each capacitor? The key to answering this question is the recognition that the wires connecting the capacitors are conductors and that in electrostatic equilibrium there can be no potential difference between any points connected to the same wire.* Thus all points connected directly together—including the top plates of each capacitor—are at the same potential. Similarly, the bottom plates and the wires connecting them are all at the same potential. Therefore, the potential differences across the two capacitors are equal. This is a very important point in the practical understanding of the electric circuits, and applies to any two circuit components that are connected in parallel:

> **The potential differences across two circuit components in parallel are equal.**

Recognizing this simple fact is essential in developing your understanding of electric circuits!

The equivalent capacitance is the ratio of the total charge on both capacitors to the voltage across the parallel combination. Solving the defining relation $C = Q/V$ for charge, we can write the charges on the two capacitors as

$$Q_1 = C_1 V \quad \text{and} \quad Q_2 = C_2 V.$$

The potential difference V is the *same* in both cases because the capacitors are connected in *parallel.* Thus the total charge is

$$Q = Q_1 + Q_2 = C_1 V + C_2 V = (C_1 + C_2)V.$$

Taking the ratio of total charge Q to the voltage V across the parallel combination gives the equivalent capacitance:

$$C = C_1 + C_2. \quad \text{(parallel capacitors)} \quad (26\text{-}9a)$$

Equation 26-9a is frequently stated as "capacitors in parallel add." You can understand this result physically by considering two parallel-plate capacitors with equal spacing. Connecting them in parallel amounts to adding their plate areas, giving a larger capacitance. Although we derived Equation 26-9a for two

---

* Even when we relax the equilibrium assumption, this conclusion will still hold in the approximation that the wires are perfect conductors.

parallel capacitors, the result that parallel capacitances add is easily extended to three or more capacitors (see Problem 56):

$$C = C_1 + C_2 + C_3 + \cdots \qquad \text{(parallel capacitors)} \qquad (26\text{-}9b)$$

What about the working voltage of the parallel combination? Both capacitors experience the full potential difference V, so the working voltage of the combination is that of whichever capacitor has the lower working voltage.

Capacitors in Series

Suppose we charge the series capacitor system of Fig. 26-16b, putting $+Q$ on the upper plate of C_1 and $-Q$ on the lower plate of C_2. The positive charge on the uppermost plate attracts $-Q$ to the lower plate of C_1 and the negative charge $-Q$ on the lowermost plate attracts $+Q$ to the upper plate of C_2 (Fig. 26-17). Note that this leaves the middle two plates together with zero net charge, which must be the case since these two plates are not connected to any external source of charge. With charge of magnitude Q on every plate, we can conclude that

| **Capacitors in series carry the same charge.**

To find the equivalent capacitance, we first solve the relation $C = Q/V$ for the voltages across the two capacitors:

$$V_1 = \frac{Q}{C_1} \quad \text{and} \quad V_2 = \frac{Q}{C_2}.$$

Here there is no need to label the Q's since series capacitors carry the same charge. But now the voltages need not be the same. Since the electric fields in the two capacitors point the same way (Fig. 26-17), the voltage across the series combination is just

$$V = V_1 + V_2.$$

Inserting our expressions for the individual potential difference gives

$$V = \frac{Q}{C_1} + \frac{Q}{C_2} = Q\left(\frac{1}{C_1} + \frac{1}{C_2}\right).$$

Dividing by Q gives V/Q, which the relation $C = Q/V$ shows is the reciprocal of the equivalent capacitance:

$$\frac{1}{C} = \frac{1}{C_1} + \frac{1}{C_2}. \qquad \text{(series capacitors)} \qquad (26\text{-}10a)$$

This result is frequently described by saying that "capacitors in series add reciprocally." The result is easily extended to three or more capacitors (see Problem 56):

$$\frac{1}{C} = \frac{1}{C_1} + \frac{1}{C_2} + \frac{1}{C_3} + \cdots \qquad \text{(series capacitors)} \qquad (24\text{-}10b)$$

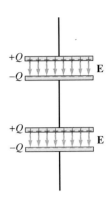

FIGURE 26-17 Capacitors in series carry the same charge.

When there are just two capacitors, combining reciprocals over a common denominator gives

$$C = \frac{C_1 C_2}{C_1 + C_2}. \qquad \text{(2 series capacitors)} \qquad \text{(26-10c)}$$

Equations 26-10 show that the combined capacitance of two series capacitors is less than the capacitance of either. You can make physical sense of this by considering parallel-plate capacitors with equal plate areas. Putting them in series effectively adds the plate separations of the two capacitors, yielding a smaller overall capacitance.

What about the voltage rating of the series combination? The full applied voltage V is the sum of the voltages across each capacitor, so each can be rated for less than the full applied voltage. The fraction of the applied voltage that appears across each capacitor depends on the relative capacitances.

● **EXAMPLE 26-7** CONNECTING CAPACITORS

(a) Find the equivalent capacitance of the combination shown in Fig. 26-18a. (b) If the maximum voltage applied between points A and B is 100 V, what should be the working voltage of C_2?

Solution
The way to handle circuit problems like this one is to reduce the circuit to a simpler one by recognizing combinations of series and parallel components, as shown in Fig. 26-18. Here C_3 and C_4 are in parallel, so they add to give an equivalent capacitance of 4.0 μF; the circuit then looks like Fig. 26-18b. This parallel combination C_{34} is in series with C_2; applying Equation 26-10c gives

$$C_{234} = \frac{C_2 C_{34}}{C_2 + C_{34}} = \frac{(12 \ \mu F)(4.0 \ \mu F)}{12 \ \mu F + 4.0 \ \mu F} = 3.0 \ \mu F.$$

Now the circuit looks like Fig. 26-18c, with C_1 and C_{234} in parallel. The equivalent capacitance of the entire circuit is their sum, or 7.0 μF (Fig. 26-18d).

To find the working voltage needed for C_2, we need to know the voltage across C_2 with 100 V across the entire combination. Since C_1 and the combination C_{234} are in parallel and connected to points A and B, both these capacitors experience the full 100 V. We can then use the defining relation $C = Q/V$ to find the charge on C_{234}:

$$Q_{234} = C_{234} V = (3.0 \ \mu F)(100 \ V) = 300 \ \mu C.$$

But C_{234} is the series combination of C_2 and C_{34} (Fig. 26-18b,c), and the charge on series capacitors is the same. So $Q_2 = 300 \ \mu C$ as well. We again use the defining relation $C = Q/V$, now to find the voltage across C_2:

$$V_2 = \frac{Q_2}{C_2} = \frac{300 \ \mu C}{12 \ \mu F} = 25 \ V.$$

This is the required working voltage for C_2.

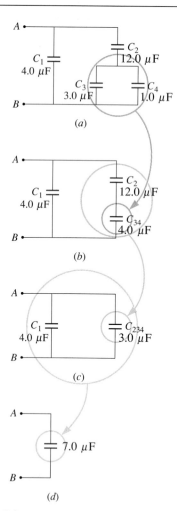

FIGURE 26-18 What is the equivalent capacitance (Example 26-7)? Analyzing the circuit involves reducing parallel and series combinations using Equations 26-9 and 26-10.

TIP Analyzing Circuits This example illustrates a very general approach to circuit analysis. You reduce the circuit to its simplest form by recognizing series and parallel combinations, then gradually build back up to find the quantity of interest. The reduction is essential, even when you're only interested in what's happening with one component. Here, for instance, we wanted to know the voltage on C_2 with 100 V across the whole combination. But to find that we had to analyze the entire circuit, and only then could we focus on C_2.

TIP Recognize Series and Parallel Parallel components have their two ends, respectively, connected *directly* together, as in Fig. 26-16a. Series components are connected in such a way that if you

imagine moving through one component, then the *only* place you can go is into the next component (Fig. 26-16b). Many connections are neither; C_2 and C_3 in Fig. 26-18a, for example, are *not* in series because after you go through C_2 the circuit splits and you could go through either C_3 or C_4. And C_1 and C_2 are *not* in parallel; even though their top plates are connected directly together, their bottom plates are not. The series and parallel formulas we've derived apply *only* to true series and parallel combinations.

EXERCISE (a) With $V_{AB} = 100$ V in Fig. 26-18, what should be C_3's working voltage? (b) What will be the charges on C_1, C_3, and C_4?

Answers: (a) 75 V; (b) $Q_1 = 400 \ \mu C$, $Q_3 = 225 \ \mu C$, $Q_4 = 75 \ \mu C$

Some problems similar to Example 26-7: 52, 54, 57, 58 ●

26-7 CAPACITORS AND DIELECTRICS

The insulating material between the plates of a capacitor serves several purposes. It maintains physical separation of the plates and minimizes charge leakage. Its molecular properties also influence the capacitance. In Chapter 23 we found that electric dipoles tend to align with an applied electric field, and we defined **dielectrics** as materials whose molecules behave as dipoles. The molecular dipole moments may be intrinsic to the molecules, or may be induced by an applied electric field, as we showed in Fig. 23-31. Essentially all insulators are dielectrics.

Suppose we have a parallel-plate capacitor charged to some voltage V_0, with air or vacuum between its plates. What happens if we insert a slab of dielectric material that fills the space between the plates? Figure 26-19 shows that molecular dipoles align with the field arising from the charge on the capacitor plates and that therefore the fields of the dipoles themselves *oppose* the capacitor field. The result is a reduction in the net electric field between the plates. How much reduction depends on details of molecular structure and molecular interactions; empirically, though, we find that a given material may be characterized by its **dielectric constant,** κ, which describes the reduction in field. If E_0 is the original field, then the field after insertion of the dielectric will be decreased by a factor $1/\kappa$: $E = E_0/\kappa$.

If the capacitor is not connected to anything, then there's no way for the charge on its plates to change. But the field between the plates has been reduced, and therefore the potential difference $V = Ed$ has decreased by the same factor $1/\kappa$; $V = V_0/\kappa$. But capacitance is the ratio of charge to voltage, so with the dielectric in place we have

$$C = \frac{Q}{V} = \frac{Q}{(V_0/\kappa)} = \frac{\kappa Q}{V_0} = \kappa C_0, \qquad (26\text{-}11)$$

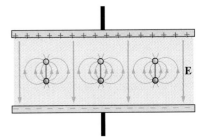

FIGURE 26-19 A capacitor with a dielectric between its plates. Molecular dipoles align with the capacitor's electric field, and their own fields point opposite to the original field. The result is a reduction in field strength.

▲ **TABLE 26-1** PROPERTIES OF SOME COMMON DIELECTRICS

DIELECTRIC MATERIAL	DIELECTRIC CONSTANT	BREAKDOWN FIELD (kV/mm)
Air	1.0006	3
Aluminum oxide	8.4	670
Glass (Pyrex)	5.6	14
Mica	5.4	100
Neoprene	6.9	12
Paper	3.5	14
Plexiglass	3.4	40
Polyethylene	2.3	50
Polystyrene	2.6	25
Quartz	3.8	8
Tantalum oxide	26	500
Teflon	2.1	60
Water	78	—

where $C_0 = Q/V_0$ is the original capacitance. Thus insertion of the dielectric increases the capacitance by a factor κ.

Capacitors are among the most difficult electronic components to miniaturize, so the ongoing revolution in microelectronics has spurred a search for suitable dielectrics with large dielectric constants, good insulating properties, and high breakdown fields. Exotic materials like tantalum oxide and strontium titanate have become widely used in recent years because of their high dielectric constants. In a few unusual applications water, with its dielectric constant of 78, is used as a dielectric. In high-energy experiments where it is necessary to store a large amount of energy for a short time, water's large dielectric constant outweighs the disadvantage of poor insulating quality. Table 26-1 lists dielectric constants and breakdown fields of selected materials.

In addition to helping build better capacitors, the relation between capacitance and dielectric constant serves as a useful probe of the structure of matter. Introducing a dielectric material between capacitor plates lowers the potential difference and therefore allows us to calculate the dielectric constant. This, in turn, gives information about the density and structure of the individual molecular dipoles. Conversely, we can use the measured dielectric constant to help identify an unknown material.

● **EXAMPLE 26-8** CAPACITORS AND DIELECTRICS

An air-insulated capacitor is charged by connecting it to a 12-V battery. The battery is then disconnected. When the space between capacitor plates is filled with an unknown plastic, the voltage between the plates drops to 4.6 V. What is the unknown material? If the plate spacing is 0.10 mm, how much voltage can the capacitor withstand with this material between its plates?

Solution
The voltage has dropped by a factor $1/\kappa = 4.6/12 = 1/2.6$. From Table 26-1, we see that a plastic with $\kappa = 2.6$ is polystyrene. With a dielectric breakdown field of 25 kV/mm,

the 0.1-mm-thick piece of polystyrene can withstand 2.5 kV. The rated working voltage would actually be lower, to allow a margin of safety.

EXERCISE An air-insulated capacitor with $C = 25\ \mu$F is connected to a 10-V battery and the battery is left connected as a quartz slab is inserted to fill the space between the plates. Find the charge on the capacitor (a) before and (b) after the slab is inserted. *Hint:* The battery maintains a fixed 10 V across the plates, but now charge can move from the battery to the plates.

Answers: (a) 250 μC; (b) 950 μC

Some problems similar to Example 26-8: 64–66 ●

What happens to the energy stored in a capacitor when a dielectric is inserted between its plates? The dielectric increases the capacitance by a factor κ, but it also decreases the potential difference by the same factor. If the energy is initially $U_0 = \frac{1}{2} C_0 V_0^2$, then after the dielectric is inserted it becomes

$$U = \frac{1}{2}(\kappa C_0)\left(\frac{V_0}{\kappa}\right)^2 = \frac{1}{2\kappa} C_0 V_0^2 = \frac{U_0}{\kappa}. \tag{26-12}$$

Since $\kappa > 1$, the energy has decreased. Where has it gone? As the dielectric moves into the capacitor, the electric field causes charge separation and rotation of the molecular dipoles. The dipoles thus gain energy from the field. In a solid the molecules interact strongly, and the energy is quickly dissipated as heat. By writing Equation 26-12 explicitly for a parallel-plate capacitor, you can show that the energy density in the presence of a dielectric is

$$u_E = \frac{1}{2}\kappa \varepsilon_0 E^2 \tag{26-13}$$

(see Problem 70). Here E is the field averaged over many molecules. The factor κ in Equation 26-13 reflects the presence of electric fields on the microscopic scale that are associated with the stretching and rotation of molecules.

That the energy of a capacitor is lower with a dielectric inserted suggests that a force acts to propel the dielectric slab into the capacitor. Figure 26-20 shows that this force originates in something we have heretofore intentionally ignored—the nonuniform fringing field beyond the plates of the capacitor. This nonuniform field acts on the dipoles in the dielectric to produce a net force toward the interior of the capacitor (see Problem 81).

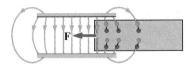

FIGURE 26-20 The nonuniform fringing field outside the capacitor is stronger nearer the plates and therefore results in a net force on the molecular dipoles in the dielectric slab. As a result, the dielectric is pulled into the capacitor.

CHAPTER SYNOPSIS

Summary

1. The work required to assemble a charge distribution is stored as the **electrostatic potential energy** of the distribution. Electrostatic potential energy resides in the electric field. Whenever an electric field is altered, energy is added to or removed from the field.
   a. The **electric field energy density** is given by

   $$u_E = \frac{1}{2}\varepsilon_0 E^2,$$

   where the SI units of u_E are J/m³. In a dielectric material, the energy density includes a factor of the dielectric constant κ.
   b. The electrostatic potential energy of a charge distribution may be determined either by computing the work required to assemble the individual charges of the distribution or, knowing the electric field of the distribution, by integrating the energy density over the volume containing the field:

   $$U = \int u_E \, dV = \frac{1}{2}\varepsilon_0 \int E^2 \, dV.$$

2. A **capacitor** is an arrangement of two conductors separated by an insulator. Transferring charge from one conductor to the other results in an electric field in the region between the conductors, and energy is stored in the field.
   a. The **capacitance** of a capacitor is the ratio of the charge to the potential difference between its conductors:

   $$C = \frac{Q}{V}.$$

   The capacitance of a parallel-plate capacitor is given by

   $$C = \frac{\varepsilon_0 A}{d},$$

   where A is the plate area and d the spacing. The capacitance of other configurations may be determined by assuming a charge, computing the associated potential difference, and taking the ratio $C = Q/V$.
   b. The energy stored in a capacitor depends on the capacitance and on the square of the potential difference:

   $$U = \frac{1}{2}CV^2.$$

c. Capacitors may be connected in **series** or **parallel**. The capacitances of parallel capacitors add:

$$C = C_1 + C_2 + C_3 + \cdots. \qquad \text{(parallel capacitors)}$$

The capacitances of series capacitors add reciprocally:

$$\frac{1}{C} = \frac{1}{C_1} + \frac{1}{C_2} + \frac{1}{C_3} + \cdots. \qquad \text{(series capacitors)}$$

d. The **working voltage** of a capacitor is the maximum potential difference that can be applied across the capacitor without risk of dielectric breakdown in the insulating material.

e. The **dielectric constant,** κ, of the insulating material used in a capacitor affects the capacitance, increasing it by a factor of the dielectric constant.

Terms You Should Understand

(Pairs are closely related terms whose distinction is important; number in parentheses is chapter section where term first appears.)

electrostatic potential energy (26-1)
energy density (26-3)
capacitor (26-4)
capacitance (26-4)
working voltage (26-6)
series, parallel (26-6)
dielectric constant (26-7)

Symbols You Should Recognize

U, u (26-2, 26-3)
C (26-4)
κ (26-7)

Problems You Should Be Able to Solve

calculating the work needed to assemble a distribution of discrete charges (26-1)
calculating electrostatic energy by integrating electric field energy density (26-3)
determining capacitance of simple capacitor configurations (26-4)
evaluating energy stored in capacitors (26-5)
analyzing parallel and series capacitor combinations (26-6)
determining the effect of dielectric materials in capacitors (26-7)

Limitations to Keep in Mind

Equation 26-6 for a parallel-plate capacitor is an approximation that neglects fringing fields at the plate edges; the approximation is good for capacitor plates much larger than their spacing.

Equation 26-10c applies *only* to *two* capacitors in series; with more capacitors Equation 26-10b must be used.

QUESTIONS

1. Two positive point charges are initially infinitely far apart. Is it possible, using only a finite amount of work, to move them until they are located a small distance d apart?

2. Two positive point charges are initially a small distance d apart. Is it possible, using a finite amount of work, to move them together until there is no separation between them?

3. The work required to assemble a certain charge distribution is exactly zero. Does this mean the assemblage of charges is in static equilibrium under the influence of the electric force alone? Explain.

4. How does the energy density a certain distance from a negative point charge compare with the energy density the same distance from a positive point charge of equal magnitude?

5. A dipole consists of two equal but opposite charges. Is the total energy stored in the field of the dipole zero? Why or why not?

6. Charge is spread over the surface of a balloon. The balloon is then allowed to expand. What happens to the energy of the electric field? If it is reduced, where does it go? If it is increased, where does the extra energy come from?

7. Why doesn't the superposition principle hold for electric field energy densities? That is, if you double the field strength at some point, why don't you simply double the energy density as well?

8. A student argues that the total energy associated with the electric field of a charged sphere must be infinite because its field extends throughout an infinite volume. Criticize this argument.

9. A capacitor is said to carry a charge Q. What is the net charge on the entire capacitor?

10. Does the capacitance of a capacitor describe the maximum amount of charge it can hold, in the same way that the capacity of a bucket describes the maximum amount of water it can hold? Explain and compare the meanings of capacitance and capacity.

11. A cylinder for storing compressed gas has a fixed volume, yet knowing this volume does not tell how much gas the cylinder can hold. Why not? How is the cylinder like a capacitor? Form analogies between quantities used in describing a capacitor and the amount of gas in the cylinder, the cylinder pressure, and the maximum pressure the cylinder can withstand.

12. A capacitor of capacitance C is charged to a potential difference V and carries charge Q. Why isn't the stored energy given simply by CV^2? After all, the work required to move a charge Q against a potential difference V is QV, and $C = Q/V$, so $Q = CV$.

13. Is a force needed to hold the plates of a charged capacitor in place? Explain.

14. Why do we say that capacitance depends only on the physical configuration of conductors making up a capacitor, not on the charge or potential difference, and yet we define capacitance as $C = Q/V$?

15. Why can't useful capacitors of arbitrarily large capacitance be made by simply reducing the spacing between parallel plates?

16. A solid conducting slab is inserted between the plates of a capacitor, as shown in Fig. 26-21. Does the capacitance increase, decrease, or remain the same?

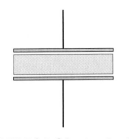

FIGURE 26-21 Question 16.

17. Why is a capacitor needed for energy storage in a camera flash? After all, the battery is the ultimate source of the flash energy, so it should be capable of supplying the needed energy.

18. Two capacitors are connected in series. Is the equivalent capacitance more or less than that of either one?

19. Two capacitors are connected in series. Could the maximum working voltage of the combination be as great as the sum of the working voltages of both capacitors? Could it be lower than this sum? Could it be lower than the working voltage of either capacitor?

20. Explain why the potential differences across parallel capacitors must be the same.

21. Two capacitors are storing equal amounts of energy, yet one has twice the capacitance of the other. How do their voltages compare?

22. Explain why inserting a dielectric between capacitor plates increases the capacitance.

23. An air-insulated parallel-plate capacitor is connected to a battery that imposes a potential difference V across the capacitor. If a dielectric slab is inserted between the capacitor plates, what happens to (a) the potential difference; (b) the capacitor charge; and (c) the capacitance?

24. An ideal dielectric would be a material whose internal dipoles did not dissipate energy as heat as they respond to an electric field. If a slab of ideal dielectric is placed part way between the plates of a capacitor, what will be its subsequent motion? Assume that the capacitor is charged but is not connected to a battery or other external circuitry.

25. A capacitor is charged and left connected to the charging battery. If you insert a dielectric slab between the capacitor plates, do you do work on the slab, or does it do work on you? Explain.

PROBLEMS

Section 26-1 Energy of a Charge Distribution

1. Three point charges, each of $+q$, are moved from infinity to the vertices of an equilateral triangle of side ℓ. How much work is required?

2. Repeat the preceding problem for the case of two charges $+q$ and one $-q$.

3. Four 50-μC charges are brought from far apart onto a line where they are spaced at 2.0-cm intervals. How much work does it take to assemble this charge distribution?

4. Repeat Example 26-1 for the case when the negative charge is $-q$ rather than $-q/2$.

5. Suppose two of the charges in Problem 1 are held in place, while the third is allowed to move freely. If this third charge has mass m, what will be its speed when it's far from the other two charges?

6. To a very crude approximation, a water molecule consists of a negatively charged oxygen atom and two "bare" pro-

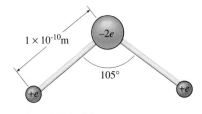

FIGURE 26-22 Problem 6.

tons, as shown in Fig. 26-22. Calculate the electrostatic energy of this configuration, which is therefore the magnitude of the energy released in forming this molecule from widely separated atoms. Your answer is an overestimate because electrons are actually "shared" among the three atoms, spending more time near the oxygen.

7. Four identical charges q, initially widely separated, are brought to the vertices of a tetrahedron of side a (Fig. 26-23). Find the electrostatic energy of this configuration.

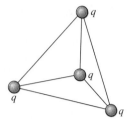

FIGURE 26-23   Problem 7 (All sides have equal length a).

8. A charge Q_0 is at the origin. A second charge, $Q_x = 2Q_0$, is brought to the point $x = a$, $y = 0$. Then a third charge Q_y is brought to the point $x = 0$, $y = a$. If it takes twice as much work to bring in Q_y as it did Q_x, what is Q_y in terms of Q_0?

Section 26-2 Two Isolated Conductors

9. Two square conducting plates 25 cm on a side and 5.0 mm apart carry charges ± 1.1 μC. Find (a) the electric field between the plates, (b) the potential difference between the plates, and (c) the stored energy.

10. Two square conducting plates measure 5.0 cm on a side. The plates are parallel, spaced 1.2 mm apart, and initially uncharged. (a) How much work is required to transfer 7.2 μC from one plate to the other? (b) How much work is required to transfer a second 7.2 μC?

11. (a) How much charge must be transferred between the initially uncharged plates of the preceding problem in order to store 15 mJ of energy? (b) What will be the potential difference between the plates?

12. Two parallel, circular metal plates of 15 cm radius are initially uncharged. It takes 6.3 J to transfer 45 μC from one plate to the other. How far apart are the plates?

13. A conducting sphere of radius a is surrounded by a concentric spherical shell of radius b. Both are initially uncharged. How much work does it take to transfer charge from one to the other until they carry charges $\pm Q$?

14. Show that the energy given by Equation 26-2 can be written as the product of the charge transferred with the *average* value of the potential during the transfer.

15. Two conducting spheres of radius a are separated by a distance $\ell \gg a$; since the distance is large, neither sphere affects the other's electric field significantly, and the fields remain spherically symmetric. (a) If the spheres carry equal but opposite charges $\pm q$, show that the potential difference between them is $2kq/a$. (b) Write an expression for the work dW involved in moving an infinitesimal charge dq from the negative to the positive sphere. (c) Integrate your expression to find the work involved in transferring a charge Q from one sphere to the other, assuming both are initially uncharged.

Section 26-3 Energy and the Electric Field

16. The energy density in a uniform electric field is 3.0 J/m³. What is the field strength?

17. A car battery stores about 4 MJ of energy. If all this energy were used to create a uniform electric field of 30 kV/m, what volume would it occupy?

18. Air undergoes dielectric breakdown at a field strength of 3 MV/m. Could you store energy in a uniform electric field in air with the same energy density as that of liquid gasoline? (See Appendix C.)

19. Find the electric field energy density at the surface of a proton, taken to be a uniformly charged sphere 1 fm in radius.

20. A pair of closely spaced square conducting plates measure 10 cm on a side. The electric field energy density between the plates is 4.5 kJ/m³. What is the charge on the plates?

21. The electric field strength as a function of position x in a certain region is given by $E = E_0(x/x_0)$, where $E_0 = 24$ kV/m and $x_0 = 6.0$ m. Find the total energy stored in a cube 1.0 m on a side, located between $x = 0$ and $x = 1.0$ m. (The field strength is independent of y and z.)

22. A sphere of radius R contains charge Q spread uniformly throughout its volume. Find an expression for the electrostatic energy contained within the sphere itself. *Hint:* Consult Example 24-1.

23. A sphere of radius R carries a total charge Q distributed over its surface. Show that the total energy stored in its electric field is $U = kQ^2/2R$.

24. A uranium-235 nucleus contains 92 protons and 143 neutrons, and has a diameter of 6.6 fm. Assuming that the proton charge is distributed uniformly throughout the nucleus, calculate the total electrostatic energy of this configuration. *Hint:* See the preceding two problems.

25. Two 4.0-mm-diameter water drops each carry 15 nC. They are initially separated by a great distance. Find the change in the electrostatic potential energy if they are brought together to form a single spherical drop. Assume all charge resides on the drops' surfaces.

26. A 2.1-mm-diameter wire carries a uniform line charge density $\lambda = 28$ μC/m. How much energy is contained in a space 1.0 m long within one wire diameter of the wire surface?

27. A long, solid rod of radius a carries uniform volume charge density ρ. Find an expression for the electrostatic energy per unit length contained *within* the rod. *Hint:* See Problem 24-31.

Sections 26-4 Capacitors

28. A capacitor's plates hold 1.3 μC when charged to 60 V. What is its capacitance?

29. The "memory" capacitor in a VCR has a capacitance of 4.0 F and is charged to 3.5 V. What is the charge on its plates?

30. What voltage is needed to put 1.6 mC on a 100-μF capacitor?

31. Figure 26-24 shows data from an experiment in which known amounts of charge are placed on a capacitor and the resulting voltage measured. Fit a line to the data, and use it to determine the capacitance.

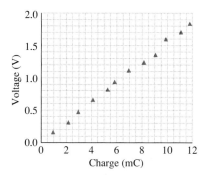

FIGURE 26-24 Problem 31 (data plot).

32. Show that the units of ε_0 may be written as F/m.

33. Find the capacitance of a parallel-plate capacitor consisting of circular plates 20 cm in radius separated by 1.5 mm.

34. A parallel-plate capacitor with 1.1-mm plate spacing has ± 2.3 μC on its plates when charged to 150 V. What is the plate area?

35. Find the capacitance of a 1.0-m-long piece of coaxial cable whose inner conductor radius is 0.80 mm and whose outer conductor radius is 2.2 mm, with air in between.

36. A capacitor consists of a conducting sphere of radius a surrounded by a concentric conducting shell of radius b. Show that its capacitance is $C = \dfrac{ab}{k(b - a)}$.

37. Figure 26-25 shows a capacitor consisting of two electrically connected plates with a third plate between them, spaced so its surfaces are a distance d from the other plates. The plates have area A. Neglecting edge effects, show that the capacitance is $2\varepsilon_0 A/d$.

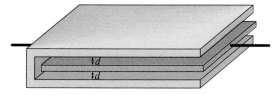

FIGURE 26-25 Problem 37.

Section 26-5 Energy Storage in Capacitors

38. The power supply of a stereo receiver contains a 2500-μF capacitor charged to 35 V. How much energy does it store?

39. Find the capacitance of a capacitor that stores 350 μJ when the potential difference across its plates is 100 V.

40. A certain capacitor stores 40 mJ of energy when charged to 100 V (a) How much would it store when charged to 25 V? (b) What is its capacitance?

41. Which can store more energy, a 1-μF capacitor rated at 250 V or a 470 pF capacitor rated at 3 kV?

42. A circuit application calls for a 10-μF capacitor that can store 12 mJ. What should be its voltage rating? The capacitors are available with voltage ratings that are multiples of 25 V.

43. A 0.01-μF, 300-V capacitor costs 25¢, a 0.1-μF, 100-V capacitor costs 35¢, and a 30-μF, 5-V capacitor costs 88¢. (a) Which can store the most charge? (b) Which can store the most energy? (c) Which is the most cost-effective energy storage device, as measured by energy stored per unit cost?

44. The charge on a capacitor is 50 mC, and it stores 2.5 J of energy. Find (a) its capacitance and (b) the voltage between its plates.

45. A camera flashtube requires 5.0 J of energy per flash. The flash duration is 1.0 ms. (a) What is the power used by the flashtube *while it is actually flashing*? (b) If the flashtube operates at 200 V, what size capacitor is needed to supply the flash energy? (c) If the flashtube is fired once every 10 s, what is its *average* power consumption?

46. The NOVA laser fusion experiment at Lawrence Livermore Laboratory in California can deliver 10^{14} W (roughly 100 times the output of all the world's power plants) of light energy when its lasers are on. But the laser pulse lasts only 10^{-9} s. (a) How much energy is delivered in one pulse? (b) The capacitor bank supplying this energy has a total capacitance of 0.26 F. Only about 0.17% (i.e., 0.0017) of the capacitor energy actually appears as light. To what voltage must the capacitor bank be charged?

47. A solid conducting slab is inserted between the plates of a charged capacitor, as shown in Fig. 26-26. The slab thickness is 60% of the plate spacing, and its area is the same as the plates'. (a) What happens to the capacitance? (b) What happens to the stored energy, assuming the capacitor is not connected to anything?

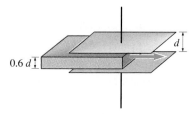

FIGURE 26-26 Problem 47.

48. Consider the two widely separated spheres of Problem 15 as a capacitor. Use energy considerations (i.e., the equation $U = \frac{1}{2}CV^2$ applies to *any* capacitor) and the answers to Problem 15 to find the capacitance.

49. The cylindrical capacitor of Example 26-5 is charged to a voltage V. Obtain an expression for the energy density as a function of radial position in the capacitor, and integrate to show explicitly that the stored energy is $\frac{1}{2}CV^2$.

Section 26-6 Connecting Capacitors Together

50. You have a 1.0-μF and a 2.0 μF capacitor. What values of capacitance could you get by connecting them in series or parallel?

51. Two capacitors are connected in series and the combination charged to 100 V. If the voltage across each capacitor is 50 V, how do their capacitances compare?

52. (a) What is the equivalent capacitance of the combination shown in Fig. 26-27? (b) If a 100-V battery is connected across the combination, what is the voltage across each capacitor? (c) What is the charge on each capacitor?

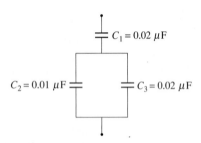

FIGURE 26-27 Problem 52.

53. You're given three capacitors: 1.0 μF, 2.0 μF, and 3.0 μF. Find (a) the maximum, (b) the minimum, and (c) two intermediate values of capacitance you could achieve with various combinations of all three capacitors.

54. What is the equivalent capacitance of the four identical capacitors in Fig. 26-28, measured between A and B?

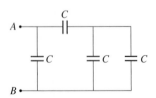

FIGURE 26-28 Problem 54.

55. You have an unlimited supply of 2.0-μF, 50-V capacitors. Describe combinations that would be equivalent to (a) a 2.0-μF, 100-V capacitor and (b) a 0.50-μF, 200-V capacitor.

56. Repeat the derivations for parallel and series capacitors, now using combinations of three capacitors.

57. What is the equivalent capacitance in Fig. 26-29?

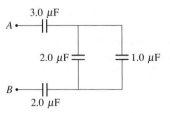

FIGURE 26-29 Problems 57, 58.

58. In Fig. 26-29, find the energy stored in the 1-μF capacitor when a 50-V battery is connected between points A and B.

59. Two capacitors C_1 and C_2 are in series, with a voltage V across the combination. Show that the voltages across the individual capacitors are

$$V_1 = \frac{C_2 V}{C_1 + C_2} \quad \text{and} \quad V_2 = \frac{C_1 V}{C_1 + C_2}.$$

60. A 0.10-μF capacitor rated at 50 V is in series with a 0.20-μF capacitor rated at 200 V. What is the maximum voltage that should be applied across the series combination? *Hint:* See the preceding problem.

61. A variable "trimmer" capacitor used to make fine adjustments has a capacitance range from 10 to 30 pF. The trimmer is in parallel with a capacitor of about 0.001 μF. Over what percentage range can the capacitance of the combination be varied?

62. Capacitors are often marked with a nominal value for the capacitance and a tolerance range within which the actual capacitance lies. For example, a 1-μF, $\pm20\%$ capacitor has capacitance between 0.8 μF and 1.2 μF. If you connect a 0.01-μF $\pm$ 20% capacitor in series with a 0.02 μF $\pm$ 30% capacitor, in what range will the resulting capacitance lie? Express as a capacitance and its associated tolerance.

63. A 5.0-μF capacitor is charged to 50 V, and a 2.0-μF capacitor is charged to 100 V. The two are disconnected from their charging batteries and connected in parallel, positive to positive. (a) What is the common voltage across each after they are connected? *Hint:* Charge is conserved. (b) Compare the total electrostatic energy before and after the capacitors are connected. Speculate on the discrepancy.

Section 26-7 Capacitors and Dielectrics

64. A parallel-plate capacitor has plates with 50 cm^2 area separated by a 25-μm layer of polyethylene. Find (a) its capacitance and (b) its working voltage.

65. A 470-pF capacitor consists of two circular plates 15 cm in radius, separated by a sheet of polystyrene. (a) What is the thickness of the sheet? (b) What is the working voltage?

66. An electrolytic capacitor is essentially a parallel-plate configuration in which aluminum plates are separated by a thin layer of aluminum oxide created by chemical action when a voltage is applied. If the effective plate area of a 2000-μF capacitor is 2.5 m^2, what are (a) the oxide layer thickness and (b) the working voltage?
67. Repeat Problem 35 for the more realistic case of a cable insulated with polyethylene.
68. An air-insulated parallel-plate capacitor has plate area 76 cm^2 and spacing 1.2 mm. It is charged to 900 V and then disconnected from the charging battery. A plexiglass sheet is then inserted to fill the space between the plates. What are (a) the capacitance, (b) the potential difference between the plates, and (c) the stored energy both before and after the plexiglass is inserted?
69. The capacitor of the preceding problem is connected to its 900-V charging battery and left connected as the plexiglass sheet is inserted, so the potential difference remains at 900 V. What are (a) the charge on the plates and (b) the stored energy both before and after the plexiglass is inserted?
70. Apply Equation 26-12 explicitly to a parallel-plate capacitor containing a dielectric, and show that the energy density between the plates is given by Equation 26-13.

Paired Problems

(Both problems in a pair involve the same principles and techniques. If you can get the first problem, you should be able to solve the second one.)
71. A pair of parallel conducting plates of area 0.025 m^2 carrying equal but opposite charges stores 1.6 J in its electric field. When the magnitude of the charge on both plates is increased by 5.0 μC, the stored energy increases to 2.4 J. Find the plate separation.
72. A capacitor stores 50 mJ of energy at voltage V_0. When the voltage is increased by 150 V, the stored energy increases to 75 mJ. Find the capacitance.
73. A 20-μF air-insulated parallel-plate capacitor is charged to 300 V. The capacitor is then disconnected from the charging battery, and its plate separation is doubled. Find the stored energy (a) before and (b) after the plate separation increases. Where does the extra energy come from?
74. Repeat the preceding problem, except that now the capacitor remains connected to the 300-V battery while the plates are separated.
75. In the capacitor network of Fig. 26-30, take $C = 6.0$ μF. Find (a) the equivalent capacitance between A and B and (b) the charge on C when 30 V is applied between A and B.
76. Take C in Fig. 26-30 as an unknown capacitance. If 100 V is applied between A and B, the network stores 5.8 mJ of energy. Find C.

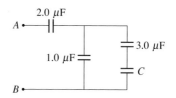

FIGURE 26-30   Problems 75, 76.

Supplementary Problems

77. A typical lightning flash transfers 30 C across a potential difference of 30 MV. Assuming such flashes occur every 5 s in the thunderstorm of Example 26-2, roughly how long could the storm continue if its electrical energy were not replenished?
78. A capacitor is constructed from a "sandwich" consisting of two long strips of aluminum foil each 2.0 cm wide and 1.6 m long, separated by two strips of 5.0-μm-thick polyethylene (Fig. 26-31). The capacitor is rolled up to make a compact cylinder. Find its capacitance. *Hint:* Because the strips are thin and closely spaced, you can treat this as a parallel-plate capacitor. But note that each foil layer in the rolled-up capacitor "sees" an oppositely charged layer on *both* sides.

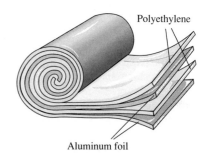

FIGURE 26-31   Problem 78.

79. Six charges $\pm q$, initially widely separated, are positioned to form a hexagon of side a, as shown in Fig. 26-32. What is the electrostatic energy of this configuration?

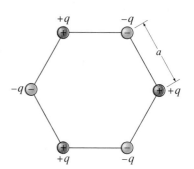

FIGURE 26-32   Problem 79.

80. Show that the result of Problem 36 reduces to that of a parallel-plate capacitor when the separation $b - a$ is much less than the radius a.

81. An air-insulated parallel-plate capacitor of capacitance C_0 is charged to voltage V_0 and then disconnected from the charging battery. A slab of material with dielectric constant κ, whose thickness is essentially equal to the capacitor spacing, is then inserted halfway into the capacitor (Fig. 26-33). Determine (a) the new capacitance, (b) the stored energy, and (c) the force on the slab in terms of C_0, V_0, κ, and the capacitor plate length L.

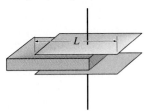

FIGURE 26-33 Problem 81.

82. Repeat parts (b) and (c) of the preceding problem, now assuming the battery remains connected while the slab is inserted.

83. We live inside a giant capacitor! Its plates are Earth's surface and the ionosphere, a conducting layer of the atmosphere beginning at about 60 km altitude. (a) What is its capacitance? *Hint:* You can treat it as either a spherical or a parallel-plate capacitor. Why? (b) The potential difference between Earth and ionosphere is about 6 MV. Find the total energy stored in this planetary capacitor.

84. Show that the result of Example 26-5 reduces to that of a parallel-plate capacitor when the separation $b - a$ is much less than the radius a. *Hint:* See Appendix A for an approximation to the logarithm.

85. Equation 26-2 gives the potential energy of a pair of oppositely charged plates. (a) Differentiate this expression with respect to the plate spacing to find the magnitude of the attractive force between the plates. (b) Compare with the answer you would get by multiplying one plate's charge by the electric field between the plates. Why do your answers differ? Which is right?

86. A solid sphere contains a uniform volume charge density. What fraction of the total electrostatic energy of this configuration is contained *within* the sphere?

87. A small dipole lies on the x axis, centered at the origin. Find an expression for the total electrostatic energy con-

tained in a thin cylindrical volume of diameter d and length ℓ, with its left end a distance ℓ from the dipole center, as shown in Fig. 26-34. Assume that ℓ is much greater than the dipole spacing. *Hint:* Since the cylinder is very thin, you can use the on-axis dipole field (Equation 23-5b) for the field throughout the cylinder.

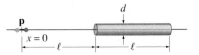

FIGURE 26-34 Problem 87.

88. A coaxial cable 15 m long consists of an inner conductor 1.0 mm in radius and an outer conductor 3.0 mm in radius, separated by polyethylene insulation. What is the electrostatic energy contained within this cable when a potential difference of 300 V is applied between its two conductors?

89. A TV antenna cable consists of two 0.50-mm-diameter wires spaced 12 mm apart. Estimate the capacitance per unit length of this cable, neglecting dielectric effects of the insulation.

90. A classical view of the electron pictures it as a purely electrical entity, whose rest mass energy mc^2 (see Section 8-7) is the energy stored in its electric field. If the electron were a sphere with charge distributed uniformly over its surface, what radius would it have to satisfy this condition? (Your answer for the electron's "size" is not consistent with modern quantum mechanics nor with experiments that suggest the electron is a true point particle.)

91. Use the fact that the static electric field is conservative to argue that there *must* be fringing field at the edges of a parallel plate capacitor. *Hint:* Remember that the plates are equipotentials, and consider the potential differences V_{AB} and V_{CD} in Fig. 26-35. What does your argument say about the strength of the fringing field relative to the field between the plates?

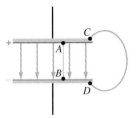

FIGURE 26-35 Problem 91.

CHAPTER 27

ELECTRIC CURRENT

Electric current in these power lines mediates the transmission of energy over long distances.

So far our discussion of electrical phenomena has been based on the assumption of electrostatic equilibrium. We now relax that assumption, and consider situations in which charges are moving. Such motion usually occurs only in materials containing free charges, so our discussion will emphasize electrical conductors. Occasionally we will also deal with charges moving in an otherwise empty region—for example, the electron beam in a TV picture tube.

27-1 ELECTRIC CURRENT

An **electric current** is a net motion of electric charge. The measure of current is the rate at which charge crosses a given area. Accordingly, its units are coulombs per second (C/s). This unit is given the special name **ampere** (A) after the French physicist André Marie Ampère:

$$1 \text{ A} = 1 \text{ C/s}.$$

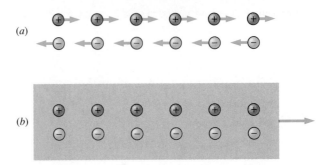

FIGURE 27-1 (a) Protons moving to the right constitute a current to the right. Since electrons are negative, *leftward* moving electrons also constitute a current to the *right*. The result in this case is a net current to the right. (b) In this bulk motion of neutral matter, the protons constitute a current to the right. But the negative electrons, also moving to the right, constitute a current to the *left*. So the net current is zero.

In electronics, biomedical applications, and many other practical situations currents are small enough that the milliampere (mA) and microampere (μA) are frequently used. When the current I is steady we write

$$I = \frac{\Delta q}{\Delta t}, \qquad \text{(steady current)} \qquad (27\text{-}1a)$$

where Δq is the net charge crossing the given area in time Δt. If the current is not steady, we consider the ratio of charge to time for arbitrarily small time intervals, giving an instantaneous current that may vary with time:

$$I = \frac{dq}{dt}. \qquad \text{(instantaneous current)} \qquad (27\text{-}1b)$$

The direction of current is the direction in which *positive* charge flows. If the moving charge is negative, as with electrons in a metal, then the current is opposite the charge motion. You can blame Benjamin Franklin for this confusing situation! It was Franklin who assigned the names "positive" and "negative" to the two kinds of electric charge. Had Franklin known that free charges in metals are electrons, he might well have reversed his terminology.

An electric current may consist of only one sign of charge in motion, or it may involve both positive and negative charge. It that case the current is determined by the *net* charge motion—that is, by the algebraic sum of the currents associated with both kinds of charge (Fig. 27-1).

Current—the rate at which charge crosses a given area—depends on the speed of the charge carriers, their density, and the charge carried by each. Consider a conductor containing n charges per unit volume, each carrying charge q and moving with speed v_d, as shown in Fig. 27-2. The quantity v_d is called the **drift speed.** In some cases—a beam of electrons in vacuum, for example—v_d is the actual particle speed. More commonly, v_d represents the time-average speed of the charge carriers, averaging out the effects of random thermal motion and collisions. If A is the conductor's cross-sectional area, then a length ℓ has

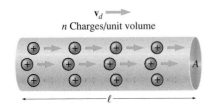

FIGURE 27-2 A conductor of cross-sectional area A containing n charge carriers per unit volume. Each moves with speed v_d and carries charge q. The total charge in a region of length ℓ is $nA\ell q$, and the current is therefore $nAqv_d$.

volume $A\ell$ and therefore contains $nA\ell$ charges. Since each carries charge q, the total charge is $\Delta Q = nA\ell q$. With drift speed v_d, the length ℓ of charge moves past a given point in time $\Delta t = \ell/v_d$, so the current is

$$I = \frac{\Delta Q}{\Delta t} = \frac{nA\ell q}{\ell/v_d} = nAqv_d. \qquad (27\text{-}2)$$

● EXAMPLE 27-1 CURRENT IN A WIRE

A copper wire with a cross-sectional area of 1.0 mm² carries a current of 5.0 A. The charge carriers in copper are electrons, and each copper atom contributes, on average, 1.3 free electrons. What is the drift speed of the electrons?

Solution
The density of copper (see inside back cover) is 8920 kg/m³, and the periodic table (also inside back cover) lists copper's atomic weight as 63.55—meaning that the mass per atom is 63.55 u. So the number density of copper atoms is

$$n = \frac{8920 \text{ kg/m}^3}{(63.55 \text{ u/atom})(1.66\times10^{-27} \text{ kg/u})}$$

$$= 8.46\times10^{28} \text{ atoms/m}^3.$$

Since each atom contributes 1.3 free electrons, the electron density is $(1.3)(8.46\times10^{28} \text{ m}^{-3}) = 1.10\times10^{29} \text{ m}^{-3}$. Solving Equation 27-2 for v_d then gives

$$v_d = \frac{I}{nAq} = \frac{5.0 \text{ A}}{(1.10\times10^{29} \text{ m}^{-3})(1.0\times10^{-6} \text{ m}^2)(1.6\times10^{-19}\text{C})}$$

$$= 0.284 \text{ mm/s}.$$

This remarkably small value is typical of drift speeds in metallic conductors.

EXERCISE A thin layer of gold, 0.10 mm wide and 3.2 μm thick, is used for connections to an integrated circuit. Find (a) the free electron density in gold and (b) the drift speed of electrons in the gold layer when it carries a current of 140 μA. The density of gold is 1.93×10^4 kg/m³, and each gold atom contributes, on average, 1.5 free electrons.

Answers: (a) 8.85×10^{28} m^{-3}; (b) 31 μm/s

Some problems similar to Example 27-1: 7–9, 55, 56 ●

How can the drift speed be so small? When you turn on a light switch, the light comes on immediately, not several thousand seconds later as the result of Example 27-1 might imply. Here it's important to distinguish between the speed of the electrons and that of the electrical signal in the wire. As soon as electrons at one end of the wire begin moving, their electric fields affect adjacent electrons, which also begin moving. This effect propagates down the wire at what is in fact nearly the speed of light, so the current begins everywhere almost simultaneously. The same thing happens when you turn on a garden hose full of water: Water comes out the far end even though water at the faucet has not had time to travel down the hose.

Current Density

In many cases electric currents are not so neatly confined as in a wire. Examples include currents in the oceans and solid Earth, in the atmosphere, in chemical solutions, and in the ionized gases that make up the stars and indeed much of the matter in the universe (Fig. 27-3). In these situations the current is spread over a rather ill-defined area, and its magnitude and direction may vary from point to point. It's useful to characterize such diffuse currents in terms of **current density,** defined as the current per unit area at a given point. Dividing Equation 27-2 by the area gives

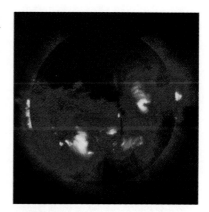

FIGURE 27-3 Strong electric currents flow in the bright yellow loops of ionized gas that arch above the Sun's surface in this image made with an orbiting x-ray telescope on the Japanese satellite Yohkoh.

$$J = \frac{I}{A} = nqv_d. \tag{27-3a}$$

The flow of charge can vary in both magnitude and direction, so current density is more generally written as a vector quantity:

$$\mathbf{J} = nq\mathbf{v}_d, \tag{27-3b}$$

where $\mathbf{v}_d$ is the drift velocity.

If current density is uniform, as in a wire, then the total current is simply the product of the current density with the wire's cross-sectional area. When the current density varies with position, then integration is necessary to calculate the total current. Problem 63 explores this case.

● EXAMPLE 27-2   AN IONIC SOLUTION

Charge carriers in salt water are positive sodium ions and negative chlorine ions, each carrying one elementary charge. A certain solution contains 6.0×10^{26} of each ion type per cubic meter. The ions drift at 2.6×10^{-5} m/s, with positive and negative ions moving in opposite directions. What is the current density in the solution? If the solution is confined to a tube of 3.0-cm^2 cross section, what is the total current?

Solution
The current density due to one type of ion—say the positive ions—is given by Equation 27-3a:

$$J = nqv_d = (6.0 \times 10^{26} \text{ m}^{-3})(1.6 \times 10^{-19} \text{ C})(2.6 \times 10^{-5} \text{ m/s})$$

$$= 2.5 \text{ kA/m}^2.$$

The negative ions drift in the opposite direction, but they also carry the opposite charge, so they provide a current density of the same magnitude and in the same direction. Therefore the net current density is 5.0 kA/m^2.

With this uniform current density through the tube's 3.0-cm^2 cross section, the total current is

$$I = JA = (5.0 \text{ kA/m}^2)(3.0 \times 10^{-4} \text{ m}^2) = 1.5 \text{ A}.$$

EXERCISE   The maximum safe current density in copper wire used in household wiring is about 6 MA/m^2. (Beyond this level the wire overheats.) What is the minimum safe wire diameter in a circuit that can carry up to 15 A?

Answer: 1.8 mm

Some problems similar to Example 27-2: 2, 6, 10, 11 ●

27-2 CONDUCTION MECHANISMS

What causes electric current? Electric charges experience forces in electric fields, so applying a field to a conductor should result in a current. You might think that it would suffice to apply the field briefly to get the charges moving; Newton's law suggests they would then keep moving. In most conductors, however, charges do not move unimpeded. They bump into things—usually ions—and quickly lose any energy they've gained from the field. To sustain a current in most materials, it is therefore necessary to maintain an electric field within the material. Having such a field does not violate our conclusion that there can be no electric field in a conductor in electrostatic equilibrium since we're explicitly considering moving charges and are therefore no longer talking about equilibrium.

Although it is generally true that electric fields result in currents, the detailed relation between current and field depends on the type of conductor. In addition to metals, important conductors include ionic solutions, plasmas (ionized gases), semiconductors, and superconductors, all of which we will consider briefly in this section.

In most materials the current density and the electric field are in the same direction, and we can therefore characterize the relation between the two using the equation

$$\mathbf{J} = \sigma\mathbf{E}, \tag{27-4a}$$

where σ is called the **conductivity** of the material. Equation 27-4a shows that conductivity is the ratio of the magnitude of the current density to the magnitude of the electric field. For many common conductors, experiment shows that this ratio is independent of field; such materials are called **ohmic,** and for them Equation 27-4 is one way of stating what is known as **Ohm's law.** (This is the *microscopic* version of Ohm's law, relating field and current density at each point. You may be familiar with the *macroscopic* version, which relates the voltage difference across a conductor to the current through it.) Ohmic materials exhibit a linear relation between current density and electric field. In **nonohmic** materials, in contrast, conductivity does depend on field, and therefore $\mathbf{J}$ and $\mathbf{E}$ are not directly proportional.

Equation 27-4a shows that the units of conductivity are A/V·m. One V/A is given the name **ohm** (symbol Ω), after the German physicist Georg Ohm (1789–1854), whose experiments clarified the relation between voltage and current. The SI unit of conductivity can therefore be written $(\Omega\cdot\text{m})^{-1}$. An equivalent way of characterizing the relation between current density and field is with the **resistivity,** ρ, defined as the inverse of the conductivity:

$$\rho = \frac{1}{\sigma},$$

so Equation 27-4a can also be written

$$\mathbf{J} = \frac{\mathbf{E}}{\rho}. \tag{27-4b}$$

The units of resistivity are $\Omega\cdot\text{m}$.

Conductivity and resistivity vary dramatically among different materials; indeed, their measurable range is one of the broadest of any physical quantity, spanning some 24 orders of magnitude. Table 27-1 lists some typical resistivities.

● EXAMPLE 27-3 HOUSEHOLD WIRING: THE ELECTRIC FIELD

A 1.8-mm-diameter copper wire carries 15 A to a household appliance. What is the electric field in the wire?

Solution
The current density is $J = I/A$; using this result in Equation 27-4b and solving for E gives

$$E = \frac{I\rho}{A} = \frac{(15\ \text{A})(1.68\times10^{-8}\ \Omega\cdot\text{m})}{(\pi)(0.90\times10^{-3}\ \text{m})^2} = 99\ \text{mV/m},$$

where we found the resistivity of copper in Table 27-1. This result is much smaller than the fields we've been discussing in electrostatic situations. Because copper is such a good conductor, only a very small field is needed to drive even a substantial current. In analyzing electric circuits we will often make the approximation that the fields and therefore potential differences in copper wires are essentially zero.

EXERCISE A uniform electric field of 0.76 V/m drives a 10-A current in an iron wire. Find the wire diameter.

Answer: 1.3 mm

Some problems similar to Example 27-3: 17–19 ●

▲ **TABLE 27-1** RESISTIVITIES

MATERIAL	RESISTIVITY ($\Omega \cdot m$)
Metallic conductors (at 20°C)	
Aluminum	2.65×10^{-8}
Copper	1.68×10^{-8}
Gold	2.24×10^{-8}
Iron	9.71×10^{-8}
Mercury	9.84×10^{-7}
Silver	1.59×10^{-8}
Ionic solutions (in water at 18°C)	
1-molar copper sulfate ($CuSO_4$)	3.9×10^{-4}
1-molar hydrochloric acid (HCl)	1.7×10^{-2}
1-molar sodium chloride (NaCl)	1.4×10^{-4}
Water, pure (H_2O)	2.6×10^{5}
Sea water (typical value)	0.22
Semiconductors (pure, at 20°C)	
Germanium	0.45
Silicon	640
Insulators	
Ceramics	10^{11}–10^{14}
Glass	10^{10}–10^{14}
Polystyrene	10^{15}–10^{17}
Rubber	10^{13}–10^{16}
Wood (dry)	10^{8}–10^{14}

Conduction in Metals

Metals contain large numbers of free electrons, which respond readily to electric fields, making these materials good conductors. We can describe conduction in a semiquantitative way by considering that these electrons are in random thermal motion and undergo collisions with the ions that are fixed in the metal's crystal structure. That description is helpful in understanding why metals obey Ohm's law, but it is not strictly correct. In fact, the major cause of resistivity involves electrons scattering from imperfections in the crystal structure—an interaction whose description necessarily involves quantum mechanics.

An electric field applied in a metal accelerates the negative electrons in the direction opposite the field. In our collision model, an electron doesn't get going very fast before it collides with an ion. It then gives up whatever energy it gained from the field, and rebounds with a random velocity. But the field is still there, and the electron again accelerates (Fig. 27-4). The result is that the electrons have a very small average velocity—the drift velocity $\mathbf{v}_d$ that we introduced earlier. Their motion therefore constitutes a net current which is itself proportional to $\mathbf{v}_d$.

The drift velocity depends on two things: the acceleration of the electrons and the rate at which they undergo collisions. The electric force $-e\mathbf{E}$ gives an acceleration $\mathbf{a} = \mathbf{F}/m = -e\mathbf{E}/m$, with m the electron mass and $-e$ its charge. If we pick a random electron and ask when it last underwent a collision, the answer will be some mean time τ called the **collision time.** During that time the electron has been accelerating with acceleration $\mathbf{a} = -e\mathbf{E}/m$, and therefore its average velocity due to the presence of the electric field—i.e., the drift velocity $\mathbf{v}_d$—will be

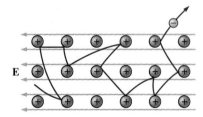

FIGURE 27-4 The path of an electron moving through a metal. The motion is almost completely random, but in the presence of an electric field there is a very slight drift antiparallel to the field.

$$\mathbf{v}_d = -\frac{e\mathbf{E}}{m}\tau.$$

Of course the electron also has its random thermal motion, but over many collisions this averages to zero while the velocity resulting from the electric field does not. Using our expression for the drift velocity in Equation 27-3b gives

$$\mathbf{J} = nq\,\mathbf{v}_d = n(-e)\mathbf{v}_d = \frac{ne^2\mathbf{E}}{m}\,\tau.$$

Comparing this expression with Equation 27-4a shows that the conductivity is

$$\sigma = \frac{ne^2}{m}\,\tau. \qquad (27\text{-}5)$$

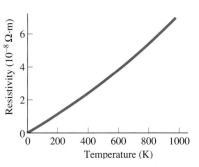

FIGURE 27-5 Resistivity of copper—the inverse of conductivity—shows a nearly linear dependence on temperature, in contrast to the classical prediction of a $\sqrt{T}$ dependence.

The collision time τ depends on how fast the electrons are moving; the faster they go, the more frequent their collisions, and the lower τ should be. The electrons have two kinds of motion: their random thermal motion and the drift velocity $\mathbf{v}_d$ acquired from the electric field. In Example 27-1 we found that a typical drift speed is on the order 1 mm/s. Thermal speeds of electrons in metals, in contrast, are around 10^6 m/s. The drift speed is therefore completely negligible in determining the collision time. That means the collision time and hence the conductivity are essentially independent of the applied electric field—and that makes Equation 27-4a a linear relation for metallic conductors. In other words, metals obey Ohm's law. We stress that this conclusion is only approximate, and that Ohm's law, unlike Gauss's law, is *not* an exact, universal statement that holds everywhere and for all materials.

Although the conductivity of a metal is independent of the applied electric field, it does depend on other factors, especially temperature. From Chapter 20 we know that classical physics gives a thermal speed proportional to the square root of the temperature, so we might expect conductivity to be proportional to $1/\sqrt{T}$ and thus resistivity to $\sqrt{T}$. Experiment, however, shows that resistivity is very nearly proportional to temperature rather than to its square root (Fig. 27-5). Here we've reached the limits of classical physics, which cannot describe fully the behavior of the free electrons in a metal. Readers of the extended version of this text will explore quantum mechanics and how it deals with electrical conduction.

● EXAMPLE 27-4 CONDUCTIVITY AND COLLISION TIME

Copper at room temperature contains 1.1×10^{29} free electrons per cubic meter. What is the collision time for these electrons?

Solution
Writing $\sigma = 1/\rho$ and solving Equation 27-5 for the collision time gives

$$\tau = \frac{m}{ne^2\rho}$$

$$= \frac{9.11\times10^{-31}\ \text{kg}}{(1.1\times10^{29}\ \text{m}^{-3})(1.6\times10^{-19}\ \text{C})^2(1.68\times10^{-8}\ \Omega\cdot\text{m})}$$

$$= 1.9\times10^{-14}\ \text{s},$$

where we found the resistivity of copper in Table 27-1. Problem 64 uses this result to estimate the mean thermal speed of the electrons.

EXERCISE Sodium contains 2.5×10^{28} free electrons per cubic meter, and the collision time is 3.4×10^{-14}s. What is the resistivity of sodium?

Answer: $4.2\times10^{-8}\ \Omega\cdot\text{m}$

Some problems similar to Example 27-4: 23, 64

●

■ **APPLICATION** NOISE IN ELECTRONIC EQUIPMENT

Although the time-average current associated with random thermal motion of charge carriers is zero, at any given instant random fluctuations may result in more charge carriers moving in one direction than in another. The result is a very small current, whose sign and magnitude fluctuate randomly. Called thermal noise, this current may disturb or even overwhelm currents of interest in sensitive electronic instruments. Thermal motion decreases at low temperatures, and sensitive circuits like the amplifiers in radio telescopes are often cooled to liquid helium temperatures (around 4 K) to reduce thermal noise (Fig. 27-6). Ultimately thermal noise limits our ability to detect and study very weak electrical signals.

FIGURE 27-6 This 12-m radio telescope at Kitt Peak National Observatory uses amplifiers cooled to 4 K with liquid helium in order to reduce electrical noise.

Ionic Solutions

An ionic solution contains positive and negative ions that respond to an electric field by moving in opposite directions, resulting in a net current (Fig. 27-7). Conductivity of the solution is limited by collisions between the ions and neutral atoms. In addition to heating the solution, some of the energy of these collisions may go into chemical reactions that store energy. Charging a car battery, for example, involves driving a current through the battery's acid solution. Electrical energy goes into reversing the chemical reactions that normally power the battery, thus building up a supply of stored energy. Conduction in ionic solutions also plays an important role in the corrosion of metals, for example those exposed to salt solutions. And the presence of an ionic solution—sweat—increases our vulnerability to electric shock. Table 27-1 includes the resistivities of some ionic solutions. Note that these solutions are poorer conductors than metals.

FIGURE 27-7 The electric eel sets up currents in the surrounding water, and can sense the presence of nearby objects by subtle variations in conductivity. It uses larger currents to kill its prey.

Plasmas

Plasma is ionized gas that conducts because it contains free electrons and positive ions. It takes substantial energy to ionize atoms, so plasmas usually occur only in high-temperature environments. The few examples of plasmas on

Earth occur in fluorescent lamps, neon signs, devices for fusion research, the ionosphere, flames, and lightning flashes (Fig. 27-8). Yet most of the matter in the universe is probably in the plasma state; the stars, in particular, are almost entirely plasma.

The electrical properties of plasma make it so different from ordinary gas that plasma is often called "the fourth state of matter." Many plasmas are so diffuse that collisions between particles are rare. These "collisionless" plasmas are sometimes far better conductors than metals, and they can sustain large electric currents with very modest electric fields.

Semiconductors

Even in the best insulators, random thermal motions occasionally dislodge electrons, giving these materials very modest conductivity. This effect increases with temperature, but is usually insignificant at normal temperatures. But a few materials—notably the element silicon—exhibit significant conductivity at room temperature. The electrical properties of these **semiconductors** make possible the microelectronic technology that plays a pervasive role in modern civilization.

Conduction in semiconductors involves not only electrons dislodged from their places in the material structure, but also the "holes" left behind by those electrons. An adjacent electron can "fall" into the hole, with the effective result that the hole has moved in the direction of the field (Fig. 27-9). Thus holes act as positive charge carriers, and a pure semiconductor like silicon contains equal numbers of negative charge carriers (free electrons) and positive charge carriers (holes).

A pure semiconductor has rather low conductivity and is useful only in a few applications. The key to semiconductor technology lies in the control of conductivity by adding very small amounts of impurities—a process called **doping.** Adding an element with five electrons in its outermost shell—as opposed to silicon's four—results in large numbers of free electrons and a much more conductive material whose charge carriers are predominantly negative. Since its charge carriers are negative, such a material is called an **N-type** semiconductor. In contrast, doping with an element containing only three outermost electrons results in a **P-type** semiconductor, whose charge carriers are predominantly holes. The wide range of semiconductor devices in use today results from carefully engineered combinations of P- and N-type material. The application below presents one of the most important such devices.

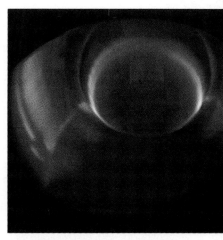

FIGURE 27-8 Plasma, or ionized gas, is an excellent electrical conductor. Photo shows glowing plasma in the Tokamak Fusion Test Reactor at the Princeton Plasma Physics Laboratory.

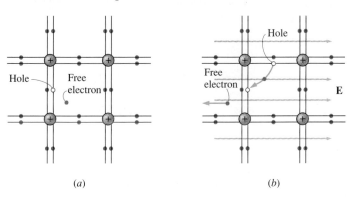

(a)   (b)

FIGURE 27-9 (a) Structure of a silicon crystal, showing each atom bound to each of its neighbors by two shared electrons. Thermal motion has dislodged one of the electrons from the crystal structure, leaving behind a hole. (b) In the presence of an electric field, the free electron drifts opposite the field direction. In addition, a nearby bound electron falls into the hole, effectively moving the hole in the direction of the field. Holes thus act as positive charge carriers.

■ **APPLICATION** TRANSISTORS AND INTEGRATED CIRCUITS

Few inventions have revolutionized human society as much as the transistor. Transistors are the basis of all modern electronic devices, from stereo amplifiers to automobile ignition systems to VCRs, laboratory instruments, and computers. Basically, a transistor is a semiconductor device in which one electrical signal controls another. The transistor shares that control function with its predecessor, the vacuum tube. But whereas vacuum tubes were large, fragile, expensive, unreliable, hot, slow, and consumed large amounts of power, transistors are small, rugged, cheap, reliable, cool, fast, and consume little power (Fig. 27-10).

Figure 27-11 shows a widely used type of transistor called the field effect transistor (FET). It consists of a slab of P-type semiconductor with two separate regions of N-type material on top, called the *drain* and *source*. Part of the structure is coated with silicon dioxide (SiO_2), an excellent insulator, and a metal *gate* coated on top. Now, a junction between P- and N-type semiconductors has the property that current flows readily from P to N but not from N to P. In the transistor of Fig. 27-11 current cannot flow between drain and source because either way an N-to-P junction is encountered. But suppose positive charge is placed on the gate, by connecting it to an appropriate potential. This charge will repel positive holes in the P-type material below the gate, and attract electrons. The result is that the "channel" between drain and source becomes temporarily N-type. Now there are no N-P junctions, and the transistor conducts. Thus the voltage applied to the gate can be used to control the drain–source current.

Varying the gate charge continuously varies the drain–source current in the same way; the transistor then functions as an *amplifier,* making a weak signal stronger. Audio and video systems all contain amplifying transistors; ultimately, for example, the weak signal from a cassette tape deck is amplified enough to drive a loudspeaker. In digital circuits like computers, in contrast, the transistor functions as a switch, with its drain–source channel either fully "on" (conducting) or "off" (nonconducting).

A transistor is fabricated from a single piece of silicon. By exposing various parts of the surface to dopant chemicals, oxygen, and metal atoms, the various N and P regions as well as the SiO_2 insulator and metallic gate are formed. The same process is used to produce entire circuits containing millions of transistors (Fig. 27-12). These **integrated circuits**—also called **chips**—make possible the complexity, miniaturization, and sophistication of modern electronic devices (Fig. 27-13).

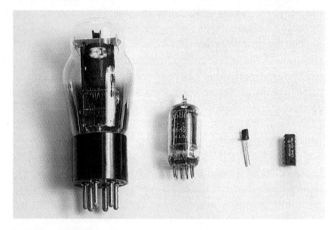

FIGURE 27-10 Vacuum tubes, a transistor, and an integrated circuit show the trend toward miniaturization in electronics. Tubes and transistor can each act as a single electronic switch, while the integrated circuit shown contains hundreds of transistors. Larger ICs now contain up to 10^8 transistors.

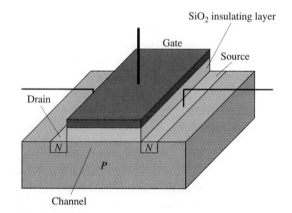

FIGURE 27-11 A field-effect transistor, or FET.

FIGURE 27-12 This PowerPC 603 microprocessor chip is at the heart of many Apple and IBM computers built in the mid-1990s. The PowerPC contains nearly 2 million transistors on a silicon chip occupying less than a square centimeter.

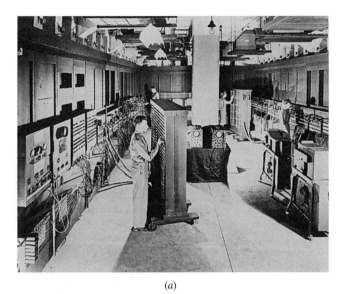
(a)

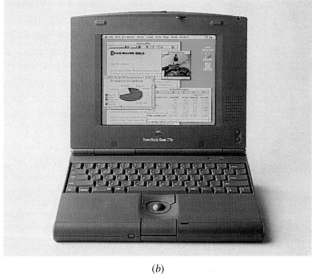
(b)

FIGURE 27-13 (a) ENIAC, one of the first computers, was built in 1946, occupied an entire room, and broke down frequently. (b) Today's laptop computers are far more powerful than ENIAC, cost orders of magnitude less, and are extremely reliable.

Superconductors

In 1911 the Dutch physicist H. Kamerlingh Onnes, studying the electrical properties of mercury at very low temperatures, found a sudden drop in resistivity at a temperature of 4.2 K. The resistivity below this temperature proved immeasurably low. Subsequent research has identified thousands of substances that become **superconductors** at sufficiently low temperatures. Currents in superconductors persist for years without any measurable decrease, suggesting that the resistivity of a superconductor is truly zero (Fig. 27-14).

For decades the known superconductors were largely metals and metal alloys that required cooling in liquid helium to achieve their superconductivity. Then, in 1986, physicists J. Georg Bednorz and K. Alex Müller of IBM's Zurich research laboratory made the stunning discovery that a class of ceramic materials becomes superconducting at temperatures around 100 K—high enough that these materials superconduct when cooled with liquid nitrogen. A flurry of research followed, soon pushing the highest superconducting temperature to well over 100 K (Fig. 27-15). Bednorz and Müller received the 1987 Nobel Prize in physics, only a year after their discovery. The search for higher temperature superconductors continues; indeed, as this book goes to press, a French team has just announced evidence for superconductivity at 250 K—about −10°F—which is warmer than a cold winter's day in the northern United States.

Superconductivity has enormous practical significance, offering loss-free transmission of electric power. Liquid-helium-cooled superconductors are used today in a number of applications, especially strong electromagnets (Fig. 27-16). Physicists and engineers are working to develop practical uses for the newer high-temperature superconductors, whose liquid nitrogen coolant is far less expensive than liquid helium. Devices based on thin films of these superconductors are already used to measure weak magnetic fields in biomedical and

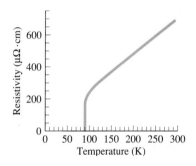

FIGURE 27-14 Resistivity versus temperature for a thin film of yttrium-barium-copper-oxide. Below the 93-K transition temperature the resistivity is truly zero.

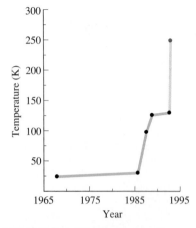

FIGURE 27-15 Since the mid-1980s, scientists have produced materials that become superconducting at increasingly high temperatures. The 250 K result is still tentative as this book goes to press.

FIGURE 27-16 Superconducting electromagnets (red structures) guide high-energy protons around the 2-km-diameter particle accelerator at Fermilab near Chicago.

geophysical applications, for measuring radiant energy, and for generating and processing microwaves. Discovery of a room-temperature superconductor would revolutionize electrical technology, and the search for such a material continues.

Although superconductivity was discovered in 1911, a satisfactory explanation of the phenomenon was not given until 1957. In that year John Bardeen, Leon Cooper, and J. Robert Schrieffer showed how superconductivity in the traditional low-temperature materials arises from a quantum-mechanical interaction among the conduction electrons. In a way that has no analog in classical physics, all the electrons move coherently through the crystal lattice, with no energy loss. Their theory earned Bardeen, Cooper, and Schrieffer the 1972 Nobel Prize in physics. A comprehensive theory of the newer, high-temperature superconducting materials has yet to be fully developed. Readers of the extended version of this text will revisit superconductivity in Chapter 41.

27-3 RESISTANCE AND OHM'S LAW

Ohm's law in the form of Equation 27-4 relates the electric field and current density within a given material. A more familiar form of Ohm's law relates voltage and current in a particular piece of material. We can relate these two—the microscopic and macroscopic forms of Ohm's law—by considering the cylindrical conductor shown in Fig. 27-17. Suppose there is a uniform electric field **E** within the conductor. Then there must be a uniform current density given by Equation 27-4b:

$$\mathbf{J} = \frac{\mathbf{E}}{\rho},$$

where ρ is the resistivity of the material. Then the total current is

$$I = JA = \frac{EA}{\rho},$$

where A is the conductor's cross-sectional area. If the conductor has length ℓ then the potential difference between its ends is

$$V = E\ell,$$

since the electric field is uniform. Note that the potential is higher (more positive) at the left, indicating that the potential drops—the energy per unit charge decreases—in the direction that charge moves through the resistor. Taking the ratio of voltage (potential difference) to current then gives

$$\frac{V}{I} = \frac{E\ell}{EA/\rho} = \frac{\rho\ell}{A}.$$

For an ohmic material, in which resistivity is independent of electric field, this equation tells us that the ratio of voltage to current depends only on the resistivity of the material and the dimensions of the particular piece. This ratio is called the electrical **resistance.** Its units are volts/ampere, or ohms (Ω). Our derivation shows that the resistance of an object with uniform cross section is just

$$R = \frac{\rho\ell}{A}. \qquad (27\text{-}6)$$

Then the ratio of voltage to current becomes

$$\frac{V}{I} = R. \qquad (27\text{-}7)$$

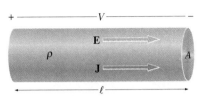

FIGURE 27-17 A cylindrical conductor made from a material with resistivity ρ. The uniform electric field **E** drives a uniform current density **J,** giving a total current $I = JA$ through the conductor's cross-sectional area A. The electric field is associated with a potential difference V across the length ℓ of the conductor.

For ohmic materials, R is constant, and Equation 27-7 is the macroscopic version of **Ohm's law.** For nonohmic materials, resistance depends on voltage, and we can consider Equation 27-7 to define the resistance at a given value of voltage and current. Again, we stress that Ohm's law is not a universal statement but an empirical law that provides a good description of ohmic materials.

Ohm's law is often written in the equivalent forms

$$V = IR \qquad \text{and} \qquad I = \frac{V}{R}.$$

This last form makes good sense, for it shows that a given voltage can push more current through a lower resistance. Two extreme cases are worth noting. An **open circuit** is a nonconducting gap with infinite resistance. No matter what the voltage across an open circuit, the current is zero. A **short circuit,** in contrast, has zero resistance. In a short circuit, currents of any magnitude can flow without requiring any potential difference or electric field. All real situations, with the exception of superconductors, lie between these two extremes.

● EXAMPLE 27-5   RESISTANCE AND OHM'S LAW

A copper wire 0.50 cm in diameter and 70 cm long is used to connect a car battery to the starter motor. What is the wire's resistance? If the starter motor draws a current of 170 A, what is the potential difference across the wire?

Solution
Table 27-1 gives 1.68×10^{-8} Ω·m for the resistivity of copper. Then Equation 27-6 gives

$$R = \frac{\rho\ell}{A} = \frac{(1.68\times10^{-8}\ \Omega\cdot\text{m})(0.70\ \text{m})}{(\pi)(0.25\times10^{-2}\ \text{m})^2} = 6.0\times10^{-4}\ \Omega.$$

This low resistance is necessary because the starter motor draws such a large current. We now apply Ohm's law to find the voltage across the wire:

$$V = IR = (170\ \text{A})(6.0\times10^{-4}\ \Omega) = 0.10\ \text{V}.$$

This is small compared with the 12 volts available from the car battery, showing that the wire is well chosen for this application (Fig. 27-18). A larger voltage difference—which would occur with a thinner wire—would mean a significant reduction in power to the motor.

FIGURE 27-18 Thick jumper cables are necessary to carry the large current used by a car's starter motor.

EXERCISE What should be the diameter of an aluminum wire that carries 15 A when the voltage across 1.0 m of the wire is 0.25 V?

Answer: 1.4 mm

Some problems similar to Example 27-5: 28, 33, 34, 36

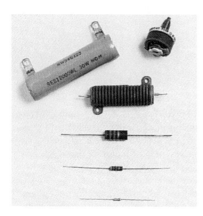

FIGURE 27-19 Typical resistors. The large unit at the upper left is a wire-wound resistor that can dissipate 30 watts. At upper right is a variable resistor, in which a rotating contact can be positioned at different points on a coil of resistance wire. Colored stripes on the smaller resistors code their resistance values. The smallest resistor shown is a carbon-film resistor that can dissipate at most $\frac{1}{4}$ watt.

Ohm's law remains valid when a conductor has nonuniform cross section, but now Equation 27-5 must be integrated to get the total resistance. Problems 69 and 70 explore this situation.

A **resistor** is a piece of conductor made to have a specific resistance. Heating elements used in electric stoves, hair dryers, irons, space heaters, and the like are all essentially resistors; so are the filaments of ordinary incandescent light bulbs. Resistors are made in a wide range of resistances for use in electronic circuits, where they help set appropriate values for voltage and current. Resistors are rated not only by their resistance but also by the maximum power they can dissipate without overheating. Figure 27-19 shows some typical resistors.

27-4 ELECTRIC POWER

We've seen that resistivity arises from collisions between electrons and ions. In these collisions, energy gained from the electric field ends up heating the conductor. Suppose a potential difference V is imposed across a conductor, driving a current I through it. The quantity V is the energy gained per unit charge as charge "falls" through the potential difference. The current I is the rate at which charge flows through the conductor. In a conductor with nonzero resistance, the energy gained from the electric field is dissipated through collisions, heating the conductor. Then the energy per unit time—that is, the power dissipated in the resistor—is the product of the energy per unit charge times the rate at which charge moves through the conductor:

$$P = IV. \quad \text{(electric power)} \tag{27-8}$$

Although we developed Equation 27-8 for power dissipated as heat in a resistance, it holds any time electrical energy is being converted to some other form. If, for example, we measure 5 V across an electric motor and 2 A through the motor, we can conclude that the motor is converting electrical to mechanical energy at the rate of 10 W (actually somewhat less because some of the power goes into heating).

Solving Ohm's law for V and putting the result in Equation 27-8 gives

$$P = I^2 R; \tag{27-9a}$$

solving instead for I gives

$$P = \frac{V^2}{R}. \tag{27-9b}$$

These are useful forms when we know the resistance and either the voltage or current. Although they may seem contradictory, both forms are equivalent and are equivalent to the more general Equation 27-8 for the case when current and voltage are related by Ohm's law.

TIP **Think About What's Constant** Equation 27-9a seems to imply that power increases with increasing resistance, while Equation 27-9b seems to suggest the opposite. Both implications are correct—*if I* in Equation 27-9a and *V* in Equation 27-9b are constants. But there's no contradiction because *I* and *V* can't both be constant while the resistance *R*—the ratio of *V* to *I*—changes. In most cases we work with sources of constant voltage—for example, the power company promises to maintain 120 V between the two contacts in a household electrical outlet—and in this case the power dissipated is inversely proportional to the resistance we connect across that voltage, as shown by Equation 27-9b. See Question 22 for more on this point.

● EXAMPLE 27-6 A LIGHT BULB

The voltage in typical household wiring is 120 V. How much current does a 100-W light bulb draw? What is the bulb's resistance under these conditions?

Solution

Solving Equation 27-8 for *I* gives

$$I = \frac{P}{V} = \frac{100 \text{ W}}{120 \text{ V}} = 0.833 \text{ A}.$$

Since we know the current, we can determine the resistance directly from Ohm's law:

$$R = \frac{V}{I} = \frac{120 \text{ V}}{0.833 \text{ A}} = 144 \text{ Ω}.$$

We could have bypassed the calculation of current and obtained *R* directly from Equation 27-9b:

$$R = \frac{V^2}{P} = \frac{(120 \text{ V})^2}{100 \text{ W}} = 144 \text{ Ω}.$$

Finally, had we known the current but not the voltage, we could have used Equation 27-9a:

$$R = \frac{P}{I^2} = \frac{100 \text{ W}}{(0.833 \text{ A})^2} = 144 \text{ Ω}.$$

The three approaches are equivalent. Use of Equation 27-9a or b merely amounts to solving symbolically for *V* or *I* before using Equation 27-8.

Because a light bulb filament undergoes a huge temperature change when turned on, its resistance is not independent of voltage and current. Our value 144 Ω holds when the light is on. When off, it is cool, and its resistance is much lower.

EXERCISE A power line has 0.20 Ω resistance per kilometer of length. If it carries 300 A of current, find (a) the voltage across 1.0 km of the wire and (b) the power dissipated in each km of wire.

Answers: (a) 60 V; (b) 18 kW

Some problems similar to Example 27-6: 42–44 ●

The Kilowatt-Hour

The SI unit of power is the watt (W), defined as 1 J/s and thus reflecting the definition of power as energy per time. We could equally well have defined the watt first, then defined the Joule as 1 W·s. The **kilowatt-hour** (kWh), a unit commonly used for electrical energy, is in fact defined in an analogous way. Just as 1 joule is the energy used by a device consuming 1 watt for 1 second, so a kilowatt-hour is the energy used by a device consuming 1 kW for 1 hour. Your household electric bill shows your electrical energy consumption in kWh; your cost for 1 kWh of electrical energy is typically in the range from 5¢ to 15¢. Since there are 3600 s in an hour, 1 kWh is equal to (1000 W)(3600 s) = 3.6 MJ. Burning a 100-W light bulb for 1 hour, for example, uses 100 watt-hours or

0.1 kWh. Although it is usually used only with electrical energy, the kWh is a perfectly good non-SI unit for describing any kind of energy; for example, it's useful to remember that the energy content of a gallon of oil or gasoline is about 40 kWh.

CHAPTER SYNOPSIS

Summary

1. **Electric current** is a net flow of electric charge, specified as the charge per unit time crossing a given area:

$$I = \frac{dq}{dt}.$$

If a material contains n free charges q per unit volume, moving with average speed v_d (called the **drift speed**), then the current through an area A perpendicular to the flow is

$$I = nqAv_d.$$

Current density is a vector specifying the current per unit area:

$$\mathbf{J} = nq\mathbf{v}_d.$$

2. **Conductivity** (symbol σ) is a property of a given material describing the ratio of electric field to current density in the material:

$$\mathbf{J} = \sigma\mathbf{E}.$$

For **ohmic** materials conductivity is independent of electric field and this relation constitutes the microscopic version of **Ohm's law.**

    Resistivity (symbol ρ) is the inverse of conductivity.

3. Conduction mechanisms vary with material, and include:
   a. **Metals,** in which the charge carriers are free electrons. Metals are ohmic materials in which resistivity arises from collisions of free electrons with ions.
   b. **Ionic solutions** are conductors because of the presence of negative and positive ions that can move through the solution.
   c. **Plasmas** are ionized gases, often with extremely high conductivity. Plasmas are rare on Earth but comprise much of the matter in the universe.
   d. **Semiconductors** conduct only poorly in their pure state, but their electrical properties can be radically altered by doping with impurities. Charge carriers in semiconductors can be electrons, positive "holes," or both. Semiconductors are the basis of modern electronic technology.
   e. **Superconductors** exhibit zero resistivity at sufficiently low temperatures, and consequently require no electric field or potential difference to drive a current.

4. **Resistance** is the ratio of voltage to current in a particular piece of material:

$$R = \frac{V}{I}.$$

Resistance depends on resistivity and physical dimensions. For an object of resistivity ρ, length ℓ, and uniform cross-sectional area A, the resistance is $R = \rho\ell/A$.

    For ohmic materials, the relation $R = V/I$ constitutes the macroscopic form of **Ohm's law.**

5. The rate at which electrical energy is converted to other forms is the product of the current I through a device and the potential difference V across it:

$$P = IV.$$

In a resistance the electrical energy is converted to heat, and the power can be written in the two equivalent forms

$$P = I^2R \quad \text{and} \quad P = \frac{V^2}{R}.$$

Terms You Should Understand

(Pairs are closely related terms whose distinction is important; number in parentheses is chapter section where term first appears.)

electric current, current density (27-1)
ampere (27-1)
drift speed, drift velocity (27-1)
conductivity, resistivity (27-2)
ohm (27-2)
ohmic, nonohmic materials (27-2)
Ohm's law (27-2, 27-3)
plasma (27-2)
semiconductor (27-2)
superconductor (27-2)
resistance (27-3)
open circuit, short circuit (27-3)

Symbols You Should Recognize

I (27-1)
v_d (27-1)
$\mathbf{J}$ (27-1)
σ, ρ (27-2)
Ω (27-2)
R (27-3)

Problems You Should Be Able to Solve

calculating current from drift speed and material properties (27-1)

relating current and current density (27-1)

relating electric field, current density, and conductivity or resistivity (27-2)

calculating resistance from resistivity and dimensions (27-3)

using Ohm's law to relate current, voltage, and resistance (27-3)

calculating electric power (27-4)

Limitations to Keep in Mind

Ohm's law is not a universal statement but an approximate empirical relation that holds with high accuracy for some materials, like metals.

QUESTIONS

1. If you physically move an electrical conductor, does this constitute a current?

2. In previous chapters we've stressed the absence of electric fields inside conductors in equilibrium. Why now do we allow such fields?

3. A wire carries a steady current. If the wire diameter decreases in the direction of the current, what happens to the current density?

4. When you talk on the telephone, your voice is heard almost immediately at the other end. Yet the drift speed of electrons in the telephone wire is on the order of millimeters per second. Explain the apparent discrepancy.

5. What is the difference between current and current density?

6. A constant electric field generally produces a constant drift velocity. How is this consistent with Newton's assertion that force results in acceleration, not velocity?

7. When caught in the open in a lightning storm, it is better to crouch low with the feet close together rather than lie flat on the ground. Why?

8. Why does the conductivity of a metal depend on the *square* of the electron charge?

9. Plasma physicists often use the approximation that there is no electric field in a plasma. How does this follow from the fact that plasma has a very large conductivity?

10. What are *P*- and *N*-type semiconductors? Does either carry a net electric charge?

11. Good conductors of electricity are often good conductors of heat. Why might this be?

12. Why can current persist forever in a superconductor with no applied voltage?

13. A plasma contains equal densities of free electrons and protons. Do you expect each to contribute equally to the net current? Explain.

14. Does an electric stove burner draw more current when it is first turned on or when it's fully hot?

15. A person and a cow are standing in a field when lightning strikes the ground nearby. Why is the cow more likely to be electrocuted?

16. You put a 1.5-V battery across a piece of material and a 100-mA current flows through the material. With a 9-V battery the current increases to 400 mA. Is the material ohmic or not?

17. The resistance of a metal increases with increasing temperature, while the resistance of a semiconductor decreases. Why the difference?

18. Macroscopic electric fields cannot exist inside a superconductor. Why not?

19. How does the fact that the drift speed of electrons in a metal is much less than their thermal speed imply that metals are ohmic conductors?

20. A 50-W and a 100-W light bulb are both designed to operate at 120 V. Which has the lower resistance?

21. A power line with a small but nonzero resistance is used to carry 450 MW of electric power from a nuclear power plant to a city. Is it most efficient to transmit this power at high voltage and low current or vice versa? Explain.

22. Equation 27-9a suggests that no power can be dissipated in a superconductor, since $R = 0$. But Equation 27-9b suggests the power should be infinite. Which is right, and why?

23. A motor made with superconducting wire and frictionless bearings is turning at constant speed and doing no mechanical work. Make an argument showing that the motor cannot be drawing current, even if it's connected to a battery. What would happen if the motor started to do mechanical work, like lifting a weight?

24. The resistivity of a pure semiconductor decreases with increasing temperature. Speculate on what might happen if a fixed voltage were applied across a piece of such material.

PROBLEMS

Section 27-1 Electric Current

1. A wire carries 1.5 A. How many electrons pass through the wire in each second?

2. In an ionic solution, 4.1×10^{15} ions, each carrying charge $+2e$, pass to the right each second; 3.6×10^{15} ions, each carrying $-e$, pass to the left in the same time. What is the net current?

3. A car battery is rated at 80 ampere-hours, meaning it can supply 80 A of current for 1 hour before it becomes discharged. If you accidentally leave the headlights on until the battery discharges, how much charge moves through the lights?

4. The electron beam that "paints" the image on a computer screen contains 5.0×10^6 electrons per cm of its length. If the electrons move toward the screen at 6.0×10^7 m/s, how much current does the beam carry? What is the direction of this current?

5. Electrons in the Stanford Linear Accelerator are accelerated to nearly the speed of light. These high-energy electrons are produced in pulses containing 5×10^{11} electrons each, lasting 1.6 μs. (a) Assuming an electron speed essentially that of light, what is the physical length of each pulse? (b) What is the peak current (i.e., the rate of charge flow while a pulse is going by)? (c) If the accelerator produces 180 pulses per second, what is the average current?

6. The National Electrical Code specifies a maximum current of 10 A in 16-gauge (0.129 cm diameter) copper wire. What is the corresponding current density?

7. Each atom in aluminum contributes about 3.5 conduction electrons. What is the drift speed in a 0.21-cm-diameter aluminum wire carrying 20 A?

8. What is the diameter of a copper wire carrying 15 A, if the drift speed is 0.86 mm/s?

9. What is the drift speed in a silver wire carrying a current density of 150 A/mm²? Each silver atom contributes 1.3 free electrons.

10. The filament of the light bulb in Example 27-6 has a diameter of 0.050 mm. What is the current density in the filament? Compare with the current density in a 12-gauge wire (diameter 0.21 cm) supplying current to the light bulb.

11. A gold film in an integrated circuit measures 2.5 μm thick by 0.18 mm wide. It carries a current density of 6.8×10^5 A/m². What is the total current?

12. A piece of copper wire joins a piece of aluminum wire whose diameter is twice that of the copper. The same current flows in both wires. The density of conduction electrons in copper is 1.1×10^{29} m⁻³; in aluminum it is 2.1×10^{29} m⁻³. Compare (a) the drift speeds and (b) the current densities in each.

13. A plasma used in fusion research contains 5.0×10^{18} electrons and an equal number of protons per cubic meter.

Under the influence of an electric field the electrons drift in one direction at 40 m/s, while the protons drift in the opposite direction at 6.5 m/s. (a) What is the current density? (b) What fraction of the current is carried by the electrons?

14. In Fig. 27-20, a 100-mA current flows through a copper wire 0.10 mm in diameter, a 1.0-cm-diameter glass tube containing a salt solution, and a vacuum tube where the current is carried by an electron beam 1.0 mm in diameter. The density of conduction electrons in copper is 1.1×10^{29} m⁻³. The current in the solution is carried equally by positive and negative ions with charges $\pm 2e$; the density of each ion species is 6.1×10^{23} m⁻³. The electron density in the beam is 2.2×10^{16} m⁻³. Find the drift speed in each region.

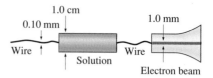

FIGURE 27-20 Problem 14.

15. The current in a wire as a function of time is given by $I(t) = 4t - 3t^2$, where t is in seconds and I in amperes. (a) Find the net charge q that has passed through the wire as a function of time, assuming $q = 0$ at $t = 0$. (b) At what time will the net charge be zero?

Section 27-2 Conduction Mechanisms

16. The electric field in an aluminum wire is 0.085 V/m. What is the current density in the wire?

17. What electric field is necessary to drive a 7.5-A current through a silver wire 0.95 mm in diameter?

18. A cylindrical tube of sea water carries a total electric current of 350 mA. If the electric field in the water is 21 V/m, what is the diameter of the tube?

19. A 1.0-cm-diameter rod carries a 50-A current when the electric field in the rod is 1.4 V/m. What is the resistivity of the rod material?

20. There is a potential difference of 2.5 V between opposite ends of a 6.0-m-long iron wire. (a) Assuming a uniform electric field in the wire, what is the current density? (b) If the wire diameter is 1.0 mm, what is the total current?

21. Use Table 27-1 to determine the conductivity of (a) copper and (b) sea water.

22. The maximum safe current in 12-gauge (0.21-cm-diameter) copper wire is 20 A. What are (a) the current density and (b) the electric field under these conditions?

23. The free-electron density in aluminum is 2.1×10^{29} m^{-3}. What is the collision time in aluminum?

24. A pure silicon crystal contains 4.9×10^{28} atoms/m^3. At room temperature, the density of electron-hole pairs is 1×10^{16} m^{-3}. In what concentration (aluminum atoms per silicon atom) must aluminum be added to give a conductivity 1000 times that of pure silicon? Assume that each aluminum atom contributes one extra hole, and that the conductivity is proportional to the density of charge carriers.

Section 27-3 Resistance and Ohm's Law

25. What is the resistance of a heating coil that draws 4.8 A when the voltage across it is 120 V?

26. What voltage does it take to drive 300 mA through a 1.2-kΩ resistance?

27. What is the current in a 47-kΩ resistor with 110 V across it?

28. The "third rail" that carries the electric power to a subway train is a rectangular iron bar whose cross section measures 10 cm $\times$ 15 cm, as shown in Fig. 27-21. What is the resistance of a 5.0-km piece of this rail?

FIGURE 27-21 A "third rail" (Problem 28).

29. What current flows when a 45-V potential difference is imposed across a 1.8-kΩ resistor?

30. A silver and an iron wire of the same length and diameter carry the same current. How do the voltages across the two compare?

31. The presence of a few ions makes air a conductor, albeit a poor one. If the total resistance between the ionosphere and Earth is 200 Ω, how much current flows as a result of a 300-kV potential difference between Earth and ionosphere?

32. A uniform wire of resistance R is stretched until its length doubles. Assuming its density and resistivity remain constant, what is its new resistance?

33. A cylindrical iron rod measures 88 cm long and 0.25 cm in diameter. (a) Find its resistance. If a 1.5-V potential difference is applied between the ends of the rod, find (b) the current, (c) the current density, and (d) the electric field in the rod.

34. You have a cylindrical piece of material 2.4 cm long and 2.0 mm in diameter. When you attach a 9-V battery to the ends of the piece, a current of 2.6 mA results. Which material from Table 27-1 do you have?

35. How must the diameters of copper and aluminum wire be related if they are to have the same resistance per unit length?

36. Extension cords are often made from 18-gauge copper wire (diameter 1.0 mm). (a) What is the resistance per unit length of this wire? (b) An electric saw that draws 7.0 A is operated at the end of an 8.0-m-long extension cord. What is the potential difference between the wall outlet and the saw?

37. Engineers call for a power line with a resistance per unit length of 50 mΩ/km. What wire diameter is required if the line is made of (a) copper or (b) aluminum? (c) If the costs of copper and aluminum wire are $1.53/kg and $1.34/kg, which material is more economical? The densities of copper and aluminum are 8.9 g/cm^3 and 2.7 g/cm^3, respectively.

38. A solid, rectangular iron bar measures 0.50 cm by 1.0 cm by 20 cm. Find the resistance between each of the three pairs of opposing faces, assuming that the faces in question are equipotentials.

39. Corrosion at battery terminals results in increased resistance, and is a frequent cause of hard starting in cars. In an effort to diagnose hard starting, a mechanic measures the voltage between the battery terminal and the wire carrying current to the starter motor. While the motor is cranking, this voltage is 4.2 V. If the motor draws 125 A, what is the resistance at the battery terminal?

40. A clear plastic trough 2.5 cm wide, 5.0 cm high, and 15 cm long has the insides of its two long sides coated with metal, as shown in Fig. 27-22. If a 60-V potential difference is applied between these sides, how much current flows when the trough contains (a) pure water and (b) sea water?

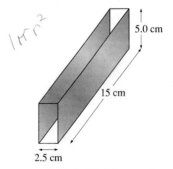

FIGURE 27-22 Problem 40.

Section 27-4 Electric Power

41. A car's starter motor draws 125 A with 11 V across its terminals. What is its power consumption?

42. A 4.5-W flashlight bulb draws 750 mA. (a) At what voltage does it operate? (b) What is its resistance?

43. A watch uses energy at the rate of 240 μW. How much current does it draw from its 1.5-V battery?

44. An electric stove burner with 35 Ω resistance consumes 1.5 kW of power. At what voltage does it operate?

45. What is the resistance of a standard 120-V, 60-W light bulb?

46. Use the numbers from Problem 31 to find the electric power dissipation in Earth's atmosphere. If we could harness this power, would it make a dent in humanity's 10^{12}-W electric power consumption?

47. If the electrons of Problem 4 are accelerated through a potential difference of 10 kV, how much power must be supplied to produce the electron beam?

48. The "instant on" feature of all the television sets in the United States requires the continuous power output of a typical large power plant—about 1000 MW. If there are 10^8 TVs in the United States, how much current does the "instant on" circuit of each draw from the 120-V power line?

49. How much total energy could the battery of Problem 3 supply?

50. During a "brown out," the power line voltage drops from 120 V to 105 V. By how much does the thermal power output of a 1500-W stove burner drop, assuming its resistance remains constant?

51. Two cylindrical resistors are made from the same material and have the same length. When connected across the same battery, one dissipates twice as much power as the other. How do their diameters compare?

52. Your author's house uses approximately 110 kWh of electrical energy each week. If that energy is supplied at 240 V, what average resistance does the house present to the power line?

53. A 2000-horsepower electric railroad locomotive gets its power from an overhead wire with 0.20 Ω/km. The potential difference between wire and track is 10 kV. Current returns through the track, whose resistance is negligible. (a) How much current does the locomotive draw? (b) How far from the power plant can the train go before 1% of the energy is lost in the wire?

54. A 100% efficient electric motor is lifting a 15-N weight at 25 cm/s. If the motor is connected to a 6.0-V battery, how much current does it draw?

Paired Problems

(Both problems in a pair involve the same principles and techniques. If you can get the first problem, you should be able to solve the second one.)

55. Electrons in a fine silver wire 20 μm in diameter drift at 0.14 mm/s. What is the current in the wire? Each silver atom contributes 1.3 free electrons.

56. A potassium slab carries a current density of 470 kA/m^2, and the drift speed of the electrons is 0.20 mm/s. If the density of potassium is 860 kg/m^3, what is the average number of free electrons contributed by each atom?

57. What is the resistance of a column of mercury 0.75 m long and 1.0 mm in diameter?

58. An integrated circuit design calls for a 470-Ω resistor. The resistor is to be 10 μm long, 1.4 μm wide, and 0.85 μm thick, with connections made at the ends 10 μm apart. Its resistivity is to be set by the appropriate amount of doping. What should that resistivity be?

59. A power plant produces 1000 MW to supply a city 40 km away. Current flows from the power plant on a single wire of resistance 0.050 Ω/km, through the city, and returns via the ground, assumed to have negligible resistance. At the power plant the voltage between the wire and ground is 115 kV. (a) What is the current in the wire? (b) What fraction of the power is lost in transmission?

60. What should be the power line voltage in the preceding problem if the transmission loss is not to exceed 2%?

61. A 240-V electric motor is 90% efficient, meaning that 90% of the energy supplied to it ends up as mechanical work. If the motor lifts a 200-N weight at 3.1 m/s, how much current does it draw?

62. An 8.5-kN elevator is powered by a 480-V electric motor that draws 24 A. If the motor is 85% efficient, how long does it take to lift the elevator 18 m?

Supplementary Problems

63. A metal bar has a rectangular cross section 5.0 cm by 10 cm, as shown in Fig. 27-23. The bar has a nonuniform conductivity, ranging from zero at the bottom to a maximum at the top. As a result, the current density increases linearly from zero at the bottom to 0.10 A/cm^2 at the top. What is the total current in the bar?

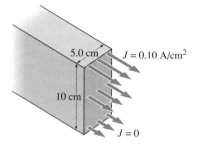

FIGURE 27-23 Problem 63.

64. Metallic copper (atomic weight 64, density 8.9 g/cm^2) forms a crystal structure with copper atoms located at the corners of cubes. (a) Use the density and atomic weight to determine the distance between copper atoms. (b) Use your result, and the collision time from Example 27-4, to

estimate the mean thermal speed of the electrons in copper. Consider τ to be the mean time between collisions.

65. The electric car of Fig. 27-24 converts 70% of its electrical energy supply into mechanical energy available at the wheels. The car weighs 640 kg and has a 96-V battery. How much current does the motor draw when the car is climbing a 10° slope at 45 km/h? Neglect friction and air resistance.

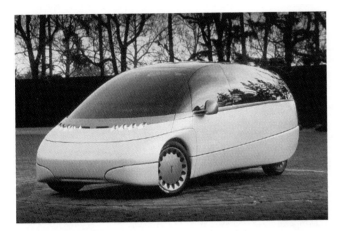

FIGURE 27-24   An electric car (Problem 65).

66. An immersion-type heating coil is connected to a 120-V outlet and immersed in a 250-ml cup of water initially at 10°C. The water comes to a boil in 85 s. What are (a) the power and (b) the resistance of the heater? Assume no heat loss, and neglect the mass of the heater.

67. A 100-Ω resistor of negligible mass is mounted inside a calorimeter. When a 12-V battery is connected for 5.0 min, the temperature inside the calorimeter rises by 26°C. What is the heat capacity of the calorimeter contents?

68. A parallel-plate capacitor has plates of 10 cm² area separated by a 1.0-mm layer of glass insulation with resistivity $\rho = 1.2 \times 10^{13}$ $\Omega \cdot$m and dielectric constant $\kappa = 5.6$. The capacitor is charged to 100 V and the charging battery disconnected. (a) What is the initial rate of discharge (i.e., the current through the insulation)? (b) At this rate, how long would it take the capacitor to discharge fully? (The rate does not remain constant; more on this in the next chapter.)

69. Figure 27-25 shows a resistor made from a truncated cone of material with uniform resistivity ρ. Consider the cone to be made of thin slices of thickness dx, like the one shown; Equation 27-6 shows that the resistance of each slab is

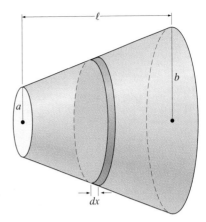

FIGURE 27-25   Problem 69.

$dR = \rho \, dx / A$. By integrating over all such slices, show that the resistance between the two flat faces is $R = \rho \ell / \pi ab$. (This method assumes the equipotentials are planes, which is only approximately true.)

70. A circular pan of radius b has a plastic bottom and metallic side wall of height h. It is filled with a solution of resistivity ρ. A metal disk of radius a and height h is placed at the center of the pan, as shown in Fig. 27-26. The side and disk are essentially perfect conductors. Show that the resistance measured from side to disk is $\rho \ln(b/a)/2\pi h$.

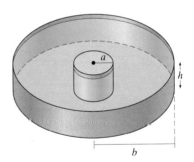

FIGURE 27-26   Problem 70.

71. At some point in a material of resistivity ρ the current density is J. Show that the power per unit volume dissipated at that point is $J^2 \rho$.

72. A thermally insulated container of sea water carries a uniform current density of 75 mA/cm². How long does it take to raise its temperature from 15°C to 20°C? Use the result of the preceding problem, and assume that both the specific heat (assumed the same as pure water) and the resistivity are constant over this range.

CHAPTER 28

ELECTRIC CIRCUITS

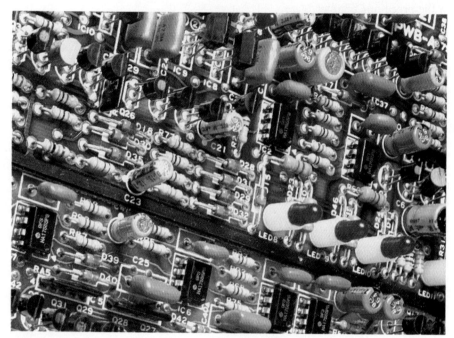

This circuit board is a complex interconnection of electronic components, including resistors, capacitors, transistors, light-emitting diodes, and integrated circuits.

28-1 CIRCUITS AND SYMBOLS

An **electric circuit** is a collection of electrical devices, called **circuit elements,** connected by conductors. A circuit usually contains a source of electrical energy, and is designed to do something useful. You are most familiar with human-made electrical circuits, which range from simple flashlights to computers, but important circuits also exist in nature. Examples include nervous systems in living organisms and Earth's global atmospheric circuit, in which thunderstorms are the batteries and the atmosphere a resistor. Your study of electric circuits should prove immensely practical, for it will help you to understand and to use effectively and safely the growing myriad of electrical and electronic devices you encounter. Basic circuit knowledge can even help you design new devices and troubleshoot old ones.

It is often helpful to represent circuits symbolically. We do so using standard symbols for circuit elements, with lines to represent the wires connecting them. We usually assume that the wires are perfect conductors, so that all points connected by wires alone are at the same potential; such points are electrically equivalent. Realizing this will greatly facilitate your interpretation of circuit diagrams! Figure 28-1 shows some common circuit symbols.

28-2 ELECTROMOTIVE FORCE

In the preceding chapter, we found that an electric field is necessary to drive a current in any conductor with nonzero resistance. But if we simply apply an electric field, say by putting excess charge on one end of the conductor, the charge will quickly redistribute itself until electrostatic equilibrium is reached and the electric field disappears. Somehow we must maintain the electric field, and with it the current, despite the tendency toward equilibrium. This requires that we compensate for the energy lost through the collisions that give the material its resistance.

What we need is a device that can maintain charge separation by converting energy from some other form into electrical energy. We call such a device a source of **electromotive force** or **emf.** (The name has historical origins; emf is not actually a force.) Most sources of emf have two electrical contacts, or **terminals,** for connection to other circuit elements. Energy conversion processes within the source move an excess of positive charge to one terminal, negative to the other, thus maintaining a potential difference between the terminals. The circuit symbol for a source of emf is shown in Fig. 28-1. The most familiar example is a battery, in which chemical energy drives electric charge to the two terminals. Other examples include electric generators, which convert mechanical energy to electrical energy; photovoltaic cells, which convert sunlight; fuel cells, which "burn" hydrogen to produce electrical energy and water; and biological cell membranes, which separate charge to control the movement of ions into the cell.

Electromotive force is quantified by the work per unit charge done by a source as it separates positive and negative charge to its two terminals. The units of emf are therefore volts, and the emf of a source is often called, loosely, its voltage. An **ideal source of emf**—one with no internal energy losses—maintains the same voltage across its terminals under all conditions. Real sources always have internal energy losses, so the terminal voltage may not equal the rated emf. We discuss this situation in the next section.

When a source of emf is not connected to any external circuit, no work is needed to maintain its terminal voltage. But current flows when the source is connected to an external circuit. This current would quickly deplete the charge at the terminals were if not for work done inside the source to separate more charge. The simple analogy of Fig. 28-2 illustrates the operation of a source of emf. The energy conversion mechanism in the emf is like the person lifting bowling balls to the table top. The balls then roll down a ramp, bumping into a series of pegs on the way down and giving up the energy they gained when lifted onto the table. Once they reach the bottom the balls roll back to where they're again lifted to repeat the cycle. An emf does the same thing: It "lifts" charge

Circuit symbols

Resistor Capacitor Source Voltmeter Switch
of emf

Ammeter Variable Variable Ground Fuse
resistor capacitor

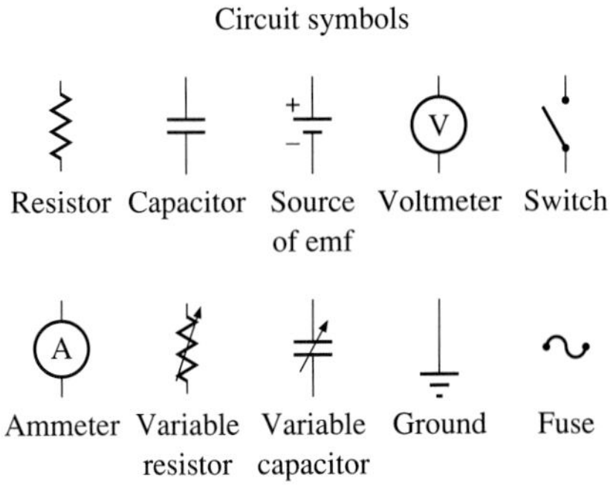

FIGURE 28-1 Common circuit symbols.

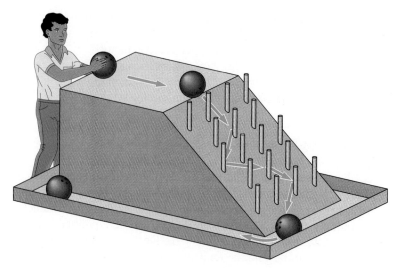

FIGURE 28-2 A gravitational analog of a source of emf. The person lifting the bowling balls represents the energy conversion mechanism in the emf, which does work on charges as it moves them against the electric field. The ramp, studded with pegs to which the balls transfer their energy, is like an external resistance connected across the emf.

against the internal electric field that points from its positive to its negative terminal. The charge then "falls" through the external circuit, dissipating its energy in the circuit resistance. When the charge returns to the source it is again "lifted," and the process continues.

28-3 SIMPLE CIRCUITS: SERIES AND PARALLEL RESISTORS

In the circuit of Fig. 28-3 a battery of emf $\mathcal{E}$ drives a current through the resistor R. How much current? The voltage across the battery is its emf $\mathcal{E}$. The battery is connected to the resistor by wires, assumed to have zero resistance. Because they have zero resistance, there is no potential difference across either wire. Therefore the voltage across the resistor is the same as the voltage across the battery, and we can immediately apply Ohm's law to find the current:

$$I = \frac{\mathcal{E}}{R}.$$

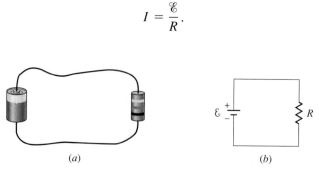

(a)   (b)

FIGURE 28-3 A circuit containing a battery and a resistor. (a) Physical circuit. (b) Schematic diagram, using the symbols of Fig. 28-1.

Energetically, this circuit is analogous to Fig. 28-2. Charge gains energy in the amount $\mathcal{E}$ joules per coulomb as it goes through the battery, then dissipates that energy in heating the resistor.

> **TIP**   **Don't Get Hung Up about Wires**   If the charge loses its energy flowing through the resistor, how does it then get back to the battery? If the wire is a perfect conductor, there's no problem because it takes no energy—and therefore no voltage—to drive current through it. In this case the current is determined entirely by the battery emf $\mathcal{E}$ and the resistance R. If you try to use Ohm's law to calculate the currents and voltages in ideal wires, you are needlessly complicating things! Of course real wires have some resistance, but if it's negligible compared with other resistances in the circuit, then the voltage across the wires is negligible, and we can approximate them as being ideal.

Series Resistors

Figure 28-4 shows a circuit containing two resistors in series. What is the current through these resistors? What is the voltage across each? Note that neither resistor is connected directly across the battery, so we can't argue that the voltage across either is the battery voltage. However, the full battery voltage does appear across the series combination, so if we knew the equivalent resistance we could solve for the current. What current? The current in *both* resistors. Since they are in series the only place current flowing through R_1 can go is through R_2. As long as there is no buildup of charge in the circuit, the current through both resistors—and, for that matter, through the battery as well—must be the same. This situation holds whenever circuit elements are in series:

| **The current through circuit elements in series is the same.**

If I is the current in the circuit of Fig. 28-4, then there must be a voltage $V_1 = IR_1$ across R_1 to drive the current through this resistor. Similarly, the voltage across R_2 is $V_2 = IR_2$. Thus, the potential difference across the two resistors together is $V_1 + V_2 = IR_1 + IR_2$. But the battery is connected directly across this series combination, so we have

$$IR_1 + IR_2 = \mathcal{E},$$

or

$$I = \frac{\mathcal{E}}{R_1 + R_2}.$$

Comparison with Ohm's law in the form $I = V/R$ shows that the two resistors in series behave like an equivalent resistance equal to the sum of their resistances. In an obvious generalization to more resistors in series, we have

$$R_{\text{series}} = R_1 + R_2 + R_3 + \cdots \qquad (28\text{-}1)$$

In other words, resistors in series add.

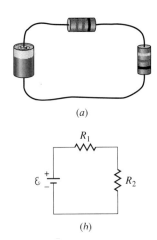

(a)

(b)

FIGURE 28-4   A battery and two resistors in series. (a) Physical circuit. (b) Schematic diagram.

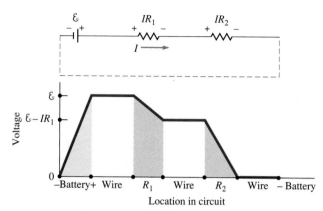

FIGURE 28-5 Voltages in the circuit of Fig. 28-4, with $V = 0$ at the negative battery terminal. Note that there is no potential difference across the wires, since they have negligible resistance, and that potential increases across the battery and decreases across the resistors. Can you tell from the graph which resistance is greater?

Given the current, we can use Ohm's law in the form $V = IR$ to solve for the voltage across each resistor:

$$V_1 = \frac{R_1}{R_1 + R_2} \mathscr{E} \tag{28-2a}$$

and

$$V_2 = \frac{R_2}{R_1 + R_2} \mathscr{E}. \tag{28-2b}$$

These expressions show that the battery voltage divides between the two resistors in proportion to their resistance. For this reason a series combination of resistors is called a **voltage divider.** Figure 28-5 depicts the voltages throughout the circuit of Fig. 28-4, and shows explicitly that the resistors divide the battery voltage.

● **EXAMPLE 28-1**   DESIGNING A VOLTAGE DIVIDER

A light bulb with a resistance (when on) of 5.0 Ω is designed to operate at a current of 600 mA. To operate this lamp from a 12-V battery, what resistance should you place in series with it?

Solution
Let R_2 be the lamp and R_1 the unknown series resistor. Since resistors in series add, the current through both resistors is $I = \mathscr{E}/(R_1 + R_2)$, which is supposed to be 600 mA or 0.60 A. Solving for R_1 gives

$$R_1 = \frac{\mathscr{E} - IR_2}{I} = \frac{12 \text{ V} - (0.60 \text{ A})(5.0 \text{ Ω})}{0.60 \text{ A}} = 15 \text{ Ω}.$$

You can also get this result by noting that the light bulb's proper operating voltage is $V = IR_2 = (0.60 \text{ A})(5.0 \text{ Ω}) = 3.0 \text{ V}$. This is one-fourth of the battery voltage, so the light bulb's 5-Ω resistance should be one-fourth of the total. That makes the total 20 Ω, leaving 15 Ω for R_1.

EXERCISE   Suppose that in Fig. 28-4 $R_1 = 470$ Ω. If the voltage across R_2 is 59% of the battery voltage, find R_2.

Answer: 676 Ω

Some problems similar to Example 28-1: 23–25

Real Batteries

What's the difference between the two 1.5-V batteries shown in Fig. 28-6? If both were ideal there would be no difference because both would maintain 1.5 V across their terminals no matter how much current was flowing. But these are real batteries. The rate at which internal chemical reactions take place limits the amount of current each can supply. Not surprisingly, the larger battery can supply more current.

We can model a real battery by considering it to be an ideal emf in series with an **internal resistance,** as shown in Fig. 28-7. Of course this is not how batteries are made since no manufacturer can make an ideal emf! The internal resistance is intrinsic to the battery, and there is no way to circumvent it. The more powerful battery is the one with lower internal resistance; it approaches more closely the ideal of zero internal resistance and can therefore supply more current.

We can understand the effect of internal resistance by considering the circuit of Fig. 28-8. This is just the series circuit of Fig. 28-4, with R_1 the internal resistance R_{int} and R_2 the external resistance R_L. R_L is called the *load resistor* because it is the thing to which we wish to deliver electric power; it is the electrical load on the battery. From Equation 28-2b we see that if R_{int} is small compared with R_L, then the voltage across the load resistance will be very close to the battery's internal emf. In this case the battery's behavior is nearly ideal, since it has essentially $\mathscr{E}$ volts across its terminals. But if we lower R_L so it becomes comparable with R_{int}, then the voltage across R_L decreases and the battery no longer seems ideal. As we lower R_L we draw more current from the battery. It takes a higher voltage to drive this current through the fixed resistance R_{int}, so more voltage drops across R_{int}, leaving less across R_L. Even if we short-circuit the battery terminals (which is not good for the battery!) we will not get infinite current—in fact, we will simply have

$$I = \frac{\mathscr{E}}{R_{int}}. \quad \text{(battery short-circuited)}$$

We conclude that a battery or other source of emf behaves more or less ideally depending on the size of its load resistance relative to its internal resistance. A calculator, for example, has a very high resistance and draws little current. It is quite happy with a small battery whose internal resistance, while relatively high, is still small compared with the calculator's resistance. A car starter motor, on the other hand, draws a large current and thus requires a battery with very low internal resistance.

FIGURE 28-6 A 1.5-V calculator battery and a 1.5-V D-cell flashlight battery have the same voltage, but the internal resistance of the calculator battery is higher.

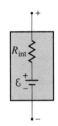

FIGURE 28-7 A real battery modeled as an ideal emf in series with an internal resistance.

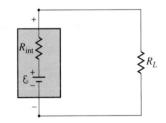

FIGURE 28-8 A real battery connected to an external load. Some voltage drops across the internal resistance, making the voltage across the battery terminals less than the battery's rated voltage.

● **EXAMPLE 28-2** STARTING A CAR

Your car's starter motor draws 125 A. The car has a 12-V battery, but while the starter motor is running the voltage across the battery terminals measures only 9.5 V. What is the internal resistance of the battery?

Solution

This circuit is just that of Fig. 28-8, with the starter being the load. With 9.5 V across the starter, there must be 2.5 V left across the internal resistance to make a total of 12 V. The current is the same throughout this series circuit, so 125 A is the current through R_{int}. Knowing current and voltage, we apply Ohm's law:

$$R = \frac{V}{I} = \frac{2.5 \text{ V}}{125 \text{ A}} = 0.020 \ \Omega.$$

A battery voltage between 9 and 11 volts is typical of a car being started. A battery voltage much below 9 volts usually indicates a weak battery, a defective starter motor, or very cold weather!

EXERCISE A 9-V battery has an internal resistance of 13 Ω. What is the maximum current that can be drawn from the battery if its terminal voltage is to remain above 8.0 V?

Answer: 77 mA

Some problems similar to Example 28-2: 12, 13, 15, 17 ●

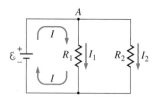

FIGURE 28-9 Parallel resistors connected across a battery. The two resistor currents sum to the battery current.

Parallel Resistors

Figure 28-9 shows two resistors in parallel, connected across an ideal battery. What is the equivalent resistance of this parallel combination? Since the two resistors are connected at top and bottom by ideal wires, the voltage across each must be the same. We made this point in Chapter 26 when we discussed parallel capacitors, and it's worth repeating here:

| The voltage across circuit elements in parallel is the same.

The parallel resistors are connected directly across the battery, so their common voltage is the battery emf $\mathscr{E}$. Applying Ohm's law then gives the current through each resistor:

$$I_1 = \frac{\mathscr{E}}{R_1}$$

and

$$I_2 = \frac{\mathscr{E}}{R_2}.$$

At the point marked A in Fig. 28-9, a current I brings in charge from the battery, while the currents I_1 and I_2 carry charge away. If charge is not to accumulate (see Problem 69), the incoming and outgoing currents must be equal; that is,

$$I = I_1 + I_2.$$

Using our expressions for the two resistor currents gives

$$I = \frac{\mathscr{E}}{R_1} + \frac{\mathscr{E}}{R_2} = \mathscr{E}\left(\frac{1}{R_1} + \frac{1}{R_2}\right).$$

Comparison with Ohm's law in the form $I = V/R$ shows that the equivalent resistance of the parallel combination is given by

$$\frac{1}{R_{\text{parallel}}} = \frac{1}{R_1} + \frac{1}{R_2}. \tag{28-3a}$$

This result is readily generalized to more than two parallel resistors:

$$\frac{1}{R_{\text{parallel}}} = \frac{1}{R_1} + \frac{1}{R_2} + \frac{1}{R_3} + \cdots. \qquad (28\text{-}3b)$$

In other words, resistors in parallel add reciprocally. Equation 28-3b shows that the resistance of a parallel combination is always lower than that of the lowest resistance in the combination. When there are only two parallel resistors, we can rewrite Equation 28-3a using a common denominator to obtain

$$R_{\text{parallel}} = \frac{R_1 R_2}{R_1 + R_2}. \qquad (28\text{-}3c)$$

Note that *parallel* resistors combine in the same way as *series* capacitors, and vice versa.

● EXAMPLE 28-3 PARALLEL AND SERIES RESISTORS

You have available three 2.0-Ω resistors. What different resistances can you make by combining all three resistors?

Solution
Figure 28-10 shows the four possible combinations. Resistors in series add, so combination (*a*) has 6.0 Ω. Resistors in parallel reciprocally add, so combination (*b*) has

$$\frac{1}{R} = \frac{1}{2.0\ \Omega} + \frac{1}{2.0\ \Omega} + \frac{1}{2.0\ \Omega} = 1.5\ \Omega^{-1},$$

for a resistance of 0.67 Ω. Combination (*c*) has two resistors in series, giving 4.0 Ω. This 4.0-Ω combination is in parallel with 2.0 Ω, so Equation 28-3c gives

$$R = \frac{(2.0\ \Omega)(4.0\ \Omega)}{2.0\ \Omega + 4.0\ \Omega} = 1.3\ \Omega.$$

Finally, combination (*d*) has two resistors in parallel, giving 1.0 Ω. This combination is in series with 2.0 Ω, for a total of 3.0 Ω. Thus you can make combinations ranging from 0.67 Ω to 6.0 Ω with these three equal resistors.

EXERCISE A 270-Ω and a 470-Ω resistor are connected in parallel, and the combination is connected in series with a 150-Ω resistor. Find the equivalent resistance of this combination.

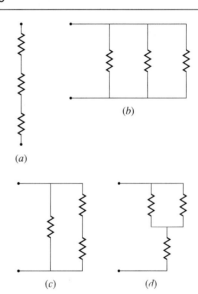

FIGURE 28-10 The four possible combinations of three equal resistors (Example 28-3).

Answer: 321 Ω

Some problems similar to Example 28-3: 16, 18, 19, 59, 60 ●

Analyzing Circuits

Many circuits contain series and parallel combinations of basic circuit elements. We can simplify these circuits by treating each series or parallel combination as a single element, often continuing the process until we can determine the voltages and currents throughout the entire circuit. Example 28-4 illustrates this procedure, which is similar to the way we dealt with capacitor combinations in Chapter 26.

● **E X A M P L E 2 8 - 4** ANALYZING A CIRCUIT

In the circuit of Fig. 28-11, what is the current through the 2-Ω resistor?

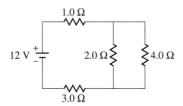

FIGURE 28-11 Circuit for Example 28-4.

Solution

We approach this problem by simplifying the circuit until we can solve for something—in this case the total current. Then we reverse the process, analyzing the circuit details until we can solve for the quantity we want. Figure 28-12 shows the steps in simplifying the circuit. We get from the original circuit, Fig. 28-12a, to Fig. 28-12b by calculating the resistance of the parallel combination of 2 Ω and 4 Ω:

$$R_{\parallel} = \frac{(2.0 \ \Omega)(4.0 \ \Omega)}{2.0 \ \Omega + 4.0 \ \Omega} = 1.33 \ \Omega.$$

Figure 28-12b shows three resistors in series, which add to get the single-resistor circuit of Fig. 28-12c. From here we can calculate the total current:

$$I = \frac{\mathscr{E}}{R} = \frac{12 \ \text{V}}{5.33 \ \Omega} = 2.25 \ \text{A}.$$

Where does this current flow? It flows from the battery through the 1-Ω resistor, then on through the parallel combination of the 2-Ω and 4-Ω resistors, then through the 3-Ω resistor and back to the battery. It does *not* all flow through the 2-Ω resistor because there are two paths the current can take when it gets to the parallel combination. However, it does all flow through the parallel combination. We already found that this combination has a resistance of 1.33 Ω, and now we know that 2.25 A flows through the combination. So the voltage across the combination is

$$V = IR = (2.25 \ \text{A})(1.33 \ \Omega) = 2.99 \ \text{V}.$$

This same voltage appears across each of the two resistors making up the parallel combination (why?) so the current through the 2-Ω resistor is

$$I = \frac{V}{R} = \frac{2.99 \ \text{V}}{2.0 \ \Omega} = 1.5 \ \text{A}.$$

In solving for this current we effectively reversed our original simplification of the circuit, first considering Fig. 28-12b to get

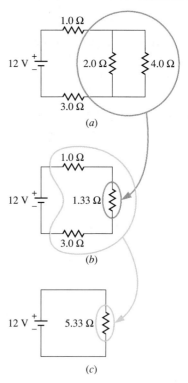

FIGURE 28-12 Simplifying the circuit by forming parallel and series combinations.

the voltage across the parallel combination, and then going to the full circuit to get the answer. At each stage we applied Ohm's law to solve for either a voltage or a current as needed.

TIP Don't Abuse Ohm's Law Ohm's law relates the voltage *across a resistor* to the current *through that resistor*. It does *not* relate arbitrary voltages and currents anywhere in a circuit. Just because there is a 12-V battery in Fig. 28-11, for example, does *not* mean that 12 V appears across the 2-Ω resistor. Be careful, too, with series and parallel combinations. As it stands, the only individual resistors in Fig. 28-11 in either series or parallel are the parallel pair of 2 Ω and 4 Ω. The 1-Ω and 2-Ω resistors are *not* in series because current flowing through the 1-Ω resistor need not all go through the 2-Ω resistor; some can go through the 4-Ω resistor instead. Equations 28-1 and 28-3 apply *only* to combinations that are strictly series or parallel, respectively.

EXERCISE   In Fig. 28-13, find (a) the current supplied by the battery and (b) the voltage across the 180-Ω resistor.

Answers: (a) 30 mA; (b) 2.9 V

Some problems similar to Example 28-4:   26–28

FIGURE 28-13   What is the voltage across the 180-Ω resistor?

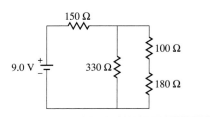

28-4 KIRCHHOFF'S LAWS AND MULTILOOP CIRCUITS

Some circuits cannot be simplified using series and parallel combinations. This often happens when there is more than one source of emf or when circuit elements are connected in complicated ways. In Fig. 28-14, for example, are resistors R_1 and R_2 in parallel? No, because R_3 separates their lower ends. Are R_1 and R_3 in series? No, because current flowing out the bottom of R_1 can go through either R_3 or R_4. Solving circuits that aren't reducible to series and parallel combinations usually requires more general techniques than we've used so far.

FIGURE 28-14   This circuit cannot be analyzed using series and parallel combinations.

Kirchhoff's Laws

In Fig. 28-5 we looked at changes in electric potential around the loop comprising a simple series circuit. The result was an increase in potential at the battery, followed by decreases at both resistors that summed to the gain in the battery. If we count the increase as a positive change and the decreases as negative changes, we can state the following:

> **The sum of the voltage changes across all the circuit elements around any closed loop is zero.**

This statement is known as **Kirchhoff's loop law,** and it applies not just to Fig. 28-5 but to *any* closed loop in a circuit. The loop law is ultimately about energy conservation; it says that charge moving around a loop gains as much energy from batteries or other sources as it loses in resistors or other energy-conversion devices.

In analyzing parallel resistors with Fig. 28-9 we noted that the current flowing into point *A* must equal the total current flowing out. This is really a statement about conservation of charge. It applies to any point in a circuit carrying steady currents since under steady-state conditions charge cannot build up or be depleted. The statement is most useful at a point where three or more wires join; such a point, like *A* in Fig. 28-9, is called a **node.** If we count currents flowing into a node as positive, and currents flowing out as negative, then the statement of charge conservation becomes the following:

> **The sum of the currents at any node is zero.**

This is **Kirchhoff's node law.**

Analyzing Multiloop Circuits

Even the most complex circuits can be analyzed using Kirchhoff's laws. Applying the laws amounts to writing equations expressing the loop law and node law for the distinct loops and nodes in the circuit. The number of equations needed is generally one less than the number of loops plus one less than the number of nodes; this is because the quantities in one loop and one node can be expressed entirely in terms of other loops and nodes, making one loop and one node equation each redundant. Here we give just two examples; electrical engineers take entire courses in circuit analysis using these laws.

● **EXAMPLE 28-5** A MULTILOOP CIRCUIT

Apply Kirchhoff's laws to find the current through R_1 in Fig. 28-15a.

Solution

We don't really need Kirchhoff's laws here; since resistors R_2 and R_3 are in parallel, we could use the methods developed earlier. But this simple circuit serves to illustrate the Kirchhoff approach. In Fig. 28-15b we've identified three loops and two nodes, and we've labeled three distinct currents. In a multiloop circuit it's not always obvious which way the currents flow, and therefore we've arbitrarily assigned directions. If an answer comes out negative, that just means the current is really flowing in the opposite direction. You can probably see in this circuit that I_2 and I_3 can't really be in opposite directions, but we will assume they are for the sake of illustration.

First we apply the loop law. Going clockwise around loop 1 starting at node B, we first encounter a voltage increase $\mathcal{E}_1$, then a drop $-I_1R_1$ in resistor R_1, then a *drop* $\mathcal{E}_2$ because we traverse the second battery going from positive to negative, and finally an increase I_2R_2 in R_2. Why an increase? Because we're going *against* the indicated direction of I_2 and, therefore, from what is purportedly the lower to the higher potential end of the resistor. Never mind if the situation is really reversed; again, the algebra will take care of it. So the loop law reads

$$\mathcal{E}_1 - I_1R_1 - \mathcal{E}_2 + I_2R_2 = 0. \quad \text{(loop 1)}$$

Our choice of going clockwise is arbitrary; going counterclockwise would give the same equation since all terms would enter with the opposite sign. Loop 2 is much easier; going clockwise from B, its equation reads

$$-I_2R_2 - I_3R_3 = 0. \quad \text{(loop 2)}$$

Here both terms are negative because we go through both resistors in the indicated current direction. What about loop 3? Physically, it consists of parts of loop 1 and loop 2. Mathematically, we could get the equation for loop 3 by subtracting our previous loop equations. In other words, the equation for loop 3 contains no new information, and we need not bother with it.

Having taken care of the loops, we apply Kirchhoff's node law to the circuit nodes. The node law says that the sum of currents at a node is zero. For node A this gives

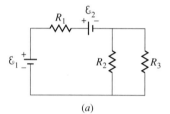

(a)

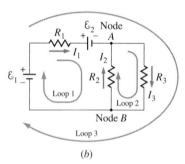

(b)

FIGURE 28-15 (a) A multiloop circuit. (b) The circuit contains 3 loops and 2 nodes. Current directions are arbitrary, and the algebraic signs of the answers will determine the actual directions (Example 28-5).

$$I_1 + I_2 - I_3 = 0, \quad \text{(node A)}$$

where I_3 enters with a minus sign since the direction we've assigned has I_3 carrying charge away from the node. Node B would give essentially the same equation. At any rate, we already have three equations in the three unknowns I_1, I_2, and I_3, and that is sufficient to solve for all three. We just need I_1, so we'll eliminate the other two. First, the node equation gives $I_3 = I_1 + I_2$; substituting in the loop 2 equation, we have $-I_2R_2 - I_1R_3 - I_2R_3 = 0$, or

$$I_2 = -\frac{I_1R_3}{R_2 + R_3}.$$

The minus sign here shows that I_1 and I_2 cannot both be in the directions indicated. Substituting this result for I_2 in the loop 1 equation and solving for I_1 then gives

$$I_1 = \frac{\mathcal{E}_1 - \mathcal{E}_2}{R_1 + \dfrac{R_2 R_3}{R_2 + R_3}}.$$

Does this result make sense? The denominator is just R_1 in series with the parallel combination of R_2 and R_3. If we inter-changed R_1 and $\mathcal{E}_2$ then we would have exactly what the equation describes: a battery of emf $\mathcal{E}_1 - \mathcal{E}_2$ connected across the resistor combination indicated by the denominator. Since the current in the series elements $\mathcal{E}_2$ and R_1 is the same, their order doesn't matter.

What about the current directions? Our answer shows that I_1 is in the direction indicated if $\mathcal{E}_1$ is the higher emf. Then I_2's direction must be opposite what we've indicated. If $\mathcal{E}_2$ is greater, these conclusions reverse. ●

● EXAMPLE 28-6 RATE THE RESISTOR

What power dissipation must resistor R_3 of Fig. 28-16a be able to tolerate?

Solution
Again, we indicate loops, nodes, and currents (Fig. 28-16b). Here we need to know I_3 to find the power dissipation in R_3. Instead of solving algebraically as in Example 28-5, we will simply write the equations directly with their numerical values. Let's go counterclockwise around loop 1; starting at A, the loop law becomes

$$6 - 2I_1 - I_3 = 0, \quad \text{(loop 1)}$$

where we've temporarily dropped the units. Starting from A and going clockwise around loop 2 (remember, the direction is arbitrary), we have

$$9 + 4I_2 - I_3 = 0. \quad \text{(loop 2)}$$

Finally, the node equation at node A reads

$$-I_1 + I_2 + I_3 = 0, \quad \text{(node } A)$$

where the first term is negative because I_1 is indicated as flowing *away* from the node. The equations for node B and loop 3 are redundant, so we're ready to solve the system. The node equation gives $I_1 = I_2 + I_3$; substituting in the loop 1 equation, we have $6 - 2I_2 - 3I_3 = 0$, or

$$I_2 = \tfrac{1}{2}(6 - 3I_3).$$

Finally, we can use this result in the loop 2 equation to get $9 + 2(6 - 3I_3) - I_3 = 0$ or

$$21 - 7I_3 = 0,$$

giving $I_3 = 3$ A. That this answer is positive indicates that I_3 is indeed upward in Fig. 28-16. The power dissipated in R_3 is therefore

$$P_3 = I_3^2 R_3 = (3 \text{ A})^2 (1 \ \Omega) = 9 \text{ W},$$

using Equation 27-9a. A 10-W resistor, the next larger size commercially available, would be just adequate.

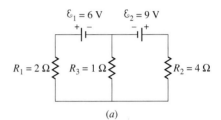

(a)

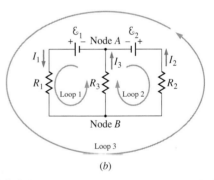

(b)

FIGURE 28-16 (a) A circuit that cannot be analyzed using series and parallel combinations. (b) There are 3 loops, 2 nodes, and 3 unknown currents whose direction assignments are, again, arbitrary (Example 28-6).

Although the circuits of Figs. 28-15 and 28-16 look very similar, that similarity is deceptive. The placement of $\mathcal{E}_2$ on the other side of node A in Fig. 28-16 means there is no way to solve this circuit using series and parallel combinations.

EXERCISE To what value should $\mathcal{E}_2$ in Fig. 28-16 be changed so that the current I_2 becomes zero?

Answer: 2.0 V

Some problems similar to Examples 28-5 and 28-6: 32–35 ●

(a)

(b)

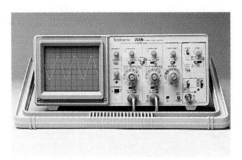

(c)

FIGURE 28-17 (a) Most voltmeters today have digital displays, although (b) analog—or moving-needle—meters are still found in older instruments. (c) An oscilloscope is essentially a pair of voltmeters in which two different voltages are indicated by the horizontal and vertical deflection of an electron beam. When the horizontal voltage varies linearly with time, the result is a plot of the vertical voltage versus time.

28-5 ELECTRICAL MEASURING INSTRUMENTS

Voltmeters

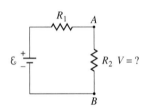

FIGURE 28-18 How to measure the voltage across R_2?

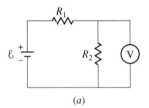

(a)

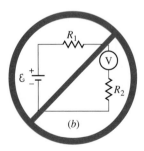

(b)

FIGURE 28-19 (a) Correct and (b) incorrect ways to measure the voltage across R_2.

A **voltmeter** is a device that indicates the potential difference across its two terminals. The indication is usually by a digital readout, although older meters use a moving needle, and oscilloscopes use the deflection of an electron beam (Fig. 28-17). Potential difference—voltage—is a property of two points, and therefore to measure the voltage between two points we connect the two terminals of the voltmeter to those points. So to measure the voltage across resistor R_2 in Fig. 28-18, we connect the voltmeter to points A and B, as shown in Fig. 28-19a. We do *not* break the circuit and insert the meter, as in Fig. 28-19b, for then we would not be measuring the voltage *across* the resistor; in fact, we would radically alter the circuit.

How good is our voltage measurement? There are two considerations here. First, how accurately does the meter indicate the voltage across its terminals? For digital meters this is usually expressed as the number of significant figures in the digital display, while for analog (moving-needle) meters, accuracy is given as a percentage of the full-scale reading. But there is a more subtle question of accuracy. Even if the meter reads exactly the voltage across its terminals, can we be sure that voltage is the same as it was before the meter was connected? The circuit of Fig. 28-20 provides the answer. There is an open circuit between points A and B, so no current flows in the resistor R. With no current through the resistor, there is no voltage across the resistor, so the voltage between points A and B is the same as the battery voltage. We could have arrived at this result more formally using our voltage divider Equation 28-2b, with R_2 set to infinity in order to represent the open circuit.

Now connect a voltmeter between points A and B. Suppose the meter itself has a resistance R_m. Now the circuit looks like Fig. 28-20b. We have a complete circuit, with current flowing from the battery through R, through the meter, and back to the battery. A voltage across R is required to drive the current, so the voltage between points A and B is now less than it was before the meter was connected. The voltmeter reading is not the same as the voltage in the absence of the meter. This discrepancy occurs not because the meter is inherently inaccurate, but for the following reason:

The instrument affects the circuit being measured.

How far off is the meter reading? That depends on the meter resistance relative to the rest of the circuit. If the meter resistance is high, it will draw little current, and its effect on the circuit will be small. Our circuit with the meter included is identical to the voltage divider of Fig. 28-4, with R_1 replaced by R and R_2 by R_m. The meter voltage is then given by Equation 28-2b:

$$V_m = \frac{R_m}{R + R_m}\mathcal{E}.$$

As R_m becomes large compared with R, the fraction $R_m/(R + R_m)$ approaches 1 and the meter voltage becomes essentially the open-circuit voltage $\mathcal{E}$. But if R_m is not large compared with R, then the meter reading will be substantially lower. We conclude that a voltmeter should have a much higher resistance than typical resistances in the circuit being measured.

How much higher? That depends on how accurate a reading we require. For 1% accuracy, a rough rule of thumb is that the meter resistance should be 100 times the circuit resistance. If we're troubleshooting a car's electrical system, where currents are large and resistances low, we can get away with a fairly low meter resistance. But if we want to meaure the voltage developed by the electrode in a chemist's pH meter, we must use a very high resistance indeed, for the pH electrode looks like a nonideal source of emf with internal resistance as high as $10^{14}\ \Omega$.

An ideal voltmeter would draw no current and so must have infinite resistance. Older moving-needle meters have difficulty approaching this ideal because they consist fundamentally of a sensitive current meter in series with a large resistor; the current that operates the meter comes from the circuit being measured. A good meter of this type that reads 10 V full scale might have a resistance of 200 kΩ. Modern digital voltmeters, in contrast, contain amplifiers that greatly reduce the current drawn from the circuit; typical meter resistances are around 10 MΩ, and much higher values can be achieved. Today's digital meters include amplifier, digital converter, and digital display in a single integrated-circuit package, making them the most economical, reliable, and accurate variety.

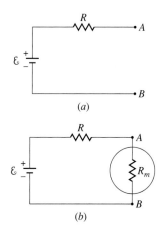

FIGURE 28-20 (a) Since no current flows, there is no voltage across the resistor, and the voltage between A and B is therefore the battery voltage. (b) A voltmeter with finite resistance R_m draws current, causing a voltage drop across R and thus lowering the voltage between A and B.

● EXAMPLE 28-7 VOLTMETERS

You wish to measure the voltage across the 40-Ω resistor of Fig. 28-21. What reading would an ideal voltmeter give? A voltmeter with a resistance of 1000 Ω?

Solution
An ideal voltmeter has infinite resistance and therefore would not alter the circuit, which is a simple voltage divider. Applying Equation 28-2b to this divider circuit gives the voltage across the 40-Ω resistor:

$$V_{40} = \frac{(40\ \Omega)(12\ \text{V})}{80\ \Omega + 40\ \Omega} = 4.0\ \text{V}.$$

Because the meter is connected in parallel with the resistor, this is also the voltage read by the meter.

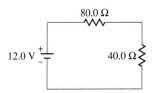

FIGURE 28-21 What is the voltage across the 40-Ω resistor (Example 28-7)?

With the nonideal voltmeter in place, the circuit becomes that of Fig. 28-22. The meter and 40-Ω resistor form a parallel combination whose resistance is given by Equation 28-3c:

$$R_\parallel = \frac{(40\ \Omega)(1000\ \Omega)}{40\ \Omega + 1000\ \Omega} = 38.5\ \Omega.$$

The circuit now looks like a voltage divider with $R_\parallel$ the lower resistor. Applying Equation 28-2b to this circuit gives

$$V_\parallel = \frac{(38.5\ \Omega)(12\ \text{V})}{80\ \Omega + 38.5\ \Omega} = 3.9\ \text{V}.$$

This $V_\parallel$ is the voltage across the parallel combination consisting of the meter and 40-Ω resistor. Since the voltage across parallel circuit elements is the same, $V_\parallel$ is both the meter reading and the voltage across the 40-Ω resistor. And this value is 0.10 V—about 2.5%—lower than the value indicated by an ideal voltmeter.

EXERCISE A 100-kΩ resistor and a 150-kΩ resistor are in series, with a 250-V potential difference across the combina-

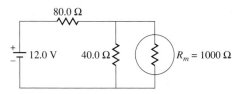

FIGURE 28-22 A nonideal voltmeter alters the circuit, lowering the overall resistance.

tion. A digital meter with a 3-significant-digit display and 1.0-MΩ resistance is used to measure the voltage across the 150-kΩ resistor. (a) What does it read? (b) By what percentage does this reading differ from that of an ideal voltmeter?

Answers: (a) 142 V; (b) 5.3%

Some problems similar to Example 28-7: 37, 38, 40, 61 ●

Ammeters

An **ammeter** measures the current flowing *through* itself. To measure the current through a circuit element it is necessary to break the circuit and insert the ammeter in series with that element (Fig. 28-23); only then will all the current through the element also go through the meter. Connecting the meter across the resistor in Fig. 28-23 would be wrong, for then the current through the resistor would not be going through the meter.

What electrical properties should the ammeter have so it doesn't alter the circuit in which it is connected? If the meter had any resistance, then the total resistance of the circuit would increase with the meter connected in series. This in turn would decrease the current, resulting in an incorrect reading. We conclude that an ideal ammeter should have zero resistance. In practice, ammeter resistance should be much lower than typical resistances in the circuit being measured.

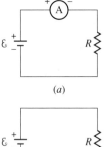

(a)

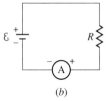

(b)

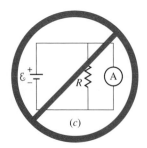

(c)

FIGURE 28-23 An ammeter goes in series with the circuit element whose current it is to measure. It doesn't matter whether the meter goes (a) before or (b) after the circuit element, in this case the resistor R. (c) Connecting it across the resistor is incorrect, since then the resistor current does not flow through the meter. In fact, this connection would probably destroy the meter!

TIP **Watch Your Language when Connecting Meters** A voltmeter measures potential difference *between* two points; hence, we connect it *across*—i.e., in parallel with—the circuit element whose voltage we wish to measure. An ammeter measures the current *through* itself; hence, we connect it in *series* with the circuit element whose current we wish to measure. If you get used to voltages appearing *across* things and currents flowing *through* them, you'll have no trouble connecting meters. But if you insist on talking about "the voltage through" something, then you'll be unable to hook up meters accurately or safely. The ways to connect meters, and the words *across* for voltage and *through* for current, go right back to the definitions of potential difference as a property of two points and of current as a flow.

Ohmmeters and Multimeters

Often we would like to measure the resistance of a particular circuit element. We can do this by connecting a source of known voltage in series with an ammeter and the unknown resistance, as in Fig. 28-24. Knowing the voltage and measuring the current then allows us to calculate the unknown resistance. A meter used for this purpose can be calibrated directly in ohms even though it is really measuring current; it is then called an **ohmmeter.**

The functions of voltmeter, ammeter, and ohmmeter are often combined in a single instrument called a **multimeter.** Multimeters include switches for selecting the quantity and range to be measured, and may be either analog or digital. Figure 28-25 shows a modern digital multimeter, or DMM.

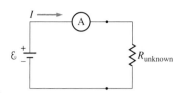

FIGURE 28-24   A simple ohmmeter consists of a known emf and an ammeter. When an unknown resistance is connected across the pair, its resistance may be determined from Ohm's law in the form $R = \mathcal{E}/I$.

Potentiometric Measurement

An elegant way to measure voltage is to compare the unknown voltage with an accurate standard, in much the same way that a pan balance weighs an unknown mass by balancing it against standard masses. Figure 28-26 shows how this scheme works. An accurately known emf $\mathcal{E}_0$ is connected across a resistor along which a sliding contact moves, forming a variable voltage divider in which the position of the sliding contact determines the voltage. The output of this voltage divider—called a **potentiometer**—is connected to the unknown voltage through a meter. The potentiometer is adjusted until the meter reads zero, at which point the potentiometer voltage—which can be read off its calibrated dial—must equal the unknown voltage. The great virtue of this method is that when the system is at null—the condition where source and unknown voltages are equal—then there is no current being drawn from the unknown regardless of the meter resistance, and therefore the method has, in principle, no effect on the circuit being measured.

In the past, potentiometric measurements were made with circuits like that of Fig. 28-26, with special batteries of precisely known emf and accurately calibrated potentiometers. Manual adjustment was used to achieve the null condition. Today, nulling is accomplished electronically or electromechanically through a process known as negative feedback. Circuits using the potentiometric technique are the basis of many powerful measurement and control devices.

FIGURE 28-25   This digital multimeter measures voltage, current, and resistance.

28-6 CIRCUITS WITH CAPACITORS

So far we have considered only circuits in which current and voltage are steady in all components. When you turn on a flashlight, for example, current starts to flow almost immediately through the bulb, batteries, and connecting metal parts. The current continues to flow steadily until you turn off the switch.

With a capacitor in a circuit, this picture changes. Circuit quantities change more gradually because of the capacitor. Why is this? Recall that a capacitor is a pair of insulated conductors that stores electrical energy when opposite charges are put on the conductors. A capacitor is characterized by its capacitance

$$C = \frac{Q}{V},$$

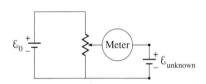

FIGURE 28-26   Potentiometric voltage measurement compares the unknown voltage with an accurately known voltage.

where Q is the magnitude of the charge on either conductor and V the voltage between the conductors. (See Section 26-4 for a review of capacitors.) Because charge and voltage are proportional in a capacitor, it is not possible to change the voltage without changing the charge. In a circuit, we change the capacitor charge by moving charge on or off the capacitor plates through wires connecting them to the rest of the circuit. This charge movement constitutes a current. The magnitude of the current is the rate at which charge is entering or leaving the capacitor. As long as the current is finite, as it is in any real circuit, the charge on the capacitor cannot change instantaneously. Because charge and voltage are proportional in a capacitor, we conclude the following:

▌ The voltage across a capacitor cannot change instantaneously.

This simple statement is the key to understanding circuits containing capacitors.

The *RC* Circuit: Charging

Consider the circuit of Fig. 28-27. The capacitor is initially uncharged, so the voltage across it is zero. What happens when we close the switch?

The switch connects the left end of the resistor to the battery's positive terminal, so the left end of the resistor goes to $\mathscr{E}$ volts (here we take the zero of potential at the battery's negative terminal). The right end of the resistor is at the same voltage as the upper capacitor plate. But the voltage across the capacitor cannot change instantaneously, and therefore remains zero just after the switch is closed. With the capacitor plates both at zero volts, the full battery voltage $\mathscr{E}$ appears across the resistor. With $\mathscr{E}$ volts across the resistor, there must be a current $I = \mathscr{E}/R$ through the resistor. This current cannot flow "through" the capacitor but serves instead to pile positive charge on the upper plate, negative charge on the lower. The same current I flows everywhere except in the insulated gap between the capacitor plates.

Now that current is flowing, charge accumulates on the capacitor, and the capacitor voltage increases in proportion to this charge. As the capacitor voltage rises, the resistor voltage falls because the voltage across the series combination of resistor and capacitor is the battery voltage $\mathscr{E}$. But the current through the resistor is proportional to the resistor voltage, so the resistor current falls as well. This in turn decreases the *rate* at which charge accumulates on the capacitor plates, lowering the rate at which the capacitor voltage increases. The voltage across the capacitor continues to increase, and the current through the resistor to decrease, but at an ever slower rate.

What happens if we wait a long time? As the capacitor voltage approaches the battery voltage, the voltage across the resistor, hence the current through the resistor, and therefore the rate of charge buildup on the capacitor, all become very small. The whole system tends more and more slowly toward a final state in which the capacitor is charged to the full battery voltage and the current in the circuit is zero. Figure 28-28 summarizes the interplay among current, charge, and voltage, while Fig. 28-29 shows the time dependence of these quantities.

We can analyze this circuit quantitatively using the loop law. Going clockwise around the loop, we first encounter a voltage increase $\mathscr{E}$ across the battery, then a drop IR across the resistor, then a drop V_C from the upper to lower capacitor plate (Fig. 28-30). But the definition of capacitance gives $V_C = Q/C$, so the loop equation becomes

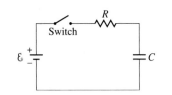

FIGURE 28-27 An *RC* circuit. The switch is closed at time $t = 0$.

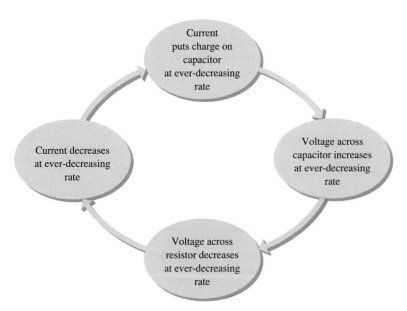

FIGURE 28-28 Interrelationships among circuit quantities in a charging *RC* circuit.

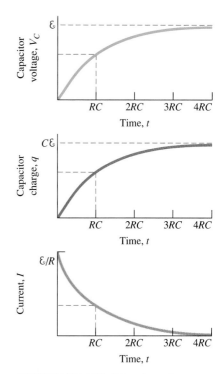

FIGURE 28-29 Time dependence of capacitor voltage, capacitor charge, and current in a charging *RC* circuit. In one time constant *RC* the capacitor voltage and charge rise to about 2/3 (actually $1 - 1/e$) of their final value, while the current drops to about 1/3 (actually $1/e$) of its initial value.

$$\mathcal{E} - IR - \frac{Q}{C} = 0.$$

This equation contains the two unknown quantities I and Q. Can we relate them? Yes—but not through a proportionality or other algebraic equation. Rather, the current is the rate at which charge is accumulating on the capacitor, or

$$I = \frac{dQ}{dt}.$$

To use this relation, we take the time derivative of the loop equation:

$$-R\frac{dI}{dt} - \frac{1}{C}\frac{dQ}{dt} = 0.$$

The battery voltage $\mathcal{E}$ does not appear in this differentiated equation because it is constant and thus its derivative is zero. Using $I = dQ/dt$ and rearranging the equation slightly gives

$$\frac{dI}{dt} = -\frac{I}{RC}. \tag{28-4}$$

This equation shows that the rate of change of current is proportional to the current itself, expressing mathematically what Fig. 28-28 shows schematically. Equations like this arise whenever the rate of change of a quantity is proportional to the quantity itself. Population growth, the increase of money in a bank account, and the decay of a radioactive element are all described by similar equations.

Like the equation for simple harmonic motion in Chapter 15, Equation 28-4 is a *differential* equation, so called because the unknown quantity I occurs in a

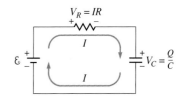

FIGURE 28-30 Voltage changes in a charging *RC* circuit. Since they sum to zero, the loop equation for this circuit is $\mathcal{E} - (IR) - (Q/C) = 0$.

derivative. The solution to a differential equation is not a single number but rather a function expressing the relation between the unknown quantity—in this case current—and the independent variable—in this case time. We can solve this particular differential equation by multiplying both sides by dt/I, in order to collect all terms involving I on one side of the equation. This gives

$$\frac{dI}{I} = -\frac{dt}{RC}.$$

We then integrate both sides, noting that RC is constant:

$$\int_{I_0}^{I} \frac{dI}{I} = -\frac{1}{RC} \int_{0}^{t} dt,$$

where $I_0 = \mathcal{E}/R$ is the initial current at the time $t = 0$ just after the switch is closed and where the integration runs from $t = 0$ to some arbitrary time t. The integral on the left is just the natural logarithm, and that on the right is just t. Then we have

$$\ln (I/I_0) = -\frac{t}{RC},$$

where we used $\ln I - \ln I_0 = \ln(I/I_0)$. To get an equation for I we exponentiate both sides, recalling that $e^{\ln x} = x$. This gives

$$\frac{I}{I_0} = e^{-t/RC},$$

or, since $I_0 = \mathcal{E}/R$,

$$I = \frac{\mathcal{E}}{R} e^{-t/RC}. \tag{28-5}$$

Thus the current in the circuit decreases exponentially with time, in agreement with our qualitative analysis.

What about the capacitor voltage? The capacitor and resistor voltages must add to the battery voltage $\mathcal{E}$, and the resistor voltage is just $V_R = IR$, or

$$V_R = \mathcal{E}e^{-t/RC}.$$

Thus the capacitor voltage is $V_C = \mathcal{E} - V_R$, or

$$V_C = \mathcal{E}(1 - e^{-t/RC}). \qquad (RC \text{ circuit, charging}) \tag{28-6}$$

Equation 28-6 shows the capacitor voltage starting at zero, and rising rapidly at first but with its rate of rise ever slowing, as it gradually approaches the battery voltage $\mathcal{E}$, again agreeing with our qualitative analysis (see Fig. 28-29, which was in fact plotted using Equations 28-5 and 28-6).

When is the capacitor fully charged? Never, according to our equations! But the rate at which it approaches full charge is determined by the quantity RC that appears in Equations 28-5 and 28-6. (Problem 45 will convince you that this quantity has the units of time.) Called the **time constant**, RC is a characteristic time for changes to occur in a circuit containing a capacitor. Equation 28-6 shows that in one time constant, the voltage rises to $\mathcal{E}(1 - 1/e)$, or to about two-thirds of the battery voltage. A practical rule of thumb says that in five time constants ($t = 5RC$) a capacitor is 99% charged (see Problem 47). The RC time constant clarifies our statement that the voltage across a capacitor cannot change instantaneously. We can now say that in times small compared with the time constant, the voltage across a capacitor cannot change appreciably. On the other hand, if we wait a long time—many time constants—we will find essentially no current flowing to the capacitor.

RC circuits with appropriate time constants are used in electronic timing applications covering microseconds to hours. In other circuits where we want voltages to change rapidly, the time constant can be annoyingly long. For example, capacitance in audio equipment can limit high-frequency response, decreasing the quality of music reproduction. You intentionally alter the response of a stereo system by adjusting bass and treble controls, which are simply variable resistors in RC circuits.

The *RC* Circuit: Discharging

Suppose we connect a charged capacitor across a resistor, as shown in Fig. 28-31. If the capacitor voltage is initially V_0, then when the circuit is connected this voltage will drive a current $I_0 = V_0/R$ through the resistor. This current transfers charge from the positive to the negative capacitor plate, lowering the charge on the capacitor. Since capacitor charge and voltage are proportional, the capacitor voltage drops, too. So, therefore, does the current, and therefore the rate at which the capacitor discharges. We therefore expect both the voltage and current in this circuit to decay toward zero. In terms of energy, that happens because the energy stored in the capacitor's electric field is gradually dissipated as heat in the resistor.

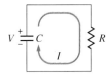

FIGURE 28-31 A discharging RC circuit.

The loop equation for Fig. 28-31 is particularly simple; going clockwise, we have

$$\frac{Q}{C} - IR = 0,$$

where the two terms are the voltage changes across the capacitor and resistor, respectively. Since we've indicated positive current in Fig. 28-31 in the direction that would *reduce* the capacitor charge Q, the rate of change dQ/dt and the current must have opposite signs: $I = -dQ/dt$. Differentiating our loop equation and substituting this expression for I gives

$$\frac{dI}{dt} = -\frac{I}{RC}.$$

This is Equation 28-4; the solution is therefore Equation 28-5, but with $I_0 = V_0/R$ instead of $\mathcal{E}/R$:

$$I = \frac{V_0}{R}e^{-t/RC}. \tag{28-7}$$

In this circuit the capacitor and resistor voltage are the same since the two are in parallel. Since the resistor voltage and current are proportional, the voltage across the capacitor and resistor is

$$V = V_0 e^{-t/RC}. \quad \text{(RC circuit, discharging)} \tag{28-8}$$

Equations 28-7 and 28-8 show that the capacitor discharges with the same characteristic time constant RC that governs its charging.

● **EXAMPLE 28-8** A CAMERA FLASH

In Chapter 26 we considered an electronic camera flash using a capacitor to store energy. A particular camera flashtube obtains its energy from a 150-μF capacitor and requires 170 V to fire. If the capacitor is charged by a 200-V battery* through a 30-kΩ resistor, how long must the photographer wait between flashes? What is the peak power drawn from the battery? Assume the capacitor is fully discharged during a flash.

Solution
The time between flashes is the time it takes the capacitor voltage to reach 170 V. To find this time, we solve Equation 28-6 for the exponential term that contains the time:

$$e^{-t/RC} = 1 - \frac{V_C}{\mathcal{E}}.$$

We then take the natural logarithm of both sides, recalling that $\ln e^x = x$, so

$$-\frac{t}{RC} = \ln\left(1 - \frac{V_C}{\mathcal{E}}\right).$$

Solving for t and setting $V_C = 170$ V, $\mathcal{E} = 200$ V, $R = 30$ kΩ, and $C = 150$ μF gives

$$t = -RC \ln\left(1 - \frac{V_C}{\mathcal{E}}\right)$$

$$= -(30\times10^3 \ \Omega)(150\times10^{-6} \ \text{F}) \ln\left(1 - \frac{170 \ \text{V}}{200 \ \text{V}}\right) = 8.5 \ \text{s}.$$

Problem 74 explores the question of power in this circuit, and shows that energy from the battery cannot all end up in the capacitor.

EXERCISE If the flash lamp in Example 28-8 has an effective resistance of 10 Ω, how long does it take the capacitor voltage to drop to 100 V as it discharges through the lamp?

Answer: 0.80 ms

Some problems similar to Example 28-8: 47–49, 51 ●

It's not always necessary to solve exponential equations in analyzing RC circuits. If we're concerned only with times short compared with the time constant, if suffices to remember that the voltage across a capacitor cannot change instantaneously. And after many time constants have passed a capacitor has essentially reached its final voltage, and there will be no current flowing to it. These two conditions are sufficient to analyze circuits on short and long time scales.

*Actually, much lower voltage batteries are used. But their voltage is increased using transistors and a transformer, working on principles described in Chapter 31.

● **EXAMPLE 28-9** LONG- AND SHORT-TIME BEHAVIOR OF AN *RC* CIRCUIT

In Fig. 28-32*a* the capacitor is initially uncharged. What is the current through R_1 the instant after the switch is closed? A long time after the switch has been closed?

Solution
The capacitor voltage cannot change instantaneously, so just after the switch is closed there can be no voltage across the capacitor and therefore none across R_2. Then the full battery voltage is across R_1, so the current in R_1 is $\mathscr{E}/R_1$. After a very long time the capacitor will be fully charged (to what voltage? —see the exercise below), and no current will flow into it. The capacitor then acts like an open circuit, and we simply have two resistors in series. The current in each is $\mathscr{E}/(R_1 + R_2)$.

How simple this example is! The uncharged capacitor has no voltage across it, so it acts instantaneously like a short circuit. The fully charged capacitor has no current into it, so it acts like an open circuit. To solve the problem we could simply redraw the circuit, once with the capacitor replaced by a wire (Fig. 28-32*b*), the second time with the capacitor simply erased from the circuit diagram (Fig. 28-32*c*). Only if we wanted to know what was happening at intermediate times would we have to resort to the solution of an equation describing the circuit.

EXERCISE When the capacitor in Example 28-9 is fully charged, what will be the voltage across it?

Answer: $\mathscr{E}R_2/(R_1 + R_2)$

Some problems similar to Example 28-9: 55–57

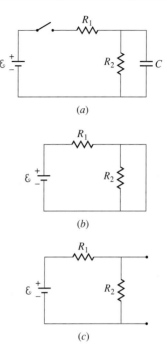

FIGURE 28-32 (*a*) A circuit with two resistors and a capacitor. (*b*) Just after the switch is closed the voltage across the capacitor is still zero, so the capacitor acts like a short circuit. (*c*) Long after the switch is closed the capacitor is fully charged. No more current flows into it, so it acts like an open circuit.

28-7 ELECTRICAL SAFETY

Whether you find yourself in a laboratory hooking up electronic equipment, or in a hospital connecting instrumentation to a patient, or on a job designing electrical devices, or simply at home plugging in appliances and tools, you should be concerned with electrical safety.

Everyone knows enough to be wary of "high voltage." People with a little more electrical sophistication are fond of saying "it isn't the voltage but the current that kills you." In fact, both points of view are partially correct. Current through the body is dangerous, but as with any resistor it takes voltage to drive that current.

Table 28-1 shows typical effects of electric currents introduced into the body through skin contact. Currents below the lethal 100 mA may result in involuntary muscle contraction that unfortunately keeps the victim in contact with the circuit. A primary danger is disturbance of the biologically generated electrical signals that pace heartbeat; this is reflected in the lethal zone of 100 to 200 mA at which currents the heart is thrown into fibrillation—uncontrolled spasms of the cardiac muscle. With electrical signals applied internally to local regions of the body, much smaller currents can be lethal. Doctors performing cardiac

▲ **TABLE 28-1** EFFECTS OF EXTERNALLY APPLIED CURRENT ON THE HUMAN ORGANISM

CURRENT RANGE	EFFECT
0.5–2 mA	Threshold of sensation
10–15 mA	Involuntary muscle contractions; can't let go
15–100 mA	Severe shock; muscle control lost; breathing difficult
100–200 mA	Fibrillation of heart; death within minutes
>200 mA	Cardiac arrest; breathing stops; severe burns

catheterization, for example, must worry about currents at the microampere level.

Above 200 mA, complete cardiac arrest may occur, breathing may stop, and there may be severe burns both internally and at the points of skin contact. Sometimes such high currents are useful: when a heart is fibrillating, doctors or emergency technicians briefly apply a high enough current to stop the heart. The heart often restarts, beating normally. The figures of Table 28-1 are rough averages, and vary widely from person to person as well as with duration of the shock and whether alternating or direct current is involved. In particular, very young children and people with heart conditions are at higher risk.

Under dry conditions, the typical human being has a resistance from one point to another on unbroken skin of about 10^5 Ω. What voltages are dangerous to such a person? At 10^5 Ω it takes

$$V = IR = (0.1 \text{ A})(10^5 \text{ Ω}) = 10,000 \text{ V}$$

to drive the fatal 100 mA. But a person who is wet or sweaty has a much lower resistance and may be electrocuted by 120-V household electricity. People have been electrocuted at voltages as low as 30 V, although such cases are rare.

It takes current to harm a person, but it takes voltage to drive that current. To be dangerous, an electric circuit must have high voltage *and* be capable of driving sufficient current. For example, a car battery can deliver 300 A, but it cannot electrocute you because its 12 volts will not drive much current through you (although you could be hurt by the energy released if you accidentally short-circuit such a battery). On the other hand the 20,000 V that runs your car's spark plugs will not electrocute you either, since the internal resistance of this high-voltage circuit is so high that it cannot deliver more than a few mA.

Because potential difference is a property of two points, receiving an electric shock requires that two parts of the body be in contact with conductors at different potentials. In typical 120-V wiring used throughout North America, one of the two wires is connected physically to the ground. This ground connection is to prevent the wiring from reaching arbitrarily high potentials with respect to the ground, as might otherwise happen in a thunderstorm or if a short circuit occurred in a power line. At the same time it means that an individual contacting the "hot" side of the circuit and any grounded conductor such as the ground, a water pipe, or a bathtub will receive a shock.

A potentially dangerous situation occurs when power tools, instruments, or appliances are used by an operator who is likely to be in contact with a grounded conductor. Examples include working outdoors with an electric drill, in a kitchen with an electric mixer, or in a laboratory with an oscilloscope. Suppose you're

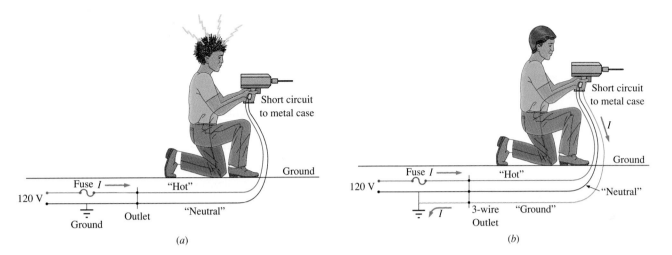

(a)   (b)

using a power tool that is plugged in through a regular two-wire cord. Normally exposed metal parts of the tool are not connected to either wire. Now suppose something goes wrong in the tool and a wire short-circuits to the metal case. If it happens to be the wire that is plugged into the grounded side of the power line there is no problem, but if it is the other wire the metal is suddenly 120 V above ground. If you are standing on the ground, or in a damp basement, or are leaning on the kitchen sink, you will receive a potentially lethal shock (Fig. 28-33a).

To avoid this danger many electrical devices are equipped with three-wire cords (Fig 28-34). The third wire runs from exposed metal parts to a grounded wire in the outlet, and normally carries no current. If a short circuit occurs the third wire provides a very low resistance path to ground (Fig. 28-33b). Large currents will flow and will blow the circuit breaker or fuse, shutting off the current. Held at ground potential by the ground wire, the operator of the device will be safe.

Because many older homes are not wired with grounded outlets, some manufacturers produce two-wire tools and other devices that are "double-insulated" to provide an extra margin of safety. Newer appliances are sometimes equipped with "polarized plugs," which can only be plugged in one way (Fig. 28-34), ensuring that exposed metal parts are most likely to end up at ground potential. Finally, electronic devices called ground-fault interrupters (Fig. 28-35) are used in kitchen, bathroom, basement, and other hazardous circuits in new homes. These devices sense a slight imbalance—5 mA or less—in current flowing in the two wires of a circuit, and shut off the circuit in less than a millisecond on the assumption that the excess current is leaking to ground—perhaps through a person. (Do ground-fault interrupters know about the node law?)

FIGURE 28-33 (a) A short circuit in an ungrounded tool could result in a lethal shock. (b) With a grounded tool, the short circuit causes a blown fuse or circuit breaker, thereby protecting the operator.

FIGURE 28-34 Plugs. (Left) grounded; (right) polarized.

FIGURE 28-35 A ground-fault interrupter protects against shock by sensing small currents leaking to ground.

CHAPTER SYNOPSIS

Summary

1. A source of **electromotive force (emf)** is a device—like a battery—that converts some form of energy into the electrical energy associated with the buildup of positive and negative charge at its two terminals. An ideal emf maintains a constant potential difference between its terminals, but energy losses in a real emf result in terminal voltage that decreases as more current is drawn from the device. When connected across a resistor, an ideal emf $\mathscr{E}$ drives a current $I = \mathscr{E}/R$ through the resistor.

2. Electric circuits may often be analyzed using series and parallel resistor combinations. Resistors in series add:

$$R_{\text{series}} = R_1 + R_2 + R_3 + \cdots,$$

while resistors in parallel add reciprocally:

$$\frac{1}{R_{\text{parallel}}} = \frac{1}{R_1} + \frac{1}{R_2} + \frac{1}{R_3} + \cdots.$$

3. More complicated circuits may be analyzed using **Kirchhoff's laws.** The **loop law** follows from conservation of energy and states that the sum of the voltage differences around any circuit loop is zero. The **node law** follows from conservation of charge and states that the sum of the currents at any circuit node is zero.

4. In using instruments to measure electrical quantities, care must be taken to ensure that the instrument does not alter the circuit being measured. An ideal **voltmeter** has infinite resistance; in practice, a voltmeter's resistance should be much higher than typical resistances in the circuit being measured. An ideal **ammeter** has zero resistance; in practice, an ammeter's resistance should be much lower than typical circuit resistances.

5. It takes time to move charge on and off the plates of a capacitor, and as a result the voltage across a capacitor cannot change instantaneously. Quantities in a resistor-capacitor (*RC*) circuit change on a characteristic time scale given by the product *RC*. When an emf $\mathscr{E}$ charges a capacitor *C* through a resistor *R*, the capacitor voltage and circuit current vary with time according to

$$V_{\text{C}} = \mathscr{E}(1 - e^{-t/RC})$$

$$I = \frac{\mathscr{E}}{R}e^{-t/RC}.$$

For a discharging capacitor, both voltage and current decrease exponentially with the same time constant *RC*.

6. **Electrical safety** is a matter of avoiding currents high enough to cause biological damage. The danger of electric shock depends on the current at which such damage occurs, the resistance of the organism, and the voltage available to drive the current.

Terms You Should Understand

(Pairs are closely related terms whose distinction is important; number in parentheses is chapter section where term first appears.)

circuit, circuit element (28-1)
electromotive force (emf) (28-2)
ideal, real emfs (28-2, 28-3)
voltage divider (28-3)
internal resistance (28-3)
Kirchhoff's loop and node laws (28-4)
voltmeter, ammeter, ohmmeter (28-5)
time constant (28-6)

Symbols You Should Recognize

electric circuit symbols (Fig. 28-1)
$\mathscr{E}$ (28-1)

Problems You Should Be Able to Solve

analyzing simple circuits with series and parallel resistors (28-3)
analyzing circuits with the loop and node laws (28-4)
determining the effects of nonideal measuring instruments (28-5)
analyzing charging and discharging *RC* circuits (28-6)
quickly determining short- and long-term behavior of *RC* circuits (28-6)
assessing electrical hazards (28-7)

Limitations to Keep in Mind

Real circuit elements deviate from their idealizations. Thus the voltage across a real battery may not be exactly its rated voltage, and voltage drops occur in real wires. Attention to circuit design will minimize these nonideal effects.

Electrical measuring instruments may affect the circuits they are measuring.

Currents may be far more dangerous than Table 28-1 implies if they are introduced beneath the skin.

QUESTIONS

1. In each of the circuits of Fig. 28-36, which, if any, of the resistors are in series? In parallel?

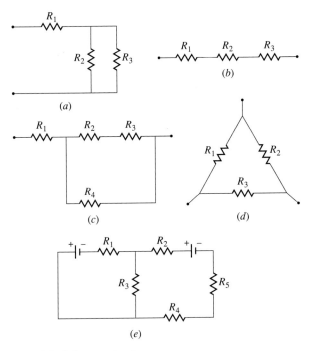

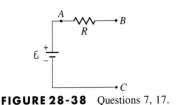

FIGURE 28-36 Question 1.

2. Are the electrical outlets in a home connected in series or parallel? How do you know?

3. In which of the circuits of Fig. 28-37 does the battery supply the same current? All the resistors have the same resistance.

4. Can the voltage across a battery's terminals differ from the rated voltage of the battery? Explain.

5. Can the voltage across a battery's terminals be higher than the rated voltage of the battery? Explain.

6. In some cities, streetlights are wired in such a way that when one light burns out, they all go out. Are the lights in series or parallel?

7. If you know the battery voltage in Fig. 28-38, can you determine the voltage between points B and C without knowing the resistance R?

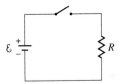

FIGURE 28-38 Questions 7, 17.

8. Must we assign zero volts to the negative terminal of a battery?

9. When the switch in Fig 28-39 is open, what is the voltage across the resistor? Across the switch?

FIGURE 28-39 Question 9.

FIGURE 28-37 Question 3.

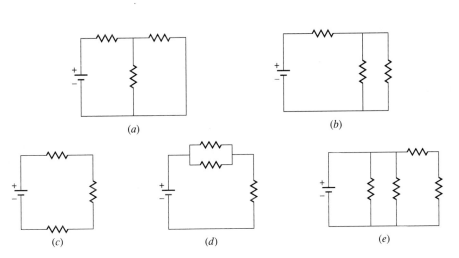

10. Two identical resistors in series dissipate equal power. How can this be, when electric charge loses energy in flowing through the first resistor?

11. What is the current through resistor R_2 in Fig. 28-40? Assume all the wires are ideal.

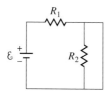

FIGURE 28-40   Question 11.

12. The resistors in Fig. 28-41 all have the same resistance. If an ideal voltmeter is connected between points A and B, what will it read?

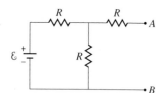

FIGURE 28-41   Question 12.

13. When a large electrical load such as a washing machine, oven, or oil burner comes on, lights throughout a house often dim. Why is this? *Hint:* Think about real wires.

14. If the node law were not obeyed at some node in an electric circuit, what would happen to the voltage at that node?

15. How would you connect a pair of equal resistors across an ideal battery in order to get the most power dissipation in the resistors?

16. You have a battery whose voltage and internal resistance are unknown. Using an ideal voltmeter and an ideal ammeter, how would you determine both these battery characteristics?

17. An ideal voltmeter is used to measure the voltages between points A and B and between C and B in the circuit of Fig. 28-38. How do the measurements compare?

18. You wish to measure the resistance R_2 in Fig. 28-42 with an ohmmeter. Can you do so while R_2 is in the circuit? Why or why not?

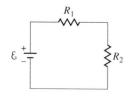

FIGURE 28-42   Question 18.

19. A student who is confused about voltage and current hooks a nearly ideal ammeter across a car battery. What happens?

20. A student who is confused about voltage and current tries to measure the voltage across a lighted light bulb by inserting a voltmeter in series with the bulb. What happens to the bulb? Explain.

21. Four identical light bulbs are connected to a battery as shown in Fig. 28-43. How does the brightness of each compare? Explain.

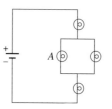

FIGURE 28-43   Questions 21, 22.

22. Suppose bulb A in Fig. 28-43 is unscrewed from its socket. How will the brightness of the three remaining bulbs change? How does the brightness of the three compare?

23. What does it mean for a capacitor to be "fully charged"?

24. Is the current into a charging capacitor in an RC circuit greatest when the capacitor voltage is greatest or when it is smallest?

25. If it takes forever to charge a capacitor fully, why is the RC time constant of any significance in describing the charging?

26. The two resistors in Fig. 28-44 have equal resistance. If the circuit has been connected for a long time, what is the voltage across the capacitor?

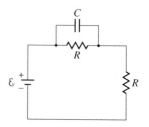

FIGURE 28-44   Question 26.

27. In one time constant, a charging capacitor reaches approximately $\frac{2}{3}$ of full charge. In one time constant, the voltage across a discharging capacitor falls to approximately $\frac{1}{3}$ of its original value. What is the origin of the approximate numerical factors $\frac{2}{3}$ and $\frac{1}{3}$ in these statements?

28. What's wrong with this news report: "A power-line worker was seriously injured when 4000 volts passed through his body"?

PROBLEMS

Section 28-1 Circuits and Symbols

1. Sketch a circuit diagram for a circuit that includes a resistor R_1 connected to the positive terminal of a battery, a pair of parallel resistors R_2 and R_3 connected to the lower-voltage end of R_1, then returned to the battery's negative terminal, and a capacitor across R_2.

2. A circuit consists of two batteries, a resistor, and a capacitor, all in series. Sketch this circuit. Does the description allow any flexibility in how you draw the circuit?

3. Resistors R_1 and R_2 are connected in series, and this series combination is in parallel with R_3. This parallel combination is connected across a battery whose internal resistance is R_{int}. Draw a diagram representing this circuit.

Section 28-2 Electromotive Force

4. What is the emf of a battery that delivers 27 J of energy as it moves 3.0 C between its terminals?

5. A 1.5-V battery stores 4.5 kJ of energy. How long can it light a flashlight bulb that draws 0.60 A?

6. If you accidentally leave your car headlights (current drain 5 A) on for an hour, how much of the 12-V battery's chemical energy is used up?

7. A battery stores 50 W·h of chemical energy. If it uses up this energy moving 3.0×10^4 C through a circuit, what is its voltage?

Section 28-3 Simple Circuits: Series and Parallel Resistors

8. A 47-kΩ resistor and a 39-kΩ resistor are in parallel, and the pair is in series with a 22-kΩ resistor. What is the resistance of the combination?

9. What resistance should be placed in parallel with a 56-kΩ resistor to make an equivalent resistance of 45 kΩ?

10. In Fig. 28-45 all resistors have the same value, R. What will be the resistance measured (a) between A and B or (b) between A and C?

11. In Fig. 28-45, take all resistors to be 1.0 Ω. If a 6.0-V battery is connected between points A and B, what will be the current in the vertical resistor?

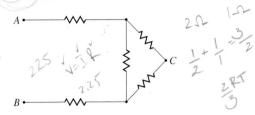

FIGURE 28-45 Problems 10, 11.

12. A defective starter motor in a car draws 300 A from the car's 12-V battery, dropping the battery terminal voltage to only 6 V. A good starter motor should draw only 100 A. What will the battery terminal voltage be with a good starter?

13. What is the internal resistance of the battery in the preceding problem?

14. Three 1.5-V batteries, with internal resistances of 0.01 Ω, 0.1 Ω, and 1 Ω, each have 1-Ω resistors connected across their terminals. To three significant figures, what is the voltage across each resistor?

15. When a 9-V battery is temporarily short-circuited, a 200-mA current flows. What is the internal resistance of the battery?

16. What possible resistance combinations can you form using three resistors whose values are 1.0 Ω, 2.0 Ω, and 3.0 Ω? (Use all three resistors.)

17. A partially discharged car battery can be modeled as a 9-V emf in series with an internal resistance of 0.08 Ω. Jumper cables are used to connect this battery to a fully charged battery, modeled as a 12-V emf in series with a 0.02-Ω internal resistance. How much current flows through the discharged battery?

18. You have a number of 50-Ω resistors, each capable of dissipating 0.50 W without overheating. How many resistors would you need, and how would you connect them, so as to make a 50-Ω combination that could be connected safely across a 12-V battery?

19. What is the equivalent resistance between A and B in each of the circuits shown in Fig. 28-46? *Hint:* In (c), think about symmetry and the current that would flow through R_2.

FIGURE 28-46 Problem 19.

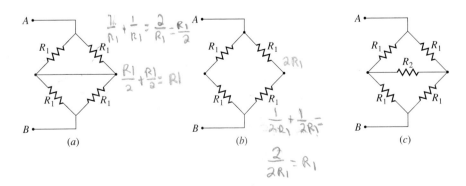

(a)

(b)

(c)

20. A 6.0-V battery has an internal resistance of 2.5 Ω. If the battery is short circuited, what is the rate of energy dissipation in its internal resistance?

21. How many 100-W, 120-V light bulbs can be connected in parallel before they blow a 20-A circuit breaker?

22. What is the current through the 3-Ω resistor in the circuit of Fig. 28-47? *Hint:* This is trivial. Can you see why?

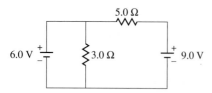

FIGURE 28-47 Problem 22.

23. Take $\mathscr{E} = 12$ V and $R_1 = 270$ Ω in the voltage divider of Fig. 28-4. (a) What should the value of R_2 in order that 4.5 V appear across R_2? (b) What will be the power dissipation in R_2?

24. A voltage divider consists of two 1.0-kΩ resistors connected in series across a 160-V emf. If a 10-kΩ resistor is connected across one of the 1.0-kΩ resistors, what will be the voltage across it?

25. In the circuit of Fig. 28-48, R_1 is a variable resistor, and the other two resistors have equal resistances R. (a) Find an expression for the voltage across R_1, and (b) sketch a graph of this quantity as a function of R_1 as R_1 varies from 0 to 10R. (c) What is the limiting value as $R_1 \rightarrow \infty$?

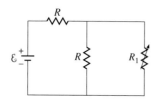

FIGURE 28-48 Problem 25.

26. In the circuit of Fig. 28-49 find (a) the current supplied by the battery and (b) the current through the 6-Ω resistor.

27. In the circuit of Fig. 28-49, how much power is being dissipated in the 4-Ω resistor?

28. A 50-Ω resistor is connected across a battery, and a 26-mA current flows. When the 50-Ω resistor is replaced with a 22-Ω resistor, a 43-mA current flows. What are the battery's voltage and internal resistance?

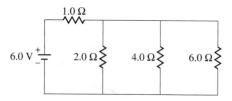

FIGURE 28-49 Problems 26, 27.

Section 28-4 Kirchhoff's Laws and Multiloop Circuits

29. In the circuit of Fig. 28-50 it makes no difference whether the switch is open or closed. What is $\mathscr{E}_3$ in terms of the other quantities shown?

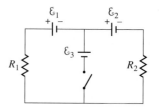

FIGURE 28-50 Problem 29.

30. What is the current through the ammeter in Fig. 28-51?

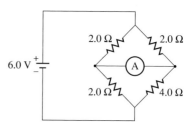

FIGURE 28-51 Problem 30.

31. In Fig. 28-52, what is the equivalent resistance measured between points A and B?

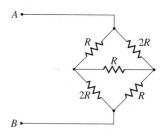

FIGURE 28-52 Problem 31.

32. Find all three currents in the circuit of Fig. 28-16, but now with $\mathscr{E}_2 = 1.0$ V.

33. Find all three currents in the circuit of Fig. 28-16 with the values given, but with battery $\mathscr{E}_2$ reversed.

34. In Fig. 28-53, take $\mathscr{E}_1 = 6.0$ V, $\mathscr{E}_2 = 1.5$ V, $\mathscr{E}_3 = 4.5$ V, $R_1 = 270$ Ω, $R_2 = 150$ Ω, $R_3 = 560$ Ω, and $R_4 = 820$ Ω. Find the current in R_3, and give its direction.

35. With all the values except $\mathscr{E}_2$ in Fig. 28-53 as given in the preceding problem, find the condition on $\mathscr{E}_2$ that will make the current in R_3 flow upward.

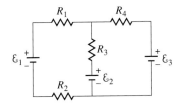

FIGURE 28-53 Problems 34, 35, 36.

36. Suppose that all resistors in Fig. 28-53 have the same value R and that $\mathcal{E}_1 = \mathcal{E}_3 = \mathcal{E}$ and $\mathcal{E}_2 = 2\mathcal{E}$. Find expressions for the currents in the four resistors, and give their directions.

Section 28-5 Electrical Measuring Instruments

37. A voltmeter with 200-kΩ resistance is used to measure the voltage across the 10-kΩ resistor in Fig. 28-54. By what percentage is the measurement in error because of the finite meter resistance?

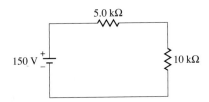

FIGURE 28-54 Problems 37, 38.

38. An ammeter with 100-Ω resistance is inserted in the circuit of Fig. 28-54. By what percentage is the measurement in error because of the nonzero meter resistance?

39. A neophyte mechanic foolishly connects an ammeter with 0.1-Ω resistance directly across a 12-V car battery whose internal resistance is 0.01 Ω. What is the power dissipation in the meter? No wonder it gets destroyed!

40. The voltage across the 30-kΩ resistor in Fig. 28-55 is measured with (a) a 50-kΩ voltmeter, (b) a 250-kΩ voltmeter, and (c) a digital meter with 10-MΩ resistance. To two significant figures, what does each read?

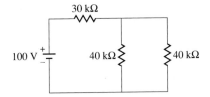

FIGURE 28-55 Problem 40.

41. You have an ammeter with 10-Ω resistance whose full-scale reading is 1.0 mA. What resistance should you put in series with it to make a voltmeter that reads 25 V full scale?

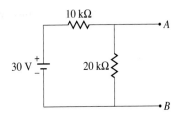

FIGURE 28-56 Problem 43.

42. Suppose you want to make a 10-A full scale ammeter out of the 1.0-mA meter of the preceding problem. What resistance should you put in parallel with it?

43. In Fig. 28-56 what are the meter readings when (a) an ideal voltmeter or (b) an ideal ammeter is connected between points A and B?

44. A resistor draws 1.00 A from an ideal 12.0-V battery. (a) If an ammeter with 0.10-Ω resistance is inserted in the circuit, what will it read? (b) If this current is used to calculate the resistance, how will the calculated value compare with the actual value?

Section 28-6 Circuits with Capacitors

45. Show that the quantity RC has the units of time (seconds).

46. If capacitance is given in μF, what will be the units of the RC time constant when resistance is given in (a) Ω, (b) kΩ, (c) MΩ? Your answers eliminate the need for tedious power-of-10 conversions.

47. Show that a capacitor is charged to approximately 99% of the applied voltage in five time constants.

48. An uncharged 10-μF capacitor and a 470-kΩ resistor are connected in series, and 250 V applied across the combination. How long does it take the capacitor voltage to reach 200 V?

49. Figure 28-57 shows the voltage across a capacitor that is charging through a 4700-Ω resistor in the circuit of Fig. 28-27. Use the graph to determine (a) the battery voltage, (b) the time constant, and (c) the capacitance.

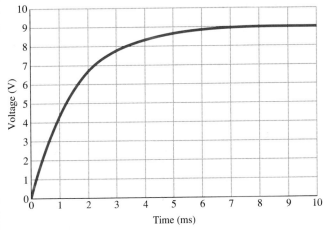

FIGURE 28-57 Problem 49.

50. The voltage across a charging capacitor in an *RC* circuit rises to $1 - 1/e$ of the battery voltage in 5.0 ms. (a) How long will it take to reach $1 - 1/e^3$ of the battery voltage? (b) If the capacitor is charging through a 22-kΩ resistor, what is its capacitance?

51. A 1.0-μF capacitor is charged to 10.0 V. It is then connected across a 500-kΩ resistor. How long does it take (a) for the capacitor voltage to reach 5.0 V and (b) for the energy stored in the capacitor to decrease to half its initial value?

52. A capacitor used to provide steady voltages in the power supply of a stereo amplifier charges rapidly to 35 V every 1/60 of a second. It must then hold that voltage to within 1.0 V for the next 1/60 s while it discharges through the amplifier circuit. If the circuit draws 1.2 A from the 35-V supply (a) what is its effective resistance and (b) what value of capacitance is needed?

53. A capacitor is charged until it holds 5.0 J of energy. It is then connected across a 10-kΩ resistor. In 8.6 ms, the resistor dissipates 2.0 J. What is the capacitance?

54. A 2.0-μF capacitor is charged to 150 V. It is then connected to an uncharged 1.0-μF capacitor through a 2.2-kΩ resistor, by closing switch S in Fig. 28-58. Find the total energy dissipated in the resistor as the circuit comes to equilibrium. *Hint:* Think about charge conservation.

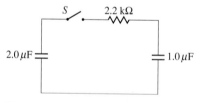

FIGURE 28-58 Problem 54.

55. For the circuit of Example 28-9, take $\mathcal{E} = 100$ V, $R_1 = 4.0$ kΩ, and $R_2 = 6.0$ kΩ, and assume the capacitor is initially uncharged. What are the currents in both resistors and the voltage across the capacitor (a) just after the switch is closed and (b) a long time after the switch is closed? Long after the switch is closed it is again opened. What are I_1, I_2, and V_C (c) just after this switch opening and (d) a long time later?

56. In the circuit of Fig. 28-59 the switch is initially open and both capacitors initially uncharged. All resistors have the same value R. Find expressions for the current in R_2 (a) just after the switch is closed and (b) a long time after the switch is closed. (c) Describe qualitatively how you expect the current in R_3 to behave after the switch is closed.

57. In the circuit of Fig. 28-60 the switch is initially open and the capacitor is uncharged. Find expressions for the current I supplied by the battery (a) just after the switch is closed and (b) a long time after the switch is closed.

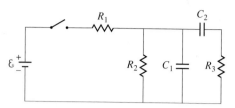

FIGURE 28-59 Problem 56.

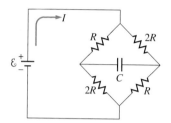

FIGURE 28-60 Problem 57.

58. Obtain an expression for the rate (dV/dt) at which the voltage across a charging capacitor increases. Evaluate your result at time $t = 0$, and show that if the capacitor continued charging steadily at this rate it would be fully charged in exactly one time constant.

Paired Problems

(Both problems in a pair involve the same principles and techniques. If you can get the first problem, you should be able to solve the second one.)

59. A 3.3-kΩ resistor and a 4.7-kΩ resistor are connected in parallel, and the pair is in series with a 1.5-kΩ resistor. What is the resistance of the combination?

60. Find the value of R in Fig. 28-61 that will make the resistance between points A and B equal to R.

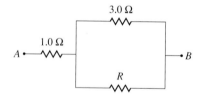

FIGURE 28-61 Problem 60.

61. A battery's voltage is measured with a voltmeter whose resistance is 1000 Ω; the result is 4.36 V. When the measurement is repeated with a 1500-Ω meter the result is 4.41 V. What are (a) the battery voltage and (b) its internal resistance?

62. An ammeter with a resistance of 1.4 Ω is connected momentarily across a battery (not the way to treat an ammeter!) and it reads 9.78 A. When the measurement is repeated with a meter whose resistance is 2.1 Ω the reading

is 7.46 A. What are (a) the battery voltage and (b) its internal resistance?

63. In Fig. 28-62, take $\mathscr{E}_1 = 12$ V, $\mathscr{E}_2 = 6.0$ V, $\mathscr{E}_3 = 3.0$ V, $R_1 = 1.0\ \Omega$, $R_2 = 2.0\ \Omega$, and $R_3 = 4.0\ \Omega$. Find the current in R_2 and give its direction.

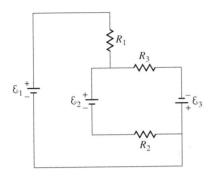

FIGURE 28-62 Problems 63, 64.

64. (a) With all values except $\mathscr{E}_2$ as given in the preceding problem, find $\mathscr{E}_2$ such that there is no current in this battery. (b) What are the currents in R_1 and R_2 under these conditions?

65. In Fig. 28-63 what are the meter readings when (a) an ideal voltmeter or (b) an ideal ammeter is connected between points A and B?

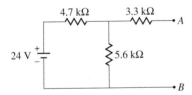

FIGURE 28-63 Problems 65, 66.

66. In Fig. 28-63 what are the meter readings when (a) a voltmeter with 50-kΩ resistance or (b) an ammeter with 150-Ω resistance is connected between points A and B?

67. An initially uncharged capacitor in an RC circuit reaches 75% of its full charge in 22.0 ms. What is the time constant?

68. Find the resistance needed in an RC circuit to bring a 20-μF capacitor from zero charge to 45% of its full charge in 140 ms.

Supplementary Problems

69. Suppose the currents into and out of a circuit node differed by 1 μA. If the node consists of a small metal sphere with diameter 1 mm, how long would it take for the electric field around the node to reach the breakdown field in air (3 MV/m)?

70. You measure the voltage across a charged 26-μF capacitor by connecting a voltmeter with 250-kΩ resistance across the capacitor. If you note the meter reading 2.0 s after connecting the meter, by what percentage does your reading differ from the initial capacitor voltage? Assume the capacitor isn't connected to any other circuitry.

71. In Fig. 28-64, what is the current in the 4-Ω resistor when each of the following circuit elements is connected between points A and B: (a) an ideal ammeter; (b) an ideal voltmeter; (c) another 4.0-Ω resistor; (d) an uncharged capacitor, right after it's connected; (e) long after the capacitor of part (d) is connected; (f) an ideal 12-V battery, with its positive terminal at A; (g) a capacitor initially charged to 12 V, right after it's connected with its positive plate at A; (h) long after the capacitor in part (g) is connected?

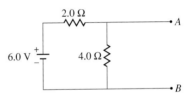

FIGURE 28-64 Problem 71.

72. A resistance R is connected across a battery with internal resistance R_{int}. Show that the maximum power dissipation in R occurs when $R = R_{\text{int}}$. *Note:* This is not the way to treat a battery! But it is the basis for matching loads in amplifiers and other devices; for example, a stereo amplifier designed to drive 8-Ω speakers has internal resistance close to 8 Ω.

73. A parallel-plate capacitor is insulated with a material of dielectric constant κ and resistivity ρ. Since the resistivity is finite, the capacitor "leaks" charge and can be modeled as an ideal capacitor in parallel with a resistor. (a) Show that the time constant of the capacitor is independent of its dimensions (provided the spacing is small enough that the usual parallel-plate approximation applies) and is given by $\varepsilon_0 \kappa \rho$. (b) If the insulating material is polystyrene ($\kappa = 2.6$, $\rho = 10^{16}\ \Omega \cdot$m), how long will it take for the stored energy in the capacitor to decrease by a factor of 2?

74. Of the total energy drawn from a battery in charging an RC circuit, show that only half ends up as stored energy in the capacitor. *Hint:* What happens to the rest of it? You will need to integrate.

75. Find a general solution for the currents in Example 28-6, in terms of the symbolic quantities $\mathscr{E}_1$, $\mathscr{E}_2$, R_1, R_2, and R_3.

FIGURE 28-65   Problem 76.

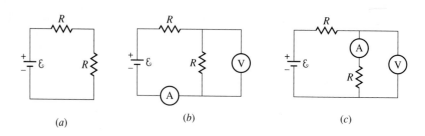

(a)   (b)   (c)

76. Two identical resistors are connected across a battery, as shown in Fig. 28-65a. Figures 28-65b and c show two ways to connect a voltmeter and ammeter in order to measure the voltage across and current through one resistor. If the voltmeter resistance is $100R$ and the ammeter resistance is $0.010R$, how would the resistances calculated from the measured voltage and current for the two connections compare with the actual resistance?

77. Write the loop and node laws for the circuit of Fig. 28-66, and show that the time constant for this circuit is $R_1 R_2 C/(R_1 + R_2)$.

78. The circuit in Fig. 28-67 extends forever to the right, and all the resistors have the same value R. Show that the equivalent resistance measured across the two terminals at left is $\frac{1}{2}R(1 + \sqrt{5})$. *Hint:* You don't need to sum an infinite series.

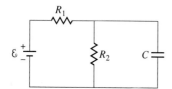

FIGURE 28-66   Problem 77.

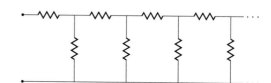

FIGURE 28-67   Problem 78.

THE MAGNETIC FIELD

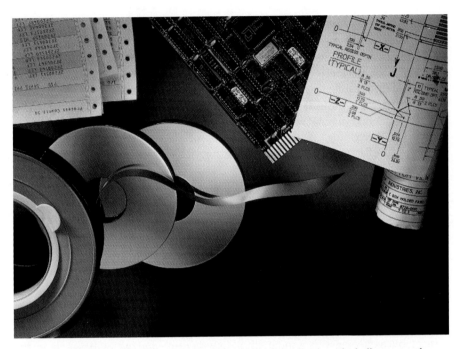

Magnetic media have replaced paper in many data storage applications, including text and graphics. An electronic circuit board controls the flow of information to and from the magnetic media.

M ost people are fascinated with magnets. Magnetism—the seemingly mysterious force you feel when you try to push two magnets together in a way they don't want to go—is always intriguing. Some uses of magnets, like holding notes on refrigerators, are mundane. But others—holding gas at a temperature of one hundred million K in a nuclear fusion reactor, or converting electrical into mechanical energy in the motors of a railroad locomotive—are more impressive. And magnetism, like electricity, is at the heart of many natural phenomena and technological devices. Video and audio tape recorders, electric motors, TV picture tubes, computer disks, and electric power plants would be impossible without magnetism. Earth's magnetism helps us find our way around, provides historical evidence for the evolution of our planet, and protects us from harmful radiation. Birds, sea turtles, and some bacteria use Earth's magnetism for navigation. Without magnetism we would not even see,

FIGURE 29-1 Iron filings align with the magnetic field, tracing out the field from a pair of bar magnets. The field is strongest near the magnetic poles.

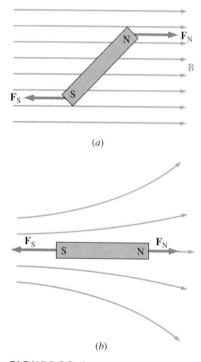

FIGURE 29-2 (a) The poles of a bar magnet experience opposite forces in a uniform magnetic field, giving rise to a torque that tends to align the magnet with the field. (b) In a nonuniform field the forces on the two poles are unequal, giving rise to a net force on the magnet.

for light itself originates in an interaction between magnetism and electricity. As with electricity, we often do not recognize the magnetic character of everyday phenomena.

In this chapter we discuss various aspects of magnetism and its relation to electricity. In subsequent chapters we will say more about the fundamental relation between magnetism and electricity, and eventually will come to understand how electricity and magnetism are manifestations of the same underlying phenomenon.

29-1 MAGNETS, POLES, AND THE MAGNETIC FIELD

An ordinary magnet has two **poles,** arbitrarily designated north and south. The interaction of magnetic poles is similar to that of electric charges: like poles repel, and unlike poles attract. This interaction may be described in terms of the **magnetic field** (symbol **B**). We say that one magnet produces a magnetic field and that a second magnet responds to the field in its immediate vicinity. As with electricity, describing magnetism in terms of fields eliminates the awkward "action at a distance" picture, in which one magnet somehow reaches across empty space to influence another.

We can trace out magnetic field lines using small bar magnets or iron filings (Fig. 29-1). The direction of the field is taken as the direction of the force on a magnetic north pole. The field is stronger—meaning it gives rise to greater forces—where field lines are closely spaced. Since magnetic poles always come in pairs, the effect of the magnetic force on a magnet is to rotate the magnet into alignment with the field (Fig. 29-2a). When the field is nonuniform the magnet experiences a force as well (Fig. 29-2b).

But what *is* a magnet? Magnets, like other matter, are composed of electrons, protons, and neutrons. The interaction of magnets and magnetic fields is fundamentally an interaction involving these particles. We begin our study of magnetism by exploring this fundamental interaction, and we will see how the phenomenon of magnetism is far more extensive than our experience with magnets would suggest. Later we'll see how magnets and magnetic materials fit into this much broader picture of magnetic phenomena.

29-2 ELECTRIC CHARGE AND THE MAGNETIC FIELD

We have mentioned several times that electricity and magnetism are closely related. One manifestation of this relation is that an *electric charge* can experience a force in a magnetic field. We defined the electric field **E** through the relation $\mathbf{F}_E = q\mathbf{E}$, where $\mathbf{F}_E$ is the electric force on a particle with charge q. Analogously, we define the magnetic field in terms of the relation between a charge q and the magnetic force $\mathbf{F}_B$ it experiences.

Consider a region in which there is no electric field, but in which a magnetic field is present. Experimentally, we find that a stationary electric charge in this region experiences no force. But if the charge is moving there may be a magnetic force on it. Experiment shows that

1. The magnetic force is always at right angles both to the velocity **v** of the charge and to the magnetic field **B** (Fig. 29-3).
2. The strength of the magnetic force is proportional to the product of the charge q, its speed v, and the magnetic field strength B.
3. The force is greatest if the charged particle is moving at right angles to the magnetic field **B** and is zero if the charge velocity **v** is parallel or antiparallel to the field. In general, the magnetic force is proportional to the sine of the angle between the vectors **v** and **B** (see Fig. 29-3).

Putting these facts together allows us to write the magnetic force compactly in terms of the vector cross product introduced in Chapter 13:

$$\mathbf{F}_B = q\mathbf{v} \times \mathbf{B}, \quad \text{(magnetic force)} \tag{29-1a}$$

where $\mathbf{F}_B$ is the magnetic force on a particle of charge q moving with velocity **v** at a point where the magnetic field is **B**. Recall from Chapter 13 that the cross product $\mathbf{A} \times \mathbf{B}$ of two vectors **A** and **B** is a vector of magnitude $AB\sin\theta$, where θ is the angle between **A** and **B**, and where the direction of $\mathbf{A} \times \mathbf{B}$ is given by the right-hand rule. The magnetic force thus has magnitude

$$F = qvB\sin\theta, \tag{29-1b}$$

while Fig. 29-4 shows how the right-hand rule applies to the vectors **v** and **B**. Because the charge q also enters Equation 29-1a, the direction of the magnetic force is that of $\mathbf{v} \times \mathbf{B}$ for a positive charge but opposite $\mathbf{v} \times \mathbf{B}$ for a negative charge.

We can regard Equation 29-1 as the definition of the magnetic field. If we put particle of charge q moving with velocity **v** in a region free of other influences (i.e., electric and gravitational fields), then the presence of a force **F** on the particle shows that there is a magnetic field **B** implied by Equation 29-1a. Equations 29-1 show that the SI units of magnetic field are N·s/C·m, a unit given the name **tesla** (T) after the Serbian-American inventor Nikola Tesla (1865–1943). One tesla is a strong magnetic field, and a smaller unit called the

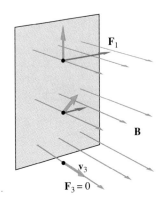

FIGURE 29-3 The magnetic force on a charged particle is perpendicular to the particle's velocity **v** and to the magnetic field **B**. The magnitude of the force depends on the angle between **v** and **B**, and is greatest when **v** and **B** are perpendicular (top). When **v** and **B** are parallel there is no magnetic force (bottom).

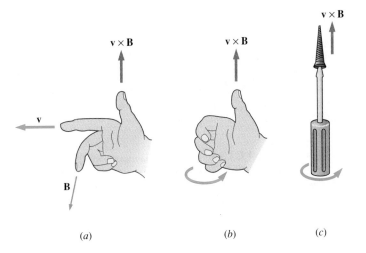

(a) (b) (c)

FIGURE 29-4 Three ways to remember the right-hand rule. In (b) and (c) the curved arrow represents a rotation taking the vector **v** onto the vector **B**.

FIGURE 29-5 This large superconducting magnet is 15 m long and produces a field of 6.6 T. It was designed to steer charged particles in the Superconducting Super Collider, a giant particle accelerator intended to probe the fundamental structure of matter.

gauss (G), equal to 10^{-4} T, is often used. Earth's magnetic field, for example, is a little under 1 G, while the field at the poles of a toy magnet may be 100 G. Strong laboratory magnets (Fig. 29-5) produce fields ranging from several T to nearly 40 T, while neutron stars—incredibly dense, rapidly rotating objects with the mass of a star compressed into a region several km in diameter—have fields up to 10^8 T.

● EXAMPLE 29-1 STEERING PROTONS

A magnetic field of 0.10 T is used to steer charged-particle beams in a nuclear physics experiment. The field points vertically upward. Three protons enter the field region, two moving horizontally and one vertically as shown in Fig. 29-6. All three are moving at 2.0×10^6 m/s. What is the magnetic force on each?

Solution

Proton 2 is moving parallel to the field. For it, $\mathbf{v} \times \mathbf{B} = \mathbf{0}$ and it experiences no magnetic force. Protons 1 and 3 are moving at right angles to the field, so $\sin\theta = 1$ and Equation 29-1b gives

$$F = qvB = (1.6 \times 10^{-19}\ \text{C})(2.0 \times 10^6\ \text{m/s})(0.10\ \text{T})$$

$$= 3.2 \times 10^{-14}\ \text{N}.$$

Since the protons carry positive charge, the direction of the force is that of the product $\mathbf{v} \times \mathbf{B}$. For proton 1, moving to the right, $\mathbf{v} \times \mathbf{B}$ is out of the page. For proton 3, moving to the left, the force is into the page. This example shows that the magnetic field alone does not determine the magnetic force. Identical particles in the same field experience different forces if their velocities are different. Had the particles been electrons the forces would have been in the opposite directions. (Why?)

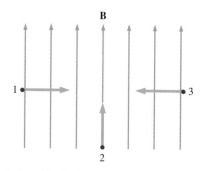

FIGURE 29-6 What is the magnetic force on each proton? (Example 29-1)

EXERCISE A 1.3-T magnetic field points in the x direction. A particle carrying 1.0 μC is moving at 20 m/s in the x-y plane and experiences a magnetic force of 1.4×10^{-5} N. What angle does the particle's velocity make with the x axis?

Answer: 33°

Some problems similar to Example 29-1: 1–11 ●

Although electricity and magnetism are related, the electric field and the magnetic field are distinct. In particular, both may be present simultaneously. In that case, a charged particle experiences both the magnetic force $\mathbf{F}_B$ of Equation 29-1 and also the electric force $q\mathbf{E}$, for a net **electromagnetic force** given by

$$F = q\mathbf{E} + q\mathbf{v} \times \mathbf{B}. \qquad \text{(electromagnetic force)} \qquad (29\text{-}2)$$

● EXAMPLE 29-2 A VELOCITY SELECTOR

A region contains uniform electric and magnetic fields of magnitude E and B, respectively, and oriented at right angles as shown in Fig. 29-7. A beam of charged particles enters the region, heading straight into the page. What speed must a particle have if it is to cross the field region undeflected?

Solution
The net force on a particle must be zero if it is to be undeflected. Consider a positively charged particle heading into the page in Fig. 29-7. The electric field points to the right, so the electric force $q\mathbf{E}$ is to the right. Applying the right-hand rule with the velocity vector $\mathbf{v}$ into the page and the magnetic field $\mathbf{B}$ downward gives a magnetic force to the left; since v and B are at right angles, the magnitude of the magnetic force is qvB. So the electric and magnetic forces are in opposite directions; for them to cancel their magnitudes must be equal:

$$qE = qvB,$$

or

$$v = \frac{E}{B}.$$

Note that this condition is independent of the charge—even of its sign, since the directions of both forces would reverse for a negative charge.

The field configuration of this example is called a **velocity selector,** since it passes undeflected only those particles with speed $v = E/B$ that are moving perpendicular to both fields.

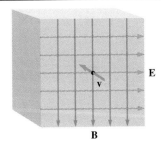

FIGURE 29-7 Crossed electric and magnetic fields give rise to oppositely-directed forces on charged particles moving perpendicular to the fields. The magnetic force depends on the particle speed, but the electric force does not; therefore, only those particles with just the right speed can traverse the field region undeflected.

Velocity selectors can be used to prepare particle beams with uniform speeds or to analyze the distribution of speeds in a charged-particle population.

EXERCISE You have a magnet that produces a 0.25-T field. What strength electric field should you apply to make a velocity selector that passes 2.8×10^5 m/s particles undeflected?

Answer: 70 kV/m

Some problems similar to Example 29-2: 12–14

29-3 THE MOTION OF CHARGED PARTICLES IN MAGNETIC FIELDS

Like any force, the magnetic force deflects a particle from the straight-line path Newton's first law says it would otherwise follow (Fig. 29-8). But unlike most other forces, the magnetic force is necessarily at right angles to the particle velocity. That means force and particle displacement are perpendicular. Since work is the dot product of force with displacement (Equation 7-5; $W = \mathbf{F} \cdot \Delta\mathbf{r}$), and the dot product of perpendicular vectors is zero, we can conclude that

▌ The magnetic force does no work on a charged particle.

Both the kinetic energy and thus the speed of a charged particle subject only to a magnetic force are therefore constant. The magnetic force changes only the

(a)   (b)

FIGURE 29-8 (a) In the absence of a magnetic field, an electron beam moves in a straight line. (b) A magnetic field causes the beam to deflect. If the electrons are moving from left to right, which way is the field pointing?

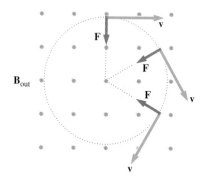

FIGURE 29-9 A charged particle moving at right angles to a uniform magnetic field describes a circular path. Note the convention that dots represent magnetic field lines emerging from the page; crosses, in contrast, would represent field lines going into the page.

direction of the particle's velocity, not its magnitude. What sort of motion results from the magnetic force? Suppose a particle with positive charge q is moving at right angles to a uniform magnetic field **B,** as shown in Fig. 29-9. At the instant the particle's velocity is to the right, the outward-pointing magnetic field gives a force $q\mathbf{v} \times \mathbf{B}$ that is downward. Since this force is at right angles to the velocity, it changes the direction but not the magnitude of the velocity. The magnetic force on the new velocity is still perpendicular to it, and since the magnitude of the velocity is unchanged, so is the magnitude of the force. What we've just described—a force of constant magnitude always at right angles to a particle's velocity—is the prescription for uniform circular motion. Thus, a charged particle moving in a plane perpendicular to a uniform magnetic field describes a circular path. The magnetic force acts exactly like the tension in a string when you tie a mass to the string and whirl it around in a circle. The tension is perpendicular to the motion and changes its direction but not its speed.

How big is the circle, and how long does it take to get around? We can answer these questions using Newton's second law. In its circular path, the charged particle undergoes an acceleration v^2/r, directed toward the center of the circle. What causes this acceleration? The magnetic force! With the field and velocity at right angles, the magnitude of the magnetic force is just

$$F = qvB,$$

and the force points toward the center of the circle. Writing Newton's law $\mathbf{F} = m\mathbf{a}$ then gives

$$qvB = m\frac{v^2}{r},$$

so

$$r = \frac{mv}{qB}. \tag{29-3}$$

This result makes sense: the larger the particle's momentum mv, the harder it is for the magnetic force to bend it out of a straight line, so the larger the radius of the orbit. On the other hand, if we make the field or charge larger, then the magnetic force increases, giving a tighter orbit.

● **EXAMPLE 29-3** A TV PICTURE TUBE

In a TV picture tube, deflection of the electron beam that "paints" the TV picture is accomplished using magnetic fields. The geometry of a certain tube requires that the electron beam be bent in a circular arc with a minimum radius of 4.5 cm (Fig. 29-10). If the electrons are accelerated from rest through a 25-kV potential difference before they enter the magnetic field region, what is the magnetic field strength required? In what direction should the field point to accomplish the deflection shown in Fig. 29-10? Assume that the field is uniform over the deflecting region and zero elsewhere.

Solution

Solving Equation 29-3 for the magnetic field strength B gives

$$B = \frac{mv}{er},$$

with e the elementary charge. "Falling" through a potential difference $V = 25$ kV, the electron of charge e acquires a kinetic energy $\frac{1}{2}mv^2 = Ve$, so its speed is

$$v = \sqrt{\frac{2Ve}{m}}.$$

Our expression for the field then becomes

$$B = \frac{m}{er}\sqrt{\frac{2Ve}{m}} = \frac{1}{r}\sqrt{\frac{2mV}{e}}$$

$$= \frac{1}{0.045 \text{ m}}\left(\frac{(2)(9.1\times10^{-31} \text{ kg})(25\times10^3 \text{ V})}{1.6\times10^{-19} \text{ C}}\right)^{1/2}$$

$$= 0.012 \text{ T}.$$

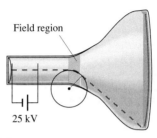

Field region

25 kV

FIGURE 29-10 A TV picture tube requiring a 4.5-cm bending radius (marked with a diagonal line) for maximum deflection of the electron beam. Which way should the magnetic field point to deflect an electron in the direction indicated (Example 29-3)?

To achieve an initially downward deflection, the force $q\mathbf{v} \times \mathbf{B}$ must be initially downward. But the electron charge is negative, so $\mathbf{v} \times \mathbf{B}$ must be upward. Application of the right-hand rule to the rightward-moving electrons shows that the magnetic field $\mathbf{B}$ must be into the page in Fig. 29-10.

EXERCISE A beam of protons moving at 2.5×10^5 m/s is deflected through a 90° turn by a 10-mT magnetic field oriented perpendicular to the beam's velocity. What is the radius of the turn?

Answer: 26 cm

Some problems similar to Example 29-3: 15, 18, 21 ●

The Cyclotron Frequency

How long does it take a charged particle to complete its circular orbit in a uniform magnetic field? The circumference of the orbit is $2\pi r$, so the period T of the circular motion is

$$T = \frac{2\pi r}{v}.$$

Using Equation 29-3 for the radius r gives

$$T = \frac{2\pi r}{v} = \frac{2\pi}{v}\frac{mv}{qB} = \frac{2\pi m}{qB}. \qquad (29\text{-}4)$$

FIGURE 29-11 The Crab Nebula, remnant of a supernova explosion observed nearly 1000 year ago. Intense radio emission occurs as electrons undergo circular motion in the Crab's magnetic field. The frequency of the emission is related to the cyclotron frequency of the electrons' motion, and its measurement therefore allows determination of the magnetic field.

This is a remarkable result, for it shows that the period of the circular motion is independent of the particle's speed and the size of the orbit. It depends only on the magnetic field and the charge-to-mass ratio of the particle. The frequency in revolutions per second is simply $1/T$, or

$$f = \frac{qB}{2\pi m}. \qquad (29\text{-}5)$$

This quantity is often called the **cyclotron frequency,** since it's the frequency at which charged particles circulate in a cyclotron particle accelerator. In astrophysics, the cyclotron frequency provides a simple way to measure the magnetic fields of distant objects (Fig. 29-11), while electrons circling in a special tube called a magnetron generate the microwaves that cook your food in a microwave oven.

■ APPLICATION THE CYCLOTRON

Physicists studying the basic structure of matter need tools that can probe the atomic nucleus and its constituent particles. The only probes sufficiently small are subatomic particles themselves, accelerated to high enough energies that they can disrupt the strong nuclear force. How is this acceleration accomplished? One way is to accelerate particles of charge q through a large potential difference V, giving each particle energy qV. But there are practical problems in the generation and handling of potential differences much over a million volts (recall Fig. 25-25). To achieve higher energies, devices are used that circumvent the need for a single large potential difference. One of the earliest and most successful such devices is the **cyclotron** (Fig. 29-12). The device consists of an evacuated chamber

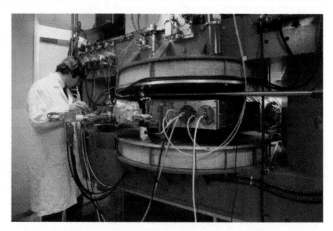

FIGURE 29-13 This cyclotron at Massachusetts General Hospital was used to produce radioisotopes for research leading to the development of the medical diagnostic procedure known as positron emission tomography (PET).

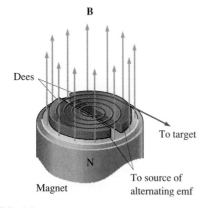

FIGURE 29-12 A cyclotron, showing one magnet pole, dees, and a typical ion trajectory. Not shown are the vacuum chamber surrounding the dees, upper pole, and frame supporting pole pieces and electromagnet windings.

between the poles of a magnet. At the center of the chamber is a source of the particles to be accelerated, usually protons or light ions. The ions undergo circular motion in the magnetic field.

Also within the evacuated chamber are two hollow conducting structures each shaped like the letter D (Fig. 29-12). A modest potential difference is applied across these two "dees," and this potential difference is made to alternate in polarity with the same frequency as the circular motion of the ions. Recall that this frequency depends only on the magnetic field strength and the charge-to-mass ratio of the ions, but not on their energy. As the ions circle around in the cyclotron, they are accelerated across the gap between the dees by the strong electric field associated with the potential difference. Because each dee is a hollow, nearly closed conducting structure, there is no electric field within the dees.

Once inside a dee, the ions simply follow a circle in the magnetic field. Halfway round they again cross the gap between dees. If the electric field were steady, the ions would be *decelerated* at this crossing. But the electric field changes direction in step with the ions' circular motion, so each time the ions cross the dee gap they are accelerated and gain more energy. They move faster and in ever larger circles, but always with the same orbital period. Eventually the ion orbits become nearly the size of the machine. At this point an electrostatic field provided by a high-voltage electrode deflects the ions out of the magnetic field and toward a target, where their interactions with target nuclei cause nuclear reactions.

In addition to providing experimental data on nuclear structure, cyclotrons are valuable in producing short-lived radioactive isotopes for a variety of purposes, particularly medical research and diagnosis. A number of large hospitals have their own cyclotrons (Fig. 29-13); in particular, the diagnostic procedure known as positron emission tomography (PET) requires cyclotron-produced radioisotopes.

At very high energies, the theory of relativity comes into play and alters our conclusion that the cyclotron frequency is independent of particle energy. As a result, the cyclotron cannot be used to achieve these relativistic energies. An alternate

FIGURE 29-14   Aerial view of Fermilab, the Fermi National Accelerator Laboratory at Batavia, Illinois. Large circle in the background is the 2-km-diameter Tevatron, a synchrotron that accelerates protons to energies of 1 TeV (10^{12} eV). Much larger synchrotrons include the Large Electron-Positron Collider at the European Center for Nuclear Research and the Superconducting Super Collider that was to have been built in Texas.

accelerator design is the **synchrotron,** in which both the magnetic field and frequency of an alternating electric field are varied to account for increasing particle energy, while the orbital radius is held constant (Fig. 29-14).

● **EXAMPLE 29-4**   DESIGNING A CYCLOTRON

A cyclotron is to accelerate protons to a kinetic energy of 5.0 MeV. If the magnetic field in the cyclotron is 2.0 T, what must be the radius of the cyclotron and the frequency at which the dee voltage is alternated?

Solution

The cyclotron frequency is given by Equation 29-5:

$$f = \frac{qB}{2\pi m} = \frac{(1.6 \times 10^{-19}\ \text{C})(2.0\ \text{T})}{(2\pi)(1.67 \times 10^{-27}\ \text{kg})} = 3.0 \times 10^7\ \text{Hz}.$$

This is the frequency required to accelerate protons at each crossing of the dee gap; incidentally, it is about the frequency of a citizens band (CB) radio transmitter.

An energy of 5.0 MeV is equal to

$$(5.0 \times 10^6\ \text{eV})(1.6 \times 10^{-19}\ \text{J/eV}) = 8.0 \times 10^{-13}\ \text{J},$$

so the proton kinetic energy is

$$K = \tfrac{1}{2}mv^2 = 8.0 \times 10^{-13}\ \text{J}.$$

Solving for the speed v gives

$$v = \sqrt{\frac{2K}{m}} = \sqrt{\frac{(2)(8.0 \times 10^{-13}\ \text{J})}{1.67 \times 10^{-27}\ \text{kg}}} = 3.1 \times 10^7\ \text{m/s}.$$

Equation 29-3 then gives the radius needed to accommodate 5-MeV protons:

$$r = \frac{mv}{qB} = \frac{(1.67 \times 10^{-27}\ \text{kg})(3.1 \times 10^7\ \text{m/s})}{(1.6 \times 10^{-19}\ \text{C})(2.0\ \text{T})} = 0.16\ \text{m}.$$

To ensure a uniform magnetic field over the particle trajectories, the radii of the magnet pole pieces would have to be somewhat larger than this value.

EXERCISE   A microwave oven uses 2.4-GHz microwaves, generated by electrons circling at this frequency in the magnetic field of a magnetron tube. (a) What is the magnetic field strength? (b) If the energy associated with an electron's circular motion is 890 eV, what is the radius of its circular path?

Answers: (a) 86 mT; (b) 1.2 mm

Some problems similar to Example 29-4:   17, 25, 26   ●

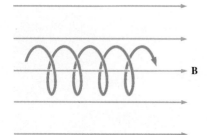

FIGURE 29-15 A particle in a uniform magnetic field describes a helical path, its motion along the field direction unaffected by the magnetic force.

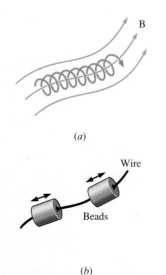

(a)

(b)

FIGURE 29-16 (a) Charged particles undergoing helical motion about the magnetic field direction are "frozen" to the field like (b) beads sliding along a wire.

FIGURE 29-17 Magnetic fields govern the behavior of matter throughout much of the universe. Here, concentrations of ionized gas trace out magnetic field loops in the Sun's atmosphere.

Particle Trajectories in Three Dimensions

What if a particle's motion is not confined to a plane perpendicular to the magnetic field? Then we can resolve its velocity into two vectors, one perpendicular and the other parallel to the magnetic field. Since the magnetic force is always perpendicular to the field, there is no component of force along the field direction—and therefore the velocity component in this direction is unaffected by the magnetic force. The force acts only on the velocity component perpendicular to the field, and here our previous analysis applies: in the plane perpendicular to the field, the particle's motion is circular with frequency given by Equation 29-5. In the absence of other forces the particle moves uniformly along the field direction while executing circular motion perpendicular to the field. The resulting trajectory is a helix, as shown in Fig. 29-15.

The absence of magnetic force along the field direction means that particles move readily along the field. But if you try to push them at right angles to the field, they simply move in larger circles about the field direction. As a result, charged particles are often described as being "frozen" to the magnetic field (Fig. 29-16). Nonuniform fields and collisions between particles make this "freezing" of particles and field less than perfect, but in many cases the particle density is low enough that the "frozen" assumption is an excellent approximation. Unlike our relatively cool planet Earth, much of the universe consists of free electrons and protons, not bound into neutral atoms. As a result, magnetic fields are a dominant influence on matter throughout much of the universe (Fig. 29-17).

An important terrestrial application of this "trapping" of particles on magnetic field lines occurs in nuclear fusion reactors. These experimental devices—whose successful development would make 1 gallon of sea water the energy equivalent of more than 300 gallons of gasoline—have to contain ionized gas (plasma) at temperatures around 100 million kelvins. They do so with "magnetic bottles," most commonly toroidal (doughnut-shaped) chambers whose circular magnetic fields never intersect the chamber walls and therefore keep the hot plasma from touching the walls (Fig. 29-18).

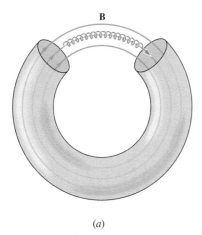

(a)

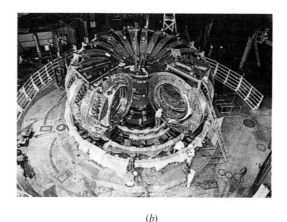

(b)

FIGURE 29-18 (a) Charged particles spiral about the circular field lines in a simplified fusion reactor. With its toroidal shape, the machine has no "ends" from which magnetic field lines emerge. The field therefore keeps charged particles away from the chamber walls. (b) The Tokamak Fusion Test Reactor at Princeton University. This view, taken during construction, shows clearly the machine's toroidal shape.

■ **APPLICATION** THE AURORA AND MAGNETIC MIRRORS

Earth itself possesses a magnetic field whose origin we will consider in the next chapter. Near the planet, the field resembles that of a bar magnet. A stream of electrons and protons, called the solar wind, flows outward from the Sun and is deflected by Earth's field, protecting us from potentially harmful radiation. But occasionally bursts of high-energy solar particles penetrate the magnetic field, eventually becoming trapped into helical motion along the field. Figure 29-19a shows that the higher field lines, where particles are most likely to be trapped, intersect Earth near the poles. High-energy particles moving along these field lines slam into the upper atmosphere near the poles, exciting oxygen and nitrogen atoms and

(b)

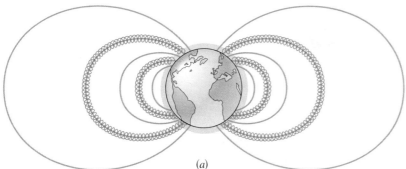

(a)

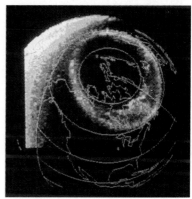

(c)

FIGURE 29-19 The aurora arises from the interaction of high-energy particles trapped in Earth's magnetic field. (a) Schematic diagram showing particles spiraling between the northern and southern polar regions along the magnetic field. (b) A colorful aurora viewed from Canada's Prince Edward Island. (c) The aurora appears as a circle surrounding Earth's north magnetic pole in this satellite image.

thereby producing the spectacular displays we call the aurora (Fig. 29-19b,c). These solar outbursts also buffet Earth's magnetic field, causing "magnetic storms" that can severely disrupt communications and electric power transmission.

Fig. 29-20 shows a particle spiraling into a region of stronger magnetic field, like the polar regions of Fig. 29-19a. Note that there is a component of magnetic force opposite to the general direction of motion. This force eventually reverses the particle's forward motion, with the result that the stronger field region acts like a **magnetic mirror,** reflecting particles that attempt to penetrate. This means that particles trapped in Earth's magnetic field bounce back and forth between northern and southern hemispheres, reflected by the magnetic mirrors of higher field strength near the poles. Auroral displays in opposite hemispheres are often strikingly similar since they are caused by the same particle populations mirroring back and forth.

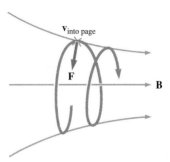

FIGURE 29-20 The magnetic mirror effect. A charged particle is spiraling into a region of stronger magnetic field. At the point shown its velocity is into the page, and the magnetic force—proportional to $\mathbf{v} \times \mathbf{B}$—has a component to the left. This force eventually reverses the rightward motion of the particle.

29-4 THE MAGNETIC FORCE ON A CURRENT

So far we have considered the magnetic force on individual charged particles. An electric current is simply a group of charged particles sharing a common motion, so we should expect a current to interact with a magnetic field.

The Force on a Straight Wire

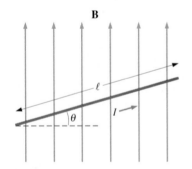

FIGURE 29-21 A straight wire carrying current I through a uniform magnetic field. The charge carriers in the wire experience a magnetic force, in this case out of the page.

Imagine a long straight wire carrying a current I through a uniform magnetic field $\mathbf{B}$, as shown in Fig. 29-21. If each charge carrier has a drift velocity $\mathbf{v}_d$ along the wire, and if each carries charge q, then the magnetic force on each charge carrier is given by Equation 29-1:

$$\mathbf{F}_q = q\mathbf{v}_d \times \mathbf{B}.$$

(Of course the charge carriers also have random thermal velocities. But these average to zero and therefore make no net contribution to the magnetic force.) If the wire has cross-sectional area A and contains n charge carriers per unit volume, then the net force on all the charge carriers in a length ℓ of the wire is

$$\mathbf{F} = nA\ell q\mathbf{v}_d \times \mathbf{B}.$$

The product $nAqv_d$ is just the current I, as we found in deriving Equation 27-2. If we define a vector ℓ whose magnitude is the length ℓ of the wire and whose direction is along the current, then we can write

$$\mathbf{F} = I\boldsymbol{\ell} \times \mathbf{B}. \qquad \text{(magnetic force on a current)} \qquad (29\text{-}6)$$

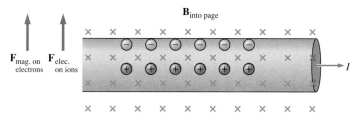

FIGURE 29-22   Electrons moving to the left in a wire are deflected upward by the magnetic force, resulting in charge separation and therefore in an upward electric force on the fixed ions. As a result the entire wire is ultimately influenced by the magnetic force on the moving electrons.

The direction of this force is at right angles to both the current and the magnetic field, or out of the page in Fig. 29-21. For a given direction of the current, the direction of the force does not depend on the sign of the charge carriers. If the current is to the right, then positive charges move to the right and the force on each is out of the page. If the charges are negative, they move to the left and the force is still out of the page because both the sign of the velocity and the sign of the charge are reversed, given the same sign for the force $q\mathbf{v} \times \mathbf{B}$.

Strictly speaking, Equation 29-6 gives the net magnetic force only on the charge carriers in the wire. But the motion of the charge carriers—typically electrons—under the influence of the magnetic force causes charge separation in the wire, and the resulting electric field exerts a force on the fixed charges—typically ions—in the wire (Fig. 29-22). Thus the entire wire experiences the force. Although its origin is not entirely magnetic, we loosely call the force given by Equation 29-6 the magnetic force on the wire.

The magnetic force on a current-carrying wire is the basis for many practical devices, including electric motors that start cars and run refrigerators, CD players, computer disk drives, subway trains, pumps, food processors, power tools, and myriad other useful instruments of modern society.

● EXAMPLE 29-5   MAGNETIC FORCE ON A POWER LINE

A power line runs along Earth's equator, where Earth's magnetic field points horizontally from south to north and has a strength of about 0.5 G. The current in the power line is 500 A, flowing from west to east. What are the magnitude and direction of the magnetic force on 1 km of the power line?

Solution

Let eastward be the x direction, northward the y direction, and upward the z direction. After we convert gauss to tesla, Equation 29-6 gives

$$\mathbf{F} = I\boldsymbol{\ell} \times \mathbf{B} = (500 \text{ A})(1000\hat{\mathbf{i}} \text{ m}) \times (0.5 \times 10^{-4}\,\hat{\mathbf{j}} \text{ T})$$

$$= 25\hat{\mathbf{k}} \text{ N}.$$

This 25-N upward force is negligible compared with the weight—on the order of 2×10^4 N—of 1 km of power line.

EXERCISE   A 2.0-m-long wire has a mass of 37 g. The wire extends horizontally, at 40° to a horizontal magnetic field of 0.075 T. What current in the wire will result in the magnetic force suspending it against gravity?

Answer: 3.8 A

Some problems similar to Example 29-5:   33, 35, 37   ●

■ APPLICATION MAGNETIC LEVITATION AND PROPULSION

Engineers throughout the world are at work on transportation systems whose operation results directly from the magnetic force on a straight current-carrying conductor. In so-called maglev vehicles, magnetic forces levitate the vehicle just a few centimeters above a conducting guideway and also provide the horizontal force that propels the vehicle (Fig. 29-23). Such vehicles should be capable of 500-km/h speeds, well above the 300 km/h maximum for high-speed rail systems. Maglev vehicles should prove especially effective for travel between cities in densely populated regions.

A related application is so-called magnetohydrodynamic (MHD) propulsion, a kind of jet propulsion in which a conducting fluid is accelerated by the magnetic force. Engineers are currently experimenting with MHD propulsion for ships. Because sea water is a reasonably good electrical conductor, passage of a current through sea water in a magnetic field results in a magnetic force on the water. Figure 29-24 shows the experimental Japanese ship Yamato-1 and its MHD propulsion system.

FIGURE 29-23 A magnetically levitated vehicle under development in Japan.

(a)

FIGURE 29-24 (a) The Japanese ship Yamato-1 uses MHD propulsion. (b) Schematic diagram of the MHD propulsion unit. Superconducting electromagnet produces a magnetic field; current flowing in the sea water between the two electrodes then experiences a magnetic force that ejects a jet of water at right. (c) One of Yamato's two MHD propulsion units.

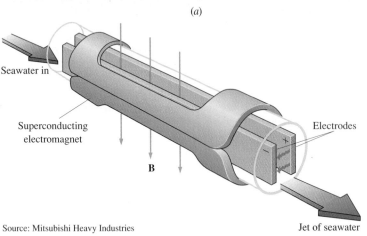

Seawater in

Superconducting electromagnet

B

Electrodes

Jet of seawater

Source: Mitsubishi Heavy Industries

(b)

(c)

■ **APPLICATION**   THE HALL EFFECT

We have seen that the force on a current-carrying conductor is independent of the sign of the charge carriers. But there is a subtle difference. Fig. 29-25 shows two conductors, each with the same current I to the right and magnetic field **B** pointing into the page. In Fig. 29-25a the current is carried by electrons moving to the *left*. The product **v** × **B** is downward, but since electrons are negative the force $q\mathbf{v} \times \mathbf{B}$ is upward. As a result, the upper edge of the conductor is negative with respect to the lower edge. In Fig. 29-25b the current is carried by protons moving to the *right*. Again the force $q\mathbf{v} \times \mathbf{B}$ is upward, so now the upper edge of the conductor is positive.

This phenomenon of charge separation is the **Hall effect**. The separated charge gives rise to an electric field and therefore to a measurable potential difference—the **Hall potential**—between opposite edges of the conductor. The sign of the Hall potential depends on the sign of the charge carriers. Charge separation will stop once the electric force on the charge carriers cancels the magnetic force. The electric force has magnitude qE and, with **B** perpendicular to the current, the magnetic force has magnitude qv_dB, with v_d the charge-carrier drift speed. Equating these magnitudes gives $E = v_dB$. In a rectangular conductor this field will be essentially uniform, so the Hall potential becomes

$$V_H = Eh = v_dBh,$$

where h is the conductor height shown in Fig. 29-25. But Equation 27-2 gives $I = nAqv_d$, with n the number density of charge carriers, A the conductor's cross-sectional area, and q the charge on each carrier. Solving for v_d and using the result in our expression for V_H gives

$$V_H = \frac{IBh}{nAq},$$

or, since $A = ht$ with t the conductor thickness shown in Fig. 29-25,

$$V_H = \frac{IB}{nqt}. \tag{29-7}$$

The quantity $1/nq$ is the **Hall coefficient**. Measurement of V_H in a sample of known thickness carrying known current in a known magnetic field yields the value of this coefficient and, thus, gives information on the nature and density of the charge carriers. Alternatively, measurement of V_H in a material with known Hall coefficient carrying a known current provides a direct measure of the magnetic field strength (Fig. 29-26). Since V_H is inversely proportional to the charge-carrier density n, Hall-effect magnetic field measurements often use semiconductors whose carrier density n is much lower than in metals. For typical magnetic fields, Hall potentials are then in millivolts (see Problem 45).

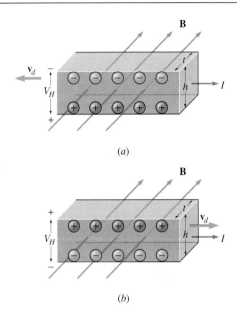

FIGURE 29-25   The Hall potential V_H arises from the deflection of charge carriers by the magnetic force when a current-carrying conductor is in a magnetic field. Although the magnetic force on the conductor is independent of the sign of the charge carriers, the sign of the Hall potential does depend on the sign of the charge carriers. In both (a) and (b) the current is to the right, carried in (a) by negative charge carriers moving to the left and in (b) by positive charge carriers moving to the right.

FIGURE 29-26   This magnetometer uses the Hall potential developed in its semiconductor probe to measure magnetic fields.

One of the most surprising discoveries of the 1980s was the so-called quantized Hall effect, where the Hall potential at low temperatures and strong magnetic fields exhibits a step-like rather than continuous increase with increasing magnetic field. This is one of the few instances where a macroscopic property manifests directly the discontinuity inherent in the quantum description of matter. Discovery of the quantized Hall effect won Klaus von Klitzing the 1985 Nobel Prize in Physics.

FIGURE 29-27 (a) A curved conductor in a nonuniform magnetic field. (b) A small segment $d\ell$ can be treated as a straight wire in a uniform field; the magnetic force on the segment is $d\mathbf{F} = I\,d\ell \times \mathbf{B}$.

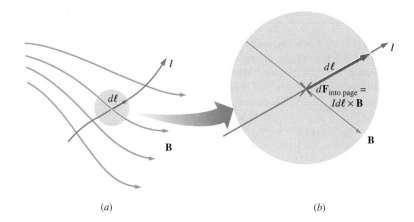

(a)　　　(b)

Nonuniform Fields and Curved Conductors

Equation 29-6 applies to a straight conductor at some arbitrary angle to a uniform magnetic field. What if the conductor bends, so its orientation relative to the field changes? Or what if the field changes in magnitude or direction over the length of the conductor? We can still make use of Equation 29-6 if we apply it only to a very small segment of the conductor, so small that it is essentially straight and the field essentially constant over its length. If the length of this small segment is $d\ell$, then Equation 29-6 becomes

$$d\mathbf{F} = I\,d\ell \times \mathbf{B}$$

for the small force on the segment (Fig. 29-27). To find the total force on the conductor, we sum the forces on all such segments. In the limit of very small segments, this sum becomes an integral, and we have

$$\mathbf{F} = \int d\mathbf{F} = \int I\,d\ell \times \mathbf{B}. \tag{29-8}$$

The integration is taken over the entire section of conductor on which we are calculating the force.

● EXAMPLE 29-6 THE FORCE ON A CURVED CONDUCTOR

A semicircular wire connects two points C and D a horizontal distance $2R$ apart. The wire carries a current I from C to D and is in a uniform magnetic field $\mathbf{B}$ pointing upward, as in Fig. 29-28a. Show that the magnetic force on this semicircular wire is the same as the force that a straight wire from C to D would experience if it carried the same current I.

Solution

Since the orientation of the wire relative to the field varies, we use the integral of Equation 29-8 to calculate the force. Figure 29-28a shows an infinitesimal segment $d\ell$ of the wire, and the angle θ that specifies its position on the semicircular arc. This

infinitesimal segment subtends an infinitesimal angle $d\theta$, as shown. A blown-up view of the segment (Fig. 29-28b) shows that the angle between the segment $d\ell$ and the magnetic field $\mathbf{B}$ is also θ. Then the force on the segment is

$$d\mathbf{F} = I\,d\ell \times \mathbf{B}.$$

The magnitude of this force element $d\mathbf{F}$ is

$$dF = I\,d\ell B\sin\theta.$$

Application of the right-hand rule shows that the direction of the force element is out of the page for all values of θ on the

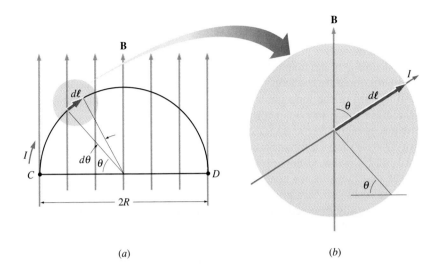

FIGURE 29-28 (*a*) A semicircular wire carries current *I* between points *C* and *D*. (*b*) Blow-up of an infinitesimal segment *dℓ* of the wire, showing that the segment makes an angle *θ* with the magnetic field.

semicircle—that is, for $0 \leq \theta \leq \pi$. Therefore, the integral in Equation 29-8 describes a sum of vectors all of which point in the same direction, so the magnitude of the net force is

$$F = \int dF = \int_0^\pi IB \sin \theta \, d\ell.$$

But *dℓ* subtends the angle *dθ*, so $d\ell = R \, d\theta$ (Fig. 29-29). Then, taking the constants *I*, *B*, and *R* outside the integral, we have

$$F = IBR \int_0^\pi \sin \theta \, d\theta = IBR \, (-\cos \theta)\Big|_0^\pi$$

$$= IBR[-(-1 - 1)] = IB(2R).$$

This is just the force we would get from Equation 29-6 for a straight wire of length 2*R* carrying current *I* perpendicular to a magnetic field **B**.

EXERCISE A 15-cm-long wire carrying 4.8 A is bent into a quarter circle and oriented like the left half of the semicircular

wire in Fig. 29-28*a* (i.e., from point *C* to the top of the semicircle). If the magnetic field shown in Fig. 29-28*a* has magnitude 0.65 T, what is the magnetic force on the wire?

Answer: 0.30 N

Some problems similar to Example 29-6: 40–43

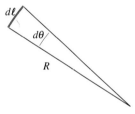

FIGURE 29-29 An angle in radians is defined as the ratio of subtended arc to radius. Here the angle *dθ* is so small that the difference between a circular arc and the straight segment *dℓ* is insignificant. Therefore $d\ell = R \, d\theta$. ●

If we closed the semicircular loop of Example 29-6, as in Fig. 29-30, the force on the straight section would have the same magnitude *IB*(2*R*) but the opposite direction from the force on the semicircle; thus, the net force on the closed loop would be zero. Problem 43 shows that this result is in fact true for *any* closed current loop in a *uniform* magnetic field.

29-5 A CURRENT LOOP IN A MAGNETIC FIELD

We've just seen that a closed, current-carrying loop experiences no net force in a uniform magnetic field. But such a loop generally does experience a torque, as we now show. Torques on current loops play important roles in a wide range of natural and technological systems.

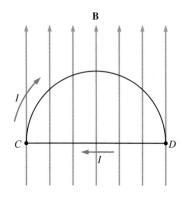

FIGURE 29-30 A closed current loop consisting of a semicircle and a straight wire. Since both carry the same current, but in opposite directions, the result of Example 29-6 shows that they experience forces of equal magnitude but opposite directions. The net force on the loop is therefore zero.

Consider a rectangular current loop in a uniform magnetic field, with the normal to the loop making an angle θ with the magnetic field, as shown in Fig. 29-31a. Since current flows in opposite directions on opposite sides and since the field is uniform, the magnetic forces on opposite sides cancel and there is no net magnetic force on the loop. The right-hand rule shows that the forces on the top and bottom of the loop point directly upward and downward, respectively, and therefore result in no net torque.

The forces on the vertical sections of the loop, however, do produce a torque. Figure 29-31b shows a top view of the loop, with the forces resulting from the upward- and downward-flowing currents in the vertical sides. From the figure, it is clear that these forces give rise to a net torque twisting the loop clockwise. Each vertical side is a straight wire of length a carrying current I at right angles to the horizontally directed magnetic field **B**, so the magnitude of the force on each side is simply

$$F_{\text{side}} = IaB.$$

The distance from the loop's central axis to the sides where this force acts is half the loop width, or $b/2$. The geometry of Fig. 29-31b shows that the angle used in calculating the torque is the same as the angle θ between the loop normal and the magnetic field. Therefore, the torque due to the force on each side is

$$\tau_{\text{side}} = F\frac{b}{2}\sin\theta = Ia\frac{b}{2}B\sin\theta.$$

Accounting for the contributions from both sides gives the magnitude of the net torque on the loop:

$$\tau = IabB\sin\theta.$$

We can express the torque in vector notation if we define a vector **A** whose magnitude is the area ab of the loop and which is perpendicular to the loop. We

FIGURE 29-31 (a) A rectangular current loop in a uniform magnetic field. The angle between the loop normal and the magnetic field is θ. (b) Top view of the current loop, showing that magnetic forces on the vertical sides result in a net torque. The angles used in the torque calculation are the same as the angle θ between the loop normal and the magnetic field.

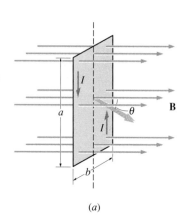

(a)

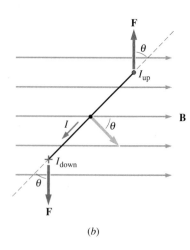

(b)

choose the direction of **A** by the right-hand rule: wrap your fingers around the loop in the direction of the current and your thumb points in the direction of **A.** Then we can write

$$\tau = I\mathbf{A} \times \mathbf{B}. \tag{29-9}$$

Although we derived this equation for a rectangular loop, it holds in fact for any current loop. The torque on a current loop depends on the current, the loop area, the magnetic field, and the orientation between loop and field.

Equation 29-9 should remind you of Equation 23-9 for the torque on an electric dipole in an electric field. There we had

$$\tau = \mathbf{p} \times \mathbf{E},$$

where **p** is the electric dipole moment and **E** the electric field. Comparison with Equation 29-9 suggests that a current loop in a magnetic field behaves analogously to an electric dipole in an electric field. As far as its response to magnetic fields is concerned, therefore,

| **A current loop constitutes a magnetic dipole.**

The quantity $I\mathbf{A}$ in Equation 29-9 plays the same role as the electric dipole moment in the equation $\tau = \mathbf{p} \times \mathbf{E}.$ We therefore call this quantity the **magnetic dipole moment, μ:**

$$\mu = I\mathbf{A}. \qquad \text{(single-turn loop)}$$

The direction of the vector μ is the same as the direction we defined for **A:** curl your right fingers in the direction of the loop current, and your thumb points in the direction of μ (Fig. 29-32). More generally, a loop may consist of N turns of conducting wire; then each contributes $I\mathbf{A}$ to give the total magnetic moment:

$$\mu = NI\mathbf{A}. \qquad \text{(magnetic dipole moment)} \tag{29-10}$$

Clearly the units of magnetic moment are A·m². Using the magnetic moment vector, Equation 29-9 then becomes

$$\tau = \mu \times \mathbf{B}, \qquad \text{(torque on a current loop)} \tag{29-11}$$

in analogy with the electric case. The magnetic moment vector of a current loop is perpendicular to the plane of the loop, and the torque given in Equation 29-11 tends to align the magnetic moment with the field (Fig. 29-33). It takes work to twist the loop's magnetic moment vector out of alignment with the field. In exact analogy with the electric-dipole case summarized in Equation 23-10, we express the associated potential energy as

$$U_{\text{magnetic}} = -\mu \cdot \mathbf{B}. \tag{29-12}$$

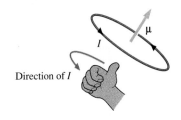

FIGURE 29-32 Using the right-hand rule to find the direction of a current loop's magnetic moment.

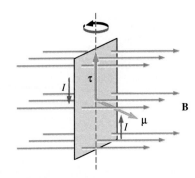

FIGURE 29-33 The torque on a current loop tends to align the loop's magnetic moment vector with the magnetic field.

● EXAMPLE 29-7 POINTING A SATELLITE

Some small satellites use the magnetic torque on current loops to point themselves in space. Earth itself provides the magnetic field, while electricity from solar panels powers coils that constitute the current loops; unlike rocket-based pointing systems, there's no fuel to run out. Three mutually perpendicular coils allow the satellite to point itself in any direction. One satellite uses 30-cm-diameter circular coils, each with 1000 turns. (a) If the satellite orbits at an altitude where Earth's magnetic field is 0.24 G, and if specifications call for a maximum torque of 1.3 mN·m from each coil, what should be the coil current? (b) How much work is done on the coil to rotate it from its maximum-torque orientation to an angle of $120°$ to the field?

Solution

(a) The maximum torque occurs when the coil's magnetic moment is perpendicular to the field; Equation 29-11 shows that this maximum torque has magnitude $\tau_{max} = \mu B$. The magnetic moment of the coil is the product of the number of turns, the current, and the coil area πR^2, as given by Equation 29-10. Thus the maximum torque is

$$\tau_{max} = NI\pi R^2 B.$$

Solving for the current gives

$$I = \frac{\tau_{max}}{N\pi R^2 B} = \frac{1.3\times10^{-3} \text{ N·m}}{(1000)(\pi)(0.15 \text{ m})^2(0.24\times10^{-4} \text{ T})}$$

$$= 0.766 \text{ A}.$$

(b) The loop's potential energy is given by Equation 29-12: $U = -\boldsymbol{\mu} \cdot \mathbf{B} = -\mu B\cos\theta$. Maximum torque occurs when $\theta = 90°$, so the work in rotating the coil—equal to the difference in its potential energies in the two orientations—is

$$W = U_{120°} - U_{90°} = -\mu B\cos 120° - (-\mu B\cos 90°)$$

$$= -\tau_{max}\cos 120° - 0 = (-1.3\times10^{-3} \text{ N·m})(\cos 120°)$$

$$= 0.65 \text{ mJ},$$

where we recognized the quantity μB as the maximum torque defined earlier.

EXERCISE An electric motor consists of a 550-turn loop with area 100 cm² in a 0.23-T magnetic field. What loop current is necessary if the motor is to develop a maximum torque of 3.7 N·m?

Answer: 2.9 A

Some problems similar to Example 29-7: 47, 48, 50–52 ●

■ APPLICATION ELECTRIC MOTORS

Electric motors are so much a part of our lives that we hardly think of them. But CD players, car starters, refrigerators, vacuum cleaners, power saws, subway trains, computer disk drives, food processors, washing machines, fans, hair dryers, water pumps, oil burners, and most industrial machinery would be difficult or impossible to build without electric motors (Fig. 29-34).

At the heart of every electric motor is a current loop in a

(a)

(b)

FIGURE 29-34 Electric motors are made in a wide range of sizes and power outputs. (a) A small motor, disassembled to show the armature (rotating coil assembly). (b) This large motor drives the world's largest fully steerable radio telescope.

magnetic field. But instead of a steady current, the loop carries a current that reverses periodically. In direct-current (DC) motors, this reversal is achieved through the electrical contacts that provide current to the loop. Figure 29-35 shows a simple motor consisting of a loop in the field of a permanent magnet. Current from an external battery reaches the loop through a set of stationary "brushes," which make contact with a pair of semicircular conductors called the "commutator." The commutator is attached rigidly to the loop, which rotates to align its dipole moment with the field. Just as it reaches alignment, however, the brushes cross the gaps in the commutator. This crossing reverses the connections of loop to battery, resulting in a reversal of the loop current. This in turn reverses the magnetic moment of the loop, so it is no longer aligned with the field. The loop then rotates another 180° to align its new magnetic moment vector with the field. Just as it reaches alignment, the current again reverses. This process repeats, resulting in continuous rotation of the loop. A rigid shaft through loop and commutator delivers mechanical work to the device powered by the motor. The source of this work is the battery or whatever supplies electrical energy to the motor. The motor itself is a device that converts electrical to mechanical energy; the magnetic field is an intermediary in this conversion.

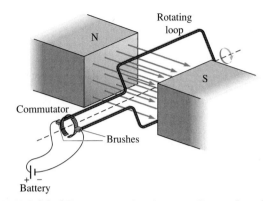

FIGURE 29-35 A simple electric motor. Current flows from the battery through the stationary brushes to the rotating commutator and loop. The direction of the loop current reverses as the two halves of the commutator rotate to contact different brushes.

What determines the rotational speed of the motor? Why doesn't it undergo a constant rotational acceleration, reaching ever greater speeds? The answer to these questions lies in another deep interaction between electricity and magnetism, an interaction that we will explore in Chapter 31.

CHAPTER SYNOPSIS

Summary

1. **Magnetism** is a fundamental interaction described in terms of **magnetic fields.** Magnetic fields interact with **moving electric charges.** A moving charged particle experiences a force that depends on its charge, its velocity, and the magnetic field. The force is at right angles to both velocity and field, and is given by

$$\mathbf{F} = q\mathbf{v} \times \mathbf{B}.$$

2. An electric current is made up of moving electric charges, so there is a net magnetic force on an electric current. The force on small current elements of length $d\ell$ is

$$d\mathbf{F} = I\,d\boldsymbol{\ell} \times \mathbf{B},$$

where I is the current, $\mathbf{B}$ the magnetic field, and $d\boldsymbol{\ell}$ an infinitesimal vector pointing in the local direction of the current. For a straight wire of length ℓ in a magnetic field, this equation becomes

$$\mathbf{F} = I\boldsymbol{\ell} \times \mathbf{B};$$

for other cases the infinitesimal force $d\mathbf{F}$ can be integrated over a current to obtain the force on the entire current.

3. A particularly important case of a current in a magnetic field is a closed current loop. Such a loop behaves like a **magnetic dipole** with magnetic dipole moment

$$\boldsymbol{\mu} = NI\mathbf{A},$$

where N is the number of turns in the loop, I the loop current, and $\mathbf{A}$ a vector perpendicular to the plane of the loop and whose magnitude is the loop area.

a. A current loop in a magnetic field experiences a torque given by

$$\boldsymbol{\tau} = \boldsymbol{\mu} \times \mathbf{B}.$$

If the field is nonuniform, the loop experiences a net force as well.

b. The potential energy associated with a current loop in a magnetic field is

$$U_B = -\boldsymbol{\mu} \cdot \mathbf{B},$$

where the zero of potential energy is taken when the loop's magnetic moment vector is perpendicular to the field.

Terms You Should Understand

(Pairs are closely related terms whose distinction is important; number in parentheses is chapter section where term first appears.)

magnetic field (29-1, 29-2)

magnetic force, electromagnetic force (29-2)

Symbols You Should Recognize

$\mathbf{B}$ (29-1, 29-2)
T (29-2)
$\boldsymbol{\mu}$ (29-5)

Problems You Should Be Able to Solve

calculating magnetic force vectors given charge, velocity, and magnetic field (29-2)

analyzing charged-particle trajectories in uniform magnetic fields (29-3)

finding magnetic forces on straight conductors in uniform fields (29-4)

using integration to determine magnetic forces with curved conductors and/or nonuniform fields (29-4)

determining magnetic dipole moments of current loops (29-5)

evaluating torque and potential energy for current loops in magnetic fields (29-5)

Limitations to Keep in Mind

Magnetic forces arise only when charges are *moving*.
The cyclotron frequency is independent of particle energy only for speeds much less than the speed of light.

QUESTIONS

1. A stationary charged particle experiences no force in a certain region. Can you conclude from this observation that there is no magnetic field in this region?

2. A charged particle moves through a region containing only a magnetic field. Under what condition will the particle experience no force?

3. An electron moving with velocity $\mathbf{v}$ through a magnetic field $\mathbf{B}$ experiences a magnetic force $\mathbf{F}$. Which of the vectors $\mathbf{F}$, $\mathbf{v}$, and $\mathbf{B}$ must be at right angles?

4. In Fig. 29-36 a high-energy gamma ray has decayed into an electron and its positively charged antiparticle, a positron. A magnetic field points out of the plane of the photograph, and the electron and positron spiral in this field. Which path belongs to which particle? Why might the paths be spirals rather than circles of constant radius?

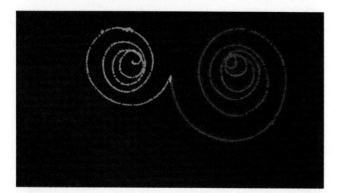

FIGURE 29-36 Creation of an electron-positron pair in the decay of a gamma ray incident from the top of the figure. Trajectories were imaged in the detector of a high-energy particle accelerator (Question 4).

5. A magnetic field points out of this page. Will a positively charged particle moving in the plane of the page circle clockwise or counterclockwise as viewed from above?

6. An electron moves through a region in a straight line at constant speed. Can you conclude that there is no magnetic field in the region? Could you so conclude if you knew that there were no electric or gravitational forces on the electron? Explain.

7. High-resolution color TV monitors sometimes have a built-in circuit that compensates for changes in the orientation of Earth's magnetic field relative to the TV picture tube as the monitor is moved from one place to another. Why is this necessary?

8. What is meant by the statement that charged particles can be "trapped" by magnetic fields?

9. An electron beam comes straight to the center of a TV screen, where it makes a spot of light. If you hold the north pole of a bar magnet on the left side of the picture tube, which way will the spot move?

10. Do particles in a cyclotron gain energy from the electric field, the magnetic field, or both? Explain.

11. A cyclotron is designed to accelerate either hydrogen or deuterium nuclei. If the magnetic field is unchanged, how must the frequency of the alternating dee voltage be changed in order to switch from hydrogen to deuterium?

12. An electron and a proton moving at the same speed enter a region containing a uniform magnetic field. Which is deflected more from its original path?

13. An electron and a proton with the same kinetic energy enter a region containing a uniform magnetic field. Which is deflected more from its original path?

14. For what orientation of electric and magnetic fields could the net force on a particle be zero?
15. Will the velocity selector of Example 29-2 work for particles coming out of the page in Fig. 29-7? Why or why not?
16. What does magnetism have to do with the fact that auroras are seen near Earth's poles?
17. In a certain region uniform electric and magnetic fields are at right angles to one another. A positively charged particle is released from rest in this region. Describe qualitatively its subsequent motion.
18. How do the period and radius of an electron's orbit in a magnetic field depend on its velocity? Assume that the electron is moving at right angles to the field.

19. Current in a certain ionic solution is carried equally by positive and negative ions. Would you expect the Hall effect to occur in this solution?
20. Two identical particles carrying equal charge are moving in opposite directions along a magnetic field, when they collide elastically head-on. Describe their subsequent motion.
21. Repeat the above question for the case when the two particles are moving instantaneously perpendicular to the field when they collide.
22. Under what conditions will a current loop in a magnetic field experience zero force? Zero torque?
23. What would happen to a motor with no commutator?

PROBLEMS

Section 29-2 Electric Charge and the Magnetic Field

1. (a) What is the minimum magnetic field needed to exert a 5.4×10^{-15}-N force on an electron moving at 2.1×10^7 m/s? (b) What magnetic field strength would be required if the field were at $45°$ to the electron's velocity?
2. An electron moving at right angles to a 0.10-T magnetic field experiences an acceleration of 6.0×10^{15} m/s^2. (a) What is the electron's speed? (b) By how much does its *speed* change in 1 ns ($= 10^{-9}$ s)?
3. What is the magnitude of the magnetic force on a proton moving at 2.5×10^5 m/s (a) at right angles; (b) at $30°$; (c) parallel to a magnetic field of 0.50 T?
4. A magnetic field of 0.10 T points in the x direction. A charged particle carrying 1.0 μC enters the field region moving at 20 m/s. What are the magnitude and direction of the force on the particle when it first enters the field region if it does so moving (a) along the x axis; (b) along the y axis; (c) along the z axis; (d) at $45°$ to both x and y axes?
5. A particle carrying a 50-μC charge moves with velocity $\mathbf{v} = 5.0\mathbf{i} + 3.2\hat{\mathbf{k}}$ m/s through a uniform magnetic field $\mathbf{B} = 9.4\hat{\mathbf{i}} + 6.7\hat{\mathbf{j}}$ T. (a) What is the force on the particle? (b) Form the dot products $\mathbf{F}\cdot\mathbf{v}$ and $\mathbf{F}\cdot\mathbf{B}$ to show explicitly that the force is perpendicular to both $\mathbf{v}$ and $\mathbf{B}$.
6. Moving in the x direction, a particle carrying 1.0 μC experiences no force. Moving with speed v at $30°$ to the x axis, the particle experiences a force of 2.0 N. What is the magnitude of the force it would experience if it moved along the y axis with speed v?
7. A proton moving with velocity $\mathbf{v}_1 = 3.6\times10^4\hat{\mathbf{j}}$ m/s experiences a force of $7.4\times10^{-16}\hat{\mathbf{i}}$ N. A second proton moving on the x axis experiences a magnetic force of $2.8\times10^{-16}\hat{\mathbf{j}}$ N. Find the magnitude and direction of the magnetic field, and the velocity of the second proton.
8. The magnitude of Earth's magnetic field is a little less than 1 G near Earth's surface. What is the maximum possible magnetic force on an electron with kinetic energy of 1 keV? Compare with the gravitational force on the same electron.

9. An alpha particle (2 protons, 2 neutrons) is moving with velocity $\mathbf{v} = 150\hat{\mathbf{i}} + 320\hat{\mathbf{j}} - 190\hat{\mathbf{k}}$ km/s at a point where the magnetic field is $\mathbf{B} = 0.66\hat{\mathbf{i}}$ T. Find the magnitude of the force on the particle.
10. An electron is moving with velocity $\mathbf{v} = 5.1\hat{\mathbf{i}} + 2.3\hat{\mathbf{j}}$ Mm/s at a point where the magnetic field is $\mathbf{B} = 0.29\hat{\mathbf{i}} - 0.62\hat{\mathbf{k}}$ T. Find the magnetic force on the electron.
11. A 1.4-μC charge moving at 185 m/s experiences a magnetic force $\mathbf{F}_B = 2.5\hat{\mathbf{i}} + 7.0\hat{\mathbf{j}}$ μN in a magnetic field $\mathbf{B} = 42\hat{\mathbf{i}} - 15\hat{\mathbf{j}}$ mT. What is the angle between the particle's velocity and the magnetic field?
12. A velocity selector uses a 60-mT magnetic field and a 24 kN/C electric field. At what speed will charged particles pass through the selector undeflected?
13. A region contains an electric field $\mathbf{E} = 7.4\hat{\mathbf{i}} + 2.8\hat{\mathbf{j}}$ kN/C and a magnetic field $\mathbf{B} = 15\hat{\mathbf{j}} + 36\hat{\mathbf{k}}$ mT. Find the electromagnetic force on (a) a stationary proton, (b) an electron moving with velocity $\mathbf{v} = 6.1\hat{\mathbf{i}}$ Mm/s.
14. A charged particle is moving at right angles to both a 1.1 kN/C electric field and a 0.75-T magnetic field. If the magnitude of the electric force on the particle is twice that of the magnetic force, what is the particle's speed?

Section 29-3 The Motion of Charged Particles in Magnetic Fields

15. What is the radius of the circular path described by a proton moving at 15 km/s in a plane perpendicular to a 400-G magnetic field?
16. How long does it take an electron to complete a circular orbit at right angles to a 1.0-G magnetic field?
17. Radio astronomers detect electromagnetic radiation at a frequency of 42 MHz from an interstellar gas cloud. If this radiation is caused by electrons spiraling in a magnetic field, what is the field strength in the gas cloud?
18. A beam of electrons moving in the x direction at 8.7×10^6 m/s enters a region where a uniform magnetic

field of 180 G points in the y direction. How far into the field region does the beam penetrate?

19. Electrons and protons with the same kinetic energy are moving at right angles to a uniform magnetic field. How do their orbital radii compare?

20. The Van Allen belts are regions in space where high-energy charged particles are trapped in Earth's magnetic field. If the field strength at the Van Allen belts is 0.10 G, what are the period and radius of the helical path described (a) by a proton with a 1.0-MeV kinetic energy? (b) by a 10-MeV proton?

21. Typical particle energies in a nuclear fusion reactor are on the order of 10 keV. If the smallest dimension of the reactor is on the order of 1 m, estimate the minimum magnetic field strength needed to ensure that protons have orbits smaller than the size of the reactor. Will this field be sufficient for electrons of the same energy?

22. Show that the orbital radius of a charged particle moving at right angles to a magnetic field B can be written

$$r = \frac{\sqrt{2Km}}{qB},$$

where K is the kinetic energy in joules, m the particle mass, and q its charge.

23. Two protons, moving in a plane perpendicular to a uniform magnetic field of 500 G, undergo an elastic head-on collision. How much time elapses before they collide again? *Hint:* Draw a picture.

24. Repeat the preceding problem for the case of a proton and an antiproton colliding head-on (a) if they have the same speed and (b) if they have different speeds. (An antiproton has the same mass as a proton, but carries the opposite charge.)

25. A cyclotron is designed to accelerate deuterium nuclei. (Deuterium has one proton and one neutron in its nucleus.) (a) If the cyclotron uses a 2.0-T magnetic field, at what frequency should the dee voltage be alternated? (b) If the vacuum chamber has a diameter of 0.90 m, what is the maximum kinetic energy of the deuterons? (c) If the magnitude of the potential difference between the dees is 1500 V, how many orbits do the deuterons complete before achieving the energy of part (b)?

26. Without changing the magnetic field, how could the cyclotron of the preceding problem be modified to accelerate (a) protons and (b) alpha particles (two protons and two neutrons)? What would be the maximum energy achievable with (c) protons and (d) alpha particles?

27. Figure 29-37 shows a simple mass spectrometer, designed to analyze and separate atomic and molecular ions with different charge-to-mass ratios. In the design shown, ions are accelerated through a potential difference V, after which they enter a region containing a uniform magnetic field. They describe semicircular paths in the magnetic field, and land on a detector a lateral distance x from where

they entered the field region, as shown. Show that x is given by

$$x = \frac{2}{B}\sqrt{\frac{2V}{(q/m)}},$$

where B is the magnetic field strength, V the accelerating potential, and q/m the charge-to-mass ratio of the ion. By counting the number of ions accumulated at different positions x, one can determine the relative abundances of different atomic or molecular species in a sample.

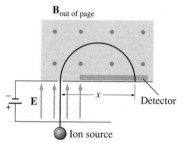

FIGURE 29-37 A mass spectrometer (Problem 27).

28. A mass spectrometer like that of the preceding problem has $V = 2000$ V and $B = 1000$ G. It is used to analyze a gas sample suspected of containing Ne, O_2, CO, SO_2, and NO_2. Ions are detected at distances of 58 cm, 68 cm, and 87 cm from the entrance to the field region. Which gases are actually present? Assume that all molecules are singly ionized.

29. A mass spectrometer is used to separate the fissionable uranium isotope U-235 from the much more abundant isotope U-238. To within what percentage must the magnetic field be held constant if there is to be no overlap of these two isotopes? Both isotopes appear as constituents of uranium hexafluoride gas (UF_6), and the gas molecules are all singly ionized.

30. An electron is moving in a uniform magnetic field of 0.25 T; its velocity components parallel and perpendicular to the field are both equal to 3.1×10^6 m/s. (a) What is the radius of the electron's helical path? (b) How far does it move along the field direction in the time it takes to complete a full orbit about the field direction?

31. An electron moving at 3.8×10^6 m/s enters a region containing a uniform magnetic field $\mathbf{B} = 18\hat{\mathbf{k}}$ mT. The electron is moving at 70° to the field direction, as shown in Fig. 29-38. Find the radius r and pitch p of its helical path, as indicated in the figure.

32. A proton in interstellar space describes a helical path about a 15-mG magnetic field, with velocity component 40 km/s perpendicular to the field. If the pitch of the helix (see Fig. 29-38) is 8.7 km, what is the proton's velocity component parallel to the field?

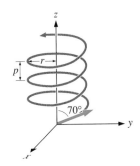

FIGURE 29-38 Problems 31, 32.

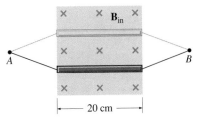

FIGURE 29-40 Problem 38.

Section 29-4 The Magnetic Force on a Current

33. What is the magnitude of the force on a 50-cm-long wire carrying 15 A at right angles to a 500-G magnetic field?

34. A wire coincides with the x axis, carrying 2.4 A in the $+x$ direction. The wire passes through a region containing a uniform magnetic field $\mathbf{B} = 0.17\hat{\mathbf{i}} + 0.32\hat{\mathbf{j}} - 0.21\hat{\mathbf{k}}$ T. Find a vector expression for the force per unit length on the wire in the magnetic field region.

35. A wire carrying 15 A makes a 25° angle with a uniform magnetic field. The magnetic force per unit length of wire is 0.31 N/m. (a) What is the magnetic field strength? (b) What is the maximum force per unit length that could be achieved by reorienting the wire in this field?

36. A wire of negligible resistance is bent into a rectangle as shown in Fig. 29-39, and a battery and resistor are connected as shown. The right-hand side of the circuit extends into a region containing a uniform magnetic field of 38 mT pointing into the page. Find the magnitude and direction of the net force on the circuit.

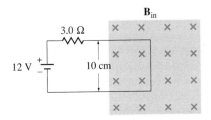

FIGURE 29-39 Problem 36.

37. In a high-magnetic-field experiment, a conducting bar carrying 7.5 kA passes through a 30-cm-long region containing a 22-T magnetic field. If the bar makes a 60° angle with the field direction, what force is necessary to hold it in place?

38. A 20-cm-long conducting rod with mass 18 g is suspended by wires of negligible mass, as shown in Fig. 29-40. The rod is in a region containing a uniform magnetic field of 0.15 T pointing horizontally into the page, as shown. An external circuit supplies current between the support points A and B. (a) What is the minimum current necessary to move the bar to the upper position shown? (b) Which direction should the current flow?

39. A piece of wire with mass per unit length 75 g/m runs horizontally at right angles to a horizontal magnetic field. A 6.2-A current in the wire results in its being suspended against gravity. What is the magnetic field strength?

40. A nonuniform magnetic field points out of the page, as shown in Fig. 29-41. The field strength increases at the rate of 2.0 mT/cm as you move to the right. A square wire loop 15 cm on a side lies in a plane perpendicular to the field, and a 2.5-A current circles the loop in the counterclockwise direction. What are the magnitude and direction of the net magnetic force on the loop?

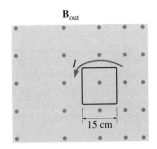

FIGURE 29-41 Problem 40.

41. A wire carrying 1.5 A passes through a region containing a 48-mT magnetic field. The wire is perpendicular to the field and makes a quarter-circle turn of radius 21 cm as it passes through the field region, as shown in Fig. 29-42. Find the magnitude and direction of the magnetic force on this section of wire.

42. A wire coincides with the x axis, and carries a current $I = 2.0$ A in the $+x$ direction. A nonuniform magnetic field points in the y direction, given by $\mathbf{B} = B_0(x/x_0)^2\hat{\mathbf{j}}$, where $B_0 = 0.22$ T, $x_0 = 1.0$ m, and x is the x coordinate. Find the force on the section of wire between $x = 1.0$ m and $x = 3.5$ m.

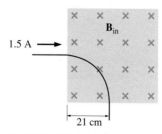

FIGURE 29-42 Problem 41.

43. Apply Equation 29-8 to a closed current loop of arbitrary shape in a *uniform* magnetic field, and show that the net force on the loop is zero. *Hint:* Both I and B are constant as you go around the loop, so you can take them out of the integral. What is the remaining vector integral?

44. A rectangular copper strip measures 1.0 mm in the direction of a uniform 2.4-T magnetic field. When the strip carries a 6.8-A current at right angles to the field, the Hall voltage across the strip is 1.2 μV. Find the number density of free electrons in the copper.

45. The probe in a Hall-effect magnetometer uses a semiconductor doped to a charge-carrier density of 7.5×10^{20} m^{-3}. The probe measures 0.35 mm thick in the direction of the magnetic field being measured, and carries a 2.5-mA current perpendicular to the field. If its Hall potential is 4.5 mV, what is the magnetic field strength?

Section 29-5 A Current Loop in a Magnetic Field

46. Show that the units of magnetic moment (A·m^2) can also be written as J/T.

47. A single-turn square wire loop 5.0 cm on a side carries a 450-mA current. (a) What is the magnetic moment of the loop? (b) If the loop is in a uniform 1.4-T magnetic field with its dipole moment vector at 40° to the field direction, what is the magnitude of the torque it experiences?

48. An electric motor contains a 250-turn circular coil 6.2 cm in diameter. If it is to develop a maximum torque of 1.2 N·m at a current of 3.3 A, what should be the magnetic field strength?

49. A bar magnet experiences a 12-mN·m torque when it is oriented at 55° to a 100-mT magnetic field. What is the magnitude of its magnetic dipole moment?

50. A single-turn wire loop 10 cm in diameter carries a 12-A current. It experiences a torque of 0.015 N·m when the normal to the loop plane makes a 25° angle with a uniform magnetic field. What is the magnetic field strength?

51. A simple electric motor like that of Fig. 29-35 consists of a 100-turn coil 3.0 cm in diameter, mounted between the poles of a magnet that produces a 0.12-T field. When a 5.0-A current flows in the coil, what are (a) its magnetic dipole moment and (b) the maximum torque developed by the motor?

52. A satellite with rotational inertia 20 kg·m^2 is in orbit at a height where Earth's magnetic field strength is 0.18 G. It

has a magnetic torquing system, as described in Example 29-7, that uses a 1000-turn coil 30 cm in diameter. What should be the current in the coil if the magnetic torque is to give the satellite a maximum angular acceleration of 0.0015 s^{-2}?

53. Nuclear magnetic resonance (NMR) is a technique for analyzing chemical structures and is also the basis of magnetic resonance imaging used for medical diagnosis. The NMR technique relies on sensitive measurements of the energy needed to flip atomic nuclei upside-down in a given magnetic field. In an NMR apparatus with a 7.0-T magnetic field, how much energy is needed to flip a proton ($\mu = 1.41 \times 10^{-26}$ A·m^2) from parallel to antiparallel to the field?

54. A wire of length ℓ carries a current I. (a) Find an expression for the magnetic dipole moment that results when the wire is wound into an N-turn circular coil. (b) For what integer value of N is this moment a maximum?

Paired Problems

(Both problems in a pair involve the same principles and techniques. If you can get the first problem, you should be able to solve the second one.)

55. Find the magnetic force on an electron moving with velocity $\mathbf{v} = 8.6 \times 10^5 \hat{\imath} - 4.1 \times 10^5 \hat{\jmath}$ m/s in a magnetic field $\mathbf{B} = 0.18 \hat{\jmath} + 0.64 \hat{k}$ T.

56. A proton moving with velocity $\mathbf{v} = 2.0 \times 10^5 \hat{\imath} + 4.0 \times 10^5 \hat{\jmath}$ m/s experiences a magnetic force $\mathbf{F} = 10\hat{\imath} - 5.0\hat{\jmath} + 21\hat{k}$ fN. What is the z component of the magnetic field?

57. Proponents of space-based particle-beam weapons have to confront the effect of Earth's magnetic field on their beams. If a beam of protons with kinetic energy 100 MeV is aimed in a straight line perpendicular to Earth's magnetic field in a region where the field strength is 48 μT, what will be the radius of the protons' circular path?

58. Electrons are accelerated through a 30-kV potential difference at the rear of a TV tube. The electron beam is initially headed straight toward the center of the tube. The TV is oriented so the beam is perpendicular to Earth's magnetic field, in a location where the field strength is 62 μT. What will be the radius of the electron beam's curved path?

59. A 170-mT magnetic field points into the page, confined to a square region as shown in Fig. 29-43. A square conduct-

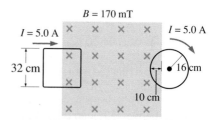

FIGURE 29-43 Problems 59, 60.

ing loop 32 cm on a side carrying a 5.0-A current in the clockwise sense extends partly into the field region, as shown. Find the magnetic force on the loop.

60. Find the force on the circular current loop shown at the right of Fig. 29-43. The loop carries 5.0 A clockwise, has radius 16 cm, and extends 10 cm into the field region.

61. An old-fashioned analog meter uses a wire coil in a magnetic field to deflect the meter needle. If the coil is 2.0 cm in diameter and consists of 500 turns of wire, what should be the magnetic field strength if the maximum torque is to be 1.6 μN·m when the current in the coil is 1.0 mA?

62. A circular wire coil 15 cm in diameter carries a 460-mA current and experiences a 0.020-N·m torque when the normal to the coil makes a 27° angle with a 42-mT magnetic field. How many turns are in the coil?

Supplementary Problems

63. Electrons in a TV picture tube are accelerated through a 30-kV potential difference and head straight for the center of the tube, 40 cm away. If the electrons are moving at right angles to Earth's 0.50-G magnetic field, by how much do they miss the screen's exact center?

64. A certain region contains both an electric field and a magnetic field. An electron moving with velocity $2.0 \times 10^4 \hat{\mathbf{j}}$ m/s experiences a force of $2.2\hat{\mathbf{i}} + 4.8\hat{\mathbf{j}}$ fN. Another electron, moving with velocity $2.0 \times 10^4 \hat{\mathbf{i}}$ m/s, experiences a force of $2.6\hat{\mathbf{j}} + 2.2\hat{\mathbf{k}}$ fN. A third electron, moving with velocity $1.0 \times 10^4 \hat{\mathbf{k}}$ m/s, experiences a force $-1.1\hat{\mathbf{i}} + 4.8\hat{\mathbf{j}}$ fN. Find a combination of uniform electric and magnetic fields that could be responsible for these forces.

65. A conducting bar with mass 15.0 g and length 22.0 cm is suspended from a spring in a region where a 0.350-T magnetic field points into the page, as shown in Fig. 29-44. With no current in the bar, the spring length is 26.0 cm. The bar is supplied with current from outside the field region, using wires of negligible mass. When a 2.00-A current flows from left to right in the bar, it rises 1.2 cm from its equilibrium position. Find (a) the spring constant and (b) the unstretched length of the spring.

66. In 2.0 μs, an electron moves 15 cm in the direction of a 0.10-T magnetic field. If the electron's velocity components perpendicular and parallel to the field are equal, (a) what is the length of its actual helical trajectory and (b) how many orbits about the field direction does it complete?

67. A square loop of side a is free to pivot about a horizontal rod passing through the centers of two sides. The loop carries a current I and is in a uniform vertical magnetic field **B**, as shown in Fig. 29-45. A mass m hangs from one side of the loop, as shown. Find an expression for the angle θ between the loop plane and the horizontal for which this system will be in equilibrium.

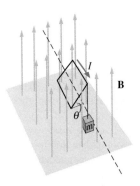

FIGURE 29-45 Problem 67.

68. A disk of radius R carries a uniform surface charge density σ and is rotating with angular frequency ω. Show that its magnetic dipole moment is $\frac{1}{4}\pi\sigma\omega R^4$. *Hint:* Divide the disk into concentric rings. Treat each as a loop carrying an infinitesimal current, and integrate over all the loops.

69. A 10-turn wire loop measuring 8.0 cm by 16 cm carrying 2.0 A lies in a horizontal plane but is free to rotate about the axis shown in Fig. 29-46. A 50-g mass hangs from one side of the loop, and a uniform magnetic field points horizontally, as shown. What magnetic field strength is required to hold the loop in its horizontal position?

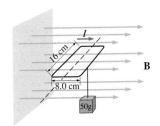

FIGURE 29-46 Problem 69.

70. A closed current loop is made from two semicircular wire arcs of radius R, joined at right angles as shown in Fig. 29-47. The loop carries a current I and is oriented with the

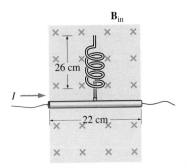

FIGURE 29-44 Problem 65.

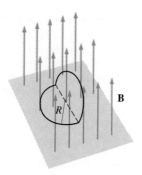

FIGURE 29-47 Problem 70.

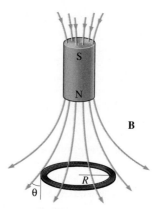

FIGURE 29-48 Problem 71.

plane of one semicircle perpendicular to a uniform magnetic field **B,** as shown. Find (a) the magnetic moment of this nonplanar loop and (b) the torque on the loop. *Hint:* You can think of the loop as a superposition of two semicircular loops, each closed along the dashed line shown (why?).

71. A circular wire loop of mass m and radius R carries a current I. The loop is hanging horizontally below a cylindrical bar magnet, suspended by the magnetic force, as shown in Fig. 29-48. If the field lines crossing the loop make an angle θ with the vertical, show that the strength of the magnet's field at the loop's position is $B = mg/2\pi RI \sin\theta$.

72. A square wire loop of mass m carries a current I. It is initially in equilibrium, with its magnetic moment vector aligned with a uniform magnetic field **B.** The loop is rotated slightly out of equilibrium about an axis through the centers of two opposite sides and then released. Show that it executes simple harmonic motion with period given by $T = 2\pi \sqrt{m/6IB}$.

73. Early models pictured the electron in a hydrogen atom as being in a circular orbit of radius 5.29×10^{-11} m about the stationary proton, held in orbit by the electric force. Find the magnetic dipole moment of such an atom. This quantity is called the *Bohr magnetron* and is typical of atomic-sized magnetic moments. *Hint:* The full electron charge passes any given point in the orbit once per orbital period. Use this fact to calculate the average current.

CHAPTER 30

SOURCES OF THE MAGNETIC FIELD

Moving electric charge produces magnetic fields. Here a large electromagnet lifts scrap metal in a recycling operation.

The preceding chapter introduced the magnetic field and its effect on matter. Here we consider the opposite question: How does matter produce magnetic fields? We've seen that magnetic fields exert forces on moving electric charges—a fact that implies a deep relation between electricity and magnetism. That relation in fact goes two ways: Not only do magnetic fields affect moving electric charges, but moving charges themselves produce magnetic fields.

30-1 THE BIOT-SAVART LAW

The first inkling of a relation between electricity and magnetism came in 1820 when the Danish scientist Hans Christian Oersted discovered that a compass needle is deflected by an electric current (Fig. 30-1). A mere month after Oersted's discovery became known in Paris, the French scientists Jean Baptiste Biot and Félix Savart had experimentally determined the form of the magnetic

FIGURE 30-1 Oersted's experiment linking electricity and magnetism is commemorated on the Oersted medal, awarded by the American Association of Physics Teachers to honor "notable contributions to the teaching of physics."

751

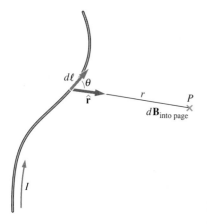

FIGURE 30-2 The Biot-Savart law gives the magnetic field $d\mathbf{B}$ at point P arising from the current I flowing along the infinitesimal vector $d\boldsymbol{\ell}$. The unit vector $\hat{\mathbf{r}}$ points from the current element I $d\boldsymbol{\ell}$ toward the field point P.

field arising from a steady current. Biot and Savart considered the contribution $d\mathbf{B}$ to the magnetic field at some point P due to a small segment $d\boldsymbol{\ell}$ in the path of a steady current I. They found, in the geometry of Fig. 30-2, that

1. The field contribution $d\mathbf{B}$ is perpendicular to both the current element $I d\boldsymbol{\ell}$ and the vector $\mathbf{r}$ from the current element to the point P.
2. The field strength dB is proportional to the current I and to the length $d\boldsymbol{\ell}$ of the current element, and inversely proportional to the square of the distance r from the current element to the field point P.
3. The field strength depends on the orientation between the vectors $d\boldsymbol{\ell}$ and $\mathbf{r}$, being proportional to the sine of the angle θ between these two vectors.

Mathematically, these results can be summarized in a compact vector equation called the **Biot-Savart law:**

$$d\mathbf{B} = \frac{\mu_0}{4\pi} \frac{I \, d\boldsymbol{\ell} \times \hat{\mathbf{r}}}{r^2}, \quad \text{(Biot-Savart law)} \qquad (30\text{-}1)$$

where $\hat{\mathbf{r}}$ is a unit vector pointing from the current element toward the field point P and where μ_0 is a constant called the permeability constant, with the exact value $4\pi \times 10^{-7}$ N/A².*

Compare the Biot-Savart law with the Coulomb expression for the electric field of an infinitesimal point charge dq:

$$d\mathbf{E} = \frac{dq}{4\pi \varepsilon_0 r^2} \hat{\mathbf{r}}.$$

Both give fields whose strengths depend on their sources—current and charge for the magnetic and electric fields, respectively. Both fields decrease as the inverse square of the distance from those sources. But here the similarity ends. Unlike electric charge, the current element $I \, d\boldsymbol{\ell}$ has direction as well as magnitude. As a result the magnetic field of a current element is not symmetric about the element but depends on the direction of the field point relative to the direction of the current element. This directional character is expressed by the cross product in Equation 30-1, which reflects Biot and Savart's discovery that the field contribution $d\mathbf{B}$ is perpendicular both to the current element $I d\boldsymbol{\ell}$ and the vector $\hat{\mathbf{r}}$ (see Fig. 30-2).

There is another distinction between the Biot-Savart law and Coulomb's law. Both describe the fields of localized structures—current elements and point charges—that are sources of the fields. It makes sense to talk about the electric field of an isolated point charge. But can we have an isolated current element? Not in a steady-state situation, where the current flowing into a current element must be the same as the current flowing out. Thus any Biot-Savart calculation necessarily involves the fields produced by many small current elements from an entire circuit. Experimentally, we find that the magnetic field obeys the superpo-

---

*That this constant has an exact value is a consequence of the definition of the ampere in terms of magnetic forces, as discussed later in this chapter.

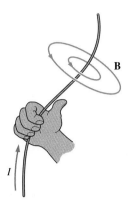

FIGURE 30-3 Magnetic field lines generally encircle a current. The field direction is given by the right-hand rule: Point your right thumb in the direction of the current, and your fingers curl in the direction of the field.

sition principle, so the net field at any point is the vector sum—or integral—of the fields of individual current elements:

$$\mathbf{B} = \int d\mathbf{B} = \int \frac{\mu_0}{4\pi} \frac{I \, d\boldsymbol{\ell} \times \hat{\mathbf{r}}}{r^2}, \qquad (30\text{-}2)$$

where the integration is taken over the entire circuit in which the current I flows. The field given by Equation 30-2 depends on the details of the current configuration, but the directionality associated with the cross product means that, quite generally, magnetic field lines encircle the current that is their source (Fig. 30-3).

● **EXAMPLE 30-1** THE FIELD OF A CURRENT LOOP

Find the magnetic field at an arbitrary point P on the axis of a circular loop of radius a carrying current I, as shown in Fig. 30-4a.

Solution

Let the loop axis be the x axis, with the origin at the loop center. Figure 30-4b shows the field contribution $d\mathbf{B}$ at P arising from a current element $I \, d\boldsymbol{\ell}$ at the top of the loop, with its direction given by the cross product $d\boldsymbol{\ell} \times \hat{\mathbf{r}}$. Figure 30-4c adds the contribution from a current element at the bottom of the loop, and shows that the net field points along the axis. To find that net field we therefore need sum only the x components of the field contributions from around the loop.

Figures 30-4b and c also show that the vectors $d\boldsymbol{\ell}$ and $\hat{\mathbf{r}}$ are perpendicular; therefore, the product $d\boldsymbol{\ell} \times \hat{\mathbf{r}}$ has magnitude $|d\boldsymbol{\ell}||\hat{\mathbf{r}}| \sin 90° = d\ell$ since $\hat{\mathbf{r}}$ is a unit vector. Then the magnitude of the field contribution $d\mathbf{B}$, as given by the Biot-Savart law, is

$$dB = \frac{\mu_0 I}{4\pi} \frac{|d\boldsymbol{\ell} \times \hat{\mathbf{r}}|}{r^2} = \frac{\mu_0 I}{4\pi} \frac{d\ell}{x^2 + a^2}.$$

To find the net field, we need to sum the x components $dB_x = dB \cos\theta$. Figure 30-4b shows that $\cos\theta = a/r = a/\sqrt{x^2 + a^2}$, so the net field becomes

$$
\begin{aligned}
B &= \int_{\text{loop}} dB \cos\theta \\
&= \int_{\text{loop}} \left(\frac{\mu_0 I}{4\pi} \frac{d\ell}{x^2 + a^2} \right) \left(\frac{a}{\sqrt{x^2 + a^2}} \right) \\
&= \frac{\mu_0 I a}{4\pi (x^2 + a^2)^{3/2}} \int_{\text{loop}} d\ell,
\end{aligned}
$$

where the integral reduces to such a simple form because the distance x is the same for all points on the loop. The remaining integral just means the sum of all the infinitesimal segments $d\ell$ around the loop—or the loop circumference $2\pi a$. Thus, the magnitude of the magnetic field on the loop axis becomes

$$B = \frac{\mu_0 I a^2}{2(x^2 + a^2)^{3/2}}. \qquad (30\text{-}3)$$

EXERCISE Earth's magnetic field at temperate latitudes is about 50 μT. You wish to produce a point where the net field is zero by orienting a 12-cm-diameter wire loop so the field at its center just cancels Earth's field. What loop current is required?

Answer: 4.8 A

Some problems similar to Example 30-1: 1–3, 5, 15, 17

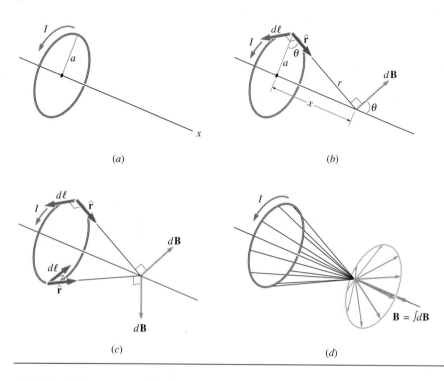

(a)

(b)

(c)

(d)

FIGURE 30-4 (*a*) A current loop whose axis is the *x* axis. (*b*) The field element *d***B** arising from the current element *I dℓ* at the top of the loop. (*c*) Field components perpendicular to the loop axis cancel in pairs. (*d*) The result is a net field along the axis.

FIGURE 30-5 The field of a current loop, as traced by iron filings. Note that field lines encircle the current-carrying wire.

We found in the preceding chapter that a current loop behaves like a magnetic dipole when it's in an external magnetic field. Does the loop also produce a dipole-like field? At points far from the loop ($x \gg a$), we can neglect a^2 compared with x^2 in the denominator of Equation 30-3, which then becomes

$$B = \frac{\mu_0 I a^2}{2x^3} = \frac{\mu_0}{2\pi}\frac{\mu}{x^3}, \qquad (x \gg a) \tag{30-4}$$

where the second equality follows by introducing the loop's magnetic dipole moment $\mu = I\pi a^2$ as defined in Equation 29-10. Equation 30-4 shows the inverse-cube behavior we found earlier for the field of an *electric* dipole. Although we derived it for a circular loop, Equation 30-4 in fact holds at large distances from a current loop of *any* shape, suggesting that the field is essentially that of a magnetic dipole. A much more difficult calculation would confirm that the distant field off the axis also shows an inverse-cube dependence and exhibits the angular dependence typical of a dipole. We conclude that, in both its response to magnetic fields and its production of magnetic fields:

| **A current loop constitutes a magnetic dipole.**

Figure 30-5 shows the field of a current loop as traced by iron filings.

■ APPLICATION   MAGNETIC FIELDS OF EARTH AND SUN

Earth, Sun, and many other astronomical objects possess magnetic fields. Reasonably close to Earth, the field approximates that of a magnetic dipole of dipole moment $\mu = 8.0 \times 10^{22}$ A·m^2 (Fig. 30-6). The direction of the dipole moment vector differs by about 11° from that of Earth's rotation axis, and this accounts for the difference between magnetic and true north. Locally, the field often deviates significantly from a pure dipole form, and these deviations provide geologists with clues to the detailed structure of the planet. The field is not constant; locally, its direction varies significantly over times as short as a few years, and the overall field reverses about every half million years or so (Fig. 30-7). At substantial distances from Earth the field is distorted from its dipole form by the solar wind, a flow of high-speed particles from the Sun.

What causes Earth's magnetic field? We know that electric currents produce magnetic fields. Deep inside Earth are a solid inner core and a liquid outer core, both rich in iron. Through an interaction not yet fully understood, the planet's rotation combined with convective motions due to internal heating produces electric currents in the liquid core. For reasons even less well understood, those currents and the resulting magnetic field undergo reversals on a time scale of approximately a million years. Problems 5 and 6 deal with Earth's magnetic field and its origin.

Earth's magnetic field is crucial to our well-being. As we saw in the preceding chapter, high-energy particles from the Sun and elsewhere are trapped in the magnetic field and have difficulty reaching Earth's surface. Thus the field protects us from harmful particulate radiation. During the field reversals, which last about 10,000 years, the magnetic field is significantly reduced and surface exposure to high-energy particulate radiation is accordingly increased. Some scientists speculate that evolutionary changes due to radiation-induced mutations may accelerate at these times.

The Sun's magnetic field probably arises in the same way as does Earth's, although the gaseous nature of the star and the intense energy flow resulting from nuclear fusion in the Sun's core make its magnetic field much more dynamic. The Sun's field reverses every 11 years, giving rise to the solar activity cycle whose best-known indicator is the count of sunspots—regions of intense magnetic field at the solar surface (Fig. 30-8).

The complex behavior of astrophysical magnetic fields is governed entirely by Newton's laws and the laws of electromagnetism. That we do not yet fully understand these magnetic fields is testimony to the rich variety of phenomena subsumed under those laws.

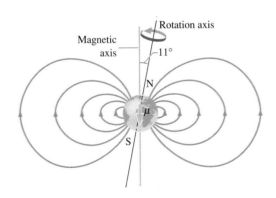

FIGURE 30-6 Earth's magnetic field approximates that of a magnetic dipole located near the center of the Earth and inclined at 11° to the rotation axis. Note from the field direction that Earth's "north" pole is really a magnetic south pole.

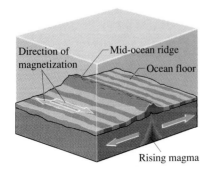

FIGURE 30-7 Magma emerges at the mid-ocean ridges, pushing older portions of the ocean floor farther apart. As the magma solidifies, it becomes magnetized in the direction of Earth's magnetic field. Analysis of the ocean floor reveals bands of alternating magnetic polarity, showing that Earth's magnetic field reverses on a time scale of 10^5 to 10^6 years. Dark bands represent regions with Earth's present magnetic polarity.

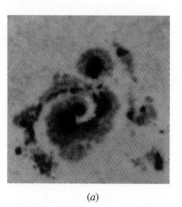
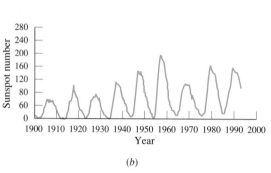

(a) (b) (c)

FIGURE 30-8 (a) A group of sunspots, each a region of intense magnetic field at the Sun's surface. (b) The solar magnetic activity cycle is clearly evident in this plot of the number of sunspots. (c) The solar corona traces the overall solar magnetic field in this photo taken during the 1991 eclipse in Hawaii.

Any steady current ultimately flows in a complete circuit, and therefore at great distances its magnetic field is dipolar. But closer in the field configuration depends on the details of the current. An important case is the field near a straight stretch of current. We can approximate this field as that of an infinite line of current, which we calculate in Example 30-2.

● EXAMPLE 30-2 THE FIELD OF A STRAIGHT WIRE

What is the magnetic field produced by an infinitely long straight wire carrying a steady current I?

Solution

Let the wire coincide with the x axis, and consider an arbitrary point P on the y axis. Figure 30-9 shows a current element $I\,d\ell$ on the wire. At point P the field $d\mathbf{B}$ of this element is perpendicular to both the wire and the vector $\hat{\mathbf{r}}$—that is, $d\mathbf{B}$ points out of the page. Clearly this is true for all current elements along the wire, so we can find the net field by integrating the field contributions dB without needing a vector integration. Using the Biot-Savart law, we have

$$dB = \frac{\mu_0 I}{4\pi}\frac{|d\boldsymbol\ell \times \hat{\mathbf{r}}|}{r^2} = \frac{\mu_0 I}{4\pi}\frac{d\ell\,\sin\theta}{r^2},$$

where we used the fact that $\hat{\mathbf{r}}$ is a unit vector making an angle θ with $d\boldsymbol\ell$. Figure 30-9 shows that $r^2 = x^2 + y^2$ and $\sin\theta = y/r = y/\sqrt{x^2+y^2}$. And, since the segment $d\ell$ lies along the x axis, $d\ell = dx$. Making these substitutions and integrating over the entire wire (from $x = -\infty$ to $x = +\infty$) gives the net field at point P:

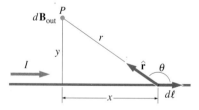

FIGURE 30-9 Geometry for calculating the field at P due to an infinite straight wire carrying a steady current I along the x axis. All current elements along the wire contribute fields at P that point out of the page.

$$B = \int dB = \int_{-\infty}^{\infty} \frac{\mu_0 I}{4\pi}\frac{y\,dx}{(x^2 + y^2)^{3/2}}$$

$$= \frac{\mu_0 I y}{4\pi}\frac{x}{y^2 \sqrt{x^2 + y^2}}\bigg|_{-\infty}^{\infty},$$

where we found the integral in the table of integrals in Appendix A. (See Problem 73 for another approach.) At the limits

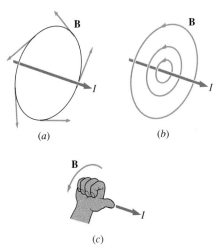

FIGURE 30-10 (a) Some magnetic field vectors associated with a straight wire carrying a steady current. (b) The corresponding magnetic field lines are circles, concentric with the wire. (c) The right-hand rule gives their direction.

$x = \pm\infty$ the expression $x/\sqrt{x^2 + y^2}$ takes on the values ± 1, so we have

$$B = \frac{\mu_0 I}{4\pi y}[1 - (-1)] = \frac{\mu_0 I}{2\pi y}. \qquad (30\text{-}5)$$

Since the wire has cylindrical symmetry this result must hold anywhere, and the result is circular field lines encircling the wire, as shown in Fig. 30-10.

The field given by Equation 30-5 falls as the inverse of the distance from the wire. This should not be surprising: we found

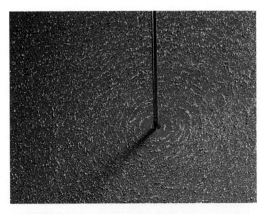

FIGURE 30-11 Iron filings trace out the circular magnetic field surrounding a current-carrying wire.

the same dependence on distance for the *electric* field of an infinitely long charged line. The field patterns differ, though, in that the electric field points radially from the charged line while the magnetic field encircles the current-carrying wire. Although Equation 30-5 applies strictly only to an infinitely long wire, it is a good approximation for a finite wire at distances small compared with the wire length (Fig. 30-11).

EXERCISE A power line carries 450 A. How close would you have to be to the line for its magnetic field to equal Earth's 50-μT field?

Answer: 1.8 m

Some problems similar to Example 30-2: 4, 11, 59, 60 ●

30-2 THE MAGNETIC FORCE BETWEEN TWO CONDUCTORS

In the preceding chapter we found the force on a straight wire of length ℓ carrying a current I through a magnetic field **B**:

$$\mathbf{F} = I\boldsymbol{\ell} \times \mathbf{B},$$

where the vector $\boldsymbol{\ell}$ is in the direction of the current flow. Through Equation 30-5 we now know the magnetic field produced by a long wire. If two long, parallel wires carry current in the same direction, as shown in Fig. 30-12, then each will experience a force arising from the other's field. We now determine that force.

If d is the distance between the wires, then at wire 2 the field magnitude B_1 due to the current I_1 is, from Equation 30-5,

$$B_1 = \frac{\mu_0 I_1}{2\pi d}.$$

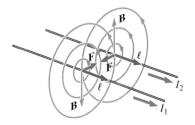

FIGURE 30-12 The magnetic force between two parallel wires carrying current in the same direction is attractive.

FIGURE 30-13 This superconducting electromagnet was torn apart by the magnetic forces from a current of 50 kA flowing in its coils. The magnetic field reached 55 T—over one million times Earth's field. Coils were made from copper/niobium composite with the strength of steel and were reinforced with Kevlar and a steel casing. The coils originally encircled the channel where a pencil has been placed to show the scale.

The field is perpendicular to wire 2, so the magnitude of the force on a length ℓ of wire 2 is

$$F_2 = I_2 \ell B_1 = \frac{\mu_0 I_1 I_2 \ell}{2\pi d}. \tag{30-6}$$

Calculating the force on a length ℓ of wire 1 would amount to interchanging the subscripts 1 and 2, giving a force of the same magnitude.

What is the direction of the force? Evaluating the cross product of ℓ and **B** in Fig. 30-12 shows that the force on wire 2 is toward wire 1, and vice versa. By using the right-hand rule, you can convince yourself that reversing one of the currents would reverse the directions of both forces. We therefore conclude the following:

> **Conductors carrying currents in the same direction experience an attractive magnetic force, while conductors carrying currents in opposite directions repel each other.**

The force between nearby conductors must be considered in the construction of electrical devices carrying large currents. In electromagnets, particularly, conductors must have enough support that magnetic forces do not destroy the device (Fig. 30-13). The hum you often hear around electrical equipment comes from the mechanical vibration of tightly wound conductors in transformers and other electrical components. This vibration results from the changing magnetic force associated with the 60-Hz alternating current.

The magnetic force between conductors is the basis for the definition of the ampere and, consequently, the coulomb. One ampere is defined as follows:

> **If two long, parallel conductors 1 meter apart carry equal currents and experience a force of 2×10^{-7} N, then the current in each is, by definition, 1 A.**

It then follows that

> **1 C is, by definition, the amount of charge passing in 1 s through a wire carrying 1 A.**

30-3 AMPÈRE'S LAW

In Chapter 23 we did several relatively cumbersome electric field calculations using Coulomb's law. For example, we found the field of an infinite line charge by integrating over all the charge elements along the line. Then in Chapter 24 we introduced Gauss's law and showed how it made electric field calculations much easier in situations with sufficient symmetry.

Can we make a statement that would enable us to calculate magnetic fields with comparable ease? Figure 30-14 shows two of the circular magnetic field lines surrounding a long wire carrying current I out of the page. Imagine moving around the inner circle. As you move a short way, form the product of the distance $d\ell$ that you travel with the component of the magnetic field in the

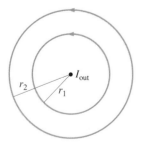

FIGURE 30-14 Two magnetic field lines surrounding a wire carrying current out of the page.

direction you're going. Here the field is entirely in the direction of your path, so that product is simply $B\,d\ell$, where B is the field magnitude; more generally, it would be the dot product $\mathbf{B} \cdot d\ell$ to account for an arbitrary angle between the field and your path. Now consider adding up all these products as you go around a complete circle. Formally, the result is the line integral $\oint \mathbf{B} \cdot d\ell$, or, in this case when the path and field coincide, just $\oint B\,d\ell$. Here the circle indicates that the integration is done around a *closed* path. The magnitude of $\mathbf{B}$ is given by Equation 30-5: $B = \dfrac{\mu_0 I}{2\pi r_1}$, where we've replaced y by the radius r_1 since Equation 30-5 applies for any orientation of the y axis perpendicular to the wire. Evaluating $\oint \mathbf{B} \cdot d\ell$ then gives

$$\oint \mathbf{B} \cdot d\ell = \frac{\mu_0 I}{2\pi r_1} \oint d\ell = \frac{\mu_0 I}{2\pi r_1}(2\pi r_1) = \mu_0 I, \qquad (30\text{-}7)$$

where we took $B = \mu_0 I/2\pi r_1$ outside the integral because r_1 is the constant radius of the circular field line. Our answer, $\mu_0 I$, is independent of the radius r_1. We would get the same answer going around the outer circle, or indeed any circular path. On a larger path the distance is greater, but the field—dropping as $1/r$—is correspondingly weaker, making the value of $\oint \mathbf{B} \cdot d\ell$ the same. Thus the line integral $\oint \mathbf{B} \cdot d\ell$ does not depend on the radius of the circular path, but only on the current I encircled by that path.

What if our path does not coincide with a field line? Figure 30-15 shows a closed loop that encircles the wire but does not coincide with a single field line. We can evaluate $\oint \mathbf{B} \cdot d\ell$ by considering each of the four segments AB, BC, CD, and DA. The radial segments BC and DA are perpendicular to the field, so here $\mathbf{B} \cdot d\ell = 0$ and these segments make no contribution to $\oint \mathbf{B} \cdot d\ell$. Now consider the arc AB that lies on our path and the arc DC that does not. We know that $\oint \mathbf{B} \cdot d\ell$ has the same value, namely $\mu_0 I$, if we go around a path coinciding with either the inner or the outer field line. Arcs AB and DC occupy the same fractions of their respective circles, and, therefore, the contribution from AB to the line integral $\oint \mathbf{B} \cdot d\ell$ is the same as we would get going instead along DC. But then since the radial segments contribute nothing, the value of $\oint \mathbf{B} \cdot d\ell$ around the irregular path $ABCDA$ is the same as its value around the outer circle—which Equation 30-7 shows to be $\mu_0 I$.

We can approximate an arbitrary loop as a sequence of concentric arcs joined by radial segments, as shown in Fig. 30-16. Applying the arguments we used with Fig. 30-15, we conclude that the line integral $\oint \mathbf{B} \cdot d\ell$ around *any* loop has the value $\mu_0 I$. Although we have considered only the field of a single straight current, this result is in fact a universal statement about the relation between current and magnetic field:

> The value of the line integral $\oint \mathbf{B} \cdot d\ell$ around any closed loop is proportional to the current encircled by that loop. Mathematically,

$$\oint \mathbf{B} \cdot d\ell = \mu_0 I_{\text{encircled}}. \qquad (30\text{-}8)$$

This statement is a form of **Ampère's law,** one of the four fundamental laws of electromagnetism. Ampère's law says that whatever arrangement of currents we

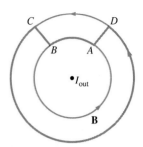

FIGURE 30-15 A closed loop that does not coincide with a single field line. The line integral of the magnetic field around this loop has the same value— $\mu_0 I$—that it has around a circular loop coinciding with a field line.

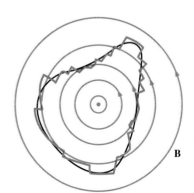

FIGURE 30-16 Approximating an irregular loop as a series of arcs and radial segments. The line integral of the magnetic field is still $\mu_0 I$, as it is for *any* closed loop.

(a)

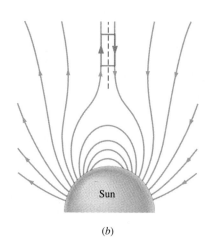

Sun

(b)

FIGURE 30-17 (a) A coronal streamer in the Sun's outer atmosphere contains oppositely directed magnetic fields in close proximity. (b) A model calculation of the coronal magnetic field. Since $\oint \mathbf{B} \cdot d\boldsymbol{\ell}$ is clearly nonzero around the loop shown, there must be current in the region encircled by the loop.

might have, and however complicated the resulting magnetic field, that field is such that the line integral $\oint \mathbf{B} \cdot d\boldsymbol{\ell}$ around any closed loop will have the value $\mu_0 I$, where I is the net current *encircled* by that loop. Compare this with Gauss's law, which says that whatever arrangement of electric charges we might have, the resulting electric field is such that the surface integral $\oint \mathbf{E} \cdot d\mathbf{A}$ has the value q/ε_0, where q is the *enclosed* charge.

Our statement of Ampère's law is true for any closed loop whatsoever, as long as the encircled current is steady—that is, never changing with time. It does not matter whether the current is in a single wire or in a number of wires or distributed throughout space; in any case, we simply add the currents to obtain the net current encircled by the loop. If there are currents flowing in opposite directions we add them with appropriate algebraic signs. A current counts as positive if, when you curl the fingers of your right hand in the direction of the loop, your right thumb points in the general direction of the current.

Although Ampère's law describes the relation between the magnetic field on a loop and the current *encircled* by that loop, we emphasize that the field **B** in Ampère's law is the *net field* from all currents, whether inside the loop or not. We found the same thing with Gauss's law, which relates the electric field **E** on a closed surface to the enclosed charge; **E** itself, however, is the *net field* arising from all sources, whether enclosed or not.

Ampère's law, like Gauss's, is a truly universal statement describing the relation between magnetic field and electric current. It holds in the electromagnetic devices we build, in atomic and molecular systems, in the interaction of fluid motion and electric charge that gives rise to Earth's magnetic field, and in distant astrophysical objects (Fig. 30-17). Although it is difficult to show mathematically, the Biot-Savart law follows logically from Ampère's law in the same sense that Coulomb's law follows from Gauss's.

● EXAMPLE 30-3   SOLAR CURRENTS

The long dimension of the rectangular loop in Fig. 30-17b is 4×10^8 m, and the magnetic field strength in the vicinity of the loop is essentially constant at 2 mT. Find the total current encircled by the rectangle. In what direction is this current flowing?

Solution

We can't insert an ammeter into the Sun's atmosphere, but we can determine the current using Ampère's law. The short ends of the rectangle in Fig. 30-17b are perpendicular to the field, so here $\mathbf{B} \cdot d\boldsymbol{\ell} = 0$ and there is no contribution to the line integral around the loop. Moving around the loop in the direction shown means going *with* the field along both long dimensions (note that this is because the field direction reverses across the dashed line). Since the field is uniform, the contribution to the line integral from each side is just $B\ell$. Thus $\oint \mathbf{B} \cdot d\boldsymbol{\ell} = 2B\ell$. Ampère's law tells us that the line integral of the magnetic field around a closed loop has the value $\mu_0 I$, so

$$2B\ell = \mu_0 I,$$

or

$$I = \frac{2B\ell}{\mu_0} = \frac{(2)(2 \times 10^{-3} \text{ T})(4 \times 10^8 \text{ m})}{4\pi \times 10^{-7} \text{ N/A}^2} = 1 \times 10^{12} \text{ A}.$$

This is a colossal current by terrestrial standards, but is typical of large-scale currents on the Sun.

Since the line integral is positive, we curl our right fingers around the loop in the direction shown by the arrows and our thumb must point in the direction of the current—that is, into the page in Fig. 30-17. In three dimensions this current flows around the Sun in approximately the equatorial plane.

TIP   Amperian Loops   The loop used with Ampère's law is truly arbitrary. It need not coincide with a field line, as Fig. 30-16 showed. Example 30-3 showed this, too; here, the rectangular loop coincided with the straight field lines over its long sides but not along its shorter ends. The loop used with Ampère's law is called an **amperian loop.** Don't confuse amperian loops with field lines!

EXERCISE   Fig. 30-18 shows a cross-sectional view of three wires carrying current perpendicular to the plane of the page. Wire A carries 20 A out of the page. $\oint \mathbf{B} \cdot d\boldsymbol{\ell}$ around loop 1 in the direction shown has the value 1.26×10^{-5} T·m, and the line integral around loop 2 has the value -6.28×10^{-6} T·m. Find the currents in wires B and C, and the directions of those currents.

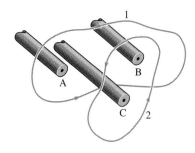

FIGURE 30-18   What are the currents in wires B and C?

Answers: B: 10 A into page; C: 5 A out of page
Some problems similar to Example 30-3:   28–30 ●

30-4 USING AMPÈRE'S LAW

For charge distributions with sufficient symmetry we used Gauss's law to solve for the electric field in a simple and elegant way. Similarly, for current distributions with sufficient symmetry we can use Ampère's law to solve for the magnetic field. Doing so requires finding suitable amperian loops over which we can evaluate the line integral of the magnetic field.

The Field of a Straight Wire

Here we use Ampère's law to determine the field of an infinite straight wire. This is the same calculation we did in Example 30-2 using the Biot-Savart law.

Figure 30-19 shows an end view of the wire, which carries current I out of the page. We know that magnetic field lines generally encircle the currents that are their sources. Here cylindrical symmetry requires that those field lines be circles

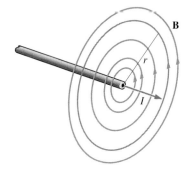

FIGURE 30-19   Circular field lines surrounding an infinitely long wire. Gray circle is an amperian loop for evaluating the line integral of the magnetic field. Although we've drawn only four field lines, the field exists everywhere, including on the amperian loop.

centered on the wire and that the magnetic field strength cannot depend on angular position around the wire. Thus B must be constant on a field line. Applying the right-hand rule—thumb along the current, curling the fingers of the right hand—shows that the magnetic field circles counterclockwise around the wire. And since the wire is infinitely long, Fig. 30-19 represents the field anywhere along the wire.

Given the symmetry, an appropriate amperian loop for evaluating the line integral is itself a circle, also shown in Fig. 30-19. The magnetic field everywhere on this amperian loop is in the same direction as the loop, so the dot product $\mathbf{B} \cdot d\boldsymbol{\ell}$ becomes simply $B \, d\ell$. Then the line integral around the loop is

$$\oint_{\text{loop}} \mathbf{B} \cdot d\boldsymbol{\ell} = \oint_{\text{loop}} B \, d\ell = B \oint_{\text{loop}} d\ell = 2\pi r B.$$

Here we can take the field magnitude B outside the integral because it's constant on the circular amperian loop; the remaining integral is then the loop circumference $2\pi r$. Ampère's law equates this line integral to μ_0 times the encircled current, so

$$2\pi r B = \mu_0 I,$$

or

$$B = \frac{\mu_0 I}{2\pi r}. \tag{30-9}$$

This result is the same as Equation 30-5, but we derived it with much less difficulty.

> **TIP**   **Symmetry Is Crucial**   Our use of Ampère's law to derive the field of a long wire depends crucially on symmetry. We cannot arbitrarily pull B outside the integral unless we know—as we do here from symmetry—that it is constant in magnitude and in direction relative to our amperian loop.

Our calculation made no assumptions about the diameter of the wire. Therefore Equation 30-9 holds for any long cylindrical wire, thick or thin, as long as we restrict ourselves to points *outside* the wire so that our amperian loop encircles the *entire* current. In fact, you can easily convince yourself that Equation 30-9 must hold *outside* any current distribution with cylindrical symmetry. To calculate the field *inside* a current distribution, however, we must be careful to use only the actual current encircled by our amperian loop. Example 30-4 illustrates this point.

● **EXAMPLE 30-4**   INSIDE A WIRE

A long, straight wire of radius R carries a current I uniformly distributed over its cross-sectional area. What is the magnetic field as a function of position within the wire?

Solution

All the symmetry arguments we used to find the field outside a wire still apply here, so the line integral around a circular

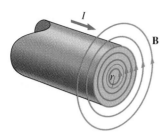

FIGURE 30-20   Cross section of a long cylindrical wire carrying current out of the page. Symmetry requires that the magnetic field be circular inside as well as outside the wire. Current is distributed uniformly over the wire; in applying Ampère's law to the amperian loop shown in gray we need the fraction of the total current encircled by the loop.

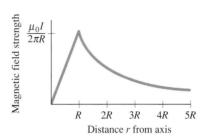

FIGURE 30-21   The magnetic field inside a wire increases linearly with radial distance, while beyond the wire radius R it drops as $1/r$.

amperian loop of radius r is still $2\pi rB$. What's different is that an amperian loop within the wire no longer encircles the entire current (Fig. 30-20). How much current is encircled? The current is distributed uniformly, giving a current density (current per unit area) $J = I/A = I/\pi R^2$. The encircled current is then the current density times the loop area, or

$$I_{\text{encircled}} = \left(\frac{I}{\pi R^2}\right)(\pi r^2) = I\frac{r^2}{R^2}.$$

Equating the line integral $2\pi rB$ to this encircled current gives

$$2\pi rB = I\frac{r^2}{R^2},$$

or
$$B = \frac{\mu_0 Ir}{2\pi R^2}. \tag{30-10}$$

Does this result make sense? The field increases linearly with distance from the axis—just as we found for the *electric* field *inside* a uniformly charged cylinder. Here the increase occurs

because we encircle more and more current—with $I_{\text{encircled}}$ growing as r^2—as long as we're inside the wire. Once we reach the surface, of course, the encircled current remains constant and the field begins to decrease inversely with distance, as described in Equation 30-9. Figure 30-21 plots the field strengths both inside and outside the wire.

This example is not merely academic; in superconducting wires, for example, strong magnetic fields can destroy superconductivity, so it's important to know the field throughout the wire.

EXERCISE   A power line 4.0 cm in diameter carries 1.5 kA. Find the magnetic field (a) 1.0 cm from the axis of the wire and (b) 10 cm from the axis.

Answers: (a) 7.5 mT; (b) 3.0 mT

Some problems similar to Example 30-4:   32, 33, 37, 38

● **EXAMPLE 30-5**   A CURRENT SHEET

An infinite flat sheet carries a current out of the page, as shown in Fig. 30-22. The current is distributed uniformly along the sheet, with the current per unit width along the sheet given by J_s. Find the magnetic field of this sheet.

Solution
What might the field of this sheet look like? Figure 30-23 suggests that we can consider the sheet to be made of many parallel wires. The vector sum of the fields of these wires gives a net field to the left above and to the right below the sheet. We can also argue from symmetry that the fields must point horizontally and with magnitude that is independent of position parallel to the sheet; since the sheet extends infinitely to the right and left, there's nothing to favor an upward or downward deflection of the field lines or any variation in field magnitude parallel to the sheet. We've drawn these horizontal field lines, and an appropriate amperian loop, in Fig. 30-22.

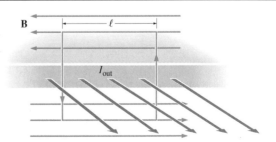

FIGURE 30-22   A current sheet (gray) extends infinitely to the left and right, as well as in and out of the page. Field lines and a rectangular amperian loop are also shown.

We already evaluated $\oint \mathbf{B} \cdot d\boldsymbol{\ell}$ for a similar geometry in Example 30-3; the result is simply $2B\ell$, with ℓ the width of the amperian rectangle. The current per unit width of the sheet is J_s, so our rectangular loop of width ℓ encircles a current

$I = J_s\ell$. Equating μ_0 times this encircled current to the line integral $\oint \mathbf{B} \cdot d\ell = 2B\ell$ gives

$$2B\ell = \mu_0 J_s \ell,$$

or
$$B = \tfrac{1}{2}\mu_0 J_s. \qquad (30\text{-}11)$$

Thus the magnetic field of an infinite current sheet is independent of distance from the sheet. We found the same result for the electric field of an infinite charged plane. Although the infinite sheet is an idealization, Equation 30-11 is a good approximation to the field near long, wide, flat conductors. Thin current sheets also form in conducting plasmas like those in fusion reactors and in the Sun's atmosphere, where they can lead to sudden dissipation of energy.

EXERCISE   Current in a printed circuit board is carried in a long copper strip 2.1 mm wide and much thinner than its width. If the current in the strip is 35 mA, spread uniformly over its cross section, find the magnetic field strength near the strip's surface.

Answer: 10 μT

Some problems similar to Example 30-5: 39–41, 44

FIGURE 30-23   A current sheet approximated by closely spaced parallel wires. The magnetic field of the sheet is the vector superposition of the fields of the individual wires. In the limit of infinitesimally spaced wires, the field lines become straight and strictly horizontal.

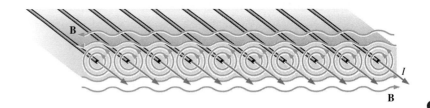

Fields of Simple Current Distributions

We have used Ampère's law to calculate the magnetic fields of two simple, symmetric current distributions. Although those magnetic fields may look quite different from the electric fields of correspondingly symmetric charge distributions, they exhibit the same dependences on distance. Table 30-1 summarizes the electric and magnetic fields of several simple charge and current distributions. Real distributions are usually more complicated, but often they can be approximated by these simple cases. Far from any current loop, for example, the magnetic field is that of a dipole. Very near *any* wire, its magnetic field is essentially that of a long straight wire. For situations that lack symmetry and in which approximations are not adequately accurate, we can always calculate the magnetic field of a steady current distribution from the Biot-Savart law, just as we can always calculate the electric field of an arbitrary charge distribution from Coulomb's law.

30-5 SOLENOIDS AND TOROIDS

We found in Chapter 26 that we can produce a uniform electric field between the two closely spaced, charged conducting plates of a capacitor. Is there an analogous device that will produce a uniform magnetic field?

Fig. 30-24 shows a coil of wire carrying current I. Close to the wire are magnetic field lines encircling the wire. We show these field lines in Fig. 30-24 at the top and bottom of the coil, where they cross the plane of the page. The net field anywhere is, of course, just the vector sum of the field contributions from all parts of the coil. You can see that inside the coil, the fields from current elements at the top and bottom all have a component to the right and so tend to

▲ **TABLE 30-1**   FIELDS OF SOME SIMPLE CHARGE AND CURRENT
DISTRIBUTIONS

DISTRIBUTION	FIELD DEPENDENCE ON DISTANCE*	ELECTRIC OR MAGNETIC FIELD
Electric dipole	$\dfrac{1}{r^3}$	
Magnetic dipole	$\dfrac{1}{r^3}$	
Spherically symmetric charge distribution	$\dfrac{1}{r^2}$	
Spherically symmetric current distribution	Impossible for steady current	
Charge distribution with line symmetry	$\dfrac{1}{r}$	
Current distribution with line symmetry	$\dfrac{1}{r}$	
Infinite flat sheet of change	Uniform field; no variation	
Current sheet	Uniform field; no variation	

* For field *outside* distribution.

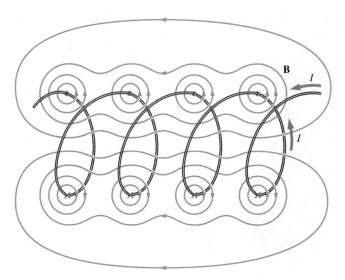

FIGURE 30-24   The magnetic field arising from the current in a loosely wound coil of wire is strongest within the coil and weaker outside. Field is shown only in the plane of the page; dots show current emerging from the page and crosses mark current going into the page.

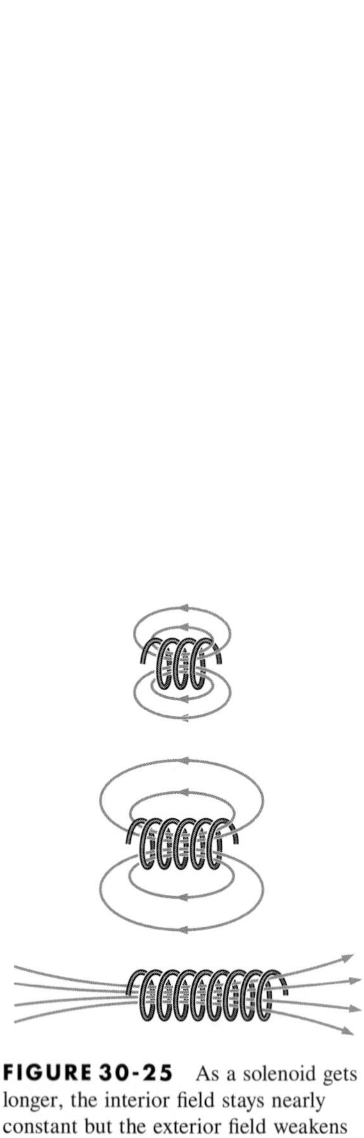

FIGURE 30-25   As a solenoid gets longer, the interior field stays nearly constant but the exterior field weakens as the field lines spread ever farther apart.

reinforce. Outside the coil, though, field components from top and bottom are in opposite directions and thus reduce the net field, which points to the left as shown in the figure.*

Imagine winding to coil more tightly and making it longer. Such a long, tightly wound coil is a **solenoid.** Figure 30-25 shows what happens as the solenoid gets longer: the interior field stays essentially the same, but exterior field lines spread into more distant regions before closing back on themselves. Therefore, the field strength outside the solenoid—as evidenced by the density of field lines—decreases. In the limit of an infinitely long solenoid, the interior field becomes perfectly straight and the exterior field in the plane of Figs. 30-24 and 30-25 goes to zero. A real solenoid approaches this ideal limit when its length is much greater than its diameter. Figure 30-26 shows the field of a solenoid as traced by iron filings.

In the long-solenoid limit, we can use Ampère's law to find the field in the solenoid. Figure 30-27 shows a cross section through a long solenoid, with a rectangular amperian loop. Since the exterior field is zero, there is no contribution to the line integral $\oint \mathbf{B} \cdot d\boldsymbol{\ell}$ from the top of the amperian loop. The loop's vertical sides are at right angles to the field, so they, too, contribute nothing to the line integral. The only contribution is from the bottom of the loop. Going counterclockwise around the loop, we move with the field on this segment. Since the field cannot vary with position along the infinite solenoid, the contribution to the line integral is just $B\ell$; since the other three sides contribute nothing, this is the value of $\oint \mathbf{B} \cdot d\boldsymbol{\ell}$.

To apply Ampère's law we need the encircled current. If there are n turns per unit length of solenoid, the amperian loop encircles $n\ell$ turns. The same current

---

* Because the coil carries a net current from right to left, there is also a weak field component outside that encircles the coil. This component plays no role in the discussion that follows, and we ignore it.

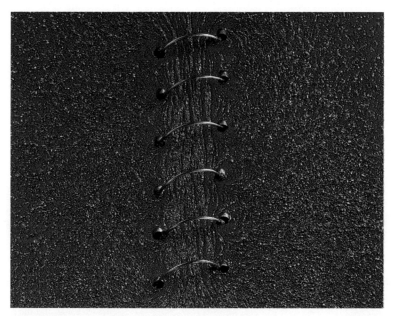

FIGURE 30-26 Iron filings trace the magnetic field of a solenoid. Note that the field is strong inside the solenoid and relatively weak outside, except near the ends.

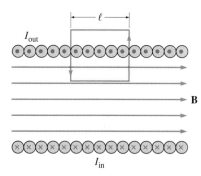

FIGURE 30-27 Cross section of a long solenoid, showing a rectangular amperian loop straddling the region where the solenoid coils emerge from the plane of the page.

I flows in each turn (why?), so the encircled current is $n\ell I$. Applying Ampère's law then gives

$$B\ell = \mu_0 n\ell I,$$

or
$$B = \mu_0 nI. \qquad \text{(solenoid)} \qquad (30\text{-}12)$$

Since the vertical dimension of the amperian rectangle never entered our calculation, this field magnitude is the same anywhere inside the solenoid. The magnetic field in the solenoid is therefore uniform. Although it looks very different, the long solenoid with its uniform magnetic field is the magnetic analog of a closely spaced parallel-plate capacitor, which produces a uniform electric field.

Although Figs. 30-24 and 30-25 depict circular coils, the derivation of Equation 30-12 is based only on Fig. 30-27, which could represent a solenoid whose coils are circular, square, or any other shape. Equation 30-12 therefore holds for coils of any shape, as long as that shape is the same over the length of the solenoid.

■ APPLICATION EVERYDAY SOLENOIDS, MEDICAL SOLENOIDS

Solenoids of all sizes are used in a wide variety of experimental and practical devices. Because a solenoid is hollow, magnetic materials like iron will be pulled into the solenoid by the nonuniform magnetic field near its ends, and solenoids are therefore used to produce straight-line motion in mechanical devices. For example, turning the key to start your car sends current through a solenoid in the car's starter motor. A magnetic field develops in the solenoid, pulling in a steel plunger. Movement of the plunger joins a set of electrical contacts allowing current to flow to the starter motor (it would be difficult to switch the roughly 100-A starter current directly from inside the car). At the same time, the rod moves a small gear to the end of the motor shaft, engaging a gear on the engine's flywheel so that the starter motor can turn the engine (Fig. 30-28).

Running a dishwasher or washing machine also involves solenoids. The valves that control the flow of water in these machines are solenoid valves, opened and closed as a steel rod moves in or out of a solenoid (Fig. 30-29).

Finally, huge solenoids with human-size interiors are used to produce the strong, uniform magnetic fields needed for magnetic resonance imaging (MRI). This high-tech medical technique images the body's interior without the need for x rays or other radiation (Fig. 30-30). The MRI solenoids use superconducting wire cooled with liquid helium to achieve high currents and correspondingly high magnetic fields.

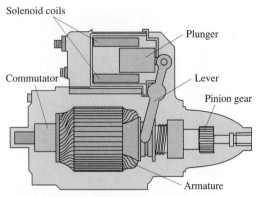

FIGURE 30-28 Cross section of a car starter motor, showing solenoid coil and mechanical linkage that engages the starter's pinion gear to turn the engine.

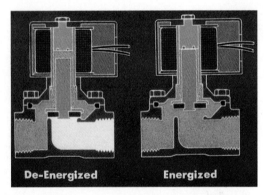

FIGURE 30-29 Cutaway diagram of a solenoid valve, widely used in washing machines, dishwashers, and many industrial applications. Solenoid coils are the red structures at top. A steel plunger extending part of the way into the coils connects to a diaphragm. At left, there is no current in the coils, and the diaphragm blocks the flow of liquid. At right, the coils are energized, producing a magnetic field. As a result, the plunger and diaphragm are pulled upward, allowing liquid to flow through the valve.

FIGURE 30-30 Patient being inserted into a superconducting solenoid used for magnetic resonance imaging (MRI), a powerful and relatively new medical diagnostic technique.

● EXAMPLE 30-6 AN MRI SOLENOID

A solenoid used in magnetic resonance imaging is 2.4 m long and 95 cm in diameter. It is wound from a niobium-titanium superconducting wire 2.0 mm in diameter, with adjacent turns separated by an insulating layer of negligible thickness. What current is necessary to produce a 1.5-T magnetic field inside the solenoid?

Solution

The magnetic field is given by Equation 30-12, $B = \mu_0 nI$, so if we know n—the number of turns per unit length—we can solve for I. Since the wire has diameter $d = 2.0$ mm, 5 turns will occupy 1 cm and therefore $n = 500$ turns/meter. Solving Equation 30-12 for I then gives

$$I = \frac{B}{\mu_0 n} = \frac{(1.5 \text{ T})}{(4\pi \times 10^{-7})(500 \text{ m}^{-1})} = 2.4 \text{ kA}.$$

This is a large current, but readily handled by the niobium-titanium superconductor.

EXERCISE Copper wire 0.40 mm in diameter is tightly wound into a long solenoid, and a 3.6-A current passed through it. What is the magnetic field in the solenoid?

Answer: 11 mT

Some problems similar to Example 30-6: 45–47, 50

Toroids

A **toroid** is a solenoid bent into a doughnut shape (Fig. 30-31). Toroidal geometry is widely used in fusion reactors, as discussed in the preceding chapter, and toroidal coils help produce the magnetic field that confines the fusion plasma.

Symmetry requires that the toroid's field lines be circular, with constant magnitude on any line. We can readily calculate the line integral of this magnetic field around a circular amperian loop, like that shown in Fig. 30-32. This loop coincides with a field line, and symmetry ensures that the magnetic field has the same magnitude over the loop. Therefore the line integral becomes

FIGURE 30-31 A toroidal coil.

$$\oint_{\text{loop}} \mathbf{B} \cdot d\boldsymbol{\ell} = \oint_{\text{loop}} B\, d\ell = B \oint_{\text{loop}} d\ell = 2\pi r B.$$

To apply Ampère's law we equate this quantity to μ_0 times the encircled current. If the toroid consists of N turns and carries a current I, then an amperian loop inside the toroid encircles a total current NI. Then Ampère's law becomes

$$2\pi r B = \mu_0 NI,$$

or
$$B = \frac{\mu_0 NI}{2\pi r}. \quad \text{(toroid)} \qquad (30\text{-}13)$$

This result holds when the amperian loop is within the toroid itself. But if it's inside the inner edge there's no encircled current and, therefore, no magnetic field. And beyond the outer edge, the amperian loop encircles equal but opposite currents, giving zero net current and again no magnetic field.* As with a solenoid, the individual turns of a toroid need not be circular.

The toroidal field of Equation 30-13 is not uniform but exhibits a $1/r$ decrease. This nonuniformity causes problems with plasma confinement in fusion reactors, where it causes a drift of particles perpendicular to the field and therefore toward the walls of the machine.

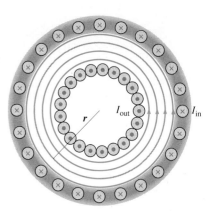

FIGURE 30-32 Cross section of a toroid showing current emerging at inner edge of coils and descending into plane of page at outer edge. Also shown are circular field lines and an amperian loop of radius r (gray) for use in calculating the field.

30-6 MAGNETIC MATTER

We began our study of magnetism in Chapter 29 with a discussion of magnets. But we quickly moved on to concentrate on the behavior of electric charges in magnetic fields. In this chapter we've further developed the relation between magnetism and electric charge, as we've seen that moving electric charge is the source of magnetic fields. What does all this business with electric charge have to do with the familiar magnets we used to introduce magnetism?

In fact the two are one and the same phenomenon. When an electron orbits an atomic nucleus, its circular motion and charge make a miniature current loop, and it therefore constitutes a magnetic dipole (Fig. 30-33). In addition the electron possesses an intrinsic magnetic dipole moment associated with an intrinsic angular momentum called "spin." Interactions among these magnetic dipole moments determine the magnetic properties of individual atoms and of

FIGURE 30-33 In the classical model of the atom, the circling electron constitutes a miniature current loop. The atom is therefore a magnetic dipole. Here the current—carried by the *negative* electron and therefore opposite the electron's motion—is counterclockwise when viewed from above, so the magnetic moment points upward.

*This neglects a weak field associated with the fact that the current has a component *around* the toroid in the plane of Fig. 30-32.

FIGURE 30-34 Magnetic domains in a thin film of ferromagnetic material. The dark and colored areas represent regions where the magnetic moments point either into or out of the plane of the figure; the different colors emphasize that regions of oppositely directed magnetic moments tend to organize into parallel structures.

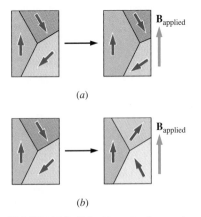

(a)

(b)

FIGURE 30-35 Domain changes in a ferromagnetic material occur through (a) domain growth and (b) domain realignment. These changes occur when a magnetic field is applied, and result in the material acquiring a net magnetic moment.

bulk matter. Although an accurate description of magnetism in matter necessarily involves quantum mechanics, we can nevertheless use our knowledge of magnetic dipoles to gain a qualitative understanding of magnetic matter.

Ferromagnetism

Experiment reveals three types of magnetic behavior in bulk matter. **Ferromagnetism,** the most familiar, is actually limited to a few substances, notably the elements iron, nickel, and cobalt, and some compounds. In a ferromagnetic material, a quantum-mechanical interaction among nearby atomic magnetic moments results in regions—called **magnetic domains**—in which all the atomic magnetic moments point in the same direction. A typical domain contains 10^{17} to 10^{21} atoms and occupies a volume on the order of 10^{-12} to 10^{-8} m^3 (Fig. 30-34). The magnetic moment of a single domain can be large since it is the sum of many atomic magnetic moment vectors all pointing in the same direction. A typical piece of ferromagnetic material, however, contains many domains with their moments in random directions and, therefore, exhibits no net magnetic moment. But when an external magnetic field is applied, a net magnetic moment develops. This occurs because domains that already happen to be aligned with the field can grow by realignment of individual atomic moments in adjacent domains (Fig. 30-35a). In addition, the magnetic moments of entire domains can rotate (Fig. 30-35b).

Since any dipole experiences a force in a nonuniform field, a piece of ferromagnetic material in such a field experiences a net force; this is why iron and other ferromagnetic materials are attracted to magnets even though they themselves are not magnets.

Removal of the applied magnetic field does not entirely destroy the overall alignment that gives the net magnetic moment in a ferromagnetic substance. The remanent magnetization is what makes permanent magnets. In so-called hard ferromagnetic materials, the remanent magnetism is strong; these materials are used specifically for making permanent magnets. Remanent magnetism is relatively weak in soft ferromagnetic materials; these are used in applications like heads for VCRs and computer disk drives, where permanent magnetization is undesirable (tapes and disks themselves use harder materials for long-term information storage).

Random thermal motions tend to disrupt the alignment of individual magnetic moments. Thus ferromagnetic effects weaken with increasing temperature. Above the so-called **Curie temperature,** ferromagnetism ceases altogether. Curie temperatures for the common ferromagnetic elements nickel, iron, and cobalt are 631, 1043, and 1395 K, respectively. The rarer ferromagnetic elements dysprosium and gadolinium have much lower Curie temperatures of 85 and 289 K, respectively. The disappearance of ferromagnetism at the Curie temperature is an example of a phase transition, analogous to the solid/liquid/gas transitions we studied in Chapter 20.

Paramagnetism

Many substances that are not ferromagnetic nevertheless consist of atoms or molecules with permanent magnetic dipole moments. What distinguishes these **paramagnetic** materials from ferromagnetic substances is the absence of a strong interaction that tends to align nearby moments. As a result, individual atomic moments are not organized into domains, but point in random directions. An applied magnetic field still brings the atomic magnetic moments into some degree of alignment, but at all but the coldest temperatures this alignment is far less complete than in a ferromagnetic material. Therefore paramagnetic materials are attracted only weakly to magnets (Fig. 30-36).

Diamagnetism

Even materials with no intrinsic magnetic moments can have magnetic moments induced when a magnet approaches or when an applied magnetic field otherwise changes. Such materials are termed **diamagnetic.** In contrast to paramagnetic and ferromagnetic materials, diamagnetic materials are repelled by magnets. We will explore the origins of diamagnetism in the next chapter.

Magnetic Susceptibility

We can characterize the effect of atomic magnetic dipoles on bulk matter just as we did in Chapter 23 for electric dipoles in dielectrics. There we found that atomic electric dipoles align with an electric field to reduce the field within the material, and we introduced the dielectric constant κ as the factor quantifying that reduction.

In a magnetic material, alignment of atomic magnetic dipoles has the opposite effect: the field in the material is increased. Figure 30-37 shows why. An electric dipole consists of two separated point charges. The strongest field associated with this dipole is the internal field pointing straight from the positive to the negative charge. When the dipole aligns with an applied electric field, this internal field is *opposite* the applied field and, therefore, *reduces* the net electric field (Fig. 30-37a). But, as we've seen, magnetic field lines always *encircle* a current. As Fig. 30-37b shows, the field inside the current loop therefore points in the *same* direction as the applied field and, therefore, *increases* the net magnetic field.

FIGURE 30-36 Liquid oxygen (O_2) is one of the more strongly paramagnetic materials. Here liquid oxygen is suspended between the poles of a magnet.

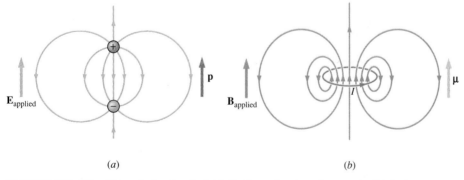

(a)   (b)

FIGURE 30-37 Although the electric fields far from electric and magnetic dipoles have the same form, the fields *within* the atomic structure giving rise to the different dipole moments have opposite directions. This is because electric dipoles consist of separated point charges, while magnetic dipoles are current loops. When external electric and magnetic fields are applied, respectively, to these two structures, the result is a reduction in the net electric field but an increase in the net magnetic field.

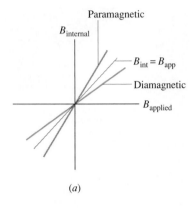

(a)

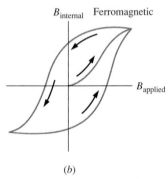

(b)

FIGURE 30-38 Internal versus applied magnetic field for different types of magnetic materials. (a) In diamagnetic and paramagnetic materials the relationship is linear; paramagnetic materials strengthen the applied field slightly, while diamagnetic materials weaken it slightly. (b) In ferromagnetic materials, the relationship depends on the strength of the applied field and on the past history of the material. In (b) field strengths along the vertical axis are much greater than along the horizontal axis, reflecting the large susceptibilities of ferromagnetic materials. Arrows indicate the direction in which fields are changed.

In analogy with the dielectric constant, we introduce the quantity κ_M, called the **relative permeability,** as the factor by which the magnetic field within a material increases as a result of the alignment of atomic magnetic dipoles. For paramagnetic materials, κ_M is slightly greater than 1; for ferromagnetic materials it is much greater than 1. In diamagnetic materials the dipoles align antiparallel to the field; this makes $\kappa_M < 1$ for these materials. Because the relative permeabilities of paramagnetic and diamagnetic materials are very close to 1, it's more convenient to work with the **magnetic susceptibility,** defined by

$$\chi_M = \kappa_M - 1.$$

The internal field in the material is then given by

$$B_{int} = \kappa_M B_{applied} = (\chi_M + 1)B_{applied}. \qquad (30\text{-}14)$$

Equation 30-14 is most useful for paramagnetic and diamagnetic materials. In a ferromagnetic material, the relative permeability and susceptibility themselves depend on the applied field, and in fact on the past history of the material's exposure to magnetic fields (Fig. 30-38). This phenomenon of **hysteresis,** in which a ferromagnetic material "remembers" past fields, is what makes permanent magnets possible. Table 30-2 lists some magnetic susceptibilities for all three types of magnetic materials. Note the entry for superconductors, which are perfectly diamagnetic. Their susceptibility of -1 implies that $\kappa_M = 0$, showing that they completely exclude magnetic fields. We will explore the reasons for this in the next chapter.

▲ **T A B L E 3 0 - 2** MAGNETIC SUSCEPTIBILITIES*

MATERIAL	MAGNETIC SUSCEPTIBILITY, χ_M
Diamagnetic materials:	
Copper	-9.6×10^{-6}
Lead	-1.6×10^{-5}
Mercury	-2.8×10^{-5}
Nitrogen (gas, 293 K)	-6.7×10^{-9}
Sodium chloride	-1.4×10^{-5}
Any superconductor	-1
Water	-9.1×10^{-6}
Paramagnetic materials	
Aluminum	2.1×10^{-5}
Chromium	3.1×10^{-4}
Oxygen (gas, 293 K)	1.9×10^{-6}
Oxygen (liquid, 90 K)	3.5×10^{-3}
Sodium	8.5×10^{-6}
Ferromagnetic materials (field and history dependent; maximum value listed)	
Iron (annealed)	5.5×10^{3}
Permalloy (55% Fe, 45% Ni)	2.5×10^{4}
Supermalloy (15.7% Fe, 79% Ni, 5.0% Mo; 0.30% Mn)	8.0×10^{5}
μ-metal (77% Ni, 16% Fe, 5% Cu, 2% Cr)	1.0×10^{5}

*At 300 K unless noted.

30-7 MAGNETIC MONOPOLES AND GAUSS'S LAW

Electric and magnetic fields, at least as we've studied them so far, have very different configurations. Electric fields begin and end on their sources—namely electric charges. Magnetic fields, in contrast, encircle their sources—namely *moving* electric charges. Magnetic fields generally form closed loops, while the electric fields we've encountered so far do not (Fig. 30-39).

Are there particles analogous to electric charges, from which magnetic field lines might originate? There might be. Symmetry arguments based on the existence of electric charge and the many similarities between electricity and magnetism have long suggested to physicists that such **magnetic monopoles**—isolated magnetic north and south poles—might exist. Furthermore, theories of elementary particles suggest that magnetic monopoles should have been created in the Big Bang event with which the universe began. However, the most recent version of the Big Bang theory suggests that monopoles should be spread so thinly as to make the chance of detecting one completely negligible.

If they existed, magnetic monopoles would be to the magnetic field what electric charges are to the electric field. Isolated monopoles would give rise to radial magnetic fields like the electric fields of point charges. Monopoles would experience forces in magnetic fields, forces given simply by the product of the monopole's "magnetic charge" and the magnetic field. And *moving* magnetic monopoles would produce electric fields—just as moving electric charges produce magnetic fields.

No one has yet found a magnetic monopole, however, although serious experimental searches continue. Every magnetic field we've ever seen has its origin in moving electric charge. The observed absence of magnetic monopoles means that the most fundamental magnetic field configuration is that of the dipole—in contrast to the simpler electric case, where the spherically symmetric point-charge field is possible. That's why Table 30-1 contains no magnetic entry under spherical geometry.

In the absence of magnetic monopoles, there is no place where magnetic field lines begin or end. That means there is no closed surface through which a nonzero net number of magnetic field lines emerges. In complete analogy with our discussion of Gauss's law in Chapter 24, we can say that the **magnetic flux**—mathematically, the surface integral $\oint \mathbf{B} \cdot d\mathbf{A}$—is zero for every closed surface. This statement is **Gauss's Law for magnetism:**

$$\oint \mathbf{B} \cdot d\mathbf{A} = 0. \qquad (30\text{-}15)$$

Equation 30-15 is another of the four fundamental laws of electromagnetism. We've now seen three of them: Gauss's law for electricity, Ampère's law (still incomplete because of its restriction to *steady* currents), and Gauss's law for magnetism. In the next chapter we'll encounter the fourth of these fundamental laws.

Although Gauss's law for magnetism has zero on its right-hand side, the law is far from being devoid of content. It says what we've already found to be the case: that magnetic fields must configure themselves so they have no beginnings or endings.

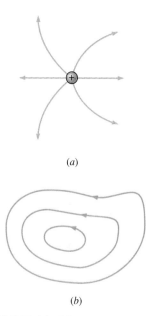

(a)

(b)

FIGURE 30-39   At this point in our study of electromagnetism, electric and magnetic field configurations are clearly distinguished. Electric fields (a) begin and end on electric charges, while magnetic fields (b) generally form closed loops.

CHAPTER SYNOPSIS

Summary

1. Magnetic fields arise from **moving electric charges.** The field at some point P arising from a steady electric current is described by the **Biot-Savart law:**

$$d\mathbf{B} = \frac{\mu_0}{4\pi} \frac{I d\boldsymbol{\ell} \times \hat{\mathbf{r}}}{r^2},$$

where $d\mathbf{B}$ is the contribution to the field from a current element flowing along the infinitesimal vector $d\boldsymbol{\ell}$ a distance r from the point P, and $\hat{\mathbf{r}}$ is a unit vector from the current element toward P. The field at P of an entire current distribution is found by integrating all its contributions $d\mathbf{B}$.

   Important special cases include a current loop, which at large distances produces a dipole field, and an infinite straight current, whose field encircles the current and decreases as $1/r$.

2. The field arising from a current exerts a magnetic force on nearby currents; as a result, parallel currents in the same direction attract and those in opposite directions repel.

3. **Ampère's law** relates the line integral of the magnetic field around an arbitrary closed loop to the current encircled by that loop:

$$\oint_{\text{loop}} \mathbf{B} \cdot d\boldsymbol{\ell} = \mu_0 I_{\text{encircled}}.$$

   This form holds for all steady currents. It may be used to calculate the magnetic field in cases with sufficient symmetry, including line symmetry (straight wires), plane symmetry (current sheets), solenoids, and toroids. The Biot-Savart law follows logically from Ampère's law.

4. Individual elementary particles and orbiting atomic electrons constitute miniature current loops, which are responsible for magnetic effects in bulk matter. The **relative permeability,** κ_M, gives the ratio of the internal magnetic field in a material to the applied field; this relation is also described by the **magnetic susceptibility,** $\chi_M = \kappa_M - 1$.

   a. In **ferromagnetic** materials, atomic magnetic moments group into domains with net magnetic moments. These domains align with an applied field, greatly increasing the field within the material. Ferromagnetic materials retain a remanent magnetization even after the applied field is removed; this phenomenon accounts for permanent magnets. $\kappa_M \gg 1$ for ferromagnetic materials, although the internal field depends on both the applied field and the material's history.

   b. In **paramagnetic** materials, individual atomic magnetic moments become partially aligned with an applied field, but there is no cooperative interaction among the individual moments. $\kappa_M > 1$ for paramagnetic materials.

   c. Diamagnetic materials have no intrinsic magnetic moments. They have $\kappa_M < 1$ and are weakly repelled from magnets.

5. No **magnetic monopoles,** or magnetic analogs of electric charge, have ever been found. Magnetic fields—originating from moving electric charge—therefore form closed loops without beginnings or ends. In the absence of magnetic monopoles, **Gauss's law for magnetism** says that the flux of the magnetic field through any closed surface is zero:

$$\oint \mathbf{B} \cdot d\mathbf{A} = 0.$$

Terms You Should Understand

(Pairs are closely related terms whose distinction is important; number in parentheses is chapter section where term first appears.)

Biot-Savart law (30-1)
Ampère's law (30-3)
solenoid, toroid (30-5)
ferromagnetism, paramagnetism, diamagnetism (30-6)
permeability, susceptibility (30-6)
magnetic monopole (30-7)
Gauss's law for magnetism (30-7)

Symbols You Should Recognize

μ_0 (30-1)
$\oint \mathbf{B} \cdot d\boldsymbol{\ell}$ (30-3)
κ_M, χ_M (30-6)

Problems You Should Be Able to Solve

calculating magnetic fields of simple current distributions by integration using the Biot-Savart law (30-1)
evaluating forces between adjacent conductors (30-2)
evaluating magnetic fields using Ampère's law in situations with sufficient symmetry (30-4)
applying results derived in the text for the fields of straight wires, current loops, solenoids, and toroids (30-1–30-5)
relating internal and applied magnetic fields given magnetic susceptibility (30-6)

Limitations to Keep in Mind

The Biot-Savart law, and Ampère's law as expressed in this chapter, are exactly valid only for *steady* currents—those that never change with time.

QUESTIONS

1. In what two senses does a current loop behave like a magnetic dipole?

2. The electric field far from a pair of equal but opposite charges has the same configuration as the magnetic field far from a current loop. Yet inside the structure giving rise to the fields, the two point in opposite directions. Why?

3. The Biot-Savart law shows that the magnetic field of a current element decreases as $1/r^2$. Could you put together a complete circuit whose field exhibits this $1/r^2$ decrease? Why or why not?

4. Do currents in the same direction attract or repel? Explain.

5. If a current is passed through an unstretched spring, will the spring contract or expand? Explain.

6. Why is it advantageous to define the ampere in terms of magnetic force rather than a standard ammeter?

7. One way to heat a plasma is to initiate a large current through it. In response the plasma compresses or "pinches" in the direction perpendicular to the current, and this adiabatic compression heats the plasma. Explain the origin of this "pinch effect."

8. The field of a long, straight wire consists of circular field lines. Does Ampère's law hold for a square loop surrounding the wire? For a circular loop not concentric with the wire? See Fig. 30-40.

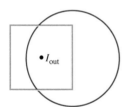

FIGURE 30-40   Question 8.

9. If the line integral around a closed loop is zero, does that mean the magnetic field on the loop is zero?

10. Must the integration path in Ampère's law coincide with a field line?

11. A short, straight length of wire has cylindrical symmetry. Why can't Ampère's law be used to find the magnetic field of a short wire? *Hint:* Think about steady currents and a short wire.

12. What must be going on inside Earth in order that it have a magnetic field?

13. Figure 30-41 shows some magnetic field lines associated with two parallel wires carrying equal currents perpendicular to the page. Are the currents in the same or opposite directions? How can you tell?

14. A solid cylinder and a hollow pipe of the same outer diameter carry the same current along their long dimensions, with the current distributed symmetrically about their

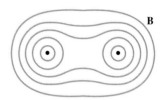

FIGURE 30-41   Question 13.

axes. How do the magnetic fields at their surfaces compare?

15. Why is a piece of iron attracted into a solenoid?

16. Would there be a net magnetic force on a piece of iron deep inside a long solenoid? Explain.

17. In what sense is a long solenoid the magnetic analog of a parallel-plate capacitor?

18. What would happen to the magnetic field inside a solenoid if you (a) doubled the solenoid length without changing the number of turns per unit length or (b) doubled the length without changing the total number of turns?

19. Identify *three* regions in the magnetic field of Fig. 30-42 where there *must* be a current. Which way is the current flowing in each region?

FIGURE 30-42   Question 19.

20. An unmagnetized piece of iron has no net magnetic dipole moment. Yet it is attracted to either pole of a bar magnet. Why?

21. How would you determine experimentally whether a substance was paramagnetic or diamagnetic?

22. Would permanent magnets be possible if the relative permeability of ferromagnetic materials were strictly constant? Explain.

23. Why do paramagnetic and ferromagnetic effects weaken with increasing temperature?

24. If magnetic monopoles existed, then a net motion of monopoles would constitute a "magnetic current." What sort of field should arise from such a "current?"

25. Figure 30-43 shows four different time-independent fields. Which are electric and which magnetic?

26. In the absence of magnetic monopoles, why do we even bother to write Gauss's law for magnetism? Does the law have any physical significance?

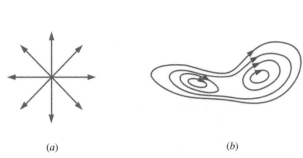

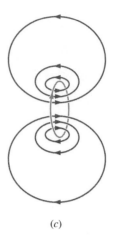

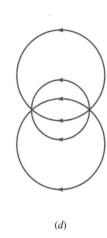

(a) (b) (c) (d)

FIGURE 30-43 Question 25.

PROBLEMS

Section 30-1 The Biot-Savart Law

1. A wire carries 15 A. You form the wire into a single-turn circular loop with magnetic field 80 μT at the loop center. What is the loop radius?

2. A single-turn wire loop is 2.0 cm in diameter and carries a 650-mA current. Find the magnetic field strength (a) at the loop center and (b) on the loop axis, 20 cm from the center.

3. A 2.2-m-long wire carrying 3.5 A is wound into a tight, loop-shaped coil 5.0 cm in diameter. What is the magnetic field at its center?

4. What is the current in a long wire if the magnetic field strength 1.2 cm from the wire's axis is 67 μT?

5. Suppose Earth's magnetic field arose from a single loop of current at the outer edge of the planet's liquid core (core radius 3000 km). How large must the current be to give the observed magnetic dipole moment of 8.0×10^{22} A·m²?

6. Earth's magnetic dipole moment is 8.0×10^{22} A·m². What is the magnetic field strength on Earth's surface at either pole?

7. A single-turn current loop carrying 25 A produces a magnetic field of 3.5 nT at a point on its axis 50 cm from the loop center. What is the loop area, assuming the loop diameter is much less than 50 cm?

8. Two identical current loops are 10 cm in diameter and carry 20-A currents. They are placed 1.0 cm apart, as shown in Fig. 30-44. Find the magnetic field strength at the center of either loop when their currents are in (a) the same and (b) opposite directions.

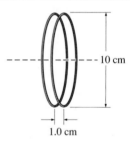

FIGURE 30-44 Problem 8.

9. You have a spool of thin wire that can handle a maximum current of 0.50 A. If you wind the wire into a loop-like coil 20 cm in diameter, how many turns should the coil have if the magnetic field at its center is to be 2.3 mT at this maximum current?

10. A single piece of wire is bent so that it includes a circular loop of radius a, as shown in Fig. 30-45. A current I flows in the direction shown. Find an expression for the magnetic field at the center of the loop.

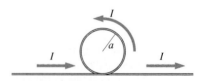

FIGURE 30-45 Problem 10.

11. Two long, parallel wires are 6.0 cm apart. One carries 5.0 A and the other 10 A, with both currents in the same direction. Where on a line perpendicular to both wires is the magnetic field zero?

12. Four long, parallel wires are located at the corners of a square 15 cm on a side. Each carries a current of 2.5 A, with the top two currents into the page in Fig. 30-46 and the bottom two out of the page. Find the magnetic field at the center of the square.

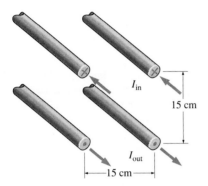

FIGURE 30-46 Problems 12, 23.

13. A power line carries a 500-A current toward magnetic north and is suspended 10 m above the ground. The horizontal component of Earth's magnetic field at the power line's latitude is 0.24 G. If a magnetic compass is placed on the ground directly below the power line, in what direction will it point?

14. An electron is moving at 3.1×10^6 m/s parallel to a 1.0-mm-diameter wire carrying 20 A. If the electron is 2.0 mm from the center of the wire, with its velocity in the same direction as the current, what are the magnitude and direction of the force it experiences?

15. Part of a long wire is bent into a semicircle of radius a, as shown in Fig. 30-47. A current I flows in the direction shown. Use the Biot-Savart law to find the magnetic field at the center of the semicircle (point P).

FIGURE 30-47 Problem 15.

16. Use the result given in Problem 59 to find the magnetic field strength at the center of a square loop of side a carrying current I.

17. Figure 30-48 shows a conducting loop formed from concentric semicircles of radii a and b. If the loop carries a current I as shown, find the magnetic field at point P, the common center.

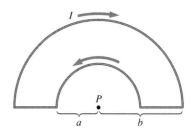

FIGURE 30-48 Problem 17.

Section 30-2 The Magnetic Force Between Two Conductors

18. In standard household wiring, parallel wires about 1 cm apart carry currents around 15 A. What is the magnitude of the force per unit length between such wires?

19. It would take a rather large apparatus to implement the definition of the ampere given in Section 30-2. Suppose you wanted to use a smaller apparatus, with wires 50 cm long separated by 2.0 cm. What force would correspond to a current of 1 A?

20. Two parallel copper rods supply power to a high-energy experiment, carrying the same current in opposite directions. The rods are held 8.0 cm apart by insulating blocks mounted every 1.5 m. If each block can tolerate a maximum tension force of 200 N, what is the maximum allowable current?

21. The structure shown in Fig. 30-49 is made from conducting rods. The upper horizontal rod is free to slide vertically on the uprights, while maintaining electrical contact with them. The upper rod has mass 22 g and length 95 cm. A battery connected across the insulating gap at the bottom of the left-hand upright drives a 66-A current through the structure. At what height h will the upper wire be in equilibrium?

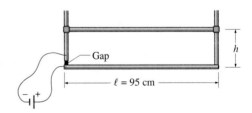

FIGURE 30-49 Problem 21.

22. Three parallel wires 4.6 m long each carry 20 A in the same direction. They are spaced at the vertices of an equilateral triangle 3.5 cm on a side. Find the magnitude of the force on each wire.

23. The wires in Fig. 30-46 carry 2.5-A currents in the directions indicated. Find the net force per unit length on the wire at lower left.

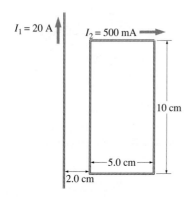

FIGURE 30-50 Problem 24.

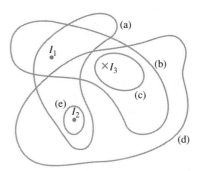

FIGURE 30-52 Problem 28.

24. A long, straight wire carries 20 A. A 5.0-cm by 10-cm rectangular wire loop carrying 500 mA is located 2.0 cm from the wire, as shown in Fig. 30-50. Find the net magnetic force on the loop.

25. A solenoid 10 cm in diameter is made with 2.1-mm-diameter copper wire wound so tightly that adjacent turns touch, separated only by enamel insulation of negligible thickness. The solenoid carries a 28-A current. In the long, straight wire approximation, what is the net force between two adjacent turns of the solenoid?

26. A long, flat conducting ribbon of width w is parallel to a long straight wire, with its near edge a distance a from the wire (Fig. 30-51). Wire and ribbon carry the same current I; this current is distributed uniformly over the ribbon. Use integration to show that the force per unit length between the two has magnitude $\dfrac{\mu_0 I^2}{2\pi w} \ln\left(\dfrac{a + w}{a}\right)$.

28. In Fig. 30-52, $I_1 = 2$ A flowing out of the page; $I_2 = 1$ A, also out of the page, and $I_3 = 2$ A, into the page. What is the line integral of the magnetic field taken counterclockwise around each loop shown?

29. The magnetic field shown in Fig. 30-53 has uniform magnitude 75 μT, but its direction reverses abruptly. How much current is encircled by the rectangular loop shown?

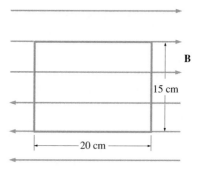

FIGURE 30-53 Problems 29, 30.

30. On the nighttime side of Earth, the solar wind draws some of the planet's magnetic field into a long "magnetotail," where field lines on opposite sides of the equatorial plane are straight but antiparallel (as in the configuration of Fig. 30-53). If the field strength is 2×10^{-4} G, what is the current per unit length flowing in the magnetotail?

31. Figure 30-54 shows a magnetic field pointing in the x direction. Its strength, however, varies with position in the

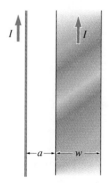

FIGURE 30-51 Problem 26.

Section 30-3 Ampère's Law

27. The line integral of the magnetic field on a closed path surrounding a wire has the value 8.8 μT·m. What is the current in the wire?

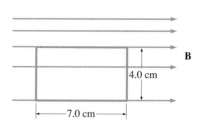

FIGURE 30-54 Problems 31, 68.

y direction. At the top and bottom of the rectangular loop shown the field strengths are 3.4 μT and 1.2 μT, respectively. How much current flows through the area encircled by the loop?

Section 30-4 Using Ampère's Law

32. (a) What is the magnetic field strength 0.10 mm from the axis of a 1.0-mm-diameter wire carrying 5.0 A distributed uniformly over its cross section? (b) What is the field strength at the surface of the wire?

33. A solid wire 2.1 mm in diameter carries a 10-A current with uniform current density. What is the magnetic field strength (a) at the axis of the wire, (b) 0.20 mm from the axis, (c) at the surface of the wire, and (d) 4.0 mm from the wire axis?

34. Show that Equations 30-9 and 30-10 give the same results when evaluated at the surface of the wire.

35. A long conducting rod of radius R carries a nonuniform current density given by $J = J_0 r/R$, where J_0 is a constant and r is the radial distance from the rod's axis. Find expressions for the magnetic field strength (a) inside and (b) outside the rod.

36. A long, hollow conducting pipe of radius R carries a uniform current I along the pipe, as shown in Fig. 30-55. Use Ampère's law to find the magnetic field strength (a) inside and (b) outside the pipe.

FIGURE 30-55   Problem 36.

37. Typically, cylindrical wires made from yttrium-barium-copper-oxide superconductor can carry a maximum current density of 6.0 MA/m² at a temperature of 77 K, as long as magnetic field at the conductor surface does not exceed 10 mT. Suppose such a wire is to carry the maximum current density. (a) At what wire diameter would the surface magnetic field equal the 10-mT limit? (b) Is this a maximum or minimum value for the diameter if the field is not to exceed the limit? (c) What current would a wire with this diameter carry?

38. A long, hollow conducting pipe of radius R and length ℓ carries a uniform current I flowing around the pipe, as shown in Fig. 30-56. Find expressions for the magnetic field (a) inside and (b) outside the pipe. *Hint:* What configuration does this pipe resemble?

39. A copper ribbon 1.0 cm wide and 0.15 mm thick is rated for a maximum safe current density of 8.8×10^6 A/m². What is the maximum magnetic field strength achievable at the surface of this ribbon?

FIGURE 30-56   Problem 38.

40. Two large, flat conducting plates lie parallel to the *x-y* plane. They carry equal currents, one in the $+x$ and the other in the $-x$ direction. In each plate the current per meter of width in the *y* direction is J_s. Find the magnetic field strength (a) between and (b) outside the plates.

41. Repeat the preceding problem for the case when one current flows in the $+x$ direction and the other in the $+y$ direction.

42. The coaxial cable shown in Fig. 30-57 consists of a solid inner conductor of radius a and a hollow outer conductor of inner radius b and thickness c. The two carry equal but opposite currents I, uniformly distributed. Find expressions for the magnetic field strength as a function of radial position r (a) within the inner conductor, (b) between the inner and outer conductors, and (c) beyond the outer conductor.

FIGURE 30-57   Problems 42, 63.

43. A hollow conducting pipe of inner radius a and outer radius b carries a current I parallel to its axis and distributed uniformly through the pipe material (Fig. 30-58). Find expressions for the magnetic field for (a) $r < a$, (b) $a < r < b$, and (c) $r > b$, where r is the radial distance from the pipe axis.

FIGURE 30-58   Problem 43.

44. A conducting slab extends infinitely in the x and y directions and has thickness h in the z direction. It carries a uniform current density $\mathbf{J} = J\hat{\mathbf{i}}$. Find the magnetic field strength (a) inside and (b) outside the slab, as functions of the distance z from the center plane of the slab.

Section 30-5 Solenoids and Toroids

45. A superconducting solenoid has 3300 turns per meter and can carry a maximum current of 4.1 kA. What is the magnetic field strength in the solenoid?

46. A solenoid used in a plasma physics experiment is 10 cm in diameter, 1.0 m long, and carries a 35-A current to produce a 100-mT magnetic field. (a) How many turns are in the solenoid? (b) If the solenoid resistance is 2.7 Ω, how much power does it dissipate?

47. You have 10 m of 0.50-mm-diameter copper wire and a battery capable of passing 15 A through the wire. What magnetic field strengths could you obtain (a) inside a 2.0-cm-diameter solenoid wound with the wire as closely spaced as possible and (b) at the center of a single circular loop made from the wire?

48. A toroidal coil of inner radius 15 cm and outer radius 17 cm is wound from 1200 turns of wire. What are (a) the minimum and (b) the maximum magnetic field strengths within the toroid when it carries a 10-A current?

49. A toroidal fusion reactor requires a magnetic field that varies by no more than 10% from its central value of 1.5 T. If the minor radius of the toroidal coil producing this field is 30 cm, what is the minimum value for the major radius of the device?

50. A long solenoid with n turns per unit length carries a current I. The current returns to its driving battery along a wire of radius R that passes through the solenoid, along its axis. Find expressions for (a) the magnetic field strength at the surface of the wire and (b) the angle the field at the wire surface makes with the solenoid axis.

51. We noted that there is a nonzero magnetic field component outside a solenoid, encircling the device, associated with the component of current flow parallel to the solenoid axis. For a long solenoid of radius R, find an expression for the ratio of this external encircling field just outside the solenoid to the field inside, and show explicitly that this ratio tends to zero as the number of turns per unit length becomes large.

52. Derive Equation 30-12 for the field of a solenoid by considering the solenoid to be made of a large number of adjacent current loops. Use Equation 30-3 for the field of a current loop, and integrate over all loops.

Section 30-6 Magnetic Matter

53. When a sample of a certain substance is placed in a 250.0-mT magnetic field, the field inside the sample is 249.6 mT. Find the magnetic susceptibility of the substance. Is it ferromagnetic, paramagnetic, or diamagnetic?

54. A container of liquid oxygen at 90 K is placed in a 500.0-G magnetic field. What is the field strength within the liquid oxygen?

55. A ferromagnetic material is placed in a 2.5-G magnetic field and the field within the material is determined to be 1.8 T. What is the magnetic susceptibility of this material?

56. Figure 30-59 shows the hysteresis curve for the ferromagnetic alloy Alnico V, commonly used in permanent magnets. Use the graph to find approximate values for the maximum field strength obtainable inside this material, (a) in the absence and (b) in the presence of an externally applied field.

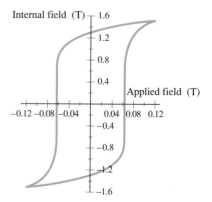

FIGURE 30-59 Problem 56.

Paired Problems

(Both problems in a pair involve the same principles and techniques. If you can get the first problem, you should be able to solve the second one.)

57. Two concentric, coplanar circular current loops have radii a and $2a$. If the magnetic field is zero at their common center, how does the current in the outer loop compare with that in the inner loop?

58. A thin conducting washer of inner radius a and outer radius $2a$ carries a current I distributed uniformly with radial position, as suggested in Fig. 30-60. Find an expression for the magnetic field strength at its center.

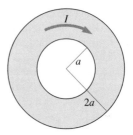

FIGURE 30-60 Problem 58.

59. Figure 30-61 shows a wire of length ℓ carrying current fed by other wires that are not shown. Point A lies on the perpendicular bisector, a distance y from the wire. Adapt the calculation of Example 30-2 to show that the magnetic field at A due to the straight wire alone has magnitude $\dfrac{\mu_0 I \ell}{2\pi y \sqrt{\ell^2 + 4y^2}}$. What is the field direction?

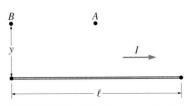

FIGURE 30-61 Problems 59, 60.

60. Point B in Fig. 30-62 lies a distance y perpendicular to the end of the wire. Show that the magnetic field at B has magnitude $\dfrac{\mu_0 I \ell}{4\pi y \sqrt{\ell^2 + y^2}}$. What is the field direction?

61. The largest lightning strikes have peak currents around 250 kA, flowing in essentially cylindrical channels of ionized air. How far from such a flash would the resulting magnetic field be equal to Earth's magnetic field strength, about 50 μT?

62. A particle-beam weapon being tested for ballistic missile defense delivers a 2.5-cm-diameter electron beam carrying a current of 10 kA at an altitude where Earth's magnetic field is 30 μT. How close to the beam would an adversary's surveillance apparatus need to be in order to detect the beam by observing a 1% change in the local magnetic field strength? Assume the beam's magnetic field is parallel to Earth's at the point of detection.

63. A coaxial cable like that shown in Fig. 30-57 consists of a 1.0-mm-diameter inner conductor and an outer conductor of inner diameter 1.0 cm and 0.20 mm thickness. A 100-mA current flows down the center conductor and back along the outer conductor. Find the magnetic field strength (a) 0.10 mm, (b) 5.0 mm, and (c) 2.0 cm from the cable axis.

64. Repeat the preceding problem if the current in the outer conductor only is increased to 200 mA.

Supplementary Problems

65. A circular wire loop of radius 15 cm and negligible thickness carries a 2.0-A current. Use suitable approximations to find the magnetic field of this loop (a) in the loop plane, 1.0 mm outside the loop, and (b) on the loop axis, 3.0 m from the loop center.

66. A long, flat conducting bar of width w carries a total current I distributed uniformly, as shown in Fig. 30-62. Use suitable approximations to write expressions for the magnetic field strength (a) near the conductor surface ($r \ll w$) and (b) very far from the conductor ($r \gg w$).

67. (a) Use the result of Problem 59 to find an expression for the magnetic field strength on the axis of a square loop of

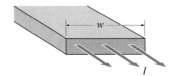

FIGURE 30-62 Problem 66.

side a carrying current I. (b) Show that your result reduces the field of a dipole—Equation 30-4—in the limit $x \gg a$.

68. The magnetic field in Fig. 30-54 is given by $\mathbf{B} = by\hat{\mathbf{i}}$, where $b = 55$ μT/m and y is the vertical coordinate in Fig. 30-54, measured in meters. Find the current density in the region.

69. A wide, flat conducting spring of spring constant $k = 20$ N/m and negligible mass consists of two 6.0-cm-diameter turns, as shown in Fig. 30-63. In its unstretched configuration the coils are nearly touching. A 10-g mass is hung from the spring, and at the same time a current I is passed through it. The spring stretches 2.0 mm. Find I, assuming the coils remain close enough to be treated as parallel wires.

10 g

FIGURE 30-63 Problem 69.

70. A solid conducting wire of radius R runs parallel to the z axis and carries a current density given by $\mathbf{J} = J_0(1 - r/R)\hat{\mathbf{k}}$, where J_0 is a constant and r the radial distance from the wire axis. Find expressions for (a) the total current in the wire, (b) the magnetic field strength for $r > R$, and (c) the magnetic field strength for $r < R$.

71. A disk of radius a carries a uniform surface charge density σ, and is rotating with angular speed ω about the central axis perpendicular to the disk. Show that the magnetic field at the disk's center is $\frac{1}{2}\mu_0 \sigma \omega a$.

72. Calculate the magnetic field of an infinite current sheet by considering the sheet to be made up of infinitesimal line currents, as suggested in Fig. 30-23, and integrating the fields of these line currents.

73. Work Example 30-2 by expressing all variables in terms of the angle θ and integrating over the appropriate range in θ.

ELECTROMAGNETIC INDUCTION

Generators use electromagnetic induction to convert mechanical energy into electral energy. These generators at Arizona's Glen Canyon Dam produce 1300 MW of electric power from the energy of falling water.

All the electric and magnetic fields we encountered in the previous chapters had their ultimate origins in electric charge, either stationary or moving. We stressed a relation between electricity and magnetism, whereby electric charge gives rise to and interacts with both the electric field and the magnetic field. The remainder of our study of electromagnetism is devoted to a much more intimate relation between electricity and magnetism, a relation in which the fields themselves interact directly. This interaction forms the basis of new electromagnetic technologies, leads toward an understanding of the nature of light, and points the way to the theory of relativity.

31-1 INDUCED CURRENTS

In 1831, the English scientist Michael Faraday and the American Joseph Henry independently carried out experiments in which electric currents arose in circuits subjected to changing magnetic fields. Figure 31-1 shows one such exper-

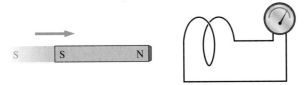

FIGURE 31-1   When a magnet is moved near a closed circuit, current flows in the circuit.

iment, in which a magnet is moved in the presence of a circuit consisting of a loop of wire and an ammeter. There is no battery or other obvious source of emf in the circuit. As long as the magnet is held still, there is no current. But while the magnet is moving, the ammeter registers a current—an **induced current.** If we modify the experiment, holding the magnet still but moving the coil, we again observe an induced current. Apparently only the relative motion matters. If we replace the magnet with another circuit in which a battery drives a current, and move the two circuits relative to each other, we get an induced current in the circuit without the battery (Fig. 31-2). If we hold the two circuits still we get no induced current. But now if we close—or open—a switch in the circuit with the battery, we find that the ammeter indicates a momentary induced current (Fig. 31-3).

The one common feature in all these experiments is a *changing magnetic field*. It does not matter whether the field changes because a magnet is moved, or because a circuit is moved near a magnet, or because the current giving rise to the field changes. In each case, an induced current appears in a circuit subjected to a changing magnetic field. We are observing here a new phenomenon—**electromagnetic induction**—whereby electrical effects arise from *changing* magnetic fields.

31-2 FARADAY'S LAW

We know from Chapter 27 that a source of electromotive force—something like a battery that supplies energy to electric charges—is needed to establish a current in a circuit. When an induced current flows in a circuit, an emf is similarly present. This **induced emf** is usually not localized at one point in the circuit, as in a battery, but may be spread throughout the conductors making up the circuit.

Experimentally, we find that the induced emf in a circuit depends on the rate of change of magnetic flux through the circuit. Before quantifying this relationship, we show how to calculate magnetic flux.

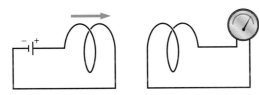

FIGURE 31-2   An induced current also arises when the magnet is replaced by a current-carrying circuit.

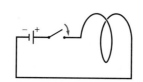 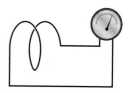

FIGURE 31-3 A current is also induced—even in the absence of any motion—when the current in an adjacent circuit changes.

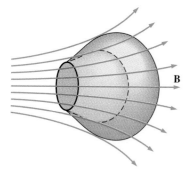

FIGURE 31-4 The flux through a circuit is the same through *any* open surface bounded by the circuit. Three such surfaces are shown here; they include the flat circular surface in the plane of the circular loop circuit, and the two bubble-like surfaces. If the fluxes through any two of these surfaces were not the same, then the flux through a *closed* surface bounded by the two would not be zero, in violation of Gauss's law for magnetism.

Magnetic Flux

We define magnetic flux in analogy with Chapter 24's definition of electric flux as the integral of the field over a surface:

$$\phi_B = \int \mathbf{B} \cdot d\mathbf{A}, \qquad (31\text{-}1)$$

where $d\mathbf{A}$ is an infinitesimal vector normal to the surface. As with electric flux, magnetic flux is proportional to the number of field lines passing through a surface. With electromagnetic induction, we're interested in the flux through an *open* surface bounded by the circuit in question. Because the magnetic flux through any *closed* surface is zero, we can in fact use *any* open surface bounded by our circuit to calculate the flux through the circuit (see Fig. 31-4).

For a flat surface in a uniform magnetic field, Equation 31-1 reduces to the simple expression $\phi_B = BA\cos\theta$, with θ the angle between the field and the normal to the surface. For curved surfaces and nonuniform fields, we add—that is, integrate—all the infinitesimal flux elements $d\phi_B = \mathbf{B} \cdot d\mathbf{A}$ over the surface to get the total flux. Note that the units of magnetic flux are T·m².

● EXAMPLE 31-1 THE MAGNETIC FLUX THROUGH A SOLENOID

A solenoid of circular cross section has radius R, consists of n turns per unit length, and carries a current I. What is the magnetic flux through each turn of the solenoid?

Solution
Away from the solenoid ends, the field is uniform and parallel to the solenoid axis, with magnitude given by Equation 30-12:

$$B = \mu_0 nI.$$

A flat surface bounded by one turn of the solenoid lies at right angles to this uniform field, so the flux is simply the product of the magnetic field and the area:

$$\phi_B = \int \mathbf{B} \cdot d\mathbf{A} = BA = \mu_0 nI\pi R^2.$$

We are being a little loose here in thinking of a single turn of the solenoid as a closed loop, but if the solenoid is tightly wound this is an excellent approximation. We will be concerned frequently with the flux through a multiturn coil, and in calculating this flux it is convenient to view each turn as an individual loop. The flux through an N-turn coil in a uniform magnetic field is just N times the flux through each turn.

EXERCISE A rectangular wire loop measuring 10 cm by 15 cm is oriented so the normal to the loop makes a 30° angle with a uniform 50-mT magnetic field. Find the magnetic flux through the loop.

Answer: 6.5×10⁻⁴ T·m²

Some problems similar to Example 31-1: 3, 4 ●

● EXAMPLE 31-2 MAGNETIC FLUX IN A NONUNIFORM FIELD

A long, straight wire carries a current I. A rectangular wire loop of dimensions ℓ by w lies with its closest edge a distance a from the wire, as shown in Fig. 31-5. What is the magnetic flux through the loop?

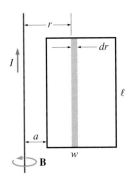

FIGURE 31-5 A rectangular loop in the magnetic field of a long wire. The field to the right of the wire points into the plane of the page, and its magnitude drops inversely with distance from the wire. Area elements for the flux calculation are strips of length ℓ and infinitesimal width dr.

Solution

The magnetic field of the wire is given by Equation 30-9:

$$B = \frac{\mu_0 I}{2\pi r},$$

where r is the distance from the wire. At the site of the loop, this field points straight into the page, perpendicular to the plane of the loop. However, the field varies with distance from the straight wire, so we cannot simply multiply the field by the loop area to get the flux. Instead, we divide the loop into thin strips of width dr and area $dA = \ell \, dr$, as shown in Fig. 31-5. With the field at right angles to each strip, $\mathbf{B} \cdot d\mathbf{A} = B \, dA$, and the flux through any strip is

$$d\phi = B \, dA = B\ell \, dr = \frac{\mu_0 I}{2\pi r} \ell \, dr.$$

Then the total flux through the loop is the integral over all such strips contained within the loop, that is, over all strips between $r = a$ and $r = a + w$:

$$\phi = \int_a^{a+w} \frac{\mu_0 I}{2\pi r} \ell \, dr = \frac{\mu_0 I \ell}{2\pi} \int_a^{a+w} \frac{dr}{r} = \frac{\mu_0 I \ell}{2\pi} \ln r \Big|_a^{a+w}$$

$$= \frac{\mu_0 I \ell}{2\pi} \ln\left(\frac{a + w}{a}\right).$$

EXERCISE A nonuniform magnetic field points in the z direction. The field strength is independent of y but varies linearly from $B = 0$ at $x = 0$ to $B = 2.0$ T at $x = 1.0$ m. A square wire loop 45 cm on a side lies in the x-y plane with one corner at the origin and two sides extending along the positive x and y axes. Find the magnetic flux through this loop.

Answer: 91 mT·m^2

Some problems similar to Example 31-2: 45, 46 ●

Flux and Induced EMF

Having quantified the notion of magnetic flux, we are now ready to state rigorously the experimental fact that changing magnetic flux induces an emf in a circuit. Our statement is a special case of **Faraday's law of induction,** which constitutes another of the four basic laws of electromagnetism:

> **The induced emf in a circuit is proportional to the rate of change of magnetic flux through any surface bounded by that circuit.**

This statement is a special case; we will later broaden its scope to include situations where no circuit is present. In SI units the proportionality constant between emf and rate of change of flux is just -1, so mathematically Faraday's law is

$$\mathcal{E} = -\frac{d\phi_B}{dt}, \tag{31-2}$$

where $\mathcal{E}$ is the emf induced in a circuit and ϕ_B the magnetic flux through any surface bounded by that circuit. Problem 1 will help convince you that the units of the rate of change of magnetic flux are indeed volts, the same as the units of emf. The minus sign in Equation 31-2 is essential, and we will soon have a great deal more to say about it. We stress that Faraday's law, Equation 31-2, applies whenever the magnetic flux through a circuit changes. We could change that flux by moving a magnet near the circuit or the circuit near a magnet. Since the flux $\mathbf{B} \cdot \mathbf{A} = BA\cos\theta$ depends on the field strength B, the area A, and the angle θ between the field and the normal to the area, we could also change the flux by changing any of these three quantities. The following examples explore these possibilities.

● EXAMPLE 31-3 A CHANGING MAGNETIC FIELD

A wire loop of radius 10 cm has a resistance of 2.0 Ω. The loop is at right angles to a uniform magnetic field **B,** as shown in Fig. 31-6. The field strength is increasing at 0.10 tesla/second. What is the magnitude of the induced current in the loop?

Solution
To find the induced current, we need to know the induced emf. Faraday's law tells us that the induced emf is related to the rate of change of magnetic flux through the circuit. With a uniform field at right angles to the loop, the flux is just the field strength times the loop area, or

$$\phi_B = \int \mathbf{B} \cdot d\mathbf{A} = \pi r^2 B.$$

Even though the magnetic field is changing with *time,* at any given instant it is uniform in *space,* which is why the integration was trivial.

We don't know B, but this doesn't matter because we are really interested in the *rate of change* of the flux, not in the flux itself. With the loop area constant, the rate of change of flux is

$$\frac{d\phi_B}{dt} = \pi r^2 \frac{dB}{dt} = (\pi)(0.10 \text{ m})^2(0.10 \text{ T/s}) = 3.14 \times 10^{-3} \text{ V}.$$

By Faraday's law, this is the magnitude of the induced emf. We then calculate the current using Ohm's law, as we would for any emf:

$$I = \frac{\mathcal{E}}{R} = \frac{3.14 \times 10^{-3} \text{ V}}{2.0 \text{ Ω}} = 1.6 \times 10^{-3} \text{ A} = 1.6 \text{ mA}.$$

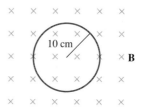

FIGURE 31-6 A circular conducting loop at right angles to a uniform magnetic field (Example 31-3).

TIP It's the Change That Counts Faraday's law relates induced emf to the *rate of change* of magnetic flux. The actual value of the flux—or of the magnetic field if that is what's changing—doesn't matter. You may be troubled by induction problems where the value of the field is not given. You don't need it; what you do need or may be asked to find is the *rate of change* of the field.

EXERCISE A square conducting loop 25 cm on a side lies at right angles to a uniform magnetic field. The loop's resistance is 8.0 Ω, and it carries a current of 14 mA. At what rate is the magnetic field changing?

Answer: 1.8 T/s

Some problems similar to Example 31-3: 4–8 ●

● EXAMPLE 31-4 A CHANGING AREA

A circuit consists of two parallel conducting rails a distance ℓ apart connected at one end by a resistance R. A conducting bar slides along the rails. The whole circuit is in a constant, uniform magnetic field **B** at right angles to the plane of the circuit, as shown in Fig. 31-7. The bar is pulled to the right with constant speed v. What is the current in the circuit? Assume the bar and rails are ideal conductors, so the total circuit resistance is R.

Solution
Here the current is driven by an induced emf arising from the change in magnetic flux that occurs as the circuit area increases. We determine the emf using Faraday's law,

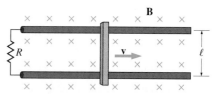

FIGURE 31-7 Pulling the bar to the right increases the circuit area, thereby increasing the magnetic flux and inducing an emf that drives a current (Example 31-4).

$\mathcal{E} = -d\phi_B/dt$. The circuit area is the rail spacing ℓ times the distance x from resistor to bar. With a uniform field perpendicular to the circuit, the flux integral $\int \mathbf{B} \cdot d\mathbf{A}$ reduces to the product of field strength with the area:

$$\phi_B = \int \mathbf{B} \cdot d\mathbf{A} = BA = B\ell x.$$

Never mind that we don't know x—we do know its rate of change and that is all we need. With the field and the rail spacing constant, the rate of change of flux is

$$\frac{d\phi_B}{dt} = B\ell \frac{dx}{dt} = B\ell v,$$

since dx/dt is just the bar velocity v. Faraday's law tells us that the magnitude of the induced emf is equal to the rate of change of flux, so

$$|\mathcal{E}| = B\ell v.$$

This emf drives a current I around the circuit:

$$|I| = \frac{|\mathcal{E}|}{R} = \frac{B\ell v}{R}.$$

EXERCISE A square wire loop 25 cm on a side moves into a region containing a uniform 1.2-T magnetic field oriented at right angles to the loop, as shown in Fig. 31-8. The loop wire is essentially resistanceless, but inserted in the loop is a 3-V flashlight bulb. How fast must the loop move to keep the bulb lit at its normal brightness?

Answer: 10 m/s

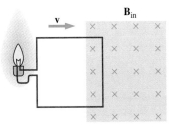

FIGURE 31-8 How fast must the loop move to light the 3-V bulb?

Some problems similar to Example 31-4: 9, 10, 16, 27, 35, 36

● **EXAMPLE 31-5** A CHANGING ORIENTATION

A circular wire loop of radius a and resistance R is initially perpendicular to a constant, uniform magnetic field B. The loop rotates with angular velocity ω about an axis through a diameter, as shown in Fig. 31-9. What is the current in the loop?

Solution
Again, we must find the rate of change of magnetic flux, from which we can get the induced emf and then the current. Here the field is uniform over the area, but the orientation of the field relative to the area is changing. The definition of magnetic flux contains a dot product to account for this orientation. Evaluating the flux, we have

$$\phi_B = \int \mathbf{B} \cdot d\mathbf{A} = \int B\,dA\cos\theta = B\cos\theta \int dA = \pi a^2 B\cos\theta,$$

where θ is the angle between the field and a perpendicular to the loop area. Here we could take the field magnitude B outside the integral because the field is uniform over the loop. And even though the orientation between loop and field changes with *time*, it does not change with *position* at a given instant, so we could also take $\cos\theta$ outside the integral.

The changing orientation is described by giving θ as a function of time. Since the loop rotates with constant angular velocity ω, we can write simply $\theta = \omega t$, where we take the zero of time when $\theta = 0$. Then the flux is

$$\phi_B = \pi a^2 B\cos\omega t,$$

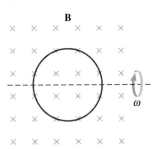

FIGURE 31-9 A wire loop rotating in a uniform magnetic field. The flux changes because of the changing orientation of the loop relative to the field, thereby inducing an emf that drives a current around the loop.

so

$$\frac{d\phi_B}{dt} = \pi a^2 B \frac{d}{dt}(\cos\omega t) = -\pi a^2 B\omega\sin\omega t.$$

By Faraday's law, the emf is then

$$\mathcal{E} = -\frac{d\phi_B}{dt} = \pi a^2 B\omega\sin\omega t,$$

giving a current

$$I = \frac{\mathcal{E}}{R} = \frac{\pi a^2 B\omega}{R}\sin\omega t.$$

(Check the units!)

Unlike the current in the previous two examples, this one changes with time. Its sinusoidal time dependence is in fact just like that of standard alternating current used for electric power—and with good reason: our rotating loop constitutes a simple alternating-current generator. Sinusoidally varying emf's and currents occur whenever conducting loops are rotated in uniform magnetic fields.

TIP Peak Values You'll often be asked for the peak value of voltage or current when either quantity is changing with time (more on this in Chapter 33). You've derived a formula like $\mathscr{E} = \pi a^2 B\omega \sin \omega t$ in Example 31-5, but you don't have a value for the time t.

It doesn't matter! The peak value occurs when $\sin \omega t = 1$, so just replace $\sin \omega t$ by 1. In Example 31-5, for example, the peak emf is just $\mathscr{E}_p = \pi a^2 B\omega$.

EXERCISE Take the radius $a = 11$ cm in Example 31-5 and the magnetic field strength $B = 0.63$ T. With what angular speed ω must the loop rotate in order to produce a peak emf of 6.0 V?

Answer: 250 s⁻¹

Some problems similar to Example: 31-5: 12, 60, 66

You might wonder about the direction of the induced emf and current in these examples since we were concerned only with magnitudes. Although that direction follows mathematically from the minus sign in Faraday's law, we will find in the next section a physical justification for the minus sign that makes it clear how to determine the direction of induced emf's and currents.

31-3 INDUCTION AND THE CONSERVATION OF ENERGY

Move a bar magnet toward a wire loop, as shown in Fig. 31-10. Without the loop present, it would take no work to move the magnet horizontally at constant velocity. But with the loop present, moving the magnet induces a current in the loop. Electrical energy is dissipated in the loop's resistance, heating the loop. Where does the energy come from?

There's only one source for the energy—the agent moving the magnet. As you move the magnet, you must be doing positive work. Otherwise you would get energy—ultimately heating the loop—for nothing. So you must have to push the magnet against some force. What force? A magnetic force, caused by the interaction of the magnet with the magnetic field produced by the induced current. Think about this! The induced current, like any current, produces a magnetic field. The magnet experiences a force in this field, and you must do work to move the magnet through that field. With the magnetic field as intermediary, the work you do ultimately ends up heating the loop.

FIGURE 31-10 As the bar magnet moves toward the conducting loop, the changing magnetic flux through the loop induces an emf that drives a current around the loop. Conservation of energy requires that the magnetic field produced by this current *oppose* the motion of the magnet, so the agent moving the magnet does work that ends up heating the loop.

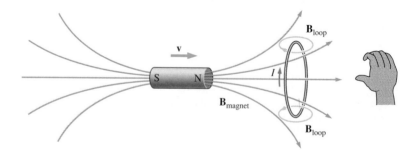

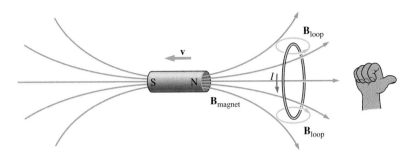

FIGURE 31-11 When the bar magnet is pulled away from the loop, the direction of the induced current is such that the magnetic force opposes the magnet's withdrawal.

As you push the magnet toward the loop, the force the loop's field exerts on the magnet must be repulsive in order that you do positive work on the magnet. In Fig. 31-10 the bar magnet's north pole is toward the loop. To provide a repulsive force the loop must therefore present a north pole to the approaching magnet; that is, field lines must emerge from the loop's interior pointing to the left. What current direction will provide such a field? Application of the right-hand rule gives the answer: Wrap the fingers of your right hand around the loop in the direction of the current, and your right thumb points in the direction of the field in the loop's interior. Thus the current direction is as shown in Fig. 31-10.

This analysis leading to the direction of the induced current was based on one simple principle: conservation of energy. For electromagnetic induction, this universal principle requires the following:

The direction of the induced emf and current is such that the magnetic field created by the induced current opposes the change in magnetic flux.

This statement, called **Lenz's law,** is represented mathematically in the minus sign on the right-hand side of Faraday's law (Equation 31-2).

What happens, for example, if you pull the bar magnet away from the loop in Fig. 31-10? Now the loop must present a south pole to the receding magnet, creating an attractive force opposing the magnet's withdrawal (Fig. 31-11). The loop current must be opposite its direction in Fig. 31-10, as again you do work to overcome the magnetic force.

Motional EMF and Lenz's Law

When a conductor moves through a magnetic field, we can understand the origin of the induced emf in terms of the magnetic force on charge carriers in the wire; this emf is called **motional emf.** In the case of motional emf we can show explicitly that Lenz's law requires energy conservation.

Consider a square conducting loop of side ℓ and resistance R being pulled with constant speed v out of a uniform magnetic field **B**, as shown in Fig. 31-12. We will show that the rate of joule heating in the loop is equal to the rate at which the agent pulling the loop does work.

In Chapter 29 we found that the magnetic force on a charge is given by $\mathbf{F} = q\mathbf{v} \times \mathbf{B}$. Pulling the loop to the right moves its free electrons through the magnetic field; the magnetic force $q\mathbf{v} \times \mathbf{B}$ on these electrons is downward in Fig. 31-12 (opposite $\mathbf{v} \times \mathbf{B}$ since electrons are negative). The resulting downward motion of the negative electrons in the left-hand side of the loop constitutes

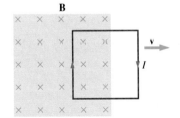

FIGURE 31-12 A conducting loop being withdrawn from a magnetic field. The magnetic force $q\mathbf{v} \times \mathbf{B}$ on charge carriers in the left side of the loop drives a current clockwise around the loop. The current direction also follows from Lenz's law: The magnetic field produced by the loop current acts to oppose the decrease in flux through the loop, and therefore points in the same direction as the original field. This requires a clockwise loop current.

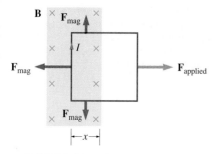

FIGURE 31-13   Forces on the loop. The total magnetic force is that on the left side alone, and the agent pulling the loop must exert an equal but opposite force to maintain constant velocity.

an upward current. This current continues clockwise around the loop, driven by an electric field associated with the separation of charge in the loop's left side.

Now we have a current I in the loop. We found in Chapter 29 that the magnetic force on a current-carrying conductor of length ℓ is $\mathbf{F} = I\boldsymbol{\ell} \times \mathbf{B}.$ Applying this expression to the conducting loop in Fig. 31-12 shows that there is no magnetic force on the right-hand side (since $\mathbf{B} = 0$ there) and that the forces on top and bottom cancel (Fig. 31-13). So the total magnetic force on the loop is that on the left side alone; since the current is upward while the magnetic field is into the page, the magnitude of this force is $I\ell B$, and the right-hand rule shows that it points to the left. Since the loop is being pulled with constant velocity, the net force on it—the sum of the magnetic force and the force applied by the agent pulling the loop—must be zero. The direction of the magnetic force, and therefore of the loop current, is consistent with this requirement.

We could equally well determine the current direction from magnetic flux considerations. As the loop leaves the field region, the flux through it decreases. The direction of the induced current is such as to oppose this decrease in flux. Therefore the magnetic field of the induced current points into the page, as the induced current tries to maintain the flux. By the right-hand rule, a field within the loop and into the page requires that the induced current flow in the clockwise direction.

To calculate the current, we must find the induced emf, which in turn is related to the rate of change of magnetic flux through the loop. With the field and loop perpendicular, and with the field uniform in the region where it is nonzero, the magnetic flux is the product of the magnetic field strength and the loop area that lies within the field:

$$\phi_B = B\ell x,$$

where x is the distance between the left edge of the loop and the right edge of the magnetic field region. The magnetic field remains constant, but as the loop moves the distance x decreases at the rate $dx/dt = -v$ (the minus sign indicates a decrease). Then the rate of change of flux is

$$\frac{d\phi_B}{dt} = \frac{d(B\ell x)}{dt} = B\ell\frac{dx}{dt} = -B\ell v,$$

so Faraday's law gives $\qquad \mathcal{E} = -\frac{d\phi_B}{dt} = B\ell v.$

This induced emf drives a current I around the loop, where

$$I = \frac{\mathcal{E}}{R} = \frac{B\ell v}{R}.$$

The rate of energy dissipation in the loop is the product of the emf and the current (Equation 27-8):

$$P = I\mathcal{E} = \frac{B\ell v}{R}B\ell v = \frac{B^2\ell^2 v^2}{R}. \qquad \left(\begin{array}{c}\text{power dissipated}\\\text{in loop}\end{array}\right)$$

We've found that the magnetic force on the loop has magnitude $I\ell B$; since the loop is moving with constant velocity, this is also the magnitude of the applied force. Equation 7-21 shows that the rate at which work is done by a force **F** acting on an object moving with velocity **v** is $P = \mathbf{F} \cdot \mathbf{v}$; here, with the applied force and velocity in the same direction, we have

$$P = Fv = I\ell Bv = \frac{B\ell v}{R}\ell Bv = \frac{B^2\ell^2 v^2}{R}, \qquad \begin{pmatrix} \text{power supplied} \\ \text{to pull loop} \end{pmatrix}$$

in agreement with our expression for the electric power dissipated in the loop. Thus, the work done by the agent pulling the loop ends up as electrical energy, which is dissipated in the resistor.

Electromagnetic induction is the principle behind many important technological devices. Induction permits us to transform mechanical into electrical energy, and provides great flexibility in the handling of electric power. The applications that follow explore some uses of induction.

■ APPLICATION ELECTRIC GENERATORS

Probably the most important technological application of induction occurs in the electric generator. The world currently uses electrical energy at the phenomenal rate of about 10^{13} watts—roughly equal to the power output of 100 billion human bodies—and virtually all of this power comes from generators. A generator is just a system of conductors in a magnetic field (Fig. 31-14). Mechanical energy is supplied to rotate the conductors, resulting in a changing magnetic flux. An emf is induced and current flows through the generator and on to whatever electrical loads are connected to it. Any source of mechanical energy can power the generator, but the most common is steam from burning fossil fuels or from nuclear fission (Fig. 31-15a). Electrical energy may be generated from the kinetic energy of water or wind. (Fig. 31-15b). A small electric generator, often called an alternator, is used to recharge the battery in an automobile while the engine is running.

Lenz's law, the conservation of energy in electromagnetic induction, is very much applicable to electric generators. Were

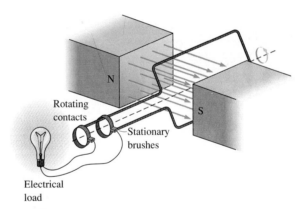

FIGURE 31-14 Simplified diagram of an electric generator. As the loop rotates in the magnetic field, the changing flux induces an emf that drives a current through the rotating contacts and stationary brushes and on through the electrical load.

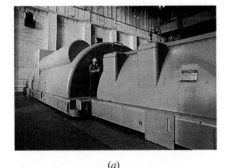

(a)

(b)

FIGURE 31-15 (a) This large power-plant generator produces over 1000 MW of electric power. (b) Wind-driven electric generators at a "wind farm" in California.

it not for Lenz's law, which requires that induced currents *oppose* the changes giving rise to them, generators would turn on their own and happily supply electricity without the need for coal, oil, or uranium. The voluminous quantities of fuel (Fig. 31-16) consumed by power plants are dramatic testimony to the minus sign appearing on the right-hand side of Equation 31-2!

An instructive introduction to Lenz's law comes about if you have access to a hand-cranked electric generator. Without any electrical load across the generator, it is easy to turn. But as you switch on increasingly heavy loads—by *lowering* the electrical load resistance—the generator gets harder to turn (Fig. 31-17). Most people find they can just sustain a 100-W light bulb with a hand generator. Think about this next time you leave a light on! You also experience Lenz's law when you turn on the headlights of a car that is idling slowly. You can hear the engine speed drop, and the car may even stall, as the car's generator gets harder for the engine to turn.

FIGURE 31-16 A 110-car trainload of coal arriving at a power plant in Texas. Some fourteen such trains arrive at the plant each week—a testimony to the minus sign in Equation 31-2!

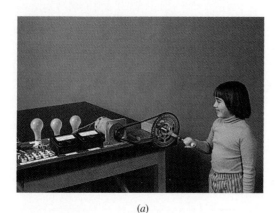

(a)

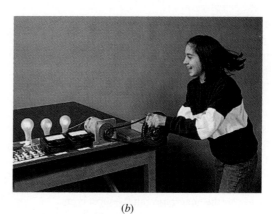

(b)

FIGURE 31-17 (a) With no electrical load, a hand-cranked generator is easy to turn. (b) With 200 watts of light bulbs connected, turning the generator becomes much more difficult. The generator emf is the same in both cases, because the generator is being turned at the same rate. But only in (b) is current flowing, giving nonzero power $P = IV$ that must be supplied by turning the generator.

● **EXAMPLE 31-6** DESIGNING A GENERATOR

An electric generator consists of a 10-turn square wire loop 50 cm on a side. The loop is turned at 60 revolutions per second, to produce standard 60-Hz alternating current like that used throughout the United States and Canada. How strong must the magnetic field be for the peak output voltage of the generator to be 170 V? (This is actually the peak voltage of standard 120-V household wiring; 120 V is an appropriate average value.)

Solution
We need to evaluate the induced emf as a function of magnetic field strength. With a uniform magnetic field, the flux through

one turn of the loop is $\int \mathbf{B} \cdot d\mathbf{A} = BA\cos\theta$, where θ is the angle between the field and the normal to the loop. But the loop rotates with angular frequency $\omega = 2\pi f$, so $\theta = 2\pi ft$. The loop area A is s^2, with s the length of the loop side, so the flux through N turns of the loop is

$$\phi_B = NBs^2\cos(2\pi ft).$$

To find the induced emf, we take the time rate of change of this flux:

$$\mathscr{E} = -\frac{d\phi_B}{dt} = -NBs^2\left[-2\pi f\sin(2\pi ft)\right]$$

$$= 2\pi NfBs^2\sin(2\pi ft).$$

The peak emf is the quantity multiplying the sine; we want this to be 170 V. Solving for the unknown magnetic field B then gives

$$B = \frac{\mathscr{E}_{\text{peak}}}{2\pi Nfs^2} = \frac{170 \text{ V}}{(2\pi)(10)(60 \text{ Hz})(0.50 \text{ m})^2} = 0.18 \text{ T}.$$

This is a typical field strength near the poles of a strong permanent magnet.

EXERCISE A generator includes a circular coil 30 cm in diameter, spinning at 3600 rpm in a uniform 0.50-T magnetic field. How many turns should it have if the peak output voltage is to be 2400 V?

Answer: 180

Some problems similar to Example 31-6: 24, 26, 51, 52, 61 ●

■ APPLICATION MAGNETIC RECORDING

Magnetic materials are widely used as information storage media. Examples include audio and video cassette tapes and computer disks of both the hard and floppy variety. Retrieving the stored information involves electromagnetic induction.

In a typical magnetic recording system the magnetic medium—usually a long plastic tape or a circular disk coated with ferromagnetic oxides—is made to move past a small coil called the **head.** To record information, current is passed through the head coil to impress a magnetization pattern on the tape or disk (Fig. 31-18). This pattern reflects the information contained in the time variation of the current supplied to the head. With analog systems, like most audio and videotapes, the current varies continuously and produces a smoothly varying magnetization pattern. (The current in audio systems may itself derive from electromagnetic induction, using a microphone in which a ferromagnetic diaphragm vibrates in response to sound waves, altering the magnetic field in a nearby coil and thus inducing a current.) With digital systems, including computer data storage and the newer digital audio tape (DAT) and video formats, coil currents of the same magnitude but opposite polarity produce regions of oppositely directed magnetization; these represent the 1's and 0's in which digital information is coded.

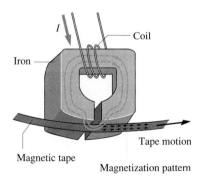

FIGURE 31-18 Recording on magnetic tape. The iron body of the head confines the magnetic field of the coil except at the gap, where a fringing field emerges to impose a magnetization pattern on the moving tape. When the tape is played back, its changing magnetization pattern imposes a changing magnetic flux at the head, inducing an emf in the head coil.

FIGURE 31-19 Mechanism of a computer disk drive includes a head (at bottom) that literally flies as close as 0.25 μm above a spinning aluminum disk coated with ferromagnetic material (red).

To retrieve the stored information, the magnetic medium is again moved past a coil, which may be the same one used for recording. The changing magnetization on the medium results in a varying magnetic flux through the coil, which induces an emf that is amplified and processed to produce images, sound, or digital data. (In audio systems, the final conversion to sound is again usually magnetic, with the amplified signal fed to the coil of a loudspeaker, where magnetic forces result in movement of the speaker cone and the generation of sound waves.)

The rate at which information can be stored and extracted depends on how densely the regions of varying magnetization can be packed without interfering and on how fast the medium moves in relation to the heads. Audio cassette tapes move past the heads at a mere 4.8 cm /s, a speed that limits these tapes' ability to record high-frequency sound faithfully. The much higher information content of video images requires higher tape-to-head speeds; in standard videocassette recorders a speed of 39.52 m /s is achieved by moving the tape past rapidly spinning heads. High-speed computer disks boast even greater speeds, allowing rapid data storage and retrieval (Fig. 31-19).

■ APPLICATION EDDY CURRENTS

Our discussion of induced currents has centered on conducting loops. But induced currents also appear in solid conductors through which magnetic flux is changing. The resistance of a solid piece of conductor is low, so the induced currents are large, resulting in substantial energy dissipation. The presence of these **eddy currents** can make it difficult to move a conductor rapidly through a magnetic field. For example, if you try to push a piece of metal—it need not be a ferrous metal like iron—between the poles of a magnet, you will find yourself working against a magnetic force.

A common demonstration of eddy currents consists of a pendulum with a metal bob that swings between the poles of a magnet (Fig. 31-20). As it swings toward the magnet, the pendulum experiences an increasing magnetic flux that induces eddy currents. The energy dissipated by these currents comes ultimately from the kinetic energy of the pendulum, which therefore stops abruptly between the magnet poles.

Eddy currents provide an alternative to friction brakes for stopping moving machinery. Rapidly rotating saw blades, for example, can be stopped abruptly by an electromagnet activated next to the blade. Similarly, eddy-current brakes are sometimes used on trains and in other applications involving rotating conductors.

In some instances eddy currents are a nuisance, acting just like friction in reducing the efficiency of machinery. To solve this problem, slots are often cut into moving conductors to make the current paths longer, thus increasing the electrical resistance and reducing the eddy currents. For example, if the solid pendulum bob in Fig. 31-20 is replaced by a slotted piece, it then swings more freely through the magnet.

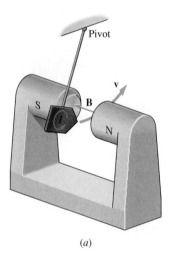

(a)

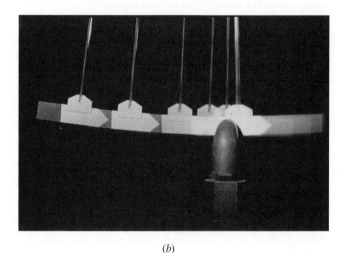

(b)

FIGURE 31-20 (a) An eddy-current pendulum consists of a conductor swinging between magnet poles. As it enters the field region, the conductor experiences a changing magnetic flux that induces a current. You should convince yourself that the current is in the direction shown as the bob begins to enter the field region. (b) The current dissipates energy in the resistance of the conductor, at the expense of its kinetic energy. Strobe photo shows rapid deceleration as the bob passes between the magnet poles.

Lenz's Law and Changing Magnetic Fields

Lenz's law—conservation of energy applied to electromagnetic induction—determines the direction of the induced current even when no motion is involved. Figure 31-21 shows a conducting loop in a magnetic field that points into the page. Suppose the field strength—and, therefore, the magnetic flux through the loop—is *decreasing*. Then the direction of the induced current must be such that the magnetic field it creates *opposes* this decrease. Therefore, the loop's field must reinforce the existing field, which means that the loop's field points into the

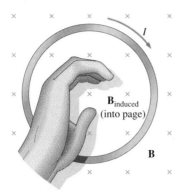

FIGURE 31-21 A uniform magnetic field is decreasing in strength, causing a decrease in magnetic flux through the conducting loop. The induced current in the loop must *oppose* this *decrease,* and therefore loop current goes clockwise in order to produce a magnetic field within the loop that reinforces the original field.

page. Applying the right-hand rule then shows that the loop current is clockwise in Fig. 31-21.

What if the magnetic field in Fig. 31-21 had been increasing in strength? Then the loop current would have been in the opposite direction, to create a magnetic field opposite the existing field in order to oppose its increase.

> **TIP** **Induction Opposes Change** Faraday's law relates the induced emf to the *rate of change* of magnetic flux. Lenz's law says that the induced emf and current are in such a direction as to oppose that *change* in flux—not the flux or the field itself. Figure 31-21 is a case in point; here the induced current is in a direction that actually reinforces the original field and flux—precisely because they are both *decreasing.* The same thing happened in Fig. 31-12, where movement of the loop decreased the flux, giving an induced current that reinforced the existing field.

● EXAMPLE 31-7 LENZ'S LAW

Two coils are arranged as shown in Fig. 31-22. If the resistance of the variable resistor is being increased, what is the direction of the induced current in the fixed resistor R?

Solution
Applying the right-hand rule, we find that the magnetic field of coil A emerges from the right side of the coil, pointing toward coil B. As the resistance increases, the current in coil A decreases, and with it the strength of coil A's magnetic field. This results in a decrease in the magnetic flux through coil B. The induced current in coil B acts to oppose this decrease in flux, so the magnetic field resulting from the induced current reinforces the field from coil A. Thus the field of coil B emerges from the

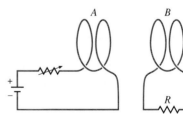

FIGURE 31-22 As the variable resistance is increased, the current in coil A decreases. This decreases the magnetic flux through coil B, resulting in an induced current whose direction is such that the magnetic field it creates in the interior of coil B reinforces the field from coil A, thereby opposing the decrease in flux through B.

right end of the coil and enters on the left end. By the right-hand rule, this requires a current from right to left in the fixed resistor.

EXERCISE A circular conducting loop lies next to a long, straight wire carrying current in the direction indicated in Fig. 31-23. In which direction is the induced current in the loop when the current in the wire is (a) increasing and (b) decreasing?

Answers: (a) counterclockwise; (b) clockwise

Some problems similar to Example 31-7: 4, 10, 19, 27, 32

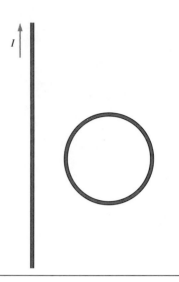

FIGURE 31-23 Which way is the current in the circular loop?

Induction in Open Circuits

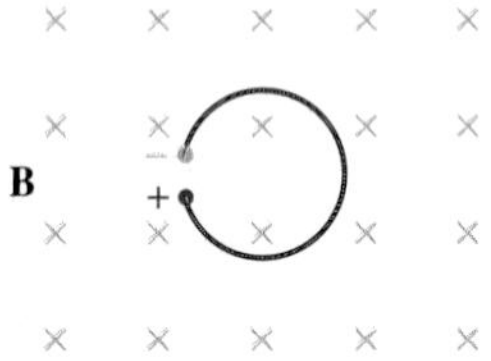

FIGURE 31-24 An open loop. In a changing magnetic field, there is an induced emf that results in charge buildup at the gap. The polarity shown results when the magnetic field strength decreases.

An induced emf also arises in an open circuit, but there it cannot drive a steady current. Figure 31-24 shows the loop and magnetic field of Fig. 31-21, but now with a small gap in the loop. Although the loop is not quite closed, there is still an induced emf. In response to this emf, charge piles up on either side of the gap, creating a voltage across it. Once the gap voltage equals the induced emf, the electric field associated with the separated charge opposes the emf's tendency to move charge, and a steady state is reached. In the open-circuit case, the entire emf implied by Equation 31-2 is available at the gap. The polarity can be determined by considering what would happen if current did flow. Since current flowed clockwise in the closed loop of Fig. 31-21, positive charge will accumulate at the bottom of the gap, as shown in Fig. 31-24.

Another example of an induced emf in an open circuit is the motional emf arising as a single conductor moves through a magnetic field. Example 31-8 treats this case quantitatively.

● EXAMPLE 31-8 THE TETHERED SATELLITE

A 1992 flight of the space shuttle Atlantis attempted to demonstrate a new method for supplying electric power to orbiting spacecraft. Once in orbit, Atlantis deployed a 520-kg satellite attached to the shuttle through a conducting tether (Fig. 31-25). With the tethered satellite flying vertically above Atlantis the two moved at approximately right angles to Earth's magnetic field, whose strength was about 30 μT at the orbital position. Find the motional emf that would have developed between the tethered satellite and the shuttle had the tether been extended its full 20-km length. (Difficulties with the tether's reel mechanism prevented full deployment from the shuttle, but a subsequent rocket-borne tether was successful.)

Solution
Figure 31-26 shows the pair flying at right angles to the magnetic field. We can imagine forming a closed circuit by letting the satellite and shuttle slide along a system of conducting rails, just like that of Example 31-4. The system is then identical to that example, and we have

$$\mathcal{E} = B\ell v = (30\times10^{-6}\text{ T})(20\times10^{3}\text{ m})(7.8\times10^{3}\text{ m/s})$$

$$= 4.7\text{ kV},$$

where we calculated the speed in low-Earth orbit in Example 9-3.

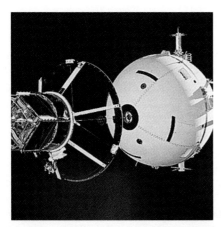

FIGURE 31-25 Tethered satellite being deployed from the space shuttle Atlantis.

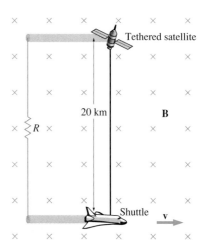

FIGURE 31-26 The tethered satellite generates electric power from the motional emf developed between the spacecraft and a satellite connected by a conducting tether (Example 31-8).

Since the magnetic flux through the rail arrangement would be increasing, the current would flow counterclockwise so that the magnetic field of the induced current would oppose this increase. That means positive charge will accumulate on the satellite and negative charge on the shuttle—a result that can also be obtained by considering the magnetic force on free charges in the tether (see Problem 57).

At the shuttle's orbital altitude there are enough free electrons from ionization of the upper atmosphere for the tenuous gas surrounding the shuttle to carry a current, completing a circuit and allowing exploitation of the induced emf for electric power generation.

EXERCISE An airplane with a wingspan of 44 m is in level flight at 300 m/s over the north pole, where Earth's magnetic field has magnitude 0.62 G and points vertically downward. Find the induced emf between the wing tips.

Answer: 0.82 V

Some problems similar to Example 31-8: 10, 13, 57, 58 ●

31-4 INDUCED ELECTRIC FIELDS

What do we really mean by an induced emf? In a circuit containing a battery, the notion of emf is clear—the emf arises in a specific device where chemical energy is converted to electrical energy associated with charge separation. This charge separation sets up an electric field that drives current in an external circuit. In the case of motional induced emf we also have a clear picture: the emf arises from the separation of charge associated with magnetic forces on the free charges in a moving conductor; again, the electric field associated with this charge separation may drive a current.

Now consider the current induced in a conducting loop by a magnetic field that changes with time. No motion is involved, yet there must be a force on the free charges in the conductor. The one force we know that acts on stationary charges is the electric force. Electric forces arise from electric fields, so there must be an electric field in a conducting loop in a changing magnetic field. This field is called an **induced electric field.** It has the same effect on charges—exerting a force $q\mathbf{E}$—as do the electric fields we considered earlier. The field itself, however, originates not in electric charges but in changing magnetic fields.

An induced electric field arises whenever a magnetic field changes with time—whether or not an electric circuit is present. When a circuit is present, then the induced field may drive a current. But the induced field, not the current, is fundamental. A single, stationary electron placed in a changing magnetic field will experience an *electric* force—clear evidence for the existence of an electric field.

We have been thinking of Faraday's law as a relation between the emf induced in a circuit and the rate of change of magnetic flux through that circuit. We now know that induced electric fields are the fundamental manifestation of changing magnetic flux, and that these fields arise whether or not circuits are present. We need to reformulate Faraday's law to describe induced electric fields without reference to circuits. The induced emf $\mathscr{E}$ that we have been writing on the left-hand side of Faraday's law means simply the work per unit charge gained by a test charge moved around a circuit. Since work is the line integral of force over distance, and electric field is the force per unit charge, we can write the emf as the line integral of the electric field. Then Faraday's law becomes

$$\oint \mathbf{E} \cdot d\boldsymbol{\ell} = -\frac{d\phi}{dt} = -\frac{d}{dt}\int \mathbf{B} \cdot d\mathbf{A}. \qquad \text{(Faraday's law)} \qquad (31\text{-}3)$$

Here the line integral on the left-hand side is taken over *any* closed loop, which need not coincide with a circuit. The flux on the right-hand side is the surface integral of the magnetic field over *any* open surface bounded by the loop on the left-hand side.

Faraday's law in the form 31-3 makes no reference to wires or other circuits. It simply describes induced electric fields, which occur whenever there are changing magnetic fields. If electric circuits are present, then induced currents occur as well—but it is the induced electric fields that are fundamental. We can state Faraday's law loosely but powerfully by saying the following:

▌ A changing magnetic field creates an electric field.

This direct interaction between the fields is the basis for numerous practical devices and, as we shall see in Chapter 34, is essential to the existence of light.

Note the similarity between Faraday's law and Ampère's law (Equation 30-8). Faraday's law gives the line integral of the electric field around a closed loop in terms of the rate of change of magnetic flux through the loop. Ampère's law gives the line integral of the magnetic field around a closed loop in terms of the current through the loop. Both give fields that *encircle* their sources—current for the source of magnetic field and changing magnetic field for the induced electric field. That means the configuration of an induced electric field is very different from that of an electric field originating in electric charge. Induced fields have no beginning or end; their field lines form closed loops encircling regions of changing magnetic field.

When a changing magnetic field has sufficient symmetry, we can evaluate the induced electric field in the same way we did the magnetic field of a symmetric current distribution. Example 31-9 illustrates this procedure.

● EXAMPLE 31-9 AN INDUCED ELECTRIC FIELD

A long solenoid has circular cross section of radius R. The current in the solenoid is increasing and, as a result, the uniform magnetic field within the solenoid increases with time; the field magnitude is given by $B = bt$, with b a constant and t the time. Find the induced electric field outside the solenoid, a distance r from the solenoid axis.

Solution

The induced electric field has no beginning or end, so the field lines must encircle the solenoid. The only field consistent with the symmetry consists of circular field lines centered on the solenoid axis, as suggested in Fig. 31-27. Since the solenoid field points into the page and is increasing in strength, the direction of the induced electric field must be such that any current it might drive would produce a magnetic field *opposing* the increase in the solenoid field. Applying the right-hand rule then shows that the induced electric field runs counterclockwise.

Faraday's law relates the line integral of the electric field to the rate of change of the encircled magnetic flux. Here a suitable integration loop is itself a circle centered on the solenoid axis. Symmetry shows that the field magnitude is constant over this loop, and if we circle the loop in the direction of the field then $\cos\theta$ in the dot product $\mathbf{E} \cdot d\boldsymbol{\ell}$ is 1. Thus

$$\oint \mathbf{E} \cdot d\boldsymbol{\ell} = \oint E\, d\ell = E \oint d\ell = 2\pi r E,$$

since $\oint d\ell$ is just the loop circumference. Faraday's law relates this quantity to the rate of change of the encircled magnetic flux. Here the flux is due to the uniform magnetic field inside the solenoid, perpendicular to our loop and confined to a circular area of radius R. (We use the long-solenoid approximation, neglecting the small magnetic field that must exist outside a solenoid of finite length.) So the encircled flux is just $\phi_B = \pi R^2 B = \pi R^2 bt$, and its rate of change is

$$\frac{d\phi_B}{dt} = \frac{d}{dt}(\pi R^2 bt) = \pi R^2 b.$$

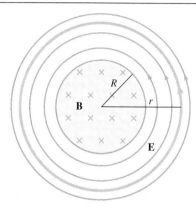

FIGURE 31-27 Cross section of a solenoid of radius R whose magnetic field is increasing with time. Field lines of the induced electric field are circles concentric with the solenoid axis. Also shown is a circular loop (gray) of radius r for evaluating the line integral in Faraday's law.

Then Faraday's law gives

$$2\pi r E = -\pi R^2 b,$$

or

$$E = -\frac{R^2 b}{2r}.$$

We've already accounted for the minus sign in arguing that the field circles counterclockwise. The $1/r$ dependence of the field strength on distance should come as no surprise; points with $r > R$ are outside a cylindrically symmetric distribution, in this case a distribution of changing magnetic flux. We found the same $1/r$ dependence for the electric and magnetic fields outside, respectively, a cylindrically symmetric charge distribution and a cylindrically symmetric current distribution.

EXERCISE Show that the electric field at points *inside* the solenoid has magnitude $br/2$.

Some problems similar to Example 31-9: 37–39, 41, 43, 44 ●

In Chapter 30 we derived the magnetic field of a solenoid using Ampère's law—which, as we've formulated it so far, applies only to steady currents. But a solenoid whose field is changing—as in Example 31-9—must have a changing current, so how can we talk about it? In fact, Ampère's law gives a good approximation to the field produced by changing current, provided that change is sufficiently slow. In the examples and problems of this chapter we assume that to be the case. In Chapter 34 we will explore what happens if we relax the assumption of slowly changing currents.

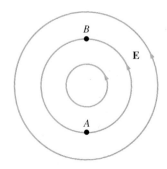

FIGURE 31-28 Two points in an induced electric field. If a positively charged particle moves counterclockwise along the field line from A to B the field does work on it. But if it moves from A to B in the clockwise direction the field does negative work; that is, an external force must do work on the charge. The work done is not path independent, and the induced electric field is not conservative.

FIGURE 31-29 This induction cooktop has no burners. Instead, coils beneath the cooktop produce time-varying magnetic fields that induce currents in the iron cookware. Resistance in the cookware then results in heating.

Conservative and Nonconservative Electric Fields

We have seen that static electric fields—those beginning and ending on stationary charge distributions—are conservative, meaning that the work required to move a charge between two points is independent of the path taken. A consequence is that it takes no work to move around a closed path in an electrostatic field; mathematically, we express this by writing

$$\oint \mathbf{E} \cdot d\boldsymbol{\ell} = 0. \quad \text{(electrostatic field)}$$

In contrast, induced electric fields generally form closed loops, and Faraday's law shows that the line integral of the electric field around a closed path in such a field is decidedly not zero. That means the induced electric field does work on a charge moved around a *closed* path and that the work done in moving between two points cannot be independent of the path taken (Fig. 31-28). The induced electric field is therefore not conservative.

The work done on charges moving around closed paths in nonconservative electric fields can be useful. A simple example is the induction cooktop (Fig. 31-29), where food cooks in a special conducting pan heated by currents in the pan itself. Those currents are driven by an induced electric field originating from a changing magnetic field in coils just below the cooktop's surface. Much more generally, any electrical device that derives its power from a generator—and that includes almost every electrical thing you use—is part of a circuit in which a nonconservative electric field drives the current.

■ APPLICATION THE TOKAMAK

In Chapter 29 we described how magnetic fields are used to confine very hot plasma in fusion reactors and showed why most reactors have a toroidal shape. The most promising of the toroidal reactor designs is the **tokamak,** in which the confining magnetic field arises in part from current within the plasma itself. Discharging a large capacitor bank through a coil produces a rapidly changing magnetic field in the "hole" of the toroidal "doughnut," resulting in an induced electric field within the torus. This field drives the plasma current that produces a magnetic field component around the minor radius of the device. Other coils around the torus provide a magnetic field component around the long dimension of the torus, giving a net magnetic field that spirals around the plasma (Fig. 31-30).

As in any conductor, particle collisions dissipate energy gained from the electric field, in this case heating the plasma. Unfortunately the resistance of a plasma drops at high temperature, making it difficult to achieve the temperature needed for fusion with resistive heating alone. Nevertheless, the induced electric field helps bring the plasma close to the fusion temperature, as well as driving currents that produce the confining magnetic field.

Since the tokamak uses an induced electric field, its operation requires a changing magnetic field. Therefore, the tokamak cannot be run continuously but operates instead in a

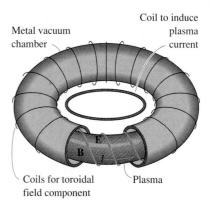

FIGURE 31-30 A tokamak fusion reactor. Current in the plasma is driven by an electric field induced by the magnetic field change associated with a rapid change in current through the central coil. Additional coils wound around the toroid give a steady magnetic field in the same direction as the plasma current. Combining with the field from the plasma current gives a net field that spirals around the plasma.

pulsed mode as current builds up rapidly and then decays in preparation for another pulse. Today's experimental tokamaks can achieve pulse durations of 10 s or more, during which they produce significant fusion energy.

■ FOR FURTHER THOUGHT NONCONSERVATIVE ELECTRIC FIELDS

The nonconservative nature of the induced electric field is strikingly demonstrated if you attempt to measure potential differences in nonconservative fields. In Chapter 25, we defined potential difference as the work required to move a unit charge between two points, and stressed that this work is independent of path for a conservative field. But when the field is nonconservative, the work is not independent of path, and the concept of potential becomes ambiguous.

Figure 31-31 shows an end view of a long solenoid surrounded by three identical resistors bent into circular arcs. If the solenoid current is increasing, an induced electric field appears in the resistors, and drives a current I in the counterclockwise direction.

Because they have the same resistance and carry the same current, the potential difference across each resistor should be the same. We could try to measure the potential difference across one resistor, for example, by connecting a voltmeter as shown in Fig. 31-31b. With current I flowing through the resis-

tance R, this meter reads IR. Since the current flows counterclockwise, we must connect the positive voltmeter terminal to point B.

But now try to measure the potential difference across the other two resistors together, as in Fig. 31-31c. With the current I flowing through the total resistance $2R$, the meter now reads $2IR$. We have two voltmeters with their terminals connected to the same points, and yet they indicate different voltages. Not only are the magnitudes of the voltages different, but even their polarities differ. How can this be? We are experiencing the nonconservative nature of the induced electric field. The two voltmeters are positioned differently with respect to the changing magnetic flux, so they sample different regions of the induced electric field. Even though the meters are connected to the same points, they measure the line integral of the induced electric field over *different* paths, and so they do not read the same voltage.*

* For a fascinating discussion of voltage measurement in induced fields, see R. Romer, "What do 'Voltmeters' Measure?: Faraday's Law in a Multiply Connected Region," *American Journal of Physics,* vol. 50, no. 12, pp. 1089–1091 (December 1982).

FIGURE 31-31 (*a*) End view of a long solenoid surrounded by three resistors in series. A changing magnetic field in the solenoid induces an emf in the resistors, and the same current *I* flows through each. (*b*) A voltmeter connected between points *A* and *B* indicates the voltage *IR* across one resistor. (*c*) A second voltmeter connected to the *same points* indicates voltage 2*IR* with the opposite sign.

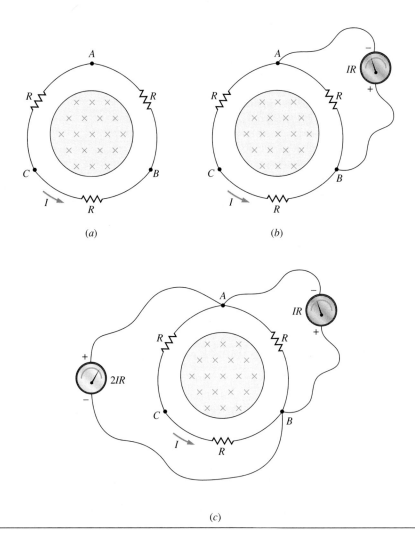

(*a*)   (*b*)

(*c*)

31-5 DIAMAGNETISM

In Section 30-6 we discussed paramagnetic and ferromagnetic materials, in which atomic magnetic dipoles align with an applied magnetic field, causing an attractive interaction between the material and a magnet. We also mentioned diamagnetic materials, in which induced magnetic dipoles align antiparallel to the applied field, causing a repulsive force. We are now ready to understand diamagnetism as a manifestation of Faraday's law at the microscopic level.

In a purely diamagnetic material, current loops associated with pairs of atomic electrons exactly cancel, leaving atoms with no intrinsic magnetic moments. Fig. 31-32*a* shows a simplified model to describe such an atom. The picture should not be taken too literally, for it uses classical physics to describe a phenomenon properly within the domain of quantum mechanics. Nevertheless, it shows qualitatively how diamagnetism is an electromagnetic induction effect.

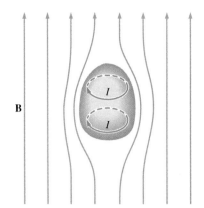

FIGURE 31-32 Simplified, classical model for a diamagnetic atom. (*a*) In the absence of an applied field, magnetic moments (cross and dot) associated with a pair of electrons exactly cancel. (*b*) The changing flux associated with an increasing magnetic field induces an electric field that speeds up one electron and slows down the other, giving rise to a net magnetic moment. (In applying the right-hand rule, remember that the electrons are negative.)

FIGURE 31-33 Induced currents in a superconductor completely cancel an applied magnetic field. Net result is that a magnetic field cannot penetrate the superconductor.

The electron orbiting clockwise in Fig. 31-32*a* has a magnetic moment pointing out of the page (apply the right-hand rule and remember that the electron carries a negative charge). In the absence of an applied magnetic field, this moment is exactly cancelled by the equal but oppositely directed moment of the other electron. But now suppose a magnetic field is applied, for example by moving the north pole of a magnet down toward the plane of Fig. 31-32. This produces a changing magnetic field in the plane of the figure, resulting in an induced electric field that alters the speeds of the orbiting electrons. The direction of this effect must be such as to oppose the increase in field. Thus the electron whose magnetic moment points out of the page has its moment increase, while the moment pointing into the page decreases. Figure 31-32*b* shows how these changes result from the alteration of electron speeds by the induced electric field.

Once the field is applied and the orbital speeds of the two electrons are no longer identical, the atom now has a net magnetic moment that points out of the page in Fig. 31-32, opposing the incoming magnet and resulting in a repulsive force. This repulsion is the distinguishing characteristic of diamagnetism. We listed a number of diamagnetic materials in Table 30-2.

A material that is entirely superconducting is perfectly diamagnetic, with magnetic susceptibility -1. This means that the magnetic field resulting from induced currents within the material completely cancels any applied field. Since these induced currents persist in the zero-resistance superconductor, the material completely excludes magnetic fields from its interior, a phenomenon known as the Meissner effect (Fig. 31-33). The repulsive force associated with the magnetic moments of a permanent magnet and a nearby superconductor results in the widely publicized phenomenon of magnetic levitation (Fig. 31-34). Readers of the extended version of this text will explore superconductivity further in Chapter 42.

FIGURE 31-34 A small magnet is levitated above a wafer of high-temperature superconductor in a bath of liquid nitrogen at 77 K.

CHAPTER SYNOPSIS

Summary

1. **Electromagnetic induction** is a fundamental phenomenon linking magnetism and electricity. Induction is described by **Faraday's law,** which states that a **changing magnetic field** produces an **induced electric field.** Unlike the conservative electrostatic field of an electric charge, this induced field is **nonconservative,** meaning it can do work on charges as they move around a closed loop. Faraday's law relates the line integral of this nonconservative electric field around an arbitrary loop to the rate of change of magnetic flux through a surface bounded by that same loop:

$$\oint \mathbf{E} \cdot d\boldsymbol{\ell} = -\frac{d\phi_B}{dt} = -\frac{d}{dt}\int \mathbf{B} \cdot d\mathbf{A}.$$

2. In order for energy to be conserved, the induced electric field is in such a direction as to oppose the change in flux that gives rise to it. This energy-conserving aspect of Faraday's law is called **Lenz's law** and is reflected mathematically in the minus sign on the right-hand side of Faraday's law.

3. When a conductor is present, the nonconservative electric field manifests itself as an **induced emf:**

$$\mathcal{E} = -\frac{d\phi_B}{dt}.$$

This emf drives an **induced current** in any circuit with finite resistance. It does not matter whether the magnetic flux is changed by moving a conductor in a magnetic field, or by moving a magnetic field near a conductor, or by altering the shape or orientation of the conductor. The generation of electric power by moving conducting loops in magnetic fields is an important technological example of induced currents.

4. **Diamagnetism** is a manifestation of electromagnetic induction on the atomic scale. Application of a magnetic field to a diamagnetic material results in induced atomic magnetic moments that cause the material to be repelled from a magnet.

Terms You Should Understand

(Pairs are closely related terms whose distinction is important; number in parentheses is chapter section where term first appears.)

electromagnetic induction (31-1)
induced current, induced emf (31-1, 31-2)
magnetic flux (31-2)
Faraday's law, Lenz's law (31-2, 31-3)
motional emf (31-3)
generator (31-3)
eddy currents (31-3)
induced electric field (31-4)

Symbols You Should Recognize

ϕ_B (31-2)
$\oint \mathbf{E} \cdot d\boldsymbol{\ell}$ (31-4)

Problems You Should Be Able to Solve

calculating magnetic flux in uniform and nonuniform fields (31-2)
calculating induced emf's from changing magnetic flux (31-2)
determining the direction of induced emf's and currents (31-3)
calculating induced electric fields in symmetric situations (31-4)

Limitations to Keep in Mind

Electromagnetic induction requires a *change* in magnetic flux. Values of magnetic field or flux alone do not matter; only the rate of change of those quantities is important.

QUESTIONS

1. A copper penny falls on a vertical path that takes it between the poles of a magnet. Does it hit the ground faster or slower than if no magnet were present?

2. A bar magnet is moved toward a conducting ring, as shown in Fig. 31-35. What is the direction of the induced current in the ring?

3. Figure 31-36 shows two concentric conducting loops, the outer connected to a battery and a switch. The switch is initially open. It is then closed, left closed for a while, then opened again. Describe the currents in the inner loop during the entire procedure.

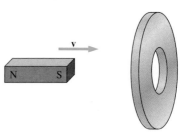

FIGURE 31-35 Question 2.

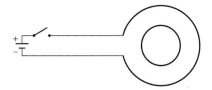

FIGURE 31-36   Question 3.

4. An electric generator is being turned at constant speed. A load resistor *R* is connected across the generator terminals. If the electrical resistance of the load is lowered, does the generator get easier or harder to turn?

5. Service manuals for cars often tell you to set the idle speed of the engine with the headlights on. Why? What does this have to do with electromagnetic induction?

6. Figure 31-37 shows two square wire loops, the first containing a battery and variable resistor. The resistor is initially at the midpoint of its resistance range. Should its resistance be lowered or raised in order to induce a clockwise current in the right-hand loop?

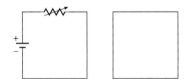

FIGURE 31-37   Question 6.

7. Consider the simple motor shown in Fig. 29-35. What happens if you connect a resistor across the motor terminals, and turn the motor by hand? What is the difference between a motor and a generator?

8. Figure 31-38 shows an open wire loop in a magnetic field. The field is changing, and charge has piled up at the loop gap with the polarity indicated. Is the magnetic field strength increasing or decreasing?

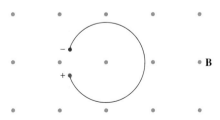

FIGURE 31-38   Question 8.

9. A constant, uniform magnetic field points into the page. A flexible, circular conducting ring lies in the plane of the page. If the ring is stretched, maintaining its circular shape, what is the direction of the induced current?

10. A student argues that it takes work to stretch the ring in the preceding question, and that therefore the ring should release all the work as energy if allowed to shrink. Is this right? Why or why not?

11. When a magnet is moved near a superconductor, the magnetic field lines never enter the superconductor. Why not?

12. Is it possible to produce an induced current that never changes? How or why not? Could you produce an induced current that was steady for some finite time? How or why not?

13. When you push a bar magnet into a conducting loop, you do work. What happens when you pull it out the other side?

14. You are turning a generator in such a way that the current it delivers remains constant. As you lower the load resistance across the generator, does the generator get easier or harder to turn?

15. Devise a way of measuring a magnetic field using Faraday's law.

FIGURE 31-39   Question 16.

16. In Fig. 31-39, a copper ring was originally resting on the wooden structure, surrounding the coil. When a rapidly changing current was applied to the coil, the ring was ejected into the air. Explain this phenomenon.

17. Fluctuations in Earth's magnetic field due to changing solar activity can wreak havoc with communications, even those using underground cables. How is this possible?

18. Why is it not possible to run a tokamak on a continuous basis?

19. Conventional brakes on a car need large surface areas to dissipate the heat of friction when the brakes are applied. Would eddy-current brakes have the same problem?

20. Which way would the eddy currents flow in Fig. 31-20*a* as the bob continues its swing and begins to emerge from the field?

21. In Chapter 29, we pointed out that a static magnetic field cannot change the energy of a charged particle. Is this true of a changing magnetic field? Discuss.

22. A long solenoid of circular cross section is oriented so that its magnetic field points out of the page, as shown in Fig. 31-40. The solenoid current is increasing. (a) What is the direction of the induced electric field at points A and B in the figure? (b) What is the magnitude of the induced electric field in the center of the solenoid? (Don't calculate! Argue from symmetry.)

23. Is the concept of electric potential (Chapter 25) useful in a nonconservative electric field? Give an example to substantiate your answer.

24. Can an induced electric field exist in the absence of a conductor?

25. Could you tell whether a given electric field arises from electric charge or from a changing magnetic field? How or why not?

26. Does a diamagnetic material experience a force in a uniform magnetic field?

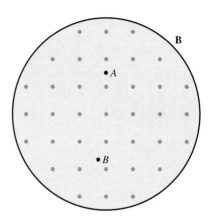

FIGURE 31-40   Question 22.

PROBLEMS

Sections 31-2 and 31-3 Faraday's Law and Induction and the Conservation of Energy

1. Show that the volt is the correct SI unit for the rate of change of magnetic flux, making Faraday's law dimensionally correct.

2. A bar magnet is moved steadily through a wire loop, as shown in Fig. 31-41. Sketch qualitatively the current and power dissipation in the loop as functions of time. Take as positive a current flowing out of the plane of the page at the top of the loop, and indicate the position of the magnet on your time axis.

3. Find the magnetic flux through a circular loop 5.0 cm in diameter oriented with the loop normal at 30° to a uniform 80-mT magnetic field.

4. A circular wire loop 40 cm in diameter has 100-Ω resistance and lies in a horizontal plane. A uniform magnetic field points vertically downward, and in 25 ms it increases linearly from 5.0 mT to 55 mT. Find the magnetic flux through the loop at (a) the beginning and (b) the end of the 25 ms period. (c) What is the loop current during this time? (d) Which way does this current flow?

5. A conducting loop of area 240 cm² and resistance 12 Ω lies at right angles to a spatially uniform magnetic field. The loop carries an induced current of 320 mA. At what rate is the magnetic field changing?

6. A conducting loop of area A and resistance R lies at right angles to a spatially uniform magnetic field. At time $t = 0$ the magnetic field and loop current are both zero. Subsequently, the current increases according to $I = bt^2$, where b is a constant with the units A/s². Find an expression for the magnetic field strength as a function of time.

7. A conducting loop with area 0.15 m² and resistance 6.0 Ω lies in the x-y plane. A spatially uniform magnetic field points in the z direction. The field varies with time according to $B_z = at^2 - b$, where $a = 2.0$ T/s² and $b = 8.0$ T. Find the loop current (a) when $t = 3.0$ s and (b) when $B_z = 0$.

8. The magnetic field inside a 20-cm-diameter solenoid is increasing at the rate of 2.4 T/s. How many turns should a coil wrapped around the outside of the solenoid have in order that the emf induced in the coil be 15 V?

9. A square wire loop of side ℓ and resistance R is pulled with constant speed v from a region of no magnetic field until

FIGURE 31-41   Problem 2.

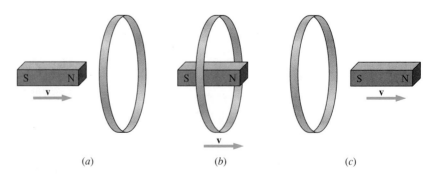

(a)   (b)   (c)

it is fully inside a region of constant, uniform magnetic field **B** perpendicular to the loop plane. The boundary of the field region is parallel to one side of the loop. Find an expression for the total work done by the agent pulling the loop.

10. A 1.8-m high runner sprints eastward at 9.5 m/s along the equator, where Earth's magnetic field points horizontally with a strength of 31 μT. (a) What is the magnitude of the emf induced between the runner's head and feet? (b) Which end is positive? *Hint:* See Fig. 30-6.

11. In Fig. 31-24 the loop radius is 15 cm, and the magnetic field is decreasing at the rate of 550 T/s. If the gap width is small compared with the loop circumference, what is the voltage across the gap?

12. A 5-turn coil 1.0 cm in diameter is rotated at 10 rev/s about an axis perpendicular to a uniform magnetic field. A voltmeter connected to the coil through rotating contacts reads a peak value of 360 μV. What is the magnetic field strength?

13. The wingspan of a 747 jetliner is 60 m. If the plane is flying at 960 km/h in a region where the vertical component of Earth's magnetic field is 0.20 G, what emf develops between the plane's wingtips?

14. A square wire loop 3.0 m on a side is perpendicular to a uniform magnetic field of 2.0 T. A 6-V light bulb is in series with the loop, as shown in Fig. 31-42. The magnetic field is reduced steadily to zero over a time Δt. (a) Find Δt such that the light will shine at full brightness during this time. (b) Which way will the loop current flow?

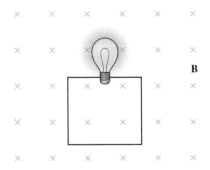

FIGURE 31-42   Problem 14.

15. In Example 31-2 take $a = 1.0$ cm, $w = 3.5$ cm, and $\ell = 6.0$ cm. Suppose the rectangular loop is a conductor with resistance 50 mΩ and that the current I in the long wire is increasing at the rate of 25 A/s. Find the induced current in the loop. In what direction does it flow?

16. A windmill with conducting blades of length ℓ is rotating with angular speed ω about a horizontal axis; the horizontal component of Earth's magnetic field at its location is B_x. Find an expression for the emf developed between the blade tips and the central axis.

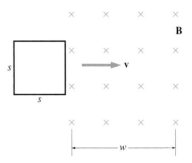

FIGURE 31-43   Problem 17.

17. A square conducting loop of side $s = 0.50$ m and resistance $R = 5.0$ Ω moves to the right with speed $v = 0.25$ m/s. At time $t = 0$ its rightmost edge enters a uniform magnetic field $B = 1.0$ T pointing into the page, as shown in Fig. 31-43. The magnetic field covers a region of width $w = 0.75$ m. Plot (a) the current and (b) the power dissipation in the loop as functions of time, taking a clockwise current as positive and covering the time until the entire loop has exited the field region.

18. A 2.0-m-long solenoid is 15 cm in diameter and consists of 2000 turns of wire. The current in the solenoid is increasing at the rate of 1.0 kA/s. (a) Find the current in a wire loop with diameter 10 cm and resistance 5.0 Ω, lying inside the solenoid in a plane perpendicular to the loop axis. (b) Repeat for a similarly oriented loop with diameter 25 cm, lying entirely outside the solenoid.

19. A solenoid 2.0 m long and 30 cm in diameter consists of 5000 turns of wire. A 5-turn coil with negligible resistance is wrapped around the solenoid and connected to a 180-Ω resistor, as shown in Fig. 31-44. The direction of the current in the solenoid is such that the solenoid's magnetic field points to the right. At time $t = 0$ the solenoid current begins to decay exponentially, being given by $I = I_0 e^{-t/\tau}$, where $I_0 = 85$ A, $\tau = 2.5$ s, and t is the time in seconds. (a) What is the direction of the current in the resistor as the solenoid current decays? What is the value of the resistor current at (b) $t = 1.0$ s and (c) $t = 5.0$ s?

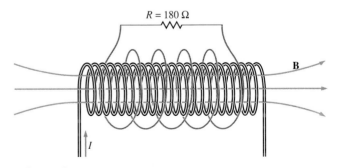

FIGURE 31-44   Problems 19–21.

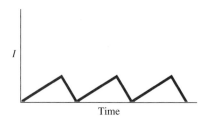

FIGURE 31-45   Problem 20.

20. Make a qualitative plot of the resistor current in the pre-ceding problem as a function of time if the solenoid current has the form shown in Fig. 31-45. Take a left-to-right resistor current as positive.

21. (a) Find an expression for the resistor current in Problem 19 if the solenoid current is given by $I = I_0 \sin \omega t$, where $I_0 = 85$ A and $\omega = 210$ s^{-1}. (b) What is the peak current in the resistor? (c) What is the resistor current when the solenoid current is a maximum?

22. A magnetic field is described by $\mathbf{B} = B_0 \sin \omega t \, \hat{\mathbf{k}}$, where $B_0 = 2.0$ T and $\omega = 10$ s^{-1}. A conducting loop with area 150 cm^2 and resistance 5.0 Ω lies in the x-y plane. Find the induced current in the loop (a) at $t = 0$ and (b) at $t = 0.10$ s.

23. In the preceding problem, what is the first time after $t = 0$ when the loop current will be zero?

24. A car alternator consists of a 250-turn coil 10 cm in diame-ter in a magnetic field of 0.10 T. If the alternator is turning at 1000 revolutions per minute, what is its peak output voltage?

25. A credit-card reader extracts information from the card's magnetic stripe as it is pulled past the reader's head. At some instant the card motion results in a magnetic field at the head that is changing at the rate of 450 μT/ms. If this field passes perpendicularly through a 5000-turn head coil 2.0 mm in diameter, what will be the induced emf?

26. A generator consists of a rectangular coil 75 cm by 1.3 m, spinning in a 0.14-T magnetic field. If it is to produce a 60-Hz alternating emf (i.e., $\mathscr{E} = \mathscr{E}_0 \sin 2\pi ft$, where $f = 60$ Hz) with peak value 6.7 kV, how many turns must it have?

27. Figure 31-46 shows a pair of parallel conducting rails a distance ℓ apart in a uniform magnetic field $\mathbf{B}$. A resistance R is connected across the rails, and a conducting bar of negligible resistance is being pulled along the rails with velocity $\mathbf{v}$ to the right. (a) What is the direction of the current in the resistor? (b) At what rate must work be done by the agent pulling the bar?

28. The resistor in the preceding problem is replaced by an ideal voltmeter. (a) To which rail should the positive meter terminal be connected? (b) At what rate must work be done by the agent pulling the bar?

29. A battery of emf $\mathscr{E}$ is inserted in series with the resistor in Fig. 31-46, with its positive terminal toward the top rail. The bar is initially at rest, and now no agent pulls it. (a) Describe the bar's subsequent motion. (b) The bar eventually reaches a constant speed. Why? (c) What is that constant speed? Express in terms of the magnetic field, the battery emf, and the rail spacing ℓ. Does the resistance R affect the final speed? If not, what role does it play?

30. A toroidal coil of square cross section has inner radius a and outer radius b. It consists of N turns of wire and carries a time-varying current $I = I_0 \sin \omega t$. A single-turn wire loop encircles the toroid, passing through its center hole as shown in Fig. 31-47. Find an expression for the peak emf induced in the loop.

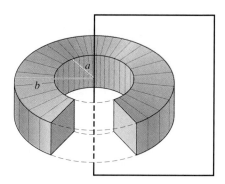

FIGURE 31-47   Problem 30.

31. A pair of parallel conducting rails 10 cm apart lie at right angles to a uniform magnetic field $\mathbf{B}$ of magnitude 2.0 T, as shown in Fig. 31-48. A 5.0-Ω and a 10-Ω resistor lie across the rails and are free to slide along them. (a) The 5-Ω resistor is held fixed, and the 10-Ω resistor is pulled to the right at 50 cm/s. What are the direction and magnitude of the induced current? (b) Now the 10-Ω resistor is held

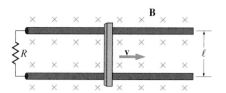

FIGURE 31-46   Problems 27–29, 33, 63.

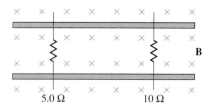

FIGURE 31-48   Problems 31, 32, and 64.

fixed, and the 5-Ω resistor is pulled to the left at 50 cm/s. What are the direction and magnitude of the induced current? (c) What is the power dissipation in the 10-Ω resistor in both cases?

32. In Fig. 31-48 the 10-Ω resistor is being moved to the right at a constant 50 cm/s. The 5-Ω resistor, initially at rest, is placed across the conducting rails. Describe qualitatively its subsequent motion, and determine its final speed.

33. In Fig. 31-46, take $\ell = 10$ cm, $B = 0.50$ T, $R = 4.0\ \Omega$, and $v = 2.0$ m/s. Find (a) the current in the resistor, (b) the magnetic force on the bar, (c) the power dissipation in the resistor, and (d) the mechanical work done by the agent pulling the bar. Compare your answers to (c) and (d).

34. A square conducting loop of side ℓ and resistance R lies in the x-y plane and is being moved with constant velocity $\mathbf{v} = v\hat{\imath}$, its sides parallel to the x and y axes. For $x < 0$ there is a uniform magnetic field $\mathbf{B} = B_0\hat{\mathbf{k}}$; for $x > 0$ the field is nonuniform and is given by $\mathbf{B} = (B_0 + bx)\hat{\mathbf{k}}$, where B_0 and b are positive constants. At time $t = 0$ the trailing side of the loop crosses the y axis, so the loop is entirely in the nonuniform field region. (a) Find an expression for the loop current for times $t \geq 0$. (b) Which way does the current flow, as viewed from the positive z axis?

35. A circular loop 40 cm in diameter is made from a flexible conductor and lies at right angles to a uniform 12-T magnetic field. At time $t = 0$ the loop starts to expand, its radius increasing at the rate of 5.0 mm/s. Find the induced emf in the loop (a) at $t = 1.0$ s and (b) at $t = 10$ s.

36. A spherical balloon initially has radius 20 cm. A conducting stripe with resistance 2.4 Ω is painted around the balloon's "equator," and a uniform 1.5-T magnetic field points perpendicular to the "equatorial plane." The balloon is deflated so its radius decreases at the constant rate of 1.0 cm/s. Find the current in the conducting stripe when the radius has been reduced to 10 cm, assuming the stripe's resistance stays constant.

Section 31-4 Induced Electric Fields

37. The induced electric field 12 cm from the axis of a solenoid with 10 cm radius is 45 V/m. Find the rate of change of the solenoid's magnetic field.

38. Find the electric force on a 50-μC charge inside the solenoid of Problem 18, if the charge is 5.0 cm from the solenoid axis.

39. Figure 31-49 shows a top view of a tokamak. The magnetic field in the center is confined to a circular area of radius 50 cm, and during a pulse it increases at the rate of 5.1 T/ms. (a) What is the magnitude of the induced electric field in the tokamak, 1.2 m from the center of the field region in Fig. 31-49? (b) What is the field direction? (c) If a proton circles the tokamak once at this radius, going with the electric field, how much energy does it gain?

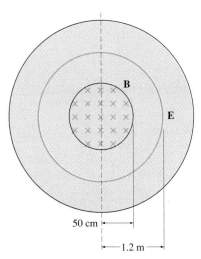

FIGURE 31-49 Top view of a tokamak (Problem 39).

40. A uniform magnetic field points into the page in Fig. 31-50. In the same region an electric field points straight up, but increases with position at the rate of 10 V/m² as you move to the right. Apply Faraday's law to a rectangular loop to show that the magnetic field must be changing with time, and calculate the rate of change.

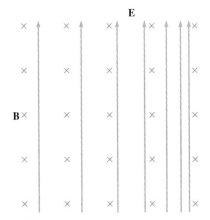

FIGURE 31-50 The magnetic field must be changing. Why? (Problem 40)

41. In Example 31-9, take the solenoid radius $R = 10$ cm and suppose the magnetic field inside the solenoid is given by $B = 0.10t^3 - 1.1t^2 + 2.8t$, with B in T and t in ms. (a) Find the electric field strength 14 cm from the solenoid axis at $t = 1.0$ ms. (b) Find a time when the induced electric field in and around the solenoid is zero.

42. Use Faraday's law to show that the electric field produced by charges on the plates of a parallel-plate capacitor cannot

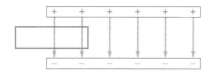

FIGURE 31-51 Problem 42.

end abruptly at the edges of the plates. *Hint:* Consider the loop shown in Fig. 31-51.

43. Figure 31-52 shows a magnetic field pointing into the page; the field is confined to a layer of thickness h in the vertical direction but extends infinitely to the left and right. The field strength is increasing with time: $B = bt$, where b is a constant. Find an expression for the electric field at all points outside the field region. *Hint:* Consult Example 30-5.

FIGURE 31-52 Problem 43.

44. The magnetic field inside a solenoid of circular cross section is given by $\mathbf{B} = bt\,\hat{\mathbf{k}}$, where $b = 2.1$ T/ms. At time $t = 0.40\ \mu$s a proton is inside the solenoid at the point $x = 5.0$ cm, $y = 0$, $z = 0$ and is moving with velocity $\mathbf{v} = 4.8 \times 10^6\,\hat{\mathbf{j}}$ m/s. Find the net electromagnetic force on the proton.

Paired Problems

(Both problems in a pair involve the same principles and techniques. If you can get the first problem, you should be able to solve the second one.)

45. A magnetic field is given by $\mathbf{B} = B_0(x/x_0)^2\,\hat{\mathbf{k}}$, where B_0 and x_0 are constants. Find an expression for the magnetic flux through a square of side $2x_0$ that lies in the x-y plane with one corner at the origin and two sides coinciding with the positive x and y axes.

46. A circular region of radius R lies in the x-y plane and contains a magnetic field given by $\mathbf{B} = B_0\dfrac{r}{R}\,\hat{\mathbf{k}}$, where r is the radial distance from the central axis of the field region. Find an expression for the magnetic flux through this region.

47. A uniform magnetic field is given by $\mathbf{B} = bt\,\hat{\mathbf{k}}$, where $b = 0.35$ T/s. Find the current in a conducting loop with area 240 cm^2 and resistance 0.20 Ω that lies in the x-y plane. In what direction is the current, as viewed from the positive z axis?

48. A uniform magnetic field is given by $\mathbf{B} = bt^3\,\hat{\mathbf{k}}$. A square conducting loop 15 cm on a side has 0.32-Ω resistance and lies in the x-y plane. At time $t = 2.5$ s, the current in the loop is 4.1 A. Find b.

49. A pair of vertical conducting rods are a distance ℓ apart and are connected at the bottom by a resistance R. A conducting bar of mass m runs horizontally between the rods and can slide freely down them while maintaining electrical contact. The whole apparatus is in a uniform magnetic field $\mathbf{B}$ pointing horizontally and perpendicular to the bar. When the bar is released from rest it soon reaches a constant speed. Find this speed.

50. A conducting bar of mass m slides down the conducting wedges shown in Fig. 31-53. The wedges are separated by a distance ℓ, connected at the top by a resistance R, and make an angle θ with the vertical. A uniform magnetic field $\mathbf{B}$ points horizontally, as shown. When released from rest the bar soon reaches a constant speed. Find an expression for this speed.

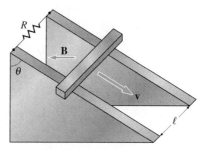

FIGURE 31-53 Problem 50.

51. Figure 31-54 shows an unusual design for a generator, consisting of a conducting bar that rotates about a central axis while making contact with a conducting ring of radius R. A uniform magnetic field is perpendicular to the ring. Wires from the axis and ring carry power to a load. Find an expression for the emf induced in this generator when the bar rotates with angular speed ω.

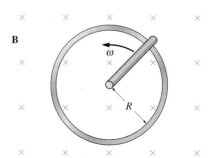

FIGURE 31-54 Problem 51.

52. A copper disk 90 cm in diameter is spinning at 3600 rpm about a conducting axle through its center, as shown in Fig. 31-55. A uniform 1.5-T magnetic field is perpendicular to the disk, as shown. A stationary conducting brush maintains contact with the disk's rim, and a voltmeter is

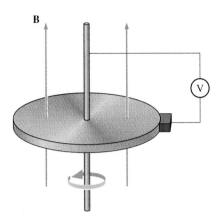

FIGURE 31-55 Problem 52.

connected between the brush and the axle. (a) What does the voltmeter read? (b) Which voltmeter lead is positive?

53. An electron is inside a solenoid, 28 cm from the solenoid axis. It experiences an electric force of magnitude 1.3 fN. At what rate is the solenoid's magnetic field changing?

54. It takes 0.43 J to push a 84-μC charge around a closed path surrounding a 1.5-m-diameter solenoid. At what rate is the solenoid's magnetic field changing?

Supplementary Problems

55. At times prior to $t = 0$, there is no current in either the solenoid or the small coil of Problem 19. Subsequently, the current in the small coil is observed to increase at 10 μA/s. What is the solenoid current as a function of time?

56. A conducting loop of area A and resistance R lies perpendicular to a uniform magnetic field B. The loop is then rotated at a uniform rate until it is upside down; this takes time Δt. Find an expression for the work done in flipping the loop.

57. A conducting rod of length ℓ moves at speed v in a plane perpendicular to a uniform magnetic field **B**, as shown in Fig. 31-56. The magnetic force on charge carriers in the rod causes charge separation, which creates an electric

field. Charge motion stops when the electric and magnetic forces on the charge carriers are equal. Show that this condition results in an electric field of magnitude vB and, therefore, in a potential difference $B\ell v$ between the rod ends, as we found from flux considerations in Example 31-8.

58. The tethered satellite system of Example 31-8 is flown in a low circular orbit about a planet of mass M and radius R. If the tether has length ℓ and generates an emf $\mathcal{E}$, find an expression for the planet's magnetic field strength.

59. Clever farmers whose lands are crossed by large power lines have been known to steal power by stringing wire near the power line and making use of the induced current—a practice that has been ruled legally to be theft. The scene of a particular crime is shown in Fig. 31-57. The power line carries 60-Hz alternating current with a peak current of 10 kA (that is, the current is given by $I = I_0 \sin \omega t$, where $I_0 = 10$ kA and $\omega = 2\pi f$, with $f = 60$ Hz). (a) If the farmer wants a peak voltage of 170 V, what should be the length ℓ of the loop shown in Fig. 31-57? (170 V is the peak of standard 120-V AC power.) (b) If all the equipment the farmer connects to the loop has an equivalent resistance of 5.0 Ω, what is the farmer's average power consumption? *Note:* The *average* power consumption is half the product of the peak voltage and peak current. (c) If the power company charges 10¢ per kWh, what is the monetary value of the energy stolen each day? (d) Without examining the farmer's lands, how, in principle, could the power company know that a crime is being committed?

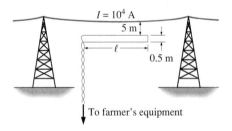

FIGURE 31-57 Problem 59.

60. A circular wire loop of resistance R and radius a lies with its plane perpendicular to a uniform magnetic field. The field strength changes from an initial value B_1 to a final value B_2. Show, by integrating the loop current over time, that the total charge that moves around the ring is

$$q = \frac{\pi a^2}{R}(B_2 - B_1).$$

Note that this result is independent of how the field changes with time.

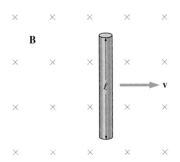

FIGURE 31-56 Problem 57.

61. A generator like that shown in Fig. 31-14 has an N-turn coil of area A spinning with angular speed ω in a uniform magnetic field B. A resistor R is connected across the generator. (a) Find an expression for the power dissipated in the resistor as a function of time. (b) Find an expression for the magnetic torque on the generator coil as a function of time. (c) Study the discussion associated with Equations 12-28a and b, and use it to show that the rate at which the agent turning the generator at constant angular speed does work is equal to the power dissipation in the resistor. Assume the coil's magnetic moment is aligned with the field a time $t = 0$.

62. A conducting disk with radius a, thickness h, and resistivity ρ is inside a solenoid of circular cross section. The disk axis coincides with the solenoid axis. The magnetic field in the solenoid is given by $B = bt$, with b a constant. Find expressions for (a) the current density in the disk as a function of the distance r from the disk center and (b) the rate of power dissipation in the entire disk. *Hint:* Consider the disk to be made up of infinitesimal conducting loops.

63. The bar in Problem 27 has mass m and is initially at rest. A constant force F is applied to the bar, pulling it to the right. (a) Formulate Newton's second law for the bar as an equation involving both v and $a = dv/dt$. (b) Use your equation to show that the bar's acceleration becomes zero when its speed reaches the value $FR/B^2\ell^2$. (c) Show by direct substitution that your equation is satisfied if v as a function of time is given by $v(t) = \dfrac{FR}{B^2\ell^2}(1 - e^{-B^2\ell^2 t/mR})$.

64. Find an expression for the speed of the left-hand resistor in Problem 32 as a function of time, in terms of its mass m, the field strength B, the speed v of the right-hand bar, the time t, and the resistance R_{left} and R_{right}.

65. A pendulum consists of a mass m suspended from two identical copper wires of negligible mass. At equilibrium the mass is a vertical distance ℓ below its supports, and the wires make 45° angles with the vertical, as shown in Fig. 31-58. A uniform magnetic field $\mathbf{B}$ points into the page. The pendulum is displaced from the plane of the page by a small angle θ_0, and at time $t = 0$ it is released. Find an expression for the voltmeter reading as a function of time.

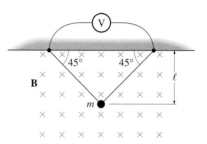

FIGURE 31-58   Problem 65.

66. A *flip coil* consists of a small coil used to measure magnetic fields. The flip coil is placed in a magnetic field with its plane perpendicular to the field, and then rotated abruptly through 180° about an axis in the plane of the coil. The coil is connected to instrumentation to measure the total charge Q that flows during this process. If the coil has N turns of area A and if its rotation axis is perpendicular to the magnetic field, show that the field strength is given by $B = QR/2NA$, where R is the coil resistance.

CHAPTER 32

INDUCTANCE AND MAGNETIC ENERGY

This transformer at an electrical substation uses mutual inductance to change the voltage at which power is transmitted.

Faraday's law implies that a changing magnetic flux through a circuit produces an induced emf in the circuit. In this chapter we consider the special case when that changing flux is itself caused by a changing current in an electric circuit. We then speak of the **inductance** of the circuit.

32-1 MUTUAL INDUCTANCE

Consider two coils arranged so that some of the magnetic flux associated with current in one coil also passes through the second coil, as in Fig. 32-1. If we change the current I_1 in the first coil, an induced emf $\mathscr{E}_2$ appears in the second. As we discussed in the preceding chapter, $\mathscr{E}_2$ depends on the rate of change of magnetic flux through the second coil. The magnetic flux depends, in turn, on the current in the first coil and on the geometrical arrangement of the two coils that determines how much flux from the first coil actually links the second. We

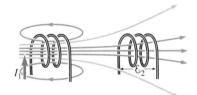

FIGURE 32-1 The mutual inductance between two coils is the ratio of magnetic flux through one coil to the current in the other coil A changing current in one coil induces an emf in the other.

characterize this geometrical arrangement by the ratio of the total magnetic flux through the second coil to the current in the first coil. This ratio defines the **mutual inductance, M,** of the two coils:

$$M = \frac{\phi_2}{I_1}. \tag{32-1}$$

Solving Equation 32-1 for ϕ_2 and differentiating, we obtain

$$\frac{d\phi_2}{dt} = M\frac{dI_1}{dt}.$$

Faraday's law says that $d\phi_2/dt$ is $-\mathscr{E}_2$, the induced emf in coil 2. So we have

$$\mathscr{E}_2 = -M\frac{dI_1}{dt}, \tag{32-2}$$

where the minus sign describes the polarity of the induced emf.

We could equally well have considered the case where a changing current in the second coil induces an emf in the first. Although it is not at all obvious, the same value of mutual inductance applies in this case even when the arrangement of the two coils is far from symmetric.

From Equation 32-2, we see that the unit of mutual inductance is the volt-second/ampere. This unit is given the name henry (H) in honor of the American scientist Joseph Henry (1797–1878), who was also the first secretary of the Smithsonian Institution. Mutual inductances found in common electronic circuits usually range from microhenrys (μH) on up to several henrys.

● **EXAMPLE 32-1** MUTUAL INDUCTANCE

A 2-turn coil is wrapped around a long solenoid with cross-sectional area $A = 26$ cm^2, wound with $n = 3500$ turns per meter of length (Fig. 32-2). Find the mutual inductance of this arrangement.

Solution
Since the solenoid field is uniform and confined to the solenoid interior, the flux through each turn of the small coil is just BA. Accounting for the two turns and using Equation 30-12 for the solenoid's magnetic field, the total magnetic flux through the small coil becomes

$$\phi_2 = 2BA = 2\mu_0 nIA,$$

where I is the solenoid current. Equation 32-1 shows that mutual inductance is the ratio of the flux in one coil to the current in the other, or

$$M = \frac{\phi_2}{I} = \frac{2\mu_0 nIA}{I} = 2\mu_0 nA$$

$$= (2)(4\pi \times 10^{-7} \text{ N/A}^2)(3500)(26 \times 10^{-4} \text{ m}^2)$$

$$= 23 \ \mu\text{H}.$$

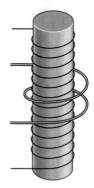

FIGURE 32-2 What is the mutual inductance of the two coils? (Example 32-1)

EXERCISE Suppose the solenoid of Example 32-1 has circular cross section, and that the smaller coil is *inside* the solenoid, with its diameter half that of the solenoid. Find the mutual inductance.

Answer: 5.7 μH

Some problems similar to Example 32-1: 6, 7, 10

● **E X A M P L E   3 2 - 2** AN IGNITION COIL

Your car's spark plugs ignite the gasoline-air mixture that burns to power the engine. The high voltage that causes sparks to jump across the plug gaps is provided by the ignition coil—actually an arrangement of two tightly wound coils (Fig. 32-3). Current from the car's 12-V battery flows through the coil with fewer turns and is interrupted periodically by a switch in the distributor. The sudden change in current induces a large emf in the coil with more turns, and this emf drives the spark. Interruption of the current is carefully timed so that the spark occurs at exactly the right point in the engine cycle. An important part of a "tune-up" is the precise adjustment of this ignition timing.

A typical ignition coil draws 3.0 A and supplies 20 kV to the spark plugs. If the current decays in 0.10 ms when the switch opens, what is the mutual inductance of the ignition coil?

Solution

The rate of change of current is

$$\frac{dI}{dt} = \frac{3.0 \text{ A}}{0.10 \times 10^{-3} \text{ s}} = 3.0 \times 10^4 \text{ A/s}.$$

Solving Equation 32-2 for M then gives

$$M = \frac{|\mathcal{E}|}{|dI/dt|} = \frac{20 \times 10^3 \text{ V}}{3.0 \times 10^4 \text{ A}} = 0.67 \text{ H}.$$

E X E R C I S E An electric toothbrush has no electrical connection to the power line (Fig. 32-4). But when the toothbrush is in its stand, a coil inside the toothbrush itself rests inside another coil in the stand, and alternating current from the power line flows in the stand coil. The current thus induced in the toothbrush coil is used to charge the batteries that power the toothbrush. Suppose the mutual inductance of this arrangement is 100 mH. At an instant when the current in the stand coil is changing at the rate of 40 A/s, what is the emf in the toothbrush coil?

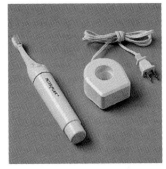

FIGURE 32-4 The batteries in this electric toothbrush are charged with energy transferred via mutual induction of coils located in the base unit, which is connected to the AC power line, and in the bottom of the brush unit. There is no direct electrical connection to the brush unit.

Answer: 4.0 V

Some problems similar to Example 32-2: 1–4 ●

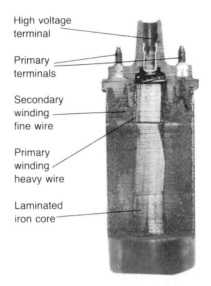

High voltage
terminal

Primary
terminals

Secondary
winding
fine wire

Primary
winding
heavy wire

Laminated
iron core

FIGURE 32-3 An automobile ignition coil uses electromagnetic induction to produce high voltage that drives sparks to ignite the gasoline.

32-2 SELF-INDUCTANCE

So far we've considered emf and current induced in a circuit by changes—like moving a magnet or varying the current in another circuit—that were external to the circuit in question. But the changing magnetic field associated with the changing current in a circuit also affects that same circuit.

Consider a circular loop carrying current I, as shown in Fig. 32-5. Magnetic field lines arising from this current loop pass through the loop, making a magnetic flux through the loop. If the current is steady, this flux is constant, and

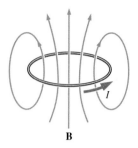

FIGURE 32-5 Magnetic flux from a circular current loop passes through the loop itself.

FIGURE 32-6 Typical inductors. At left is a 20-H unit, wound on an iron core for increased inductance. The others have inductances in the mH range.

there is no induced electric field. But if we change the loop current, then the flux changes, and an induced electric field arises. In order to conserve energy this field opposes the change that causes it—in this case the change in loop current. If the current is counterclockwise and we increase its strength, an induced electric field will appear in the clockwise direction to oppose the current increase. If we decrease the current, the induced electric field will have the opposite sense, now trying to maintain the current. The induced electric field therefore makes it difficult to change the current in the circuit.

This property of a circuit whereby its own magnetic field opposes changes in current is termed **self-inductance.** All circuits possess self-inductance, but this inductance is important only when the circuit encircles a great many of its own magnetic field lines or when current changes very rapidly. A simple piece of wire exhibits very little opposition to current changes in the 60-Hz alternating current used in electric power systems. But in a TV set or computer, where currents change on time scales of billionths of a second, self-inductance of the wires themselves must be taken into account.

An **inductor** is a device designed specifically to exhibit self-inductance. A typical inductor consists of a coil of wire, constructed so that a great deal of its own magnetic flux is encircled. Some inductors are wound on iron cores to promote flux concentration (Fig. 32-6). Ideally, the only electrical property of an inductor is its inductance. But real inductors are made from wire, so they have resistance as well.

As long as the current in an inductor is steady, the inductor acts just like a piece of wire. But when the current changes, the changing magnetic flux induces an emf that opposes the change in current. The more rapidly the current changes, the greater the rate of change of flux and so the greater the emf. We characterize the inductor by its self-inductance, L, defined as the ratio of magnetic flux through the inductor to current in the inductor:

$$L = \frac{\phi_B}{I}.$$

(32-3)

The unit of self-inductance, like that of mutual inductance, is the henry. Inductance is a constant determined by the physical design of an inductor. In principle, we can calculate the inductance of any inductor, but in practice this is difficult unless the geometry is particularly simple. Inductors for use in electronic circuits are available commercially in a wide range of inductance values.

● **EXAMPLE 32-3** INDUCTANCE OF A SOLENOID

A long solenoid of cross-sectional area A and length ℓ has n turns per unit length. What is its self-inductance?

Solution

Equation 32-3 shows that self-inductance is the ratio of magnetic flux to current. In Chapter 30 we used Ampère's law to find the magnetic field of a long solenoid: $B = \mu_0 nI$. Since the field is uniform and perpendicular to the solenoid coils, the flux

through each turn is just BA. With n turns per unit length, our solenoid has $N = n\ell$ turns. The total flux through the solenoid is then

$$\phi_B = NBA = (n\ell)(\mu_0 nI)(A) = \mu_0 n^2 A\ell I.$$

Then Equation 32-3 gives

$$L = \frac{\phi_B}{I} = \mu_0 n^2 A\ell.$$

(32-4)

Does this result make sense? As the area increases, so does the flux and therefore the inductance. As the length increases so does the number of turns, and therefore the flux and the inductance increase. Equation 32-4 reflects these trends. Can you see why the inductance should be proportional to the *square* of the number of turns per unit length?

EXERCISE What is the self-inductance of the MRI solenoid in Example 30-6 (page 768)?

Answer: 0.53 H

Some problems similar to Example 32-3: 11, 15, 20

The induced emf in an inductor is determined by Faraday's law, which relates the emf to the rate of change of magnetic flux:

$$\mathcal{E} = -\frac{d\phi_B}{dt}.$$

Differentiating Equation 32-3, the definition of inductance, gives

$$\frac{d\phi_B}{dt} = L\frac{dI}{dt}.$$

Then Faraday's law becomes

$$\mathcal{E} = -L\frac{dI}{dt}. \qquad (32\text{-}5)$$

This equation gives the emf $\mathcal{E}$ induced in an inductor L when the current in the inductor is changing at the rate dI/dt. The minus sign again tells us that the emf *opposes* the change in current. For this reason the inductor emf is often called a **back emf;** it works *against* changes brought about by an externally applied emf.

When the current in an inductor is steady, then $dI/dt = 0$ and there is no emf in the inductor. In this case, the inductor acts just like a piece of wire. But when the current changes the inductor responds by producing a back emf that opposes the change in current. Now the inductor acts very much like a battery, with the magnitude of its emf dependent on how fast the current changes. If we try to start or stop current suddenly, dI/dt is very large and a very large back emf appears. This is not merely mathematics! Rapid switching of inductive devices such as solenoids, solenoid valves, or motors can result in destruction of delicate electronic devices by induced currents. And people have been killed opening switches in circuits containing large inductors.

● **EXAMPLE 32-4** A DANGEROUS INDUCTOR

A current of 5.0 A is flowing in a 2.0-H inductor. The current is reduced steadily to zero in 1.0 ms. What is the magnitude of the emf in the inductor while the current is being turned off?

Solution

Because the current changes steadily, its time rate of change has magnitude

$$\frac{dI}{dt} = \frac{5.0 \text{ A}}{1.0 \text{ ms}} = 5000 \text{ A/s},$$

so

$$|\mathcal{E}| = L\frac{dI}{dt} = (2.0 \text{ H})(5000 \text{ A/s}) = 10{,}000 \text{ V},$$

enough to produce a lethal shock. Note that this voltage is quite unrelated to the voltage of the battery or whatever else was supplying the inductor current. We could have a 6-volt battery and still be electrocuted trying to open the circuit rapidly when a large inductance is present.

EXERCISE A neon lamp that glows only when the voltage across it exceeds 90 V is connected across a 1.2-H inductor carrying 750 mA. When the current is interrupted the lamp flashes. Find the maximum time over which the current could have dropped to zero, assuming a steady decrease.

Answer: 10 ms

Some problems similar to Example 32-4: 12–14 ●

32-3 INDUCTORS IN CIRCUITS

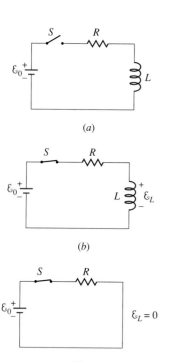

(a)

(b)

(c)

FIGURE 32-7 (a) An *RL* circuit. With the switch open no current flows. (b) Immediately after the switch is closed there is still no current. The inductor produces a back emf equal but opposite the battery emf, and therefore the rate of change of current is not zero. (c) After a long time, the rate of change of current and therefore the inductor emf approach zero. The inductor then acts like a piece of wire.

Here we examine circuits containing batteries, resistors, and inductors, analogous to the *RC* circuits of Chapter 28. In the qualitative analysis of *RC* circuits we found a useful guiding principle: The voltage across a capacitor cannot change instantaneously. We can make an analogous statement for inductors. Because the inductor emf is proportional to the rate of change of current in the inductor and because an infinite emf is physically impossible, we conclude the following:

▍ The current through an inductor cannot change instantaneously.

Thus, the effect an inductor has on current is analogous to the effect a capacitor has on voltage. Much of our understanding of capacitors can be applied to inductors if we interchange the words *voltage* and *current*.

Building Up the Current

Figure 32-7a shows a circuit containing a battery, switch, resistor, and inductor (symbol ᙁᙁᙁ). What happens when we close the switch? Initially the inductor current is zero; since it can't change instantaneously, it must remain zero immediately after the switch is closed. But this is a series circuit, so the inductor and resistor currents are equal. With zero current immediately after the switch is closed, there must be no voltage across the resistor. Therefore the inductor must produce a back emf equal to that of the battery, with the polarity shown in Fig. 32-7b. Even though there is, at this instant, no current in the inductor, the presence of an emf indicates that the *rate of change* of the current is not zero. Going around the circuit in the direction that the battery would drive a current, the inductor polarity indicated in Fig. 32-7b shows that the inductor emf is *negative*—and Equation 32-5 therefore shows that dI/dt is *positive*—i.e., the current is *increasing*.

As the current rises, so does the voltage across the resistor (since $V_R = IR$). Since the battery emf is constant, that means the inductor emf goes *down*—and that means the rate of change of current goes down. Thus, the current in the circuit builds up, but at an ever-decreasing rate. Concurrently, the inductor emf goes down. Eventually the current reaches a steady value, at which point dI/dt and, therefore, the inductor emf are zero. In this ultimate steady state the inductor acts like a piece of wire, and the circuit looks like Fig. 32-7c. The steady-state current is just $\mathcal{E}_0/R$, where $\mathcal{E}_0$ is the battery emf. Figure 32-8 summarizes graphically this qualitative analysis of the *RL* circuit.

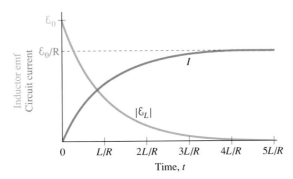

FIGURE 32-8 Inductor current and magnitude of the inductor emf as functions of time. Vertical axis represents either current or voltage.

To analyze the *RL* circuit quantitatively, we apply the loop law. Going clockwise from the negative terminal of the battery, we first encounter a voltage increase $\mathcal{E}_0$ due to the battery. Then the voltage decreases by *IR* in the resistor. Finally, there is a voltage change $\mathcal{E}_L$ in the inductor. This change is actually a decrease, because the inductor emf opposes the increase in current. However, we will simply call the inductor emf $\mathcal{E}_L$ and let Equation 32-5 take care of the signs. Then the loop law becomes

$$\mathcal{E}_0 - IR + \mathcal{E}_L = 0. \tag{32-6a}$$

If we differentiate this equation with respect to time, the battery emf $\mathcal{E}_0$ drops out because it is constant, giving

$$\frac{d\mathcal{E}_L}{dt} = R\frac{dI}{dt}.$$

Equation 32-5 shows that $dI/dt = -\mathcal{E}_L/L$, so our differentiated loop equation becomes

$$\frac{d\mathcal{E}_L}{dt} = -R\frac{\mathcal{E}_L}{L}. \tag{32-6b}$$

This differential equation describes a quantity—$\mathcal{E}_L$—whose rate of change is proportional to itself. We discussed such equations in Chapter 28 when we considered the *RC* circuit. Equation 32-6b is similar to Equation 28-4, but with current *I* replaced by the inductor emf $\mathcal{E}_L$, capacitance *C* by *L*, and *R* by $1/R$. The solution to Equation 32-6b is that of Equation 28-4, provided we make the appropriate substitutions for *I*, *C*, and *R*:

$$\mathcal{E}_L = -\mathcal{E}_0 e^{-Rt/L}. \tag{32-7}$$

This equation shows that the inductor emf decays exponentially to zero, starting from an initial value of $-\mathcal{E}_0$ (negative because the inductor emf *opposes* the battery emf). We can now solve for the current using Equation 32-6a:

$$I = \frac{\mathcal{E}_0 + \mathcal{E}_L}{R} = \frac{\mathcal{E}_0 + (-\mathcal{E}_0 e^{-Rt/L})}{R} = \frac{\mathcal{E}_0}{R}(1 - e^{-Rt/L}). \tag{32-8}$$

With a capacitor, we characterized the exponentially changing quantities in terms of the capacitive time constant RC. With an inductor, we have an **inductive time constant** L/R. Significant changes in current cannot occur on time scales much shorter than L/R. On the other hand, an RL circuit will approach a steady state, with zero $\mathscr{E}_L$, only after many inductive time constants.

Why is the inductive time constant a quotient of L and R rather than a product, as in the capacitor case? In Problem 23 you will convince yourself mathematically that L/R does indeed have the units of seconds. But you can also understand this physically. The larger L, the larger the back emf and the longer it takes the current to build up. The larger R, the smaller the final current and so, all else being equal, the smaller the rate of change of current, and, therefore, the smaller the inductive effects.

● EXAMPLE 32-5   FIRING UP AN ELECTROMAGNET

A large electromagnet used for lifting scrap metal has a self-inductance of 56 H. It is connected through a switch to a constant 440-V power source; the total resistance of the circuit is 2.8 Ω. When the switch is closed, how long does it take to bring the magnet current to 75% of its final value?

Solution

Letting $t \to \infty$ in Equation 32-8 shows that the final current is $I_{final} = \mathscr{E}_0/R$, as we argued in our qualitative analysis. Setting the current I in Equation 32-8 to $0.75\mathscr{E}_0/R$ gives

$$0.75 = 1 - e^{-Rt/L},$$

or

$$e^{-Rt/L} = 0.25.$$

Taking the natural logarithm of both sides (recall that $\ln e^x = x$) gives

$$-Rt/L = \ln(0.25),$$

or

$$t = -\frac{L}{R}\ln(0.25) = -\frac{56\ \text{H}}{2.8\ \Omega}\ln(0.25) = 28\ \text{s}.$$

(The minus sign canceled since the logarithm of a number less than 1 is negative.) Our answer is approximately one time constant ($L/R = 20$ s). This should not be surprising since we found with RC circuits that quantities following equations like Equation 28-6 (and, therefore, its analog, Equation 32-8) reach $1 - 1/e$, or about two-thirds, of their final value in one time constant.

EXERCISE   A 1.0-kΩ resistor is in series with an inductor, and a 12-V battery is connected across the pair. The current rises to 8.5 mA in 21 μs. Find the inductance.

Answer: 61 mH

Some problems similar to Example 32-5: 27–29, 32 ●

The Current Decays

Figure 32-9a shows an RL circuit with a two-way switch. Throwing the switch to position A allows current to build up in the inductor as we've already described. Then, at time $t = 0$, we throw it to position B. This disconnects the battery, leaving a circuit electrically equivalent to Fig. 32-9b. Just prior to $t = 0$ there was some current I_0 flowing downward in the inductor. Since the inductor current cannot change instantaneously, that current must continue just after the switch is closed, as shown in Fig. 32-9b. To drive this current, the inductor must develop an emf in the direction shown. Now the inductor emf is positive, so Equation 32-5 shows that the current is *decreasing,* as we might well expect since the battery has been disconnected. As the current decreases, so does the voltage across the resistor. So, therefore, does the inductor emf and, therefore, the rate of change of current. We thus expect both the current and the inductor emf to decrease, but at an ever-decreasing rate.

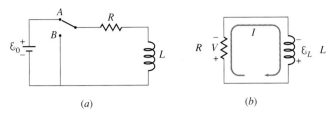

(a)   (b)

FIGURE 32-9 (a) Throwing the switch to position A allows current to build up in the inductor. (b) Throwing it to position B gives a circuit containing only the inductor and resistor. The current then decays exponentially.

Note that the inductor emf here is like any other induced emf: it *opposes* the change giving rise to it. In this case that change is the decrease in current caused by disconnecting the battery. The inductor responds with an emf in such a direction as to keep that current flowing.

Applying the loop law to Fig. 32-9b gives

$$\mathcal{E}_L - IR = 0.$$

Using $\mathcal{E}_L = -L\, dI/dt$ from Equation 32-5, the loop equation becomes

$$\frac{dI}{dt} = -\frac{R}{L}I.$$

This is just like Equation 32-6b, but with I replacing $\mathcal{E}_L$. The solution follows by analogy with Equation 32-7:

$$I = I_0 e^{-Rt/L}, \qquad\qquad (32\text{-}9)$$

where I_0 is the inductor current when the switch is thrown from A to B. Equation 32-9 shows that the current decays with the same exponential time constant L/R that described its buildup (Fig. 32-10). The resistor voltage IR and therefore the inductor emf also decay in the same way.

It is not always necessary to use Equations 32-8 and 32-9 in describing RL circuits. For times very short compared with the time constant L/R, it suffices to use the fact that inductor currents cannot change instantaneously. And after many time constants, inductors in a circuit containing only steady sources will act like wires. Example 32-6 explores this situation.

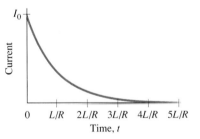

FIGURE 32-10 Exponential decay of the current in the circuit of Fig. 32-9b.

● **EXAMPLE 32-6** SHORT TIMES, LONG TIMES

In the circuit of Fig. 32-11a, the switch is initially open. What is the current in resistor R_2 immediately after the switch is closed? A long time after the switch is closed? Long after the switch is closed, it is again opened. What is the current in R_2 just after it is opened? A long time after?

Solution

Just before we close the switch, the current in the inductor is zero. The current cannot change instantaneously, so it remains

zero just after the switch is closed. At this instant the inductor might as well be an open circuit, giving the circuit shown in Fig. 32-11b. Then all the current from R_1 flows through R_2, so

$$I = \frac{\mathcal{E}_0}{R_1 + R_2}.$$

If we wait long enough, the circuit will reach a steady state in which $dI/dt = 0$. So then there is no inductor emf, and the

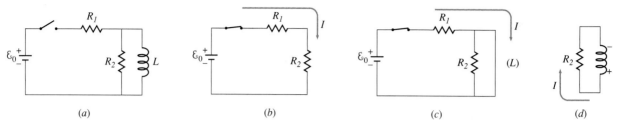

FIGURE 32-11   (a) Circuit for Example 32-6. (b) Just after the switch is closed the inductor acts like an open circuit. (c) A long time later the inductor emf is zero, so it acts like a wire. (d) When the switch opens again, current continues to flow in the inductor and on through R_2.

inductor acts like a wire. We can redraw the circuit as Fig. 32-11c, in which all the current from R_1 goes through L, and none through R_2. The resulting current in R_1 and L is just $\mathcal{E}_0/R_1$.

Now the switch is opened again. Current in R_1 stops abruptly since there is no way charge can get through the open switch. But the current through the inductor, which was $\mathcal{E}_0/R_1$ just before the switch was opened, remains $\mathcal{E}_0/R_1$ the instant after the switch is opened. There is only one place this current can go—through R_2, from bottom to top (Fig. 32-11d). So just after the switch is opened, the current in R_2 is $\mathcal{E}_0/R_1$. Notice that the value of R_2 has no effect on this current, which is determined entirely by the battery emf and the resistance R_1.

What about the voltage across R_2? This is given by Ohm's law:

$$V_2 = I_2 R_2 = \frac{\mathcal{E}_0 R_2}{R_1}.$$

The larger R_2, the larger the voltage that appears when the switch is opened. If R_2 has infinite resistance, or is not in the circuit, the voltage will be arbitrarily large as the inductor seeks at all cost to keep the current flowing. This dangerous situation can result in arcing and vaporization of circuit conductors, and even in electric shock. In circuits with large inductance, resistors are often placed in parallel with inductors to alleviate these dangers.

Finally, the current in Fig. 32-11d decays exponentially to zero. Plots of the currents in R_2 and L as functions of time are shown in Fig. 32-12.

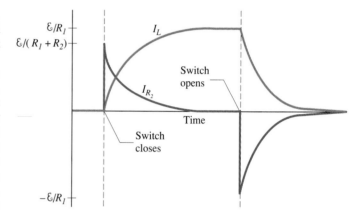

FIGURE 32-12   Currents in R_2 and L for Example 32-6. Note that the time constants before and after the switch opens are different; this reflects the fact that R_1 is out of the circuit after the switch opens.

EXERCISE   In Fig. 32-11 take $\mathcal{E}_0 = 12$ V, $R_1 = 56\ \Omega$, $L = 48$ mH, and suppose the switch has been closed for a long time. What is the maximum value of R_2 for which the inductor emf will not exceed 100 V when the switch is opened?

Answer: 467 Ω

Some problems similar to Example 32-6:   36, 37, 61, 62   ●

32-4 MAGNETIC ENERGY

In the situations we considered in Figs. 32-9b and 32-11d, current flows in circuits containing only a resistor and an inductor. Energy is dissipated, heating the resistor. Where does this energy come from?

Because there is a current in the inductor, there is also a magnetic field. The change in that magnetic field is what produces the emf that drives current around the circuit. As the current decreases, so does the inductor's magnetic field. Eventually the circuit reaches a state where there is no current, no magnetic field—and a hot resistor. So where did the resistor's thermal energy come from? It came from the magnetic field.

Like the electric field, the magnetic field contains stored energy. Our decaying *RL* circuit is analogous to a discharging *RC* circuit, in which the electric field between the capacitor plates disappears as thermal energy appears in the resistor. As in the electric case, magnetic energy is not limited to circuits. *Any* magnetic field contains energy. Release of magnetic energy drives a number of practical devices and also powers violent events throughout the universe (Fig. 32-13).

We can reinterpret the *RL* circuit of Fig. 32-9 in terms of energy. With the switch in position *A*, the battery supplies energy to the resistor and inductor. In the resistor the energy is dissipated as heat, but in the inductor it goes into the magnetic field. When the switch is thrown to position *B* the battery—the ultimate energy source for the circuit—is disconnected. Energy dissipated in the resistor now comes from the decaying magnetic field of the inductor. That energy came originally from the battery but was stored temporarily in the magnetic field. Figure 32-14 outlines these energy transfers.

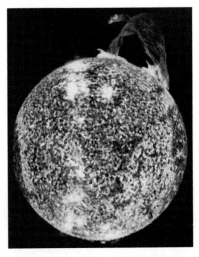

FIGURE 32-13 This eruption of a huge prominence from the Sun's surface involves the release of energy stored in magnetic fields.

Magnetic Energy in an Inductor

How much energy is stored in an inductor's magnetic field? We can answer this by considering first the *rate* of energy storage. If we multiply Equation 32-6a—the loop law for an *RL* circuit—by the current *I*, the result is

$$I\mathscr{E}_0 - I^2R + I\mathscr{E}_L = 0,$$

or, using Equation 32-5 for $\mathscr{E}_L$,

$$I\mathscr{E}_0 - I^2R - LI\frac{dI}{dt} = 0.$$

What do the terms in this equation mean? The first is the product of the battery's current and emf—a product we know gives electrical power. That this term is positive means the battery supplies energy *to* the circuit at the rate $I\mathscr{E}_0$. The second term, I^2R, is the rate of energy dissipation in the resistor (recall Equation 27-9). The negative sign means that this energy is taken *from* the circuit. The third term is also negative (since the current is increasing, dI/dt is positive) and represents the rate at which the inductor takes energy from the circuit. But the

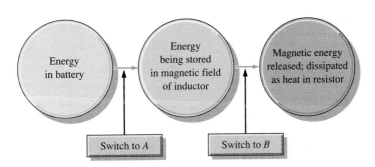

Energy in battery → Energy being stored in magnetic field of inductor → Magnetic energy released; dissipated as heat in resistor

Switch to *A*   Switch to *B*

FIGURE 32-14 Energy transfers in the circuit of Fig. 32-9.

inductor does not dissipate this energy; instead, it stores the energy in its growing magnetic field. The rate at which the inductor stores energy is thus

$$P = LI\frac{dI}{dt}.$$

Suppose we increase the current in an inductor by some small amount dI over a small time interval dt. Since the power is the rate of energy storage, the energy dU stored during this time is thus

$$dU = P\,dt = LI\,dI$$

We find the total energy stored in bringing the inductor current from zero to some final value I by summing—that is, integrating—all the dU's:

$$U = \int dU = \int P\,dt = \int_0^I LI\,dI = \tfrac{1}{2}LI^2\big|_0^I$$

Evaluating at the two limits then gives the stored energy:

$$U = \tfrac{1}{2}LI^2. \tag{32-10}$$

This much energy is therefore released when the magnetic field decays.

● EXAMPLE 32-7 QUENCHING A SUPERCONDUCTING MAGNET

Loss of coolant is a danger in superconducting electromagnets. The current is suddenly left without its zero-resistance path, and energy stored in the magnetic field is rapidly released. To prevent explosive energy release, copper wire is incorporated into the conducting system to lengthen the time constant L/R in the event of such a "quench." The superconducting MRI solenoid of Example 30-6 carries a 2.4-kA current and has 0.53-H inductance. In its nonsuperconducting state, the total resistance is 31 mΩ. (a) How much energy is stored in the solenoid's magnetic field? (b) If the coils suddenly lose their superconductivity, what is the initial rate of energy release?

Solution
(a) Equation 32-10 gives the magnetic energy:

$$U = \tfrac{1}{2}LI^2 = (\tfrac{1}{2})(0.53\text{ H})(2.4\times10^3\text{ A})^2 = 1.53\text{ MJ}.$$

(b) When superconducting ceases, the current decays according to Equation 32-9. At the instant the decay starts the current still has its original value. Since the magnetic energy released is dissipated in the resistor, the rate of energy release is just I^2R, or

$$P = I^2R = (2.4\times10^3\text{ A})^2(0.031\text{ Ω}) = 180\text{ kW}.$$

This is a substantial power; equivalent to 1,800 100-W light bulbs burning in the space of this roughly human-size device. The following exercise explores the duration of this power surge.

EXERCISE How long is it before 90% of the magnetic energy in Example 32-7 has been dissipated?

Answer: 20 s

Some problems similar to Example 32-7: 42, 44, 47 ●

Magnetic Energy Density

A long solenoid is a particularly simple inductor in which the magnetic field is essentially uniform. We can readily evaluate the energy density using this magnetic field, just as we found the electric field energy density using a parallel-plate capacitor.

In Example 32-3 we found that the inductance of a long solenoid of length ℓ and cross-sectional area A is $L = \mu_0 n^2 A \ell$, with n the number of turns per unit length. Equation 32-10 then gives the magnetic energy stored in the solenoid when it carries current I:

$$U = \tfrac{1}{2} L I^2 = \tfrac{1}{2} \mu_0 n^2 A \ell I^2 = \frac{1}{2\mu_0} (\mu_0 n I)^2 A \ell = \frac{B^2}{2\mu_0} A \ell,$$

where we recognized the quantity $\mu_0 n I$ as B, the magnetic field in the solenoid (Equation 30-12). The quantity $A \ell$ is the volume containing this field, so the energy per unit volume—the **magnetic energy density**—is

$$u_B = \frac{B^2}{2\mu_0}. \qquad \text{magnetic energy density} \qquad (32\text{-}11)$$

Although we derived this expression for the field of a solenoid, it is, in fact, a universal expression for the local magnetic energy density. Wherever there is a magnetic field, there is stored energy.

Compare Equation 32-11 with Equation 26-3 for the energy density in an electric field:

$$u_E = \tfrac{1}{2} \varepsilon_0 E^2.$$

The expressions for electric and magnetic energy densities are similar. Each is proportional to the *square* of the field strength, and each contains the appropriate constant. That the constant appears in the numerator in one case and the denominator in the other has no deep significance; it is merely a consequence of the way SI units are defined.

● **EXAMPLE 32-8** ENERGY IN EARTH'S MAGNETIC FIELD

The magnetic field strength near Earth's surface is about 50 μT. (a) How much energy is contained in 1 cubic kilometer of this field? (b) How does this compare with the electrical energy in the same volume, given a fair-weather electric field of 100 V/m?

Solution
(a) The magnetic field is essentially constant over this volume, so the total energy is the product of the energy density with the volume V:

$$U = u_B V = \frac{B^2}{2\mu_0} V = \frac{(50 \times 10^{-6} \text{ T})^2}{(2)(4\pi \times 10^{-7} \text{ N/A}^2)} (1 \times 10^3 \text{ m})^3$$

$$= 1 \text{ MJ}.$$

This is not a particularly large energy; a mere gallon of gasoline, for example, stores about 100 times as much.

(b) Since we're considering equal volumes, it suffices to compare energy densities. Then

$$\frac{u_B}{u_E} = \frac{B^2/2\mu_0}{\tfrac{1}{2}\varepsilon_0 E^2} = \frac{B^2}{\mu_0 \varepsilon_0 E^2}$$

$$= \frac{(50 \times 10^{-6} \text{ T})^2}{(4\pi \times 10^{-7} \text{ N/A}^2)(8.85 \times 10^{-12} \text{ F/m})(100 \text{ V/m})^2}$$

$$= 2 \times 10^4$$

Thus the electrical energy is even smaller, by a factor of 20,000. The energy stored in both of Earth's electromagnetic fields is very small compared, for example, with the gravitational energy of an equivalent volume of air or the amount of solar energy incident each day on a square kilometer. In other systems, though, magnetic energy may dominate.

EXERCISE A typical sunspot is about 50,000 km in diameter, and extends into the Sun as a cylinder about 30,000 km long. Its magnetic field is about 0.2 T. What is the magnetic energy in such a spot?

Answer: 9×10^{26} J

Some problems similar to Example 32-8: 51,52 ●

When the magnetic field varies with position it is necessary to divide the volume into infinitesimal elements and integrate to find the total magnetic energy. Example 32-9 illustrates this process:

● **EXAMPLE 32-9** MAGNETIC ENERGY IN A TOROID

A toroidal coil of square cross section has inner radius R and side ℓ as shown in Fig. 32-15. The coil consists of N turns, and carries a current I. What is the total magnetic energy stored in the toroid?

Solution
In Chapter 30, we found that the magnetic field in a toroid is given by

$$ B = \frac{\mu_0 NI}{2\pi r}, $$

with r the distance from the central axis of the toroid. The magnetic energy density in the toroid is then

$$ u_B = \frac{B^2}{2\mu_0} = \frac{\mu_0 N^2 I^2}{8\pi^2 r^2}. $$

To calculate the total energy, consider a thin ring of thickness dr and height ℓ located at a distance r from the axis of the toroid, as shown in Fig. 32-15. The volume of this ring is

$$ dV = 2\pi r \ell \, dr , $$

so the magnetic energy in the ring is

$$ dU = u_B \, dV = \frac{\mu_0 N^2 I^2}{8\pi^2 r^2} 2\pi r \ell \, dr = \frac{\mu_0 N^2 I^2 \ell}{4\pi r} dr. $$

To find the total energy, we integrate over all such rings within the toroid; that is, from $r = R$ to $r = R + \ell$:

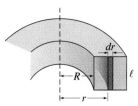

FIGURE 32-15 Cross section of a toroidal coil (Example 32-9). Also shown is a part of a volume element, in the shape of a ring of thickness dr.

$$ U = \int_R^{R+\ell} dU = \frac{\mu_0 N^2 I^2 \ell}{4\pi} \int_R^{R+\ell} \frac{dr}{r} = \frac{\mu_0 N^2 I^2 \ell}{4\pi} \ln\left(\frac{R+\ell}{R}\right). $$

Comparing this result with the expression $U = \frac{1}{2}LI^2$ would allow us to calculate the inductance of the toroid (see Problem 65).

EXERCISE A long straight wire of radius R carries current I. Find an expression for the magnetic energy stored in a region of length ℓ, between the wire surface and a distance $2R$ from the wire's axis.

Answer: $(\mu_0 I^2 \ell \ln 2)/4\pi$

Some problems similar to Example 32-9: 54–56, 63, 64 ●

CHAPTER SYNOPSIS

Summary

1. The **mutual inductance** of a pair of coils is defined as the ratio of the total flux in the second coil to the current in the first:

$$ M = \frac{\phi_2}{I_1}. $$

By Faraday's law, the emf in the second coil is proportional to the rate of change of current in the first:

$$ \mathcal{E}_2 = -M\frac{dI_1}{dt}. $$

The same mutual inductance M describes the emf developed in the first coil as a result of changing current in the second coil.

2. A changing current in a coil or circuit gives rise to a changing magnetic flux through that same circuit. This changing flux, in turn, gives rise to an induced electric field that opposes the original change in current. A device constructed to exploit this property of **self-inductance** is called an **inductor.** The self-inductance L of an inductor is the ratio of magnetic flux to current:

$$L = \frac{\phi}{I}.$$

Faraday's law relates the emf in an inductor to the rate of change of current:

$$\mathcal{E} = -L\frac{dI}{dt}.$$

The direction of the emf is such as to oppose changes in the inductor current. Self-inductance prevents the inductor current from changing instantaneously.

3. In a circuit containing a resistor R and inductor L, changes occur exponentially with an **inductive time constant** L/R. The rising current in a series RL circuit is given by

$$I = \frac{\mathcal{E}_0}{R}(1 - e^{-Rt/L}),$$

where $\mathcal{E}_0$ is the battery emf and t the time. If the current is subsequently allowed to decay, it goes exponentially to zero:

$$I = I_0 e^{-Rt/L},$$

where t is measured from the start of the decay.

4. Electrical energy supplied to an inductor ends up stored in the inductor's magnetic field. When an inductance L carries current I, the stored magnetic energy is

$$U = \tfrac{1}{2}LI^2.$$

5. All magnetic fields, not only those of inductors, contain stored energy, with the local **magnetic energy density** given by

$$u_B = \frac{B^2}{2\mu_0}.$$

Terms You Should Understand

(Pairs are closely related terms whose distinction is important; number in parentheses is chapter section where term first appears.)

mutual inductance, self-inductance (32-1, 32-2)
inductor (32-2)
back emf (32-2)
inductive time constant (32-3)
magnetic energy, magnetic energy density (32-4)

Symbols You Should Recognize

M, L (32-1, 32-2)
u_B (32-4)

Problems You Should Be Able to Solve

calculating mutual inductance for simple coil configurations (32-1)
finding induced emf in a second coil given mutual inductance and rate of change of current in the first coil (32-1)
calculating self-inductance for simple configurations (32-2)
calculating back emf given self-inductance and rate of change of current (32-2)
analyzing time-dependent behavior of RL circuits (32-3)
calculating magnetic energy (32-4)

Limitations to Keep in Mind

Real inductors, other than those made from superconductors, have resistance as well as inductance.

QUESTIONS

1. Figure 32-16 shows two pairs of identical coils in different geometrical arrangements. For which arrangement is the mutual inductance greatest? Why?

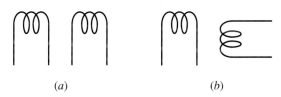

(a)   (b)

FIGURE 32-16 Question 1.

2. A car battery has an emf of 12 V, yet energy from the battery provides the 20,000-V spark that ignites the gasoline. How is this possible?

3. When two coils are connected in series but are physically far apart, they behave as a single inductor whose inductance is the sum of the individual inductances. Why might this not be true if they are close together?

4. You have a fixed length of wire to wind into an inductor. Will you get more inductance if you wind a short coil with large diameter, or a long coil with small diameter?

5. You have a fixed length of wire of resistance R. You want to wind the wire into a small space and use it as a resistor. How would you wind it so as to minimize its self-inductance?

6. In wiring circuits that operate at high frequencies, like TV sets or computers, it is important to avoid extraneous loops in wires. Why?

7. In a popular demonstration of induced emf, a light bulb is connected across a large inductor in an LR circuit, as shown in Fig. 32-17. When the switch is opened, the bulb flashes brightly and burns out. Why?

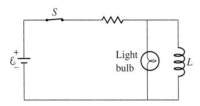

FIGURE 32-17   Question 7.

8. In the RL circuit of Fig. 32-7a, can the inductor emf exceed the battery emf (a) when the switch is first closed? (b) When the switch is opened after being closed for a long time?

9. What is the time constant for an inductive circuit made entirely from a superconductor?

10. Does it take more or less than one time constant for current in an RL circuit to build up to half its steady-state value?

11. If you increase the resistance in an RL circuit, what effect does this have on the inductive time constant?

12. Speculate on what would happen if you connected an ideal battery directly across an ideal inductor, with no resistance anywhere in the circuit.

13. How could you modify the simple RL circuit of Fig. 32-7a to prevent dangerous voltages from developing when the switch is opened?

14. List some similarities and differences between inductors and capacitors.

15. A 1-H inductor carries 10 A, and a 10-H inductor carries 1 A. Which inductor contains more stored energy?

16. Does the energy density in a magnetic field depend on the direction of the field?

17. The field of a magnetic dipole extends to infinity. Is there an infinite amount of energy stored in the dipole field? Why or why not?

18. It takes work to push two bar magnets together with like poles facing each other. Where does this energy go?

PROBLEMS

Section 32-1 Mutual Inductance

1. Two coils have a mutual inductance of 2.0 H. If current in the first coil is changing at the rate of 60 A/s, what is the emf in the second coil?

2. A 500-V emf appears in a coil when the current in an adjacent coil changes at the rate of 3.5 A/ms. What is the mutual inductance of the coils?

3. The current in one coil is given by $I = I_p \sin 2\pi ft$, where $I_p = 75$ mA, $f = 60$ Hz, and $t =$ time. Find the peak emf in a second coil if the mutual inductance between the coils is 440 mH.

4. Two coils have a mutual inductance of 580 mH. One coil is supplied with a current given by $I = 3t^2 - 2t + 4$, where I is in amperes and t in seconds. What is the induced emf in the other coil at time $t = 2.5$ s?

5. An alternating current given by $I_p \sin 2\pi ft$ is supplied to one of two coils whose mutual inductance is M. (a) Find an expression for the emf in the second coil. (b) When $I_p = 1.0$ A and $f = 60$ Hz, the peak emf in the second coil is measured at 50 V. What is the mutual inductance?

6. Find the mutual inductance of the two-coil system described in Problem 19 of Chapter 31.

7. Two long solenoids of length ℓ both have n turns per unit length. They have circular cross sections with radii R and $2R$, respectively. The smaller solenoid is mounted inside the larger one, with their axes coinciding. Find the mutual inductance of this arrangement, neglecting any nonuniformity in the magnetic field near the ends.

8. Coils A and B have mutual inductance 25 mH. At time $t = 0$ the current in coil A is zero. Subsequently a time-varying current is supplied to A, and the induced emf in coil B is given by $\mathcal{E} = 50 + 0.2t$, with $\mathcal{E}$ in V and t in ms. Find an expression for the time-varying current in coil A.

9. A rectangular loop of length ℓ and width w is located a distance a from a long, straight wire, as shown in Fig. 32-18. What is the mutual inductance of this arrangement?

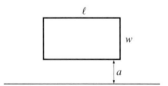

FIGURE 32-18   Problem 9.

10. Two wire loops of radii a and b lie in the same plane and have a common center. Find the mutual inductance of this arrangement, assuming $b \gg a$. *Hint:* With $b \gg a$ the magnetic field will be essentially uniform over the smaller loop. See Example 30-1.

Section 32-2 Self-Inductance

11. What is the self-inductance of a solenoid 50 cm long and 4.0 cm in diameter that contains 1,000 turns of wire?

12. The current in an inductor is changing at the rate of 100 A/s, and the inductor emf is 40 V. What is its self-inductance?

13. A 2.0-A current is flowing in a 20-H inductor. A switch is opened, interrupting the current in 1.0 ms. What emf is induced in the inductor?

14. A 60-mA current is flowing in a 100-mH inductor. Over a period of 1.0 ms the current is reversed, going steadily to 60 mA in the opposite direction. What is the inductor emf during this time?

15. A cardboard tube measures 15 cm long by 2.2 cm in diameter. How many turns of wire must be wound on the full length of the tube to make a 5.8-mH inductor?

16. The current in a 2.0-H inductor is given by $I = 3t^2 + 15t + 8$, where t is in seconds and I in amperes. Find an expression for the magnitude of the inductor emf.

17. The emf in a 50-mH inductor has magnitude $|\mathscr{E}| = 0.020t$, with t in seconds and $\mathscr{E}$ in volts. At $t = 0$ the inductor current is 300 mA. (a) If the current is increasing, what will be its value at $t = 3.0$ s? (b) Repeat for the case when the current is decreasing.

18. The current in a 40-mH inductor is given by $I = I_0 e^{-bt}$, where $I_0 = 10$ A and $b = 20$ s^{-1}. What is the magnitude of the inductor emf at (a) $t = 0$ (b) $t = 25$ ms, and (c) $t = 50$ ms?

19. A 2,000-turn solenoid is 65 cm long and has cross-sectional area 30 cm^2. What rate of change of current will produce a 600-V emf in this solenoid?

20. You have a plastic rod 20 cm long and 1.5 cm in diameter. What inductance will you get if you wind the entire rod with a single layer of (a) 22-gauge (0.64-mm-diameter) and (b) 34-gauge (0.16-mm-diameter) wire? Assume adjacent turns are touching, separated only by a negligible thickness of enamel insulation.

21. The emf in a 50-mH inductor is given by $\mathscr{E} = \mathscr{E}_p \sin \omega t$, where $\mathscr{E}_p = 75$ V and $\omega = 140$ s^{-1}. What is the peak current in the inductor? (Assume the current swings symmetrically about zero.)

22. A coaxial cable consists of an inner conductor of radius a and outer conductor of radius b, as shown in Fig. 32-19. Current flows along one conductor and back along the other. Show that the inductance per unit length of the cable is $\frac{\mu_0}{2\pi} \ln(b/a)$.

FIGURE 32-19 Problems 22, 68.

Section 32-3 Inductors in Circuits

23. Show that the inductive time constant has the units of seconds.

24. What inductance should you put in series with a 100-Ω resistor to give a time constant of 2.2 ms?

25. The current in a series RL circuit rises to 20% of its final value in 3.1 μs. If $L = 1.8$ mH, what is the resistance R?

26. The current in a series RL circuit rises to half its final value in 7.6 s. What is the time constant?

27. A 10-H inductor is wound of wire with resistance 2.0 Ω. If the inductor is connected across an ideal 12-V battery, how long will it take the current to reach 95% of its final value?

28. In a series RL circuit like Fig. 32-7a, $\mathscr{E}_0 = 45$ V, $R = 3.3$ Ω, and $L = 2.1$ H. If the current is 9.5 A, how long has the switch been closed?

29. In Fig. 32-7a, take $R = 2.5$ kΩ and $\mathscr{E}_0 = 50$ V. When the switch is closed, the current through the inductor rises to 10 mA in 30 μs. (a) What is the inductance? (b) What will be the current in the circuit after many time constants?

30. A series RL circuit like Fig. 32-7a has $\mathscr{E}_0 = 60$ V, $R = 22$ Ω, and $L = 1.5$ H. Find the rate of change of the current (a) immediately after the switch is closed and (b) 0.10 s later.

31. In Fig. 32-7a, take $R = 100$ Ω, $L = 2.0$ H, and $\mathscr{E}_0 = 12$ V. At 20 ms after the switch is closed, what are (a) the circuit current, (b) the inductor emf, (c) the resistor voltage, (d) the rate of change of the circuit current, and (e) the power dissipation in the resistor?

32. Show that a series RL circuit reaches 99% of its final current in approximately 5 time constants.

33. Resistor R_2 in Fig. 32-20 is to limit the emf that develops when the switch is opened. What should be its value in order that the inductor emf not exceed 100 V?

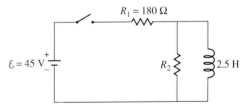

FIGURE 32-20 Problem 33.

34. In Fig. 32-9a take $\mathscr{E}_0 = 12$ V, $R = 2.7$ Ω, and $L = 20$ H. Initially the switch is in position B and there is no current anywhere. At $t = 0$ the switch is thrown to position A, and at $t = 10$ s it is thrown back to position B. Find the inductor current at (a) $t = 5.0$ s and (b) $t = 15$ s.

35. A 5.0-A current is flowing through a nonideal inductor with $L = 500$ mH. If the inductor is suddenly short circuited, the inductor current drops to 2.5 A in 6.9 ms. What is the resistance of the inductor?

36. In Fig. 32-21, take $\mathscr{E}_0 = 12$ V, $R_1 = 4.0$ Ω, $R_2 = 8.0$ Ω, $R_3 = 2.0$ Ω, and $L = 2.0$ H. What is the current I_2

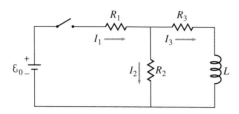

FIGURE 32-21 Problem 36.

(a) immediately after the switch is first closed and (b) a long time after the switch is closed? (c) After a long time the switch is again opened. Now what is I_2?

37. In Fig. 32-22, take $\mathcal{E}_0 = 20$ V, $R_1 = 10$ Ω, $R_2 = 5.0$ Ω, and assume the switch has been open for a long time. (a) What is the inductor current immediately after the switch is closed? (b) What is the inductor current a long time after the switch is closed? (c) If after a long time the switch is again opened, what will be the voltage across R_1 immediately afterwards?

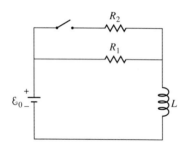

FIGURE 32-22 Problem 37.

Section 32-4 Magnetic Energy

38. How much energy is stored in a 5.0-H inductor carrying 35 A?

39. What is the current in a 10-mH inductor when the stored energy is 50 μJ?

40. A 220-mH inductor carries 350 mA. How much energy must be supplied to the inductor in raising the current to 800 mA?

41. A 12-V battery, 5.0-Ω resistor, and 18-H inductor are connected in series and allowed to reach a steady state. (a) What is the energy stored in the inductor? (b) Once in the steady state, over what time interval is the energy dissipated in the resistor equal to that stored in the inductor?

42. A battery, switch, resistor, and inductor are connected in series. When the switch is closed the current rises to half its steady-state value in 1.0 ms. How long does it take for the magnetic energy in the inductor to rise to half its steady-state value?

43. The current in a 2.0-H inductor is decreased linearly from 5.0 A to zero over 10 ms. (a) What is the average rate at

which energy is being extracted from the inductor during this time? (b) Is the instantaneous rate constant?

44. When a nonideal 1.0-H inductor is short-circuited, its magnetic energy drops to one-fourth of its original value in 3.6 s. What is its resistance?

45. The current in a 2.0-H inductor is increasing. At some instant, the current is 3.0 A and the inductor emf is 5.0 V. At what rate is the inductor's magnetic energy increasing at this instant?

46. A 500-turn solenoid is 23 cm long, 1.5 cm in diameter, and carries 65 mA. How much magnetic energy does it contain?

47. A superconducting solenoid with inductance $L = 3.5$ H carries 1.8 kA. Copper is embedded in the coils to carry the current in the event of a quench (see Example 32-7). (a) What is the magnetic energy in the solenoid? (b) What is the maximum resistance of the copper that will limit the power dissipation to 100 kW immediately after a loss of superconductivity? (c) With this resistance, how long will it take the power to drop to 50 kW?

48. Show that the quantity $B^2/2\mu_0$ has the units of energy density (J/m³).

49. The Alcator fusion experiment at MIT has a 50-T magnetic field. What is the magnetic energy density in Alcator?

50. What is the magnetic field strength in a region where the magnetic energy density is 7.8 J/cm³?

51. The magnetic field of a neutron star is about 10^8 T. How does the energy density in this field compare with the energy density stored in (a) gasoline and (b) pure uranium-235 (mass density 19×10^3 kg/m³)? Consult Appendix C.

52. A loop of magnetic field arches above the Sun's surface, forming a tube approximately 10^5 km long and 10^4 km in diameter. If the magnetic field strength in the tube is 50 G, what is its magnetic energy content?

53. A single-turn loop of radius R carries current I. How does the magnetic energy density at the loop center compare with that of a long solenoid of the same radius, carrying the same current, and consisting of n turns per unit length?

54. A magnetic field is given by $\mathbf{B} = B_0(x/a)^2\hat{\mathbf{j}}$, where B_0 and a are constants. Find an expression for the magnetic energy in a cube of side a with one corner at the origin and sides extending along the coordinate axes.

55. Estimate the total energy in Earth's magnetic field by integrating the magnetic energy density over the entire volume outside the planet, assuming that the field strength at Earth's surface is a constant 0.5 G and drops as $1/r^3$, with r the distance from Earth's center. (This estimate neglects the factor-of-two variation from pole to equator, any field inside the planet, and distortions of Earth's magnetic field by the solar wind.)

56. A toroid of inner radius 1.5 m and square cross section is wound with 2,500 turns. What must be the length ℓ of its cross-sectional square if the toroid contains 80 J of magnetic energy at a current of 63 A?

Paired Problems

(Both problems in a pair involve the same principles and techniques. If you can get the first problem, you should be able to solve the second one.)

57. Two coils have mutual inductance M. The current supplied to coil A is given by $I = bt^2$. Find an expression for the magnitude of the induced emf in coil B.

58. Two coils have mutual inductance M, and a time-varying current is supplied to coil A; at time $t = 0$ that current is zero. The magnitude of the induced emf in coil B is given by $\mathcal{E} = b\sqrt{t}$. Find an expression for the current in coil A.

59. In the circuit of Fig. 32-7a, take $\mathcal{E}_0 = 5.0$ V and $R = 1.8\ \Omega$. At 2.5 s after the switch is closed, the circuit current is 250 mA. Find the inductance.

60. In the circuit of Fig. 32-7a, take $\mathcal{E}_0 = 5.0$ V and $R = 1.8\ \Omega$. At 2.5 s after the switch is closed, the inductor emf is 2.1 V. Find the inductance.

61. In Fig. 32-11a, take $\mathcal{E}_0 = 25$ V, $R_1 = 1.5\ \Omega$, and $R_2 = 4.2\ \Omega$. What is the voltage across R_2 (a) immediately after the switch is first closed and (b) a long time after the switch is closed? (c) Long after the switch is closed it is again opened. Now what is the voltage across R_2?

62. In Fig. 32-11, take $R_2 = 5R_1$. If the maximum possible value for the inductor emf in this circuit is 300 V, what is the battery emf $\mathcal{E}_0$?

63. A wire of radius R carries a current I distributed uniformly over its cross section. Find an expression for the magnetic energy per unit length in the region from R to $100R$.

64. A wire of radius R carries a current I distributed uniformly over its cross section. Find an expression for the total magnetic energy per unit length *within* the wire.

Supplementary Problems

65. (a) Use the result of Example 32-9 to determine the inductance of a toroid. (b) Show that your result reduces to the inductance of a long solenoid when $R \gg \ell$.

66. Two long, flat parallel bars of width w and spacing d carry equal but opposite currents I, as shown in Fig. 32-23. (a) Use Ampère's law to find the magnetic field between the bars. Take $d \ll w$ so you can neglect fringing fields. (b) Use your result to find the magnetic energy per unit length stored between the bars. (c) Compare your result in (b) with the expression $U = \frac{1}{2}LI^2$ to find an expression for the inductance per unit length of the bars.

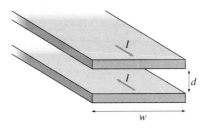

FIGURE 32-23 Problem 66.

67. (a) Use Equation 32-9 to write an expression for the power dissipation in the resistor as a function of time, and (b) integrate from $t = 0$ to $t = \infty$ to show that the total energy dissipated is equal to the energy initially stored in the inductor, namely, $\frac{1}{2}LI_0^2$.

68. (a) Find the magnetic energy density as a function of radial distance for the coaxial cable of Problem 22. (b) Integrate over the volume between the cable conductors to show that the total energy per unit length of the cable is

$$U = \frac{\mu_0}{4\pi}I^2\ln(b/a).$$

Hint: Your volume element should be a cylindrical shell of radius r, thickness dr, and length ℓ. What is its volume dV? (c) Use the expression $U = \frac{1}{2}LI^2$ to find the inductance per unit length, and show that your result agrees with that of Problem 22.

69. An electric field and a magnetic field have the same energy density. Obtain an expression for the ratio E/B, and evaluate this ratio numerically. What are its units? Is your answer close to any of the fundamental constants listed inside the front cover?

70. Two long, straight, parallel wires are a distance d apart. The wires have radius a, where $a \ll d$. Current flows down one wire and back along the other. Find the inductance per unit length of the parallel wires. Assume the wire radius is so small that you can neglect the magnetic flux within the wires themselves.

71. The switch in the circuit of Fig. 32-24 is closed at time $t = 0$, at which instant the inductor current is zero. Write the loop and node laws for this circuit, and show that they are satisfied if the inductor current is given by $I = (\mathcal{E}_0/R_1)(1 - e^{-R_\parallel t/L})$, where $R_\parallel$ is the resistance of R_1 and R_2 were they connected in parallel.

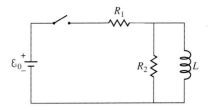

FIGURE 32-24 Problem 71.

72. Earth's magnetic field ends abruptly on the sunward side at approximately the point where the magnetic energy density has dropped to the same value as the kinetic energy density in the solar wind. Near Earth, the solar wind contains about 5 protons and 5 electrons per cubic centimeter, and flows at 400 km/s. Treating Earth's field as that of a dipole with dipole moment 8×10^{22} J/T, estimate the distance to the point above the equator where the field ends.

CHAPTER 33

ALTERNATING-CURRENT CIRCUITS

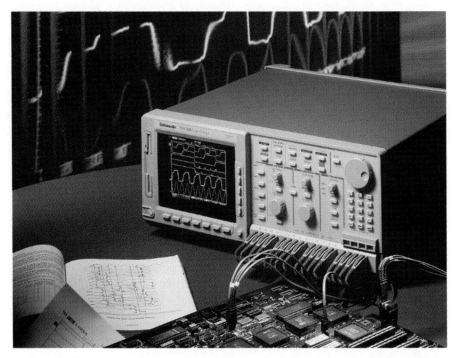

Time-varying voltages and currents are essential in electronic circuits and in electric power transmission. This oscilloscope displays time-varying voltages measured in a computer circuit.

Chapter 28 considered direct-current (DC) circuits, in which the source of electrical energy is a battery or other device whose emf does not change with time. When we turn on a circuit containing only resistors and a DC emf, currrent starts to flow immediately and remains steady until the circuit is turned off. Even when we add capacitance, as in Section 28-6, or inductance, as in Section 32-3, all currents and voltages eventually reach steady values.

We now turn our attention to alternating-current (AC) circuits, in which sources of electrical energy vary with time. A familiar AC circuit is standard household wiring. Alternating current with a frequency of 60 Hz is used almost universally for electric power generation and transmission, for reasons we will discuss in Section 33-6. Devices such as stereos, TVs, radios, and microwave ovens involve more rapidly varying alternating currents.

33-1 ALTERNATING CURRENT

In describing time-varying electrical quantities, we will consider only sinusoidal variations. More complicated variations can be analyzed as superpositions of sinusoidal functions, as we described in Section 16-5. A sinusoidal AC voltage or current is characterized by its amplitude, frequency, and phase constant—the same quantities we developed in Chapter 15 to describe simple harmonic motion. Amplitude is specified by giving the peak value (V_p, I_p) or the **root-mean-square** value (V_{rms}, I_{rms}). The rms value is an average obtained by squaring the signal, taking the time average, and then taking the square root. This procedure is used because the direct average of an AC signal is zero since it spends as much time below zero as above. Use of rms values also facilitates the calculation of power in AC circuits. For a sine wave, rms and peak values are related by

$$V_{rms} = \frac{V_p}{\sqrt{2}} \quad \text{and} \quad I_{rms} = \frac{I_p}{\sqrt{2}}, \tag{33-1}$$

as you can show in Problem 7. When we speak of 120-V household wiring, for example, we are giving the rms voltage.

In practical and engineering situations we usually describe frequency f in cycles per second, or hertz (Hz). In mathematical analysis of alternating current, it is usually more convenient to use the angular frequency ω, measured in radians per second or, equivalently, inverse seconds (s^{-1}). The relation between the two,

$$\omega = 2\pi f, \tag{33-2}$$

is the same as for rotational and simple harmonic motion, and for the same reason: a full cycle contains 2π radians.

Sometimes we are interested in the phase constant ϕ of an AC signal, which describes when the sine curve crosses zero. A full mathematical description of an AC voltage or current then includes its amplitude (V_p, I_p), frequency (ω), and phase constant (ϕ):

$$V = V_p \sin(\omega t + \phi_V) \quad \text{and} \quad I = I_p \sin(\omega t + \phi_I). \tag{33-3}$$

Here we've labeled the phase constants with subscripts V and I to indicate that voltage and current—even in the same circuit element—need not have the same phase constant. We will often take one phase constant to be zero; then the other describes the *phase difference* between the voltage and current. Phase difference is an important quantity in AC circuits.

Figure 33-1 plots a typical AC voltage, showing the relation between peak and rms amplitudes.

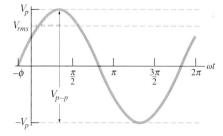

FIGURE 33-1 A sinusoidally varying AC voltage, showing peak, rms, and peak-to-peak amplitudes; the latter is just the difference between the extreme values, here equal to $2V_p$. The waveform completes a full cycle as ωt increases by 2π. The phase constant ϕ is $\pi/6$ or 30°; note that a positive phase constant shifts the curve to the left.

● EXAMPLE 33-1 HOUSEHOLD VOLTAGE

Standard household wiring in the United States supplies 120 V rms at 60 Hz. Express this voltage mathematically in the form of Equation 33-3, assuming that the voltage is rising through zero at time $t = 0$.

Solution

The rms and peak voltages are related by Equation 33-1, so

$$V_p = \sqrt{2}V_{\text{rms}} = (\sqrt{2})(120 \text{ V}) = 170 \text{ V}.$$

The angular frequency is 2π times the frequency in Hz, so

$$\omega = 2\pi f = (2\pi)(60 \text{ Hz}) = 377 \text{ s}^{-1}.$$

When $t = 0$ we want Equation 33-3 to give $V = 0$; that requires $\phi = 0$ or $\phi = \pi$. But only with $\phi = 0$ will the curve be *rising* at $t = 0$, so Equation 33-3 becomes

$$V = 170\sin(377t),$$

with V in volts and t in seconds. We will frequently take the phase constant to be zero as in this example; only when we're comparing signals with different phase does the value of ϕ become significant.

EXERCISE A 1.0-kHz sinusoidal current with rms amplitude 1.5 A drives a loudspeaker in a test of a stereo system. Express this current in the form of Equation 33-3, assuming zero phase constant.

Answer: $I = 2.12\sin(6.28\times10^3 t)$, with I in A and t in s.

Some problems similar to Example 33-1: 1, 4, 5 ●

33-2 CIRCUIT ELEMENTS IN AC CIRCUITS

Here we examine separately the AC behavior of resistors, capacitors, and inductors so we can subsequently understand what happens when we combine these elements in AC circuits.

Resistors

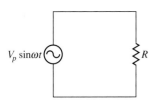

FIGURE 33-2 A resistor connected across an AC generator (symbol ⊝).

An ideal resistor is a device whose current and voltage are always proportional:

$$I = \frac{V}{R}.$$

Figure 33-2 shows a resistor R connected across an AC generator, making the voltage across the resistor equal to the generator voltage. The generator voltage is described by Equation 33-3, where we take $\phi_V = 0$. Then the current is

$$I = \frac{V}{R} = \frac{V_p\sin\omega t}{R} = \frac{V_p}{R}\sin\omega t.$$

The current has the same frequency as the voltage, and, since its phase constant is also zero, the voltage and current are *in phase*—that is, they peak at the same time. The peak current is simply the peak voltage divided by the resistance: $I_p = V_p/R$. Because voltage and current are both sinusoidal, their rms values are in the same ratio as their peak values; thus $I_{\text{rms}} = V_{\text{rms}}/R$.

Capacitors

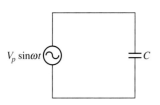

FIGURE 33-3 A capacitor across an AC generator.

Figure 33-3 shows a capacitor connected across an AC generator. In Chapter 26, we defined a capacitor as a device in which voltage and charge are directly proportional:

$$q = CV.$$

Differentiating this relation gives

$$\frac{dq}{dt} = C\frac{dV}{dt}.$$

But dq/dt is the capacitor current I, so

$$I = C\frac{dV}{dt}.$$

The generator voltage $V_p \sin \omega t$ appears directly across the capacitor, so the current is

$$I = C\frac{d}{dt}(V_p \sin \omega t)$$

$$= \omega C V_p \cos \omega t = \omega C V_p \sin(\omega t + \pi/2). \qquad (33\text{-}4)$$

This equation shows clearly the phase and amplitude relations between current and voltage in a capacitor. Because the cosine curve is just a sine curve shifted left by $\pi/2$ or 90°, Equation 33-4 tells us that:

| **The current in a capacitor leads the voltage by 90°.**

Figure 33-4 shows graphically this relation between current and voltage in a capacitor.

The term $\omega C V_p$ multiplying the cosine in Equation 33-4 is the amplitude of the current, so we can write

$$I_p = \omega C V_p,$$

or, in a form resembling Ohm's law,

$$I_p = \frac{V_p}{1/(\omega C)} = \frac{V_p}{X_C}, \qquad (33\text{-}5)$$

where we have defined $X_C = 1/\omega C$.

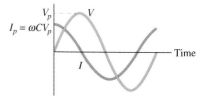

FIGURE 33-4 Current in a capacitor leads the voltage by $\pi/2$ or 90°.

Equation 33-5 shows that the capacitor acts somewhat like a resistor of resistance $X_C = 1/\omega C$. But not quite! This "resistance" does give the relation between the peak voltage and peak current, but it doesn't tell the whole story. The capacitor also introduces a phase difference between voltage and current. This phase difference reflects a fundamental physical difference between resistors and capacitors. A resistor dissipates electrical energy as heat. A capacitor stores and releases electrical energy. Over a complete cycle, the agent turning the generator in Fig. 33-3 does no net work, while the agent turning the generator with the resistive load of Fig. 33-2 continually does work that gets dissipated as heat in the resistor. Because the quantity X_C in Equation 33-5 does not act quite like a resistance, we give it the special name **capacitive reactance**. Like resistance, reactance is measured in ohms (Ω).

Does it make sense that X_C depends on frequency? Yes—as frequency goes to zero, X_C goes to infinity. At zero frequency nothing is changing, there is no need to move charge on or off the plates, so no current flows, and the capacitor might as well be an open circuit. As frequency increases, larger currents flow to move charge on and off the capacitor in ever shorter times, so the capacitor looks increasingly like a short circuit. We often summarize this behavior qualitatively by saying that a capacitor at low frequencies acts like an open circuit, while at high frequencies it acts like a short circuit.

Why does the capacitor current *lead* the voltage? Because the capacitor voltage is proportional to its charge, and it takes current to move charge onto the capacitor plates. Therefore current must flow *before* the voltage can change significantly. We found the same thing with the *RC* circuit of Chapter 28; there current flowed as soon as the switch was closed, but it took time to build up the capacitor voltage.

Inductors

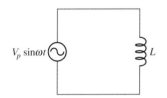

FIGURE 33-5 An inductor across an AC generator.

Figure 33-5 shows an inductor connected across an AC generator. The loop law for this circuit is

$$V_p \sin \omega t + \mathscr{E}_L = 0.$$

From the preceding chapter, we know that the inductor emf is given by

$$\mathscr{E}_L = -L\frac{dI}{dt},$$

so the loop law becomes $\quad V_p \sin \omega t = L\frac{dI}{dt}.$

To obtain a relation involving the current *I* rather than its derivative, we integrate this equation:

$$\int V_p \sin \omega t \, dt = \int L\frac{dI}{dt} dt = \int L \, dI,$$

or $\qquad -\dfrac{V_p}{\omega}\cos \omega t = LI.$

Here we have set the integration constants to zero because nonzero values would represent a DC emf and current that are absent in this circuit. Solving for *I* then gives

$$I = -\frac{V_p}{\omega L}\cos \omega t = \frac{V_p}{\omega L}\sin(\omega t - \pi/2), \qquad (33\text{-}6)$$

where the last step follows because $\sin(\alpha - \pi/2) = -\cos\alpha$ for any α.

Equation 33-6 shows that the current in the inductor lags the applied voltage by $\pi/2$ or 90° (i.e., Equation 33-6 is Equation 33-3 with $\phi = -\pi/2$). Equivalently:

▎ The voltage across an inductor leads the inductor current by 90°.

Figure 33-6 plots this phase relation.

Equation 33-6 also shows that the peak current is

$$I_p = \frac{V_p}{\omega L} = \frac{V_p}{X_L}. \tag{33-7}$$

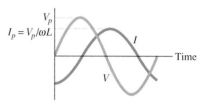

FIGURE 33-6 Voltage across an inductor leads the current by $\pi/2$ or 90°.

Again, this equation resembles Ohm's law, with a "resistance" $X_L = \omega L$. But as with the capacitor, no power is dissipated in the inductor. Instead, energy is alternately stored and released as the inductor's magnetic field builds up, then decays. To distinguish it from dissipative resistance, we call X_L the **inductive reactance.** Inductive reactance, too, is measured in ohms.

Does it make sense that the inductive reactance increases with increasing ω and increasing L? An inductor is a device that, through its induced back emf, opposes changes in current. The greater the inductance, the greater the opposition to changing current. And the more rapidly the current is changing, the more vigorously the inductor opposes the change, so the inductive reactance increases at high frequencies. In the extreme case of very high frequencies, an inductor looks like an open circuit. But at very low frequencies it looks more and more like a short circuit, until with direct current (zero frequency), an inductor exhibits zero reactance because there is no change in current.

Why does the inductor voltage *lead* the current? Because a changing current in an inductor induces an emf. *Before* the current can build up significantly there must first, therefore, be voltage across the inductor.

Table 33-1 summarizes the phase and amplitude relations in resistors, capacitors, and inductors.

● **EXAMPLE 33-2** INDUCTORS AND CAPACITORS

A capacitor is connected across the 60-Hz, 120-V rms power line, and an rms current of 200 mA flows. What is the capacitance? What inductance would have to be connected across the power line for the same current to flow? Would there be anything different about the circuit containing the inductor?

Solution

The peak current and voltage are related through Equation 33-5:

so

$$I_p = \frac{V_p}{1/\omega C},$$

$$C = \frac{I_p}{\omega V_p}.$$

We are given the rms voltage and current, but since only the ratio of these quantities appears in our equation, it doesn't

▲ **TABLE 33-1** PHASE AND AMPLITUDE RELATIONS IN CIRCUIT ELEMENTS

CIRCUIT ELEMENT	PEAK CURRENT/VOLTAGE	PHASE RELATION
Resistor	$I_p = \dfrac{V_p}{R}$	V, I in phase
Capacitor	$I_p = \dfrac{V_p}{X_C} = \dfrac{V_p}{1/\omega C}$	I leads V by 90°
Inductor	$I_p = \dfrac{V_p}{X_L} = \dfrac{V_p}{\omega L}$	V leads I by 90°

matter whether we use rms or peak values. With $f = 60$ Hz, $\omega = 2\pi f$ or 377 s^{-1}, so

$$C = \frac{I}{\omega V} = \frac{0.20 \text{ A}}{(377 \text{ s}^{-1})(120 \text{ V})} = 4.4 \text{ } \mu\text{F}.$$

An inductor that passes the same current must have the same reactance, so

$$\omega L = \frac{1}{\omega C},$$

or $$L = \frac{1}{\omega^2 C} = \frac{1}{(377 \text{ s}^{-1})^2 (4.4 \times 10^{-6} \text{ F})} = 1.6 \text{ H}.$$

Although the currents are the same, the two situations differ in that current leads voltage by 90° in the capacitor and lags by 90° in the inductor.

EXERCISE Inductors (called "ballast") are often used to limit the current in fluorescent lamps. In a particular lamp operating at 60 Hz, 80 V (rms) appears across a 0.53-H ballast inductor. (a) What is the rms inductor current? (b) What capacitance could be used in place of the inductor to provide the same current?

Answers: (a) 400 mA; (b) 13 μF

Some problems similar to Example 33-2: 13, 14, 17, 20 ●

Phasor Diagrams

Phase and amplitude relations in AC circuits may be summarized graphically in **phasor diagrams.** A phasor is an arrow whose fixed length represents the amplitude of an AC voltage or current. The phasor rotates counterclockwise about the origin with the angular frequency ω of the AC quantity. The component of the phasor on the vertical axis then represents the sinusoidally varying AC signal. Figure 33-7a shows phasors for the current and voltage in a resistor. The lengths of the phasors are related by Ohm's law, $V_p = I_p R$. The current and voltage phasors always point in the same direction, showing that current and voltage in the resistor are in phase. Figures 33-7b and 33-7c show phasor diagrams for a capacitor and an inductor. In each, the lengths of the phasors are related by the appropriate reactance, so $V_p = I_p X$. As the phasors rotate, they remain at right angles, indicating the phase relation between current and voltage in these reactive circuit elements. You should convince yourself that all the relationships of Table 33-1 are correctly described by the phasor diagrams of Fig. 33-7. Although phasor diagrams do not add much to our understanding of AC circuits containing only one circuit element, they will greatly simplify the analysis of more complicated circuits.

FIGURE 33-7 Phasor diagrams showing voltage and current in (*a*) a resistor, (*b*) a capacitor, and (*c*) an inductor. Lengths of phasors correspond to amplitudes V_p and I_p, while projection on the vertical axis gives the instantaneous value. Angle between voltage and current phasors gives their relative phase.

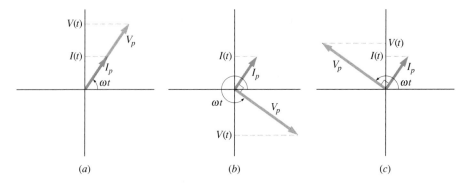

▲ **T A B L E   3 3 - 2**   CAPACITORS AND INDUCTORS

	CAPACITOR	**INDUCTOR**
Defining relation	$C = \dfrac{q}{V}$	$L = \dfrac{\phi_B}{I}$
Defining relation, differential form	$I = C\dfrac{dV}{dt}$	$\mathcal{E} = -L\dfrac{dI}{dt}$
Opposes changes in	Voltage	Current
Energy storage	In electric field $U = \frac{1}{2}CV^2$	In magnetic field $U = \frac{1}{2}LI^2$
Behavior in low-frequency limit	Open circuit	Short circuit
Behavior in high-frequency limit	Short circuit	Open circuit
Reactance	$X_C = 1/\omega C$	$X_L = \omega L$
Phase	Current leads by 90°	Voltage leads by 90°

Capacitors and Inductors: A Comparison

Here and in previous chapters, we have considered separately the behavior of capacitors and inductors. Many of the properties of these devices are analogous. A capacitor opposes instantaneous changes in voltage, while an inductor opposes instantaneous changes in current. In an *RC* circuit with a DC emf, voltage builds up exponentially across the capacitor, with time constant *RC*. In the analogous *RL* circuit, current builds up exponentially in the inductor, with time constant L/R. A capacitor stores electrical energy given by $\frac{1}{2}CV^2$. An inductor stores magnetic energy given by $\frac{1}{2}LI^2$. A capacitor acts like an open circuit at low frequencies, an inductor like a short circuit at low frequencies. Each exhibits the opposite behavior at high frequencies.

Capacitors and inductors are complementary devices, reflecting a deeper complementarity between electric and magnetic fields. Any verbal description of a capacitor applies to an inductor if we replace the words "capacitor" with "inductor," "electric" with "magnetic," and "voltage" with "current." Table 33-2 summarizes the complementary aspects of capacitors and inductors.

■ **A P P L I C A T I O N**   LOUDSPEAKER SYSTEMS

Loudspeakers in high-quality audio systems invariably contain two or more individual *drivers*—devices for converting electrical energy to sound—within the same enclosure (Fig. 33-8a). The most common form of driver includes a wire coil attached to a diaphragm and suspended in the field of a permanent magnet (Fig. 33-8b). Varying current in the coil then leads to a varying force that drives the diaphragm back and forth to produce sound.

Faithful reproduction of low-frequency sound requires a large driver, called a *woofer*. The woofer is large both because much of the sound power in music lies in the low-frequency range (think of a drum versus a flute!) and because the human

ear is much less sensitive to low frequencies (recall Fig. 17-4). But the woofer's large size gives it a large mechanical inertia, which means it cannot respond effectively with the rapid movements needed to reproduce high-frequency sound. Consequently, a much smaller driver, the *tweeter,* is used for high frequencies.

Power from an amplifier does not "know" about the mechanical properties of the drivers, so connecting both drivers directly to the same amplifier would result in low-frequency power being dissipated ineffectually in the tweeter and high-frequency power in the woofer. To prevent this inefficiency, a speaker system contains a *crossover network* to "steer" power to the appropriate drivers. Figure 33-8c shows a simple crossover network that exploits the frequency-dependent behavior of capacitors and inductors. At low frequencies the inductor reactance ωL is low, and current flows readily through the inductor to the woofer coil. But at high frequencies the inductor reactance is high, and little high-frequency power reaches the woofer. The capacitor's behavior is the opposite: It blocks low-frequency power from reaching the tweeter, while passing high-frequency power. Many speaker systems also employ a *midrange* driver, with a capacitor and inductor in series to block power at both high- and low-frequency extremes. Example 33-4, later in this chapter, explores quantitatively the behavior of the midrange circuitry.

The circuits of Fig. 33-8c are examples of *filters,* widely used in electronic systems to pass preferentially a range of frequencies. Problem 74 explores a simple filter.

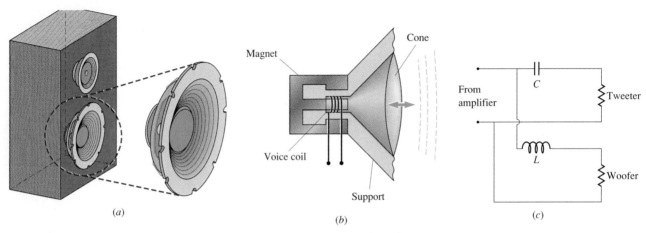

FIGURE 33-8 (a) A loudspeaker system, showing the woofer and tweeter drivers for low- and high-frequency sound production. (b) Cutaway view of a moving-coil driver. (c) The crossover network "steers" power at the appropriate frequencies to the woofer and tweeter. Resistances represent the two drivers.

33-3 *LC* CIRCUITS

In this section we consider circuits containing both inductors and capacitors. The properties of these circuits reflect directly the complementary nature of the two devices.

LC Oscillations

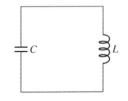

FIGURE 33-9 An *LC* circuit.

Figure 33-9 shows a circuit with a capacitor C and inductor L. Suppose the capacitor is initially charged to some voltage V_p and corresponding charge q_p, then connected to the inductor. What happens?

Initially, the capacitor is fully charged, while the inductor current is zero. There is electrical energy stored in the capacitor, but no energy in the inductor. This initial state is shown in Fig. 33-10a. Then the capacitor begins to discharge through the inductor. It cannot do so all at once, because the inductor opposes changes in current. So current in the inductor rises gradually, and with it the magnetic energy stored in the inductor. At the same time the capacitor voltage, charge, and stored electrical energy decrease. Some time later, the initial energy is divided equally between the capacitor and inductor, as in Fig. 33-10b. But the capacitor keeps discharging, eventually reaching zero charge, as in Fig. 33-10c. Now there is no voltage across the capacitor and no stored electrical energy. All the energy that was initially in the electric field of the capacitor is in the magnetic field of the inductor.

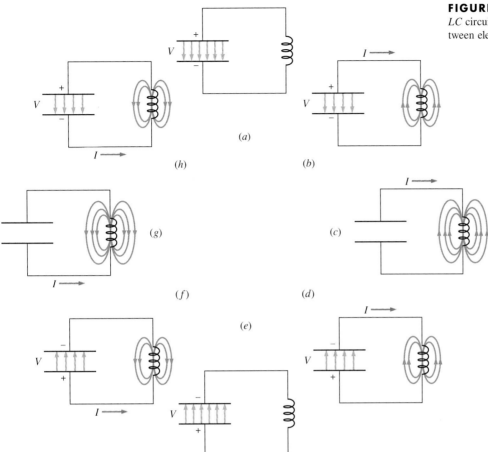

FIGURE 33-10 Oscillation in an *LC* circuit, showing energy transfer between electric and magnetic fields.

Does everything stop at this point? No, because current is flowing in the inductor. Current in an inductor cannot change abruptly, so the current keeps flowing and piles positive charge on the bottom plate of the capacitor. Stored electrical energy increases as the capacitor charges, and the inductor current and stored magnetic energy decrease. Eventually (Fig. 33-10*e*), the capacitor is fully charged in the opposite direction from its initial state. Again all the energy is in the capacitor. Again the capacitor begins to discharge, and the process repeats, now with a counterclockwise current. All the energy is transferred to the inductor (Fig. 33-10*g*), and then back to the capacitor, which again attains its initial state (Fig. 33-10*a*). Provided there is no energy loss, the oscillation repeats indefinitely.

This *LC* oscillation should remind you of the mass-spring system we studied in Chapter 15. There, energy was transferred back and forth between kinetic energy of the mass and potential energy of the spring. Here, energy is transferred back and forth between magnetic energy of the inductor and electrical energy of the capacitor. The mass-spring system oscillates with frequency determined by the mass *m* and spring constant *k*. Similarly, the *LC* circuit oscillates with frequency determined by the inductance *L* and capacitance *C*. Table 33-3 shows some analogies between mass-spring systems and *LC* circuits. We will develop these analogies more rigorously in the next section.

Analogies with *LC* circuits are so useful that engineers sometimes simulate complicated systems, such as bridges, automobile suspensions, or world energy

▲ **TABLE 33-3** *LC* CIRCUITS AND MASS-SPRING SYSTEMS

LC CIRCUIT	MASS-SPRING
Charge q	Displacement x
Current $I = dq/dt$	Velocity $v = dx/dt$
Inductance L	Mass m
Capacitance C	$1/k$ (k = spring constant)
Magnetic energy $U_B = \frac{1}{2}LI^2$	Kinetic energy $U_K = \frac{1}{2}mv^2$
Electric energy $U_E = \frac{1}{2}(1/C)q^2$	Potential energy $U = \frac{1}{2}kx^2$
Frequency $\omega = 1/\sqrt{LC}$	Frequency $\omega = \sqrt{k/m}$
Resistance	Friction

usage, with networks of *LC* circuits. Such a network is called an analog computer because its behavior is analogous to that of the system under study.

Analyzing the *LC* Circuit

We described the *LC* circuit qualitatively in terms of transfer between electric and magnetic energy. This description suggests a way to analyze the circuit quantitatively. The total energy in the circuit is the sum of the magnetic and electric energy:

$$U = U_B + U_E = \frac{1}{2}LI^2 + \frac{1}{2}\frac{q^2}{C}.$$

The time derivative of this equation is

$$\frac{dU}{dt} = \frac{d}{dt}\left(\frac{1}{2}LI^2 + \frac{1}{2}\frac{q^2}{C}\right).$$

But since the total energy does not change, $dU/dt = 0$. Carrying out the differentiations in our expression for dU/dt, we then have

$$LI\frac{dI}{dt} + \frac{q}{C}\frac{dq}{dt} = 0. \tag{33-8}$$

Substituting $I = dq/dt$ and $dI/dt = d^2q/dt^2$ gives

$$L\frac{d^2q}{dt^2} + \frac{1}{C}q = 0. \tag{33-9}$$

Equation 33-9 is a differential equation describing the capacitor charge as a function of time. We encountered a similar equation in Chapter 15 when we studied the mass-spring system:

$$m\frac{d^2x}{dt^2} + kx = 0. \qquad (15\text{-}4)$$

We found that Equation 15-4 could be satisfied by a sinusoidal function of time, with frequency given by

$$\omega = \sqrt{k/m}. \qquad (15\text{-}11)$$

Equation 33-9 is identical to Equation 15-4 except that q replaces x, L replaces m, and $1/C$ replaces k. Therefore the solution of Equation 33-9 is a sinusoidal oscillation whose frequency is given by Equation 15-11 with L replacing m and $1/C$ replacing k:

$$q = q_p\cos\omega t, \qquad (33\text{-}10)$$

where

$$\omega = \frac{1}{\sqrt{LC}}. \qquad (33\text{-}11)$$

Here we chose cosine rather than sine since our qualitative description (Fig. 33-10) started with the capacitor charge at its peak value. We could equally well have used a sine function with phase constant $\pi/2$. Differentiating the charge (Equation 33-10) gives the current:

$$I = \frac{dq}{dt} = \frac{d}{dt}(q_p\cos\omega t) = -\omega q_p\sin\omega t. \qquad (33\text{-}12)$$

All other circuit quantities follow from Equations 33-10 and 33-12. The capacitor voltage, obtained from the relation $q = CV$, is

$$V_C = \frac{q}{C} = \frac{q_p}{C}\cos\omega t.$$

The electrical energy in the capacitor is, therefore,

$$U_E = \tfrac{1}{2}CV^2 + (\tfrac{1}{2}C)\left(\frac{q_p}{C}\cos\omega t\right)^2 = \frac{q_p^2}{2C}\cos^2\omega t,$$

while the magnetic energy in the inductor is

$$U_B = \tfrac{1}{2}LI^2 = \tfrac{1}{2}L(-\omega q_p\sin\omega t)^2 = \tfrac{1}{2}L\omega^2 q_p^2\sin^2\omega t.$$

We can verify that our solution conserves energy by adding the electric and magnetic energies:

$$U_{\text{total}} = U_E + U_B = \frac{q_p^2}{2C}\cos^2\omega t + \tfrac{1}{2}L\omega^2 q_p^2\sin^2\omega t.$$

But Equation 33-11 shows that $\omega^2 = 1/LC$, so we have

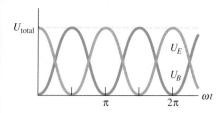

FIGURE 33-11 Electric and magnetic energies in an LC circuit. Their sum is constant.

$$U_{total} = \frac{q_p^2}{2C} \cos^2 \omega t + \frac{1}{2} \frac{L}{LC} q_p^2 \sin^2 \omega t$$

$$= \frac{q_p^2}{2C} (\cos^2 \omega t + \sin^2 \omega t) = \frac{q_p^2}{2C},$$

since $\cos^2 \omega t + \sin^2 \omega t = 1$. Thus the total energy is independent of time and is equal to the initial energy stored in the capacitor. Figure 33-11 is a plot of the electric and magnetic energies as functions of time, showing that the two always sum to a constant.

● EXAMPLE 33-3 A PIANO TUNER

You wish to make an LC circuit oscillate at 440 Hz (A above middle C) to assist in tuning pianos. You have available a 2.0-H inductor. What value of capacitance should you use? If you initially charge the capacitor to 5.0 V, what will be the peak charge on the capacitor and the peak current in the circuit?

Solution

The oscillation frequency is given by Equation 33-11. Solving for C gives

$$C = \frac{1}{\omega^2 L} = \frac{1}{(2\pi f)^2 L} = \frac{1}{[(2\pi)(440 \text{ Hz})]^2 (2.0 \text{ H})} = 65.4 \text{ nF}.$$

The capacitor charge and voltage are related through $C = q/V$, so

$$q_p = CV_p = (65.4 \text{ nF})(5.0 \text{ V}) = 327 \text{ nC}.$$

Equation 33-12 then shows that the peak current is

$$I_p = \omega q_p = 2\pi f q_p = (2\pi)(440 \text{ Hz})(327 \text{ nC}) = 0.90 \text{ mA}.$$

Problem 38 shows how the current also follows from energy considerations.

EXERCISE An FM radio transmitter requires an LC circuit that oscillates at the 89.5-MHz transmitter frequency. What should be the inductance if the circuit capacitance is 47 pF?

Answer: 67 nH

Some problems similar to Example 33-3: 27, 28, 31, 32 ●

Resistance in *LC* Circuits—Damping

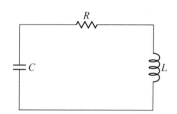

FIGURE 33-12 An *RLC* circuit.

Real inductors, capacitors, and wires have resistance. Both this intrinsic resistance and any external resistance are represented by the resistor R in Fig. 33-12. What happens in such a resistive LC circuit?

Provided the resistance is small—small enough that only a small fraction of the energy is lost in one cycle—then our analysis of the preceding section applies. The circuit oscillates at a frequency given very nearly by Equation 33-11. But as current flows back and forth through the resistor, energy is dissipated as heat. On each successive cycle, the total energy decreases. Consequently, the amplitude of the oscillations decreases with time.

We can analyze this RLC circuit by starting with Equation 33-8, but now setting dU/dt equal to the rate of energy dissipation in the resistor:

$$LI\frac{dI}{dt} + \frac{q}{C}\frac{dq}{dt} = -I^2 R,$$

where the minus sign indicates that energy is *lost* in the resistor. Writing $I = dq/dt$ as we did in the preceding section leads to

$$L\frac{d^2q}{dt^2} + R\frac{dq}{dt} + \frac{q}{C} = 0.$$

This equation is mathematically identical to Equation 15-22 for damped simple harmonic motion, showing that our analogies of Table 33-3 continue to hold when resistance is present. Using Equation 15-23, which is the solution to Equation 15-22, and the appropriate analogies from Table 33-3, we can construct the solution for our decaying *RLC* circuit:

$$q = q_p e^{-Rt/2L} \cos \omega t. \qquad (33\text{-}13)$$

Other quantities show similar behavior, with oscillation amplitude decaying exponentially with time constant $2L/R$. Figure 33-13 shows an oscilloscope trace of the capacitor voltage in a circuit undergoing damped oscillations.

Equations 15-23 and 33-13 are correct only when the energy dissipation is small. As the electrical resistance increases, the oscillations decay more rapidly and the frequency of oscillation decreases. Finally, when the exponential time constant $2L/R$ equals the inverse of the natural frequency given by Equation 33-11, much of the energy is lost in the time of one undamped oscillation period. This situation is termed **critical damping,** and at this value of R circuit quantities decay exponentially to zero, in analogy with a critically damped mechanical system (Section 15-6). For greater values of R, the circuit is **overdamped** and also exhibits no oscillation.

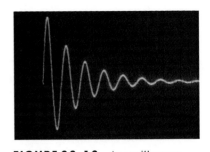

FIGURE 33-13 An oscilloscope displays the capacitor voltage in a damped *RLC* circuit. Note the exponential decline in amplitude of the oscillations.

33-4 DRIVEN *RLC* CIRCUITS AND RESONANCE

What happens if we connect an *RLC* circuit to an AC generator, as shown in Fig. 33-14? Because the circuit is analogous to a mass-spring system, we might expect it to exhibit resonant behavior analogous to the mechanical resonance discussed in Section 15-7.

Resonance in the *RLC* Circuit

Suppose we vary the generator frequency in Fig. 33-14 while keeping the peak voltage constant. How much current flows in the *RLC* circuit? At low frequencies the capacitor acts almost like an open circuit (its reactance $X_C = 1/\omega C$ is large), so little current flows. At high frequencies the inductor acts almost like an open circuit (its reactance $X_L = \omega L$ is large), so little current flows. At some intermediate frequency the current must be a maximum. We now show that this **resonant frequency,** ω_r, is in fact the undamped natural frequency $1/\sqrt{LC}$.

Figure 33-14 is a series circuit, so the *same* current flows through all the components. We know that the voltage in a capacitor lags the current by 90°,

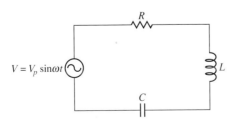

FIGURE 33-14 A series *RLC* circuit driven by an AC generator.

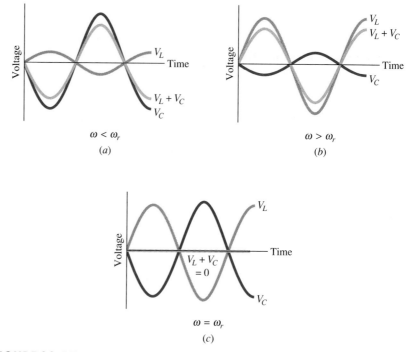

FIGURE 33-15 In a series *RLC* circuit, capacitor and inductor voltages are 180° out of phase at all frequencies. (*a*) At low frequencies, the voltage is greatest across the capacitor. (*b*) At high frequencies, the voltage is greatest across the inductor. (*c*) At resonance, capacitor and inductor voltages have equal amplitude, and therefore they cancel completely.

while the voltage in an inductor leads by 90°. Since the same current flows through the inductor and capacitor, the inductor and capacitor voltages are therefore 180° out of phase. This phase relation holds at *any* frequency. At low frequencies, however, the capacitor's reactance is greatest and the capacitor voltage is therefore greater than the inductor voltage (Fig. 33-15*a*). At high frequencies the opposite is true, with the inductor voltage being greatest (Fig. 33-15*b*). Since the two voltages are 180° out of phase, they tend to cancel, but at high or low frequencies Figs. 33-15*a* and *b* show that their different amplitudes mean this cancellation is not complete.

Is there a frequency at which the capacitor and inductor voltages exactly cancel? The peak current and voltage in these two components are related by Equations 33-5 and 33-7, respectively. Since the current is the same, comparison of these equations shows that the voltages will be the same when the capacitive reactance $1/\omega C$ is equal to the inductive reactance ωL, which gives

$$\omega_r^2 = \frac{1}{LC}.$$

This is precisely Equation 33-11's condition for the undamped natural frequency.

At resonance, then, the capacitor and inductor voltages completely cancel. The voltage across the pair together is zero, and—at the resonant frequency only—they might as well be replaced by a wire. The circuit current at resonance is determined entirely by the resistance. At any other frequency the effects of capacitance and inductance do not cancel completely, and the current is lower.

Frequency Response of the *RLC* Circuit

Here we derive a general expression for the current as a function of frequency in the series *RLC* circuit, using the phasor diagrams introduced in Section 33-2. Since the same current flows through all components of the series circuit, we represent this current by a single phasor of length I_p in Fig. 33-16. Also shown are phasors for the resistor, capacitor, and inductor voltages. The resistor voltage phasor is in the same direction as the current because these two are in phase, but the capacitor and inductor voltages are at 90° relative to the current, representing the phase relations in these components.

Applying the loop law to Fig. 33-14 gives $V - V_R - V_L - V_C = 0$; thus, the applied voltage V is the sum of the voltages across the other three circuit elements. This statement is true for the instantaneous values of the voltages at any time. But because they have different phase constants, the voltages V_R, V_L, and V_C peak at different times—and, thus, the peak applied voltage is *not* the sum of the peak voltages across the resistor, inductor, and capacitor. We can find the relation among the peak voltages by adding their phasors *vectorially*—a process that accounts for both magnitude and phase. Figure 33-16*b* shows the result of this phasor addition. Applying the Pythagorean theorem, we see that the magnitude of the applied voltage phasor—that is, the peak applied voltage—is given by

$$V_p^2 = V_{Rp}^2 + (V_{Cp} - V_{Lp})^2.$$

Expressing this in terms of the current and the resistances and reactances gives

$$V_p^2 = I_p^2 R^2 + (I_p X_c - I_p X_L)^2,$$

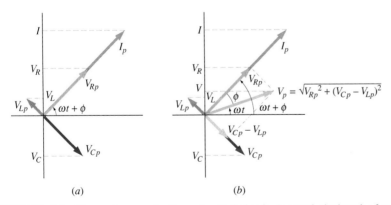

(a)   (b)

FIGURE 33-16 Phasor diagrams for the series *RLC* circuit. As usual, the length of each phasor indicates the peak value (subscript *p*) of the associated quantity. (*a*) The current in the three circuit elements is the same, but the voltages are out of phase. The whole system rotates with angular speed ω, and instantaneous values of voltages and current are given by the projections on the vertical axis. (*b*) The three voltage phasors sum vectorially to the applied voltage. Note that for this case the current leads the applied voltage; the phase difference ϕ is given by Equation 33-16.

or

$$I_p = \frac{V_p}{\sqrt{R^2 + (X_C - X_L)^2}} = \frac{V_p}{Z},$$ (33-14)

where we have defined

$$Z = \sqrt{R^2 + (X_C - X_L)^2}.$$

Equation 33-14 has the form of Ohm's law, with Z playing the role of resistance. We call Z the **impedance** of the circuit. Impedance is a generalization of resistance to include the frequency-dependent effects of capacitance and inductance. Putting in our expressions for the reactances gives

$$Z = \sqrt{R^2 + \left(\frac{1}{\omega C} - \omega L\right)^2}.$$ (33-15)

In agreement with our earlier analysis, this equation shows that the circuit impedance becomes very large at high and low frequencies, and has its lowest value, R, at resonance.

Figure 33-17 is a plot of Equation 33-14, showing peak current versus frequency for several values of resistance. As we lower the resistance, the peak current at resonance rises. Although the current at other frequencies rises, too, it does so to a much lower extent than at resonance. This is because the impedance at resonance depends only on the resistance, but includes reactive effects at other frequencies. As a result, the resonance curve becomes more sharply peaked as the resistance drops. For a circuit with very low resistance, the current at resonance is dramatically different from that at even a slightly different frequency. Such a circuit, called a **high-Q** (for high-quality) circuit, does a good job of distinguishing its resonant frequency from nearby frequencies. A rigorous definition of Q can be given in terms of the width of the resonance curve (see Problem 81). When you tune an older radio, you are adjusting the variable capacitor in a high-Q LC circuit that selects the desired station from among the hodgepodge of radio signals reaching the antenna (Fig. 33-18).

Equation 33-14 relates the peak current and the applied voltage in the RLC circuit, but it does not tell the whole story. In general, current and voltage are

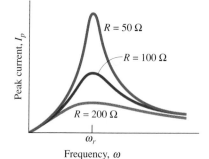

FIGURE 33-17   Resonance curves for an RLC circuit with $L = 5.0$ mH, $C = 0.22$ μF, and several resistance values.

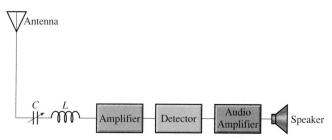

FIGURE 33-18   Simplified diagram of an older radio, using an LC circuit with a variable capacitor to select the desired station frequency. The signal is subsequently amplified, the audio information is extracted in the detector, and the amplified audio signal drives the loudspeaker. Modern radios often use sophisticated digital circuitry in place of the LC frequency selector.

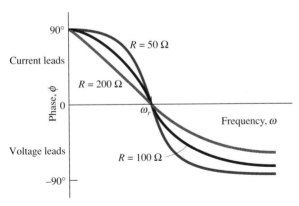

FIGURE 33-19   Phase relations for the *RLC* circuits whose resonance curves are shown in Fig. 33-17.

not in phase. Figure 33-16*b* shows that the phase difference ϕ between the two is

$$\tan \phi = \frac{V_{Cp} - V_{Lp}}{V_{Rp}}.$$

Because the voltages are proportional to the reactances and resistance, this expression may be written

$$\tan \phi = \frac{X_C - X_L}{R} = \frac{1/\omega C - \omega L}{R}. \tag{33-16}$$

Does this equation make sense? At resonance, $X_L = X_C$ and the phase difference is zero. Here the effects of capacitance and inductance cancel, and the circuit behaves like a pure resistance. At frequencies below resonance, capacitive reactance is greatest and the phase difference is positive, indicating that current leads voltage—as we expect in a circuit dominated by capacitance. Above resonance $X_L > X_C$ and the phase difference is negative, indicating that voltage leads current—as we expect in a circuit dominated by inductance. At high- and low-frequency extremes, $\tan \phi$ becomes arbitrarily large and the phase differences approach 90°, as shown in Fig. 33-19.

TIP   **Phase Matters**   You can't analyze an AC circuit by treating resistors, capacitors, and inductors all as "resistors" with resistances R, X_C, and X_L. Why not? Because associated with each component is a different phase relation between current and voltage. Only a phasor diagram—or mathematical analysis that includes phase relations—correctly characterizes the circuit. The phase relations in a series *RLC* circuit show up in the minus sign joining the capacitive and inductive reactance, and in the Pythagorean-like addition of the resistance and reactances in Equations 33-14 and 33-15.

● **EXAMPLE 33-4** DESIGNING A LOUDSPEAKER SYSTEM

Current flows to the midrange speaker in a loudspeaker system through a 2.2-mH inductor and a capacitor, in series. (See *Application: Loudspeaker Systems,* earlier in this chapter.) (a) What should be the capacitance in order that a given applied voltage produces the maximum current at a frequency of 1.0 kHz? (b) What should be the speaker's resistance in order that the same voltage produces a current with half the maximum value at 600 Hz? (c) If the peak output voltage of the amplifier is 20 V, what should be the capacitor's peak voltage rating?

Solution

(a) Inductor, capacitor, and resistor comprise a series RLC circuit, with resonant frequency $f_r = \omega_r/2\pi = 1/2\pi\sqrt{LC}$. Solving for C gives

$$C = \frac{1}{4\pi^2 f_r^2 L} = \frac{1}{(4\pi^2)(1.0\times10^3\text{ Hz})^2(2.2\times10^{-3}\text{ H})} = 11.5\,\mu\text{F}.$$

(b) Equation 33-14 gives the current in an RLC circuit. The denominator in this equation—the impedance $Z = \sqrt{R^2 + (X_C - X_L)^2}$—has the value R at resonance, when $X_L = X_C$. Thus the current will have half its maximum value where the $Z = 2R$. Using $X_C = 1/\omega C$ and $X_L = \omega L$, we want R such that

$$Z = \sqrt{R^2 + \left(\frac{1}{\omega C} - \omega L\right)^2} = 2R$$

when $f = \omega/2\pi = 600$ Hz, or $\omega = 3.77\times10^3$ s^{-1}. Squaring and solving for R gives

$$R = \frac{1}{\sqrt{3}}\left(\frac{1}{\omega C} - \omega L\right) = 8.53\,\Omega,$$

where we used the 11.5-μF answer to part (a) for the capacitance C. This resistance value is typical for a loudspeaker. The exercise below shows that the current also has half its maximum value at 1.7 kHz, so the midrange speaker gets at least half its maximum current in the range from about 600 to 1700 Hz.

(c) The maximum current flows at the resonant frequency, so the Ohm's-law-like Equation 33-5 shows that the maximum capacitor voltage also occurs at this frequency and is given by

$$V_{Cp} = I_p X_C = \left(\frac{V_p}{R}\right)\left(\frac{1}{\omega C}\right)$$

$$= \left(\frac{20\text{ V}}{8.53\,\Omega}\right)\left(\frac{1}{(2\pi)(1000\text{ Hz})(11.5\times10^{-6}\text{ F})}\right)$$

$$= 32\text{ V}.$$

How can this value be *greater* than the peak applied voltage? Remember that there is another source of emf in the circuit—the inductor, whose emf depends on the rate of change of current and may therefore exceed the applied voltage. In this relatively low-Q circuit the peak capacitor voltage is not too much greater than the peak applied voltage, but in high-Q circuits like those used in radio transmitters capacitors may have to withstand voltages hundreds of times those supplied to the circuit.

EXERCISE Find a second frequency where the speaker current in Example 33-4 has half its maximum value. *Hint:* Square Equation 33-15, isolate the frequency-dependent terms, take a square root, and solve a quadratic equation. Or use graphical techniques.

Answer: 1.7 kHz

Some problems similar to Example 33-4: 43, 44, 46, 69, 70, 76 ●

33-5 POWER IN AC CIRCUITS

In Section 33-2, we noted that average power dissipation is zero in a circuit containing only a capacitor or an inductor. We can understand this physically because the reactive element alternately stores and releases energy rather than dissipating it as heat. Mathematically, we can see this from Fig. 33-20a, which shows the current, voltage, and instantaneous power in a capacitor. The power is the product of the current and voltage. Because these two are out of phase, the power is positive half the time, and negative half the time. When the power is positive, the capacitor is absorbing energy from the source of emf that drives the current. When the power is negative, the capacitor is returning energy to the driving source. The net energy transferred to the capacitor over one cycle is $\int P\,dt$, or the area under the power versus time curve, and is zero in this case.

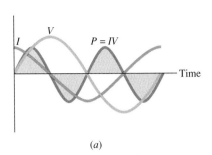

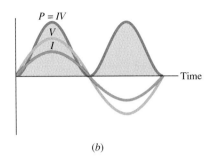

(a)

(b)

FIGURE 33-20 Current I, voltage V, and power consumption IV. (a) Current and voltage in a capacitor are out of phase by 90°. The power is alternately positive (absorbing energy from the source of emf) and negative (returning energy to the source), and the net energy consumption (shaded area) over one cycle is zero. (b) Current and voltage are in phase in a resistor and therefore the power is always positive, meaning that the resistor is a net energy consumer.

Figure 33-20b, in contrast, shows current, voltage, and instantaneous power in a resistor. Since current and voltage are always in phase, the power is always positive, and the resistor always takes energy from the source. Comparison of Figs. 33-20a, b suggests that the phase difference between current and voltage is important in determining the average power consumption of an AC circuit. We can see this more clearly if we imagine slipping the current and voltage just slightly out of phase, as in Fig. 33-21. Then there are narrow regions where the power is negative, so the average power over one cycle is slightly less than in the resistor case. As the phase difference increases, so does the time that the power is negative, until at 90° phase difference, the time-average power is zero.

We can develop a general expression for power in AC circuits by considering the time-average product of voltage and current with arbitrary phase difference ϕ:

$$\langle P \rangle = \langle [I_p \sin(\omega t + \phi)][V_p \sin \omega t] \rangle,$$

where $\langle \rangle$ indicates a time average over one cycle. Expanding the current term using a trig identity (see Appendix A) gives

$$\langle P \rangle = I_p V_p \langle (\sin^2 \omega t)(\cos \phi) + (\sin \omega t)(\cos \omega t)(\sin \phi) \rangle.$$

The average of $(\sin \omega t)(\cos \omega t)$ is zero, as we've just shown for two signals 90° out of phase. The quantity $\sin^2 \omega t$ swings from 0 to 1, and is symmetric about $\frac{1}{2}$, so its average value is $\frac{1}{2}$ (Fig. 33-22). Then we have

$$\langle P \rangle = \tfrac{1}{2} I_p V_p \cos \phi.$$

Writing the peak values as $\sqrt{2}$ times the rms values gives

$$\langle P \rangle = \tfrac{1}{2}\sqrt{2} I_{rms} \sqrt{2} V_{rms} \cos \phi = I_{rms} V_{rms} \cos \phi. \qquad (33\text{-}17)$$

This equation confirms our earlier graphical arguments. When the voltage and current are in phase, the average power is just the product $I_{rms} V_{rms}$. (This, in fact, is a principal reason for using rms values: with them, the expression for average

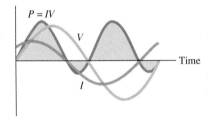

FIGURE 33-21 Average power consumption decreases as current and voltage go out of phase.

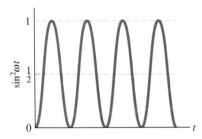

FIGURE 33-22 The average of $\sin^2 \omega t$ is $\frac{1}{2}$.

power is the same as in the DC case.) But with current and voltage out of phase, the average power is smaller; at 90° phase difference it is zero.

The factor cos ϕ is called the **power factor.** A purely resistive circuit has a power factor of 1, while a circuit with only inductance and capacitance has a power factor of zero. In circuits containing resistance along with capacitance and/or inductance, the power factor generally depends on frequency; in the series *RLC* circuit, for example, it is 1 at resonance but lower at other frequencies.

33-6 TRANSFORMERS AND POWER SUPPLIES

A **transformer** is a pair of wire coils in close proximity. Supplying an alternating current to one coil—called the **primary**—results in a changing magnetic flux through the other coil, or **secondary.** According to Faraday's law, an induced emf then appears in the secondary coil. This emf can drive a current in circuitry connected to the secondary coil. Thus, the transformer transfers electrical energy from the primary circuit to the secondary circuit, even though there is no direct electrical connection between the two.

Practical transformers are often wound on iron cores that concentrate flux, ensuring that essentially all the magnetic flux produced by the primary coil goes through the secondary. Figure 33-23 shows a simplified diagram of a transformer, along with its circuit symbol.

The transformer in Fig. 33-23*a* has two turns in its primary and four in its secondary. Since the same changing flux passes through each turn of each coil, the total emf induced the secondary must be twice that of the primary. The transformer is therefore a **step-up-transformer;** it steps up the voltage by a factor of 2. Interchanging primary and secondary would give a **step-down** transformer, with the secondary voltage half that of the primary. In general, the ratio of the peak (or rms) primary voltage V_1 to the peak (or rms) secondary voltage V_2 is the same as the ratio of the numbers of turns in the two coils:

$$V_2 = \frac{N_2}{N_1} V_1.\tag{33-18}$$

Aren't we getting something for nothing with a step-up transformer? No—a step-up transformer increases voltage but not power. In an ideal transformer, all the power supplied to the primary is transferred to the secondary; therefore,

$$I_1 V_1 = I_2 V_2.\tag{33-19}$$

If the voltage goes up, the current goes down, and vice versa. Real transformers have losses associated with resistance in their windings and heating of their iron cores, but good engineering holds these losses to a few per cent of the total power.

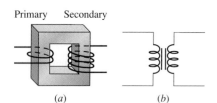

Primary   Secondary

(a)   *(b)*

FIGURE 33-23 (*a*) A transformer, consisting of two coils wound on an iron core. (*b*) Transformer circuit symbol.

58,7.5
71,5.02
0

■ APPLICATION ELECTRIC POWER DISTRIBUTION

The power loss in a resistance R carrying current I is given by I^2R. That means it's most efficient to transmit electric power at high voltage and low current. Lower voltages, on the other hand, are safer and easier to handle. How can we satisfy both these considerations in practical power systems? With transformers!

Power-plant generators operate at about 20 kV. Transformers at the power plant then step this up to several hundred kilovolts for long-distance transmission. At a city or town the voltage is dropped to several kilovolts for distribution within the municipality. Transformers near each building reduce the voltage further, for example, to 120 V and 240 V for household use. Individual electrical devices within a building use transformers to meet their particular voltage requirements. Figure 33-24 outlines the voltage transformations in power transmission.

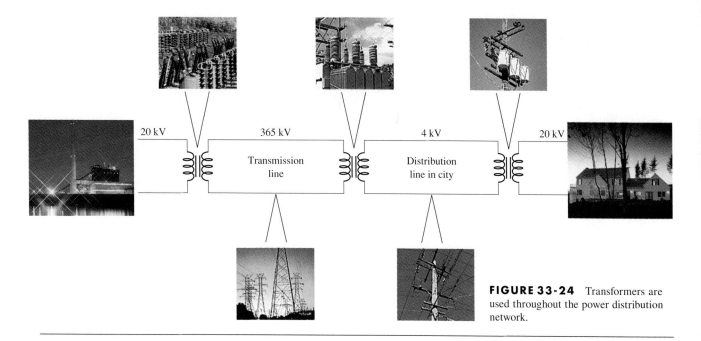

FIGURE 33-24 Transformers are used throughout the power distribution network.

Transformers are widely used to provide voltages different from those of available sources. Most transistor circuits, for example, require a few tens of volts or less; step-down transformers bring the 120-V AC power line down to this level. Television tubes and computer monitors require tens of kV, calling for step-up transformers.

Transformers exploit electromagnetic induction, and therefore, they work only with time-varying current—that is, AC. A transformer cannot be used to step up the DC voltage from a battery, unless current from the battery is somehow interrupted periodically—as it is, for example, by the distributor in an automobile ignition system, whose ignition coil is really a transformer (see Example 32-2). In fact, one of the main reasons for using AC power is that it is readily transformed from one voltage level to another.

FIGURE 33-25 Circuit symbol for a diode, with preferred direction of current indicated.

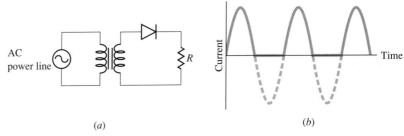

(a)   (b)

FIGURE 33-26 (a) A simple DC power supply consists of a transformer and diode, which supply current to the load resistance. The diode passes current in only one direction, cutting off the negative half of each cycle and giving the load current shown in (b).

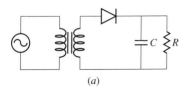

(a)

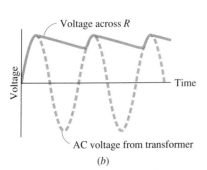

(b)

FIGURE 33-27 (a) Adding a capacitor reduces variations in the load voltage and therefore in the load current. This circuit is a complete DC power supply. (b) Load voltage as a function of time.

DC Power Supplies

Devices like light bulbs and electric heaters work equally well on AC or DC. But others, especially electronic circuits, require DC power. How do we get DC from the AC power line?

In Chapter 27's discussion of semiconductors, we mentioned that a junction between *P*-type and *N*-type semiconductors passes current in one direction but not the other. A **diode** is a *PN* junction designed specifically to be such a "one-way valve" for electric current. An ideal diode acts like a short circuit to current flowing in the preferred direction and like an open circuit in the opposite direction (Fig. 33-25).

Figure 33-26a shows a diode and resistor connected to the secondary of a transformer. The resistor represents the load—that is, whatever circuit we wish to supply with DC power. The transformer brings the 120-V AC power line to a level appropriate to the load. The diode passes current in only one direction, resulting in the load current shown in Fig. 33-26b.

Although the load current in Fig. 33-26 flows in one direction, it still varies drastically with time. If the load were a stereo amplifier, for example, its speakers would emit a loud hum. So we smooth, or **filter,** the output of the diode. The simplest filter is a capacitor, as shown in Fig. 33-27a. As the voltage on the left side of the diode rises, the capacitor voltage rises rapidly because of the short time constant associated with the low resistance of the diode in its preferred direction. But then the AC voltage begins to fall, and the diode "turns off." The capacitor cannot discharge from right to left through the diode but only through the load resistance. If the time constant *RC* is long enough, the capacitor voltage will drop only slightly before the AC voltage again rises and sends a new surge of current through the diode to bring the capacitor to its maximum charge (Fig. 33-27b). By making the capacitance large enough—so large that the time constant is much longer than the period of the AC power—we can make load current and voltage arbitrarily smooth. Large capacitors are expensive, so in practice additional filters—often using transistors or integrated circuits—are added to the simple capacitive filter in critical applications. High-quality power supplies for audio equipment or electronic instrumentation achieve **ripple factors**—the ratio of the fluctuation amplitude to the DC level—of 10^{-5} or better. More complicated diode circuits can also be used to produce signals requiring less filtering (see Problem 80).

CHAPTER SYNOPSIS

Summary

1. **Alternating current** varies with time. A sinusoidal AC signal is characterized by its peak amplitude, frequency, and phase constant; for example:

$$I = I_p \sin(\omega t + \phi).$$

2. In a resistor, the ratio of voltage to current is always constant, so

$$I_p = \frac{V_p}{R},$$

and the current and voltage are in phase.

3. In a capacitor the ratio of peak voltage and current is determined by the **capacitive reactance:**

$$I_p = \frac{V_p}{X_C}, \quad \text{where} \quad X_C = \frac{1}{\omega C}.$$

The current in the capacitor **leads** the voltage by $90°$.

4. In an inductor the ratio of peak voltage and current is determined by the **inductive reactance:**

$$I_p = \frac{V_p}{X_L}, \quad \text{where} \quad X_L = \omega L.$$

The current in the inductor **lags** the voltage by $90°$.

5. **Phasors** are vector-like arrows showing the amplitude and phase of AC signals, and are useful in analyzing AC circuits.

6. In an undriven LC circuit, energy oscillates between electric and magnetic forms at the **resonant frequency** $\omega_r = 1/\sqrt{LC}$. The amplitude of the oscillation decays exponentially as energy is dissipated in the circuit resistance.

7. In a series RLC circuit, the effects of inductance and capacitance exactly cancel at the resonant frequency. At this frequency the circuit exhibits the minimum **impedance,** and therefore passes the maximum current. At resonance the current and voltage are in phase. At lower frequencies the capacitor dominates and current leads voltage, while at higher frequencies the inductor dominates and voltage leads current. The impedance of a series RLC circuit is

$$Z = \sqrt{R^2 + (X_C - X_L)^2},$$

while the phase difference between current and voltage is given by

$$\cos \phi = \frac{R}{Z}.$$

8. The power dissipated in an AC circuit depends on the relative effects of resistance and reactance. In a purely reactive circuit, current and voltage are $90°$ out of phase, and no power is dissipated. In a purely resistive circuit, the average power dissipation is $I_{rms} V_{rms}$. When both resistance and reactance are present, the power dissipation depends on the **power factor,** $\cos \phi$, where ϕ is the phase difference between current and voltage. In general, the time-average power consumed in an AC circuit is

$$P = I_{rms} V_{rms} \cos \phi.$$

9. A **transformer** uses electromagnetic induction to transfer electric power between two circuits. The ratio of the peak or rms voltages in the transformer's two windings is the same as the ratio of the numbers of turns in the windings:

$$V_2 = \frac{N_2}{N_1} V_1,$$

while the power VI is the same in each winding of an ideal transformer.

10. DC **power supplies** use **diodes** to change alternating to direct power. Capacitors then smooth out the remaining time variation to produce the steady voltages and currents needed to power electronic equipment.

Terms You Should Understand

(Pairs are closely related terms whose distinction is important; number in parentheses is chapter section where term first appears.)

AC, DC (introduction)
peak, rms amplitudes (33-1)
frequency, angular frequency (33-1)
capacitive reactance, inductive reactance (33-2)
phasor (33-2)
resonant frequency (33-4)
power factor (33-5)
transformer (33-6)
diode (33-6)

Symbols You Should Recognize

I_p, V_p; I_{rms}, V_{rms} (33-1)
f, ω (33-1)
ϕ (33-1)
X_C, X_L (33-2)
Z (33-4)

Problems You Should Be Able to Solve

calculating current and voltage in AC circuits involving individual resistors, capacitors, and inductors (33-2)
analyzing oscillating LC circuits (33-3)
analyzing driven RLC circuits (33-4)

calculating power in AC circuits (33-5)
designing simple power supplies using transformers, diodes, and capacitive filters (33-6)

Limitations to Keep in Mind

This chapter considers only sinusoidally varying AC signals. More complex time variations must first be analyzed into sums of sinusoidal terms, using techniques mentioned in Chapter 16.

Care must be taken in adding AC signals with different phases. Phasor diagrams provide a convenient way of doing this.

QUESTIONS

1. Two AC signals have the same amplitude but different frequencies. Are their rms amplitudes the same?
2. Does it make sense to talk about the phase difference between two AC signals of different frequencies? Sketch a diagram to confirm your answer.
3. What is meant by the statement "a capacitor is a DC open circuit"?
4. How can current keep flowing in an AC circuit containing a capacitor? After all, a capacitor contains a gap between two conductors, and no charge can cross this gap.
5. Why does it make sense that inductive reactance increases with frequency?
6. The same AC voltage appears across a capacitor and a resistor, and the same rms current flows in each. Is the power dissipation the same in each?
7. When a particular inductor and capacitor are connected across the same AC voltage, the current in the inductor is larger than in the capacitor. Will this be true at all frequencies?
8. An inductor and capacitor are connected in series across an AC generator, and the rms voltage across the inductor is found to be larger than that across the capacitor. Is the generator frequency above or below resonance?
9. When the capacitor voltage in an undriven LC circuit reaches zero, why don't the oscillations stop?
10. Why is the quantity ωL not called the resistance of an inductor?
11. Why is Equation 33-5 not a full description of the relation between voltage and current in a capacitor? What equation does give the full relation?
12. If you double both the capacitance and inductance in an LC circuit, what effect does this have on the resonant frequency?
13. In a series RLC circuit, the applied voltage lags the current. Is the frequency above or below resonance?
14. In a series RLC circuit, the applied voltage leads the current. Is the peak voltage greater across the capacitor or the inductor?
15. At a certain frequency, the impedance of a series RLC circuit is twice the resistance of the circuit. Can you tell whether the frequency is above or below resonance? Which is it, or why can't you tell?
16. What does it mean to say that the capacitor in an RLC circuit dominates at low frequencies?
17. The voltage across two circuit elements in series is zero. Is it possible that the voltages across the individual elements are nevertheless not zero? Give an example.
18. If you measure the rms voltages across the resistor, capacitor, and inductor in a series RLC circuit, will they add to the rms value of the generator voltage? Reconcile your answer with the loop law. (See also Problem 47.)
19. In a fluorescent light fixture an inductor, called the ballast, is used to limit current to the lamp. Why is an inductor preferable to a resistor?
20. What is the power factor in a circuit containing only a resistor? Does this power factor change with frequency?
21. To save electrical energy, should you strive for a large or small power factor?
22. When an AC motor runs with no significant mechanical load, its power factor is very nearly zero. What must happen when the motor begins doing mechanical work?
23. A step-up transformer increases voltage, or energy per unit charge. How is this possible without violating conservation of energy?
24. The iron cores of transformers are often made by laminating together thin sheets of iron separated by an insulating substance. Why is this preferable to making the cores out of solid pieces of iron? *Hint:* Think about eddy currents.
25. A battery charger runs off the 120-V AC power line. It supplies up to 30 A to recharge a 12-V car battery, yet it can be plugged into a 15-A circuit without blowing the circuit breaker. How is this possible?
26. Manuals for electronic instruments that run off the AC power line, including stereo amplifiers, often caution against connecting the instrument to DC power sources. Why? *Hint:* What would happen if a DC voltage were imposed across the transformer in such an instrument?

PROBLEMS

Section 33-1 Alternating Current

1. Much of Europe uses AC power at 230 V rms and 50 Hz. Express this AC voltage in the form of Equation 33-3, taking $\phi = 0$.
2. An rms voltmeter connected across the filament of a TV picture tube reads 6.3 V. What is the peak voltage across the filament?
3. An oscilloscope displays a sinusoidal signal whose peak-to-peak voltage (see Fig. 33-1) is 28 V. What is the rms voltage?
4. An industrial electric motor runs at 208 V rms and 400 Hz. What are (a) the peak voltage and (b) the angular frequency?
5. An AC current is given by $I = 495 \sin(9.43t)$, with I in milliamperes and t in milliseconds. Find (a) the rms current and (b) the frequency in Hz.
6. What are the phase constants for each of the signals shown in Fig. 33-28?

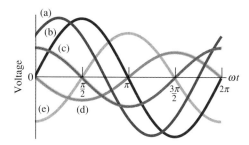

FIGURE 33-28   Problem 6.

7. The rms amplitude is defined as the square root of the average of the square of the signal. For a periodic function the time average is the integral over one period, divided by the period. For a sinusoidal voltage given by $V = V_p \sin \omega t$, show explicitly that $V_{\text{rms}} = V_p/\sqrt{2}$.
8. How are the rms and peak voltages related for the square wave shown in Fig. 33-29? See Problem 7.

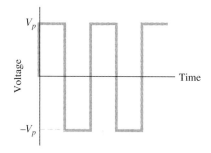

FIGURE 33-29   A square wave (Problem 8).

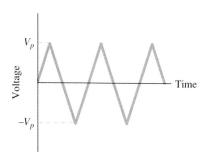

FIGURE 33-30   A triangle wave (Problem 9).

9. How are the rms and peak voltages related for the triangle wave in Fig. 33-30? See Problem 7.
10. A 10-V rms AC signal with frequency 1.0 kHz is 90° ahead of another signal of the same frequency but with peak voltage 7.1 V. On the same graph, plot voltage versus time for both these signals. Choose scales appropriate to display one full cycle.
11. The most general expression for a sinusoidal AC current may be written either $I = I_1 \sin \omega t + I_2 \cos \omega t$ or $I = I_p \sin(\omega t + \phi)$. Find relations between I_1, I_2, I_p, and ϕ that make these expressions equivalent. (See Appendix A for trig identities.)

Section 33-2 Circuit Elements in AC Circuits

12. Show that the unit of both capacitive and inductive reactance is the ohm.
13. What is the rms current in a 1.0-μF capacitor connected across the 120-V rms, 60-Hz AC line?
14. A 470-Ω resistor, 10-μF capacitor, and 750-mH inductor are each connected across 6.3-V rms, 60-Hz AC power sources. Find the rms current in each.
15. Find the reactance of a 3.3-μF capacitor at (a) 60 Hz, (b) 1.0 kHz, and (c) 20 kHz.
16. A 15-μF capacitor carries an rms current of 1.4 A. What is the minimum safe voltage rating for the capacitor if the frequency is (a) 60 Hz or (b) 1.0 kHz?
17. A capacitor and a 1.8-kΩ resistor pass the same current when each is separately connected across a 60-Hz power line. What is the capacitance?
18. (a) A 2.2-H inductor is connected across the 120-V rms, 60-Hz power line. What is the rms inductor current? (b) Repeat if the same inductor is connected across the 230-V rms, 50-Hz power line commonly used in Europe.
19. A 50-mH inductor is connected across a 10-V rms AC generator, and an rms current of 2.0 mA flows. What is the generator frequency?
20. A 2.0-μF capacitor has a capacitive reactance of 1.0 kΩ. (a) What is the frequency of the applied voltage? (b) What

inductance would give the same value for inductive reactance at this frequency? (c) How would the two reactances compare if the frequency were doubled?

21. A 1.2-μF capacitor is connected across a generator whose output is given by $V = V_p \sin 2\pi ft$, where $V_p = 22$ V, $f = 60$ Hz, and t is in seconds. (a) What is the peak current? (b) What are the magnitudes of the voltage and (c) the current at $t = 6.5$ ms?

22. A voltage $V = V_p \sin \omega t$ is applied across a capacitor. What is the minimum frequency for which the current will be zero at time $t = 20$ μs?

23. What is the maximum charge on the plates of a 16-μF capacitor connected across the 120-V rms, 60-Hz AC power line?

24. At 10 kHz an inductor has 10 times the reactance of a capacitor. At what frequency will their reactances be equal?

25. A 0.75-H inductor is in series with a fluorescent lamp, and the series combination is across the 120-V rms, 60-Hz power line. If the rms inductor voltage is 90 V, what is the rms lamp current?

26. A 2.2-nF capacitor and a capacitor of unknown capacitance are connected in parallel across a 10-V rms sinewave generator. At 1.0 kHz, the generator supplies a total current of 3.4 mA rms. The generator frequency is then decreased until the rms current has dropped to 1.2 mA. Find (a) the unknown capacitance and (b) the lower frequency.

Section 33-3 *LC* Circuits

27. Find the resonant frequency of an *LC* circuit consisting of a 0.22-μF capacitor and a 1.7-mH inductor.

28. An *LC* circuit with $C = 18$ mF undergoes *LC* oscillations with period 2.4 s. What is the inductance?

29. You have a 2.0-mH inductor and wish to make an *LC* circuit whose resonant frequency spans the AM radio band (550 kHz to 1600 kHz). What range of capacitance should your variable capacitor cover?

30. The FM radio band covers the frequency range from 88 MHz to 108 MHz. If the variable capacitor in an FM receiver ranges from 10.9 pF to 16.4 pF, what inductor should be used to make an *LC* circuit whose resonant frequency spans the FM band?

31. You want to use an *LC* circuit in a timing application. The circuit is to start with the capacitor fully charged, and the voltage should drop to zero in 15 s. You have available a 25-H inductor. What capacitance should you use?

32. An *LC* circuit includes a 20-μF capacitor and has a period of 5.0 ms. The peak current is 25 mA. Find (a) the inductance and (b) the peak voltage.

33. An *LC* circuit includes a 0.025-μF capacitor and a 340-μH inductor. (a) If the peak voltage on the capacitor is 190 V, what is the peak current in the inductor? (b) How long after the voltage peak does the current peak occur?

34. If the capacitor in the preceding problem is rated at 600 V, what is the maximum inductor current that will keep the capacitor within its rating?

35. At the instant when the electric and magnetic energies are equal in the *LC* circuit of Problem 33, the current is 540 mA. (a) What is the instantaneous voltage? Find (b) the peak voltage, (c) the peak current, and (d) the total energy.

36. In an *LC* circuit, what fraction of a cycle passes before the energy in the capacitor falls to one-fourth of its peak value?

37. One-eighth of a cycle after the capacitor in an *LC* circuit is fully charged, what are each of the following as fractions of their peak values: (a) capacitor charge, (b) energy in the capacitor, (c) inductor current, (d) energy in the inductor?

38. Show from conservation of energy that the peak voltage and current in an *LC* circuit are related by $I_p = V_p \sqrt{\dfrac{C}{L}}$.

39. The 2000-μF capacitor in Fig. 33-31 is initially charged to 200 V. Describe how you would manipulate switches A and B to charge the 500-μF capacitor to 400 V. Include the times you would throw the switches.

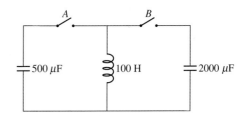

FIGURE 33-31   Problem 39.

40. A damped *LC* circuit consists of a 0.15-μF capacitor and a 20-mH inductor with resistance 1.6 Ω. How many cycles will the circuit oscillate before the peak voltage on the capacitor drops to half its initial value?

41. A damped *RLC* circuit includes a 5.0-Ω resistor and a 100-mH inductor. If half the initial energy is lost after 15 cycles, what is the capacitance?

Section 33-4 Driven *RLC* Circuits and Resonance

42. A series *RLC* circuit has $R = 75$ Ω, $L = 20$ mH, and a resonant frequency of 4.0 kHz. (a) What is the capacitance? (b) What is the impedance of the circuit at resonance? (c) What is the impedance at 3.0 kHz?

43. If the speaker system of Example 33-4 is driven by a 10-V peak, 1.0-kHz sine wave, what will be the peak voltage across the capacitor?

44. An *RLC* circuit includes a 1.5-H inductor and a 250-μF capacitor rated at 400 V. The circuit is connected across a sine-wave generator whose peak voltage is 32 V. What minimum resistance must the circuit have to ensure that the

capacitor voltage does not exceed its rated value when the generator is at the resonant frequency?

45. TV channel 2 occupies the frequency range from 54 MHz to 60 MHz. A series *RLC* tuning circuit in a TV receiver includes an 18-pF capacitor and resonates in the middle of the channel 2 band. (a) What is the inductance? (b) To let the whole signal in, the resonance curve must be broad enough that the current throughout the band be no less than 70% of the current at the resonant frequency. What constraint does this place on the circuit resistance?

46. An *RLC* circuit includes a 10-Ω resistor, 1.5-μF capacitor, and 50-mH inductor. The capacitor is rated at 1200 V. The circuit is driven by an AC source whose peak voltage is 100 V. (a) What would be the peak capacitor voltage at resonance? (b) Make a graph of the peak capacitor voltage as a function of frequency, and from it determine the frequency range that should be avoided for the capacitor to stay within its voltage rating.

47. A 2.0-H inductor and a 3.52-μF capacitor are connected in series with a 50-Ω resistor, and the combination is connected to an AC generator supplying 24 V peak at 60 Hz. (a) At the instant the generator voltage is at its peak, what is the instantaneous voltage across each circuit element? Show explicitly that these sum to the generator voltage. (b) If rms voltmeters are connected across each of the three components, what will they read? Do their readings sum to the rms generator voltage? Does this contradict the loop law?

48. Show that the impedance of an *RLC* circuit driven at frequency ω can be written

$$Z = \sqrt{R^2 + \omega^2 L^2 (1 - \omega_r^2/\omega^2)^2},$$

where ω_r is the resonant frequency.

49. Figure 33-32 shows the phasor diagram for an *RLC* circuit. (a) Is the driving frequency above or below resonance? (b) Complete the diagram by adding the applied voltage phasor, and from your diagram determine the phase difference between applied voltage and current.

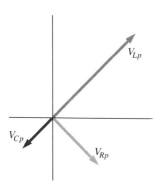

FIGURE 33-32 Problem 49.

50. An AC voltage of fixed amplitude is applied across a series *RLC* circuit. The component values are such that the cur-

rent at half the resonant frequency is half the current at resonance. (a) Show that the current at twice the resonant frequency is also half that at resonance. (b) Sketch phasor diagrams for both off-resonance cases.

51. For the circuit of Problem 46, find the phase relation between applied voltage and current at frequencies of (a) 550 Hz and (b) 700 kHz.

52. For the circuit of Problem 46, find two frequencies at which the voltage and current will be 60° out of phase. Which leads in each case?

Section 33-5 Power in AC Circuits

53. An electric drill draws 4.6 A rms at 120 V rms. If the current lags the voltage by 25°, what is the drill's power consumption?

54. A series *RLC* circuit has resistance 100 Ω and impedance 300 Ω. (a) What is the power factor? (b) If the rms current is 200 mA, what is the power dissipation?

55. A series *RLC* circuit has power factor 0.80 and impedance 100 Ω at 60 Hz. (a) What is the circuit resistance? (b) If the inductance is 0.10 H, what is the resonant frequency?

56. An *RLC* circuit with $R = 10 \ \Omega$, $C = 2.0 \ \mu$F, and $L = 500$ mH is connected to an AC generator supplying 80 V rms. (a) Find the power factor when the generator frequency is half the resonant frequency. Find the power dissipation (b) at half the resonant frequency and (c) at the resonant frequency.

57. A power plant produces 60-Hz power at 365 kV rms and 200 A rms. The plant is connected to a small city by a transmission line with total resistance 100 Ω. What fraction of the power is lost in transmission if the city's power factor is (a) 1.0 or (b) 0.60? (c) Is it more economical for the power company if the load has a large power factor or a small one? Explain.

Section 33-6 Transformers and Power Supplies

58. A rural power line carries 2.3 A rms at 4000 V. A step-down transformer reduces this to 235 V rms to supply a house. Find (a) the turns ratio of the transformer and (b) the current in the 235-V line to the house.

59. A transformer steps up the 120-V rms AC power line voltage to 23 kV rms for a TV picture tube. If the rms current in the primary is 1.0 A, and the transformer is 95% efficient, what is the secondary current?

60. A car battery charger runs off the 120-V rms AC power line, and supplies 10 A DC at 14 V. (a) If the charger is 80% efficient in converting the line power to the DC power it supplies to the battery, how much current does it draw from the AC line? (b) If electricity costs 9.5¢/kWh, how much does it cost to run the charger for 10 hours? Assume the power factor is 1.

61. The transformer in the power supply of Fig. 33-27*a* has an output voltage of 6.3 V rms at 60 Hz, and the capacitance

is 1200 μF. (a) With an infinite load resistance, what would be the output voltage of the power supply? (b) What is the minimum load resistance for which the output would not drop more than 1% from this value? Assume that the discharge time in Fig. 33-27b is essentially a full cycle.

62. A power supply like that in Fig. 33-27a is supposed to deliver 22 V DC at a maximum current of 150 mA. The transformer's peak output voltage is appropriate to charge the capacitor to a full 22 V, and the primary is supplied with 60 Hz AC. What value of capacitance will ensure that the output voltage stays within 3% of the rated 22 V?

Paired Problems

(Both problems in a pair involve the same principles and techniques. If you can get the first problem, you should be able to solve the second one.)

63. A sine-wave generator delivers a signal whose peak voltage is independent of frequency. Two identical capacitors are connected in parallel across the generator, and the generator supplies a peak current I_p at frequency f_1. The capacitors are then connected in series across the generator. To what frequency should the generator be tuned to bring the current back to I_p?

64. A 1.0-μF capacitor and a 2.0-μF capacitor are connected in parallel across a sine-wave generator, and the generator supplies a total rms current of 25 mA at 1.0 kHz. Assuming the generator voltage is independent of frequency, at what frequency will it supply the same current when the capacitors are in series?

65. The peak current in an oscillating LC circuit is 850 mA. If $L = 1.2$ mH and $C = 5.0$ μF, what is the peak voltage?

66. An LC circuit with $C = 470$ pF is oscillating at 7.3 MHz. If the peak voltage is 95 V, what is the peak current?

67. An RLC circuit includes a 3.3-μF capacitor and a 27-mH inductor. The capacitor is charged to 35 V, and the circuit begins oscillating. Ten full cycles later the capacitor voltage peaks at 28 V. What is the resistance?

68. An RLC circuit with $R = 1.2$ Ω and $C = 10$ μF loses 2% of its initial energy in one oscillation cycle. What is its inductance?

69. A series RLC circuit with $R = 5.5$ Ω, $L = 180$ mH, and $C = 0.12$ μF is connected across a sine-wave generator. If the inductor can handle a maximum current of 1.5 A, what is the maximum safe value for the generator's peak output voltage when it is tuned to resonance?

70. A series RLC circuit with $R = 1.3$ Ω, $L = 27$ mH, and $C = 0.33$ μF is connected across a sine-wave generator. If the capacitor's peak voltage rating is 600 V, what is the maximum safe value for the generator's peak output voltage when it is tuned to resonance?

Supplementary Problems

71. Two capacitors are connected in parallel across a 10-V rms, 10-kHz sine-wave generator, and the generator sup-

plies a total rms current of 30 mA. When the capacitors are rewired in series, the rms generator current drops to 5.5 mA. Find the values of the two capacitances.

72. An LC circuit starts at $t = 0$ with its 2000-μF capacitor at its peak voltage of 14 V. At $t = 35$ ms the voltage has dropped to 8.5 V. (a) What will be the peak current? (b) When will the peak current occur?

73. An undriven RLC circuit with inductance L and resistance R starts oscillating with total energy U_0. After N cycles the energy is U_1. Find an expression for the capacitance, assuming the circuit is not heavily damped.

74. Figure 33-33 shows a **low-pass filter.** When an alternating voltage is applied at the V_{in} terminals, the output voltage V_{out} depends on frequency. (a) Show that $V_{out} = V_{in}/\sqrt{1 + (RC\omega)^2}$, where the voltages are either peak or rms values (there is also a phase difference). (b) At what frequency is the output voltage down from the input voltage by a factor of $1/\sqrt{2}$? (This is called the *half-power point* since the power—proportional to V^2—is down by a factor of $\frac{1}{2}$.) *Hint:* You can repeat the phasor analysis of Section 33-4, but without the inductor. Or you can start from Equation 33-14, with X_L set to zero.

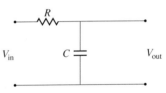

FIGURE 33-33   Problem 74.

75. You wish to make a "black box" with two input connections and two output connections, as shown in Fig. 33-34. When you put a 12-V rms, 60-Hz sine wave across the input, a 6.0-V, 60-Hz signal should appear at the output, with the output voltage leading the input voltage by 45°. Design a circuit that could be used in the "black box."

FIGURE 33-34   Problem 75.

76. A series RLC circuit with $R = 47$ Ω, $L = 250$ mH, and $C = 4.0$ μF is connected across a sine-wave generator whose peak output voltage is independent of frequency. Find the frequency range over which the peak current will exceed half its value at resonance. *Hint:* You can solve this problem graphically or, with appropriate algebraic manipulations, using quadratic equations.

77. A sine-wave generator with peak output voltage of 20 V is applied across a series RLC circuit. At the resonant fre-

quency of 2.0 kHz the peak current is 50 mA, while at 1.0 kHz it is 15 mA. Find R, L, and C.

78. Use phasor analysis to show that the parallel RLC circuit of Fig. 33-35 has impedance $Z = \left[\dfrac{1}{R^2} + \left(\dfrac{1}{X_C} - \dfrac{1}{X_L}\right)^2\right]^{-1/2}$

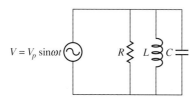

FIGURE 33-35 Problem 78.

79. A 2.5-H inductor is connected across a 1500-μF capacitor. A 5.0-kg mass is connected to a spring. What should be the spring constant if the mechanical and electrical systems have the same resonant frequency?

80. Figure 33-36 shows a diode circuit called a *full-wave bridge,* whose output requires less filtering than the single-diode circuit of Fig. 33-26. (a) Sketch a graph of the resistor current as a function of time, covering two full cycles of the AC generator, and (b) explain why less filtering is needed.

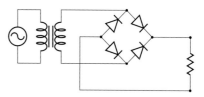

FIGURE 33-36 A full-wave diode bridge (Problem 80).

81. For RLC circuits in which the resistance is not too large, the Q factor may be defined as the ratio of the resonant frequency to the difference between the two frequencies where the power dissipated in the circuit is half that dissipated at resonance. Show, using suitable approximations, that this definition leads to the expression $Q = \omega_r L/R$, with ω_r the resonant frequency.

82. Consider a series circuit containing an AC generator, a resistor, and a capacitor. Construct a phasor diagram, and derive expressions for the circuit impedance and the phase angle between the applied voltage and the current. Show that the current always leads the voltage.

83. Consider a series circuit containing an AC generator, a resistor, and an inductor. Construct a phasor diagram, and derive expressions for the circuit impedance and the phase angle between the applied voltage and the current. Show that the voltage always leads the current.

CHAPTER 34

MAXWELL'S EQUATIONS AND ELECTROMAGNETIC WAVES

Electromagnetic waves include light and the radio waves essential to modern communications. This antenna communicates with orbiting satellites.

At this point we have introduced the four fundamental laws of electromagnetism—Gauss's law for electricity, Gauss's law for magnetism, Ampère's law, and Faraday's law—that govern the behavior of electric and magnetic fields throughout the universe. We have seen how these laws describe the electric and magnetic interactions that make matter act as it does and have explored many practical devices that exploit the laws of electromagnetism. Here we extend the fundamental laws to their most general form and show how they predict the existence of electromagnetic waves.

34-1 THE FOUR LAWS OF ELECTROMAGNETISM

Table 34-1 summarizes the four laws as we introduced them in earlier chapters. As you look at these four laws together, you can't help noticing some strong similarities. On the left-hand sides of the equations, the two laws of Gauss are identical but for the interchanging of **E** and **B.** Similarly, the laws of Ampère and Faraday have left-hand sides that differ only in the interchange of **E** and **B.**

On the right-hand sides, things are more different. Gauss's law for electricity involves the charge enclosed by the surface of integration, while Gauss's law for magnetism has zero on the right-hand side. Actually, though, these laws are similar. Since we have no experimental evidence for the existence of isolated magnetic charge, the enclosed magnetic charge on the right-hand side of Gauss's law for magnetism is zero. If and when magnetic monopoles are discovered, then the right-hand side of Gauss's law for magnetism would be nonzero for any surface enclosing net magnetic charge.

The right-hand sides of Ampère's and Faraday's laws are distinctly different. In Ampère's law we find the current—the flow of electric charge—as a source of magnetic field. We can understand the absence of a similar term in Faraday's law because we have never observed a flow of magnetic monopoles. If we had such a flow, then we would expect this magnetic current to produce an electric field encircling the magnetic current.

Two of the differences among the four laws of electromagnetism would be resolved if we knew for sure that magnetic monopoles exist. That current theories of elementary particles suggest the existence of monopoles is a tantalizing hint that there may be a fuller symmetry between electric and magnetic phenomena. The search for symmetry, based not on logic or experimental evidence but on an intuitive sense that nature should be simple, has motivated some of the most important discoveries in physics.

34-2 AMBIGUITY IN AMPÈRE'S LAW

There remains one difference between the equations of electricity and magnetism that would not be resolved by the discovery of magnetic monopoles. On the right-hand side of Faraday's law we find the term $d\phi_B/dt$ that describes

▲ **TABLE 34-1** FOUR LAWS OF ELECTROMAGNETISM (STILL INCOMPLETE)

LAW	MATHEMATICAL STATEMENT	WHAT IT SAYS
Gauss for **E**	$\oint \mathbf{E} \cdot d\mathbf{A} = \dfrac{q}{\varepsilon_0}$	How charges produce electric field; field lines begin and end on charges
Gauss for **B**	$\oint \mathbf{B} \cdot d\mathbf{A} = 0$	No magnetic charge; magnetic field lines do not begin or end
Faraday	$\oint \mathbf{E} \cdot d\boldsymbol{\ell} = -\dfrac{d\phi_B}{dt}$	Changing magnetic flux produces electric field
Ampère (steady currents only)	$\oint \mathbf{B} \cdot d\boldsymbol{\ell} = \mu_0 I$	Electric current produces magnetic field

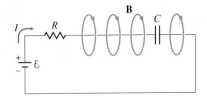

FIGURE 34-1 A charging *RC* circuit, showing some magnetic field lines surrounding the current-carrying wire.

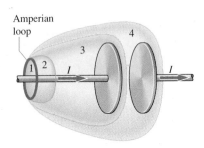

FIGURE 34-2 Ampère's law relates the line integral around the loop to the current through *any* surface bounded by the loop. There is no current through surface 3, so Ampère's law is ambiguous.

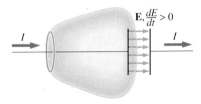

FIGURE 34-3 There is a changing electric field in the charging capacitor and, therefore, a changing electric flux through surface 3 of Fig. 34-2.

changing magnetic flux as a source of electric field. We find no comparable term in Ampère's law. Are we missing something? Is it possible that a changing electric flux produces a magnetic field? So far, we have described no experimental evidence for such a conjecture. It is suggested only by our sense that the near symmetry between electricity and magnetism is not a coincidence. If a changing electric flux did produce a magnetic field, just as a changing magnetic flux produces an electric field, then we would expect a term $d\phi_E/dt$ on the right-hand side of Ampère's law.

In our statement of Ampère's law in Chapter 30, we emphasized that the law applied only to steady currents—those that never change. The reason for this restriction is suggested in Fig. 34-1, which shows a simple *RC* circuit. While the capacitor charges, there is a current in the circuit that decreases with time. This current produces a magnetic field, as suggested in the figure. Let us apply Ampère's law to calculate this field.

Ampère's law tells us that the line integral around any closed loop is proportional to the encircled current:

$$\oint \mathbf{B} \cdot d\ell = \mu_0 I.$$

By the encircled current, we mean the current through *any* surface bounded by the loop. Figure 34-2 shows four such surfaces. The same current flows through surfaces 1, 2, and 4. But there is no current through surface 3, which passes between the capacitor plates. So Ampère's law is ambiguous in that surface 3 would not give the same answer for the magnetic field.

This ambiguity does not arise with steady currents. In an *RC* circuit the steady-state current is everywhere zero, and thus the right-hand side of Ampère's law is zero for *any* surface. It's only when currents are changing with time that there may be situations like that of Fig. 34-2 where Ampère's law becomes ambiguous.* That is why the form of Ampère's law we have used until now is strictly valid only for steady currents.

Can we salvage Ampère's law, extending it to cover unsteady currents without affecting its validity in the steady case? Symmetry between Ampère's and Faraday's laws has already suggested that a changing electric flux might produce a magnetic field. Between the plates of a charging capacitor there is an electric field whose magnitude is increasing (Fig. 34-3). That means there is a changing electric flux through surface 3 of Figure 34-2.

It was the Scottish physicist James Clerk Maxwell who, about 1860, suggested that a changing electric flux should give rise to a magnetic field. Since that time many experiments, including direct measurement of the magnetic field inside a charging capacitor, have confirmed Maxwell's remarkable insight. Maxwell quantified his idea by introducing a new term into Ampère's law:

$$\oint \mathbf{B} \cdot d\ell = \mu_0 I + \mu_0 \varepsilon_0 \frac{d\phi_E}{dt}. \tag{34-1}$$

---

* You might argue that we could produce a nonzero steady current in the *RC* circuit by steadily increasing the applied emf, again making Ampère's law ambiguous. But we could not do so forever since that would require infinite energy. Steady current means current that *never* changes. A current that is steady over a finite time interval must be started and stopped, so it isn't really steady.

Now there is no ambiguity. The integral is taken around any loop, I is the current through *any* surface bounded by the loop, and ϕ_E is the electric flux through that surface. With our charging capacitor, Equation 34-1 gives the same magnetic field no matter which surface we choose. For surfaces 1, 2, and 4 of Fig. 34-3, the current I makes all the contribution to the right-hand side of the equation (here we assume that the electric field outside the capacitor is zero). For surface 3, through which no current flows, the right-hand side of Equation 34-1 comes entirely from the changing electric flux. You can readily verify that the term $\varepsilon_0 d\phi_E/dt$ has the units of current, and that, for the charging capacitor, this term is numerically equal to the current I (see Problem 3). Although the changing electric flux is not an electric current, it has the same effect as a current in producing a magnetic field. For this reason Maxwell called the term $\varepsilon_0 d\phi_E/dt$ the **displacement current.** The word "displacement" has historical roots that do not provide much physical insight. But the word "current" is meaningful in that the effect of displacement current is indistinguishable from that of real current in producing magnetic fields.

● **EXAMPLE 34-1** DISPLACEMENT CURRENT PRODUCES MAGNETIC FIELD

A parallel-plate capacitor with circular plates a distance d apart is charged through long, straight wires as shown in Fig. 34-4. The potential difference between the plates is increasing at the rate dV/dt. Find an expression for the magnetic field as a function of position between the plates.

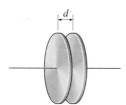

FIGURE 34-4 A circular capacitor (Example 34-1).

Solution

With long, straight feed wires, the situation has cylindrical symmetry. The only magnetic field with this symmetry has circular field lines and a magnitude that depends only on the radial distance r from the symmetry axis, as shown in Fig. 34-5. A magnetic field line within the capacitor encircles no conduction current—no flow of charge—but it does encircle a changing electric field and therefore a displacement current. If the field line has radius r, the encircled electric flux is

$$\phi_E = \int \mathbf{E} \cdot d\mathbf{A} = \pi r^2 E = \pi r^2 \frac{V}{d},$$

where the uniformity of the field allows us to calculate the field as the ratio of potential difference to plate spacing and the flux as a simple product of field and area. Then the displacement current is

$$I_D = \varepsilon_0 \frac{d\phi_E}{dt} = \frac{\varepsilon_0 \pi r^2}{d} \frac{dV}{dt}.$$

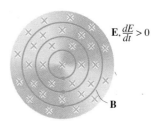

FIGURE 34-5 Electric and magnetic fields between the circular capacitor plates. The electric field strength is increasing, so the displacement current is in the same direction as the electric field. Pointing the right thumb in this direction then shows that the magnetic field circles clockwise.

With cylindrical symmetry, the line integral on the left-hand side of Ampère's law becomes

$$\oint \mathbf{B} \cdot d\boldsymbol{\ell} = 2\pi r B.$$

Equating this quantity to μ_0 times the encircled displacement current gives

$$2\pi r B = \frac{\mu_0 \varepsilon_0 \pi r^2}{d} \frac{dV}{dt},$$

so

$$B = \frac{\mu_0 \varepsilon_0 r}{2d} \frac{dV}{dt}.$$

This field, with its magnitude increasing linearly with r, should remind you of the magnetic field inside a cylindrical wire (see Example 30-4). Problem 4 extends this calculation to the field outside the capacitor.

We can find the direction of the induced magnetic field just as we did for the fields of ordinary conduction currents: Point your right thumb in the direction of the current and your right fingers curl in the direction of the magnetic field. But which way does the displacement current go? In this example the electric field strength is increasing, so $d\phi_E/dt$ is positive, and the displacement current is in the direction of the electric field (see Fig. 34-5). If the electric field strength were decreasing, the displacement current would be opposite the field.

The induced magnetic field in a practical capacitor is minuscule, as the following exercise illustrates. We'll soon see, however, that the significance of displacement-current-induced magnetic fields is vastly greater than in this simple example.

EXERCISE In 1984 D. F. Bartlett and T. R. Corle of the University of Colorado first measured the magnetic field inside a charging capacitor using a sensitive magnetometer called a superconducting quantum interference detector (SQUID). They used a capacitor with circular plates spaced 1.22 cm, connected across a 340-V peak sine-wave generator operating at 1.25 kHz. What was the peak magnetic field strength 3.0 cm from the capacitor axis?

Answer: 3.65×10^{-11} T, less than one millionth of Earth's magnetic field

Some problems similar to Example 34-1: 4–6 ●

34-3 MAXWELL'S EQUATIONS

It was Maxwell's genius to recognize that Ampère's law should be modified to reflect the symmetry suggested by Faraday's law. The consequences of Maxwell's discovery go far beyond anything he could have imagined. To honor Maxwell, the four complete laws of electromagnetism are given the collective name **Maxwell's equations.** This full and complete set of equations, first published in 1864, governs the behavior of electric and magnetic fields everywhere. Table 34-2 summarizes Maxwell's equations.

These four simple, compact statements are all it takes to describe classical electromagnetic phenomena. Everything electric or magnetic that we have considered and will consider—from polar molecules to electric current; resistors, capacitors, inductors, and transistors; solar flares and cell membranes; electric generators and thunderstorms; computers and TV sets; the northern lights and fusion reactors—all these can be described using Maxwell's equations. And despite this wealth of phenomena, we have yet to discuss a most important manifestation of electromagnetic fields.

Maxwell's Equations in Vacuum

Consider Maxwell's equations in a region free of any matter—in vacuum. We have learned enough about electromagnetism to anticipate that the fields them-

▲ **TABLE 34-2** MAXWELL'S EQUATIONS

LAW	MATHEMATICAL STATEMENT	WHAT IT SAYS	EQUATION NUMBER
Gauss for **E**	$\oint \mathbf{E} \cdot d\mathbf{A} = \dfrac{q}{\varepsilon_0}$	How charges produce electric field; field lines begin and end on charges	(34-2)
Gauss for **B**	$\oint \mathbf{B} \cdot d\mathbf{A} = 0$	No magnetic charge; magnetic field lines do not begin or end	(34-3)
Faraday	$\oint \mathbf{E} \cdot d\boldsymbol{\ell} = -\dfrac{d\phi_B}{dt}$	Changing magnetic flux produces electric field	(34-4)
Ampère	$\oint \mathbf{E} \cdot d\boldsymbol{\ell} = \mu_0 I + \varepsilon_0 \mu_0 \dfrac{d\phi_E}{dt}$	Electric current and changing electric flux produce magnetic field	(34-5)

selves will still be able to interact, change, and carry energy even in the absence of matter. To express Maxwell's equations in vacuum, we simply remove all reference to matter—that is, to electric charge:

$$\oint \mathbf{E} \cdot d\mathbf{A} = 0 \tag{34-6}$$

$$\oint \mathbf{B} \cdot d\mathbf{A} = 0 \tag{34-7}$$

$$\oint \mathbf{E} \cdot d\boldsymbol{\ell} = -\frac{d\phi_B}{dt} \tag{34-8}$$

$$\oint \mathbf{B} \cdot d\boldsymbol{\ell} = \mu_0 \varepsilon_0 \frac{d\phi_E}{dt}. \tag{34-9}$$

In vacuum the symmetry is complete, in that electric and magnetic fields enter on an equal footing.* The only source of each field is a change in the other field.

34-4 ELECTROMAGNETIC WAVES

Faraday's law—Equation 34-8—shows that a changing magnetic field induces an electric field. In general, this induced electric field is itself changing. Ampère's law—Equation 34-9—shows that a changing electric field induces a magnetic field, which itself may be changing. The two laws together suggest the possibility of **electromagnetic waves,** in which each type of field continually produces the other in an electromagnetic structure that propagates through space. We will now show, directly from Maxwell's equations, that such waves are indeed possible. In the process we will discover the properties of electromagnetic waves and will come to a deep understanding of the nature of light.

In Chapter 16 we found that a sinusoidal wave propagating in the x direction can be represented by a function of the form $A \sin(kx - \omega t)$, where A is the wave amplitude, k the wave number, and ω the angular frequency. We now demonstrate that electric and magnetic fields of this form satisfy Maxwell's equations. Our demonstration is highly mathematical, but its physical significance is profound, for it reveals a rich new realm of electromagnetic phenomena that play a vital role in the workings of our universe.

We propose an electromagnetic wave consisting of electric and magnetic fields oriented at right angles to each other and to the direction of wave motion, as suggested in Fig. 34-6. The electric field points in the y direction, the magnetic field points in the z direction, and the wave travels in the x direction. In Fig. 34-6b the field lines are pictured only on the surface of a rectangular box. We must imagine these lines continuing straight forever in the y and z directions, giving a wave whose properties do not vary with y or z. Such a wave, whose

---

* The appearance of the constants ε_0 and μ_0 in Ampère's law but not in Faraday's law is an accident of our choice of units. That Faraday's law contains a minus sign, while Ampère's does not, is actually a symmetry, reflecting the complementary way in which electric and magnetic fields produce each other.

FIGURE 34-6 Fields of a plane electromagnetic wave. (*a*) Field vectors (*not* field lines), showing sinusoidal variation that indicates that this is a simple harmonic wave. The electric and magnetic fields are perpendicular and in phase. (*b*) Fields in the plane wave do not depend on *y* or *z*, and thus the fields actually extend forever in the *y* and *z* directions. This figure shows some field lines in a rectangular slab of the wave. Note that their spacing reflects the orientation and the sinusoidal variation shown in (*a*).

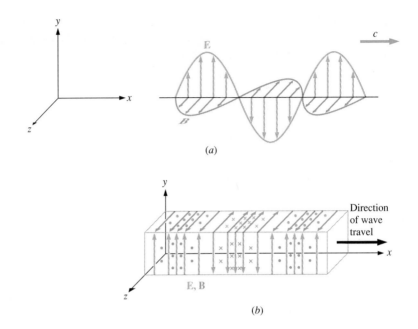

properties are independent of position in planes perpendicular to the propagation direction, is called a **plane wave.** Mathematically, our plane wave fields are described by

$$\mathbf{E}(x, t) = E_p \sin(kx - \omega t)\hat{\mathbf{j}} \qquad (34\text{-}10)$$

and

$$\mathbf{B}(x, t) = B_p \sin(kx - \omega t)\hat{\mathbf{k}}, \qquad (34\text{-}11)$$

where the peak amplitudes E_p and B_p are constants and where $\hat{\mathbf{j}}$ and $\hat{\mathbf{k}}$ are unit vectors in the *y* and *z* directions. In Fig. 34-6*b* the variation in spacing of the field lines, and their reversal from one region of densely spaced lines to another, reflect the sinusoidal dependence of the wave fields on position in the *x* direction. We chose a sinusoidal wave shape because we are familiar with such waves from Chapter 16 and because a specific mathematical form will make our derivation more concrete. But we emphasize here, as in Chapter 16, that *any* functions of $x \pm vt$ or, equivalently, $kx \pm \omega t$, are admissable waveforms (see Problem 74 here and Problem 54 of Chapter 16).

We now show that the electric and magnetic fields pictured in Fig. 34-6 and described by Equations 34-10 and 34-11 do indeed satisfy Maxwell's equations. Note first that our field lines continue forever, with no beginnings or ends; therefore Gauss's laws for electricity and magnetism in vacuum (Equations 34-6 and 34-7) are satisfied.

Faraday's Law

To see that Faraday's law is satisfied, consider an observer looking directly toward the *x-y* plane in Fig. 34-6. Such an observer would see electric field lines going up and down and magnetic field lines coming straight in and out, as shown in Fig. 34-7. Consider the small rectangular loop of height h and infinitesimal width dx shown in the figure. Evaluating the line integral of the electric field **E** around this loop, we get no contribution from the short ends because they are at right angles to the field. Going around counterclockwise, we get a contribution $-Eh$ as we go down the left side against the field direction. Then we get a positive contribution going up the right side. Because of the variation in field strength with position, the field strength on the right side of the loop is different from that on the left. Let the change in field be dE, so the field on the right side of the loop is $E + dE$, giving a contribution of $(E + dE)h$ to the line integral. Then the line integral of **E** around the loop is

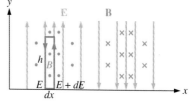

FIGURE 34-7 Cross section of Fig. 34-6 in the x-y plane. Also shown is a rectangular loop for evaluating the line integral in Faraday's law.

$$\oint \mathbf{E} \cdot d\boldsymbol{\ell} = -Eh + (E + dE)h = h\,dE.$$

Physically, this nonzero line integral means that we are dealing with an induced electric field. Induced by what? By a changing magnetic flux through the loop. The electric field of the wave arises because of the changing magnetic field of the wave. The area of the loop is $h\,dx$, and the magnetic field **B** is at right angles to this area, so the magnetic flux through the loop is just

$$\phi_B = Bh\,dx.$$

The rate of change of flux through the loop is then

$$\frac{d\phi_B}{dt} = h\,dx\,\frac{dB}{dt}.$$

Faraday's law relates the line integral of the electric field to the rate of change of flux:

$$\oint \mathbf{E} \cdot d\boldsymbol{\ell} = -\frac{d\phi_B}{dt},$$

or, using our expressions for the line integral and the rate of change of flux,

$$h\,dE = -h\,dx\,\frac{dB}{dt}.$$

Dividing through by $h\,dx$, we have

$$\frac{dE}{dx} = -\frac{dB}{dt}. \qquad (34\text{-}12a)$$

In deriving this equation, we considered changes in E with position at a fixed instant of time, as pictured in Fig. 34-7, so our derivative dE/dx means the rate of change of E with position while time is held fixed. Similarly, in evaluating the derivative of magnetic flux, we were concerned only with the time rate of change at the fixed position of our loop. Both our derivatives represent rates of change with respect to one variable while the other variable is held fixed, and are therefore partial derivatives (see Tip box on p. 383 if you're not familiar with partial derivatives). Equation 34-12a should then be written more properly:

$$\frac{\partial E}{\partial x} = -\frac{\partial B}{\partial t}. \qquad (34\text{-}12\text{b})$$

Equation 34-12b—which is just Faraday's law applied to our electromagnetic wave—tells us that the rate at which the electric field changes with *position* is related to the rate at which the magnetic field changes with *time*.

Ampère's Law

Now imagine an observer looking down on Fig. 34-6 from above. This observer sees the magnetic field lines lying in the *x-z* plane, and electric field lines emerging perpendicular to the *x-z* plane as shown in Fig. 34-8. We can apply Ampère's law (Equation 34-9) to the infinitesimal rectangle shown, just as we applied Faraday's law to a similar rectangle in the *x-y* plane. Going counterclockwise around the rectangle, we get no contribution to the line integral of the magnetic field on the short sides, since they lie perpendicular to the field. Going down the left side, we get a contribution Bh to the line integral. Going up the right side, against the field, we get a negative contribution $-(B + dB)h$, where dB is the change in B from one side of the rectangle to the other. So the line integral in Ampère's law is

$$\oint \mathbf{B} \cdot d\boldsymbol{\ell} = Bh - (B + dB)h = -h\, dB.$$

The electric flux through the rectangle is simply $Eh\, dx$, so the rate of change of electric flux is

$$\frac{d\phi_E}{dt} = h\, dx\, \frac{dE}{dt}.$$

Ampère's law relates the line integral of the magnetic field to this time derivative of the electric flux, giving

$$-h\, dB = \varepsilon_0 \mu_0 h\, dx\, \frac{dE}{dt}.$$

Dividing through by $h\, dx$ and noting again that we are really dealing with partial derivatives, we have

$$\frac{\partial B}{\partial x} = -\varepsilon_0 \mu_0 \frac{\partial E}{\partial t}. \qquad (34\text{-}13)$$

Equations 34-12 and 34-13—derived from Faraday's and Ampère's laws—express fully the requirements that Maxwell's universal laws of electromagnetism pose on the field structure postulated in Fig. 34-6. The two equations are remarkable in that each describes an induced field that arises from the changing

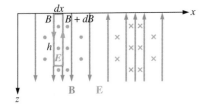

FIGURE 34-8 Cross section of Fig. 34-6 in the *x-z* plane, showing a rectangular loop for evaluating the line integral in Ampère's law.

of the other field. That other field, in turn, arises from the changing of the first field. Thus we have a self-perpetuating electromagnetic structure, whose fields exist and change without the need for charged matter. If Equations 34-10 and 34-11, which describe the fields in Fig. 34-6, can be made consistent with Equations 34-12 and 34-13, then we will have demonstrated that our electromagnetic wave does indeed satisfy Maxwell's equations and is thus a possible configuration of electric and magnetic fields.

Faraday, Ampère, and the Wave Fields

To see that Equation 34-12 is satisfied, we differentiate the electric field of Equation 34-10 with respect to x, and the magnetic field of Equation 34-11 with respect to t:

$$\frac{\partial E}{\partial x} = \frac{\partial}{\partial x}[E_p \sin(kx - \omega t)] = kE_p \cos(kx - \omega t)$$

and

$$\frac{\partial B}{\partial t} = \frac{\partial}{\partial t}[B_p \sin(kx - \omega t)] = -\omega B_p \cos(kx - \omega t).$$

Putting these expressions in for the derivatives in Equation 34-12 gives

$$kE_p \cos(kx - \omega t) = -[-\omega B_p \cos(kx - \omega t)].$$

The cosine term cancels from this equation, showing that the equation holds if

$$kE_p = \omega B_p. \tag{34-14}$$

To see that Equation 34-13 is also satisfied, we differentiate the magnetic field of Equation 34-11 with respect to x and the electric field of Equation 34-10 with respect to t:

$$\frac{\partial B}{\partial x} = kB_p \cos(kx - \omega t)$$

and

$$\frac{\partial E}{\partial t} = -\omega E_p \cos(kx - \omega t).$$

Using these expressions in Equation 34-13 then gives

$$kB_p \cos(kx - \omega t) = -\varepsilon_0 \mu_0 [-\omega E_p \cos(kx - \omega t)].$$

Again, the cosine term cancels, showing that this equation is satisfied if

$$kB_p = \varepsilon_0 \mu_0 \omega E_p. \tag{34-15}$$

Our analysis has shown that electromagnetic waves whose form is given by Fig. 34-6 and Equations 34-10 and 34-11 can exist, provided that the amplitudes E_p and B_p, and the frequency ω and wave number k, are related by Equations 34-14 and 34-15. Physically, the existence of these waves is possible because a change in either kind of field—electric or magnetic—induces the other kind of field, giving rise to a self-perpetuating electromagnetic field structure. Maxwell's theory thus leads to the prediction of an entirely new phenomenon—the electromagnetic wave. We will now explore some properties of these waves.

34-5 THE SPEED OF ELECTROMAGNETIC WAVES

In Chapter 16 we found that the speed of a sinusoidal wave is given by the ratio of the angular frequency and wave number:

$$\text{wave speed} = \frac{\omega}{k}.$$

To determine the speed of our electromagnetic wave, we solve Equation 34-14 for E_p:

$$E_p = \frac{\omega B_p}{k},$$

and use this expression in Equation 34-15:

$$kB_p = \varepsilon_0\mu_0\omega E_p = \frac{\varepsilon_0\mu_0\omega^2 B_p}{k}.$$

Solving for the wave speed ω/k then gives

$$\text{wave speed} = \frac{\omega}{k} = \frac{1}{\sqrt{\varepsilon_0\mu_0}}. \tag{34-16}$$

This remarkably simple result shows that the speed of an electromagnetic wave in vacuum depends only on the electric and magnetic constants ε_0 and μ_0. All such electromagnetic waves, regardless of frequency or amplitude, share this speed. Although we derived this result for sinusoidal waves, superposition considerations of Section 16-5 show that it holds for any wave shape.

We can easily calculate the speed given in Equation 34-16:

$$\frac{1}{\sqrt{\varepsilon_0\mu_0}} = \frac{1}{[(8.85\times10^{-12}\ \text{F/m})(4\pi\times10^{-7}\ \text{H/m})]^{1/2}} = 3.00\times10^8\ \text{m/s}.$$

But this is precisely the speed of light! As early as 1600, Galileo had tried to measure the speed of light by uncovering lanterns on different mountain tops. He was able to conclude only that "If not instantaneous, it is extraordinarily rapid." In 1728, James Bradley, in England, used changes in the apparent positions of stars resulting from Earth's orbital motion to calculate a value of 2.95×10^8 m/s for the speed of light. Bradley's result differs by less than 2% from the value 2.99792458×10^8 m/s used in the 1983 definition of the meter in terms of the speed of light. Furthermore, the Dutch physicist Christian Huygens had suggested in 1678—again, about 200 years before Maxwell—that light is a wave. A substantial body of optical experiments had confirmed Huygens' theory, although neither theory nor experiment could say what sort of wave light might be. Now, in the 1860s, came Maxwell. Using a theory developed from laboratory experiments involving electricity and magnetism, with no reference whatever to optics or light, Maxwell showed how the interplay of electric and magnetic fields could result in electromagnetic waves. The speed of those waves—calculated from the quantities ε_0 and μ_0 that were determined in laboratory experiments having nothing to do with light—was precisely the known speed of light. Maxwell was led inescapably to the conclusion that light is an electromagnetic wave.

Maxwell's identification of light as an electromagnetic phenomenon is a classic example of the unification of knowledge toward which science is ever striving. With one simple calculation, Maxwell brought the entire science of optics under the umbrella of electromagnetism. Maxwell's work stands as a crowning intellectual triumph, an achievement whose implications are still expanding our view of the universe.

34-6 PROPERTIES OF ELECTROMAGNETIC WAVES

Our demonstration that electromagnetic waves satisfy Maxwell's equations places definite constraints on the properties of those waves. The wave frequency ω and wave number k are not both arbitrary, but must be related through

$$\frac{\omega}{k} = c, \tag{34-17a}$$

where $c = 1/\sqrt{\varepsilon_0 \mu_0}$ is the speed of light. In Chapter 16 we related the angular frequency ω and wave number k to the more familiar frequency f and wavelength λ through the equations $\omega = 2\pi f$ and $k = 2\pi/\lambda$. Therefore we can also write Equation 34-17a in the form

$$f\lambda = c. \tag{34-17b}$$

Furthermore, Equation 34-14 shows that

$$E = \frac{\omega}{k}B = cB. \tag{34-18}$$

Thus, the field magnitudes in the wave are not independent but are in the ratio of the speed of light. Also, equations 34-10 and 34-11 require that the electric and magnetic fields be in phase. (This is why we wrote E and B in Equation 34-18 even though Equation 34-14 relates only the peak values E_p and B_p.) Finally, Fig. 34-6 has the electric and magnetic fields perpendicular to each other and to the direction of wave propagation. Although it is not clear that we had to start with a wave of this form, it is in fact the case that only waves with **E** and **B** in phase and with **E, B,** and the propagation direction all perpendicular to each other can satisfy Maxwell's equations in vacuum (see Problem 11).

● **EXAMPLE 34-2** LASER LIGHT

A laser beam with wavelength 633 nm is propagating in the $+z$ direction. Its electric field is parallel to the x axis and has amplitude 6.0 kV/m. Find the wave frequency, and the direction and amplitude of the magnetic field.

Solution

Equation 34-17b relates the wavelength and frequency to the speed of light. Solving for f gives

$$f = \frac{c}{\lambda} = \frac{3.00 \times 10^8 \text{ m/s}}{633 \times 10^{-9} \text{ m}} = 4.74 \times 10^{14} \text{ Hz}.$$

If we imagine reorienting the wave of Fig. 34-6 so it propagates along the z direction, then rotate it about the z direction so the electric field is parallel to the x axis, we find that the magnetic field is parallel to the y axis. The magnetic field amplitude follows from Equation 34-18:

$$B_p = \frac{E_p}{c} = \frac{6.0 \times 10^3 \text{ V/m}}{3.00 \times 10^8 \text{ m/s}} = 2.0 \times 10^{-5} \text{ T}.$$

EXERCISE An electromagnetic wave is propagating in the $-y$ direction, with its magnetic field parallel to the x axis. The magnetic field amplitude is 8.0 μT. Write an expression for the wave's electric field vector at the point where the magnetic field points in the $+x$ direction and is at its peak value.

Answer: $\mathbf{E} = -2.4\hat{\mathbf{k}}$ kV/m

Some problems similar to Example 34-2: 24–27

●

34-7 THE ELECTROMAGNETIC SPECTRUM

Although an electromagnetic wave's frequency and wavelength must be related by Equation 34-17b, one or the other of these quantities is completely arbitrary. That means we can have electromagnetic waves of any frequency, or, equivalently, any wavelength. Direct measurement shows that visible light occupies a wavelength range from about 400 nm to 700 nm, corresponding to frequencies from 7.5×10^{14} Hz to 4.3×10^{14} Hz. The different wavelengths or frequencies correspond to different colors, with red at the long-wavelength, low-frequency end of the visible region and the blue at the short-wavelength, high-frequency end (Fig. 34-9).

The range of frequencies occupied by visible light is rather limited. What about electromagnetic waves whose frequencies lie above and below the visible range? Such invisible electromagnetic waves were unknown in Maxwell's time. A brilliant confirmation of Maxwell's theory came in 1888, when the German physicist Heinrich Hertz succeeded in generating and detecting electromagnetic

FIGURE 34-9 Visible light occupies a wavelength range from about 400 nm to 700 nm. But electromagnetic waves come in a much broader range of wavelengths.

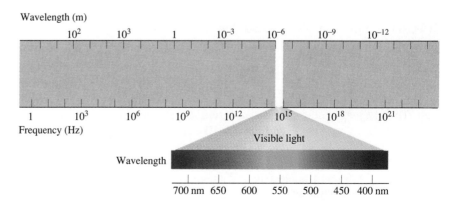

waves of much lower frequency than visible light. Hertz intended his work only to verify Maxwell's modification of Ampère's law, but the practical consequences have proven enormous. In 1896, the Italian scientist Guglielmo Marconi demonstrated that he could generate and detect the so-called "Hertzian waves." In 1901, he transmitted electromagnetic waves across the Atlantic Ocean, creating a public sensation. From the pioneering work of Hertz and Marconi, spurred by the theoretical efforts of Maxwell, came the entire technology of radio, television, and microwaves that so dominates modern society. We now consider all electromagnetic waves in the frequency range from a few Hz to about 3×10^{11} Hz as radio waves, with ordinary AM radio at about 10^6 Hz, FM radio at 10^8 Hz, and microwaves used for radar, cooking, and satellite communications at 10^9 Hz and above.

Between radio waves and visible light lies the infrared frequency range. Electromagnetic waves in this region are emitted by warm objects, even when they are not hot enough to glow visibly. For this reason, infrared detectors are used to determine subtle body temperature differences in medical diagnosis, to examine buildings for heat loss, and to study the birth of stars in clouds of interstellar gas and dust (Fig. 34-10).

Beyond the visible region are the ultraviolet rays responsible for sunburn, then the highly penetrating x rays, and finally the gamma rays whose primary terrestrial source is radioactive decay. All these phenomena, from radio to gamma rays, are fundamentally the same: They are all electromagnetic waves, differing only in frequency and wavelength. All travel with speed c, and all consist of electric and magnetic fields produced from each other through the induction processes described by Faraday's and Ampère's laws. Naming the different types of electromagnetic waves is just a matter of convenience; there are no gaps in the continuous range of allowed frequencies and wavelengths. Practical differences arise because waves of different wavelengths interact differently with matter; in particular, shorter wavelengths tend to be generated and absorbed most efficiently by smaller systems. Figure 34-11 shows the range of electromagnetic waves—the **electromagnetic spectrum**—displayed in a single diagram.

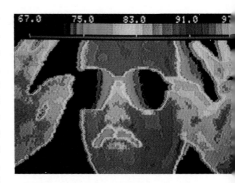

FIGURE 34-10 Subtle variations in body temperature are color coded in this infrared image of a human face.

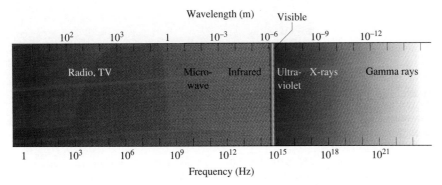

FIGURE 34-11 The electromagnetic spectrum ranges from radio waves to gamma rays. Note the logarithmic scale, in which equal intervals on the diagram correspond to factors-of-10 changes in frequency and wavelength.

■ **APPLICATION** THE NEW ASTRONOMY

FIGURE 34-12 Earth's atmosphere is opaque to most electromagnetic waves, although "windows" of transparency exist in several wavelength ranges—especially the visible. Bottom of diagram represents Earth's surface, with top of the gray curve the height to which waves of a given wavelength penetrate from outer space.

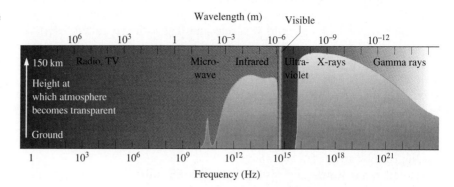

A glance at Fig. 34-11 shows that visible light occupies a small fraction of the electromagnetic spectrum. For centuries our only information about the universe beyond Earth—except for an occasional meteorite—came from visible light. Processes like those occurring on the visible surface of our Sun and many other stars produce predominantly visible light. Optical astronomy, utilizing visible light, gave a good picture of the universe to the extent that it consists of objects not too different from the visible part of the Sun. The restriction to optical astronomy was in part imposed by Earth's atmosphere. Transparent to visible light, the atmosphere is largely opaque to other forms of electromagnetic radiation, although "windows" of relative transparency exist in parts of the radio and infrared bands (Fig. 34-12). The discovery by Bell Telephone Laboratories electrical engineer Karl Jansky in 1931 that radio waves from outer space can be detected on Earth led to the development of radio astronomy. For decades, radio astronomy has given a picture of the universe that complements the optical view, showing phenomena that are simply not detectable by optical means (Fig. 34-13).

The onset of the space age in the late 1950s finally opened the entire electromagnetic spectrum to astronomers. Before this time there were surprisingly few suggestions that anything interesting might be found beyond the visible range. But satellites carrying infrared, ultraviolet, x-ray, and gamma-ray detectors have literally revolutionized our view of the universe (Fig. 34-14). Exotic objects like neutron stars—with the mass of the Sun crushed to a diameter of a few kilometers—and black holes, whose gravity is so strong that not even light can escape—are now objects of astronomical study. The opening of the entire electromagnetic spectrum has brought a new richness to astronomy, showing that our universe contains some of the most unusual objects that the laws of physics permit. Phenomena that were once bizarre conjectures of theoreticians are now observed regularly. Closer to home, observations of the Sun with ultraviolet and x-ray instruments have brought new understandings of the star that sustains us. And by turning space-borne infrared detectors toward Earth, we have learned much about the structure and resources of our own planet (Fig. 34-15).

(a)

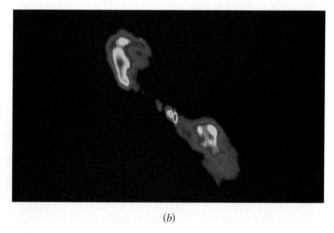

(b)

FIGURE 34-13 (a) Some of the 27 dish antennas that comprise the Very Large Array (VLA) radio telescope in New Mexico. (b) The galaxy Centaurus A, a powerful radio emitter, imaged with the VLA. The two lobes are jets of material ejected from the galaxy's central core. The VLA was tuned to a wavelength of 20 cm for this observation.

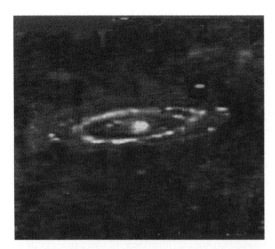

FIGURE 34-14 This false-color infrared image of the Andromeda galaxy was taken with the Infrared Astronomical Satellite. Yellow indicates regions of brightest infrared emission, corresponding to places where new stars are probably forming.

FIGURE 34-15 Infrared image of the New York City area, taken from a Landsat satellite. Areas of vegetation show reddish in this false-color image; note Central Park on Manhattan Island. Landsat images resolve objects as small as 30 m; here you can see individual ships near the Verrazano Bridge at lower center.

34-8 POLARIZATION

The fields of an electromagnetic wave in vacuum (and in most materials) are perpendicular to the propagation direction, but within the plane perpendicular to the wave propagation the orientation of one field is still arbitrary. **Polarization** is a wave property that specifies the electric field direction; since the two fields are perpendicular, polarization also determines the magnetic field direction (Fig. 34-16).

Electromagnetic waves used in radio, TV, and radar are generated in such a way that they have a definite polarization. So are the light waves produced by most lasers. In contrast, visible light from hot sources like the Sun or a light bulb is **unpolarized,** consisting of a mixture of electromagnetic waves with random field orientations.

Unpolarized light may be polarized either by reflection off surfaces or when it passes through substances whose molecular or crystal structure has a preferred direction called the **transmission axis.** Many crystals and synthetic materials like the plastic Polaroid have this property. For example, sunlight reflecting off the hood of a car becomes partially polarized in the horizontal direction. Polaroid sunglasses, with their transmission axis vertical, block this reflected glare without significantly reducing overall light intensity.

FIGURE 34-16 Some field vectors for an electromagnetic wave polarized in the y direction.

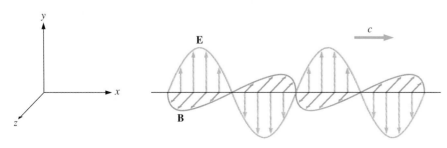

FIGURE 34-17 Two pairs of Polaroid sunglasses with their transmission axes at right angles. Where they overlap, no light can get through.

A polarizing material passes unattenuated only that component of the wave's electric field that lies along its preferred direction. If θ is the angle between the field and the polarizer's preferred direction, then the field component in the preferred direction is $E\cos\theta$. As we will show shortly, the intensity S of an electromagnetic wave is proportional to the square of the field strength; as a result a wave of intensity S_0 incident on a polarizer emerges with intensity given by the so-called **Law of Malus:**

$$S = S_0 \cos^2\theta. \qquad (34\text{-}19)$$

This equation shows that electromagnetic waves will be blocked completely if $\theta = 90°$, a situation that occurs when unpolarized light passes through one polarizer to give it a definite polarization, then through another oriented at $90°$ to the first (Fig. 34-17).

When unpolarized light passes through a polarizer its intensity is cut in half. You can see this from Equation 34-19 because the unpolarized light includes a mix of waves with random polarization angles θ. Averaging over all possible angles in Equation 34-19 amounts to taking the average of $\cos^2\theta$ over a full cycle. We've seen on several occasions that the average of the square of a sinusoidal function is $\frac{1}{2}$, so a polarizer does indeed cut the intensity of unpolarized light in half.

● **EXAMPLE 34-3** MULTIPLE POLARIZERS

Unpolarized light with intensity S_0 is incident on a "stack" of three polarizers. The first has its polarization axis vertical, the second is at $25°$ to the vertical, and the third is at $70°$ to the vertical (Fig. 34-18). What is the intensity of light emerging from this stack?

Solution

We've just seen that the first polarizer cuts the unpolarized intensity in half, giving $\frac{1}{2}S_0$ for the intensity incident on the second polarizer. Equation 34-19 shows that the second and third polarizers each reduce the intensity by a factor $\cos^2\theta$, where θ is the angle between the incident polarization direction—established by one polarizer—and the next polarizer's axis. For the second polarizer this angle is $25°$; for the third it is $70° - 25° = 45°$. Thus light emerges from the stack with intensity

$$S = (\tfrac{1}{2})(\cos^2 25°)(\cos^2 45°)S_0 = 0.205S_0.$$

Interestingly, this is greater than the intensity we would get passing light through a vertical polarizer followed by a single polarizer at $70°$. And, as the exercise below shows, it is much

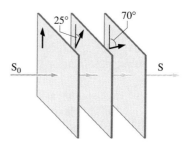

FIGURE 34-18 A stack of polarizers. Arrows on the sheets indicate directions of the polarization axes.

greater than what we would get by interchanging the second and third polarizers. Can you see why?

EXERCISE Rework Example 34-3 with the second and third polarizers interchanged.

Answer: $S = 0.029S_0$

Some problems similar to Example 35-3: 35, 36, 72 ●

Polarization can tell us much about sources of electromagnetic waves or about materials through which the waves travel. Many astrophysical processes produce polarized waves; measuring the polarization then gives clues to the mechanisms operating in distant objects. Polarization of light as it passes

FIGURE 34-19 Photomicrograph of a thin section of rock placed between crossed polarizers. Individual mineral crystals within the rock rotate the light's electric field, altering the transmitted light intensity.

FIGURE 34-20 Plastic model of a Gothic cathedral, photographed between polarizing sheets. The resulting patterns reveal stresses, helping architects and engineers understand the response of the building to wind and weight loading.

through materials helps geologists to understand the composition and formation of rocks (Fig. 34-19) and helps engineers to locate stresses in structures (Fig. 34-20).

■ APPLICATION ELECTRO-OPTIC MODULATION

The atomic structure of some materials causes them to rotate the polarization direction of incident light, an effect demonstrated in Figs. 34-19 and 34-20. Applying an electric field may alter the structure and, therefore, the material's effect on light passing through it. This is called the **electro-optic** effect.

An **electro-optic modulator** (EOM) is a device that uses the electro-optic effect to control the intensity of light. Figure 34-21 shows an EOM consisting of an electro-optic crystal between crossed polarizers. Electrodes coated on the ends of the crystal allow application of a voltage that produces an

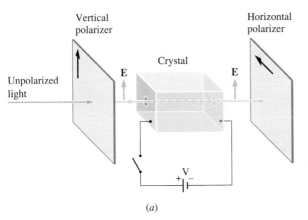

(a)

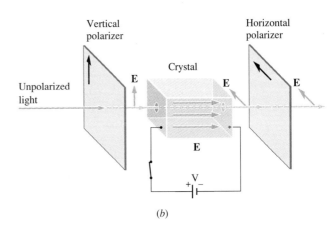

(b)

FIGURE 34-21 An electro-optic modulator consists of an electro-optic crystal between crossed polarizers. Electrodes permit application of a voltage across the crystal, and "windows" in the electrodes let light through. (a) With no applied voltage there is no rotation of the polarization, and the crossed polarizers therefore prevent light transmission through the device. (b) An applied voltage produces an electric field in the crystal. This alters the crystal structure and causes it to rotate the polarization of the incident light. A 90° rotation, shown here, gives maximum light transmission.

electric field in the crystal. With the proper design and orientation, the crystal in Fig. 34-21 has no effect on the polarization direction of light when there is no applied voltage. Since the polarizers are at right angles, light polarized by the first is then blocked by the second and so no light gets through the device (Fig. 34-21a).

When a voltage is applied across the crystal the polarization direction of the incident light is now rotated. At some value of the voltage—typically a kilovolt or more for most crystals—the polarization rotates through 90°. Then, light emerging from the crystal is polarized in the direction of the second polarizer and passes unattenuated through that polarizer. Thus the entire EOM passes the maximum light at this particular voltage.

Applying a time-varying voltage causes the EOM to vary, or *modulate,* the light intensity. Voice or music, for example, can

be transmitted on a laser beam by applying the amplified signal from a microphone to an EOM and using a light-sensitive detector to extract the audio information at the other end.

Turning on and off the voltage that gives maximum light transmission makes the EOM into an electronic shutter. Typical EOMs achieve on/off times on the order of a nanosecond, making them much faster than mechanical or even conventional electronic shutters and flash lamps. Prior to the 1983 redefinition of the meter in terms of the speed of light, EOM-based systems provided some of the most accurate measurements of *c*. Today EOM's enjoy a wide variety of scientific and commercial applications, especially those requiring rapid control of laser beams. An increasingly important use is the conversion of electrical signals to light pulses on the highest speed fiber optic communications systems.

34-9 PRODUCING ELECTROMAGNETIC WAVES

We have shown that electromagnetic waves can exist, and have explored some of their properties. But how do these waves originate?

All that's necessary is to produce a changing electric or magnetic field. Once a changing field of either type exists, Faraday's and Ampère's laws ensure the production of the other type—and so on, to give a propagating electromagnetic wave. Ultimately, changing fields of either type occur when we alter the motion of electric charge. Therefore:

❙ *Accelerated* electric charge is the source of electromagnetic waves.

In a radio transmitter, the accelerated charges are electrons moving back and forth in an antenna, driven by an alternating voltage from an *LC* circuit (Fig. 34-22). In an x-ray tube, high-energy electrons decelerate rapidly as they slam into a target; their deceleration is the source of the electromagnetic waves, now in the x-ray region of the spectrum. In the hot plasma of a fusion reactor, high-speed electrons veer around ions under the influence of the attractive electric force; their change in direction represents an acceleration that is responsible for the production of electromagnetic waves—a detrimental effect because the waves sap the plasma's energy, thereby cooling it. In the magnetron tube of a microwave oven, electrons circle in a magnetic field; their centripetal acceleration is the source of the microwaves that cook your food. And the altered movement of electrons in atoms—although described accurately only by quantum mechanics—is the source of most visible light. If the motion of the accelerated charges is periodic, then the wave frequency is that of the motion; more generally, systems are most efficient at producing (and receiving) electromagnetic waves whose wavelength is comparable to the size of the system. That's why TV antennas are on the order of 1 m in size, while nuclei—some 10^{-15} m in diameter—produce gamma rays.

Calculation of electromagnetic waves emitted by accelerated charges presents challenging but important problems for physicists and communications engineers. Figure 34-23a shows a "snapshot" of the electric field produced by a single point charge undergoing simple harmonic motion, while Fig. 34-23b

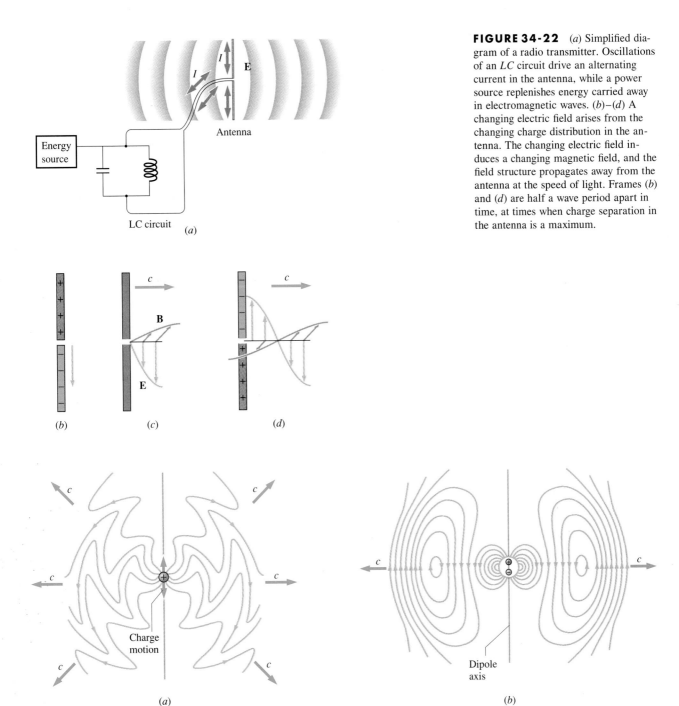

FIGURE 34-22 (a) Simplified diagram of a radio transmitter. Oscillations of an *LC* circuit drive an alternating current in the antenna, while a power source replenishes energy carried away in electromagnetic waves. (b)–(d) A changing electric field arises from the changing charge distribution in the antenna. The changing electric field induces a changing magnetic field, and the field structure propagates away from the antenna at the speed of light. Frames (b) and (d) are half a wave period apart in time, at times when charge separation in the antenna is a maximum.

FIGURE 34-23 "Snapshots" showing the electric fields of oscillating charge distributions. (a) A single point charge executing simple harmonic motion in the horizontal plane. Note that the field close to the charge approximates the radial field of a stationary point charge, but that farther out the "kinks" in the field—arising from the accelerated motion—are more prominent and become essentially perpendicular to the radial direction. These transverse kinks move outward at the speed of light, and constitute the wave fields. (b) The electric field of an oscillating dipole. Note that the field forms closed loops, detached from the dipole. These are the outward propagating wave fields. The larger field loops shown were formed when the oscillating dipole had the opposite orientation. Not shown in either figure are the equally important magnetic fields.

shows the field of an oscillating dipole—a configuration approximated by many systems from antennas to atoms and molecules. Both figures show that the waves are strongest in the direction at right angles to the acceleration of the charge distribution and that there is no radiation in the direction of the acceleration. This accounts for, among other phenomena, the directionality of radio and TV antennas, which transmit and receive most effectively perpendicular to the long direction of the antenna.

The fields shown in Fig. 34-23 seem to bear little resemblance to the plane-wave fields of Fig. 34-7 that we used to demonstrate the possibility of electromagnetic waves. We could produce true plane waves only with an infinite sheet of accelerated charge—an obvious impossibility. But far from the source, the curved field lines evident in Fig. 34-23b, for example, would appear straight, and the wave would begin to approximate a plane wave. So our plane-wave analysis is a valid approximation at great distances—typically many wavelengths—from a localized wave source. Closer to the source more complicated expressions for the wave fields apply, but these, too, satisfy Maxwell's equations.

34-10 ENERGY IN ELECTROMAGNETIC WAVES

We showed in previous chapters that electric and magnetic fields contain energy. Here we have considered electromagnetic waves, in which a combination of electric and magnetic fields propagates through space. As the wave moves, it must transport the energy contained in those fields.

We define the **intensity, S,** as the rate at which an electromagnetic wave transports energy across a unit area. This is the same definition we used for wave intensity in Chapter 16, and the units are also the same: power per unit area, or W/m². We can calculate the intensity of a plane electromagnetic wave by considering a rectangular box of thickness dx and cross-sectional area A with its face perpendicular to the wave propagation (Fig. 34-24). Within this box are wave fields $\mathbf{E}$ and $\mathbf{B}$ whose energy densities are given by Equations 26-3 and 32-11:

$$u_E = \tfrac{1}{2}\varepsilon_0 E^2$$

$$u_B = \frac{B^2}{2\mu_0}.$$

If dx is sufficiently small, the fields don't change much over the box, so the total energy in the box is just the sum of the electric and magnetic energy densities multiplied by the box volume $A\,dx$:

$$dU = (u_E + u_B)\,A\,dx = \frac{1}{2}\left(\varepsilon_0 E^2 + \frac{B^2}{\mu_0}\right)A\,dx.$$

This energy moves with speed c, so all the energy contained in the box length dx moves out of the box in a time $dt = dx/c$. The rate at which energy moves through the cross-sectional area A is then

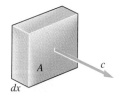

FIGURE 34-24 A box of length dx and cross-sectional area A at right angles to the propagation of an electromagnetic wave.

$$\frac{dU}{dt} = \frac{1}{2}\left(\varepsilon_0 E^2 + \frac{B^2}{\mu_0}\right)\frac{A\,dx}{dx/c} = \frac{c}{2}\left(\varepsilon_0 E^2 + \frac{B^2}{\mu_0}\right)A.$$

So the intensity S, or rate of energy flow per unit area, is

$$S = \frac{c}{2}\left(\varepsilon_0 E^2 + \frac{B^2}{\mu_0}\right).$$

We can recast this equation in simpler form by noting that, for an electromagnetic wave, $E = cB$ and $B = E/c$. Using these expressions to replace one of the E's in the term E^2 with B and similarly one of the B's in the term B^2 with E, we have

$$S = \frac{c}{2}\left(\varepsilon_0 cEB + \frac{EB}{\mu_0 c}\right) = \frac{1}{2\mu_0}(\varepsilon_0\mu_0 c^2 + 1)\,EB.$$

But $c = 1/\sqrt{\varepsilon_0\mu_0}$, so $\varepsilon_0\mu_0 c^2 = 1$, giving

$$S = \frac{EB}{\mu_0}. \tag{34-19a}$$

Although we derived Equation 34-19a for an electromagnetic wave, it is in fact a special case of the more general result that nonparallel electric and magnetic fields are accompanied by a flow of electromagnetic energy. In general, the rate of energy flow per unit area is given by

$$\mathbf{S} = \frac{\mathbf{E} \times \mathbf{B}}{\mu_0}. \tag{34-19b}$$

Here a vector $\mathbf{S}$ is used to signify not only the magnitude of the energy flow, but also its direction. For an electromagnetic wave in vacuum, in which $\mathbf{E}$ and $\mathbf{B}$ must be at right angles, Equation 34-19b reduces to Equation 34-19a, with the direction of energy flow the same as the direction of wave travel. The vector intensity $\mathbf{S}$ is called the **Poynting vector** after the English physicist J. H. Poynting, who suggested it in 1884. Poynting's name is especially fortuitous, for the Poynting vector points in the direction of energy flow. Problem 76 explores an important application of the Poynting vector to fields that do not constitute an electromagnetic wave.

Equations 34-19 give the intensity at the instant when the fields have magnitudes E and B. In an electromagnetic wave the fields oscillate, and so does the intensity. We're usually not interested in this rapid oscillation. For example, an engineer designing a solar collector doesn't care that sunlight intensity oscillates at about 10^{14} Hz. What she really wants is the *average* intensity, $\bar{S}$. Because the instantaneous intensity of Equation 34-19a contains a product of sinusoidally varying terms, which are in phase, the average intensity is just half the peak intensity:

$$\bar{S} = \frac{\overline{EB}}{\mu_0} = \frac{E_p B_p}{2\mu_0} \quad \text{(average intensity)} \tag{34-20a}$$

(This follows because, as we've seen on several occasions, the average of $\sin^2 \omega t$ over one cycle is $\frac{1}{2}$.) We wrote Equation 34-20a in terms of both the electric and magnetic fields, but we can use the wave condition $E = cB$ to eliminate either field in terms of the other:

$$\bar{S} = \frac{E_p{}^2}{2\mu_0 c} \tag{34-20b}$$

and

$$\bar{S} = \frac{cB_p{}^2}{2\mu_0}. \tag{34-20c}$$

● **EXAMPLE 34-4** SOLAR ENERGY

The average intensity of sunlight on a clear day at noon is about 1 kW/m². What are the electric and magnetic fields in sunlight? How many solar collectors would you need to replace a 4.8-kW electric water heater in noonday sun, if each collector has an area of 2.0 m² and converts 40% of the incident sunlight to heat?

Solution

Solving Equation 34-20b for the electric field gives

$$E_p = \sqrt{2\mu_0 c \bar{S}}$$

$$= [(2)(4\pi \times 10^{-7} \text{ H/m})(3.0 \times 10^8 \text{ m/s})(1 \times 10^3 \text{ W/m}^2)]^{1/2}$$

$$= 0.87 \text{ kV/m}.$$

The peak magnetic field is then given by $B_p = E_p/c$, so

$$B_p = \frac{E_p}{c} = \frac{870 \text{ V/m}}{3.0 \times 10^8 \text{ m/s}} = 3 \times 10^{-6} \text{ T}.$$

At 1 kW/m², we would then need 4.8 m² of collector area if the collectors were 100% efficient. At 40% efficiency, we therefore need 4.8 m²/0.40 = 12 m², for a total of 6 collectors.

EXERCISE A laser produces an average power of 7.0 W in a light beam 1.0 mm in diameter. Find (a) the average intensity and (b) the peak electric field of the laser light.

Answers: (a) 8.9 MW/m²; (b) 82 kV/m

Some problems similar to Example 34-4: 38–41, 43

●

Waves from Localized Sources

As an electromagnetic wave propagates through empty space, its total energy does not change. With plane waves the intensity—power per unit area—does not change either. But when a wave originates in a localized source like an atom, a radio transmitting antenna, a light bulb, or a star, its wavefronts are not planes but expanding spheres (recall Fig. 16-18). The wave's total energy remains the same, but as it expands that energy is spread over the area of an ever larger sphere—whose area increases as the square of the distance from the source. Therefore, as we found in Chapter 16, the power per unit area—the intensity—decreases as the inverse square of the distance:

$$S = \frac{P}{4\pi r^2}. \tag{34-21}$$

Here S and P can be either the peak or average intensity and power, respectively, and r is the distance from a localized source. This intensity decrease occurs not because electromagnetic waves "weaken" and lose energy but because that energy gets spread ever more thinly.

Because the intensity of an electromagnetic wave is proportional to the *square* of the field strengths (Equations 34-20), Equation 34-21 shows that the *fields* of a spherical wave decrease as $1/r$. Contrast that with the $1/r^2$ decrease in the electric field of a stationary point charge, and you can see why the electromagnetic wave fields associated with an accelerated charge dominate in all but the immediate vicinity of the charge (see Fig. 34-23a).

● **EXAMPLE 34-5** A GARAGE-DOOR OPENER

A radio-activated garage-door opener responds to signals with average intensity as weak as 20 μW/m². If the transmitter unit produces a 240-mW signal, broadcast in all directions, what is the maximum distance at which the transmitter will activate the door opener? What is the minimum value for the peak electric field to which the unit responds?

Solution
Since the waves spread out in all the directions, Equation 34-21 applies. Solving for r gives

$$r = \sqrt{\frac{P}{4\pi S}} = \sqrt{\frac{240\times10^{-3}\text{ W}}{(4\pi)(20\times10^{-6}\text{ W/m}^2)}} = 31\text{ m}.$$

Solving Equation 34-20b gives the electric field corresponding to the unit's 20-μW/m² sensitivity:

$$E_p = \sqrt{2\mu_0 c \overline{S}}$$
$$= \sqrt{(2)(4\pi\times10^{-7}\text{ H/m})(3.00\times10^8\text{ m/s})(20\times10^{-6}\text{ W/m}^2)}$$
$$= 0.12\text{ V/m}.$$

The sensitivity of radio receiving equipment is often expressed in terms of the minimum electric field strength.

EXERCISE A stereo receiver's AM tuner section has a rated sensitivity of 2.1 mV/m. What is the maximum distance at which this unit can receive broadcasts from a radio station's 5.0-kW transmitter, assuming the signal is broadcast in all directions?

Answer: 261 km

Some problems similar to Example 34-5: 49, 51, 52, 61, 62   ●

34-11 WAVE MOMENTUM AND RADIATION PRESSURE

We know from mechanics that moving objects carry both energy and momentum. The same is true for electromagnetic waves. Maxwell showed that the wave energy U and momentum p are related by

$$p = \frac{U}{c}. \tag{34-22}$$

If an electromagnetic wave is incident on an object and the object absorbs the wave energy (as, for example, a black object exposed to sunlight), then the object also absorbs the momentum given by Equation 34-22. If the wave's average intensity is $\overline{S}$, then it carries energy per unit area at the average rate $\overline{S}$ J/s/m². According to Equation 34-22 it therefore carries momentum per unit area at the rate $\overline{S}/c$. Newton's law in its general form $\mathbf{F} = d\mathbf{p}/dt$ tells us that the rate of change of an object's momentum is equal to the net force on the object. Therefore, if an object absorbs electromagnetic wave momentum $\overline{S}/c$ per unit area per unit time, it experiences a force per unit area of this magnitude. Since force per unit area is pressure, we call this quantity the **radiation pressure:**

$$P_{\text{rad}} = \frac{\overline{S}}{c}. \tag{34-23}$$

The radiation pressure is doubled if an object reflects electromagnetic waves, in the same way that bouncing a basketball off a backboard changes the ball's momentum by $2mv$ and, therefore, delivers momentum $2mv$ to the backboard.

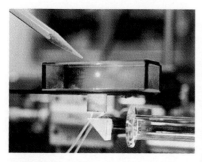

FIGURE 34-25 The star-like image is a 20-micron particle levitated by a laser beam reflected upward by the prism shown at the bottom. The star-like rays are due to diffraction (see Chapter 37) inside the camera.

FIGURE 34-26 Two-stage sailing spacecraft proposed for interstellar travel. At the target star, laser light from the solar system would reflect from the large sail to the smaller one, bringing the latter to a stop. The diameter of the large sail is 1000 km.

The pressure exerted by ordinary light is very small, but Dartmouth College physicists E. F. Nichols and G. F. Hull demonstrated its existence in a sensitive experiment performed in 1903. With high-energy laser light or with objects of low mass and large area, radiation pressure can be appreciable. Lasers exert enough light pressure to levitate small particles (Fig. 34-25), and the pressure of sunlight has been suggested as a means of driving interplanetary "sailing ships" (Fig. 34-26). Finally, the idea that electromagnetic waves carry momentum played a crucial role in Einstein's development of his famous equation $E = mc^2$.

● EXAMPLE 34-6 STAR WARS

A proposed ballistic missile defense system calls for a laser that can focus 25 MW of light on an attacking warhead. The weapon works by heating the warhead to the point of destruction, but it also delivers momentum that alters the warhead's trajectory. If the beam dwells on a 200-kg warhead for 15 s, what velocity change does it impart to the warhead? Estimate the distance by which the warhead will be knocked off course over its remaining 30 minutes of flight.

Solution

The energy delivered in a 25-MW beam acting for 15 s is (25 MW)(15 s) = 375 MJ. According to Equation 34-22, the associated momentum is

$$p = \frac{U}{c} = \frac{375 \times 10^6 \text{ J}}{3.00 \times 10^8 \text{ m/s}} = 1.25 \text{ kg} \cdot \text{m/s}.$$

The change in the warhead's velocity is given by $m\Delta v = \Delta p$, where Δp is the change in its momentum. Assuming the warhead absorbs all the beam's momentum, we then have

$$\Delta v = \frac{\Delta p}{m} = \frac{1.25 \text{ kg} \cdot \text{m/s}}{200 \text{ kg}} = 0.00625 \text{ m/s}.$$

This is insignificant compared with a typical warhead speed of 7 km/s. Even though we don't know the direction of the velocity change, we can estimate crudely the error Δx in the impact point by multiplying this change by the flight time:

$$\Delta x = \Delta v\, t = (0.00625 \text{ m/s})(30 \text{ min})(60 \text{ s/min}) = 11 \text{ m}.$$

This is totally negligible, especially for a nuclear warhead. Even with this enormously powerful laser, radiation pressure has an insignificant effect.

EXERCISE A laser delivers 5.0 MW/m². If the beam is directed upward, what is the maximum mass for a 100-μm-diameter particle to be suspended in the beam?

Answer: 1.3×10^{-11} kg

Some problems similar to Example 34-6: 56–59 ●

CHAPTER SYNOPSIS

Summary

1. Maxwell's modification of Ampère's law adds a **displacement current** term $\varepsilon_0 d\phi_E/dt$, showing that changing electric flux is a source of magnetic field. This modified law completes the set of **Maxwell's equations**—the four equations that govern the behavior of electromagnetic fields.

2. The interplay of electric and magnetic fields described by Faraday's and Ampère's laws gives rise to **electromagnetic waves.** When they propagate through vacuum, these waves
   (a) Travel at the speed of light, $c = 1/\sqrt{\varepsilon_0\mu_0} = 3.00 \times 10^8$ m/s.
   (b) Have their electric and magnetic fields at right angles to each other and to the direction of wave propagation.
   (c) Have their field magnitudes related by $E = cB$.
   (d) Can have any frequency or wavelength, provided the two are related by the equivalent expressions $f\lambda = c$ or $\omega/k = c$.

3. Radio waves, television, microwaves, infrared, visible light, ultraviolet, x rays, and gamma rays are all forms of electromagnetic radiation. They differ only in frequency and wavelength, and together comprise the **electromagnetic spectrum.**

4. **Polarization** describes the orientation of a wave's electric field in the plane perpendicular to the propagation direction. When a polarized wave passes through a polarizing material, its intensity is reduced by a factor $\cos^2\theta$, where θ is the angle between the wave polarization and the preferred axis of the material.

5. Electromagnetic waves are produced by accelerated electric charges, as in the alternating current of a radio antenna.

6. Electromagnetic waves carry energy. The rate at which energy is transported per unit area is the wave **intensity.** The **Poynting vector,**

$$\mathbf{S} = \frac{\mathbf{E} \times \mathbf{B}}{\mu_0},$$

describes this energy transport for any configuration of electromagnetic fields. The average intensity has half the peak value, or

$$\bar{S} = \frac{E_p B_p}{2\mu_0}.$$

The intensity of a plane wave remains constant, while the intensity from a localized source decreases as the inverse square of the distance from the source.

7. An electromagnetic wave with energy U also carries momentum $p = U/c$. As a result, it exerts a **radiation pressure** $P_{\text{rad}} = \bar{S}/c$ on an object that absorbs the wave and twice this pressure on a reflecting object.

Terms You Should Understand

(Pairs are closely related terms whose distinction is important; number in parentheses is chapter section where term first appears.)

displacement current (34-2)
Maxwell's equations (34-3)
electromagnetic wave (34-4)
electromagnetic spectrum (34-7)
polarization (34-8)
intensity, Poynting vector (34-10)
radiation pressure (34-11)

Symbols You Should Recognize

$\varepsilon_0 d\phi_E/dt$ (34-2) $S, \mathbf{S}, \bar{S}$ (34-10)
c (34-5) P_{rad} (34-11)

Problems You Should Be Able to Solve

evaluating induced magnetic fields in symmetric situations (34-2)
relating frequency and wavelength of electromagnetic waves (34-6)
relating electric and magnetic field strengths in electromagnetic waves (34-6)
calculating light intensity emerging from one or more polarizers (34-8)
relating wave intensity and fields (34-10)
evaluating wave intensity and fields as a function of distance from localized sources (34-10)
calculating radiation pressure and its effects (34-11)

Limitations to Keep in Mind

The description of electromagnetic waves developed in this chapter applies strictly only in vacuum.

QUESTIONS

1. Why is Maxwell's modification of Ampère's law essential to the existence of electromagnetic waves?
2. The presence of magnetic monopoles would require modification of Gauss's law for magnetism. Which of the other Maxwell equations would also need modification?
3. There is displacement current between the plates of a charging capacitor, yet no charge is moving between the plates. In what sense is the word "current" appropriate here?
4. Is there displacement current in an electromagnetic wave? Is there ordinary conduction current?
5. List some similarities and differences between electromagnetic waves and sound waves.
6. What aspect of the electromagnetic wave considered in Section 34-4 ensures that Gauss's laws for electricity and magnetism are satisfied?
7. Explain why parallel electric and magnetic fields in vacuum could not constitute an electromagnetic wave.
8. The speed of an electromagnetic wave is given by $c = \lambda f$. How does the speed depend on frequency? On wavelength?
9. When astronomers observe a supernova explosion in a distant galaxy, they see a sudden, simultaneous rise in visible light and other forms of electromagnetic radiation. How is this evidence that the speed of light is independent of frequency?
10. Turning a TV antenna so its rods point vertically may change the quality of your TV reception. Why? Think about polarization.
11. Unpolarized light is incident on two sheets of Polaroid with their polarization directions at right angles, and no light gets through. A third sheet is inserted between the other two, and now some light gets through. How can this be?
12. Why is it not possible to define exactly where the visible region of the spectrum ends?
13. Why did the field of x-ray astronomy flourish only after the advent of space flight?
14. The Sun emits most of its electromagnetic wave energy in the visible region of the spectrum, with the peak in the yellow-green. Our eyes are sensitive to the same range, with peak sensitivity in the yellow-green. Is this a coincidence?
15. Suppose your eyes were sensitive to radio waves rather than light. What things would look bright?
16. An LC circuit is made entirely from superconducting materials, yet its oscillations eventually damp out. Why?
17. If you double the field strength in an electromagnetic wave, what happens to the intensity?
18. The intensity of light falls off as the inverse square of the distance from the source. Does this mean that electromagnetic wave energy is lost? Explain.
19. When your picture is taken with a flash camera, why doesn't the momentum of the light flash knock you over?
20. Some long-distance power transmission lines use DC rather than AC, despite the need to convert between DC and AC at either end. Why might this be? What energy loss mechanism occurs with AC but not DC?
21. Electromagnetic waves do not readily penetrate metals. Why might this be?

PROBLEMS

Section 34-2 Ambiguity in Ampère's Law

1. A uniform electric field is increasing at the rate of 1.5 V/m·μs. What is the displacement current through an area of 1.0 cm² at right angles to the field?
2. A parallel-plate capacitor has square plates 10 cm on a side and 0.50 cm apart. If the voltage across the plates is increasing at the rate of 220 V/ms, what is the displacement current in the capacitor?
3. A parallel-plate capacitor of plate area A and spacing d is charging at the rate dV/dt. Show that the displacement current in the capacitor is equal to the conduction current flowing in the wires feeding the capacitor.
4. A capacitor with circular plates is fed with long, straight wires along the axis of the plates. Show that the magnetic field *outside* the capacitor, in a plane that passes through the interior of the capacitor and is perpendicular to the

axis, is given by $B = \dfrac{\mu_0 \varepsilon_0 R^2}{2rd} \dfrac{dV}{dt}$. Here R is the plate radius, d the spacing, dV/dt the rate of change of the capacitor voltage, and r the distance from the axis.
5. A parallel-plate capacitor has circular plates with radius 50 cm and spacing 1.0 mm. A uniform electric field between the plates is changing at the rate 1.0 MV/m·s. What is the magnetic field between the plates (a) on the symmetry axis, (b) 15 cm from the axis, and (c) 150 cm from the axis?
6. An electric field points into the page and occupies a circular region of radius 1.0 m, as shown in Fig. 34-27. There are no electric charges in the region, but there is a magnetic field forming closed loops pointing clockwise, as shown. The magnetic field strength 50 cm from the center of the region is 2.0 μT. (a) What is the rate of change of the electric field? (b) Is the electric field increasing or decreasing?

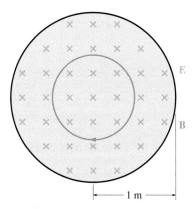

FIGURE 34-27 Problem 6.

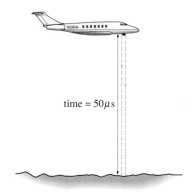

time = $50\mu s$

FIGURE 34-28 Problem 14.

Section 34-4 Electromagnetic Waves

7. At a particular point the instantaneous electric field of an electromagnetic wave points in the $+y$ direction, while the magnetic field points in the $-z$ direction. In what direction is the wave propagating?

8. The fields of an electromagnetic wave are $\mathbf{E} = E_p \sin(kz + \omega t)\hat{\mathbf{j}}$ and $\mathbf{B} = B_p \sin(kz + \omega t)\hat{\mathbf{i}}$. Give a unit vector in the direction of propagation.

9. The electric field of a radio wave is given by $\mathbf{E} = E\sin(kz - \omega t)(\hat{\mathbf{i}} + \hat{\mathbf{j}})$. (a) What is the peak amplitude of the electric field? (b) Give a unit vector in the direction of the magnetic field at a place and time where $\sin(kz - \omega t)$ is positive.

10. Show by differentiation and substitution that Equations 34-12b and 34-13 can be satisfied by fields of the form $E(x, t) = E_p f(kx \pm \omega t)$ and $B(x, t) = B_p f(kx \pm \omega t)$, where f is any function of the argument $kx \pm \omega t$.

11. Show that it is impossible for an electromagnetic wave in vacuum to have a time-varying component of its electric field in the direction of its magnetic field. *Hint:* Assume $\mathbf{E}$ does have such a component, and show that you cannot satisfy both Gauss and Faraday.

Section 34-5 The Speed of Electromagnetic Waves

12. A light-minute is the distance light travels in one minute. Show that the Sun is about 8 light-minutes from Earth.

13. Your intercontinental telephone call is carried by electromagnetic waves routed via a satellite in geosynchronous orbit at an altitude of 36,000 km. Approximately how long does it take before your voice is heard at the other end?

14. An airplane's radar altimeter works by bouncing radio waves off the ground and measuring the round-trip travel time (Fig. 34-28). If that time is 50 μs, what is the altitude?

15. Roughly how long does it take light to go 1 foot?

16. If you speak via radio from Earth to an astronaut on the moon, how long is it before you can get a reply?

17. "Ghosts" on a TV screen occur when part of the signal goes directly from transmitter to receiver, while part takes a longer route, reflecting off mountains or buildings (Fig. 34-29). The electron beam in a 50-cm-wide TV tube "paints" the picture by scanning the beam from left to right across the screen in about 10^{-4} s. If a "ghost" image appears displaced about 1 cm from the main image, what is the difference in path lengths of the direct and indirect signals?

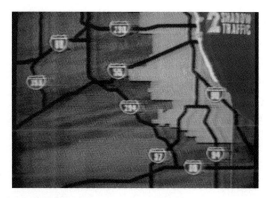

FIGURE 34-29 Ghost images of highways appear on this TV traffic report (Problem 17).

18. A computer can fetch information from its memory in 3.0 ns, a process that involves sending a signal from the central processing unit (CPU) to memory and awaiting the return of the information. If signals in the computer's wiring travel at $0.60c$, what is the maximum distance between the CPU and the memory? Your answer shows why high-speed computers are necessarily compact.

19. The speed of electromagnetic waves in a transparent dielectric is given by $1/\sqrt{\kappa\varepsilon_0\mu_0}$, where κ is the dielectric

constant described in Chapter 26 (see Problem 69). An experimental measurement gives 1.97×10^8 m/s for the speed of light in a piece of glass. What is the dielectric constant of this glass at optical frequencies?

Section 34-6 Properties of Electromagnetic Waves

20. What are the wavelengths of (a) a 100-MHz FM radio wave, (b) a 3.0-GHz radar wave, (c) a 6.0×10^{14}-Hz light wave, and (d) a 1.0×10^{18}-Hz x ray?
21. A 60-Hz power line emits electromagnetic radiation. What is the wavelength?
22. Antennas for transmitting and receiving electromagnetic radiation usually have typical dimensions on the order of half a wavelength. Look at a TV antenna, and estimate the wavelength and frequency of a TV signal.
23. A CB radio antenna is a vertical rod 2.75 m high. If this length is one-fourth of the CB wavelength, what is the CB frequency?
24. A microwave oven operates at 2.4 GHz. What is the distance between wave crests in the oven?
25. What would be the electric field strength in an electromagnetic wave whose magnetic field equalled that of Earth, about 50 μT?
26. Dielectric breakdown in air occurs at an electric field strength of about 3×10^6 V/m. What would be the peak magnetic field in an electromagnetic wave with this value for its peak electric field?
27. A radio receiver can detect signals with electric fields as low as 320 μV/m. What is the corresponding magnetic field?

Section 34-8 Polarization

28. An electromagnetic wave is propagating in the z direction. What is its polarization direction, if its magnetic field is in the y direction?
29. Polarized light is incident on a sheet of polarizing material, and only 20% of the light gets through. What is the angle between the electric field and the polarization axis of the material?
30. Vertically polarized light passes through a polarizer whose polarization axis is oriented at 70° to the vertical. What fraction of the incident intensity emerges from the polarizer?
31. A polarizer blocks 75% of a polarized light beam. What is the angle between the beam's polarization and the polarizer's axis?
32. An electro-optic modulator (see Fig. 34-21) is supposed to switch a laser beam between fully off and fully on, as its crystal rotates the beam polarization by 90° when a voltage is applied. But a power-supply failure results in only enough voltage for a 72° beam rotation. What fraction of the laser light is transmitted when it is supposed to be fully on?

33. Unpolarized light of intensity S_0 passes first through a polarizer with its polarization axis vertical, then through one with its axis at 35° to the vertical. What is the light intensity after the second polarizer?
34. Vertically polarized light passes through two polarizers, the first at 60° to the vertical and the second at 90° to the vertical. What fraction of the light gets through?
35. Unpolarized light with intensity S_0 passes through a stack of five polarizing sheets, each with its axis rotated 20° with respect to the previous one. What is the intensity of the light emerging from the stack?
36. Unpolarized light of intensity S_0 is incident on a "sandwich" of three polarizers. The outer two have their transmission axes perpendicular, while the middle one has its axis at 45° to the others. What is the light intensity emerging from this "sandwich?"
37. Polarized light with average intensity S_0 passes through a sheet of polarizing material which is rotating at 10 rev/s. At time $t = 0$ the polarization axis is aligned with the incident polarization. Write an expression for the transmitted intensity as a function of time.

Section 34-10 Energy in Electromagnetic Waves

38. A typical laboratory electric field is 1000 V/m. What is the average intensity of an electromagnetic wave with this value for its peak field?
39. What would be the average intensity of a laser beam so strong that its electric field produced dielectric breakdown of air (which requires $E_p = 3 \times 10^6$ V/m)?
40. Estimate the peak electric field inside a 625-W microwave oven under the simplifying approximation that the microwaves propagate as a plane wave through the oven's 750-cm^2 cross-sectional area.
41. A radio receiver can pick up signals with peak electric fields as low as 450 μV/m. What is the average intensity of such a signal?
42. Show that the electric and magnetic energy densities in an electromagnetic wave are equal.
43. A laser blackboard pointer delivers 0.10 mW average power in a beam 0.90 mm in diameter. Find (a) the average intensity, (b) the peak electric field, and (c) the peak magnetic field.
44. The laser of Example 34-6 produces a spot 80 cm in diameter at its target. What are the rms electric and magnetic fields at the target?
45. The United States' safety standard for continuous exposure to microwave radiation is 10 mW/cm^2. The glass door of a microwave oven measures 40 cm by 17 cm and is covered with a metal screen that blocks microwaves. What fraction of the oven's 625-W microwave power can leak through the door window without exceeding the safe exposure to someone right outside the door? Assume the power leaks uniformly through the window area.

46. A 1.0-kW radio transmitter broadcasts uniformly in all directions. What is the intensity of its signal at a distance of 5.0 km from the transmitter?

47. Use the fact that sunlight intensity at Earth's orbit is 1368 W/m^2 to calculate the Sun's total power output.

48. About two-thirds of the solar energy at Earth's orbit reaches the planet's surface. At what rate is solar energy incident on the entire Earth? See the previous problem, and compare your result with the roughly 10^{13} W rate at which humanity consumes energy.

49. During its 1989 encounter with Neptune, the Voyager 2 spacecraft was 4.5×10^9 km from Earth (Fig. 34-30). Its images of Neptune were broadcast by a radio transmitter with a mere 21-W average power output. What would be (a) the average intensity and (b) the peak electric field received at Earth if the transmitter broadcast equally in all directions? (The received signal was actually somewhat stronger because Voyager used a directional antenna.)

FIGURE 34-30 Neptune, photographed by the Voyager II spacecraft when it was 4.5×10^9 km from Earth. How long did it take the radio signal carrying this image to reach Earth? (Problem 49)

50. A quasar 10 billion light-years from Earth appears the same brightness as a star 50,000 light-years away. How does the power output of the quasar compare with that of the star?

51. At 1.5 km from the transmitter, the peak electric field of a radio wave is 350 mV/m. (a) What is the transmitter's power output, assuming it broadcasts uniformly in all directions? (b) What is the peak electric field 10 km from the transmitter?

52. The peak electric field at a point 25 m from a point source of electromagnetic waves is 4.2 kV/m. What is the peak magnetic field 1.0 m from the source?

53. A typical fluorescent lamp is a little over 1 m long and a few cm in diameter. How do you expect the light intensity to vary with distance (a) near the lamp but not near either end and (b) far from the lamp?

Section 34-11 Wave Momentum and Radiation Pressure

54. A camera flash delivers 2.5 kW of light power for 1.0 ms (Fig. 34-31). Find (a) the total energy and (b) the total momentum carried by the flash.

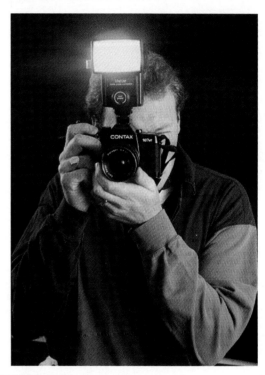

FIGURE 34-31 How much energy and momentum are in light from the camera flash? (Problem 54)

55. What is the radiation pressure exerted on a light-absorbing surface by a laser beam whose intensity is 180 W/cm^2?

56. A laser beam shines vertically upward. What laser power is necessary for this beam to support a flat piece of aluminum foil with mass 30 μg and diameter equal to that of the beam? Assume the foil reflects all the light.

57. The average intensity of noonday sunlight is about 1 kW/m^2. What is the radiation force on a solar collector measuring 60 cm by 2.5 m if it is oriented at right angles to the incident light and absorbs all the light?

58. Serious proposals have been made to "sail" spacecraft to the outer solar system using the pressure of sunlight. How much sail area must a 1000-kg spacecraft have if its accel-

eration at Earth's orbit is to be 1 m/s²? Assume the sails are made from reflecting material. *Hint:* Can you neglect the Sun's gravity?

59. A 65-kg astronaut is floating in empty space. If the astronaut shines a 1.0-W flashlight in a fixed direction, how long will it take the astronaut to accelerate to a speed of 10 m/s?

60. A "photon rocket" emits a beam of light instead of the hot gas of an ordinary rocket. How powerful a light source would be needed for a photon rocket with thrust equal to that of a space shuttle (35 MN)? Compare your answer with humanity's total electric power generating capability, about 10^{12} W.

Paired Problems

(Both problems in a pair involve the same principles and techniques. If you can get the first problem, you should be able to solve the second one.)

61. Find the peak electric and magnetic fields 1.5 m from a 60-W light bulb that radiates equally in all directions.

62. At 4.6 km from a radio transmitter, the peak electric field in the radio wave measures 380 mV/m. What is the transmitter's power, assuming it broadcasts equally in all directions?

63. Unpolarized light is incident on two polarizers with their axes at 45°. What fraction of the incident light gets through?

64. Find the angle between two polarizers if unpolarized light incident on the pair emerges with 10% of its incident intensity.

65. What is the radiation force on the door of a microwave oven if 625 W of microwave power hits the door at right angles and is reflected?

66. What is the power output of a laser whose beam exerts a 55-mN force on an absorbing object oriented at right angles to the beam? The object is larger than the beam's cross section.

67. A 60-W light bulb is 6.0 cm in diameter. What is the radiation pressure on an opaque object at the bulb's surface?

68. A white dwarf star is approximately the size of Earth but radiates about as much energy as the Sun. Estimate the radiation pressure on an absorbing object at the white dwarf's surface.

Supplementary Problems

69. Maxwell's equations in a dielectric resemble those in vacuum (Equations 34-6 through 34-9), but with ϕ_E in Ampère's law replaced by $\kappa\phi_E$, where κ is the dielectric con-

stant introduced in Chapter 26. Show that the speed of electromagnetic waves in such a dielectric is $c/\sqrt{\kappa}$.

70. Use appropriate data from Appendix E to calculate the radiation pressure on a light-absorbing object at the Sun's surface.

71. A radar system produces pulses consisting of 100 full cycles of a sinusoidal 70-GHz electromagnetic wave. The average power while the transmitter is on is 45 MW, and the waves are confined to a beam 20 cm in diameter. Find (a) the peak electric field, (b) the wavelength, (c) the total energy in a pulse, and (d) the total momentum in a pulse. (e) If the transmitter produces 1000 pulses per second, what is its average power output?

72. In a stack of polarizing sheets, each sheet has its polarization axis rotated 14° with respect to the preceding sheet. If the stack passes 37% of the incident, unpolarized light, how many sheets does it contain?

73. The peak electric field measured at 8.0 cm from a light source is 150 W/m², while at 12 cm it measures 122 W/m². Describe the shape of the source.

74. Show that Equations 34-12b and 34-13 may be combined to yield a wave equation like Equation 16-13. *Hint:* Take the partial derivative of one equation with respect to x and of the other with respect to t, and use the fact that

$$\frac{\partial}{\partial x}\left(\frac{\partial f}{\partial t}\right) = \frac{\partial}{\partial t}\left(\frac{\partial f}{\partial x}\right)$$

for any well-behaved function $f(x, t)$.

75. Studies of the origin of the solar system suggest that sufficiently small particles might be blown out of the solar system by the force of sunlight. To see how small such particles must be, compare the force of sunlight with the force of gravity, and solve for the particle radius at which the two are equal. Assume the particles are spherical and have density 2 g/cm³. Why do you not need to worry about the distance from the Sun?

76. A cylindrical resistor of length ℓ, radius a, and resistance R carries a current I. Calculate the electric and magnetic fields at the surface of the resistor, assuming the electric field is uniform throughout, including at the surface. Calculate the Poynting vector, and show that it points into the resistor. Calculate the flux of the Poynting vector (that is, $\int \mathbf{S} \cdot d\mathbf{A}$) over the surface of the resistor to get the rate of electromagnetic energy flow into the resistor, and show that the result is just I^2R. Your result shows that the energy heating the resistor comes from the fields surrounding it. These fields are sustained by the source of electrical energy that drives the current.

P A R T   4   CUMULATIVE PROBLEMS

These problems combine material from chapters throughout the entire part or, in addition, from chapters in earlier parts, or they present special challenges.

1. An air-insulated parallel-plate capacitor has plate area 100 cm² and spacing 0.50 cm. The capacitor is charged to a certain voltage and then disconnected from the charging battery. A thin-walled, nonconducting box of the same dimensions as the capacitor is filled with water at 20.00°C. The box is released at the edge of the capacitor and moves without friction into the capacitor (Fig. 1). When it reaches equilibrium the water temperature is 21.50°. What was the original voltage on the capacitor?

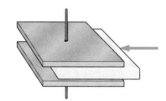

FIGURE 1   Cumulative Problem 1.

2. A wire of length ℓ and resistance R is formed into a closed rectangular loop twice as long as it is wide. It is mounted on a nonconducting horizontal axle parallel to its longer dimension, as shown in Fig. 2. A uniform magnetic field **B** points into the page, as shown. A string of negligible mass is wrapped around a drum of radius a attached to the axle, and a mass m is attached to the string. The string is many times longer than the drum

circumference; thus, many turns are wrapped around the drum. When the mass is released it falls and eventually reaches a speed that, averaged over one cycle of the loop's rotation, is constant from one rotation to the next. Find an expression for that average terminal speed.

3. Five wires of equal length 25 cm and resistance 10 Ω are connected to form two equilateral triangles that share a common side, as shown in Fig. 3. Two solenoids are perpendicular to the plane of the figure, as shown. Both solenoids have diameter 10 cm, and both extend a long way in and out of the page. The magnetic fields of both solenoids point out of the page; the field strength in the left-hand solenoid is increasing at 50 T/s while that in the right-hand solenoid is decreasing at 30 T/s. Find the current in the resistance wire shared by both triangles. Which way does the current flow?

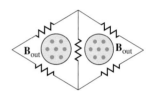

FIGURE 3   Cumulative Problem 3.

4. A long solenoid of length ℓ and radius R has a total of N turns. The solenoid current is increasing linearly with time: $I(t) = bt$, where b is a constant. (a) Find an expression for the rate at which the magnetic energy in the solenoid is increasing. (b) Find an expression for the induced electric field at the inner edge of the solenoid coils. (c) Evaluate the Poynting vector at the inner edge of the coils, and show by integration that electromagnetic energy is flowing into the solenoid at a rate equal to the buildup of magnetic energy. Use the long-solenoid approximation throughout, neglecting any variations along the direction of the solenoid axis.

5. A coaxial cable consists of an inner conductor of radius a and an outer conductor of radius b; the space between the conductors is filled with insulation of dielectric constant κ (Fig. 4). The cable's axis is the z axis. The cable is used to carry electromagnetic energy from a radio transmitter to a broadcasting antenna. The electric field between the conductors points radially from the axis,

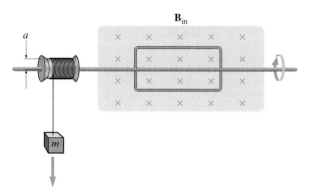

FIGURE 2   Cumulative Problem 2.

and is given by $E = E_0 \dfrac{a}{r} \cos(kz - \omega t)$. The magnetic field encircles the axis, and is given by $B = B_0 \dfrac{a}{r} \cos(kz - \omega t)$. Here E_0, B_0, k, and ω are constants. (a) Show, using appropriate closed surfaces and loops, that these fields satisfy Maxwell's equations. Your result shows that the cable acts as a "waveguide," confining an electromagnetic wave to the space between the conductors. (b) Find an expression for the speed at which the wave propagates along the cable.

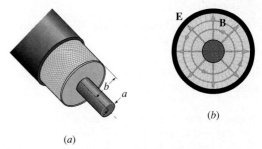

(a)

(b)

FIGURE 4   Cumulative Problem 5. (a) A coaxial cable. (b) Cross section, showing the electric and magnetic fields. The fields also vary with position z along the cable axis, according to the equations given.

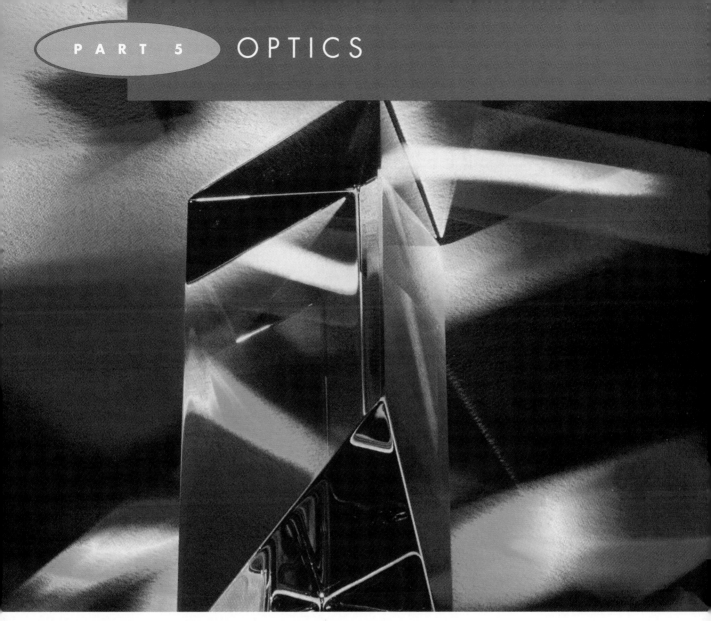

In "Prismatic Abstract," multiple light beams converge on a prism to create a dazzlingly color-ful effect. Photographic artist Pete Saloutous exploited the principles of optics in creating this work.

CHAPTER 35

REFLECTION AND REFRACTION

Reflection and refraction guide light in a wide range of natural and technological applications. Here the image of a bee's head appears in a large-scale version of the optical fibers that are essential to modern communications.

Maxwell's brilliant work shows that the phenomena of **optics**—that is, the behavior of light—are manifestations of the laws of electromagnetism. Except in the atomic realm, where it is necessary to use quantum mechanics, all optical phenomena can be understood in terms of electromagnetic wave fields as described by Maxwell's equations. But in many cases we need not resort to the full electromagnetic or even wave description of light to understand optics. When the objects with which light—or, for that matter, any other wave—interacts are much larger than the wavelength, then to a good approximation the light travels in straight lines called **rays** (Fig. 35-1), except when it actually hits something. **Geometrical optics** is the study of light under conditions when the ray approximation is valid. In this chapter we show how geometrical optics describes the behavior of light at interfaces between two different materials. In the next chapter we will show how geometrical optics explains optical systems including mirrors, lenses, the human eye, and many optical instruments.

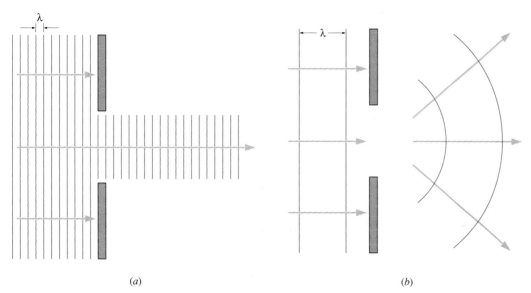

(a) (b)

FIGURE 35-1 Light waves incident on an opaque barrier with a hole in it. (*a*) When the hole diameter is much greater than the wavelength λ, light emerges from the hole in essentially a straight line. In this case it is appropriate to treat the light as traveling in straight rays except where it actually hits something. (*b*) When the hole diameter is comparable to or smaller than the wavelength, then interference causes the wavefronts to bend, and the ray approximation is no longer adequate.

35-1 REFLECTION AND TRANSMISSION

When a light ray propagating in one medium strikes the interface with a second medium, some light may be reflected back into the original medium and some transmitted into the new medium. Properties of the two materials, and the propagation direction of the incident light, determine the details of the reflection and transmission. It is also possible for light energy to be absorbed as it propagates through a medium, but here we will neglect this absorption.

Some materials—notably metals—reflect nearly all the light incident on them. It's no coincidence that these materials are also good electrical conductors. The oscillating electric field of a light wave drives the metal's free electrons into oscillatory motion, and this accelerated motion, in turn, produces electromagnetic waves. The effect of all the oscillating electrons together is simply to reradiate the wave back into the original medium. Ultimately, that's why metals appear shiny.

In other materials, the atomic and molecular configurations are such that the material absorbs most of the incident light energy. Such materials therefore appear dark and opaque. Since opaque materials basically destroy light by converting it to other forms of energy, they will play little role in our exploration of optics.

Still other materials are essentially transparent, allowing light to propagate with little energy loss. These materials are generally insulators whose electrons, bound to individual atoms, cannot respond freely to the fields of an incident wave. The electrons do respond in a more limited way, however, effectively producing oscillating molecular dipoles. Although a microscopic description of

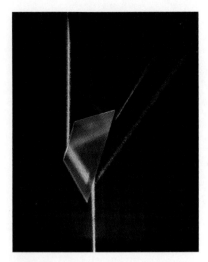

FIGURE 35-2 A laser beam incident on a glass prism shows partial reflection and partial transmission at the interfaces between air and glass.

this process is complicated, the net effect in most simple dielectric materials is a reduction in the propagation speed of electromagnetic waves, as wave energy is absorbed and reradiated by the molecular dipoles.

In this chapter we explore the processes that occur at the interfaces between transparent and reflective materials and between different transparent materials. Reflection is the significant process in the former case, while in the latter both reflection and transmission generally occur (Fig. 35-2).

35-2 REFLECTION

Reflection returns some or all of the light incident on an interface to its original medium. Whether the reflection is essentially complete, as from a metal, or partial, as from a transparent material, it satisfies the same geometrical condition. Experiment—as well as analysis based on Maxwell's equations—shows that the incident ray, the reflected ray, and the normal to the interface between the two materials all lie in the same plane, and that the angle θ_1' that the reflected ray makes with the normal is equal to the angle θ_1 made by the incident ray (Fig. 35-3a). That is,

$$\theta_1' = \theta_1, \tag{35-1}$$

where the subscript 1 designates angles in the first medium. These angles are designated the **angle of reflection** and **angle of incidence,** respectively.

When a beam of parallel light rays reflects off a smooth surface, each ray in the beam reflects at the same angle, and the entire beam is thus reflected without distortion (Fig 35-3b). This process is called **specular reflection.** But if the surface is rough, then individual rays, while still obeying Equation 35-1, reflect from differently oriented pieces of the surface. As a result the reflection is **diffuse,** with the original beam spreading in all directions (Fig. 35-3c). White wall paint is a good example of a diffuse reflector, while the aluminum or silver coating of a mirror is an excellent specular reflector. When we speak of reflection we generally mean specular reflection.

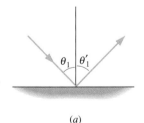

(a)

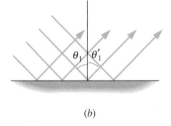

(b)

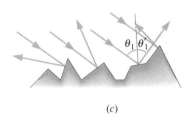

(c)

FIGURE 35-3 (a) The angles of reflection and incidence are equal. (b) In specular reflection, a smooth surface reflects a light beam undistorted. (c) A rough surface results in diffuse reflection. The law of reflection still holds for each individual ray.

● EXAMPLE 35-1 THE CORNER REFLECTOR

Two mirrors stand vertically, at right angles to each other as shown in Fig. 35-4a. Show that any light ray incident in the horizontal plane will return antiparallel to its incident direction.

Solution

Figure 35-4b shows a top view of the two mirrors. The law of reflection ensures that the two angles marked θ are equal, as are the two marked ϕ. But the mirrors make a 90° angle, and therefore $\phi = 90° - \theta$. The incident and first reflected ray make an angle 2θ, while the first reflected ray and the outgoing ray make an angle $2\phi = 180° - 2\theta$. Thus these two angles sum to 180°, which shows that the incident and outgoing rays are antiparallel. This result holds regardless of the actual value of the incidence angle θ.

Adding a third mirror at right angles to the other two allows the corner reflector to return any beam in the direction from which it came—regardless of the orientation of the beam or the reflector. Corner reflectors, often made with prisms rather than mirrors, are widely used in optics. A corner reflector left on the moon allows laser-based measurements of the moon's distance to within about 15 cm (Fig. 35-5).

EXERCISE Two mirrors make an angle of 135°. A light ray is incident on the first mirror at an angle of 60° to the mirror's normal. Find (a) the angle of incidence on the second mirror and (b) the angle between the original incident beam and the final outgoing beam.

Answers: (a) 75°; (b) 90°

Some problems similar to Example 35-1: 2, 4, 5

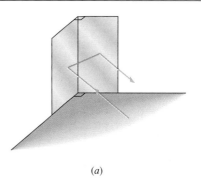

(a)

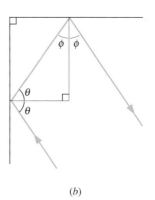

(b)

FIGURE 35-4 (a) Two-dimensional corner reflector, made from two mirrors at right angles. (b) Any ray incident in the horizontal plane returns parallel to its original direction.

FIGURE 35-5 Astronauts left this array of corner reflectors on the moon. Reflecting laser beams off the array allows determination of the moon's position to within 15 cm.

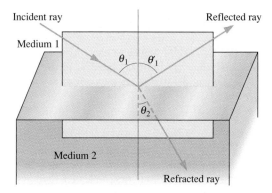

FIGURE 35-6 Reflection and refraction at an interface. This case, where the refracted ray is bent *toward* the normal, occurs when medium 2 has a higher refractive index. Angles θ_1, θ'_1, and θ_2 are, respectively, the angles of incidence, reflection, and refraction. Light-colored surface is a plane perpendicular to the interface.

35-3 REFRACTION

Figures 35-2 and 35-6 show that light passing between two transparent media is partially reflected and partially transmitted at their interface. In addition, the transmitted beam changes direction at the interface—a phenomenon known as **refraction.**

Physically, refraction occurs because the wave speeds in the two media are different. That difference, in turn, results from delays introduced by the interaction of the electromagnetic wave fields with atomic electrons. But the wave frequency f—and thus the wave period $T = 1/f$—must be the same on both sides of the interface; otherwise, on one side of the interface more wave crests would pass in a given time than on the other side, implying the creation or destruction of waves at the interface. In Fig. 35-7a waves in medium 1 travel at speed v_1, and the wavelength—the distance between wave crests—is therefore $v_1 T$. In medium 2 the speed has some lower value v_2, and the wavelength $v_2 T$ is correspondingly shorter. The shaded triangles in Fig. 35-7a are right triangles with a common hypotenuse, and the ratio of the opposite side to the length of this hypotenuse defines the sines of the angles θ_1 and θ_2. Equating expressions for the hypotenuse length (opposite side divided by sine) in terms of these two angles gives

$$\frac{v_1 T}{\sin \theta_1} = \frac{v_2 T}{\sin \theta_2},$$

or

$$\frac{\sin \theta_2}{\sin \theta_1} = \frac{v_2}{v_1}.$$

We characterize the effect of a transparent medium on light through its **index of refraction**, n, defined as the ratio of the speed c of light in vacuum to the speed v of light in the medium:

$$n = \frac{c}{v}. \qquad \text{(index of refraction)} \qquad (35\text{-}2)$$

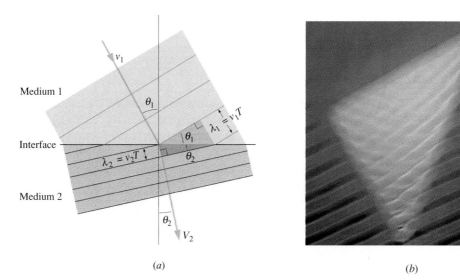

(a)

(b)

FIGURE 35-7 (a) Waves refract at an interface because wave speeds—and therefore the wavelengths—are different in the two media. Two shaded triangles have a common hypotenuse, whose length is $vT/\sin\theta$, where v and θ are taken on either side of the interface. (b) Water waves refract as they pass over a triangular patch of shallower water. Note the decreased wavelength and different orientations of the waves within the shallow region. Reflection at the interfaces has resulted in waves propagating in several directions.

(Again there's an electromagnetic connection: the index of refraction is just the square root of the dielectric constant that we introduced in Chapter 26. But don't try to calculate refractive indices from the dielectric constants of Table 26-1; the dielectric constant is frequency dependent, and its value at optical frequencies is significantly different from the DC values of Table 26-1.) Table 35-1 lists some indices of refraction. Note that for gases the index of refraction is essentially 1, the same as that of vacuum.

▲ **TABLE 35-1** INDICES OF REFRACTION*

SUBSTANCE	INDEX OF REFRACTION, n
Gases	
Air	1.000293
Carbon dioxide	1.00045
Liquids	
Water	1.333
Ethyl alcohol	1.361
Glycerine	1.473
Benzene	1.501
Diiodomethane	1.738
Solids	
Ice (H_2O)	1.309
Fused quartz (SiO_2)	1.458
Polystyrene	1.49
Glass (crown)	1.52
Sodium chloride (NaCl)	1.544
Glass (flint)	1.6–1.9
Diamond (C)	2.419
Rutile (TiO_2)	2.62

*At 1 atm pressure and temperatures ranging from 0°C to 20°C, measured at a wavelength of 589 nm (the yellow line of sodium).

Using the definition 35-2 in our ratio of the sines of the angles of refraction and incidence, and then cross multiplying, gives **Snell's law:**

$$n_1 \sin\theta_1 = n_2 \sin\theta_2. \quad \text{(Snell's law)} \quad (35\text{-}3)$$

This law was developed geometrically in 1621 by Willebrord van Roijen Snell of the Netherlands and described analytically in the 1630s by René Descartes of France, where it is known to this day as Descartes' law. It allows us to predict what will happen to light at an interface, provided we know the refractive indices of the two media.

● **EXAMPLE 35-2** CD MUSIC

The laser beam that "reads" information from a compact disc is 0.737 mm wide at the point where it strikes the underside of the disc and forms a converging cone with half-angle 27°, as shown in Fig. 35-8. It then travels through a 1.2-mm-thick layer of transparent plastic with refractive index 1.55 before reaching the very thin, reflective information layer near the disc's top surface. What is the beam diameter (d in Fig. 35-8) at the information layer?

Solution
Figure 35-8 shows that $d = D - 2x$, where $D = 0.737$ mm is the beam diameter as it hits the disc. From the figure we also see that $x = t \tan\theta_2$, where $t = 1.2$ mm is the thickness of the plastic. Finally, Snell's law gives $\theta_2 = \sin^{-1}\left(\dfrac{\sin\theta_1}{n}\right)$, where we set the refractive index of air to 1. Putting these relations together gives

$$d = D - 2t \tan\left[\sin^{-1}\left(\frac{\sin\theta_1}{n}\right)\right]$$

$$= 737 \ \mu\text{m} - (2)(1200 \ \mu\text{m}) \tan\left[\sin^{-1}\left(\frac{\sin 27°}{1.55}\right)\right]$$

$$= 1.8 \ \mu\text{m},$$

which is a little larger than the "pits" cut into the CD to store its information. This narrowing of the beam plays a crucial role in keeping CDs noise free. The tiniest dust speck would blot out information at the μm-scale information layer, but at the point where the laser beam actually enters the disc—the closest dust can get to the information layer—it would take mm-size dust to cause problems.

EXERCISE Figure 35-9 shows a polystyrene cylinder whose height is equal to its diameter. What is the maximum incidence angle θ_1 at which a light ray striking the top center of the cylinder will emerge through the bottom without first striking the side? The medium surrounding the cylinder is air.

Answer: 42°

Some problems similar to Example 35-2: 14, 15, 17, 18

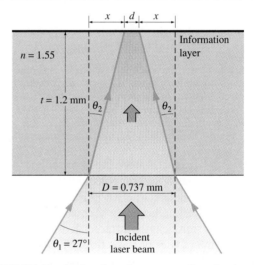

FIGURE 35-8 Section through a compact disc, showing convergence of the laser beam to a narrow spot at the information layer. All CDs share a common refractive index of 1.55 (Example 35-2).

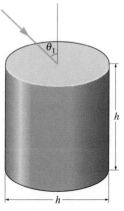

FIGURE 35-9 What is the maximum θ_1 for which the beam will emerge through the bottom of the polystyrene cylinder?

Snell's law applies whether light goes from a medium of lower to higher refractive index or the reverse, as you can see by reversing the path of the light in Fig. 35-7. In the former case the refracted ray bends *toward* the normal, while in the latter case it bends *away* from the normal.

● **EXAMPLE 35-3** IN AND OUT: PARALLEL RAYS

In Fig. 35-10, a light ray propagating in air strikes a transparent slab of thickness d and refractive index n with incidence angle θ_1. Show that the ray emerges from the slab propagating parallel to its original direction.

Solution
Applying Snell's law to the upper interface gives

$$\sin\theta_2 = \frac{\sin\theta_1}{n},$$

where we've taken $n_1 = 1$ for air and $n_2 = n$ for the slab. At the lower interface, with the light going from the slab to air, we have $n_1 = n$ and $n_2 = 1$, so Snell's law gives

$$\sin\theta_4 = n\sin\theta_3.$$

But the slab faces are parallel so, as the figure suggests, $\theta_3 = \theta_2$. Combining our two versions of Snell's law then gives

$$\sin\theta_4 = n\left(\frac{\sin\theta_1}{n}\right) = \sin\theta_1,$$

showing that the incident and outgoing rays are indeed parallel. They are, however, displaced by the distance x shown in the figure. You can find that displacement in Problem 60.

EXERCISE A piece of crown glass is immersed in di-iodomethane. A light ray strikes the glass with incidence angle

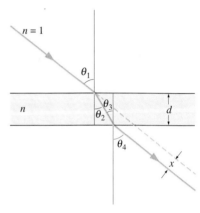

FIGURE 35-10 A light beam incident on a transparent slab emerges with its original direction, but is displaced from its original path.

40°. Find the angle the refracted beam in the glass makes with the normal to the interface.

Answer: 47.3°

Some problems similar to Example 35-3: 11, 13, 60

●

We derived Snell's law using Fig. 35-7*a*, which shows the wavelength change that occurs as the wave speed changes between two media. We argued that the frequencies in the two media must be the same, and since wavelength and frequency are related by $\lambda f = v$, Equation 35-2 then shows that the wavelengths in the two media are related by

$$\frac{\lambda_2}{\lambda_1} = \frac{n_1}{n_2}. \qquad (35\text{-}4)$$

● **EXAMPLE 35-4** LIGHT AND DIAMOND

Light with wavelength 589 nm in vacuum enters a diamond. Find the light speed and wavelength in the diamond.

Solution
Equation 35-2 gives

$$v = \frac{c}{n} = \frac{3.00\times10^8 \text{ m/s}}{2.419} = 1.24\times10^8 \text{ m/s}.$$

The refractive index of vacuum is 1, so Equation 35-4 gives

$$\lambda_{\text{diamond}} = \frac{\lambda_{\text{vacuum}}}{n} = \frac{589 \text{ nm}}{2.419} = 243 \text{ nm}.$$

The wavelength and speed are reduced by the same factor, as they must be to keep the wave frequency unchanged.

EXERCISE Microwaves propagate through glass at about 1.4×10^8 m/s. What is the refractive index of glass at microwave frequencies?

Answer: 2.14

Some problems similar to Example 35-4: 8, 9, 19

●

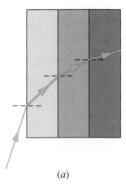

(a)

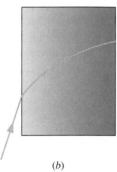

(b)

FIGURE 35-11 (a) Light propagating through a series of slabs with increasing refractive indices. At each interface the light bends more toward the normal. (b) Light follows a curved path in a medium with a continuously increasing refractive index.

Multiple Refractions

Engineered optical systems often use several layers of refractive material to minimize reflective losses and certain types of distortion. To describe the path of a light ray in such a system, we need only apply Snell's law at each of the interfaces using the appropriate pair of refractive indices (Fig. 35-11a). When the layers are parallel the angular deflection of a light ray is the same as if it had gone through a single interface from the first medium to the last (see Problem 54). In many natural situations, including the human eye and Earth's atmosphere, the index of refraction is a continuously varying function of position (Fig. 35-11b). We can approximate such a case as a sequence of thin layers, each with a different refractive index. Going toward the limit of infinitely many infinitesimal layers gets us arbitrarily close to the exact solution, in which the light follows a curved path. The mirage shown in Fig. 35-12 results from the fact that the refractive index of air is temperature dependent, causing the path of light rays to bend continuously.

(a)

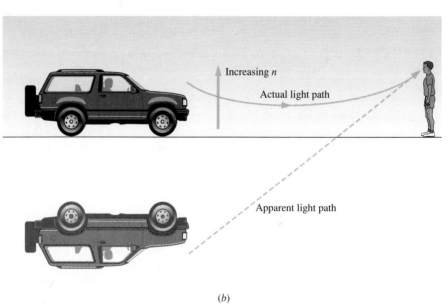

(b)

FIGURE 35-12 (a) Mirages on a hot road. (b) The mirage occurs because hot air near the highway surface has a lower refractive index, resulting in a curved path for the light. The apparent position of the vehicle is not its actual position. Can you see why it appears upside down?

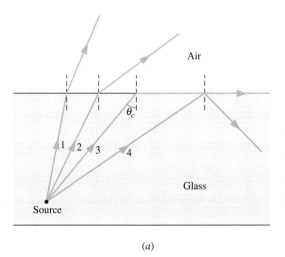

(a)

(b)

FIGURE 35-13 (a) Light propagating inside a glass block is refracted away from the normal at the glass-air interface. Ray 3, incident at the critical angle, just skims along the interface. At higher incidence angles, as with ray 4, the light undergoes total internal reflection at the interface. Not shown are weak reflected rays at angles less than the critical angle. (b) Three light beams strike a lens and are refracted at the air-glass interface. The blue beam strikes the subsequent glass-air interface nearly perpendicularly, and essentially all of it is transmitted. The yellow beam strikes more obliquely but is still largely transmitted. The red beam strikes at an incidence angle greater than the critical angle and undergoes total internal reflection.

35-4 TOTAL INTERNAL REFLECTION

Figure 35-13 shows light propagating inside a glass block and striking the interface with the surrounding air. Since air's refractive index is lower than that of glass, rays are bent *away* from the normal as they leave the glass. As the incidence angle increases, so does the angle of refraction—and Snell's law shows that the latter is always greater than the incidence angle. So at some incidence angle (see ray 3 in Fig. 35-13) the angle of refraction reaches 90°. What then?

If the incidence angle is increased further, we find that the light is *totally* reflected at the interface. (There is always *some* reflection back into the glass, but now it's *all* reflected.) This phenomenon is called **total internal reflection,** and the incidence angle at which it first occurs is the **critical angle,** θ_c. We can find θ_c by setting $\theta_2 = 90°$ (i.e., $\sin\theta_2 = 1$) in Snell's law (Equation 35-3). The critical angle is then θ_1, and we have

$$\sin\theta_c = \frac{n_2}{n_1}. \qquad (35\text{-}5)$$

Since the sine of an angle cannot exceed 1, we must have $n_2 \leq n_1$ in order for this equation to have a solution. Thus, total internal reflection occurs only when light propagating in one medium strikes an interface with a medium of lower refractive index, and it occurs whenever the incidence angle exceeds the critical angle.

Total internal reflection makes uncoated glass an excellent reflector when it's oriented appropriately (Fig. 35-14). Corner reflectors (Example 35-1) actually use total internal reflection in solid cubes of glass, rather than individual mirrors at right angles. For an observer inside a medium of higher refractive index, the existence of the critical angle affects the view of the outside world, as the example below shows. Finally, total internal reflection is the basis of the optical fibers now widely used in communications.

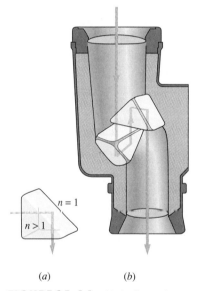

(a)   (b)

FIGURE 35-14 (a) A glass prism redirects light through total internal reflection. (b) Binoculars use a pair of prisms to "fold" the light path, allowing a more compact design.

● EXAMPLE 35-5 WHALE WATCH

Planeloads of whale watchers fly above the ocean, as shown in Fig. 35-15. A whale looks upward, watching the planes. Within what range of viewing angles can the whale see the planes?

Solution

If the whale emitted light, that light would emerge from the ocean surface only for incidence angles less than the critical angle—that is, those angles within the cone shown in Fig. 35-15. Since the path of light is reversible, the whale can only see objects above the surface when it looks within this cone. For the water-air interface Equation 35-5 gives

$$\theta_c = \sin^{-1}\left(\frac{1}{1.333}\right) = 48.6°,$$

where we found water's refractive index in Table 35-1. The geometry of Fig. 35-15 then shows that this is also the half-angle of the cone, so the whale must look within 48.6° of the vertical to see out.

Can the whale see a plane only if it's *actually* within this angle? No. As the strongly refracted ray from the right-hand plane suggests, the entire outside world appears to the whale compressed into a cone of half-angle θ_c. If the whale looks beyond this cone, it will see instead reflections of objects below the surface.

EXERCISE A diamond is submerged in water. What is the critical angle at the diamond-water interface?

Answer: 33.4°

Some problems similar to Example 35-5: 22, 23, 26, 28, 33

FIGURE 35-15 The whale sees the entire world above the surface in a cone of half-angle θ_c. Looking beyond this cone, it sees reflections of objects below the surface.

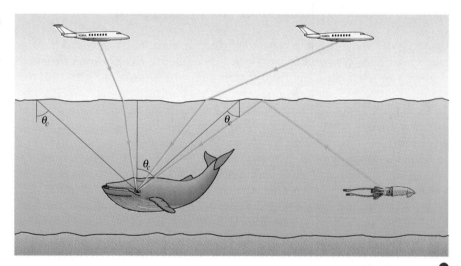

■ APPLICATION OPTICAL FIBERS AND LIGHTWAVE COMMUNICATION

Refraction and total internal reflection are the basis for **optical fibers,** which revolutionized communications in the 1980s. Some three *billion* km of optical fiber were installed in the United States alone during that decade, and today they carry telephone conversations, television signals, computer data, and other information. Undersea fibers link the entire planet in a fiber-optic network.

A typical fiber consists of a glass core only 8 μm in diameter, surrounded by a so-called cladding consisting of glass with a lower refractive index than the core. The propagation of light in the fiber is a process whereby the core-cladding interface guides the light along the fiber. In some fibers this takes place by abrupt internal reflection at the interface (Fig. 35-16), while in others gradual refraction in the cladding guides the light. The glass used in optical fibers is so pure that a 1-km thick slab would appear as transparent as an ordinary window pane. Today's fibers use infrared light at 1.31 μm wavelength, although the anticipated development of 1.55-μm systems should result in lower losses. And fibers are under study that actually regenerate signals as they travel, extracting energy from an additional light beam.

Optical fibers offer several advantages over copper wire or open-air transmission of electromagnetic waves. The main advantage is their very high **bandwidth**—the rate at which they

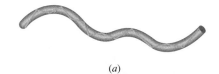

(a)

(b)

FIGURE 35-16 (*a*) A beam of light undergoes a series of total internal reflections that guide it along an optical fiber. (*b*) A bundle of actual fibers.

FIGURE 35-17 The thin fiber-optic cable and the much bulkier cable of copper wires carry information at the same rate.

can carry information. This bandwidth arises because the information carried on fiber is encoded on infrared radiation with a frequency on the order of 10^{14} Hz—much higher than the microwave frequencies used in conventional communication systems. A single fiber, for example, can carry tens of thousands of telephone conversations (Fig. 35-17).

Fibers are lighter and more rugged than copper cables and are less easy to "tap" illicitly. Because they are insulators, optical fibers are also less susceptible to electrical "noise" and are therefore used to carry information in electrically noisy environments like power plants or high-energy physics laboratories.

Optical fibers play a key role in the network that today links nearly all the world's computers. Computers on one floor or within a building may be connected in a small, relatively low-bandwidth network with copper wire. But connections between buildings are nearly always made with optical fibers, linking individual local networks into larger institution-wide networks whose optical links can transfer large amounts of information with reasonable speed. Entire institutions—corporations, universities, government laboratories, and the like—are then linked by very high speed fiber-optic "superhighways" that carry enormous loads of information "traffic."

Optical fiber technology requires more than just fiber: appropriate opto-electronic devices are needed at each end to convert information from light to electrical forms and vice versa. Such devices include miniature lasers made from semiconductors, and high-speed light-sensitive diodes. As the cost of these devices drops, high-speed optical fiber will soon connect to individual homes and offices, allowing nearly instant access to telecommunications, video resources, libraries, and databases.

35-5 DISPERSION

Refraction ultimately occurs because of the interaction of electromagnetic wave fields with atomic electrons. Although the details of that interaction require a quantum description, it should not be surprising that the index of refraction depends on frequency. After all, as we found in Chapter 15, the response of a system (here an atom) generally depends on the frequency of the driving force (here the oscillating fields of a wave).

Because the index of refraction depends on frequency, different frequencies—different colors for visible light—will be refracted through different angles. A light beam containing many colors will therefore be spread into its constituent colors, a process called **dispersion** (Fig. 35-18). The classic example of dispersion is Newton's demonstration that white light is really a mixture of all the colors in the visible spectrum. Newton not only broke white light into its

FIGURE 35-18 Dispersion of white light by a glass prism. The refractive index of the glass is greater at higher frequencies—shorter wavelengths—and therefore results in greater refraction at the blue end of the spectrum. The incident beam is at bottom left, and a white reflected beam leaves the prism going straight downward. This beam has undergone an additional refraction that has again combined its colors.

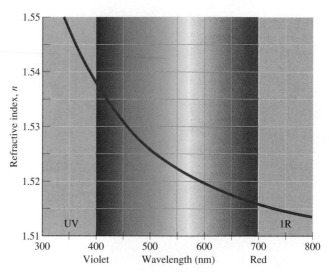

FIGURE 35-19 Index of refraction as a function of wavelength for high-dispersion crown glass.

constituent colors, as in Fig. 35-18, but he also recombined the colors to produce the original white light.

In most materials in most frequency ranges, the index of refraction increases with increasing frequency and therefore decreases with increasing wavelength (Fig. 35-19). That means colors toward the violet end of the spectrum are refracted through the greatest angles, as is evident in Fig. 35-18.

Dispersion is the basis for the widely used technique of **spectral analysis** or **spectroscopy,** in which substances or processes are characterized by analyzing the frequencies of electromagnetic radiation they emit, transmit, or absorb. Hot, dense objects, for example, emit a continuous band of radiation, while diffuse gases radiate at only a few specific wavelengths (Fig. 35-20). The existence of those discrete wavelengths provided some of the strongest evidence for the nature of the atom, and today spectral analysis allows astronomers to identify

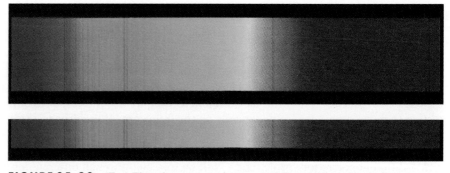

FIGURE 35-20 (Top) The solar spectrum is an essentially continuous band of wavelengths, produced by the hot, dense gases of the Sun's visible surface. Dark lines are discrete wavelengths absorbed by overlying gases. (Bottom) Spectrum of a diffuse gas—in this case hydrogen—consists of light at discrete wavelengths. The pattern of lines allows identification of the emitting material.

and measure the abundances of elements in distant astrophysical objects. Geologists use spectral analysis to identify minerals, and chemists use infrared spectra to study molecules. Spectral analysis is a powerful tool in nearly every branch of science. Although early spectroscopes used prisms, most modern instruments use instead a device called a diffraction grating, whose operation we describe in Chapter 37.

Dispersion can be a nuisance in optical systems. Glass lenses, for example, focus different colors at different points, resulting in a distortion known as chromatic aberration. This effect can be minimized by making composite lenses of materials with different refractive indices.

● EXAMPLE 35-6 HOW MUCH DISPERSION?

White light strikes the prism in Fig. 35-21 normal to one surface. The prism is made of the glass whose refractive index is plotted in Fig. 35-19. Find the angle between outgoing red (700 nm) and violet (400 nm) light.

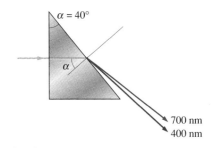

FIGURE 35-21 Example 35-6.

Solution
There is no refraction at the air-glass interface since the incident ray is normal to the surface. At the second interface medium 2 is air with $n_2 = 1$, and medium 1 is the glass whose indices at the two wavelengths we can read from Fig. 35-19: $n_{400} = 1.538, n_{700} = 1.516$. The geometry of Fig. 35-21 shows that the incidence angle at the second interface is the angle $\alpha = 40°$ at the top of the prism. The angles of refraction are then given by solving Snell's law for the two values of θ_2:

$$\theta_{400} = \sin^{-1}(n_{400} \sin \alpha) = \sin^{-1}[(1.538)(\sin 40°)] = 81.34°,$$

and

$$\theta_{700} = \sin^{-1}(n_{700} \sin \alpha) = \sin^{-1}[(1.516)(\sin 40°)] = 77.02°.$$

The angle between the two outgoing beams is therefore $\theta_{400} - \theta_{700} = 4.32°$, with the violet beam (400 nm) experiencing the greatest deflection. This 4.32° spread is called the *angular dispersion* of the beam.

EXERCISE If the angle α in Fig. 35-21 is increased, what is the maximum angular dispersion that can be achieved by refracting both the 400 nm and 700 nm beams? At what α does this occur?

Answer: $\Delta\theta = 9.7°$ at $\alpha = \sin^{-1}(1/n_{400}) = 40.556°$

Some problems similar to Example 35-6: 34–38 ●

■ APPLICATION THE RAINBOW

Nature provides a beautiful application of dispersion and internal reflection in the rainbow (Fig. 35-22), which occurs when sunlight strikes rain or other water droplets in the air. An observer standing between the Sun and the rain then sees the circular arc of colored bands. Figure 35-23 shows that the center of that arc lies on the line joining the Sun to the observer's eye. That means each observer sees a different rainbow! Furthermore, the rainbow's arc always subtends an angle of approximately 42°. How does the rainbow form, and why does it have this geometry?

Theories of the rainbow date back many centuries. By 1635 Descartes had produced a nearly complete explanation of the

FIGURE 35-22 Reflection, refraction, and dispersion all act to produce a rainbow.

rainbow's shape and apparent location, but because he did not know about dispersion he could not account for the colors. Some years later Newton, in his *Optics,* produced a full explanation.

Figure 35-24 shows a light ray passing through a spherical raindrop. The incidence angle θ in Fig. 35-24 is arbitrary, and parallel light rays striking the curved surface of the drop will experience a range of values for θ. There will, therefore, be a range of angles ϕ between the incident and outgoing rays. As Fig. 35-25 shows, however, there is a maximum angle* ϕ_{max} of about 42° and more light is returned at angles close to ϕ_{max} than at other angles. That is why a bright band—the rainbow—appears in the sky at an angle of about 42° to the direction of the Sun's rays. Problems 64 and 65 show how to find ϕ_{max}.

The "bunching" of light rays near ϕ_{max} explains why a bright band should appear, but why should it be colored? Because the refractive index varies with wavelength, so does ϕ_{max}. Each color will therefore appear brightest at a slightly different angle. For water, the refractive index in the visible region ranges from $n_{red} = 1.330$ to $n_{violet} = 1.342$. Using these values with the results of Problems 64 and 65 yields $\phi_{red} = 42.53°$ and $\phi_{violet} = 40.78°$. Thus the rainbow is seen as a band, itself subtending an angle of about 1.75°, with red at the top.

One occasionally sees a fainter and larger arc above the primary rainbow. This secondary rainbow results from two internal reflections, and as a result the order of its colors is reversed. Problem 66 explores the secondary rainbow.

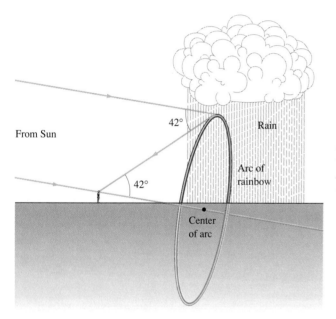

FIGURE 35-23 The rainbow is a circular arc located at 42° from the line that includes the Sun, the observer, and the center of the arc. The Sun is so far away that its rays are essentially parallel. Here part of the rainbow is blocked by the ground; however, observers in aircraft can sometimes see the entire circle.

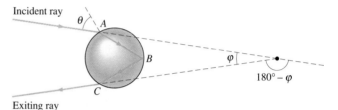

FIGURE 35-24 A light ray passing through a spherical raindrop undergoes refraction at each interface and internal reflection at the back of the drop. (Not shown are the reflected rays at interfaces A and C and the transmitted ray at interface B.) The ray undergoes an overall deflection of $180° - \phi$.

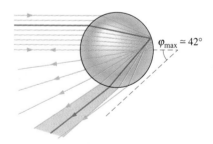

FIGURE 35-25 Parallel rays striking a water drop. The minimum angle through which a ray is deflected is about 138°, corresponding to a maximum of about 42° in the angle ϕ of Fig. 35-24. Furthermore, rays tend to "bunch" at this angle, as suggested by the broad outgoing beam defined by three nearly parallel rays. This "bunching" causes an observer to see a bright arc at 42° from the antisolar direction in Fig. 35-23. The thicker ray undergoes the minimum deviation and has, correspondingly, the maximum ϕ.

---

* This angle actually corresponds to a *minimum* deflection in which the ray is turned through an angle of $180° - 42° = 138°$. Other rays suffer more deflection; one hitting normal to the drop surface, for example, is turned through 180°, corresponding to $\phi = 0$.

35-6 REFLECTION COEFFICIENTS AND THE POLARIZING ANGLE

The laws of geometrical optics describe the paths of light rays at interfaces between different media, but they cannot tell how much of the light is reflected and how much transmitted at an interface. But application of Ampère's and Faraday's laws to electromagnetic wave fields at an interface does give that information. Such a calculation is beyond the scope of this book; here we look briefly at the results.

Normal Incidence

When light is incident normally on an interface, application of Ampère's and Faraday's laws yields a fairly simple result:

$$R = \left(\frac{n_2 - n_1}{n_2 + n_1}\right)^2, \tag{35-6}$$

where n_1 and n_2 are the refractive indices of the two media, and the **reflection coefficient** R is the fraction of the incident intensity that is reflected back into the medium of incidence. Since the difference $n_2 - n_1$ appears squared in Equation 35-6, the fraction of light reflected is the same for light crossing the interface in either direction. Energy conservation requires that light not reflected must be transmitted into the second medium, so the **transmission coefficient** is therefore

$$T = 1 - R = \frac{4n_1 n_2}{(n_1 + n_2)^2}. \tag{35-7}$$

Note that these results give $R = 0$ and $T = 1$ for $n_1 = n_2$, as we would expect for this case when there is really no interface. Equations 35-6 and 35-7 are closely related to Equations 17-7 and 17-8 for the reflection of waves on strings, although there we deal with amplitudes and here with intensities.

● **EXAMPLE 35-7** A TRIPLE-GLAZED WINDOW

In an effort to cut heat loss, a homeowner replaces single-glazed windows with triple-glazed windows. Both use the same type of glass, whose refractive index is 1.5. Unfortunately the new windows not only conserve heat, but they also let in less light. Find the reduction in intensity for light incident normally on the window as a result of the change from single to triple glazing.

Solution

Equation 35-7 shows that the fraction of light transmitted at a single air-glass or glass-air interface is

$$T = \frac{4n_1 n_2}{(n_1 + n_2)^2} = \frac{(4)(1)(1.5)}{(1 + 1.5)^2} = 0.96.$$

A single sheet of glass has two such interfaces, each cutting the intensity by this factor 0.96. So the transmitted intensity is down by a factor $0.96^2 = 0.922$ from its incident value. A triple-glazed window, on the other hand, has a total of six interfaces, cutting the intensity to $0.96^6 = 0.783$ of its incident value. Comparing the two results gives an intensity ratio of $0.783/0.992 = 0.79$ for the triple as compared with the single glazing.

Our calculation here is approximate for a subtle reason. Both reflection and transmission occur at each interface. Of the light entering a slab of glass, some is reflected back into the glass at the exit side, and of that a small fraction is reflected again at the first side, and most of that eventually joins the first exiting beam. "Higher order" effects—associated with more reflections—may also affect the outgoing beam. In our window example these effects are small. But by carefully matching the thicknesses of transparent slabs to appropriate multiples of the light wavelength, it is possible to use wave interference to enhance or diminish the transmitted or reflected intensity. Antireflection coatings on lenses use this effect, which we will study in Chapter 37.

EXERCISE A light beam propagating through air is incident normally on a transparent material, and 17.2% of the incident intensity is reflected. Find the refractive index and use Table 35-1 to identify the material.

Answer: 2.42, diamond

Some problems similar to Example 35-7: 40–42

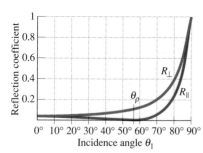

FIGURE 35-26 Reflection coefficients as a function of incidence angle for the two cases of light polarized with its electric field parallel and perpendicular to the plane defined by the incident and reflected rays, for an interface between air and glass with $n = 1.5$. At 56° incidence the parallel polarization is entirely transmitted ($R_\parallel = 0$), so an unpolarized incident beam will become polarized on reflection.

FIGURE 35-27 Reflection in a plate-glass window is especially prominent at oblique incidence angles.

Oblique Incidence

It should be no surprise that the reflection and transmission coefficients for light incident on an interface at an oblique angle depend on that angle. Less obvious is the fact that they also depend on the polarization of the light. The equations for these coefficients are quite complicated, and we give only their graphical representation (Fig. 35-26). The graph shows two curves for the reflection coefficients $R_\perp$ and $R_\parallel$ at an air-glass interface. These correspond to polarizations with the wave electric field perpendicular and parallel, respectively, to the plane of Fig. 35-6. Light of arbitrary polarization can be expressed as a combination of these two.

At normal incidence ($\theta_1 = 0$), both curves start from the value $R = 0.04$ implied in our calculation giving $T = 0.96$ in Example 35-7. At large incidence angles both coefficients approach 1. You've undoubtedly experienced this effect when standing in front of a large plate-glass window; as you look along the window at an oblique angle, the reflection of objects on your side of the window becomes more prominent (Fig. 35-27).

The Polarizing Angle

Figure 35-26 shows that light with the perpendicular polarization reflects more strongly than does light with the parallel polarization, except at normal and grazing ($\theta_1 = 90°$) incidence. That means unpolarized light reflecting from the surface will come off partially polarized. At one particular angle, marked θ_p in Fig. 35-26, the parallel polarization does not reflect at all. Unpolarized light incident at this special angle—called the **polarizing angle** or the **Brewster angle**—comes off entirely polarized. Orienting a piece of glass or other transparent material at its polarizing angle therefore offers one way of producing light polarized perpendicular to the plane of the incident and reflected rays. Alternatively, successive transmissions at the polarizing angle can enhance the parallel component of polarization.

The reason there is no reflection for the parallel polarization at a particular angle follows from the fact that electromagnetic radiation ultimately arises from accelerated charges and that, as we saw in Fig. 34-23, there is no radiation in the direction of the acceleration. Figure 35-28 shows what happens when light with parallel polarization is incident at the polarizing angle, 56.3° in glass with $n = 1.5$. Snell's law gives an angle of refraction $\theta_2 = \sin^{-1}(\sin\theta_p/n) = 33.7°$. Now $56.3° + 33.7° = 90°$ and that means, as Fig. 35-28 shows, that any reflected ray would be perpendicular to the refracted ray. But the reflected ray really arises from the acceleration of the electrons in the glass—acceleration which, according to Newton's law of motion, is in the direction of the driving force. But here the driving force comes from the wave's electric field, and with polarization in the plane of Fig. 35-28, that field is itself parallel to the direction the reflected ray would take. But we've seen that there is no radiation in the direction of the acceleration, and hence for this special geometrical condition there is no reflected ray.

Figure 35-28 shows that, quite generally, this condition of no reflection of the parallel polarization will be met when the angles of incidence and refraction sum to 90°: $\theta_p + \theta_2 = 90°$, where we've set the incidence angle equal to the polarizing angle. This, in turn, implies that $\sin\theta_2 = \cos\theta_p$. But with the incidence angle $\theta_1 = \theta_p$, Snell's law gives $\sin\theta_2 = (n_1/n_2)\sin\theta_p$. Applying our condition $\sin\theta_2 = \cos\theta_p$, we then have $\cos\theta_p = (n_1/n_2)\sin\theta_p$. Multiplying both sides by $n_2/n_1 \cos\theta_p$ then gives

$$\tan\theta_p = \frac{n_2}{n_1}. \tag{35-8}$$

For the air-glass interface whose reflection coefficients are plotted in Fig. 35-26, with $n_1 = 1$ and $n_2 = 1.5$, this equation indeed gives $\theta_p = 56.3°$.

A phenomenon similar to the polarizing angle holds for reflection from metals and other surfaces, although here the parallel polarization is minimized but not completely eliminated. Reduction of glare using Polaroid sunglasses works by eliminating the polarized light reflected off surfaces (Fig. 35-29).

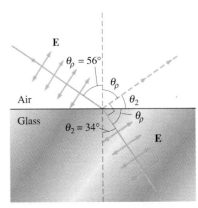

FIGURE 35-28 An electromagnetic wave polarized in the plane of the figure is incident on an interface. When the angles of incidence and refraction sum to 90°, the reflected ray would be in the direction of the electric field in the lower medium. Since there is no radiation in the direction of charged-particle acceleration, there is no reflected ray in this special case. Numbers shown are for an air-glass interface.

(a)   (b)

FIGURE 35-29 Identical views of a storefront, photographed (a) without and (b) with a polarizing filter. Light reflecting obliquely from the glass is partially polarized. The polarizing filter eliminates this component, allowing the inside of the store to show much more clearly.

CHAPTER SYNOPSIS

Summary

1. **Geometrical optics** treats the behavior of light under the approximation that it travels in straight lines—**rays**—except at interfaces between different materials. This approximation is valid whenever waves interact with systems that are much larger than their wavelength.

2. **Reflection** occurs to some extent at nearly all interfaces. When a light ray reflects, the **angle of incidence** and **angle of reflection**—both measured from the normal to the interface—are equal. **Specular reflection** occurs from smooth surfaces, while rough surfaces produce **diffuse reflection.**

3. **Refraction** is the bending of light at an interface between transparent materials. The directions of the incident and refracted rays are related by **Snell's law:**

$$n_1 \sin\theta_1 = n_2 \sin\theta_2,$$

where θ_1 and θ_2 are the angles of incidence and refraction, respectively. The **indices of refraction** are n_1 and n_2, defined as the ratios of the speed of light in vacuum to the speed in a given medium.

4. **Total internal reflection** occurs when light propagating in a medium with refractive index n_1 is incident on an interface at an angle greater than the **critical angle** θ_c given by

$$\sin\theta_c = \frac{n_2}{n_1}.$$

Total internal reflection is possible only when $n_2 < n_1$.

5. The index of refraction generally depends on frequency. This results in **dispersion** of the different colors as they refract through different angles. Dispersion is the basis of **spectroscopy.**

6. For normal incidence, the fraction of light intensity reflected at an interface is given by the **reflection coefficient:**

$$R = \left(\frac{n_2 - n_1}{n_2 + n_1}\right)^2.$$

The **transmission coefficient** is $T = 1 - R$. For oblique incidence the reflection and transmission coefficients depend on both the incidence angle and the polarization. When the angle of incidence is equal to the **polarizing angle** (also called the **Brewster angle**) there is no reflection of light polarized in the plane of the incident and refracted rays.

Terms You Should Understand

(Pairs are closely related terms whose distinction is important; number in parentheses is chapter section where term first appears.)

geometrical optics (introduction)
ray (introduction)
angles of incidence, reflection, refraction (35-2, 35-3)
specular reflection, diffuse reflection (35-2)
refraction (35-3)
index of refraction (35-3)
Snell's law (35-3)
total internal reflection (35-4)
critical angle (35-4)
optical fiber (35-4)
dispersion (35-5)
spectroscopy (35-5)
reflection coefficient, transmission coefficient (35-6)
polarizing angle (35-6)

Symbols You Should Recognize

θ_1, θ_1', θ_2 (35-2, 35-3)
n (35-3)
θ_c (35-4)
θ_p (35-6)

Problems You Should Be Able to Solve

determining the directions of reflected and refracted rays at interfaces (35-2 and 35-3)
evaluating conditions for total internal reflection (35-4)
analyzing dispersion (35-5)
determining reflected and transmitted intensities at normal incidence (35-6)
evaluating the polarizing angle (35-6)

Limitations to Keep in Mind

Geometrical optics is an approximation valid only when light interacts with systems much larger than its wavelength.
Characterizing the optical properties of a material by a single number, the index of refraction, is a simplification. In many materials optical properties depend on direction, and include absorption as well as transmission.

QUESTIONS

1. Are light rays real? Discuss.
2. It's usually inappropriate to consider low-frequency sound waves as traveling in rays. Why? Why is it more appropriate for high-frequency sound and for light?
3. Describe why a spoon appears bent when it's in a glass of water.
4. Why do a diamond and an identically shaped piece of glass sparkle differently?
5. Specular reflection occurs with "smooth" surfaces. But on the microscopic scale all surfaces are rough. What do you suppose should be the criterion for "smoothness" in dealing with reflection?
6. White light goes from air through a glass slab with parallel surfaces. Will its colors be dispersed when it emerges from the glass?
7. Would light behave as in Fig. 35-14a if the refractive index of the prism were 1.25? Explain.
8. You send white light through two identical glass prisms, oriented as shown in Fig. 35-30. Describe the beam that emerges from the right-hand prism.

White light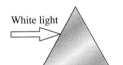

FIGURE 35-30   Question 8.

9. Why can optical fibers carry much more information than copper wires?
10. What would happen if you scratched the outside of the reflecting surface of a prism used for total internal reflection?
11. Lightning produces a sudden burst of static in a nearby radio receiver. The static comprises a broad band of radio frequencies. But in a very distant receiver the "noise" from the flash arrives over an extended time. What does this say about the refractive index of Earth's atmosphere for radio waves?
12. What does a fish see as it looks around in directions above the horizontal? Explain.
13. In glass, which end of the visible spectrum has the lowest critical angle for total internal reflection?
14. Looking out the window of a lighted room at night, you see clear reflections of the room's interior. In the daytime those reflections would be much less obvious, yet the reflection coefficient has not changed. Explain.
15. Why can't you walk to the end of the rainbow?
16. What is wrong with the painting in Fig. 35-31? *Hint:* The rainbow subtends a half-angle of 42°.

FIGURE 35-31   (What's wrong with this painting (*Niapra*, by Harry Fenn)? (Question 16.)

17. Suppose the refractive index of water were not frequency dependent. Would anything like the rainbow occur?
18. Figure 35-32 shows a ball inside a transparent sphere inside an aquarium tank. The transparent sphere and the tank can each be filled with water. Explain which combination is occurring in each frame.

FIGURE 35-32   Question 18.

19. Why are polarizing sunglasses better than glasses that simply cut down on the total amount of light?
20. Does the transmitted intensity always exceed the reflected intensity at an air-glass interface?
21. Under what conditions will the polarizing angle be less than 45°?

PROBLEMS

Section 35-2 Reflection

1. Through what angle should you rotate a mirror in order that a reflected ray rotate through 30°?
2. The mirrors in Fig. 35-33 make a 60° angle. A light ray enters parallel to the symmetry axis, as shown. (a) How many reflections does it make? (b) Where and in what direction does it exit the mirror system?

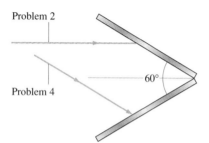

FIGURE 35-33 Problems 2, 4, 5.

3. To what angular accuracy must two ostensibly perpendicular mirrors be aligned in order that an incident ray returns on a path within 1° of its incident direction?
4. If a light ray enters the mirror system of Fig. 35-33 propagating in the plane of the page and parallel to one mirror, through what angle will it be turned?
5. Suppose the angle in Fig. 35-33 is changed to 75°. A ray enters the mirror system parallel to the axis. (a) How many reflections does it make? (b) Through what angle is it turned when it exits the system?
6. Two plane mirrors make an angle ϕ. For what value of ϕ will a light ray reflecting once off each mirror be turned through the same angle ϕ?
7. Two plane mirrors make an angle ϕ. A light ray enters the system and is reflected once off each mirror. Show that the ray is turned through an angle $360° - 2\phi$.

Section 35-3 Refraction

8. In which substance in Table 35-1 does the speed of light have the value 2.292×10^8 m/s?
9. Information in a compact disc is stored in "pits" whose depth is essentially one-fourth of the wavelength of the laser light used to "read" the information. That wavelength is 780 nm in air, but the wavelength on which the pit depth is based is measured in the $n = 1.55$ plastic that makes up most of the disc. Find the pit depth.
10. Light is incident on an air-glass interface, and the refracted light in the glass makes a 40° angle with the normal to the interface. The glass has refractive index 1.52. Find the incidence angle.
11. A light ray propagates in a transparent material at 15° to the normal to the surface. When it emerges into the sur-

rounding air, it makes a 24° angle with the normal. What is the refractive index of the material?
12. Light propagating in the glass ($n = 1.52$) wall of an aquarium tank strikes the interior edge of the wall with incidence angle 12.4°. What is the angle of refraction in the water?
13. A block of glass with $n = 1.52$ is submerged in one of the liquids listed in Table 35-1. For a ray striking the glass with incidence angle 31.5°, the angle of refraction is 27.9°. What is the liquid?
14. A meter stick lies on the bottom of the rectangular trough in Fig. 35-34, with its zero mark at the left edge of the trough. You look into the long dimension of the trough at a 45° angle, with your line of sight just grazing the top edge of the tank, as shown. What mark on the meter stick do you see if the trough is (a) empty, (b) half full of water, and (c) full of water?

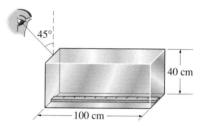

FIGURE 35-34 Problem 14.

15. You look at the center of one face of a solid cube of glass, on a line of sight making a 55° angle with the normal to the cube face. What is the minimum refractive index of the glass for which you will see through the opposite face of the cube?
16. The cylindrical tank in a public aquarium is 10 m deep, 11 m in diameter, and is full to the brim with water. If a flashlight shines on the tank from above, what is the minimum angle its beam can make with the horizontal if it is to illuminate part of the tank bottom?
17. You're standing 2.3 m horizontally from the edge of a 4.5-m-deep lake, with your eyes 1.7 m above the water surface. A diver holding a flashlight at the lake bottom shines the light so you can see it. If the light in the water makes a 42° angle with the vertical, at what horizontal distance is the diver from the edge of the lake?
18. You've dropped your car keys at night off the end of a dock into water 1.6 m deep. A flashlight held directly above the dock edge and 0.50 m above the water illuminates the keys when it's pointed at 40° to the vertical, as shown in Fig. 35-35. What is the horizontal distance x from the edge of the dock to the keys?
19. A light ray is propagating in a crystal where its wavelength is 540 nm. It strikes the interior surface of the crystal with an incidence angle of 34° and emerges into the surrounding

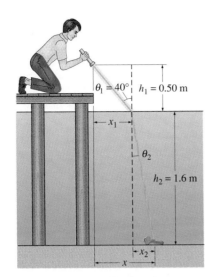

FIGURE 35-35 What is the horizontal distance x from the edge of the dock to the lost keys? (Problem 18)

air at 76° to the surface normal. Find (a) the light's frequency and (b) its wavelength in air.

20. The prism in Fig. 35-36 has $n = 1.52$, $\alpha = 60°$, and is surrounded by air. A light beam is incident at $\theta_1 = 37°$. Find the angle δ through which the beam is deflected.

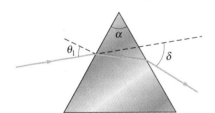

FIGURE 35-36 Problems 20, 57, 61.

Section 35-4 Total Internal Reflection

21. Find the critical angle for total internal refraction in (a) ice, (b) polystyrene, and (c) rutile. Assume the surrounding medium is air.

22. A drop of water is trapped in a block of ice. What is the critical angle for total internal reflection at the water-ice interface?

23. What is the critical angle for light propagating in crown glass when the glass is immersed in (a) water, (b) benzene, and (c) diiodomethane?

24. Total internal reflection occurs at an interface between a plastic and air at incidence angles greater than 37°. What is the refractive index of the plastic?

25. Light propagating in a medium with refractive index n_1 encounters a parallel-sided slab with index n_2. On the other side is a third medium with index $n_3 < n_1$. Show that the condition for avoiding internal reflection at *both* interfaces

is that the incidence angle at the n_1-n_2 interface be less than the critical angle for an n_1-n_3 interface. In other words, the index of the intermediate material doesn't matter.

26. An aquarium measures 30 cm front to back, as shown in Fig. 35-37. It is made of glass with thickness much less than the size of the aquarium and is full of water with $n = 1.333$. You put your eye right up to the center of the aquarium's front wall and can still see the entire back wall. What is the maximum value of the aquarium's width w? *Hint:* You can ignore the glass; see the preceding problem.

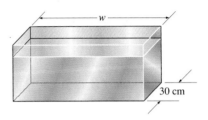

FIGURE 35-37 Problem 26.

27. What is the minimum refractive index for which total internal reflection will occur as shown in Fig. 35-14a? Assume the surrounding medium is air and that the prism is an isosceles right triangle.

28. Where and in what direction would the main beam emerge if the prism in Fig. 35-14a were made of ice?

29. What is the speed of light in a material for which the critical angle at an interface with air is 61°?

30. The prism of Fig. 35-14a has $n = 1.52$. When it is immersed in a liquid, a beam incident as shown in the figure ceases to undergo total reflection. What is the minimum value for the liquid's refractive index?

31. A compound lens is made from crown glass ($n = 1.52$) bonded to flint glass ($n = 1.89$). What is the critical angle for light incident on the flint-crown interface?

32. Find a simple expression for the speed of light in a material in terms of the critical angle at an interface between the material and vacuum.

33. A scuba diver sets off a camera flash a distance h below the surface of water with refractive index n. Show that light emerges from the water surface through a circle of diameter $2h/\sqrt{n^2 - 1}$.

Section 35-5 Dispersion

34. Laser beams with wavelengths 650 nm (red) and 410 nm (blue) strike an air-glass interface with incidence angle 50°. If the glass has refractive indices of 1.680 and 1.621 for the blue and red light, respectively, what will be the angle between the two beams in the glass?

35. Suppose the red and blue beams of the preceding problem are now propagating in the same direction *inside* the glass. For what range of incidence angles on the glass-air interface will one beam be totally reflected and the other not?

36. White light propagating in air is incident at 45° on the equilateral prism of Fig. 35-38. Find the angular dispersion γ of the outgoing beam, if the prism has refractive indices $n_{red} = 1.582$, $n_{violet} = 1.633$.

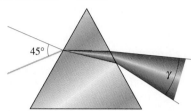

FIGURE 35-38 Problem 36 (angles of dispersed rays are not accurate.)

37. Two of the prominent spectral lines—discrete wavelengths of light—emitted by glowing hydrogen are hydrogen-α at 656.3 nm and hydrogen-β at 486.1 nm. Light from glowing hydrogen passes through a prism like that of Fig. 35-21, then falls on a screen 1.0 m from the prism. How far apart will these two spectral lines be? Use Fig. 35-19 for the refractive index.

38. Light from glowing sodium contains the two discrete wavelengths 589.0 nm and 589.6 nm. This light is passed through a prism like that of Fig. 35-21 and then allowed to fall on a screen 2.0 m distant. For wavelengths near 600 nm, the refractive index of the prism is $n = 1.546 - 4.47 \times 10^{-5} \lambda$, with λ the wavelength in nm. What must be the prism's apex angle α in order that the two sodium wavelengths be separated on the screen by 1.5 mm?

Section 35-6 Reflection Coefficients and the Polarizing Angle

39. Light is normally incident on ice. What fraction of the intensity is transmitted?

40. An aquarium's walls are made from crown glass with $n = 1.52$. What fraction of the intensity is reflected for light incident normally (a) from inside the water-filled aquarium and (b) from outside?

41. What is the refractive index of a material that transmits 92.4% of the light normally incident on it from air?

42. Light is incident normally on the outside of a glass-walled aquarium. The refractive indices of the glass and water are 1.52 and 1.333, respectively. What fraction of the incident intensity is transmitted into the water?

43. When a crystal is submerged in water ($n = 1.333$) the transmission coefficient for light incident normally on the crystal increases by 10% over its value in air. What is the refractive index of the crystal?

44. What would be the refractive index of a material for which normally incident light was half reflected, half transmitted? Assume the light is incident from air.

45. The reflection coefficient for normally incident light is the same when a block of plastic is submerged in water and in diiodomethane. What is the refractive index of the plastic?

46. Find the polarizing angle for diamond when light is incident from air.

47. What is the refractive index of a material for which the polarizing angle in air is 62°?

48. What is the polarizing angle for light incident from below on the surface of a pond?

Paired Problems

(Both problems in a pair involve the same principles and techniques. If you can get the first problem, you should be able to solve the second one.)

49. Light propagating in air strikes a transparent crystal at incidence angle 35°. If the angle of refraction is 22°, what is the speed of light in the crystal?

50. A laser beam with wavelength 633 nm is propagating in air when it strikes a transparent material at incidence angle 50°. If the angle of refraction is 27°, what is the wavelength in the material?

51. A cylindrical tank 2.4 m deep is full to the brim with water. Sunlight first hits part of the tank bottom when the rising Sun makes a 22° angle with the horizon. Find the tank's diameter.

52. For what diameter tank in the preceding problem will sunlight strike some part of the tank bottom whenever the Sun is above the horizon?

53. Light is incident from air on the flat wall of a polystyrene water tank. If the incidence angle is 40°, what angle does the light make with the tank normal in the water?

54. A parallel-sided slab with refractive index n_2 separates two media with indices n_1 and n_3. Show that a light ray incident on the n_1-n_2 interface then enters the third medium with the same angle of refraction it would have had if the slab had not been present. (Assume total internal reflection does not occur at either interface.)

55. Light strikes a right-angled glass prism ($n = 1.52$) in a direction parallel to the prism's base, as shown in Fig. 35-39. The point of incidence is high enough that the refracted ray hits the opposite sloping side. (a) Through which side of the prism does the beam emerge? (b) Through what angle has it been deflected?

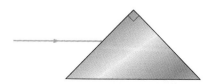

FIGURE 35-39 Problems 55, 56.

56. Repeat the preceding problem if the prism has refractive index 1.15.
57. Repeat Problem 20 for the case $n = 1.75$, $\alpha = 40°$, and $\theta_1 = 25°$.
58. The surfaces of a glass sheet are not quite parallel, but rather make an angle of 10°. The glass has refractive index $n = 1.52$. A light beam strikes one side of the sheet at incidence angle 35°, coming in on the thicker side of the normal. Through what angle is the beam direction changed when it emerges on the opposite side of the glass sheet?

Supplementary Problems

59. A cubical block is made from two equal-size slabs of materials with different refractive indices, as shown in Fig. 35-40. Find the index of the right-hand slab if a light ray is incident on the center of the left-hand slab and then describes the path shown.

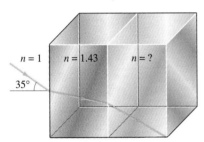

FIGURE 35-40 Problem 59.

60. Find an expression for the displacement x in Fig. 35-10, in terms of θ_1, d, and n.
61. Light is incident with incidence angle θ_1 on a prism with apex angle α and refractive index n, as shown in Fig. 35-36. Show that the angle δ through which the outgoing beam deviates from the incident beam is given by
$$\delta = \theta_1 - \alpha + \sin^{-1}\left\{ n \sin\left[\alpha - \sin^{-1}\left(\frac{\sin\theta_1}{n}\right) \right] \right\}.$$
Assume the surrounding medium has $n = 1$.
62. Taking $n = 1.5$ and $\alpha = 60°$, plot the deviation δ of the preceding problem over the range $45° < \theta_1 < 50°$, and use your plot to find the incidence angle for minimum deviation. Trace the incident beam for this value of θ_1. Your result should be the symmetric path shown in Fig. 35-41; in fact, the minimum deviation always occurs with the incidence angle that gives this path, for any n and α.
63. Show that a three-dimensional corner reflector (three mirrors in three mutually perpendicular planes, or a solid cube in which total internal reflection occurs) turns an incident light ray through 180°, so it returns in the direction from which it came. *Hint:* Let $\mathbf{q} = q_x\hat{\mathbf{i}} + q_y\hat{\mathbf{j}} + q_z\hat{\mathbf{k}}$ be a vector in the direction of propagation. How does this vector get

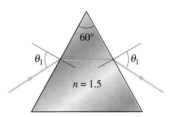

FIGURE 35-41 Minimum deviation through an equilateral prism occurs when the path is symmetric (Problem 62).

changed on reflection by a mirror in a plane defined by two of the coordinate axes?
64. Show that the angle ϕ that appears in Fig. 35-24 is given by $\phi = 4 \sin^{-1}\left(\dfrac{\sin\theta}{n}\right) - 2\theta$, where θ is the angle of incidence.
65. (a) Differentiate the result of the preceding problem to show that the maximum value of ϕ occurs when the incidence angle θ is given by $\cos^2 \theta = \frac{1}{3}(n^2 - 1)$. (b) Use this result and that of the preceding problem to find ϕ_{max} in water with $n = 1.333$.
66. Figure 35-42 shows the approximate path of a light ray that undergoes internal reflection twice in a spherical water drop. Find the maximum angle ϕ for this case, taking $n = 1.333$. This is the angle at which the secondary rainbow appears.

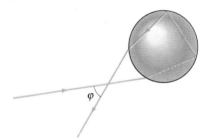

FIGURE 35-42 Problem 66.

67. *Fermat's principle* states that the path of a light ray between two points is such that the time to traverse that path is an extremum (either a minimum or a maximum) when compared with the times for nearby paths. Consider two points A and B on the same side of a reflecting surface, and show that a light ray traveling from A to B via a point on the reflecting surface will take the least time if its path obeys the law of reflection. Thus, the law of reflection (Equation 35-1) follows from Fermat's principle.
68. Use Fermat's principle (see preceding problem) to show that a light ray going from point A in one medium to point B in a second medium will take the least time if its path obeys Snell's law. Thus, Snell's law follows from Fermat's principle.

CHAPTER 36

IMAGE FORMATION AND OPTICAL INSTRUMENTS

This mirror directs sunlight into the McMath–Pierce Solar Telescope in Arizona, which uses optical technology to form a detailed image of the Sun's surface.

Reflection and refraction alter the direction of light propagation, following laws we developed in the preceding chapter. A wide variety of natural and technological systems use this effect to form images. In this chapter we study image formation using the approximation of geometrical optics—an approximation that remains valid as long as we consider only length scales much greater than the wavelength of light.

When we view an object directly, light comes to our eyes straight from the object. When we view an object with an optical system, our eyes perceive light that seems to come straight from the object but whose path has actually been altered. As a result we see an **image** that may be different in size, orientation, or apparent position from the actual object. In some cases light actually comes from the image to our eyes; the image is then called a **real image.** In other cases light only apparently comes from the image location; the image is then called a **virtual image.**

36-1 PLANE MIRRORS

When you look at yourself in a flat mirror, you see an image that appears to be behind the mirror by the same distance that you are in front of it. The image is upright and the same size as you are, but appears reversed. Why?

Figure 36-1 shows how the image in a plane mirror comes about. In Fig. 36-1a we concentrate on a small object, in this case an arrowhead. (We'll frequently use arrows to represent objects in image-forming situations because they're both simple and sufficiently asymmetric that we can see whether images are inverted or upright.) We have drawn three light rays that leave the object, reflect off the mirror, and enter the observer's eye. The rays reflect at the mirror with equal angles of incidence and reflection. As Fig. 36-1a shows, light looks to the observer like it's coming from a point behind the mirror. That point is the location of the arrowhead's image. In this case the image is virtual because no light actually comes from behind the mirror.

Since two nonparallel lines define a point, we need only two rays to locate the arrowhead in Fig. 36-1a. We've repeated this image-location process in Fig. 36-1b, using as one of the rays the ray that reflects normally. The same procedure also locates the bottom of the arrow, and we could obviously fill in additional points to locate the entire arrow; the resulting image is shown in Fig. 36-1b.

Note that the triangles OPQ and $O'PQ$ in Fig. 36-1b share a common side and that the angles OPQ and $O'PQ$ are both right angles. And because the angles of incidence and reflection are equal, so are the angles OQP and $O'QP$. Therefore triangles OPQ and $O'PQ$ are congruent, showing that the distance PQ' that the arrowhead's image lies behind the mirror is equal to the distance OP from the actual arrowhead to the mirror. A similar analysis applies to the rays from the bottom of the arrow, so we can conclude that the image appears as far behind the mirror as the object is in front. Furthermore, since extensions of the rays from the top and bottom of the arrow normal to the mirror pass through the top and bottom of the image, respectively, the image must be the same height as the object itself.

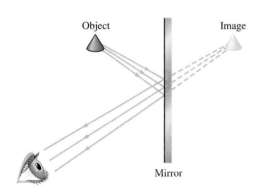

(a)

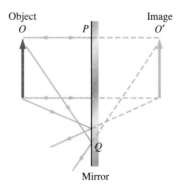

(b)

FIGURE 36-1 Image formation in a plane mirror. (a) Light reflected off the mirror seems to come from an image behind the mirror. Since there is really no light coming from behind the mirror, this is a virtual image. (b) Two rays from each point on an object serve to locate that point's image. Triangles OPQ and $O'PQ$ are congruent, showing that object and image are equidistant from the mirror.

● EXAMPLE 36-1 WHAT SIZE MIRROR TO BUY?

You want the smallest mirror that will show your full image. How tall must it be?

Solution

Figure 36-2 shows the situation. Because the angles of incidence and reflection are equal, light from your foot reflects from the mirror at a point that's vertically halfway between your eye and the floor. Similarly, light from the top of your head reflects midway between that point and your eye. The total distance from top to bottom of the mirror then needs to be half your eye-foot distance plus half the distance from your eye to the top of your head—for a total of half your height. Note that this result does not depend on how far from the mirror you stand. It does, however, require that you fix the mirror to the wall at just the right height.

EXERCISE You buy a mirror that's half your height, but you affix it to the wall 2 cm below its optimum location. How much of your image will be cut off?

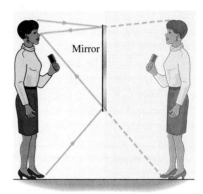

FIGURE 36-2 A mirror half your height shows your entire image.

Answer: Top 4 cm

Some problems similar to Example 36-1: 1–3 ●

Image Reversal

FIGURE 36-3 The image of a right hand is the image's left hand—but it's still the image of the *right* hand.

Figure 36-3 shows that the image in a plane mirror appears reversed left to right. So why isn't it inverted top to bottom as well?

The image does indeed appear reversed, but in the sense that its left hand is the image of the object's *right* hand. It's not that the image of the object's left hand appears opposite the right hand, any more than the image of the head appears opposite the feet. A more accurate description is that the mirror reverses front to back. Objects lying parallel to the mirror are not altered at all, but an object pointing perpendicular to the mirror is reversed. In Fig. 36-4 the effect is to alter only one of the three coordinate axes, and that alters handedness, rotation, and all other phenomena connected with the right-hand rules we've been using.

Although image formation in a single plane mirror is straightforward, multiple reflections with more than one mirror can produce many—in some cases infinitely many—images; see Fig. 36-5 and Problem 2.

36-2 CURVED MIRRORS

In contrast to plane mirrors, curved mirrors form images that may be upright or inverted, virtual or real, and whose sizes need not be those of the original objects.

Parabolic Mirrors

A parabola—the curve generated by a quadratic function like $y = x^2$—has an interesting geometrical property. For any point on the parabola, a line drawn parallel to the parabola's axis makes the same angle to the normal as does a

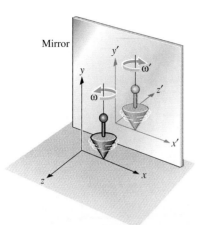

FIGURE 36-4 Actually, the mirror reverses front to back. Here the x and y axes of a normal right-handed coordinate system are unchanged in the mirror, but the image of the z-axis—which runs perpendicular to the mirror—is reversed. As a result the image is a left-handed coordinate system. The rotation of the spinning top, for example, is reversed in the mirror. But its angular velocity vector still points upward because a left-hand rule applies in the mirror world.

FIGURE 36-5 Multiple images formed by several plane mirrors.

second line drawn to a special point called the **focus** or **focal point** (Fig. 36-6a). That means a concave mirror of parabolic shape will reflect rays parallel to the parabola's axis so they converge at the focus (Fig. 36-6b). This effect can be used to concentrate light to very high intensities (Fig. 36-7a). Conversely, a point source of light at the focus will emerge from the mirror in a beam of parallel rays (Fig. 36-7b).

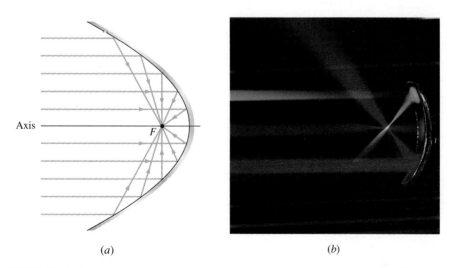

(a)   (b)

FIGURE 36-6 Drawing and photo show that a parabolic mirror reflects rays parallel to its axis to a common focus. (a) Ray diagram. (b) Photo shows rays reflecting from an actual mirror.

(a)

(b)

FIGURE 36-7 (a) This solar furnace uses a huge parabolic reflector to concentrate sunlight. (b) A flashlight has a light bulb near the focus of a parabolic mirror and thus produces a beam of nearly parallel light rays.

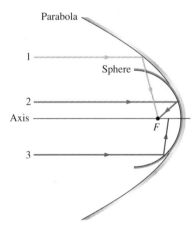

FIGURE 36-8 A sphere and parabola essentially coincide over a limited area. Rays 1 and 2 strike the parabolic mirror and reflect to the focus F, and 2 essentially reaches F from the spherical mirror as well. But ray 3, striking the spherical mirror farther from the axis, does not.

Spherical Concave Mirrors

It's much easier to make a spherical mirror than a parabolic one. Figure 36-8 shows that a parabola and a sphere essentially coincide over a limited range of the sphere's surface, and therefore over this range we can expect a sphere, too, to focus parallel light with a high degree of accuracy. Most mirrors used for focusing are therefore spherical in shape. Because a spherical mirror is not exactly parabolic, parallel rays do not converge at exactly the same point. The resulting distortion, called **spherical aberration,** is minimized by making the actual mirror only a tiny fraction of the entire sphere.

■ APPLICATION HUBBLE TROUBLE

The Hubble Space Telescope was launched in 1990 as the flagship of a new generation of space-based astronomical observatories. Although Hubble is smaller than many ground-based telescopes, its vantage point above the atmosphere was to make it optically superior in resolving astronomical objects. Furthermore, Hubble can observe in infrared, visible, and ultraviolet wavelengths.

In the months following Hubble's launch, engineers and scientists checking the telescope were frustrated by their inability to achieve a clear focus. They came reluctantly to the conclusion that Hubble's 2.4-m-diameter primary mirror was flawed. Subsequent investigation showed that an error of 1.3 mm in the placement of instruments used during manufacture of the mirror had resulted in its being ground to the wrong curvature. The mirror itself is off by only 2.3 μm at its edge, but this is an enormous flaw in an optical system designed to be accurate to a small fraction of the wavelength of light. Hubble specifications called for 70% of the light energy from a point source to fall within an angular diameter of 0.1 second of arc; because of its spherical aberration, at best 16% of the light fell in the prescribed zone.

Although the Hubble mirror could not be replaced or repaired, in 1993 astronauts managed to install corrective lenses that restored Hubble's optical system to better than its original design specifications. The result is a superb astronomical instrument with image quality greatly superior to that of ground-based telescopes (Fig. 36-9).

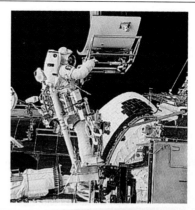

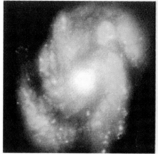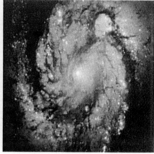

FIGURE 36-9 (Top) An astronaut installing new optical equipment during the 1993 repair of the Hubble Space Telescope. (Bottom) Hubble images of M-100, a galaxy 50 million light-years from Earth, taken before and after the repair.

We can see how spherical mirrors form images by tracing two rays from each of several points on the object, just as we did for plane mirrors. There are some special rays that make this process much simpler. First is any ray parallel to the axis; under the assumption that the spherical mirror approximates a parabola, we know that this ray reflects through the focal point *F*. A second useful ray lies along the line from the focus to the point on the object; under the parabolic approximation we know that this ray will emerge parallel to the axis. Two other useful rays are the ray that strikes the very center of the mirror, which reflects with equal angles on either side of the mirror axis, and the ray through the center of curvature, which, because it strikes normal to the surface of the spherical mirror, returns on itself. Any two of these rays suffice to locate the image.

Figure 36-10 shows the results of these ray tracings, using the two rays involving the focus, to find the image location in three cases. In all three cases symmetry ensures that the bottom of the image arrow will be on the axis, so we haven't bothered to trace it. In the more general case where the object isn't sitting on the axis we would also trace to find the bottom of the image arrow. In Fig. 36-10*a* we see that a distant object—beyond the mirror's center of curvature (*C*)—forms a smaller, inverted image. Light actually emerges from this image, so it's a *real* image. If you looked from the left in Fig. 36-10*a* you would actually see the image in space in front of the mirror (Fig. 36-11).

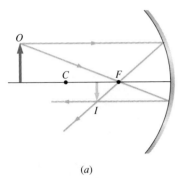

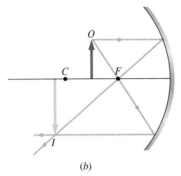

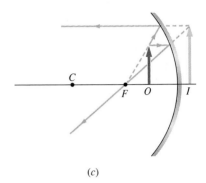

(a)   (b)   (c)

FIGURE 36-10   Image formation in a concave spherical mirror, found by exploiting the fact that light from the focus reflects paraxially from the mirror, and vice versa. *O* denotes the object and *I* its image. (*a*) With the object beyond the mirror's center of curvature *C*, the mirror forms a real image that is inverted and reduced in size. (*b*) With the object between the center of curvature and focus, the image is real, inverted, and magnified. (*c*) When the object is closer than the focus, the image is enlarged, upright, and virtual. Note, in this case, to get the ray including the object and the focus we had to extend the actual light path back toward the focus.

FIGURE 36-11   A bear meets its real image, formed by the concave mirror at rear. Note that the bear and its image are both in *front* of the mirror. From the fact that they're about the same size, what can you conclude about their position relative to the focal point?

As the object moves closer to the mirror the real image grows; with the object between the center of curvature and the focus, the image is larger than the object, and farther from the mirror (Fig. 36-10*b*). As the object moves toward the focus the image grows larger and moves rapidly away from the mirror. With the object right at the focus the rays emerge in a parallel beam and there is no image. Finally, rays from an object closer to the mirror than the focus diverge after reflection. To an observer they appear to come from a point behind the mirror. Thus there is a virtual image, in this case upright and enlarged (Fig. 36-10*c*).

Convex Mirrors

A convex mirror reflects on the outside of its spherical curvature, causing light to diverge rather than focus. Therefore, there is no possibility of forming a real image with a convex mirror. But Fig. 36-12 shows that the mirror can still form a virtual image. Although the focus has less obvious physical significance in this case, its location still controls the geometry of reflected rays. As Fig. 36-12 shows, we can still draw a ray parallel to the axis and another ray that would go through the focus if the mirror weren't in its way. By tracing these rays through the mirror we can see the directions in which they reflect. The reflected rays appear to diverge from a common point to the right, showing that there is a virtual image, upright and reduced in size, to the right of the mirror. You can convince yourself that for a convex mirror the image always has these characteristics. Convex mirrors are widely used where an image of a broad region needs to be captured in a small space (Fig. 36-13).

Curved and Plane Mirrors

You can understand curved mirrors qualitatively by thinking about bending a plane mirror. A plane mirror produces a virtual image of the same size as the object. If you bend it so it becomes slightly convex, the rays diverge more and the virtual image shrinks a little. If you bend it so it becomes slightly concave, the rays converge more and the virtual image grows. Figure 36-14 shows the effect of both these changes. Making the mirror concave also moves the focus inward from infinitely far away. If you bend it enough that the focus moves closer to the mirror than the object, then you're in a whole new realm where the mirror produces a real image, as in Figs. 36-10a and b.

The Mirror Equation

Drawing ray diagrams is a useful way to get an intuitive feel for image formation with curved mirrors. However, more precise image locations are obtained from the **mirror equation,** which we now derive.

Figure 36-15 shows a portion of a concave spherical mirror whose **focal length**—the distance from the mirror to the focal point—is f. As usual, we assume that the mirror is such a small portion of a sphere that it's essentially parabolic. Mathematically, that means the mirror's extent in the horizontal direction of Fig. 36-15 is much smaller than its vertical extent. As a result, we can approximate horizontal distances to points on the mirror as being measured to a vertical line through the mirror's apex, point A in Fig. 36-15. Under this approximation the base of the green triangle is *approximately* the focal length f. Using this triangle we can then write

$$\tan \alpha = \frac{h}{f},$$

with h the object height. Using the yellow triangle, we can equally well write

$$\tan \alpha = \frac{-h'}{\ell' - f},$$

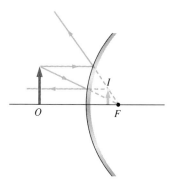

FIGURE 36-12 Image formation using a convex mirror. The image is always virtual, upright, and reduced in size.

FIGURE 36-13 Reflecting spheres produce reduced, upright, virtual images.

FIGURE 36-14 Funhouse mirrors are alternately convex and concave, producing stretched and shrunken images.

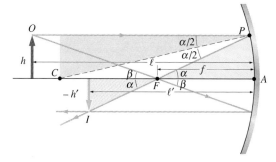

FIGURE 36-15 Ray diagram for derivation of the mirror equation. Horizontal distances are approximated as being measured to the horizontal position of the mirror apex A.

where ℓ' is the distance from the image to the mirror. We write $-h'$ so h' can be a *negative* number to indicate an *inverted* image. Equating our two expressions for $\tan\alpha$ and rearranging slightly gives

$$\frac{h'}{h} = -\frac{\ell'-f}{f} \tag{36-1}$$

We can do a similar analysis for the two triangles with vertex angles β, giving $\tan\beta = -h'/f$ and $\tan\beta = h/(\ell-f)$, where ℓ is the distance from mirror to object. Equating these two expressions gives

$$\frac{h'}{h} = -\frac{f}{\ell-f}. \tag{36-2}$$

Equating the right-hand sides of Equations 36-1 and 36-2 and carrying out some algebra then gives the mirror equation:

$$\frac{1}{\ell} + \frac{1}{\ell'} = \frac{1}{f}. \qquad \text{mirror equation} \tag{36-3}$$

We define the ratio h'/h as the mirror's **magnification,** M. Given the focal length and the image or object distance, M can be found from Equation 36-1 or 36-2, respectively. Or, by solving Equation 36-3 for f, the magnification is minus the ratio of image to object distance:

$$M = \frac{h'}{h} = -\frac{\ell'}{\ell}. \tag{36-4}$$

We can extract one more useful fact from Fig. 36-15 by considering a line from the point P where the parallel ray hits the mirror to point C, the mirror's center of curvature. Since it's a radius, this line is normal to the mirror, so the law of reflection therefore ensures that it bisects the angle between two rays at P, which is equal to the angle α. By considering the light gray triangle we can then write $\tan(\alpha/2) = h/R$, where R is the mirror's curvature radius. Now the angle α in Fig. 36-15 must be very small under our assumption that the mirror curves only slightly. We can therefore use the approximations $\tan\alpha \simeq \alpha$ and $\tan(\alpha/2) \simeq \alpha/2$. Rewriting the expressions $\tan\alpha = h/f$ and $\tan(\alpha/2) = h/R$ using this small-angle approximation then gives $h/R = h/2f$, or

$$f = \frac{R}{2}. \tag{36-5}$$

Thus, the focal length of a spherical mirror is half its curvature radius.

We emphasize that the formulas derived in this section are approximations based on the assumption that the spherical mirror is a good approximation to a parabola. Equivalent ways of stating this same approximation are that the curvature radius and focal length of the mirror are large compared with the mirror's actual diameter, and that all rays reflecting from the mirror make only small angles with the mirror axis.

● EXAMPLE 36-2 OPTICAL ILLUSION

A popular "optical illusion" gadget consists of a tiny model car that seems to "float" in the top of a bowl-shaped depression (Fig. 36-16). The "floating" car is actually the real image of a solid model mounted just above a concave mirror at the bottom of the device. The mirror's focal length is 2.0 cm. If the image is 11 cm above the mirror and 3.2 cm long, where and what size is the actual model car?

Solution
We solve Equation 36-3 for the object distance:

$$\ell = \left(\frac{1}{f} - \frac{1}{\ell'}\right)^{-1} = \left(\frac{1}{2.0 \text{ cm}} - \frac{1}{11 \text{ cm}}\right)^{-1} = 2.44 \text{ cm}.$$

Then the object size follows from Equation 36-4:

$$h = -\frac{\ell h'}{\ell'} = -\frac{(2.44 \text{ cm})(-3.2 \text{ cm})}{11 \text{ cm}} = 0.71 \text{ cm},$$

where we use a negative value for h' because the real image is inverted. We could also have obtained this result from Equation 36-1.

FIGURE 36-16 Optical illusion: A real image of the model car floats above the bottom of the bowl.

EXERCISE You stand 7.8 m from a concave spherical mirror, and, looking toward the mirror, you see an upside-down image of yourself with half your height. (a) How far apart are you and your image? (b) What is the mirror's curvature radius?

Answers: (a) 3.9 m; (b) 5.2 m

Some problems similar to Example 36-2: 4–8 ●

Although we derived the mirror equation in a situation involving a real image, the equation applies as well to virtual-image formation if we adopt the convention that a *negative* image distance puts the image *behind* the mirror—in the realm where we find virtual images. And we can handle *convex* mirrors as well if we take the focal length and therefore the curvature radius as *negative* quantities.

> **TIP** **Sign Conventions—Positive and Real, Negative and Virtual** The sign conventions for mirrors make sense: Objects and real images have positive distances. Images that aren't real (i.e., virtual images) are more absurd: they have negative distances. Mirrors that can make real images have positive focal lengths; mirrors that cannot, have negative focal lengths. On the other hand, real images may have negative height—but that's understandable because it simply means they're inverted.

● EXAMPLE 36-3 SIZING UP HUBBLE

Figure 36-17 shows a technician standing in front of the Hubble Space Telescope's primary mirror, whose focal length is 5.52 m. Use the photo to estimate the locations of the technician and his image relative to the mirror.

Solution
Direct measurement of the photo shows that the image of the technician's head appears about 3.3 times the size of his actual head. (This is an underestimate because the image is farther from the camera and thus appears smaller in relation to the technician than it really is.) The image is upright, so its height h' is positive. With $h'/h = 3.3$, Equation 36-4 then gives $\ell' = -3.3\ell$. The negative sign here indicates that the image is *behind* the mirror and is, therefore, virtual. Using this result in the mirror equation 36-3 gives

FIGURE 36-17 A technician standing in front of the Hubble Space Telescope mirror. How far are the technician and his image from the mirror (Example 36-3)?

$$\frac{1}{\ell} + \frac{1}{-3.3\ell} = \frac{1}{f}.$$

Solving for ℓ we have

$$\ell = f\left(1 - \frac{1}{3.3}\right) = 0.697f = (0.697)(5.52 \text{ m}) = 3.85 \text{ m}.$$

Thus, the technician is 3.85 m in front of the mirror, and his image is 3.3 times this, or about 13 m, behind the mirror.

Does this result make sense? The 3.8-m object distance is *less* than the focal length, as Fig. 36-10 showed it must be for the concave mirror to make a virtual image.

EXERCISE You're scrutinizing your nose using a hand-held concave vanity mirror with curvature radius 2.2 m. How far from your face should you hold the mirror to see your nose doubled in size?

Answer: 55 cm

Some problems similar to Example 36-3: 9–11 ●

■ APPLICATION NONIMAGING OPTICS AND THE QUEST FOR SOLAR ENERGY

Sometimes we want to concentrate light energy without necessarily forming an image. In solar energy systems producing heat, for example, concentrated sunlight leads to higher temperature and thus to higher efficiency; with photovoltaic systems inexpensive reflectors can be used to concentrate solar energy onto more expensive photovoltaic cells, reducing the cell area needed.

Nonimaging optics is a new field, and the design of concentrators is an ongoing challenge for optical engineers. Figure 36-18 shows a popular concentrator design, called a compound parabolic concentrator. Unlike a parabolic mirror designed for image formation, this concentrator is made from parabolic segments that do not form a single complete parabola. As Fig. 36-18a shows, all light entering the concentrator from a broad range of angles ends up at the narrow exit end. The light doesn't converge to form an image, but its intensity has nevertheless been increased because its energy is spread over a smaller area. Thus, the concentrator acts as a "funnel" for light.

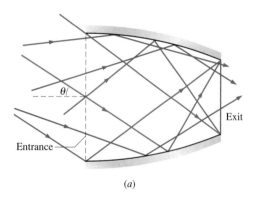

(a)

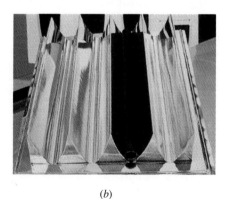

(b)

FIGURE 36-18 (a) A compound parabolic concentrator. Rays (blue) entering at the angle θ to the concentrator axis are reflected to the edge of the exit aperture. Rays incident at smaller angles (red) pass through the exit aperture after at most one reflection; thus all light entering over the angular range 2θ is concentrated at the exit. (b) A section of parabolic concentrators used to concentrate solar energy on a black tube carrying a heat-transfer fluid. A piece of pipe lies at the exit of one concentrator; since light paths are reversible, the fact that the entire concentrator appears black confirms that the path of an incident light ray ends up on the black tube.

Figure 36-18*b* shows a section of a compound parabolic reflector used to concentrate light onto a tube carrying heat-transfer fluid in a solar thermal power plant. Because the concentrator accepts light from a broad angular range, it works without the need for expensive tracking equipment that would keep it facing the Sun.

Nonimaging concentrators have other applications as well. The first compound parabolic concentrator, built in 1965, was designed to collect light from a high-energy physics experiment. Nonimaging optics in the infrared were used on the Cosmic Background Explorer satellite that, in 1992, provided crucial evidence in support of the Big Bang theory of the origin of the universe. And under development are solar-powered lasers that use nonimaging concentrators as the "pump" that supplies energy to run the laser.

36-3 LENSES

A **lens** is a piece of transparent material shaped in such a way that parallel light rays are refracted toward a single point, again called the **focus** or **focal point.** The focus may be real or only apparent—i.e., virtual— depending on whether the lens is convex (**converging lens;** see Fig. 36-19) or concave (**diverging lens;** see Fig. 36-20). We will see in the next section that a lens made with spherical surfaces approximates this focusing behavior and will show how the focal length is related to the curvatures of the lens surfaces and to its index of refraction. (An alternative lens design—the **graded-index lens**—is flat but has a refractive index that varies with position.) For now we assume we have working lenses and show how they form images.

We will use the **thin-lens** approximation, in which a lens's thickness is much less than the curvature radii of its surfaces, its focal length, and the distances to any objects and their images. Unlike a mirror, a lens works in both directions and, therefore, has two focal points. We will show, again in the next section, that the two focal lengths of a thin lens are the same.

A lens works by refracting light as it enters the first lens surface, and again as it exits the second. Because the two surfaces are not parallel, there is a net change in the direction of a light ray. We will consider the details of those refractions in the next section. In the thin-lens approximation, however, it suffices to consider that light simply bends when it crosses the center plane of the lens. Although that isn't really what happens, the lens is so thin compared with other lengths of interest that the distance between its two surfaces is negligible. Under this approximation a light ray crossing the lens at its axis suffers no effect whatsoever since for it the refractive effects of the opposite surfaces cancel (recall Example 35-3).

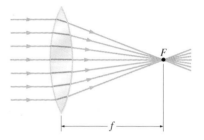

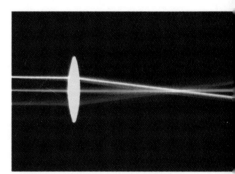

FIGURE 36-19 A convex lens brings parallel light rays to a real focus *F*. The focal length *f* is the distance from the lens to the focal point.

Lens Images by Ray Tracing

We can use ray tracing to find lens images, just as we did with mirrors. Again, two rays from any point serve to fix its image. It's convenient with lenses to use the ray parallel to the lens axis—which gets refracted through the focus—and the ray that passes undeviated through the lens center. Figure 36-21 shows the results for different object placements in relation to a converging lens. In Fig. 36-21*a* we see that an object farther out than two focal lengths produces a smaller, inverted real image on the other side of the lens. Since light really emanates from this image, you could see it without actually looking through the lens. As the object moves in, the image grows until, when the object lies between one and two focal lengths, the image is farther than 2*f* from the lens and is larger

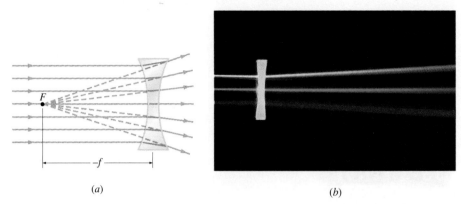

FIGURE 36-20 A concave lens has a virtual focus, since light rays only appear to meet at the focus. The focal length is negative, indicating that the lens causes light to diverge.

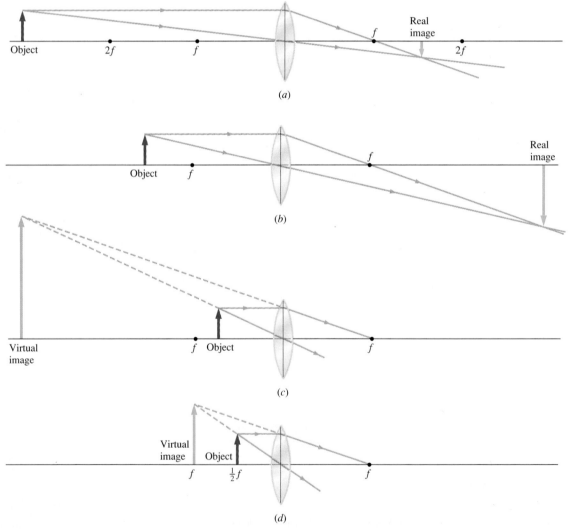

FIGURE 36-21 Image formation with a converging lens. Marks on axis are distances from the lens center, in units of the focal length (a) and (b), an object distance greater than the focal length results in an inverted real image. In (b) the image is enlarged if the object e is between one and two focal lengths. In (c) and (d), a virtual image forms when the object is within the focal length. When the noves within half the focal length the image moves within the focal length, as shown in (d).

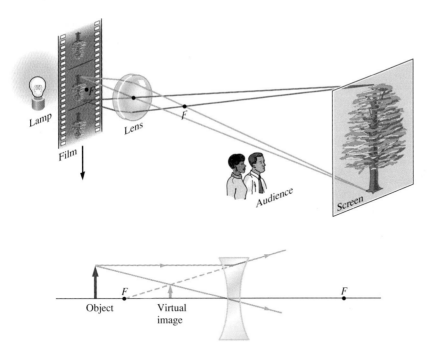

FIGURE 36-22 Simplified diagram of a movie projector. The film lies just beyond the focal length, and thus the lens produces an enlarged real image on the distant screen. The audience normally views the image from the projector side of the screen. Note that the film is upside down.

FIGURE 36-23 A diverging lens always forms a reduced, upright, virtual image, visible only through the lens.

than the object. The image on a movie screen is formed in this way (Fig. 36-22). Moving the object closer than the focal point produces an enlarged virtual image that can be seen only by an observer looking *through* the lens (Fig. 36-21c). Moving the object still closer causes the virtual image to move inward and shrink; when the object is within half the focal length of the lens, then the image is within one focal length (Fig. 36-21d). The virtual image remains always larger than the object.

Figure 36-23 shows ray tracings for a diverging lens. Like a convex mirror, this lens produces only virtual images that are upright and reduced in size. Like the virtual images of Fig. 36-21c and d, these virtual images are visible only through the lens. You should convince yourself that the basic geometry of Fig. 36-23 does not change even if the object moves within the focal length

TIP **Understanding Lenses** Before going on to the mathematics of lenses, try answering the following simple questions about the image shown in Fig. 36-24. Then check your answers against those in the footnote below.*
(a) What would happen if the bottom half of the lens were covered?
(b) What would happen if the lens were removed?
(c) What would happen if the screen's distance from the lens were doubled?
(d) Would there be an image in the absence of the screen?

*Answers: (a) You might think that half the image would disappear. But Fig. 36-21 shows complete images forming from rays that travel only through the top half of the lens. We could equally well have formed the images with rays from the bottom half, or any other portion of the lens. Covering part of the lens—any part—diminishes the intensity of the image but does not block *any* of it. (b) Did you think the image would now appear upright? No! The lens is *essential* to the image formation; it's what brings rays leaving each point of the candle flame together again at a single point. Absent the lens, the screen would appear diffusely lit by the candle's light, but there would (*continued*)

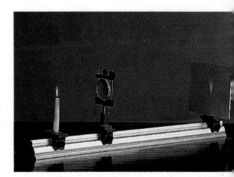

FIGURE 36-24 A simple optical system, consisting of an object (the candle), a converging lens, and a white screen on which the image appears. This system is a realization of Fig. 36-21b.

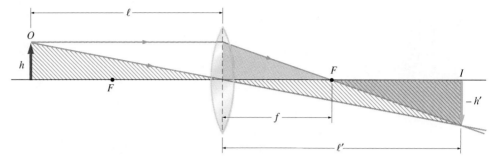

FIGURE 36-25 Ray diagram for deriving the lens equation. The two colored triangles are similar, as are the hatched triangles.

A Lens Equation

We can locate lens images quantitatively by deriving an equation like we did for mirrors, now using Fig. 36-25. Note that the two hatched (////) triangles in Fig. 36-25 are similar, and therefore the magnification is

$$M = \frac{h'}{h} = -\frac{\ell'}{\ell},$$

(36-6)

where ℓ and ℓ' are the object and image distances, respectively, and where we again take a negative height to signify an inverted image. Thus, the magnitude of the magnification is the ratio of object to image distance, as it was with mirrors. The two colored triangles are also similar, and therefore

$$\frac{-h'}{\ell' - f} = \frac{h}{f}.$$

Combining this result with Equation 36-6 and doing some algebra then gives

$$\frac{1}{\ell} + \frac{1}{\ell'} = \frac{1}{f}, \qquad \text{lens equation}$$

(36-7)

which is identical to the mirror equation 36-3. Note that putting the object at infinity ($\ell = \infty$) gives $\ell' = f$ and that putting the object at the focus ($\ell = f$)

be no image. (c) Did you think the image size would double? No! The image exists only at the one place where rays from each point on the candle converge. Again, the screen would be diffusely lit. Moving it very slightly from its position, in contrast, would result in an blurred rendition of the candle. (Moving the lens also would allow refocusing the image at the new screen location.) (d) The screen helps make the image visible to observers looking from various directions. But it plays no role in the actual image formation; absent the screen, there would still be an image at the point where the screen is now located. An observer looking toward the lens from the right, into the cone of light diverging from the image, would see an upside-down candle flame apparently in "thin air."

For more intuition on image formation, see Fred M. Goldberg and Lillian C. McDermott, "An Investigation of Student Understanding of the Real Image Formed by a Converging Lens or Concave Mirror," *American Journal of Physics,* vol. 55, no. 2, pp. 108–119 (February 1987).

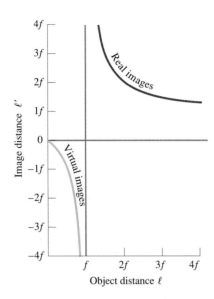

FIGURE 36-26 Image versus object distance for converging lenses. Object distances less than the focal length f correspond to virtual images (negative ℓ'). The image moves toward infinity as the object distance approaches the focal length from either direction, and concurrently the image grows arbitrarily large.

gives $\ell' = \infty$. Thus, Equation 36-7 reflects our ray-tracing diagrams that show parallel rays (such as those from an object at a great distance) converging at the focal point, and vice versa.

Although we derived Equation 36-7 for the case of a real image, the equation holds for virtual images if we consider the image distance negative; in that case the image is on the same side of the lens as is the object. And it holds for diverging lenses if we consider the focal length negative. Again these conventions make sense, as suggested in the tip before Example 36-4. Figure 36-26 is a graphical summary of the lens equation for a converging lens.

> **TIP** **Sign Conventions** As with mirrors, the sign conventions for lenses make sense. Objects and real images have positive distances; virtual images "aren't really there" and have negative distances. Converging lenses—those that can form real images—have positive focal lengths; diverging lenses can form only virtual images, and they have negative focal lengths.

● **EXAMPLE 36-4** SLIDE SHOW

You're projecting slides onto a wall 2.6 m from a slide projector whose single lens has focal length 12.0 cm. (a) How far should the slides be from the lens? (b) How big will be the image of a 35-mm slide?

Solution
We want the image focused on the screen, so the image distance ℓ' is 2.6 m. Solving Equation 36-7 for the object distance ℓ then gives

$$\ell = \left(\frac{1}{f} - \frac{1}{\ell'}\right)^{-1} = \left(\frac{1}{12\ \text{cm}} - \frac{1}{260\ \text{cm}}\right)^{-1} = 12.58\ \text{cm}.$$

This is just beyond the focal length, as Figs. 36-21 and 36-26 suggest it should be to get an enlarged, distant, real image.

Equation 36-6 then gives the height h' of the image formed of a 35-mm slide:

$$h' = -h\frac{\ell'}{\ell} = -(3.5\ \text{cm})\left(\frac{260\ \text{cm}}{12.58\ \text{cm}}\right) = -72\ \text{cm}.$$

The minus sign indicates inversion, showing that the slide must go in the projector upside down to get an upright image.

EXERCISE If the slide in Example 36-4 moves 1 mm farther from the lens, what will happen to the position of the image?

Answer: moves 36 cm toward projector

Some problems similar to Example 36-4: 16–18 ●

● **EXAMPLE 36-5** FINE PRINT

You're using a magnifying glass (a converging lens) with 30 cm focal length to read a telephone book (Fig. 36-27). How far from the page should you hold the lens in order to see the print enlarged 3 times?

Solution

Here the image is virtual, so ℓ' is negative and $3\times$ magnification then corresponds to $\ell' = -3\ell$. So Equation 36-7 becomes

$$\frac{1}{\ell} - \frac{1}{3\ell} = \frac{2}{3\ell} = \frac{1}{f} = \frac{1}{30 \text{ cm}},$$

whence $\ell = (2)(30 \text{ cm})/3 = 20$ cm. Figure 36-27 confirms that the image appears further from the lens than the actual page.

EXERCISE A magnifying glass enlarges print by 50% when it's held 9.0 cm from a page. What is its focal length?

FIGURE 36-27 A magnifying glass is a converging lens used in its virtual-image mode.

Answer: 27 cm

Some problems similar to Example 36-5: 19, 25 ●

36-4 THE LENSMAKER'S FORMULA

We now examine in detail the refraction effects that cause lenses to form images. Every lens has two refracting surfaces, at least one, and usually both, of which are curved. We therefore look first at image formation due to refraction at a *single* curved interface. To understand the behavior of a lens we will then consider two such interfaces, using the image formed by the first as the object for the second.

Refraction at a Curved Surface

Figure 36-28 shows part of a spherically curved piece of material (typically glass or plastic) with curvature radius R and refractive index n_2, surrounded by a medium with index n_1 (typically air). A point-like object O lies on an axis through the center of curvature of the refracting surface. We consider only light rays from O that are nearly parallel to the axis. Under this restriction the labeled angles in Fig. 36-28 are all small, and we can use the approximation $\sin x \simeq \tan x \simeq x$, where $x \ll 1$ with x an angle measured in radians.

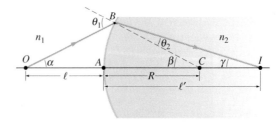

FIGURE 36-28 Refraction at a curved surface produces an image at I of the object at O.

In Fig. 36-28 a light ray from O propagating along the axis suffers no refraction because it is normal to the interface. An image of O should therefore form at a point where other rays from O intersect the axis. The figure shows one such ray, which hits the interface at B with some small incidence angle θ_1, and propagates in the second medium at angle θ_2 to the normal. Under the small-angle approximation, Snell's law—$n_1 \sin\theta_1 = n_2 \sin\theta_2$—becomes simply

$$n_1\theta_1 = n_2\theta_2.$$

From triangles BCI and OBC, we see that $\theta_2 = \beta - \gamma$ and $\theta_1 = \alpha + \beta$. Using these results in our small-angle version of Snell's law gives

$$n_1(\alpha + \beta) = n_2(\beta - \gamma).$$

Furthermore, in the small-angle approximation the arc BA is so close to a straight line that we can write $\alpha \simeq \tan\alpha \simeq BA/\ell$, with $\ell = OA$ the object's distance from the refracting surface. Similarly, $\beta \simeq BA/R$, and $\gamma \simeq BA/\ell'$. Thus, our expression of Snell's law becomes

$$n_1\left(\frac{BA}{\ell} + \frac{BA}{R}\right) = n_2\left(\frac{BA}{R} - \frac{BA}{\ell'}\right)$$

or, on canceling the arc length BA and rearranging,

$$\frac{n_1}{\ell} + \frac{n_2}{\ell'} = \frac{n_2 - n_1}{R}. \tag{36-8}$$

Notice that the angle α does not appear in this equation. Therefore, this relation between the object and image distances holds for *all* rays as long as they satisfy the small-angle approximation. That means all such rays come to a common focus at I in Fig. 36-28, so an image must appear at this point.

   Although we derived Equation 36-8 for a case of real image formation, as usual it holds for virtual images as well if we consider the image distance negative. It also works for surfaces that are concave toward the object—the opposite of Fig. 36-28—if we take R as a negative number.

● **EXAMPLE 36-6** A CYLINDRICAL AQUARIUM

An aquarium is made from a thin-walled tube of transparent plastic 70 cm in diameter (Fig. 36-29a). For a cat looking directly into the aquarium, what is the apparent distance to a fish 15 cm from the aquarium wall?

Solution

We can neglect the aquarium wall; since it's thin and has essentially parallel faces, light suffers no net deflection in passing through the wall. The surface is concave toward the object, so $R = -35$ cm. With $\ell = 15$ cm, $n_1 = 1.333$ for water, and $n_2 = 1$ for air, we can solve Equation 36-8 for the image distance ℓ':

$$\ell' = \frac{n_2}{\left(\dfrac{n_2 - n_1}{R} - \dfrac{n_1}{\ell}\right)} = \frac{1}{\left(\dfrac{1 - 1.333}{-35 \text{ cm}} - \dfrac{1.333}{15 \text{ cm}}\right)}$$

$$= -12.6 \text{ cm}.$$

The negative answer indicates that we have a virtual image, as shown in Fig. 36-29b.

   A special case of Equation 36-8 is a flat surface, for which $R = \infty$. The right-hand side is then zero, but the equation still relates object and image distances. Looking down into a swimming pool from above, for example, you see virtual images of objects in the pool. The objects appear closer than they actually are, as Fig. 36-30 and the following exercise illustrate.

EXERCISE The bottom of a swimming pool looks to be 1.5 m below the surface. What is the pool's actual depth?

Answer: 2.0 m

Some problems similar to Example 36-6: 29, 30, 32–34

(a)

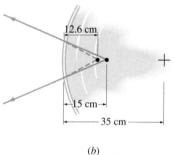

(b)

FIGURE 36-29 (a) A cylindrical aquarium. (b) Top view, showing formation of a virtual image of a fish that is actually 15 cm from the aquarium wall.

FIGURE 36-30 Refraction at a flat surface produces an image I that appears closer than the object O. Objects at the bottom of a swimming pool, for example, appear closer than they really are.

Lenses, Thick and Thin

Figure 36-31 shows a lens of arbitrary thickness t made from material with refractive index n. For simplicity we will consider the lens surrounded by air or vacuum, with refractive index equal to 1. An object O_1 lies a distance ℓ_1 to the left of the left-hand surface. This surface focuses light from O_1 to form image I_1. Light from this image impinges on the right-hand surface, forming a second image I_2. We want to relate the original object O_1 and final image I_2.

We developed Equation 36-8, our description of refraction at a single curved surface, using Fig. 36-28. In that figure we considered formation of a real image, but we quickly noted that Equation 36-8 applies to virtual and real images alike. In considering a two-surface lens it proves more convenient to place the object where its image in the first surface is a virtual one, as in Fig. 36-31. Again, our final result will apply whatever the object placement.

At the left-hand surface in Fig. 36-31, the quantities in Equation 36-8 are $\ell = \ell_1$, $\ell' = \ell'_1$, $n_1 = 1$, $n_2 = n$, and $R = R_1$, so Equation 36-8 becomes

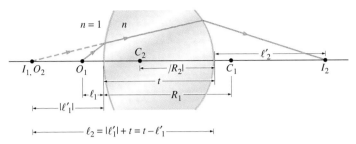

FIGURE 36-31 A thick lens. The left-hand surface forms the virtual image I_1, which is also labelled O_2 because it forms the object for the right-hand surface. Together, the two surfaces then produce the final real image I_2. Absolute value signs are used on distances whose algebraic signs in Equation 36-8 are negative—i.e., the distance ℓ_1' to the virtual image and the curvature radius R_2 of the right-hand surface, which is concave toward the incident light.

$$\frac{1}{\ell_1} + \frac{n}{\ell_1'} = \frac{n-1}{R_1}. \qquad \text{(left-hand surface)}$$

The "object" O_2 for the right-hand surface is the image I_1 since light incident on that surface looks like it's coming from I_1. The distance ℓ in Equation 36-8 for this "object" is $\ell = \ell_2 = t - \ell_1'$, with t the lens thickness. Why minus? Because I_1 is a *virtual* image, so its image distance ℓ_1' is a *negative* quantity. The physical distance between the image and the lens is the positive quantity $-\ell_1'$, and that quantity adds to the lens thickness to give the distance from this intermediate image to the right-hand surface. Also at the right-hand surface, $\ell' = \ell_2'$, $n_1 = n$, $n_2 = 1$, and $R = R_2$. Thus, at the right-hand surface Equation 36-8 reads

$$\frac{n}{t - \ell_1'} + \frac{1}{\ell_2'} = \frac{1-n}{R_2}. \qquad \text{(right-hand surface)}$$

Now we'll let the lens become arbitrarily thin, taking the limit $t \to 0$. Then the first term in our right-surface equation becomes $-n/\ell_1'$. But the term n/ℓ_1' also occurs in the left-surface equation. So we can add the two equations to eliminate the intermediate-image distance ℓ_1'. Since we'll be left with only one object distance, ℓ_1, and one image distance, ℓ_2', we can drop the subscripts 1 and 2 on these quantities. The result is

$$\frac{1}{\ell} + \frac{1}{\ell'} = (n-1)\left(\frac{1}{R_1} - \frac{1}{R_2}\right). \qquad (36\text{-}9)$$

The left-hand side of this equation is identical to the left-hand side of Equation 36-7. So we can equate the right-hand sides of the two equations to get an expression for the focal length of our lens:

$$\frac{1}{f} = (n-1)\left(\frac{1}{R_1} - \frac{1}{R_2}\right). \qquad \text{(lensmaker's formula)} \qquad (36\text{-}10)$$

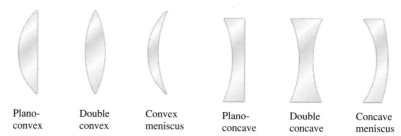

| Plano-convex | Double convex | Convex meniscus | Plano-concave | Double concave | Concave meniscus |

FIGURE 36-32 Common lens types.

We emphasize that the radii R_1 and R_2 in Equation 36-10 may be positive or negative, depending on the lens curvature. In Fig. 36-31, for example, R_1 is positive because the left-hand surface is convex toward the incident light but R_2 is negative because the right-hand surface is concave toward the light. One of the radii can also be infinite if the corresponding surface is flat.

Lenses are made in a variety of shapes, as shown in Fig. 36-32. Those that are thicker at the center are converging lenses, for which Equation 36-10 gives a positive focal length. Those that are thinner at the center are diverging lenses, with negative focal length. These characterizations of lens behavior reverse if the medium surrounding the lens has a higher refractive index than the lens itself (see Problem 71, which generalizes Equation 36-10 for an arbitrary refractive index in the surrounding medium).

● EXAMPLE 36-7 A PLANO-CONVEX LENS

The plano-convex lens in Fig. 36-32 is made from material with refractive index n. Given that the curved surface has curvature radius R, find an expression for the focal length of the lens. Show that the focal length is the same whether light is incident on the lens from right or left.

Solution

With light incident from the left in Fig. 36-32, $R_1 = R$. A flat surface has infinite curvature, so $R_2 = \infty$. Then Equation 36-10 gives

$$f = \left[(n-1)\left(\frac{1}{R} - \frac{1}{\infty} \right) \right]^{-1} = \frac{R}{n-1},$$

since $1/\infty = 0$. If light is incident from the right, then $R_1 = \infty$ and, since the curved surface is now concave toward the light, $R_2 = -R$. Then

$$f = \left[(n-1)\left(\frac{1}{\infty} - \frac{1}{-R} \right) \right]^{-1} = \frac{R}{n-1},$$

showing that the two focal points are indeed equidistant from the lens.

EXERCISE A double-convex lens (see Fig. 36-32) is made of glass with $n = 1.75$. If the curvature radii of the two sides are equal, what should that radius be in order to give the lens a 25-cm focal length?

Answer: 37.5 cm

Some problems similar to Example 36-7: 36–38 ●

Lens Aberrations

Lenses exhibit several types of optical defects. We've already described **spherical aberration** in connection with mirrors; this defect occurs because a spherical mirror only approximates the ideally focusing shape of a parabola. Lenses

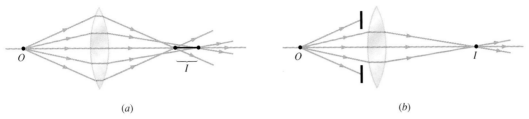

(a)   (b)

FIGURE 36-33 (a) Spherical aberration occurs because rays striking the lens farther from the axis and at greater angles focus closer to the lens than do near-axial rays. The result is a "smearing" of the image. (b) This defect can be minimized by using only the central portion of the lens, at the expense of a dimmer image.

FIGURE 36-34 Chromatic aberration occurs because the refractive index varies with wavelength, causing different colors to focus at different points.

with spherical surfaces also exhibit spherical aberration (Fig. 36-33a). In our analysis of image formation by a curved refracting surface we had to make the small-angle approximation; otherwise, we would not have reached the conclusion that all rays from an object come to a common focus. Spherical aberration can be minimized by ensuring that incident rays are as close and as parallel as possible to the lens axis. That's easy for distant objects but harder with objects close to the lens. Using only the central portion of the lens eliminates those rays at larger angles (Fig. 36-33b), leading to a sharper focus. That's why a camera focuses over a wider range when it is "stopped down," the outer part of its lens covered by an adjustable diaphragm. The tradeoff, of course, is that there is less light available.

Because the refractive index varies with wavelength, Equation 36-10 predicts different focal lengths for different colors. The result is **chromatic aberration** (Fig. 36-34). High-quality optical systems minimize this defect by using composite lenses made from different materials whose differing refractive indices allow several colors to focus at the same point. Chromatic aberration is unique to lenses since mirrors reflect light of all colors in exactly the same way. Figure 36-35 shows both spherical and chromatic aberration in a simple magnifying lens.

Figure 36-36 shows several other common lens defects. Modern lenses are often made with complicated nonspherical shapes to limit aberrations, giving undistorted images over wider ranges than the spherical lenses we have been discussing. Optical engineers still use ray tracing, but now they do so with computers.

FIGURE 36-35 This magnifying lens shows both spherical aberration (distortion of the lines) and chromatic aberration (colors).

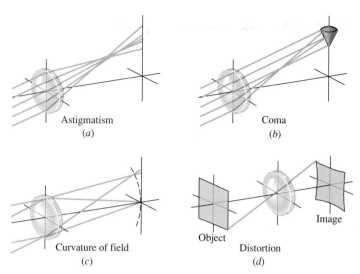

FIGURE 36-36 Lens aberrations. (*a*) Astigmatism results when the focusing quality of the lens varies with angular position, smearing images more in some directions. (*b*) Coma, caused by poor focusing of off-axis objects, produces images with comet-like tails. (*c*) Curvature of field describes the fact that extended objects focus on a curve rather than a plane. (*d*) Distortion changes straight lines into curves.

■ **APPLICATION** FRESNEL LENSES

Large-diameter lenses are impracticably thick. Lighthouses, in particular, cannot use single lenses to concentrate their intense beams because too much light would be absorbed in the glass. To solve this problem the French scientist Augustin Fresnel, in 1822, devised a way of making large lenses thin. A Fresnel lens consists of concentric rings shaped to approximate segments of an ordinary lens surface, with step-like jumps to keep the overall structure thin (Fig. 36-37*a*). Fresnel lenses are still used in lighthouses (Fig. 36-37*b*), while Fresnel lenses made from plastic sheets appear in home, office, and car windows to provide magnification or wide-angle views (Fig. 36-37*c*). They are also used in some overhead projectors.

FIGURE 36-37 (*a*) A Fresnel lens approximates a normal curved lens by a series of short segments. (*b*) A giant Fresnel lens from a nineteenth-century lighthouse. (*c*) Fresnel lens made from sheet plastic mounted on the rear window of a van provides a wide-angle view of what's behind the van.

36-5 OPTICAL INSTRUMENTS

Numerous optical instruments make use of mirrors and/or lenses. We have already considered some simple applications using a single lens or mirror, including film and slide projectors, magnifying glasses, and vanity mirrors. More complicated systems use several optical elements (lenses or mirrors), but the principles we have developed still apply. In particular, we trace light through a sequence of optical elements by using the image formed by one element as the object for the next.

The Eye

Our eyes are our primary optical instruments. Each eye is a complex optical system with several refracting surfaces and mechanisms to vary both the focal length and the amount of light admitted. Figure 36-38 shows that the eye is essentially a fluid-filled ball about 2.3 cm in diameter. Light enters through the hard, transparent cornea and passes through a fluid called the aqueous humor before entering the lens. On exiting the lens, light traverses the vitreous humor that fills the main body of the eyeball and finally strikes the retina. The retina is covered with light-sensitive cells of two types. Cells called cones are sensitive to different colors but require moderate amounts of light, while rod cells function at low intensity but lack color discrimination. Both types of cells produce electrochemical signals that carry information to the brain.

A properly functioning eye produces a well-focused real image on the retina. Contrary to general belief, most of the refraction is provided not by the lens but by the cornea. The lens, however, is the adjustable element. Ciliary muscles pulling on the lens alter its focal length, allowing fine adjustment of the eye's focus—a process known as accommodation. Other muscles automatically adjust the iris, enlarging or contracting the pupil opening to compensate for different light levels.

When the ciliary muscles are relaxed, the lens is relatively flat, with focal length about 1.7 cm. But for nearsighted (myopic) people, the image still forms in front of the retina, causing distant objects to appear blurred. Diverging

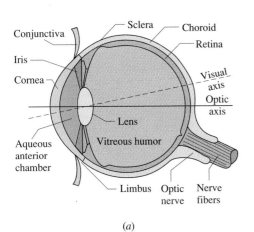

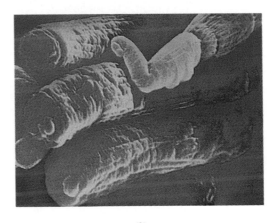

(a)   (b)

FIGURE 36-38 (a) The human eye. (b) Scanning-electron-microscope image showing rods and a cone (blue structure) in a human retina.

FIGURE 36-39 (*a*) A myopic eye focuses light from distant objects in front of the retina. (*b*) A diverging lens corrects the problem, creating a virtual image closer to the eye. The eye's image of this virtual image then falls on the retina.

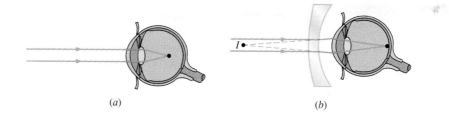

(*a*)

(*b*)

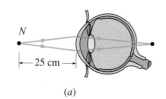

(*a*)

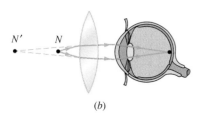

(*b*)

FIGURE 36-40 (*a*) A farsighted eye cannot focus light from the standard 25-cm near point at *N*. (*b*) A converging corrective lens makes light from an object at the near point appear to come from the point *N'* which is at the minimum distance where this eye can focus.

corrective lenses produce closer intermediate images that the myopic eye can then focus (Fig. 36-39). In farsighted (hyperopic) eyes, the image of nearby objects would focus behind the retina, and converging corrective lenses are used (Fig. 36-40). Actually, everyone is farsighted, in the sense that there is a minimum distance, called the **near point,** below which the eye cannot focus sharply. In the typical human the near point is at about 25 cm, and this value is taken as a standard. By the time one is 50, however, the typical near point is at about 40 cm, and it can easily be several meters.

Prescriptions for corrective lenses usually specify the corrective power, *P*, in **diopters,** where the diopter measure of a lens is the inverse of its focal length in meters. Thus, a 1-diopter lens has $f = 1$ m, while a 2-diopter lens has $f = 0.5$ m and is more powerful in that it refracts light more sharply.

The human eye is a complex biological system, in many ways a direct extension of the brain. Although we understand its optical behavior, we understand much less clearly how visual perception actually occurs. Recently biologists and computer scientists have teamed to produce an electronic "silicon retina" whose behavior mimics many subtle aspects of the natural eye and may therefore shed light on the process of perception.

● EXAMPLE 36-8 AN AUTHOR'S EYES

Writing this book has prematurely strained your author's eyes, which now cannot achieve clear focus at distances less than 1.2 m. What diopter lenses will correct this problem?

Solution

"Correct" in this context means to bring the range for clear focus to the standard 25-cm near point. According to Fig. 36-40*b*, that means we want a lens that will produce a virtual image at 1.2 m of an object at 25 cm. The required power follows directly from Equation 36-7:

$$P = \frac{1}{f} = \frac{1}{\ell} + \frac{1}{\ell'} = \frac{1}{0.25 \text{ m}} + \frac{1}{-1.2 \text{ m}} = 3.17 \text{ diopters},$$

where the image distance is negative because the image is virtual.

EXERCISE A nearsighted person cannot see clearly beyond 80 cm. Prescribe a lens power that will image the most distant objects at 80 cm, giving clear vision at all distances.

Answer: −1.25 diopters, the minus sign signifying a diverging lens

Some problems similar to Example 36-8: 46–48 ●

Cameras

A camera is much like the eye, except that it focuses its image on film or a light-sensitive electronic device (Fig. 36-41). Today most still cameras use the long-established technology of film based on light-sensitive chemicals, but the development of silicon chips known as CCDs (charge-coupled devices) has already made film obsolete in astronomical imaging and may someday have the

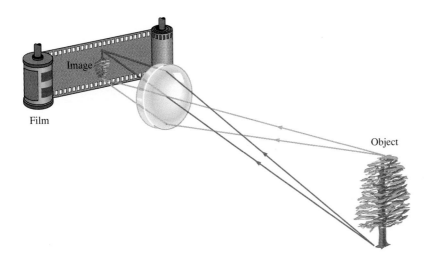

FIGURE 36-41 Optical system of a simple camera.

same effect on conventional photography. Video cameras also use CCDs.

Films and other light-sensitive media require that the amount of light admitted to a camera be regulated for optimum image quality. Too much light, and the image will be overexposed and "washed out"; too little and it will be too dark. Obviously, a larger lens admits more light. But the image brightness also depends on how much the lens concentrates the light. A lens with a short focal length bends light more, and will thus produce a brighter—but smaller—image than a lens of equal diameter that has a longer focal length. Suppose that light of a given intensity is incident on a lens of diameter d. Since intensity is power per unit area, the rate at which light energy enters the camera will be proportional to d^2. For distant objects, Equations 36-6 and 36-7 show that the image size scales with the focal length, and therefore, the image area is proportional to f^2. Thus, the intensity at the image will be proportional to $(d/f)^2$. The quantity d/f is called the **speed** of the lens. A fast lens—larger d/f—gives a brighter image and, therefore, can capture that image with a shorter exposure time.

Photographers and camera manufacturers usually use the inverse of the lens speed, called the focal ratio, **f-ratio,** or f-stop, equal to f/d. Increasing the f-ratio decreases the image intensity. Since that intensity scales as $(d/f)^2$, an increase in the f-ratio by a factor of $\sqrt{2}$ results in a halving of the image intensity. That's why adjustable cameras are commonly marked with f-ratios of 1.4, 2, 2.8, 4, 5.6, 8, 11, . . . ; these numbers are the rounded square roots of 2, 4, 8, 16, 32, 64, 128, ...

In a camera with a fixed-focus lens, changing the f-ratio is accomplished with an adjustable iris that covers part of the lens (Fig. 36-42). We've already seen that lenses focus best when only the central part of the lens is used. This is the cause of a photographer's common dilemma: to use a larger lens aperture to gather more light, or a smaller aperture to get a greater range of distances in focus.

Many of today's cameras automatically adjust both the focus and the f-ratio to optimize image quality, although professional photographers usually prefer to retain control over these settings. Increasingly, too, cameras incorporate zoom lenses whose focal lengths can be altered by physically moving the lens elements (Fig. 36-43). A lens of fixed diameter is necessarily slower at its telephoto (larger focal length) settings, as Problem 49 shows.

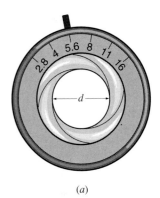

(a)

(b)

FIGURE 36-42 An adjustable diaphragm "stops down" the lens, decreasing its effective diameter and increasing the f-ratio.

FIGURE 36-43 Photos taken from the same point with (a) short and (b) long focal length settings of an adjustable zoom lens.

(a)

(b)

Magnifiers and Microscopes

We wouldn't need optical instruments to examine small objects if we could bring our eyes arbitrarily close to the objects. But we've seen that the average human eye cannot focus much closer than about 25 cm. We therefore use lenses to put enlarged images of small objects at distances where our eyes can focus.

What matters is not so much the actual size of the image, but how much of our field of view it occupies. Consequently we define the **angular magnification** m as the ratio of the angle subtended by an object seen through a lens to the angle subtended as seen by the naked eye when the object is at the standard 25-cm near point. Figure 36-44a shows that the former angle, α, is given by $\alpha = h/25$ cm, where h is the object height and α is in radians. The maximum magnification would occur with the image itself at the near point (see Problem 51), but it's more comfortable to view a very distant image. We therefore place the object at just inside the focal length, forming an enlarged virtual image at a great distance. Figure 36-44b shows this geometry, from which we see that the angle β subtended by the image is essentially h/f. Then the magnification is

$$m = \frac{\beta}{\alpha} = \frac{h/f}{h/25 \text{ cm}} = \frac{25 \text{ cm}}{f} \qquad \text{(simple magnifier)}. \qquad (36\text{-}11)$$

This angular magnification is achieved only with the eye very close to the lens, which is not the way we normally hold an ordinary magnifying glass. But it is the way we place our eyes on many common instruments that use simple magnifying lenses for their eyepieces.

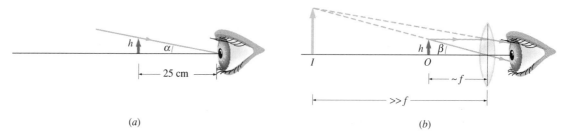

(a)

(b)

FIGURE 36-44 (a) An object of height h subtends a small angle $\alpha \simeq h/25$ cm at the standard 25-cm near point. (b) Putting the object near the focus of a converging lens gives an image that subtends an angle $\beta \simeq h/f$. The angular magnification is the ratio $m = \beta/\alpha$.

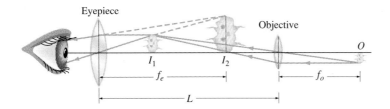

FIGURE 36-45 Image formation in a compound microscope. Figure is not to scale; distance L should be much greater than either focal length, and the image I_1 should be very near the eyepiece's focus, resulting in greater magnification.

Single lenses can produce angular magnifications of about 4 before aberrations compromise the image quality. Higher power magnification therefore requires more than one lens. A **compound microscope** is a two-lens system in which a lens of short focal length called the **objective** forms a magnified real image. This image is then viewed with a second lens, the **eyepiece,** positioned as a simple magnifier (Fig. 36-45). The object being viewed is positioned just beyond the focal length of the objective lens, and its image falls just inside the focal length of the eyepiece. If both focal lengths are small compared with the distance between the lenses, then the object distance for the objective lens is approximately the objective focal length f_o, and the resulting image distance is approximately the lens spacing L. The real image formed by the objective lens is larger than the object by the ratio of the image to object distance, or $-L/f_o$. The eyepiece enlarges the image further, by a factor of its angular magnification 25 cm$/f_e$. Then the overall magnification of the microscope is

$$M = M_o m_e = -\frac{L}{f_o}\left(\frac{25 \text{ cm}}{f_e}\right), \qquad \text{(compound microscope)} \quad (36\text{-}12)$$

where, as usual, the minus sign signifies an inverted image.

Optical microscopes work well as long as the approximation of geometrical optics holds—that is, when the object being viewed is much larger than the wavelength of light. Viewing smaller objects requires waves of shorter wavelength. In the widely used electron microscope, those "waves" are electrons, whose wavelike nature we will examine in Chapter 39.

Telescopes

A telescope collects light from distant objects, either forming an image or supplying light to instruments for analysis. Telescopes are classified as **refracting** or **reflecting,** depending on whether the main light-gathering element is a lens or mirror. Small hand-held telescopes and binoculars are refracting instruments, as are telephoto camera lenses and some older astronomical telescopes. Modern astronomical telescopes are almost invariably reflectors.

The simplest refracting telescope consists of a single lens that images distant objects at essentially its focal point. Film or a CCD placed at that point captures the image, or an eyepiece is used to view the image. The world's largest refracting telescope, at Yerkes Observatory in Wisconsin, has a 1-m-diameter lens with 12-m focal length. Photographers can think of it as a 12,000-mm telephoto lens. Figure 36-46 shows the imaging process in an astronomical refracting telescope. The focal points of the objective and eyepiece lenses are nearly coincident, so the real image of a distant object that forms at the objective's focus is then seen through the eyepiece as a greatly enlarged virtual image. The angular

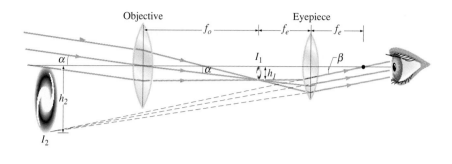

FIGURE 36-46 Image formation in a refracting telescope. A distant object, in this case a galaxy, is imaged first at the focus of the objective lens (image I_1). An eyepiece with its focus at nearly the same point then gives an enlarged virtual image (I_2). The angles α and β are given by $\alpha = h_1/f_o$ and $\beta = h_1/f_e$ in the small-angle approximation, leading to Equation 36-13 for the angular magnification.

FIGURE 36-47 Hexagonal mirror of the Keck Telescope on Hawaii's Mauna Kea shows in this photo taken through the slit in the telescope's dome. The mirror is 10 m across.

magnification is the ratio of the angle β subtended by the final image to the angle α subtended by the actual object; Fig. 36-46 shows that this ratio is

$$m = \frac{\beta}{\alpha} = \frac{f_o}{f_e}. \qquad \text{(refracting telescope)} \qquad (36\text{-}13)$$

Since a real image is inverted and a virtual image is upright, a two-lens refracting telescope gives an inverted image. This is fine for astronomical work, but telescopes designed for terrestrial use have an extra lens, a diverging eyepiece (see Problem 70) or a set of reflecting prisms (as in binoculars; see Fig. 35-14b) to produce an upright image.

Reflecting telescopes offer many advantages over refractors. Mirrors have reflective coatings on their front surfaces, eliminating chromatic aberration because light need not pass through glass. Having only one optically active surface allows much larger reflectors to be built since the mirrors can be supported across their entire back surfaces—unlike lenses, which can be supported only at the edges. Where the largest refracting telescope ever built has a 1-m-diameter lens, the newest reflectors boast diameters in the 10-m range (Fig. 36-47). These designs incorporate segmented and/or flexible mirrors whose shape can be adjusted under computer control for optimum focusing; with so-called adaptive optics, such systems may adjust rapidly enough to compensate for the atmospheric turbulence that has traditionally limited the resolution of ground-based telescopes.

The simplest reflecting telescope is a curved mirror with a CCD or film at its focus. Superb image quality results, in principle limited only by wave effects we will discuss in the next chapter. More often the telescope is used as a "light bucket," collecting light from stars or other distant sources too small to image even with today's large optical telescopes. Then a secondary mirror sends light to a focus at a convenient spot for telescope-mounted instrumentation. Optical fibers may also be used to bring light collected by the primary mirror to fixed instruments. Alternatively, an eyepiece can be mounted to examine the image visually. Figure 36-48 shows several common designs for reflecting telescopes.

Magnification is not a particularly important quantity in astronomical telescopes, which are used more for spectral and other analysis than for direct imaging. More important is the light-gathering power of the instrument, which is determined simply by the area of its objective lens or primary mirror. The 10-m Keck telescope, for instance, has 100 times the light-gathering power of the 1-m Yerkes refractor and more than 17 times the power of the 2.4-m Hubble Space Telescope.

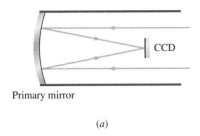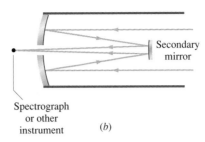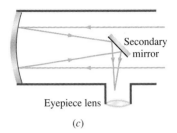

Primary mirror

(a)

Spectrograph or other instrument *(b)*

Eyepiece lens *(c)*

FIGURE 36-48 Common arrangements for reflecting telescopes. (*a*) Placing a CCD or film at the prime focus gives the best image quality. (*b*) In the Cassegrain design light reflects from a secondary mirror and passes through a hole in the primary. The secondary may be either plane or convex. This design is widely used in large telescopes, with the added option of placing detectors directly at the prime focus as in (*a*). (*c*) The Newtonian design, used primarily in small telescopes, has an angled secondary mirror to direct light to the eyepiece.

CHAPTER SYNOPSIS

Summary

1. Reflection and refraction can result in **images,** which are **real** or **virtual** depending on whether or not light actually comes from the image location.
2. Plane mirrors form only virtual images, equal in size to the object being imaged. Concave parabolic mirrors have a **focus** at which parallel rays meet; such mirrors form upright, magnified virtual images of objects located closer than the focus and inverted real images of objects beyond the focus. Concave spherical mirrors approximate the focusing behavior of parabolic mirrors. Convex curved mirrors form only virtual images, upright and reduced in size. The **mirror equation** quantifies the relation between object and image distances ℓ and ℓ':

$$\frac{1}{\ell} + \frac{1}{\ell'} = \frac{1}{f},$$

where the **focal length** f is the distance from the mirror's apex to its focus. The **magnification** is the ratio $M = -\ell'/\ell$, where a negative ℓ' corresponds to a virtual image and therefore to a positive M and so to an upright image.

3. **Lenses,** made from transparent materials, are shaped in such a way that they refract light to a focus. In the **thin-lens** approximation, the curvature radii of the lens surfaces are much greater than the thickness of the lens. Such a lens has two foci, located equal distances f on either side of the lens. The object and image distances and magnification then obey equations identical to those for mirrors. The focal length of a thin lens is given by the **lensmaker's equation:**

$$\frac{1}{f} = (n - 1)\left(\frac{1}{R_1} - \frac{1}{R_2}\right),$$

where R_1 and R_2 are the curvature radii of the lens surfaces, taken as positive when the surface is convex toward the incident light. A positive focal length f corresponds to a **converging lens** and a negative f to a **diverging lens.** Lens **aberrations** are focusing defects resulting from imperfect lens shapes and the variation of refractive index with wavelength.

4. Optical instruments use lenses and/or mirrors to form useful images. The power of instrument lenses is often given in **diopters**—the inverse of the focal length in meters. Image brightness depends on the lens diameter d in relation to the focal length f and is characterized by the **f-ratio** f/d.

Terms You Should Understand

(Pairs are closely related terms whose distinction is important; number in parentheses is chapter section where term first appears.)

real image, virtual image (introduction)
focus, focal length (36-2)
magnification (36-2)
thin lens (36-3)
converging lens, diverging lens (36-3)
spherical aberration, chromatic aberration (36-2, 36-3)
near point (36-5)
diopter (36-5)
f-ratio (36-5)
angular magnification (36-5)
compound microscope (36-5)
reflecting telescope, refracting telescope (36-5)

Symbols You Should Recognize

ℓ, ℓ' (36-2, 36-3)
f (36-2, 36-3)
M, m (36-2—36-4)

Problems You Should be Able to Solve

finding mirror images and their sizes using ray tracing and algebraic techniques (36-2)
finding lens images and their sizes using ray tracing and algebraic techniques (36-3)
analyzing image formation at a single refracting surface (36-4)

designing lenses and analyzing their focal properties (36-4)
designing and analyzing simple optical instruments (36-5)

Limitations to Keep in Mind

Our description of imaging is based on the approximation of geometrical optics, valid only when objects and optical components are much larger than the wavelength of light.

Formulas for image formation in spherical mirrors and lenses are approximations, valid under the conditions that the curved surfaces used form only a small portion of a full sphere, and that all light rays are nearly parallel to the mirror or lens axis.

QUESTIONS

1. How can you see a virtual image, when it really "isn't there"?
2. You're trying to photograph yourself in a mirror, using an autofocus camera that sets its focus by bouncing ultrasound waves off the subject at which the camera is pointed. Why might the photo come out blurred?
3. You lay a magnifying glass (which is just a converging lens) on a printed page. Looking toward the glass, you move it away from the page. Explain the changes in what you see, especially as you move the lens beyond its focal length.
4. Under what circumstances will the image in a concave mirror be the same size as the object?
5. Describe the shapes of the mirrors making the images in Fig. 36-49.

FIGURE 36-49 Question 5.

6. If you're handed a converging lens, what can you do to estimate quickly its focal length?

7. What is the meaning of a negative object distance? Of a negative focal length?
8. A diverging lens always makes a reduced image. Could you use such a lens to start a fire by focusing sunlight? Explain.
9. Is there any limit to the temperature you can achieve by focusing sunlight? Think about the second law of thermodynamics.
10. Can a concave mirror make a reduced real image? A reduced virtual image? An enlarged real image? An enlarged virtual image? Give conditions for each image that is possible.
11. If you placed a screen at the location of a virtual image, would you see the image on the screen? Why or why not?
12. Where must the object be placed to form a reduced real image with a concave mirror or converging lens?
13. Where should you place a flashlight bulb in relation to the focus of its reflector?
14. Is the image on a movie screen real or virtual? How do you know?
15. Does a fish in a spherical bowl appear larger or smaller than it actually is?
16. A block of ice contains a hollow, air-filled space in the shape of a double-convex lens. Describe the optical behavior of this space.
17. The refractive index of the human cornea is about 1.38. If you can see clearly in air, why can't you see clearly underwater? Why do goggles help?
18. Why does "stopping down" a lens or mirror allow it to focus more sharply?
19. If you increase the f-stop setting on your camera from 4 to 5.6, what happens to the amount of light admitted through the lens?
20. The compound microscope and the refracting telescope are both two-lens optical systems. How do their designs reflect their different uses?
21. Cheap binoculars sometimes show blurred images with different colors evident across the blurred region. Why?

22. Do you want a long or short focal length for the objective lens of a telescope? Of a microscope?
23. Give several reasons reflecting telescopes are superior to refractors.
24. Given that star images cannot be resolved even with telescopes, why are large telescopes superior to ordinary cameras for studying stars?

PROBLEMS

Sections 36-1 and 36-2 Plane and Curved Mirrors

1. A shoe store uses small floor-level mirrors to let customers view prospective purchases. At what angle should such a mirror be inclined so that a person standing 50 cm from the mirror with eyes 140 cm off the floor can see her feet?
2. Two plane mirrors occupy the first four meters of the positive x and y axes, as shown in Fig. 36-50. Find the locations of all images of an object at $x = 2$ m, $y = 1$ m.

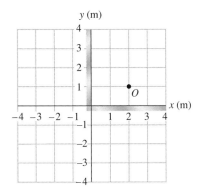

FIGURE 36-50 Problems 2, 3.

3. Suppose a stick were placed in Fig. 36-50 with its left end at O and extending parallel to the x axis. How many units long could it be for an observer at $x = 1$ m, $y = 1$ m to be able to see its entire length reflected in the x-axis mirror, which ends at $x = 4$ m.
4. A candle is 36 cm from a concave mirror with focal length 15 cm and on the mirror axis. (a) Where is its image? (b) How do the image and object sizes compare? (c) Is the image real or virtual?
5. An object is five focal lengths from a concave mirror. (a) How do the object and image heights compare? (b) Is the image upright or inverted?
6. The McMath-Pierce solar telescope at Kitt Peak National Observatory (Fig. 36-51) uses a single concave mirror to produce a real solar image 80 cm in diameter. (a) What is the focal length of the mirror? (b) How far is the image from the mirror? *Hint*: Consult Appendix E.
7. A virtual image is located 40 cm behind a concave mirror with focal length 18 cm. (a) Where is the object? (b) By how much is the image magnified?

FIGURE 36-51 The McMath-Pierce solar telescope (Problem 6).

8. (a) Where on the axis of a concave mirror would you place an object in order to produce a full-size image? (b) Will the image be real or virtual?
9. A 12-mm-high object is 10 cm from a concave mirror with focal length 17 cm. (a) Where, (b) how high, and (c) what type is its image?
10. An object's image in a 27-cm-focal-length concave mirror is upright and magnified by a factor of 3. Where is the object?
11. What is the curvature radius of a mirror that produces a 9.5-cm-high virtual image of a 5.7-cm-high object, when the object is located 22 cm from the mirror?
12. When viewed from Earth, the moon subtends an angle of 0.5° in the sky. How large an image of the moon will be formed by the 3.6-m-diameter mirror of the Canada-France-Hawaii telescope, which has a focal length of 8.5 m?
13. At what two distances could you place an object from a 45-cm-focal-length concave mirror in order to get an image 1.5 times the object's size?
14. Very distant objects are imaged close to the focus of a concave mirror. If a birdwatcher's telescope has a concave mirror with focal length 85 cm, what is the minimum distance at which a bird's image will be within 5 mm of the focal plane? (The focal plane is the plane through the focus and perpendicular to the mirror axis.)

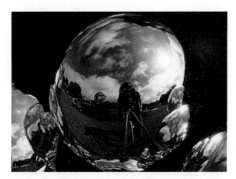

FIGURE 36-52   Problem 15.

FIGURE 36-53   The Yerkes Observatory boasts the world's largest refracting telescope (Problem 22).

15. You look into a reflecting sphere 80 cm in diameter and see an image of your face at one-third its normal size (Fig. 36-52). How far are you from the sphere's surface?

Section 36-3 Lenses

16. By what factor is the image magnified when an object is placed 1.5 focal lengths from a converging lens? Is the image upright or inverted?
17. A lens with 50-cm focal length produces a real image the same size as the object. How far from the lens are image and object?
18. A real image is 4 times as far from a lens as is the object. What is the object distance, measured in focal lengths?
19. How far from a page should you hold a lens with 32-cm focal length in order to see the print magnified 1.6 times?
20. A converging lens has a focal length of 4.0 cm. A 1.0-cm-high arrow is located 7.0 cm from the lens with its lowest point 5.0 mm above the lens axis. Make a full-scale drawing of the situation, and use ray tracing to locate the image. Confirm using the lens equation.
21. A simple camera uses a single converging lens to focus an image on its film. If the focal length of the lens is 45 mm, what should be the lens-to-film distance for the camera to focus on an object 80 cm from the lens?
22. The largest refracting telescope in the world, at Yerkes Observatory, has a 1-m-diameter lens with focal length 12 m (Fig. 36-53). If an airplane flew 1 km above the telescope, where would its image occur in relation to the images of the very distant stars?
23. A small magnifying glass can focus sunlight sufficiently to ignite paper. Calculate the actual size of the solar image produced by a magnifying glass consisting of an optically perfect lens with 25-cm focal length. *Hint:* You'll need to consult Appendix E.
24. By holding a magnifying glass 25 cm from my desk lamp, I find I can focus an image of the lamp's bulb on a wall 1.6 m from the lamp. What is the focal length of my magnifying glass?

25. A lens has focal length $f = 35$ cm. Find the type and height of the image produced when a 2.2-cm-high object is placed at distances (a) $f + 10$ cm and (b) $f - 10$ cm.
26. How far apart are object and image produced by a converging lens with 35-cm focal length when the object is (a) 40 cm and (b) 30 cm from the lens?
27. An object and its lens-produced real image are 90 cm apart. If the lens has 20-cm focal length, what are the possible values for the object distance and magnification?
28. An object is placed two focal lengths from a diverging lens. (a) What type of image forms, (b) what is the magnification, and (c) where is the image?

Section 36-4 The Lensmaker's Formula

29. You're standing in a wading pool and your feet appear to be 30 cm below the surface. How deep is the pool?
30. A tiny insect is trapped 1.0 mm from the center of a spherical dew drop 4.0 mm in diameter. As you look straight into the drop at the insect, what is its apparent distance from the edge of the drop?
31. Use Equation 36-8 to show that an object at the center of a glass sphere will appear to be its actual distance—one radius—from the edge. Draw a ray diagram showing why this result makes sense.
32. You're underwater, looking through a spherical air bubble (Fig. 36-54). What is its actual diameter if it appears, along your line of sight, to be 1.5 cm in diameter?
33. Rework Example 36-6 for a fish 15 cm from the *far* wall of the tank.
34. Consider the inverse of Example 36-6: You're inside a 70-cm-diameter hollow tube containing air, and the tip of your nose is 15 cm from the wall of the tube. The tube is

FIGURE 36-54 Problem 32.

immersed in water, and a fish looks in. To the fish, what is the apparent distance from your nose to the tube wall?

35. Two specks of dirt are trapped in a crystal ball, one at the center and the other halfway to the surface. If you peer into the ball on a line joining the two specks, the outer one appears to be only one-third of the way to the other. What is the refractive index of the ball?

36. My magnifying glass is a double convex lens with equal curvature radii of 32 cm. If the lens glass has $n = 1.52$, what is its focal length?

37. Two lenses made of the same material have the same focal length. One is plano-convex, the other double convex with both curvatures the same. How do the curvature radii of the two lenses compare?

38. For what refractive index would the focal length of a plano-convex lens be equal to the curvature radius of its one curved surface?

39. An object is 28 cm from a double convex lens with $n = 1.5$ and curvature radii 35 cm and 55 cm. Where and what type is the image?

40. A double convex lens has equal curvature radii of 35 cm. An object placed 30 cm from the lens forms a real image at 128 cm. What is the refractive index of the lens?

41. A plano-convex lens has curvature radius 20 cm and is made from glass with $n = 1.5$. Use the generalized lens-maker's formula given in problem 71 to find the focal length when the lens is (a) in air, (b) submerged in water ($n = 1.333$) and (c) embedded in glass with $n = 1.7$. Comment on the sign of your answer to (c).

42. A slide projector has a double convex lens of focal length 104 mm. (a) What should be the slide-to-focal point distance to focus the image on a screen 4.5 m distant? What will be the magnification? (b) Repeat for the case when the lens is replaced by a 78-mm version.

43. Two plano-convex lenses are geometrically identical, but one is made from crown glass ($n = 1.52$), the other from flint glass. An object at 45 cm from the lens focuses to a real image at 85 cm with the crown-glass lens and at 53 cm with the flint-glass lens. Find (a) the curvature radius (common to both lenses) and (b) the refractive index of the flint glass.

44. A double convex lens with equal 38-cm curvature radii is made from glass with refractive indices $n_{red} = 1.51$, $n_{blue} = 1.54$ at the edges of the visible spectrum. If a point source of white light is placed on the lens axis at 95 cm

from the lens, over what range will its visible image be smeared?

45. An object placed 15 cm from a plano-convex lens made of crown glass focuses to a virtual image twice the size of the object. If the lens is replaced with an identically shaped one made from diamond, what type of image will appear and what will be its magnification? See Table 35-1.

Section 36-5 Optical Instruments

46. You find that you have to hold a book 55 cm from your eyes for the print to be in sharp focus (Fig. 36-55). What power lens is needed to correct your farsightedness?

55 cm

FIGURE 36-55 Problem 46.

47. My grandmother's new reading glasses have 3.8-diopter lenses to provide full correction of her farsightedness. Her old glasses were 2.5 diopters. (a) Where is the near point for her unaided eyes? (b) Where will be the near point if she wears her old glasses?

48. A particular eye has a focal length of 2.0 cm instead of the 2.2 cm that would be required for a sharply focused image on the retina. (a) Is this eye nearsighted or farsighted? (b) What power of corrective lens is needed?

49. A camera's zoom lens covers the focal length range from 38 mm to 110 mm. (a) You point the camera at a distant object and photograph it first at 38 mm and then with the camera zoomed out to 110 mm. Compare the sizes of its images on the two photos. (b) If the camera's lowest f-ratio is 3.8 at 38 mm, what is it at 110 mm? Assume the effective lens area doesn't change.

50. A camera can normally focus as close as 60 cm, but it has provisions for mounting additional lenses at the outer end of the main lens to provide closeup capability. What type and power of auxiliary lens will allow the camera to focus as close as 20 cm? The distance between the two lenses is negligible.

51. The maximum magnification of a simple magnifier occurs with the image at the 25-cm near point. Show that the angular magnification is then given by $m = 1 + \dfrac{25 \text{ cm}}{f}$, where f is the focal length.

52. A compound microscope has objective and eyepiece focal lengths of 6.1 mm and 1.7 cm, respectively. If the lenses are 8.3 cm apart, what is the magnification of the instrument?

53. A 300-power compound microscope has a 4.5-mm-focal-length objective lens. If the distance from objective to eyepiece is 10 cm, what should be the focal length of the eyepiece?

54. To the unaided eye, the planet Jupiter has an angular diameter of 50 arc seconds. What will be its angular size when viewed through a 1-m-focal-length refracting telescope with an eyepiece whose focal length is 40 mm?

55. A Cassegrain telescope like that shown in Fig. 36-48b has 1.0-m focal length, and the convex secondary mirror is located 0.85 m from the primary. What should be the focal length of the secondary in order to put the final image 0.12 m behind the front surface of the primary mirror?

56. The Hubble Space Telescope is essentially a Cassegrain reflector like that shown in Fig. 36-48b. The focal lengths of the concave primary and convex secondary mirrors are 5520.00 mm and −679.00 mm, respectively. The secondary is located at 4906.071 mm from the apex of the primary. Draw a ray diagram showing the path of initially parallel incident rays through the telescope, and use appropriate equations to determine where in Fig. 36-48b such rays are finally focused.

Paired Problems

(Both problems in a pair involve the same principles and techniques. If you can get the first problem, you should be able to solve the second one.)

57. (a) How far from a 1.2-m-focal length concave mirror should you place an object in order to get an inverted image 1.5 times the size of the object? (b) Where will the image be?

58. (a) How far from a 48-cm-focal length concave mirror should you place an object in order to get an upright image 1.5 times the size of the object? (b) Where will the image be?

59. Find the focal length of a concave mirror if an object 15 cm from the mirror has a virtual image 2.5 times the object's actual size.

60. An object is held 6.0 cm from the surface of a reflecting ball, and its image appears three-quarters full size. What is the ball's diameter?

61. How far from a 1.6-m focal length concave mirror should you place an object to get an upright image magnified by a factor of 2.5?

62. How far from a 25-cm-focal-length lens should you place an object to get an upright image magnified by a factor of 1.8?

63. An object and its lens-produced real image are 2.4 m apart. If the lens has 55-cm focal length, what are the possible values for the object distance and magnification?

64. An object and its converging-lens-produced virtual image are 2.4 m apart. If the lens has 55-cm focal length, what are the possible values for the object distance and magnification?

65. An object is 68 cm from a plano-convex lens whose curved side has curvature radius 26 cm. The refractive index of the lens is 1.62. Where and of what type is the image?

66. Both surfaces of a double concave lens have curvature radii of 18 cm, and the refractive index is 1.5. If a virtual image appears 14 cm from the lens, where is the object?

Supplementary Problems

67. A distant object subtends an angle α at a lens of focal length f, as shown in Fig. 36-56. Take the object distance $\ell \gg f$, and assume the angle α is very small. In this approximation, find an expression for the size of the real image formed by the lens. *Hint:* For $\ell \gg f$, what is the approximate image distance?

68. Show that the powers of closely-spaced lenses add; that is, placing a 1-diopter lens in front of a 2-diopter lens gives the equivalent of a single 3-diopter lens.

69. Show that identical objects placed equal distances on either side of the focal point of a concave mirror or converging lens produce images of equal size. Are the images of the same type?

70. Galileo's first telescope used the arrangement shown in Fig 36-57, with a double convex eyepiece placed slightly before the focus of the objective lens. Use ray tracing to

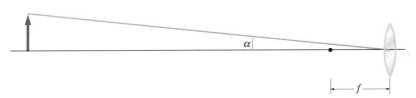

FIGURE 36-56 Problem 67.

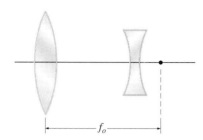

FIGURE 36-57 Problem 70.

show that this system gives an upright image, which makes the design useful for terrestrial observing.

71. Generalize the derivation of the lensmaker's formula (Equation 36-10) to show a lens of refractive index n_{lens} in an external medium with index n_{ext} has focal length given by

$$\frac{1}{f} = \left(\frac{n_{lens}}{n_{ext}} - 1\right)\left(\frac{1}{R_1} - \frac{1}{R_2}\right).$$

72. An object is located 40 cm from a lens made from glass with $n = 1.5$. In air, the resulting image is real and is located 32 cm from the lens. When the entire system is immersed in a liquid, the image becomes virtual and is reduced to half the object's size. Use the result given in the preceding problem to find the refractive index of the liquid.

73. A Newtonian telescope like that of Fig. 36-48c has a primary mirror with 20-cm diameter and 1.2-m focal length. (a) Where should the flat diagonal mirror be placed to put the focus at the edge of the telescope tube? (b) What shape should the flat mirror have to minimize blockage of light to the primary?

74. A parabola is described by the equation $y = x^2/4f$, where f is the distance from vertex to focus. (a) Find the equation for the circle (which would be a sphere in three dimensions) that closely approximates the parabola $y = x^2$. *Hint:* How are the focal length and curvature radius of a spherical mirror related? (b) Solve the circle's equation for y, and use the binomial theorem to show that the equation reduces to that of the parabola for $x \ll 1$.

75. Just before Equation 36-9 are two equations describing refraction at the two surfaces of a lens with thickness t. Combine these equations to show that the object distance ℓ and image distance ℓ' for such a lens are related by

$$\frac{1}{\ell} + \frac{1}{\ell'} - \frac{[(n-1)\ell - R_1]^2 t}{\ell R_1[t(\ell + R_1) + n\ell(R_1 - t)]}$$
$$= (n - 1)\left(\frac{1}{R_1} - \frac{1}{R_2}\right).$$

76. Use the result of the preceding problem to find an expression for the focal length of a transparent sphere with radius R and refractive index n, as measured from the exit surface.

CHAPTER 37

INTERFERENCE AND DIFFRACTION

The colors seen in a soap bubble result from constructive interference of light waves reflecting off the inner and outer surfaces of the thin film that forms the bubble wall. Note that the far end of the bubble appears dark; this is because it is too thin for constructive interference at any visible wavelength.

The preceding chapters considered reflection and refraction, and their application in image formation, from the approximation of geometrical optics—an approximation valid when we can ignore the wave nature of light. We now turn to **physical optics,** which treats optical phenomena in which the wave nature of light plays an essential role. Two related phenomena, interference and diffraction, are central to much of physical optics.

37-1 INTERFERENCE

We considered wave **interference** in Chapter 16, showing how this phenomenon arises when two or more waves meet at the same place, their amplitudes adding in a manner that may be constructive or destructive depending on their relative phases. Interference of electromagnetic waves—including light—occurs in the same way; since both the electric and magnetic fields obey the superposition principle, the net fields at a point are the vector sums of the fields of individual waves. Depending on their directions, the superposition of those fields may result in an increase (constructive interference) or a decrease (destructive interference) in the overall field strength.

Coherence

Although interference occurs any time two waves interact, we only get steady interference patterns when the interacting waves are **coherent,** meaning that they maintain a constant phase relation (Fig. 37-1a,b). Coherence also requires that the two waves have exactly the same well-defined wavelength. (Waves of a single, sharply defined wavelength are called **monochromatic** because they correspond to a single color of light.) Very slight deviations from exactly equal wavelengths may still permit observable interference, the optical equivalent of beats. Even with lasers, whose light output is highly monochromatic and coherent, it is difficult to keep two different sources in exactly the same phase relation. But when light from a single source is split into two or more beams then, even if the source is itself incoherent, the beams will always be in the same phase relation and will therefore produce an interference pattern (Fig. 37-1c).

Interference in Thin Films

We know that light is partially reflected and partially transmitted at an interface between two transparent media. Figure 37-2a shows what happens when light propagating in air hits a thin film of transparent material. Partial reflection at

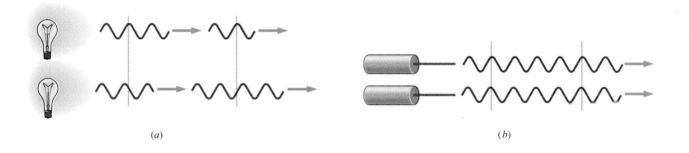

(a)   (b)

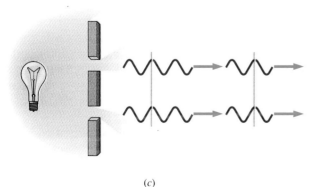

(c)

FIGURE 37-1 (a) Light bulbs emit incoherent light, consisting of wavetrains whose phases are related in a random way. Wavetrains from two different light bulbs would have random phase relations, and superposing them would not result in a steady interference pattern. (b) Laser light is coherent over much longer distances and therefore two lasers could be made to interfere if their wavelengths were exactly the same. (c) Two separate beams derived from a single point source maintain a constant phase relation even when the source itself is incoherent. Here light from a single bulb has been split by passing it through two holes in an opaque barrier. In this case the two beams are exactly in phase; making one beam traverse a longer path would introduce a constant phase difference. A single beam could also be split by partial reflection.

FIGURE 37-2 Reflection and refraction at a thin layer of transparent material. (*a*) Reflected beams 1 and 2 emerging at left are coherent and should therefore interfere. (*b*) The reflected and incident beams are 180° out of phase when they reflect at an interface that goes from a lower to a higher refractive index. (*c*) There is no phase difference on reflection when the incident medium has a higher refractive index.

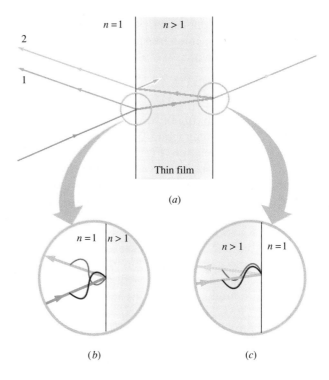

both surfaces results in two beams propagating back into the original medium. Since these beams derive from the same source they are coherent, and they should therefore interfere. (Additional reflections actually give rise to multiple beams, each successively fainter; here we consider only the first two reflected beams.)

Before we can analyze this interference, we need to understand exactly what happens when light reflects. In Section 17-5 we showed that waves on a string are partially reflected and partially transmitted at a junction between two strings with different mass per unit length. When the wave met a string with *greater* mass per unit length, it was reflected *with a 180° phase change.* When it met a string with *lower* mass per unit length, it was reflected *with no phase change.* The physical reason for these phase relations was shown in Fig. 17-7 for the extreme cases of infinite mass per unit length and zero mass per unit length. Although we will not prove it, Maxwell's equations show that light behaves in the same way, with the refractive index playing the role of mass per unit length. That is, light propagating in a material with refractive index n_1 reflects with a 180° phase change at the interface with a material of index n_2 if $n_1 < n_2$. If, on the other hand, $n_2 < n_1$, then there is no phase change. Figures 37-2*b* and *c* show these phase relations.

Now suppose the film in Fig. 37-2 were negligibly thin compared with the wavelength of light. Then the reflected beams 1 and 2 would be 180° out of phase since beam 1 reflects with a 180° phase change and beam 2 with no phase change. If the film has thickness d, then there will also be a phase difference because beam 2 travels an additional distance. For simplicity we assume that light is incident on the film at nearly normal incidence, so the extra distance traveled by beam 2 is essentially $2d$. What phase difference does this introduce? An extra half wavelength corresponds to a phase change of 180° (Fig. 37-3*a*), while a full wavelength separates the two waves by a full cycle, putting them back in phase (Fig. 37-3*b*). The 180° phase change would also occur if the path lengths differed by $\frac{3}{2}\lambda$, $\frac{5}{2}\lambda$, and so forth, and there would be no change if they differed by any integer number of wavelengths. There's also the 180° phase change introduced by the different reflections, which combines with the phase change due to the path-length difference to put beams 1 and 2 *in phase* if the path-length difference is $\frac{1}{2}\lambda$, $\frac{3}{2}\lambda$, $\frac{5}{2}\lambda$, and so forth. Conversely, beams 1 and 2 are out of phase by 180° if the path length is an integer multiple of the wavelength.

If beams 1 and 2 are in phase they interfere constructively, making the combined reflected beam brighter than it otherwise would be. We've just seen that the condition for this constructive interference is that the path-length difference $2d$ be an odd integer multiple of a half wavelength. We can express this mathematically by writing

$$2d = (m + \tfrac{1}{2})\lambda_n,$$

where m is a nonnegative integer and where λ_n is the wavelength *in the transparent film* since that's where the extra path length occurs. We saw in Chapter 35 that the wavelength in a material with refractive index n is reduced by a factor $1/n$ from its value in air or vacuum. If λ is the wavelength in air, then $\lambda_n = \lambda/n$, and our condition for constructive interference becomes

$$2nd = (m + \tfrac{1}{2})\lambda. \qquad (37\text{-}1a)$$

Conversely, destructive interference occurs when

$$2nd = m\lambda. \qquad (37\text{-}1b)$$

The destructive interference is not complete, however, since the two beams do not have the same intensity.

Suppose a film is illuminated with light of a single color—that is, a single wavelength λ. If the thickness of a film varies with position, then constructive interference described by Equation 37-1a will occur at different places. Thus there will be bright bands separated by darker areas in which destructive interference occurs. If, on the other hand, the film is illuminated with white light, then the interference conditions will be satisfied at different places for the different colors that comprise white light. The bands of color you see in a soap bubble or oil slick result from such thin-film interference (Fig. 37-4). If part of the film is too thin to meet the constructive interference condition of Equation 37-1a for *any* visible wavelength, then that part of the film will appear dark. You can see this at the top of the soap film in Fig. 37-4 and also in the photo at the beginning of this chapter, where one end of the soap bubble appears dark.

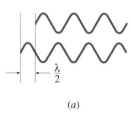

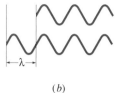

(a)

(b)

FIGURE 37-3 (*a*) Two waves begin to propagate at the same instant, but the lower one travels half a wavelength farther. When they come together they are 180° out of phase. (*b*) When the lower wave travels a full wavelength farther, they come together in phase.

FIGURE 37-4 Interference patterns formed in a soap film illuminated with white light. The film's thickness increases toward the bottom, resulting in constructive interference for a given wavelength occurring at several different vertical positions. The film is too thin at the top for constructive interference at any visible wavelength, so it appears dark.

● **EXAMPLE 37-1** A SOAP FILM

A rectangular wire loop 20 cm high is dipped into a soap solution and then held vertically, producing a soap film whose thickness varies linearly from essentially zero at the top to 1.0 μm at the bottom (Fig. 37-5). (a) If the film is illuminated with 650-nm light, how many bright bands (as shown in Fig. 37-4) will appear? (b) What region of the film would be dark if it were illuminated with white light? The refractive index of the film is that of water, $n = 1.33$.

Solution

The film thickness at the position of a bright band is given by Equation 37-1a. To find the number of bright bands, we can solve this equation for the value of m at which the thickness d would equal the maximum film thickness:

$$m = \frac{2nd}{\lambda} - \frac{1}{2} = \frac{(2)(1.33)(1.0 \ \mu m)}{0.65 \ \mu m} - \frac{1}{2} = 3.6.$$

Since m must be an integer, the largest m corresponding to a thickness within the 1.0-μm maximum is 3. Since m ranges from zero to 3, there are a total of 4 bright bands.

At a given wavelength the minimum thickness for constructive interference occurs at $m = 0$. Taking the minimum wavelength for visible light at 400 nm = 0.40 μm, Equation 37-1a with $m = 0$ then gives

$$d = \frac{\lambda}{4n} = \frac{0.40 \ \mu m}{(4)(1.33)} = 0.0752 \ \mu m.$$

Since the film goes linearly from zero thickness to a maximum of 1.0 μm over its 20-cm height, this thickness occurs 0.0752 of the way from the top edge, or at $(0.0752)(20 \ cm) = 1.5 \ cm$.

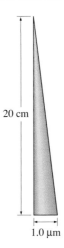

FIGURE 37-5 Cross section through a vertical soap film. The angle between the surfaces is greatly exaggerated.

Above this level the film is too thin for constructive interference in visible light, so the top 1.5 cm of film appears darker.

EXERCISE At what distances from the top of the film will bright bands appear if the film in Example 37-1 is illuminated with 450-nm light?

Answer: 1.69 cm, 5.08 cm, 8.46 cm, 11.8 cm, 15.2 cm, 18.6 cm

Some problems similar to Example 37-1: 1–3, 5 ●

We derived Equations 37-1 on the assumption that our thin film was surrounded by air. Frequently, though, a thin film is sandwiched between two different media—as when an oil film floats on water. If the intermediate medium has a higher refractive index than the medium at its rear (i.e., at the right in Fig. 37-2), then our results still apply. But if the rear medium has a greater refractive index than the thin film, then there will be an additional 180° phase change at the rear interface, and therefore, our interference conditions will reverse (see the application below, as well as Problems 63 and 64).

■ **APPLICATION** ANTIREFLECTION COATINGS

Partial reflection limits the amount of light that can be transmitted from one transparent material into another. That limitation, described quantitatively in Equation 35-7 and Figure 35-26, reduces the light-gathering power of cameras, binoculars, and other lens-based instruments and of nonimaging devices like windows and solar collectors.

Coating lenses, photovoltaic cells, and other critical light-gathering components with thin layers of appropriate materials can reduce reflection through the use of destructive interference. Since energy is conserved, the result is necessarily greater transmission into the light-gathering device. Normally such **antireflection coatings** have refractive indices between those of air and glass; consequently, there is a 180° phase change at *each* interface, and thus, Equation 37-1a rather than 37-1b gives the condition for *destructive* interference. Thus, the minimum thickness for an antireflection coating, given by Equation 37-1a with $m = 0$, is $d = \lambda/4n$. Since this result depends on wavelength, antireflection coatings are not equally effective for all colors. With composite lenses and multiple layers, however, reduced reflection is possible over the visible spectrum.

A perfect antireflection coating might seem impossible, even in principle, since the reflected rays 1 and 2 in Fig. 37-2 do not have the same intensity. But the situation is actually more complicated. There are additional rays resulting from multiple reflections inside the transparent material, and these also interfere. A full solution of the problem involving the application of Maxwell's equations at both interfaces shows that the reflection becomes exactly zero when the antireflection layer has not only the right thickness but also refractive index $n_2 = \sqrt{n_1 n_3}$, where n_1 and n_3 are the indices of the two media separated by the antireflection coating. A widely used antireflection coating, magnesium fluoride (MgF_2), has $n = 1.38$, a value that approximates the zero-reflection condition for high-index flint glasses in air.

Interference in thin layers is the basis of some very sensitive optical measuring techniques. The shape of a lens, for example, can be checked for accuracy to within a fraction of the wavelength of light using **Newton's rings**—interference patterns formed by interference of light reflected between the lens and a perfectly flat glass plate (Fig. 37-6; see also Problems 73-74).

The Michelson Interferometer

A number of optical instruments use interference for precise measurement of small distances. Among the simplest and most important of these is the **Michelson interferometer,** invented by the American physicist Albert Michelson and used in the 1880s by Michelson and his colleague Edward W. Morley in a famous experiment that paved the way for the theory of relativity. We discuss the Michelson-Morley experiment in the next chapter; here we describe the interferometer, which is still used for precision measurements.

Figure 37-7 shows the basic design of the Michelson interferometer. The key idea is that light from a monochromatic source is split into two beams by a half-silvered mirror called a **beam splitter.** The beam splitter is set at a 45° angle, so the reflected and transmitted beams travel perpendicular paths. Each then reflects off a flat mirror and returns to the beam splitter. The beam splitter again transmits and reflects half the light incident on it, with the result that some light from the originally separated beams is recombined. Since this light shares a common source, it is coherent and therefore interferes. The interference pattern is observed with a viewing lens located at the bottom of Fig. 37-7.

Suppose the path lengths in the two perpendicular arms of the interferometer were exactly the same or differed by a multiple of the wavelength. When the beams recombined they would then undergo fully constructive interference. If the path-length difference were a multiple of a half wavelength, on the other hand, the beams would interfere destructively. In reality, however, the mirrors are never exactly perpendicular, and therefore light reflecting from different parts of the mirrors experiences slightly different path lengths and thus recombines with different phase lags. To the observer, the result is a series of alternating light and dark fringes corresponding to constructive and destructive interference.

Now suppose one mirror is moved slightly. The path-length differences change and therefore, the interference pattern shifts. Moving the mirror a mere quarter wavelength, for example, adds an extra half wavelength to the round-trip

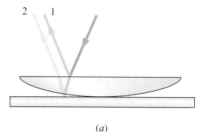

(a)

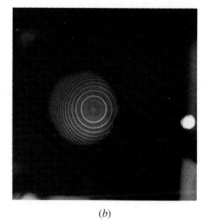

(b)

FIGURE 37-6 (a) Newton's rings arise from the difference in path lengths between rays like 1 and 2 shown here. (b) Ring pattern produced in a test of a telescope's optical system.

FIGURE 37-7 (*a*) Schematic diagram of a Michelson interferometer. The beam splitter splits incident light into reflected and transmitted beams of equal intensity. (*b*) The observer looking through the viewer sees interference fringes arising from differing optical path lengths.

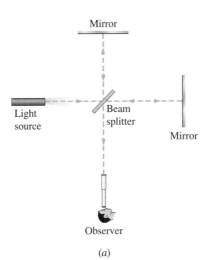

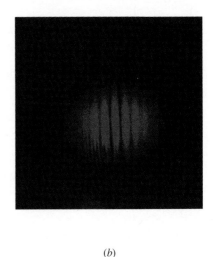

(*a*)

(*b*)

FIGURE 37-8 A modern interferometer, assembled from common laboratory optical components. The light source is a helium-neon laser producing 633-nm red light, and the lens at the right is used to project the interference fringes onto a screen. The fringes shown in Fig. 37-7*b* were produced with this interferometer.

path. That results in an additional 180° phase shift, moving dark fringes to where light ones were, and vice versa. An observer looking through the viewer as the mirror moves sees the fringes shift to their new positions. Shifts of a fraction of the distance between fringes are readily detected, allowing measurement of mirror displacements to within a small fraction of the wavelength.

A similar fringe shift occurs if a transparent material is placed in one path, retarding the beam because of its refractive index. This approach allows accurate measurements of the refractive indices of gases, which are so close to one that conventional methods would not be useful (see Problem 16).

Measurements with the Michelson interferometer depend not on the interference pattern itself but on *changes* in that pattern. Therefore, it doesn't really matter whether or not the path lengths are exactly equal. Whatever those lengths, there will be an interference pattern, and changes in that pattern will reflect changes in the relative optical paths. What is important is that components of the interferometer be mounted securely so their positions cannot change by even a fraction of the wavelength of light. The Michelson-Morley interferometer was constructed on a huge stone slab floated in mercury. Today, interferometers using lasers and precision optical components are easily assembled on optical tables designed for stability and isolation from vibration (Fig. 37-8).

● EXAMPLE 37-2 SANDSTORM

A sandstorm has pitted the aluminum mirrors of a desert solar energy installation, and engineers want to know the depths of these pits so they can estimate the mirrors' useful lifetimes. They construct a Michelson interferometer with a sample from one of the pitted mirrors in place of one of the flat mirrors. With a 633-nm laser as the light source, the interference pattern shown in Fig. 37-9 results. What is the approximate depth of the pit?

Solution

The "bumps" in the interference pattern correspond to light that has traveled the extra distance into the bottom of the pit and back. A shift of one full cycle in the interference pattern would correspond to a full wavelength change in the path. As Fig. 37-9 shows, the pit causes a fringe shift corresponding to about 0.2 wavelengths. Since light makes a round trip through the pit, its actual depth is therefore 0.1λ, or 63 nm. Try measuring that with a meter stick!

EXERCISE A Michelson interferometer uses sodium light with wavelength 589.6 nm. As one mirror is moved, a photocell connected to a counter "watches" a fixed point through the viewer. If the counter records 4878 fringes passing that point, how far did the mirror move?

Answer: 1.438 mm

Some problems similar to Example 37-2: 14–16

FIGURE 37-9 Fringe pattern resulting from a pitted mirror. The size of the bump is a measure of the pit depth.

37-2 HUYGENS' PRINCIPLE AND DIFFRACTION

The Dutch scientist Christian Huygens was the first to suggest that light might be a wave, and in 1678 he stated a principle that remains useful for describing how light waves propagate. Although Huygens worked nearly two centuries before Maxwell, his principle, like all results in classical optics, can be derived ultimately from Maxwell's equations of electromagnetism. **Huygens' principle** states the following:

> **All points on a wavefront act as point sources of spherically propagating "wavelets" that travel at the speed of light appropriate to the medium. At a short time Δt later, the new wavefront is the unique surface tangent to all the forward-propagating wavelets.**

Figure 37-10 shows how Huygens' principle accounts for the propagation of plane and spherical (or cylindrical) waves. It is also possible to derive the laws of reflection and refraction from Huygens' principle.

Diffraction

Diffraction is the bending of light or other waves as they pass by objects. Figure 37-11 shows plane waves incident on an opaque barrier containing a hole. Since the waves are blocked by the barrier, Huygens' wavelets produced near the barrier edge cause the wavefronts to bend at the barrier (see blowups in Fig. 37-11). When the width of the hole is much greater than the wavelength, as in Fig. 37-11a, this diffraction is of little consequence, and the waves effectively propagate straight through the hole in a beam defined by the hole size. But when the hole size and wavelength are comparable, then wavefronts emerging from the hole spread in a broad pattern (Fig. 37-11b). Here diffraction is the dominant effect on the propagation. Thus diffraction, although it always occurs when waves pass an object, is significant only on length scales comparable to or

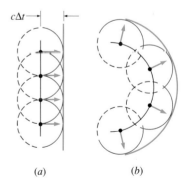

(a)   (b)

FIGURE 37-10 Application of Huygens' principle to (a) plane and (b) spherical waves. In each case the wavefront acts like a set of point sources that emit circular waves. A short time Δt later the wavelets have expanded to radius cΔt, and the new wavefront is the surface tangent to the wavelets. To follow the propagation further one could draw new wavelets originating on the new wavefront.

FIGURE 37-11 Plane waves incident on an opaque barrier with a hole in it are diffracted at the edges of the hole, as shown by Huygens' wavelet constructions. (*a*) When the hole size is much larger than the wavelength this diffraction is negligible, and the waves essentially propagate in a straight beam defined by the hole. (*b*) When the hole size and wavelength are comparable, the emerging wavefronts spread in a broad beam, approaching circular wavefronts as the hole size becomes negligibly small.

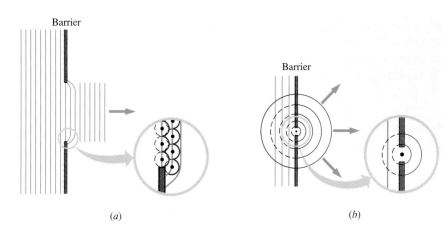

(*a*)   (*b*)

FIGURE 37-12 Diffraction of water waves as they pass through a gap in a breakwater results in circular wavefronts.

smaller than the wavelength. That's why we could ignore diffraction and assume that light always travels through a single medium in straight lines, when we considered optical systems with dimensions much larger than the wavelength of light.

Diffraction is not limited to light. When a stereo is playing in an adjacent room, you can hear the bass more clearly than the high notes. That's because the higher frequency sound does not diffract as much since its wavelength is shorter, and therefore, the situation is more like that of Fig. 37-11*a*. Lower notes, on the other hand, may have wavelengths comparable to the size of doorways or rooms, so they diffract readily. Similarly, diffraction of water waves occurs when their wavelength is comparable to the size of obstacles they encounter (Fig. 37-12).

In the remainder of this chapter we discuss interference and diffraction phenomena. Although we characterize phenomena by one or the other of these names, it's important to recognize that both interference and diffraction result from the wave nature of light and that both can play important roles in the same optical system. Often the wavefronts that result from diffraction subsequently interfere to produce characteristic patterns (Fig. 37-13). Ultimately this effect limits our ability to image small objects, as we will explore quantitatively later in this chapter.

37-3 DOUBLE-SLIT INTERFERENCE

In Chapter 16 we looked briefly at the interference patterns produced by a pair of coherent sources. With light, such a source pair can be made by passing light through a pair of narrow slits. In 1802 Thomas Young used this approach in an historic experiment that confirmed the wave nature of light. Young admitted sunlight to his laboratory through a small hole, then passed the light through a pair of closely spaced slits, after which it illuminated a screen. Waves diffract at each slit, resulting in cylindrical wavefronts that interfere in the region between slits and screen (Fig. 37-14*a*). Figure 37-14*b* shows how this interference should produce alternating regions of light and dark on the screen, while Fig. 37-14*c* shows that that is indeed what is observed. These light and dark regions are called interference **fringes.**

(a)

FIGURE 37-13 (a) Light waves diffract as they pass by the straight edge of an opaque barrier. Subsequent interference of the diffracted waves produces the interference fringes shown. (b) Diffraction of light passing by a screw again shows interference fringes and thus results in a fuzzy shadow of the screw. (c) Diffraction of light by a pair of crosshairs in a circular aperture. Diffraction effects from both the crosshairs (straight fringes) and the aperture itself (circular fringes) are evident.

(b)　　　　　(c)

The bright fringes in Fig. 37-14 occur at points where paths from the two slits are either the same length (for the central fringe) or differ by an integer number of wavelengths, causing the two waves to arrive in phase. Dark regions correspond to path lengths differing by an odd integer number of half wavelengths, making the waves 180° out of phase. We can relate the spacing of the interference fringes to the wavelength and slit spacing if we assume that the distance L

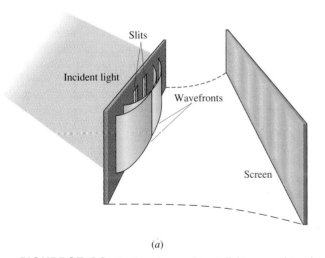

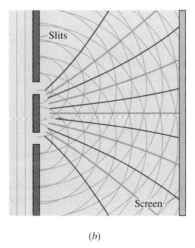

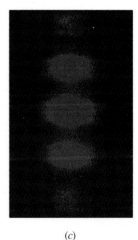

(a)　　　　　(b)　　　　　(c)

FIGURE 37-14 In Young's experiment, light passes through a pair of closely spaced slits. Light from the two slits subsequently interferes, producing a series of light and dark fringes on the screen. (a) Diagram showing the slits and one pair of wavefronts. Where the two wavefronts intersect the interference is constructive. Also shown is the screen, which in practice would be much farther from the slits. (b) Top view of the system, showing how interference between wavefronts should produce a pattern of light and dark fringes on the screen. (c) Photo produced by using film in place of the screen shows the expected fringe pattern.

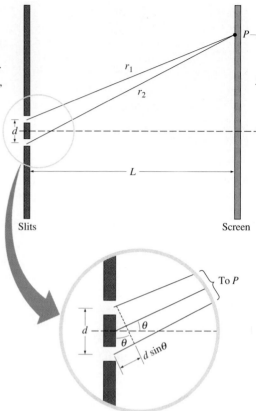

FIGURE 37-15 (a) Geometry for finding locations of the interference fringes. A bright fringe occurs when the path-length difference $r_2 - r_1$ is an integer multiple of the wavelength. (b) For $L \gg d$ the paths to P are nearly parallel, and the difference in their lengths is $d \sin\theta$. Can you find similar triangles to prove that the angles labeled θ are indeed the same?

from slits to screen is much greater than the slit spacing d (Fig. 37-15). Then the two paths r_1 and r_2 from the slits to a point P on the screen are very nearly parallel, and the angle each makes with the horizontal in Fig. 37-15 is essentially the same as the angle θ made by a line to P from the point midway between the slits. Figure 37-15b then shows that the length difference between these paths is essentially $d \sin\theta$. The condition that a bright fringe be located at point P is that this difference be an integer number of wavelengths, or

$$d \sin\theta = m\lambda. \qquad \text{(bright fringes)} \qquad (37\text{-}2a)$$

The integer m is the **order** of the fringe, with the central bright fringe being the zeroth order fringe and with higher order fringes on either side.

Waves interfere destructively when they arrive at the screen 180° out of phase, which occurs when their path lengths differ by an odd integer multiple of a half wavelength:

$$d \sin\theta = (m + \tfrac{1}{2})\lambda, \qquad \text{(dark fringes)} \qquad (37\text{-}2b)$$

where m is any integer.

In a typical double-slit experiment, L may be on the order of 1 m, d a fraction of 1 mm, and λ the sub-μm wavelength of visible light. Then we have the additional condition that $\lambda \ll d$. This makes the fringes very closely spaced on the screen, so the angle θ in Fig. 37-15b is small even for large orders m. Then

$\sin\theta \simeq \tan\theta = y/L$, and Equations 37-2 and the geometry of Fig. 37-15 show that a fringe's position y on the screen, measured from the central maximum, is given by

$$y_{\text{bright}} = m\frac{\lambda L}{d} \quad \text{and} \quad y_{\text{dark}} = (m + \tfrac{1}{2})\frac{\lambda L}{d}. \quad \text{(37-3a,b)}$$

These equations show that the fringe spacing depends on wavelength, as confirmed experimentally in Fig. 37-16. Measurement of fringe spacing enabled Young to determine the wavelength of light.

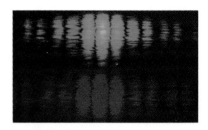

FIGURE 37-16 Interference patterns made with the same double-slit system, using two different wavelengths of light. The different fringe spacings reflect the wavelength dependence in Equations 37-3.

● EXAMPLE 37-3 LASER WAVELENGTH

A pair of narrow slits are 0.075 mm apart and are located 1.5 m from a screen. Laser light shining through the slits produces an interference pattern whose third-order bright fringe is 3.8 cm from the screen center. Find the wavelength of the light.

Solution
Here we have $m = 3$, $L = 1.5$ m, and $d = 0.075$ mm, so we can solve Equation 37-3a for the wavelength λ:

$$\lambda = \frac{y_{\text{bright}}d}{mL} = \frac{(0.038 \text{ m})(0.075\times10^{-3} \text{ m})}{(3)(1.5 \text{ m})} = 633 \text{ nm}.$$

This is in fact the wavelength of the red light from low-power helium-neon lasers commonly used in physics demonstrations.

EXERCISE What slit spacing will produce bright fringes 1.8 cm apart on a screen 85 cm from the slits, if the slits are illuminated with 589-nm light?

Answer: 27.8 μm

Some problems similar to Example 37-3: 20–23, 28

Intensity in the Interference Pattern

Geometric arguments allowed us to find the positions of the maxima and minima of a two-slit interference pattern. We can find the actual intensity variation by algebraically superposing the interfering waves. You might think this could be done by adding the wave intensities. But no! It's the electric and magnetic fields that obey the superposition principle, not the wave intensities. Intensity is proportional to the *square* of either field; if we added intensities we could never get the cancellation that occurs in destructive interference.

Consider again a point P on the screen of a double-slit apparatus (Fig. 37-17). Since light from the two slits reaches P over paths of different lengths, we expect that the expanding waves will have different intensities when they reach P. In the approximation $d \ll L$, however, the path-length difference $d\sin\theta$ is so small that we can neglect this effect. So we consider that the two waves arriving at P have electric fields of equal amplitude, given by expressions like Equation 34-10. Since we're considering a fixed point P, it is convenient to take the origin at P to eliminate the kx term in Equation 34-10. Then we can express the electric fields of the two waves in the form

$$E_1 = E_p \sin\omega t \quad \text{and} \quad E_2 = E_p \sin(\omega t + \phi),$$

where E_p is the common amplitude, ω the common frequency, and ϕ the phase difference that occurs because of the different path lengths. We have not bothered with vector notation because both waves, since they originate in the same source, are polarized in the same direction and thus their electric fields will simply add algebraically. Then net electric field at P is

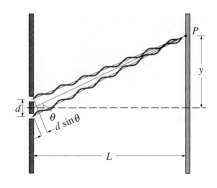

FIGURE 37-17   Waves leaving the slits with the same phase arrive at P displaced by the path-length difference $d\sin\theta$. This corresponds to phase difference $\phi = d\sin\theta(2\pi/\lambda)$.

$$E = E_1 + E_2 = E_p[\sin\omega t + \sin(\omega t + \phi)].$$

In Appendix A we find the trigonometric identity $\sin\alpha + \sin\beta = 2\sin[(\alpha+\beta)/2]\cos[(\alpha-\beta)/2]$, which, with $\alpha = \omega t$ and $\beta = \omega t + \phi$, transforms our expression for E into

$$E = 2E_p \sin\left(\omega t + \frac{\phi}{2}\right)\cos\left(\frac{\phi}{2}\right).$$

We also used $\cos(-x) = \cos x$ to eliminate a minus sign in the argument of the cosine. Thus, the electric field at P oscillates with the wave frequency ω, and its overall amplitude is scaled by the factor $\cos(\phi/2)$.

What is the phase difference ϕ? We've already seen that the path-length difference is $d\sin\theta$, with d the slit spacing and θ the angle subtended at the slits between P and the slit centerline. If this difference is one-half wavelength λ, then we have a phase difference of 180° or π radians. More generally, the phase difference in radians is whatever fraction of the full cycle (2π radians) the path difference $d\sin\theta$ is of the wavelength:

$$\phi = 2\pi\frac{d\sin\theta}{\lambda}.$$

Using this result in our expression for the electric field at P gives

$$E = 2E_p \sin\left(\omega t + \frac{\phi}{2}\right)\cos\left(\frac{\pi d\sin\theta}{\lambda}\right).$$

This equation describes a field whose peak amplitude is not E_p but $2E_p\cos(\pi d\sin\theta/\lambda)$. The average intensity then follows from Equation 34-20b:

$$\bar{S} = \frac{[2E_p\cos(\pi d\sin\theta/\lambda)]^2}{2\mu_0 c} = 4\bar{S}_0\cos^2\left(\frac{\pi d\sin\theta}{\lambda}\right), \tag{37-4}$$

where $\bar{S}_0 = E_p^2/2\mu_0 c$ is the average intensity of either wave alone. Since the cosine function varies between -1 and 1, Equation 37-4 shows that the intensity varies between zero and $4\bar{S}_0$ as the angular position changes. We can also write Equation 37-4 in terms of position y on the screen. Under the approximation $d \gg \lambda$ even high-order fringes will occur at small angles θ, so we can write $\sin\theta \simeq \tan\theta \simeq y/L$, giving

$$\bar{S} = 4\bar{S}_0\cos^2\left(\frac{\pi d}{\lambda L}y\right). \tag{37-5}$$

Now $\cos^2\alpha$ has its maximum value, 1, when its argument is an integer multiple of π. Thus, the maxima of Equation 37-5 occur when $dy/\lambda L$ is an integer m, or when $y = m\lambda L/d$. This is just the condition of Equation 37-3a, showing that our intensity calculation is fully consistent with the simpler geometrical analysis. But the intensity calculation tells us more: it gives not only the positions of bright and dark fringes, but also the intensity variation in between, as shown in Fig. 37-18.

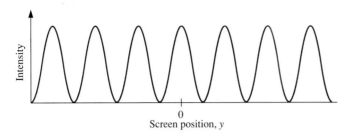

FIGURE 37-18 Intensity as a function of position y on the screen of a double-slit system.

37-4 MULTIPLE-SLIT INTERFERENCE AND DIFFRACTION GRATINGS

Systems with multiple slits play a crucial role in optical instrumentation and in the analysis of materials. As we will soon see, gratings manufactured with several thousand slits per centimeter make possible high-resolution spectroscopic analysis. At a much smaller scale the regularly spaced rows of atoms in a crystal act much like a multiple-slit system for x rays, and the resulting x-ray patterns reveal the crystal structure.

Figure 37-19 shows waves from three evenly spaced slits interfering at a screen. Maximum intensity requires that all three waves either be in phase, or differ in phase by an integer number of wavelengths. Our criterion for the maximum in a two-slit pattern, $d \sin \theta = m\lambda$, ensures that waves from two adjacent slits will add constructively. Since the slits are evenly spaced with distance d between each pair, waves coming through a third slit will be in phase with the other two if this criterion is met. So the criterion for a maximum in an N-slit system is still Equation 37-2a:

$$d \sin \theta = m\lambda. \quad \text{(maxima in multi-slit interference)}$$

With more than two waves, however, the criterion for destructive interference is more complicated. Somehow all the waves need to sum to zero. Figure 37-20 shows that this happens for three waves when each is out of phase with the others by one-third of a cycle. Thus, the path-length difference $d \sin \theta$ must be either $(m + \frac{1}{3})\lambda$ or $(m + \frac{2}{3})\lambda$, where m is an integer. The case $(m + \frac{3}{3})\lambda$ is excluded because then the path lengths differ by a full wavelength, giving constructive interference and thus a maximum in the interference pattern. More generally we can write

$$d \sin \theta = \frac{m}{N}\lambda, \quad (37\text{-}6)$$

for destructive interference in an N-slit system, where m is an integer *but not an integer multiple of N*. The reason for the exclusion is that when m is an integer multiple of N then m/N is an integer; then the path-length difference is an integer number of wavelengths, resulting in constructive rather than destructive interference. Mathematically, when m/N is an integer, Equation 37-6 becomes equivalent to Equation 37-2a that gives the condition for constructive interference.

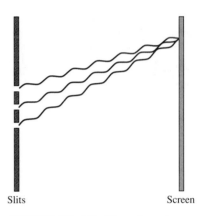

FIGURE 37-19 Waves from three evenly spaced slits interfere constructively when they arrive at the screen in phase.

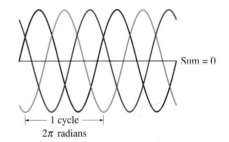

FIGURE 37-20 Waves from three slits must be out of phase by one-third of a cycle in order to interfere destructively. With this phase relationship the three waves sum to zero at every point.

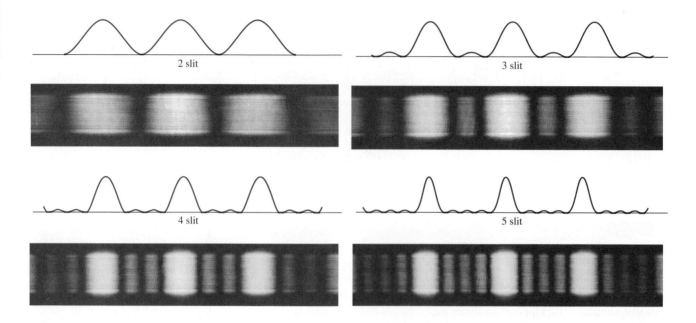

Figure 37-21 shows interference patterns and intensity plots from some multiple-slit systems. Note that the bright, or *primary,* maxima are separated by several minima and fainter, or *secondary,* maxima. Why this complex pattern? Our analysis of the three-slit system shows two minima between every pair of primary maxima; for example, we considered the minima at $d\sin\theta$ equal to $(m + \frac{1}{3})\lambda$ or $(m + \frac{2}{3})\lambda$, which lie between the maxima at $d\sin\theta$ equal to $m\lambda$ and $(m + 1)\lambda$. More generally, Equation 37-6 shows that there are $N - 1$ minima between each pair of primary maxima given by Equation 37-2a. The secondary maxima that lie between these minima result from interference that is neither fully destructive nor fully constructive. The figure shows that the primary maxima become much brighter and narrower as the number of slits increases, while the secondary maxima become relatively less bright. With a large number N of slits, then, we should expect a pattern of bright but narrow primary maxima, with broad, essentially dark regions in between.

Diffraction Gratings

A set of many very closely spaced slits is called a **diffraction grating** and proves very useful in the spectroscopic analysis of light. Diffraction gratings commonly measure several cm across and have several thousand slits—usually called lines—per cm. Gratings are made by photoreducing images of parallel lines or by ruling with a diamond stylus on aluminum-plated glass (Fig. 37-22). Gratings like the slit systems we have been discussing are **transmission gratings** since light passes through the slits. **Reflection gratings** produce similar interference effects by reflecting incident light.

We've seen that the maxima of the multi-slit interference pattern are given by the same criterion, $d\sin\theta = m\lambda$, that applies to a two-slit system. For $m = 0$ this equation implies that all wavelengths peak together at the central maximum, but for larger values of m the angular position of the maximum depends on wavelength. Thus, a diffraction grating can be used in place of a prism to

FIGURE 37-22 A diffraction grating disperses white light into its constituent colors. The enlarged spectrum at rear results from reflecting the diffracted beam off a concave mirror.

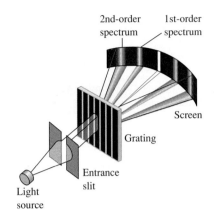

2nd-order spectrum
1st-order spectrum
Screen
Grating
Entrance slit
Light source

FIGURE 37-23 Essential elements of a grating spectrometer. The entrance slit regulates the amount of light entering the instrument. The grating disperses the light into its component wavelengths, which reach the screen to produce spectra of the different orders. An electronic detector would normally be used in place of the screen, and in most spectrometers the grating rotates to put different wavelengths on the detector.

disperse light into its component wavelengths, and the integer m is therefore called the **order** of the dispersion. Figure 37-23 shows a grating spectrometer working on this principle. Because the maxima in N-slit interference are very sharp for large N (recall Fig. 37-21), a grating with many slits diffracts individual wavelengths to very precise locations.

● **EXAMPLE 37-4** A GRATING SPECTROMETER

Light from glowing hydrogen contains discrete wavelengths ("spectral lines") called hydrogen-α and hydrogen-β, at 656.3 nm and 486.1 nm, respectively. Find the angular separation between these two wavelengths in a spectrometer using a grating with 6000 slits per cm. Consider both the first-order ($m = 1$) and second order ($m = 2$) dispersion.

Solution
With 6000 slits/cm, the slit spacing is $d = (1/6000)$ cm $= 1.667\ \mu$m. Applying the criterion $d\sin\theta = m\lambda$ for the first-order spectrum, we then have

$$\theta_{1\alpha} = \sin^{-1}\left(\frac{\lambda}{d}\right) = \sin^{-1}\left(\frac{0.6563\ \mu\text{m}}{1.667\ \mu\text{m}}\right) = 23.2°$$

and

$$\theta_{1\beta} = \sin^{-1}\left(\frac{\lambda}{d}\right) = \sin^{-1}\left(\frac{0.4861\ \mu\text{m}}{1.667\ \mu\text{m}}\right) = 17.0°.$$

Thus, the angular separation is 6.2°. Repeating the same calculation with $m = 2$ gives $\theta_{2\alpha} = 51.9°$ and $\theta_{2\beta} = 35.7°$, for an angular spread of 16.2°. This wider spacing is characteristic of higher order dispersion.

EXERCISE The bright yellow light emitted by glowing sodium vapor actually consists of two spectral lines at 589.0 nm and 589.6 nm. Find the angular separation of these lines in a second-order spectrum taken with a 4800-slit/cm grating.

Answer: 0.04°

Some problems similar to Example 37-4: 35–37 ●

In Example 37-4 the two lines are near the ends of the visible spectrum, and the values calculated for the angular positions show that there is no overlap between the first- and second-order visible spectra. But higher order spectra do overlap, a fact that astronomers use in high-resolution spectroscopy when they wish to observe two different spectral lines simultaneously. Problem 39 explores this situation, while Fig. 37-24 shows the relative positions of the different orders.

Resolving Power

The detailed shapes and wavelengths of spectral lines contain a wealth of information about the systems in which light originates (Fig. 37-25). Studying

FIGURE 37-24 Positions of the different orders in a grating spectrum. The vertical separation between orders has been introduced for clarity; in actuality they would overlap. The central line of each spectrum is at 550 nm. Note the increased dispersion of the higher orders.

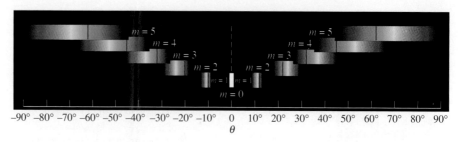

these details requires a high dispersion in order to separate nearby spectral lines or to analyze the intensity versus wavelength profile of a single line. Suppose we pass light containing two spectral lines of nearly equal wavelengths λ and λ' through a grating. Looking back at Fig. 37-21, you can see that the grating produces not just a single maximum for each wavelength in each order, but a

FIGURE 37-25 The interaction of a magnetic field with atomic electrons causes spectral lines to split. Plots show intensity versus wavelength for the 435.8-nm spectral line of mercury, with and without a 25-kG magnetic field. A high-resolution spectrometer is needed to resolve these lines, which are only 0.027 nm apart.

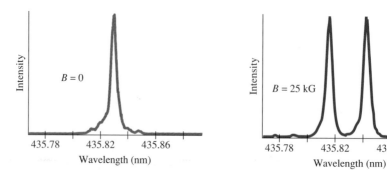

series of minima and lesser maxima in between. We will just be able to distinguish two spectral lines if the peak of one line corresponds to the first minimum of the other; any closer and the lines will blur together (Fig. 37-26). Suppose wavelength λ has its mth-order maximum at angular position θ. The criterion for this maximum is, as usual, $d \sin \theta_{\max} = m\lambda$. We can equally well write this as $d \sin \theta_{\max} = \dfrac{mN}{N}\lambda$, with N the number of slits in the grating. Equation 37-6 then shows that we get an adjacent minimum if we add 1 to the numerator mN. Thus, there is an adjacent minimum whose position satisfies

$$d \sin \theta_{\min} = \frac{mN + 1}{N}\lambda.$$

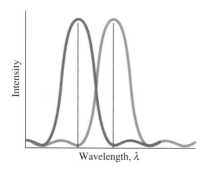

FIGURE 37-26 Two wavelengths are just distinguishable if the maximum of one falls on the first minimum of the other; any closer and they blur together.

Our criterion that the two wavelengths λ and λ' be just distinguishable is that the maximum for λ' fall at the location of this minimum for λ. But the maximum for λ' satisfies $d \sin \theta'_{\max} = m\lambda' = \dfrac{mN}{N}\lambda'$, so for $\theta'_{\max} = \theta_{\min}$ we must have

$$(mN + 1)\lambda = mN\lambda'.$$

It's convenient to express this result in terms of the wavelength difference $\Delta\lambda = \lambda' - \lambda$. Solving our equation relating λ and λ' for λ' and using the result to write $\Delta\lambda$ in terms of λ alone then gives

$$\frac{\lambda}{\Delta\lambda} = mN. \quad \text{(resolving power)} \tag{37-7}$$

The quantity $\lambda/\Delta\lambda$ is the **resolving power** of the grating, a measure of its ability to distinguish closely spaced wavelengths. The higher the resolving power, the smaller the wavelength difference $\Delta\lambda$ that can be distinguished in the spectrum. That the resolving power increases with order is not surprising since we found that higher order spectra are more dispersed. Resolving power also depends on the number N of lines in the grating, although if the entire grating is not illuminated then N in Equation 37-7 becomes the number of lines actually illuminated.

● **EXAMPLE 37-5**   "SEEING" A DOUBLE STAR

A certain double star system consists of a massive star essentially at rest, with a smaller companion in circular orbit. The stars are far too close to each other, and the system far too distant from Earth, for the pair to appear as anything but a single point in even the largest telescopes. Yet astronomers can "see" the companion star through the Doppler shift in the wavelengths of its spectral lines that occurs because of the star's motion. In the system under observation the hydrogen-α line from the companion star, at 656.272 nm when the source is at rest, is shifted to 656.215 nm (this corresponds to an orbital speed of about 26 km/s) when the companion is moving toward Earth. If the telescope's spectrometer grating has 2000 lines/cm and measures 2.5 cm across, what order spectrum will resolve the hydrogen-α lines from the stationary star and its orbiting companion?

Solution

Here $N = (2000 \text{ lines/cm})(2.5 \text{ cm}) = 5000$ lines. Solving Equation 37-7 for m then gives

$$m = \frac{\lambda}{N\,\Delta\lambda} = \frac{656.272 \text{ nm}}{(5{,}000)(656.272 \text{ nm} - 656.215 \text{ nm})} = 2.3.$$

Since the order number must be an integer, the observation must be made in third order.

EXERCISE   To the nearest thousand, how many lines are needed to resolve the magnetically split spectral lines of Fig. 37-25 in first order, given a splitting of 0.027 nm at 435.8 nm?

Answer: 16,000

Some problems similar to Example 37-5: 44–47 ●

X-Ray Diffraction

The wavelengths of x rays, around 0.1 nm, are far too short to be dispersed with diffraction gratings produced mechanically or photographically. The regular spacing of atoms in a crystal, however, provides a "grating" of the appropriate scale. In the classical electromagnetic description, the reflection of electromagnetic waves occurs as electrons in each atom are set into oscillation by the electric field of the incident x-ray beam. The oscillating electrons then reradiate the energy they have absorbed, and the combined beam from all the atoms obeys the law of reflection. When the atomic spacing is regular, and comparable to the wavelength, however, interference enhances the reflected radiation at certain angles. Figure 37-27 shows an x-ray beam interacting with the layers of atoms in a crystal. In Fig. 37-27b we see that waves reflecting at one layer travel a total distance $2d\sin\theta$ farther than those reflecting at the layer above, where θ is the angle between the incident beams and the atomic planes. The outgoing beams will interfere constructively when this difference is an integer multiple of the wavelength:

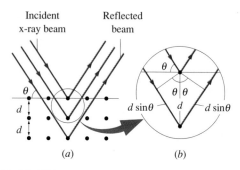

FIGURE 37-27 (a) X rays reflecting off the planes of atoms in a crystal. (b) X rays striking a lower plane travel an extra distance $2d \sin \theta$. The outgoing beam is enhanced by constructive interference when this distance is an integer multiple of the x-ray wavelength.

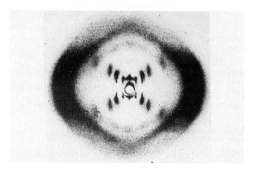

FIGURE 37-28 This x-ray diffraction image of DNA, made by the British scientist Rosalind Franklin in 1952, was crucial in establishing the structure of the DNA molecule.

$$2d \sin \theta = m\lambda. \quad \text{(Bragg condition)} \qquad (37\text{-}8)$$

First derived by W. L. Bragg, this **Bragg condition** allows one to use a crystal with known spacing as a diffraction grating at x-ray wavelengths. More important is the converse: much of what we know about crystal structure comes from probing crystals with x rays and using the resulting diffraction patterns to deduce the positions of their atoms (Fig. 37-28).

■ APPLICATION ACOUSTO-OPTIC MODULATION

Anything containing regularly spaced structures—like the lines in a diffraction grating or the atoms in a crystal—can act as a diffraction grating for waves of suitable wavelength. A recent and very useful technological development, the **acousto-optic modulator** (AOM), uses diffraction from the periodic variations in the structure of a quartz crystal caused by the presence of a sound wave in the crystal.

Figure 37-29 shows a diagram of an AOM. Attached to the crystal is a transducer—a loudspeaker-like device that converts an alternating voltage (usually in the ultrasound frequency range) into sound waves that propagate through the quartz. The regular spacing of the acoustic wavefronts constitutes a diffraction grating whose spacing may be varied by changing the driving frequency. Light incident on the quartz is diffracted at preferred angles given by Equation 37-8, just as are x rays by regularly spaced atomic planes. Changing the sound frequency allows the diffracted beam to be scanned in direction; turning the sound on and off switches the diffracted beam on and off. These effects are widely used in lightwave communication and other optical technologies; most laser printers, for example, use AOMs to control the laser beam that

"paints" a picture of the printed page on a light-sensitive surface. Setting up the AOM's "diffraction grating" requires propagating sound waves into the crystal. Since this takes time, the AOM is not as fast a lightwave modulator as is the electro-optic modulator discussed in Chapter 34. The AOM is, however, much less expensive and less demanding of the associated electronic circuitry.

Acoustic-optic modulators have another useful property. The diffraction grating established by the propagating sound waves is *moving* through the crystal, and the diffracted light is therefore Doppler shifted in frequency. A detailed analysis shows that the light frequency is shifted up or down by an amount equal to the sound frequency. Thus, the AOM may be used to encode information by altering slightly the frequency of light—an optical version of the process used to encode audio signals on radio waves in FM radio. Alternatively, AOMs can be used to produce light signals whose frequencies differ only slightly; bringing these signals together then produces the optical equivalent of the beat phenomenon we discussed in Section 16-5.

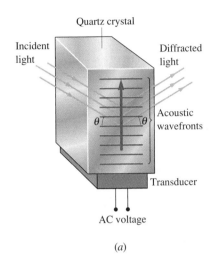

FIGURE 37-29 (*a*) Schematic diagram of an acousto-optic modulator. The transducer drives sound waves into the crystal, and the acoustic wavefronts act like a moving diffraction grating to incident light. (*b*) An acousto-optic modulator. A laser beam enters the transparent crystal while electrical signals supplied by the cable in the foreground generate sound waves that act as a diffraction grating to modify the beam's propagation direction and frequency. Such modulators are used in laser printers and many other applications.

37-5 SINGLE-SLIT DIFFRACTION

Why are we saving the single-slit case for last? Shouldn't that be the easiest? It would be—except that up to now we've neglected the slit width. According to Huygens' principle, each point in the slit acts as a source of cylindrical waves, and all these waves should interfere. Thus, a single slit is really like a system with infinitely many slits! Only when the slit width is very small can we make the approximation, implicit in our earlier analysis, that the slit behaves as a single source.

Figure 37-30 shows light incident on a slit of width a. Each point in the slit acts as a source of spherical wavelets propagating in all directions to the right of the slit. In Fig. 37-30 we focus on a particular direction described by the angle θ and will look at interference of light from the five points shown. Figure 37-30b concentrates on the points from which rays 1, 2, and 3 originate and shows that the path lengths for rays 1 and 3 differ by $\frac{1}{2}a\sin\theta$. These two beams will interfere destructively if this distance is half the wavelength; that is, if $\frac{1}{2}a\sin\theta = \frac{1}{2}\lambda$ or $a\sin\theta = \lambda$. But if rays 1 and 3 interfere destructively, so do rays 3 and 5, which have the same geometry, and so do rays 2 and 4, for the same reason. In fact, a ray leaving *any* point in the lower half of the slit will interfere destructively with the point located a distance $a/2$ above it. Therefore, an observer viewing the slit system at the angle θ satisfying $a\sin\theta = \lambda$ will see no light.

Similarly, the sources for rays 1 and 2 are $a/4$ apart and will therefore interfere destructively if $\frac{1}{4}a\sin\theta = \frac{1}{2}\lambda$, or $a\sin\theta = 2\lambda$. But then so will rays 2 and 3, and rays 3 and 4; in fact, any ray from a point in the lower three quarters of the slit will interfere destructively with a ray from the point $a/4$ above it, and therefore, an observer looking at an angle satisfying $a\sin\theta = 2\lambda$ will see no light.

We could equally well have divided the slit into six sections with seven evenly spaced points; we would then have found destructive interference if $\frac{1}{6}a\sin\theta = \frac{1}{2}\lambda$, or $a\sin\theta = 3\lambda$. We could obviously continue this process for any number of points in the slit, and therefore, we conclude that destructive interference occurs for all angles satisfying

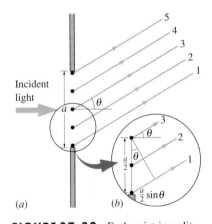

FIGURE 37-30 Each point in a slit acts as a source of Huygens' wavelets, which interfere in the region to the right of the slit. (*a*) Here we consider 5 points within the slit, and the relative phases of their light as viewed at an angle θ. (*b*) The path-length difference for two slits is their separation times $\sin\theta$.

$$a \sin \theta = m\lambda, \quad \text{(destructive interference, single-slit diffraction)} \quad (37\text{-}9)$$

with m any nonzero integer and a the slit width. Note that the case $m = 0$ is excluded; it produces not destructive interference but a central maximum in which all waves are in phase.

> **TIP** **Interference and Diffraction** Equation 37-9 for the *minima* of a single-slit diffraction pattern looks just like Equation 37-2a for the *maxima* of a multi-slit interference pattern, except that the slit width a replaces the slit spacing d. What's going on? Why does the same equation give the minima in one case and the maxima in another? Because we're dealing with two distinct but related phenomena. In the multi-slit case leading to Equation 37-2a, we considered each slit to be so narrow that it could be considered a single source, thus neglecting the interference of waves originating within the same slit. In the single-slit case leading to Equation 37-9, we do not neglect the slit width, and the resulting diffraction pattern occurs precisely because of the interference of waves from different points within the same slit.

● EXAMPLE 37-6 DIFFRACTION: A NARROW SLIT

For what slit width, in terms of wavelength, will the first minimum lie at an angular position of 45°?

Solution
With $m = 1$, Equation 37-9 gives

$$a = \frac{\lambda}{\sin \theta} = \frac{\lambda}{\sin 45°} = \sqrt{2}\lambda.$$

Here, with the slit width nearly equal to the wavelength, diffraction dominates the wave propagation. The incident light is

therefore spread over a wide angular range. The exercise below shows that the beam spreading decreases as the slit width grows.

EXERCISE Light is incident on a slit whose width is 20 times its wavelength. Find the angular spread of the beam, taken as the angle between the first minima on either side of the central maximum.

Answer: 5.7°

Some problems similar to Example 37-6: 49–51 ●

Intensity in Single-Slit Diffraction

We could try to find the intensity of the single-slit diffraction pattern by adding the electric fields of the individual beams, as we did for two-slit interference. Since there are infinitely many points along the slit, each constituting a point source of light, we would have to set up and evaluate a complicated integral. Instead we use the graphical method of phasors, introduced in Chapter 33 to deal with alternating currents and voltages of different phases. We have a similar situation here, as we want to combine waves with different phases that arise from different points on the slit. The phasor method is entirely equivalent to integrating over the electric field contributions from the individual sources in the slit.

We use phasors to represent the electric fields at the screen of the individual waves originating from points in the slit. The length of a phasor gives the field amplitude, and its direction gives its phase. Consider N such points, spaced evenly across a slit of width a, like the five points in Fig. 37-30. These divide the slit into $N - 1$ sections, each of width $a/(N - 1)$. At some angle θ, the phase difference in radians of light coming from two adjacent points will be the ratio of their path-length difference to the wavelength, times the 2π radians that comprise a complete cycle. As usual, the path-length difference is the point spacing times $\sin\theta$ (see Fig. 37-30b). Here the point spacing is $a/(N - 1)$, so we have

$$\Delta\phi = \left(\frac{a}{N - 1}\sin\theta\right)\left(\frac{2\pi}{\lambda}\right) \tag{37-10}$$

for the phase difference between waves from adjacent points. All the waves have essentially the same amplitude, so we now need to add N equal-length phasors each of which differs in phase by $\Delta\phi$ from its nearest neighbors.

Consider first the case $\theta = 0$. Then Equation 37-10 gives $\Delta\phi = 0$ so, as Fig. 37-31a shows, the phasors are all in the same direction and sum to give a large total amplitude. This is the central maximum of the diffraction pattern. If, on the other hand, $\Delta\phi$ is such that the N phasors sum to a closed path, then the total amplitude—the net displacement from the beginning to the end of the phasor diagram—is zero (Fig. 37-31b). Here we have a point of zero intensity in the interference pattern. More generally, the phasors sum to give an amplitude E_θ that is neither zero nor as great as a simple sum of the individual amplitudes, as shown in Fig. 37-31c.

Now each point in the slit differs in phase by $\Delta\phi$ from its immediate neighbors. Thus, the second point differs by $\Delta\phi$ from the first point, the third point by $2\Delta\phi$ from the first point, and so forth until the Nth point differs in phase by $(N - 1)\Delta\phi$ from the first point. But this last quantity $(N - 1)\Delta\phi$ is just the total phase difference ϕ from one end of the slit to the other. Using Equation 37-10, we can write this phase difference as

$$\phi = (N - 1)\left(\frac{a}{N - 1}\sin\theta\right)\left(\frac{2\pi}{\lambda}\right) = \frac{2\pi}{\lambda}a\sin\theta. \tag{37-11}$$

To consider all points in the slit, we take the limit as $N \to \infty$; then the end-to-end chain of phasors in Fig. 37-31c becomes a circular arc of radius R, as shown in Fig. 37-32. If we stretched this arc into a line, its length would be the amplitude E_0 of Fig. 37-31a; therefore, E_0 is also the length of the arc. From the geometry of Fig. 37-32 we see that

$$\sin(\phi/2) = \frac{E_\theta/2}{R}.$$

We can also write the angle $\phi/2$ using the definition of an angle in radians as the ratio of arc length to radius. Here $\phi/2$ subtends an arc of length $E_0/2$, so we have

$$\phi/2 = \frac{E_0/2}{R}.$$

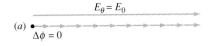
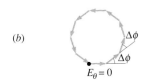
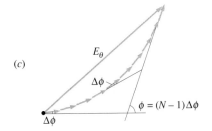

FIGURE 37-31 Phasor addition to find the amplitude in single-slit diffraction. (a) When $\theta = 0$ in Fig. 37-30 all waves have the same phase, and their amplitudes add to produce the central maximum. (b) When the phasor sum is a closed loop, the net phasor displacement is zero and so is the wave amplitude. (c) In general, the phasors add to produce an amplitude that is neither zero nor as great as in the central maximum. The angle between the first and last phasor is $N\Delta\phi$. Here $N = 10$ in all three frames, and in general we designate the resultant amplitude by E_θ. When the phasors all have the same phase, E_0 is the resultant amplitude.

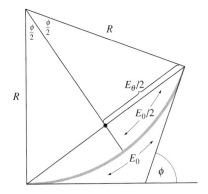

FIGURE 37-32 In the limit $N \to \infty$ the chain of phasors becomes a circular arc of length E_0.

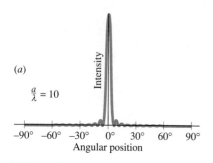

(a)

$\frac{a}{\lambda} = 10$

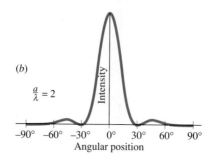

(b)

$\frac{a}{\lambda} = 2$

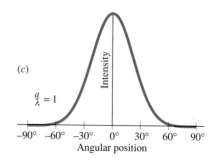

(c)

$\frac{a}{\lambda} = 1$

FIGURE 37-33 Intensity in single-slit diffraction, as a function of the angle θ from the centerline, for three values of of slit width a, expressed in units of wavelength λ. When the slit is much wider than the wavelength, most of the light is concentrated in a narrow central peak. But as the slit width approaches the wavelength, the diffracted beam becomes very wide.

Dividing these two equations gives the ratio of the amplitudes:

$$\frac{E_\theta}{E_0} = \frac{\sin(\phi/2)}{\phi/2}.$$

But the wave intensity $\bar{S}$ is proportional to the square of the amplitude, so

$$\bar{S}_\theta = \bar{S}_0 \left[\frac{\sin(\phi/2)}{\phi/2} \right]^2,$$

or, substituting ϕ from Equation 37-11,

$$\bar{S}_\theta = \bar{S}_0 \left[\frac{\sin\left(\frac{\pi a}{\lambda} \sin\theta\right)}{\frac{\pi a}{\lambda} \sin\theta} \right]^2 \tag{37-12}$$

for the intensity in the single-slit diffraction pattern in terms of its value at the central maximum.

Figure 37-33 shows plots of Equation 37-12 for several values of the slit width a in relation to the wavelength λ. Note that for wide slits—large a/λ—the central peak is very narrow and the secondary peaks are much smaller. Here diffraction is negligible, and the beam essentially propagates through the slit in the ray approximation of geometrical optics. But as the slit narrows the diffracted beam spreads, until, with $a = \lambda$, it covers an angular width of some 120°.

The intensity given by Equation 37-12 will be zero when the numerator on the right-hand side is zero—that is, when the argument of the outermost sine function is an integer multiple of π. That occurs when $\frac{\pi a}{\lambda} \sin\theta = m\pi$, or when $a \sin\theta = m\lambda$. Thus, we recover our earlier result of Equation 37-9 for the angular positions where destructive interference gives zero intensity.

● **EXAMPLE 37-7** OTHER PEAKS

Figure 37-33 shows that secondary maxima lie approximately midway between the minima of the diffraction pattern. Use this approximate location to find the intensity at the first of these secondary maxima, in terms of the central-peak intensity $\bar{S}_0$.

Solution

The first and second minima are at positions given by $a \sin\theta_1 = \lambda$ and $a \sin\theta_2 = 2\lambda$, respectively. With the first of the secondary maxima midway between them, its angular position is given by $a \sin \theta = \frac{3}{2}\lambda$, or $\frac{a}{\lambda} \sin \theta = \frac{3}{2}$. (Here we assume θ is small, so $\sin \theta \simeq \theta$, and midway in angular position θ is therefore also midway in $\sin \theta$.) Using this result in Equation 37-12 then gives

$$\bar{S}_\theta = \bar{S}_0 \left[\frac{\sin\left(\dfrac{\pi a}{\lambda} \sin \theta\right)}{\dfrac{\pi a}{\lambda} \sin \theta} \right]^2 = \bar{S}_0 \left[\frac{\sin(3\pi/2)}{3\pi/2} \right]^2$$

$$= \frac{4\bar{S}_0}{9\pi^2} = 0.045\bar{S}_0.$$

Thus, the intensity at the first secondary maximum is only about 4.5% of the central-peak intensity. Note that this result is independent of the slit width, provided the secondary peak exists (which it may not; see Fig. 37-33c).

EXERCISE Show by direct substitution that the intensity in single-slit diffraction has half its maximum value when $\sin \theta = 1.3916 \dfrac{\lambda}{\pi a}$.

Some problems similar to Example 37-7: 52–54 ●

In analyzing single-slit diffraction we considered only parallel rays in the diffraction region. To make the diffraction pattern actually appear on a screen, we would have to focus those rays with a lens between slit and screen (Fig. 37-34). Diffraction associated with parallel rays is called **Fraunhofer diffraction.** In general, Fraunhofer diffraction occurs when the distance from the diffracting system is large compared with the wavelength. Fraunhofer diffraction is an approximation to the more general case of **Fresnel diffraction,** whose analysis is more complicated because it also accounts for nonparallel rays that exist near a diffracting system.

Multiple Slits, Revisited

We neglected the width of slits in treating interference effects in multiple-slit systems. In effect, we were assuming the slits were so small that the central diffraction peak spread into the entire space beyond the slit system. When the slit width is not negligible, then the waves from each slit are not of uniform intensity, but instead exhibit a single-slit diffraction pattern. The superposition of waves from two or more slits then results in a pattern that combines single-slit diffraction with multiple-slit interference (Fig. 37-35).

37-6 THE DIFFRACTION LIMIT

Diffraction imposes a fundamental limit on the ability of optical systems to distinguish closely spaced objects. Consider two point sources of light illuminating a slit. The sources are so far from the slit that waves reaching the slit are essentially plane waves, but the different source positions mean the waves reach

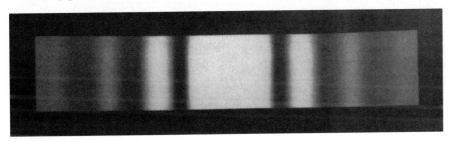

FIGURE 37-34 Single-slit diffraction pattern produced by focusing light from a single slit onto a screen. Note the bright central peak and secondary maxima separated by points of destructive interference.

FIGURE 37-35 When the slit width is not negligible, the double-slit pattern shows the regular variations of double-slit interference confined to an "envelope" in the shape of a single-slit diffraction pattern. (Top) A plot of the intensity measured at each point on the screen. (Bottom) Interference/diffraction pattern as viewed on a screen. Compare with Fig. 37-33 for a single slit; note that the two patterns are essentially identical except that here a characteristic double-slit pattern of alternating light and dark bands appears within each bright band of the single-slit pattern.

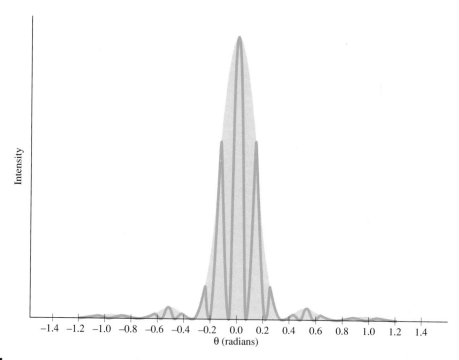

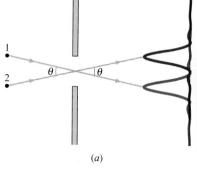

(a)

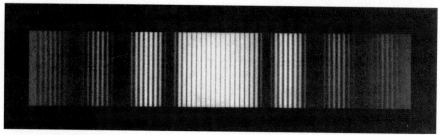

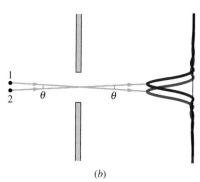

(b)

FIGURE 37-36 Two distant sources at different angular positions produce diffraction patterns whose central peaks have the same angular separation θ as the sources. (a) For sufficiently large θ the central peaks are clearly distinct, but (b) at small angular separations they overlap.

the slit at different angles. We assume the sources are incoherent, so there is no interference between the two. Then light diffracting at the slit produces two single-slit diffraction patterns, one for each source. Because the sources are at different angular positions, the central maxima of these patterns do not coincide, as shown in Fig. 37-36.

If the angular separation between the sources is great enough, then the central maxima of the two diffraction patterns will be entirely distinct. In that case the eye or any other optical system examining the diffraction pattern can clearly distinguish the two sources (Fig. 37-36a). But as the sources get closer the central maxima begin to overlap (Fig. 37-36b). The two sources remain distinguishable as long as the total intensity pattern shows two peaks. Since the sources are incoherent, they don't interfere, and the total intensity is just the sum of the individual intensities. Figure 37-37 shows how that sum loses its two-peak structure as the diffraction patterns merge. In general, two peaks are barely distinguishable if the central maximum of one coincides with the first minimum of the other. This condition is called the **Rayleigh criterion,** and when it is met we say that the two sources are just barely **resolved.**

What does all this have to do with optical instruments and images? Simply this: All optical systems are analogous to the single slit we've been considering. Every system has an aperture of finite size through which light enters the system. That aperture may be an actual slit or hole, like the diaphragm that "stops down" a camera lens, or it may be the full size of a lens or mirror. So all optical systems ultimately suffer loss of resolution if two sources—or two parts of the same extended object—have too small an angular separation. Thus, diffraction fundamentally limits our ability to probe the structure of objects that are either very small or very distant.

Figure 37-36 shows that the angular separation between the diffraction peaks is equal to the angular separation between the sources themselves. Then the Rayleigh criterion is just met if the angular separation between the two sources is equal to the angular separation between a central peak and the first minimum. We found earlier that the first minimum in single-slit diffraction occurs at angular position given by

$$\sin\theta = \frac{\lambda}{a},$$

with a the slit width and with θ measured from the central peak. In most optical systems the wavelength is much less than the size of any apertures, so we can use the small-angle approximation $\sin\theta \simeq \theta$ with θ in radians. Then the Rayleigh criterion—the condition that two sources be just resolvable—for single-slit diffraction becomes

$$\theta_{\min} = \frac{\lambda}{a}. \qquad \text{(Rayleigh criterion, slit)} \qquad (37\text{-}13a)$$

Most optical systems have circular apertures rather than slits. The diffraction pattern from such an aperture is a series of concentric rings (Fig. 37-38). A more involved mathematical analysis shows that the angular position of the first ring relative to the central peak, and therefore, the minimum resolvable source separation for a circular aperture, is

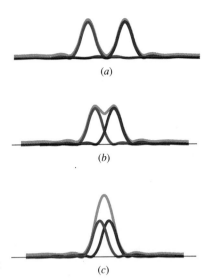

FIGURE 37-37 Since the two sources are incoherent, the total intensity is just the sum of the intensities of the two diffraction patterns. Here we see the intensities in two patterns and their sum (light blue). (a) Fully resolved; (b) barely resolved (Rayleigh criterion); (c) unresolved.

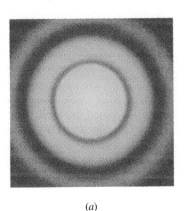

(a)

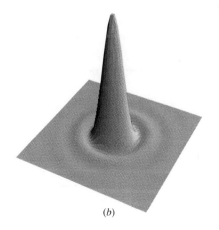

(b)

FIGURE 37-38 (a) Diffraction pattern produced by a circular aperture. (b) Three-dimensional plot of intensity versus position.

FIGURE 37-39 Diffraction patterns produced by a pair of point sources imaged through a circular hole. The angular separation of the sources decreases going from left to right, and when the sources are too close they cannot be resolved.

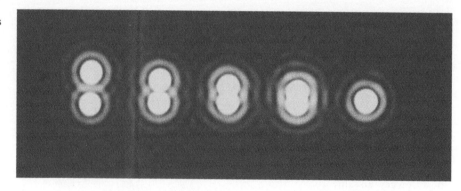

$$\theta_{min} = \frac{1.22\lambda}{D}, \qquad \text{(Rayleigh criterion, circular aperture)} \quad (37\text{-}13b)$$

with D the aperture diameter. In most optical systems the angle θ is very small, and Equations 37-13a and b hold for this case. For larger angles, θ_{min} in both equations should be replaced with $\sin\theta_{min}$. Figure 37-39 shows the loss of resolution as two nearby sources are imaged through a circular aperture.

Equations 37-13 show that increasing the aperture size allows smaller angular differences to be resolved. In optical instrument design, that means larger mirrors, lenses, and other components. An alternative is to decrease the wavelength used, which may or may not be an option depending on the source. In high-quality optical systems diffraction is often the limiting factor preventing perfectly sharp image formation; such systems are said to be **diffraction limited.** Among exceptions are large ground-based telescopes, whose imaging quality is limited by atmospheric turbulence. From its vantage point above the atmosphere, the Hubble Space Telescope is the first large diffraction-limited astronomical telescope.

● **EXAMPLE 37-8** ASTEROID ALERT

An asteroid appears on a collision course with Earth, at a distance of 20×10^6 km. What is the minimum size asteroid that the 2.4-m-diameter diffraction-limited Hubble Space Telescope could resolve at this distance, using 550-nm reflected sunlight?

Solution

Resolving the asteroid means being able to distinguish its opposite edges in the telescope image. Suppose the asteroid's long dimension (it might not be spherical) is ℓ. Then at a distance $L \gg \ell$ it subtends an angle given very nearly by $\theta = \ell/L$. Using this result in Equation 37-13b with the mirror diameter and wavelength given, we have

$$\frac{\ell}{L} = \frac{1.22\lambda}{D},$$

or

$$\ell = \frac{1.22\lambda L}{D} = \frac{(1.22)(550\times10^{-9}\ \text{m})(20\times10^9\ \text{m})}{2.4\ \text{m}} = 5.6\ \text{km}.$$

This is a potentially dangerous object, comparable in size to the asteroid that some scientists believe caused the extinction of the dinosaurs, and somewhat larger than the comet fragments that slammed into Jupiter in 1994, causing Earth-sized disturbances on the giant planet.

●

● EXAMPLE 37-9 STAR WARS: THE DIFFRACTION CHALLENGE

A system once proposed for defense against ballistic missiles calls for focusing high-power laser beams onto attacking missiles. A particular design specifies infrared laser light with 2.8-μm wavelength focused to a spot 50 cm in diameter on a missile 2500 km distant. What is the minimum diameter for a concave mirror that can achieve this spot size?

Solution

Figure 37-40a shows a parallel beam from the laser striking the mirror, then focusing on the distant missile. We assume the focal length is equal to the missile's distance, giving the best possible focus. Nevertheless, the mirror acts like a circular aperture that produces a central spot whose angular radius is given by the Rayleigh criterion of Equation 37-13b. We want the angular radius of that spot to correspond to a 25-cm-radius spot at a distance of 2500 km. Clearly the small-angle approximation applies, and Fig. 37-40b shows that we can write

$$\theta = \frac{25 \text{ cm}}{2500 \text{ km}} = 1.0 \times 10^{-7} \text{ radians}$$

for the angular radius of the spot. Solving Equation 37-13b for the minimum mirror diameter then gives

$$D = \frac{1.22\lambda}{\theta} = \frac{(1.22)(2.8 \times 10^{-6} \text{ m})}{1.0 \times 10^{-7}} = 34 \text{ m}.$$

This enormous mirror size is one reason many scientists have found "star wars" weapons designs implausible. A related problem is the long focal length; with $f = 2500$ km, a 34-m mirror is only about 40 μm from being perfectly flat, and it has to retain its exact shape despite the thermal stresses imposed by a 20-MW laser beam on a space-based mirror. An alternative approach is to use so-called phased arrays, in which a number of smaller sources with precisely related phases simulate the effect of the beam from a large mirror.

EXERCISE Two ants stand 1 cm apart. Assuming diffraction limited vision, at what distance could a human eye, with an iris aperture of 4 mm, tell that two are indeed separate creatures? Assume a wavelength of 550 nm, near the middle of the visible

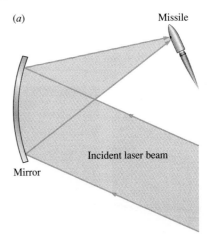

(a)

Missile

Incident laser beam

Mirror

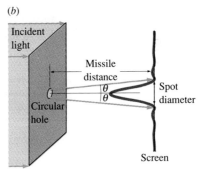

(b)

Incident light

Missile distance

Circular hole

θ
θ

Spot diameter

Screen

FIGURE 37-40 (a) In this laser-based antimissile system, the mirror acts like a circular aperture to limit the minimum size of the focused beam. (b) The equivalent single-hole diffraction system. The angle θ_{min} in the Rayleigh criterion is the ratio of spot radius to the missile distance.

spectrum. (Other limitations of the human eye prevent its realizing this diffraction limit.)

Answer: 60 m

Some problems similar to Examples 37-8 and 37-9: 55–59   ●

■ APPLICATION LONG-PLAYING CDs

Standard compact discs have a maximum playing time of 74 minutes—a value that is in fact set by the optical diffraction limit. Information on a CD is encoded in a sequence of "pits" cut into the CD surface (Fig. 37-41). The pit depth is approximately one-quarter wavelength of the laser light used to "read" the information. The laser shines from below, where it "sees" each pit as a "bump" that introduces a round-trip change of

one-half wavelength in the light path, causing destructive interference between incident and reflected beams (Fig. 37-42). As the disc rotates, the interference is alternately constructive and destructive, and the interfering light falling on a detector produces an electrical signal corresponding to the information encoded in the CD's pits.

The pits on a standard CD are about 0.5 μm wide, up to 3 μm long, and lie in spiral tracks spaced 1.6 μm apart. These are close to the minimum dimensions that can be used before diffraction effects prevent resolution of the individual pits. Thus, the diffraction limit determines the total number of pits, and therefore the total information stored on the disc. Each

second of music recording requires on the order 10^6 pits, and the result is the 74-minute maximum recording time.

Today's CD players use 780-nm infrared laser light. The reason is economic: inexpensive semiconductor lasers operating at this wavelength are readily available. The diffraction limit on compact-disc playing time has spurred a vigorous research competition among electronics corporations, in the hope of making shorter wavelength semiconductor lasers. By the early 1990s several corporations had developed blue or blue-green lasers, with a Sony unit at 447 nm having the shortest wavelength. Expect longer playing CDs, thanks to the less stringent diffraction limit, when these new short-wavelength lasers become commercially viable.

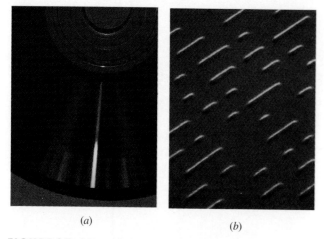

(a)   (b)

FIGURE 37-41 A compact disc encodes information in billions of microscopic pits arranged in spiral tracks 1.6 μm apart. (a) The pits are too small to see, but their spiral tracks act as a diffraction grating, producing the rainbow of colors seen here. (b) Electron micrograph showing individual pits in adjacent tracks. These CD pits are among the smallest manufactured structures. See also Fig. 1-11.

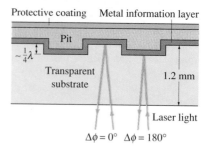

FIGURE 37-42 Schematic diagram showing a laser beam "reading" from the underside of a compact disk. Destructive interference caused by the 180° phase difference occurs when the beam reflects from one of the pits. Diagram is not to scale; the metal layer containing the pit-encoded information is less than 100 nm thick, and the protective coating is 10–30 μm thick. See also Example 35-2.

■ APPLICATION INTERFEROMETRY

At a given wavelength, Equations 37-13 show that resolving small angular separations requires a large aperture. To see detail in distant astronomical objects, in particular, requires impractically large apertures. A technique called **interferometry** provides a way around this problem. Used most commonly at radio wavelengths, interferometry employs two or more individual apertures (i.e., radio telescopes), usually separated by a considerable distance. Signals from each are combined with their phases intact, so that they interfere. A two-telescope interferometer works like a two-slit system in reverse: the system

is most sensitive to radiation from angular positions that produce constructive interference (Fig. 37-43). The resolution of an interferometer is approximately that of a single slit-like aperture whose width is the spacing between the telescopes.

Today's interferometers include installations with many radio telescopes at a single site (Fig. 37-44), as well as coordinated instruments spread across the globe. The first interferometric arrays of optical telescopes began operation in the 1990s, and astronomers look forward to radio interferometers using the Earth-moon distance as their "aperture."

FIGURE 37-43 A radio interferometer consists of two or more radio telescopes at different locations. The interferometer can detect very slight changes in phase of the incoming wavefronts and can therefore resolve very small differences in angular position of objects in the sky.

(a)

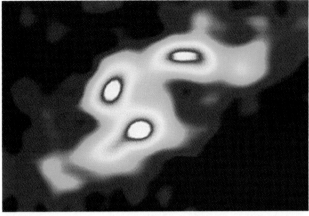

(b)

FIGURE 37-44 (a) Dish antennas of the IRAM interferometer in Spain, a joint French-German-Spanish instrument used extensively to study radio emission from interstellar molecules. (b) IRAM image of the spiral galaxy IC342 shows the distribution of interstellar hydrogen cyanide in the galaxy.

CHAPTER SYNOPSIS

Summary

1. **Interference** and **diffraction** are wave phenomena and constitute the subject of **physical optics.**
2. Interference effects require **coherent** beams of light, whose phases are related in a fixed way.
   a. Interference occurs in thin transparent films when light beams reflected from the front and back surfaces of the film recombine. For a film whose refractive index is larger than that of its surroundings, the interference is constructive when the film thickness d, refractive index n, and the light wavelength λ are related by $2nd = (m + \frac{1}{2})\lambda$, where m is an integer.
   b. The Michelson interferometer utilizes the interference of light traveling on two perpendicular paths to make precise distance measurements.
3. **Huygens' principle** treats each point on a wavefront as a source of expanding spherical wavelets. Superposition of the wavelets describes the propagation of the wavefront and shows that waves undergo **diffraction** when passing by an opaque edge.
4. When coherent light of wavelength λ diffracts on passing through a pair of narrow slits with spacing d, the resulting interference pattern shows bright fringes at angular positions given by

$$d \sin \theta = m\lambda,$$

where m is an integer. The intensity in the interference pattern is given by

$$\bar{S} = 4\bar{S}_0 \cos^2\left(\frac{\pi d \sin \theta}{\lambda}\right),$$

where $\bar{S}_0$ is the intensity incident on the slit system.
5. A multiple-slit system has primary interference maxima in the same position as a double-slit system, with multiple minima and secondary maxima in between. For large numbers of slits the primary maxima are narrow and bright and the region between them relatively dark. A system with many slits constitutes a **diffraction grating,** which disperses light into its component wavelengths. A grating produces spectra of many **orders,** corresponding to the integer m in the equation that locates the interference maxima. The **resolving power** of an N-slit grating is given by

$$\frac{\lambda}{\Delta\lambda} = mN,$$

where $\Delta\lambda$ is the minimum separation between wavelengths that can be distinguished in the mth-order spectrum. Other periodic systems, like the layers of atoms in a crystal, can serve as diffraction gratings for appropriate wavelengths.
6. When the width of a single slit is not negligible compared with the wavelength, interference of light passing through different parts of the slit produces a **diffraction pattern** with a central maximum surrounded by lesser maxima separated by points of zero intensity. The intensity as a function of angular position is given by

$$\bar{S}_\theta = \bar{S}_0 \left[\frac{\sin\left(\dfrac{\pi a}{\lambda}\sin\theta\right)}{\dfrac{\pi a}{\lambda}\sin\theta} \right]^2 .$$

7. The **diffraction limit** in optical systems arises because diffraction effects merge the images of two objects when their angular separation is sufficiently small. The **Rayleigh criterion** gives the minimum angular separation that can be resolved by a given aperture. For a circular aperture of diameter D, the Rayleigh criterion is

$$\theta_{min} = \frac{1.22\lambda}{D},$$

where θ_{min} is in radians. Large apertures are therefore necessary to resolve objects with small angular separation.

Terms You Should Understand

(Pairs are closely related terms whose distinction is important; number in parentheses is chapter section where term first appears.)

physical optics, geometrical optics (introduction and preceding chapters)
interference (37-1)
coherence (37-1)
monochromatic light (37-1)
antireflection coating (37-1)
Michelson interferometer (37-1)
beam splitter (37-1)
Huygens' principle (37-2)
diffraction (37-2)
order (37-3, 37-4)

diffraction grating (37-4)
resolving power (37-4)
Bragg condition (37-4)
diffraction limit (37-6)
Rayleigh criterion (37-6)

Symbols You Should Recognize

m (37-1, 37-3, 37-4)
θ_{min} (37-4, 37-6)

Problems You Should Be Able to Solve

determining conditions for constructive and destructive interference in thin films (37-1)
analyzing experiments using Michelson interferometers (37-1)
analyzing double-slit interference patterns (37-3)
finding locations of maxima and minima in multiple-slit interference (37-4)
analyzing dispersion of light with diffraction gratings (37-4)
finding minima and beam widths in single-slit diffraction (37-5)
finding the diffraction limit in optical systems (37-6)

Limitations to Keep in Mind

This chapter's analysis of double-slit interference assumes slit widths much smaller than the wavelength; otherwise the pattern combines diffraction and interference.
Our analysis of diffraction applies only in the Fraunhofer limit, in which the incident light is essentially parallel and the pattern is formed at large distances from the diffracting aperture.

QUESTIONS

1. A prism bends blue light more than red. Is the same true of a diffraction grating?
2. Why does an oil slick show colored bands?
3. Would a Michelson interferometer work even if its two arms were not exactly the same length?
4. Why does a soap bubble turn colorless just before it dries up and breaks?
5. Why don't you see interference effects between the front and back of your eyeglasses?
6. Hold two fingers very close together while looking through them at a source of light. Explain what you see.
7. Figure 37-45 shows the shadow cast by a ball bearing on the end of a needle, under monochromatic laser light. Explain the existence of the fringes shown and of the white spot at the center of the ball bearing's shadow.

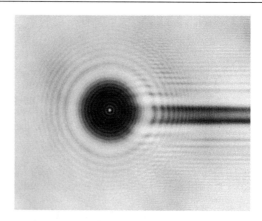

FIGURE 37-45 Question 7.

8. Why don't you see interference effects when you look out a window through venetian blinds?

9. You can hear around corners but you can't see around corners. Why the difference?

10. In deriving the intensity in double-slit interference, we could not simply add the intensities from the two slits. Why not?

11. In sketching the intensity pattern for two sources imaged through a single slit, we could simply add the intensities from the two sources. Why?

12. Explain the roles of diffraction and interference in the working of a diffraction grating.

13. The primary maxima in multiple-slit interference are in the same angular positions as those in double-slit interference. Why, then, do diffraction gratings have thousands of slits instead of just two?

14. In what way is a widely separated pair of small radio telescopes superior to a single large one? In what way is it inferior?

15. Describe the change in the diffraction pattern of a single slit as the slit is narrowed.

16. What pattern would result from passing red and blue light simultaneously through a double-slit system?

17. When the moon passes in front of a star, the intensity of the starlight fluctuates instead of dropping abruptly. Explain.

18. Why might it be desirable to observe a double star system in blue light rather than red?

19. In analyzing crystal structure using x-ray diffraction, it is necessary to take data with the crystal in several different orientations. Why?

20. Microwaves emerge from a rectangular horn, wider in the horizontal direction than in the vertical. Will the resulting beam be wider in the horizontal or in the vertical direction? Explain.

21. A double-slit system has one slit much narrower than the wavelength of the incident light, the other much wider. Describe the resulting intensity pattern.

22. Sketch roughly the diffraction pattern you would expect for light passing through a square hole a few wavelengths wide.

PROBLEMS

Section 37-1 Interference

1. Find the minimum thickness of a soap film ($n = 1.33$) in which 550-nm light will undergo constructive interference.

2. Light of unknown wavelength shines on a precisely machined wedge of glass with refractive index 1.52. The closest point to the apex of the wedge where reflection is enhanced occurs where the wedge is 98 nm thick. Find the wavelength.

3. Monochromatic light shines on a glass wedge with refractive index 1.65, and enhanced reflection occurs where the wedge is 450 nm thick. Find all possible values for the wavelength in the visible range.

4. White light shines on 100-nm-thick sliver of fluorite ($n = 1.43$). What wavelength is most strongly reflected?

5. As a soap bubble ($n = 1.33$) evaporates and thins, the reflected colors gradually disappear. (a) What is its thickness just as the last vestige of color vanishes? (b) What is the last color seen?

6. An oil film ($n = 1.25$) floats on water, and a soap film ($n = 1.33$) is suspended in air. Find the minimum thickness for each that will result in constructive interference with 500-nm green light.

7. Light reflected from a thin film of acetone ($n = 1.36$) on a glass plate ($n = 1.5$) shows maximum reflection at 500 nm and minimum at 400 nm. Find the minimum possible film thickness.

8. What minimum thickness of a coating with refractive index 1.35 should you use on a glass lens to minimize reflection at 500 nm, the approximate center of the visible spectrum?

9. An oil film with refractive index 1.25 floats on water. The film thickness varies from 0.80 μm to 2.1 μm. If 630-nm light is incident normally on the film, at how many locations will it undergo enhanced reflection?

10 Microwave ovens operate at a frequency of 2.4 GHz. What is the minimum thickness for a plastic tray with refractive index 1.45 that will cause enhanced reflection of microwaves incident normal to the plate?

11. Two perfectly flat glass plates are separated at one end by a piece of paper 0.065 mm thick. A source of 550-nm light illuminates the plates from above, as shown in Fig. 37-46. How many bright bands appear to an observer looking down on the plates?

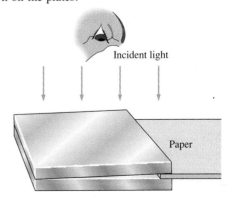

FIGURE 37-46   An air wedge (Problems 11, 12, 13, 72).

12. An air wedge like that shown in Fig. 37-46 shows N bright bands when illuminated from above. Find an expression for the number of bands that will appear if the air is replaced by a liquid of refractive index n different from that of the glass.

13. You apply a slight pressure with your finger to the upper of a pair of glass plates forming an air wedge as in Fig. 37-46. The wedge is illuminated from above with 500-nm light, and you place your finger where, initially, there is a dark band. If you push gently so the band becomes light, then dark, then light again, by how much have you deflected the plate?

14. A Michelson interferometer uses light from glowing hydrogen at 486.1 nm. As you move one mirror, 530 bright fringes pass a fixed point in the viewer. How far did the mirror move?

15. What is the wavelength of light used in a Michelson interferometer if 550 bright fringes go by a fixed point when the mirror moves 0.150 mm?

16. One arm of a Michelson interferometer is 42.5 cm long and is enclosed in a box that can be evacuated. The box initially contains air, which is gradually pumped out to create a vacuum. In the process, 388 bright fringes pass a fixed point in the viewer. If the interferometer uses light with wavelength 641.6 nm, what is the refractive index of the air?

17. The evacuated box of the previous problem is filled with chlorine gas, whose refractive index is 1.000772. How many bright fringes pass a fixed point as the tube fills?

18. Roughly how many cycles should be in the wavetrains of Fig. 37-1a, given that the wavetrains last about 10^{-8} s? Assume a visible wavelength of about 600 nm.

19. Your personal stereo is in a dead spot caused by direct reception from an FM radio station at 89.5 MHz interfering with the signal reflecting off a wall behind you. How much farther from the wall should you move in order that the interference be fully constructive?

Section 37-3 Double-Slit Interference

20. A double-slit experiment with $d = 0.025$ mm and $L = 75$ cm uses 550-nm light. What is the spacing between adjacent bright fringes?

21. A double-slit experiment has slit spacing 0.12 mm. (a) What should be the slit-to-screen distance L if the bright fringes are to be 5.0 mm apart when the slits are illuminated with 633-nm laser light? (b) What will be the fringe spacing with 480-nm light?

22. With two slits separated by 0.37 mm, the interference pattern has bright fringes with angular spacing 0.065°. What is the wavelength of the light illuminating the slits?

23. The green line of gaseous mercury at 546 nm falls on a double-slit apparatus. If the fifth dark fringe is at 0.113° from the centerline, what is the slit separation?

24. What is the angular position θ of the second-order bright fringe in a double-slit system with 1.5-μm slit spacing if the light has wavelength (a) 400 nm or (b) 700 nm?

25. Light shines on a pair of slits whose spacing is three times the wavelength. Find the locations of the first- and second-order bright fringes on a screen 50 cm from the slits. *Hint:* Do Equations 37-2 apply?

26. A double-slit experiment has slit spacing 0.035 mm, slit-to-screen distance 1.5 m, and wavelength 500 nm. What is the phase difference between two waves arriving at a point 0.56 cm from the center line?

27. For a double-slit experiment with slit spacing 0.25 mm and wavelength 600 nm, at what angular position is the path difference equal to one-fourth of the wavelength?

28. A screen 1.0 m wide is located 2.0 m from a pair of slits illuminated by 633-nm laser light, with its center on the centerline of the slits. Find the highest order bright fringe that will appear on the screen if the slit spacing is (a) 0.10 mm; (b) 10 μm.

29. Laser light at 633 nm falls on a double-slit apparatus with slit separation 6.5 μm. Find the separation between (a) the first and second and (b) the third and fourth bright fringes, as seen on a screen 1.7 m from the slits.

30. A tube of glowing gas emits light at 550 nm and 400 nm. In a double-slit apparatus, what is the lowest order 550-nm bright fringe that will fall on a 400-nm dark fringe, and what are the corresponding orders?

Section 37-4 Multiple-Slit Interference and Diffraction Gratings

31. In a 5-slit system, how many minima lie between the zeroth-order and first-order maxima?

32. In a 3-slit system the first minimum occurs at an angular position of 5°. Where is the first maximum?

33. A 5-slit system with 7.5-μm slit spacing is illuminated with 633-nm light. Find the angular positions of (a) the first 2 maxima and (b) the 3rd and 6th minima.

34. On a screen 1.25 m from a multiple-slit system, the interference pattern shows bright maxima separated by 0.86° and 7 minima between each bright maximum. (a) How many slits are there? (b) What is the slit separation if the incident light has wavelength 656.3 nm?

35. Green light at 520 nm is diffracted by a grating with 3000 lines per cm. Through what angle is the light diffracted in (a) first and (b) fifth order?

36. Find the angular separation between the red hydrogen-α spectral line at 656 nm and the yellow sodium line at 589 nm if the two are observed in 3rd order with a 3500 line/cm grating spectrometer (Fig. 37-47).

37. Light is incident normally on a grating with 10,000 lines per cm. What is the maximum order in which (a) 450-nm and (b) 650-nm light will be visible?

FIGURE 37-47 Spectral lines of hydrogen-α and sodium. What would be their angular separation if they appeared on a single spectrum? (Problem 36)

38. Visible light has wavelengths between 400 nm and 700 nm. What is the lowest pair of consecutive orders for which there will be some overlap between the visible spectra as dispersed by a grating?

39. A solar astronomer is studying the Sun's 589-nm sodium spectral line with a 2500 line/cm grating spectrometer whose fourth order dispersion puts the wavelength range from 575 nm to 625 nm on a detector. The astronomer is interested in observing simultaneously the so-called calcium-K line, at 393 nm. What order dispersion will put this line also on the detector?

40. (a) What portions of the 4th and 5th order visible spectra overlap in a 3000 lines/cm grating spectrometer? (b) How would your answer change for a 1000 lines/cm grating? (c) For a 10,000 lines/cm grating?

41. Estimate the number of lines per cm in the grating used to produce Fig. 37-24.

42. A grating spectrometer's detector covers an angular range of 10° and can be swung at any angle within ±75° of the normal to the grating. If the grating has 1200 lines/cm, in what orders can the hydrogen-α spectral line at 656 nm and the calcium-K line at 393 nm both fall on the detector? (b) What will be the angular position of the K line under these conditions?

43. When viewed in 6th order, the 486.1-nm hydrogen-β spectral line is flanked by another line that appears at the position of 484.3 nm in the 6th order spectrum. Actually the line is from a different order of the spectrum. What are the possible visible wavelengths of this line?

44. (a) Find the resolving power of a grating needed to separate the sodium-D spectral lines, which are at 589.0 nm and 589.6 nm. (b) How many lines must the grating have to achieve this resolution in first order? (This is very low resolution by present-day spectroscopic standards.)

45. Echelle spectroscopy uses relatively course gratings in high order. Compare the resolving power of an 80 line/mm echelle grating used in 12th order with a 600 line/mm grating used in 1st order, assuming the two have the same width.

46. The International Ultraviolet Explorer satellite carries a spectrometer with a 2.0-cm-wide grating ruled at 102 lines/mm. What is the minimum wavelength difference it can resolve in 12th order when observing in the ultraviolet at 155 nm?

47. You wish to resolve the calcium-H line at 396.85 nm from the hydrogen-ε line at 397.05 nm in a 1st-order spectrum. To the nearest hundred, how many lines should your grating have?

48. X-ray diffraction in calcium chloride (KCl) results in a first-order maximum when the x rays graze the crystal plane at 8.5°. If the x-ray wavelength is 97 pm, what is the spacing between crystal planes?

Sections 37-5 and 37-6 Single-Slit Diffraction and the Diffraction Limit

49. For what ratio of slit width to wavelength will the first minima of a single-slit diffraction pattern occur at ±90°?

50. Light with wavelength 633 nm is incident on a 2.5 μm wide slit. Find the angular width of the central peak in the diffraction pattern, taken as the angular separation between the first minima.

51. A beam of parallel rays from a 29-MHz citizen's band radio transmitter passes between two electrically conducting (hence opaque to radio waves) buildings located 45 m apart. What is the angular width of the beam when it emerges from between the buildings?

52. Use trial-and-error with a calculator, or a more sophisticated root-finding method, to verify the number 1.3916 in the exercise following Example 37-7.

53. Find the intensity as a fraction of the central peak intensity for the second secondary maximum in single-slit diffraction, assuming the peak lies midway between the second and third minima.

54. The width of a peak is often given in terms of the *full width at half maximum* (FWHM), meaning the width measured where the peak has half its maximum value (Fig. 37-48). Use the number mentioned in Problem 52 to find an expression for the angular FWHM of the central peak in single-slit diffraction, in terms of the wavelength λ and slit width a.

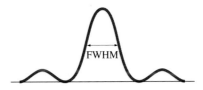

FIGURE 37-48 A peak's full width at half maximum is the width measured at half the maximum height (Problem 54).

FIGURE 37-49 CIA agents using satellite imaging to identify terrorists in the film *Patriot Games*. How big a mirror or lens must the satellite's optical system have? (Problem 55)

FIGURE 37-50 Jupiter, photographed with ground-based telescope (left) and from space (right). Atmospheric turbulence limits the ground-based image quality while diffraction limits the space-based image (Problem 61).

55. The movie *Patriot Games* has a scene in which CIA agents use spy satellites to identify individuals in a terrorist camp (Fig. 37-49). Suppose that a minimum resolution for distinguishing human features is about 5 cm. If the spy satellite's optical system is diffraction limited, what diameter mirror or lens is needed to achieve this resolution from an altitude of 100 km? Assume a wavelength of 550 nm.

56. Suppose the 10-m-diameter Keck telescope in Hawaii could be trained on San Francisco, 3400 km away. Would it be possible to read a (a) newspaper or (b) a billboard sign at this distance? Justify your answers by giving the minimum separation resolvable with Keck at 3400 km, assuming 550-nm light.

57. What is the minimum spot diameter to which a camera set at f-ratio of 16 can focus parallel light with 650-nm wavelength? *Hint:* Equation 37-13b gives the minimum angular spacing between the central maximum and first minimum; here you want the angular spread of the circle that marks the minimum.

58. The distance from the center of a circular diffraction pattern to the first minimum on a screen 0.85 m distant from the diffracting aperture is 15,000 wavelengths. What is the aperture diameter?

59. While driving at night, your eyes' irises have dilated to 3.1-mm diameter. If your vision were diffraction limited, what would be the greatest distance at which you could see as distinct the two headlights of an oncoming car, which are spaced 1.5 m apart? Take $\lambda = 550$ nm.

60. Two stars are 4.0 light-years apart, in a galaxy 20×10^6 light-years from Earth. What minimum separation of two radio telescopes, acting together as an interferometer, is needed to resolve them? The telescopes operate at 2.1 GHz and are pointing straight upward; the stars are directly overhead.

61. Under the best conditions, atmospheric turbulence limits the resolution of ground-based telescopes to about 1 arc second (1/3600 of a degree) as shown in Fig. 37-50. For what aperture sizes is this limitation more severe than that of diffraction at 550 nm? Your answer shows why large ground-based telescopes do not produce better images than small ones, although they do gather more light.

62. Two objects are separated by approximately one wavelength of the light with which an observer is attempting to resolve them. Show that the Rayleigh criterion then requires that the distance to the objects be less than the diameter of the observing aperture, and thus the Fraunhofer diffraction approximation is violated. It is in fact impossible to resolve two objects as close as one wavelength.

Paired Problems

(Both problems in a pair involve the same principles and techniques. If you can get the first problem, you should be able to solve the second one.)

63. A thin film of toluene ($n = 1.49$) floats on water. What is the minimum film thickness if the most strongly reflected light has wavelength 460 nm?

64. Oil with refractive index 1.38 forms a 210-nm-thick film on a piece of glass. What color of visible light is most strongly reflected?

65. Find the total number of lines in a 2.5-cm-wide diffraction grating whose third-order spectrum has the 656-nm hydrogen-α spectral line at an angular position of 37°.

66. Light is diffracted by a 5000 line/cm grating, and a detector sensitive only to visible light finds a maximum intensity at 28° from the central maximum. (a) What is the wavelength of the light? (b) In what order is it seen at the 28° position?

67. A 400 line/mm diffraction grating is 3.5 cm wide. Two spectral lines whose wavelengths average to 560 nm are just barely resolved in the 4th-order spectrum of this grating. What is the difference between their wavelengths?

68. What order is necessary to resolve wavelengths of 647.98 nm and 648.07 nm using a 4500-line grating?

69. What diameter optical telescope would be needed to re-solve a Sun-sized star 10 light-years from Earth? Take $\lambda = 550$ nm. Your answer shows why stars appear as point sources in optical astronomy.

70. Could the 305-m radio telescope at Arecibo in Puerto Rico resolve the star in the preceding problem, if it were observing at a wavelength of 1.0 cm?

Supplementary Problems

71. White light shines on a 250-nm-thick layer of diamond ($n = 2.42$). What wavelength *of visible light* is most strongly reflected?

72. An air wedge like that of Fig. 37-46 displays 10,003 bright bands when illuminated from above. If the region between the plates is then evacuated, the number of bands drops to 10,000. Calculate the refractive index of the air.

73. In Fig. 37-6 the mth Newton's ring appears a distance r from the center of the lens. Show that the curvature radius of the lens is given approximately by $R = r^2/(m + \frac{1}{2})\lambda$, where the approximation holds when the thickness of the air space is much less than the curvature radius.

74. Given the result of the preceding problem, how many bright Newton's rings would be seen with a 2.5-cm-di-ameter glass lens with curvature radius 7.5 cm, when illuminated with 500-nm light?

75. How many rings would be seen if the system of the preceding problem were immersed in water ($n = 1.33$)?

76. A thin-walled glass tube of length L containing a gas of unknown refractive index is placed in one arm of a Michelson interferometer using light of wavelength λ. The tube is then evacuated, and during the process m bright fringes pass a fixed point as seen in the viewer. Find an expression for the refractive index of the gas.

77. The signal from a 103.9-MHz FM radio station reflects off a building 400 m away, effectively producing two sources of the same signal. You're driving at 60 km/h along a road parallel to a line between the station's antenna and the building and located a perpendicular distance of 6.5 km from them. How often does the signal appear to fade when you're driving roughly opposite the transmitter and building?

78. A satellite dish antenna 1.5 m in diameter receives TV broadcasts at a frequency of about 6 GHz (Fig. 37-51). (a) What is the angular size of its beam, defined as the full width of the central diffraction peak? (b) How many communications satellites could fit in geosynchronous orbit above Earth's equator if antennas like this one were not to pick up signals from more than one satellite? Your answer shows why geosynchronous orbit is such valuable "real estate."

FIGURE 37-51 A satellite dish (Problem 78).

79. The component of a star's velocity in the radial direction relative to Earth is to be measured using the doppler shift in the hydrogen-β spectral line, which appears at 486.1 nm when the source is stationary relative to the observer. What is the minimum speed that can be detected by observing in 1st order with a 10,000 line/cm grating 5.0 cm across? *Hint:* See Equation 17-11.

80. Light is incident on a dispersion grating at an angle α to the normal. Show that the condition for maximum light intensity becomes $d(\sin \theta \pm \sin \alpha) = m\lambda$.

81. In a double-slit experiment, a thin glass plate with refractive index 1.56 is placed over one of the slits. The fifth bright fringe now appears where the second dark fringe previously appeared. How thick is the plate if the incident light has wavelength 480 nm?

82. An arrangement known as Lloyd's mirror (Fig. 37-52) allows interference between a direct and a reflected beam from the same source. Find an expression for the separation of bright fringes on the screen, given the distances d and D in Fig. 37-53, and the wavelength λ of the light. *Hint:* Think of two sources, one the virtual image of the other.

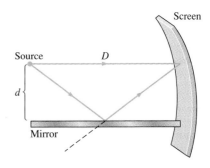

FIGURE 37-52 Lloyd's mirror (Problem 82).

P A R T 5   CUMULATIVE PROBLEMS

These problems combine material from chapters throughout the entire part or, in addition, from chapters in earlier parts, or they present special challenges.

1. A *grism* is a grating ruled onto a prism, as shown in Fig. 1. The grism is designed to transmit undeviated one wavelength of the spectrum in a given order, as refraction in the prism compensates for the deviation at the grating. Find an equation relating the separation d of the grooves that constitute the grating, the wedge angle α of the prism, the refractive index n, the undeviated wavelength λ_0, and order m_0.

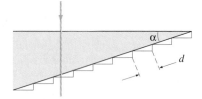

FIGURE 1   A grism (Cumulative Problem 1).

2. A double-slit system consists of two slits each of width a, with separation d between the slit centers ($d > a$). Light of intensity S_0 and wavelength λ is incident on the system, perpendicular to the plane containing the slits. Find an expression for the outgoing intensity as a function of angular position θ, taking into account both the slit width and the separation. Plot your result for the case $d = 4a$, and compare with Figure 37-35.

3. A closed cylindrical tube whose glass walls have negligible thickness measures 5.0 cm long by 5.0 mm in diameter. It is filled with water, initially at 15°C, and placed with its long dimension in one arm of a Michelson interferometer. The water is not perfectly transparent, and it absorbs 3.2% of the light energy incident on it. The laser power incident on the water is 50 mW, and the wavelength is 633 nm. The refractive index of water in the vicinity of 15°C is given approximately by $n = 1.335 - 8.4 \times 10^{-5}T$, where T is the temperature in °C. As the water absorbs light energy, how long does it take the interference pattern to shift by one whole fringe?

4. A radio antenna broadcasts a 6.5-MHz signal in all directions. The antenna is on top of a tower 300 m above sea level, and the tower is located 2.0 km from the shore. An airplane is flying toward the tower, at twice the altitude of the tower top, as shown in Fig. 2. Radio waves from the tower reflect off the ocean surface with a 180° phase change, but there is no significant reflection off the land. As the plane heads toward the tower, at what horizontal distance or distances from the tower will its radio equipment detect a minimum in the radio signal amplitude due to destructive interference of the direct and reflected radio waves?

5. In one type of optical fiber, called a *graded-index fiber*, the refractive index varies in a way that results in light rays

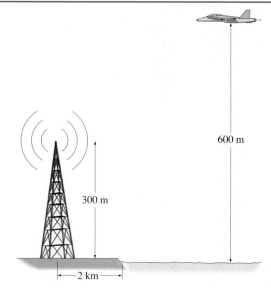

FIGURE 2   Cumulative Problem 4 (figure is not to scale).

being guided along the fiber on curved trajectories, rather than undergoing abrupt reflections. Figure 3 shows a simple model that demonstrates this effect; it also describes the basic optical effect in mirages. A slab of transparent material has refractive index $n(y)$ that varies with position y perpendicular to the slab face. A light ray enters the slab at $x = 0$, $y = 0$, making an angle θ_0 with the normal just inside the slab. The refractive index at this point is $n(y = 0) = n_0$. (a) By writing $\sin\theta$ in Snell's law in terms of the components dx and dy of the ray path, show that that path (written in the form of x as a function of y) is given by

$$ x = \int_0^y \frac{n_0 \sin\theta_0}{\sqrt{[n(y)]^2 - n_0^2 \sin^2\theta_0}}\, dy. $$

(b) Suppose $n(y) = n_0(1 - ay)$, where $n_0 = 1.5$ and $a = 1.0\ \text{mm}^{-1}$. If $\theta_0 = 60°$, find an explicit expression for x as a function of y, and plot your result to give the actual ray path. Explain the shape of your curve in terms of what happens when the ray reaches a point where $n(y) = n_0 \sin\theta_0$. What happens beyond this point?

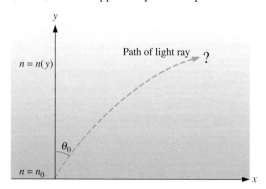

FIGURE 3   Cumulative Problem 5.

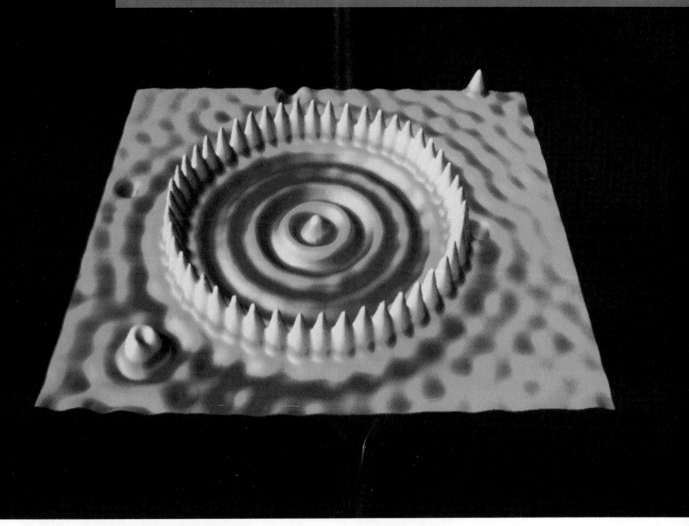

A "quantum corral" of 48 iron atoms on a copper surface encloses circular ripples associated with the wave-like nature of electrons on the surface. The image was taken with a scanning-tunneling microscope, which utilizes the principles of quantum mechanics. The ideas of modern physics allow scientists to not only understand but also, increasingly, to image and even manipulate matter at the atomic scale.

CHAPTER 38

THE THEORY OF RELATIVITY

Albert Einstein's theory of relativity revolutionized our understanding of space and time. Here Einstein explains his theory at a 1931 lecture in California.

By the last quarter of the nineteenth century, the basic laws of electromagnetism had been formulated. Maxwell's theory had demonstrated the electromagnetic nature of light, whose practical consequences we explored in the three preceding chapters. In the same era the work of Samuel Morse (telegraph), Alexander Graham Bell (telephone), Hertz and Marconi (radio), Thomas A. Edison (electric light, phonograph), and many others laid the foundation of electromagnetic technology. Yet at the same time the insights of Maxwell led to baffling questions and contradictions that shook the roots of physical understanding and even of common sense.

From the resolution of these contradictions arose the theory of relativity, a theory that radically altered the philosophical basis for our understanding of the

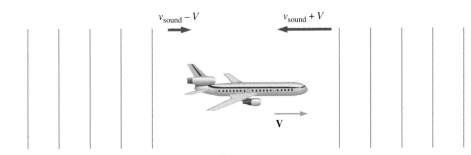

physical world, and whose influence spilled over into all areas of twentieth-century thought. The theory of relativity stands as a monument of human intellect and imagination. It transcends the everyday world of common sense, and shows us a universe whose richness is almost beyond imagination.

38-1   SPEED *c* RELATIVE TO WHAT?

Maxwell's equations show that electromagnetic waves can exist, and that all such waves travel with speed c. Speed c relative to what? When we described the mechanics of a taut string in Chapter 16, we encountered a wave equation showing that waves could propagate along the string with a certain speed. A certain speed relative to what? Clearly, to the string. Similarly, sound waves propagate through the air with a certain speed relative to the air. If you move through the air, the speed of sound *relative to you* will not be the same as its speed relative to the air (Fig. 38-1).

The Ether Concept

Each type of wave has a characteristic speed *relative to the medium in which it propagates.* What about light? What is the medium through which it propagates? Light reaches us from the most distant galaxies, traveling through seemingly empty space. Yet all our other experience with waves suggests that there should be a medium—and that the speed c in Maxwell's equations should be the speed relative to that medium. Nineteenth-century scientists, supposing that light waves were like mechanical waves, believed that light propagated through a tenuous substance called the **ether.** The ether permeated the entire universe, filling the smallest voids and permitting light to go anywhere. Electric and magnetic fields were visualized as stresses and strains in the ether. This mechanical view—that electromagnetic phenomena, including light, were disturbances of some substance—was deeply ingrained in the nineteenth-century scientists because of their previous experience that Newton's laws explained all known physical phenomena.

The ether had to have some unusual properties. First, it must be tenuous and without significant viscosity, or it could not creep into every corner of the universe. And it must offer no resistance to the motion of material bodies, or the planets would soon lose their energy and spiral into the Sun. At the same time the ether must be very stiff, for the speed of light is large. (If you make a spring stiffer, waves travel more quickly along it.) Indeed, the constants ε_0 and μ_0 must describe the mechanical properties of the ether that account for the high speed of light. These and other mechanical requirements make the ether a rather improbable substance, but without the ether it seemed there could be no waves, and the question "speed c relative to what?" would leave us floundering for an answer.

The existence of electromagnetic waves traveling at speed c follows from Maxwell's equations. But this result could be true only in a frame of reference fixed with respect to the ether, for if we move relative to the ether we should expect light to travel at a different speed relative to us. Therefore Maxwell's equations—that is, our description of electromagnetism—were presumably correct only in the ether's frame of reference.

This situation put electromagnetism in a rather different position from mechanics. In mechanics, the concept of absolute motion is meaningless. You can eat your dinner, or throw a ball, or do any mechanical experiment, as well on an airplane moving steadily at 1000 km/h as you can when the airplane is standing still on the ground. You need not take the uniform motion of the plane into account. This is the principle of **Galilean relativity,** which states that the laws of mechanics are valid in all frames of reference in uniform motion (see Section 3-8). But the laws of electromagnetism could only be valid in the ether's frame of reference, for it seemed that only in this frame could the prediction of electromagnetic waves moving at speed c be correct.

38-2 MATTER, MOTION, AND THE ETHER

Given the existence of the ether, it is natural to ask about Earth's motion relative to it. If Earth is moving through the ether, we should expect light to travel faster relative to us when it comes from the direction toward which Earth is moving. On the other hand, Earth might be at rest relative to the ether. Because other planets, stars, and galaxies move with respect to Earth, it is hard to imagine that ether is everywhere fixed with respect to Earth alone, for this violates the Copernican view that Earth does not occupy a privileged spot in the universe. But maybe Earth drags with it the ether in its immediate vicinity. If this "ether drag" occurs, then the speed of light must be independent of direction, but if ether drag does not occur then the speed of light measured on Earth must depend on direction. Through observation and experiment, nineteenth-century physicists sought to resolve the question of Earth's motion through the ether.

Aberration of Starlight

Imagine standing in a rainstorm with rain falling vertically. To keep dry, you hold your umbrella with its shaft straight up, as shown in Fig. 38-2a. But if you run, as in Fig. 38-2b, you will keep driest if you tilt your umbrella forward. Why? Because then the direction of rainfall *relative to you* is not straight down but at an angle, as shown in Fig. 38-2c. This argument presupposes that you do not drag with you a large volume of air. If such an "air drag" occurred, raindrops entering the region around you would be accelerated quickly in the horizontal direction by the air moving with you, so they would now fall vertically relative to you, as in Fig. 38-2d. No matter which way you ran, as long as you dragged air with you, you would point your umbrella vertically upward to stay dry.

FIGURE 38-2 (*a*) Standing still in vertically falling rain, you hold your umbrella overhead to keep dry. (*b*) Running, you tilt the umbrella to compensate for the rain's motion relative to you. (*c*) The situation of (*b*) seen from the runner's frame of reference. (*d*) If you dragged a volume of air with you, you would hold the umbrella overhead whatever your state of motion.

This umbrella example is exactly analogous to the observation of light from stars, with the rain being starlight and the umbrella a telescope. If Earth does not drag ether with it, the direction from which starlight comes will depend on the motion of Earth relative to the ether. But if "ether drag" occurs in analogy with Fig. 38-2*d*, then light from a particular star will always come from the same direction.

In fact we do observe a change in the direction of starlight. As Earth swings around in its orbit, we must first point a telescope one way to see a particular star. Then, six months later, Earth's orbital motion is in exactly the opposite direction, and we must point the telescope in a slightly different direction. This phenomenon is called **aberration of starlight** and shows that *Earth does not drag the ether*.

The Michelson-Morley Experiment

If we reject the pre-Copernican notion that Earth alone is at rest relative to the ether, then aberration of starlight forces us to conclude that Earth moves through the ether. Furthermore, the relative velocity of the motion must change throughout the year, as Earth orbits the Sun.

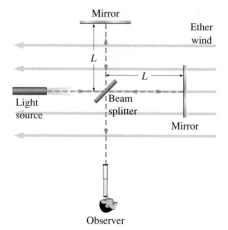

FIGURE 38-3   Simplified diagram of a Michelson interferometer. An "ether wind" blowing in the direction shown should result in a longer time for the light beam on the horizontal arm.

FIGURE 38-4   Vector diagram showing resultant velocity **u** of light moving at right angles to an ether wind with speed v. The speed of light relative to the ether is c.

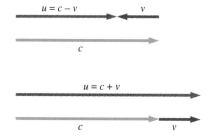

FIGURE 38-5   Vector diagrams showing the resultant velocities **u** for light moving with and against an ether wind whose speed is v.

In a series of experiments done in 1881–1887, the American scientists Albert A. Michelson and Edward W. Morley attempted to determine Earth's velocity relative to the ether. They used Michelson's interferometer (Fig. 38-3), whose operation we described in the preceding chapter. Recall that the interferometer produces a pattern of interference fringes that shifts if the round-trip travel time for light on one of its two perpendicular arms changes. The interference pattern reflects, among other things, possible differences in travel times that arise from differences in the speed of light in different directions—differences that should result from Earth's motion through the ether. Rotating the apparatus through 90° would interchange the directions of the arms and should therefore shift the interference pattern.

Now suppose that Earth moves at speed v relative to the ether. Then from the viewpoint of an observer on Earth, there is an "ether wind" blowing past Earth. Suppose that the Michelson-Morley apparatus is oriented with one light path parallel to the wind, the other perpendicular. Consider a light beam moving the distance L at right angles to the ether wind. The beam must be aimed slightly upwind, in order that it will actually move perpendicular to the wind. The light moves in this direction at speed c relative to the ether, but the ether wind sweeps it back so its path in the Michelson-Morley apparatus is at right angles to the wind. From Fig. 38-4, we see that its speed relative to the apparatus is

$$u = \sqrt{c^2 - v^2},$$

so the round-trip travel time is

$$t_{\text{perpendicular}} = \frac{2L}{u} = \frac{2L}{\sqrt{c^2 - v^2}}. \tag{38-1}$$

Light sent a distance L "upstream"—against the ether wind—travels at speed c relative to the ether but at speed $c - v$ relative to Earth. It therefore takes a time

$$t_{\text{upstream}} = \frac{L}{c - v}.$$

Returning, the light moves at $c + v$ relative to Earth, taking

$$t_{\text{downstream}} = \frac{L}{c + v}$$

(Fig. 38-5). So the round-trip time parallel to the wind is

$$t_{\text{parallel}} = \frac{L}{c - v} + \frac{L}{c + v} = \frac{2cL}{c^2 - v^2}. \tag{38-2}$$

The two round-trip travel times differ, with the trip parallel to the ether wind always taking longer (see Problems 1, 2, and 4). Light on the parallel trip is slowed when it moves against the ether wind, then speeds up when it moves with the wind. But slowing always dominates, because the light spends more time moving against the wind than with it.

The Michelson-Morley experiment of 1887 was sensitive enough to detect differences in the speed of light at least an order of magnitude smaller than Earth's orbital speed. The experiment was repeated with the apparatus oriented in different directions, and at different times throughout the year, and the same simple but striking result always emerged: there was never any difference in the travel times for the two light beams. In terms of the ether concept, the Michelson-Morley experiment showed that *Earth does not move relative to the ether*.

A Contradiction in Physics

Aberration of starlight shows that Earth does not drag ether with it. Earth must therefore move relative to the ether. But the Michelson-Morley experiment shows that it does not. This contradiction is a deep one, rooted in the fundamental laws of electromagnetism and in the analogy between mechanical waves and electromagnetic waves. The contradiction arises directly in trying to answer the simple question "with respect to what does light move at speed c?"

Physicists at the end of the nineteenth century made many ingenious attempts to resolve the dilemma of light and the ether, but their explanations were either inconsistent with experiment or lacked sound conceptual bases.

38-3 SPECIAL RELATIVITY

In 1905, at the age of 26, Albert Einstein (Fig. 38-6) proposed a theory that resolved the dilemma and at the same time altered the very foundation of physical thought. Einstein declared simply that the ether is a fiction. But then with respect to what does light move at speed c? With respect, Einstein declared, to anyone who cares to observe it. This statement is at once simple, radical, and conservative. Simple, because its meaning is clear and obvious. Anyone who measures the speed of light will get the value $c = 3.0 \times 10^8$ m/s. Radical, because it alters our commonsense notions of space and time. Conservative, because it asserts for electromagnetism what had long been true in mechanics: that the laws of physics do not depend on the motion of the observer. Einstein summarized his new ideas in the **special theory of relativity,** which is expressed in this simple sentence:

❙ The laws of physics are the same in all inertial frames of reference.

Recall that inertial frames are those which are not accelerated—i.e., those in which the laws of *mechanics* were already valid. Einstein's statement encompasses *all* laws of physics, including mechanics and electromagnetism. The prediction that electromagnetic waves move at speed c must, then, be a universal prediction that holds in *all* inertial frames of reference. The *special* theory of relativity is special because it is valid only for the special case of inertial frames. Later we will discuss the general theory of relativity, in which this restriction is removed.

Einstein's relativity readily explains the result of the Michelson-Morley experiment, for no matter what the speed of Earth relative to anything, an observer on Earth should measure the same speed for light in all directions. But at the same time, we will see that relativity flagrantly violates our common-sense notions of space and time.

FIGURE 38-6 Albert Einstein was a young father when, at age 26, he formulated the special theory of relativity.

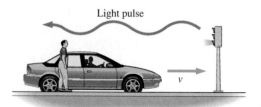

FIGURE 38-7 As the car passes the pedestrian, a light pulse goes by. Both driver and pedestrian measure the same speed c for the light, even though they are in relative motion.

38-4 SPACE AND TIME IN RELATIVITY

Consider a car driving past a pedestrian standing by the roadside (Fig. 38-7). Driver and pedestrian each measure the speed of the light from a blinking traffic signal. Relativity says they will get the same answer, $c = 3.0 \times 10^8$ m/s, even though the car is moving toward the source of light. How can this be? Consider how each observer might make the measurement. Let each be equipped with a meter stick and an accurate, high-speed electronic stopwatch. Suppose that a light pulse passes the front ends of both meter sticks just as they coincide. Each observer measures the time it takes the light pulse to cross the meter stick, then divides the distance (one meter) by the measured time to get the speed of light. Since the stick on the car is moving toward the light source, common sense suggests that the light will pass the far end of this stick first, and therefore that the time on the car's stopwatch will be shorter. But this violates relativity, for if both observers use the same path length for their speed-of-light measurement, and if they get different times, then they will have measured different speeds for light. In fact, both stopwatches will read the same time, even though common sense tells us that the light passes the end of the "moving" meter stick "earlier."

How can this be? It follows logically from the statement of special relativity, which in turn is consistent with physical experiments. But how can it be? Something must be "wrong" with someone's meter stick or stopwatch. Maybe the motion of the car somehow affects the stopwatch on the car. But no: this suggestion violates the spirit of relativity, which says that steady, uniform motion is undetectable—that it makes no more sense to say that the car is moving and the pedestrian is at rest than to say the opposite. That is the whole point of relativity—any frame of reference in uniform motion is as good as any other for doing physics. So there can be nothing wrong with the clocks and meter sticks.

The only things left to go "wrong" are time and space. Time and space—the seemingly passive, universal backgrounds in which all physical events take place—must themselves depend on the observer. Two observers in different frames of reference, moving uniformly relative to each other, are measuring different quantities when they use clocks to record the passage of time and meter sticks to determine distances in space. In relativity it is the laws of physics—not measures of time and space—that must be the same for everyone. Time and space are altered in ways that allow the laws of physics, including Maxwell's theory and its prediction that light waves travel at speed c, to be the same for all observers in uniform motion.

In exploring space and time in relativity, we'll often speak of events. An **event** is an occurrence specified by giving its position—three spatial coordinates—and its time. Your birth, for example, is an event: it occurred at a certain place and a certain time, and both are needed to specify it fully.

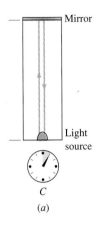

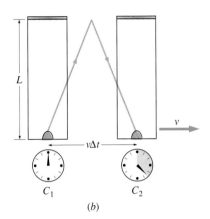

FIGURE 38-8 A "light-box clock" seen from two frames of reference. Light leaves the source, reflects off the mirror, and returns to the source. Clock C is attached to the box, while the box moves between clocks C_1 and C_2 at speed v. (a) In a frame of reference at rest with respect to the box, light travels a distance $2L$ at speed c. (b) In a frame of reference in which the box is moving, light travels a greater distance. But its speed c is the same in all frames of reference, so the time interval is longer in this frame. Part (b) shows the box both when the light is emitted and again when it returns to the source.

Time Dilation

To see how time is altered, consider the simple device shown in Fig. 38-8. It consists of a box of length L with a light source at one end and a mirror at the other. Let a flash of light leave the source, travel to the mirror, and return to the source. We want to know the time interval between two distinct events: the light flash leaving the source and the flash returning to the source.

In Fig. 38-8a we consider the experiment in a frame of reference S' at rest with respect to the box. The light travels a distance $2L$ in this frame, giving a round-trip travel time of

$$\Delta t' = \frac{2L}{c}.$$

Now consider the *same* experiment, viewed from a frame of reference S in which the box is moving to the right with speed v. In this frame there are two clocks along the path of the box. These clocks are synchronized, and are in the same frame of reference, so they both measure the same quantity. The box passes the first clock just as the light flash goes off, and at this instant the clock reads zero. Just as the light flash returns to the source, the box passes the second clock. The time this clock reads at the instant the box passes is the time, measured in frame S, between the emission and return of the light flash.

We can calculate this time interval Δt in frame S, just as we did in S', by figuring the total distance traveled by the light and dividing by its speed. Figure 38-8b shows the situation. In frame S, the box moves to the right a distance $v\,\Delta t$ in the time Δt between emission and return of the light flash. Meanwhile the light takes a diagonal path up to the mirror of the moving box, then back down to the source. The total length of this path is twice the diagonal from source to mirror, or, using the Pythagorean theorem, $2\sqrt{L^2 + (v\,\Delta t/2)^2}$. The time required for light to go this distance is just the distance divided by the speed of light, or

$$\Delta t = \frac{2\sqrt{L^2 + (v\,\Delta t/2)^2}}{c}. \tag{38-3}$$

Notice that we explicitly used the theory of relativity in writing Equation 38-3. We did not vectorially add the horizontal speed of the box to the vertical speed of light to get a new speed of light in frame S, for relativity says that the speed of light is the same in all frames of reference. Had we altered the speed, we would have had an increased path length in S, but an increased speed of light as well, and would have found that the time intervals in both frames were the same. But no! Relativity requires that we use the same speed c in both frames, even though the path lengths differ. That is why we get different answers for the time.

The unknown time Δt appears on both sides of Equation 38-3. Multiplying through by c and squaring gives

$$c^2(\Delta t)^2 = 4L^2 + v^2(\Delta t)^2.$$

We then solve for $(\Delta t)^2$ to get

$$(\Delta t)^2 = \frac{4L^2}{c^2 - v^2} = \frac{4L^2}{c^2}\left(\frac{1}{1 - v^2/c^2}\right).$$

Taking the square root of both sides, and noting that $2L/c$ is just the time $\Delta t'$ measured in the frame S' at rest with respect to the box, we have

$$\Delta t = \frac{\Delta t'}{\sqrt{1 - v^2/c^2}} \qquad \text{or} \qquad \Delta t' = \Delta t\sqrt{1 - v^2/c^2}. \qquad (38\text{-}4)$$

Equation 38-4 describes the phenomenon of **time dilation,** in which the time interval between two events is always shortest in a frame of reference in which the two events occur at the same place. (The time measured in this frame is called the **proper time,** although relativity precludes our considering it any "better" a measure of time than that made in any other frame of reference.) In our example, the two events are the emission and the return of the light flash, and they occur at the same place—the bottom of the box—in the box frame S', but at different places in S. Thus $\Delta t'$—the time interval measured in S', or the proper time—is shorter than Δt, as you can see from Equation 38-4.

Time dilation is sometimes characterized by saying that "moving clocks run slow," but this statement is not strictly correct because relativity rules out our saying that one clock is moving and another not. What the statement means is what we've just seen: that the time interval between two events is shortest in a frame of reference where the two events occur at the same place (Fig. 38-9).

We have illustrated time dilation with a very special device—a "light-box clock." But the phenomenon would occur with any other timing device, for it is not that something unusual happens to the clock, but to time itself. If we take away the light box in Fig. 38-8, giving Fig. 38-9, the clocks will still show the same discrepancy. There is no use searching for a physical mechanism that slows things down. All manifestations of time—the oscillations of the quartz crystal in a digital watch, the swing of a pendulum clock, the period of vibration of atoms in an atomic clock, biological rhythms, and human lifetimes—all are affected in the same way.

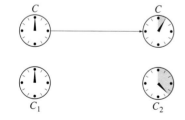

FIGURE 38-9 Clock C moves between clocks C_1 and C_2, which are at rest relative to each other and synchronized in their rest frame. Time between the event of C passing C_1 and the event of C passing C_2 is shorter in C's frame of reference.

● EXAMPLE 38-1 STAR TREK

Solution

A spaceship leaves Earth on a one-way trip that Earthbound observers judge will take 25 years. If the ship travels at 0.95c relative to Earth, how long is the journey as measured by a clock on board the ship?

Solution

The spaceship is like the light box we used in deriving Equation 38-4 for time dilation, in that in its frame the beginning and end of the journey occur at the same place—namely, on the ship. So the time interval on the ship clock is the $\Delta t'$ of Equation 38-4. For the observers on Earth, the beginning and end of the journey occur at different places, so their 25-year time interval is Δt. Applying Equation 38-4 then gives

$$\Delta t' = \Delta t \sqrt{1 - v^2/c^2} = (25 \text{ y})(\sqrt{1 - (0.95c)^2/c^2})$$

$$= 7.8 \text{ y}.$$

Thus at this high relative speed, the ship's time is considerably shorter. We'll soon explore what happens if the ship turns around and returns to Earth.

EXERCISE An extraterrestrial spacecraft whizzes through the solar system at 0.80c. How long does it take to go the 8.3 light-minute distance from Earth to the Sun (a) according to an observer on Earth and (b) according to an alien aboard the ship?

Answers: (a) 10.4 min; (b) 6.23 min

Some problems similar to Example 38-1: 7, 8, 16
●

Why don't we notice time dilation as we travel about in our everyday lives? Because the factor v^2/c^2 in Equation 38-4 is so small for any velocities we have relative to Earth. Even in a jet airplane, we are moving at 1000 km/h or only about $10^{-6} c$. Then time in the airplane is different from that on Earth by only about 1 part in $(10^6)^2$, which amounts to a few milliseconds per century. This illustrates an important point: any results predicted by relativity should agree with our common sense, Newtonian ideas when relative velocities are small compared with the speed of light. Only at substantial fractions of c do relativistic effects become obvious. Since our intuitions and common sense are built on experience at low relative velocities, it is not surprising that effects at high velocities seem counter to common sense.

■ APPLICATION MOUNTAINS AND MUONS: CONFIRMING TIME DILATION

Time dilation is clearly illustrated in experiments with subatomic particles moving, relative to us, at speeds near that of light. In a classic experiment, the "clocks" are the lifetimes of unstable particles called muons, which are created by the interaction of cosmic rays with Earth's upper atmosphere. The experiment consists in counting the number of muons incident each hour on the top of a mountain—Mt. Washington in New Hampshire, about 2000 m above sea level. The measurement is then repeated at sea level (Fig. 38-10).

Using a detector that records only those muons moving at about 0.994c at the mountaintop altitude, the experiment shows an average of about 560 muons with this speed are incident on the mountaintop each hour. If the mountain weren't there, the muons would travel from the mountaintop altitude to sea level in a time given by

$$\Delta t = \frac{2000 \text{ m}}{(0.994)(3.0 \times 10^8 \text{ m/s})} = 6.7 \text{ } \mu s.$$

The muon's decay rate is such that one should expect only about 25 of the original 560 muons to remain after this 6.7-μs interval, so that's approximately the number we might expect to detect each hour at sea level. More muons would survive a shorter time interval; in particular, we should expect 414 muons surviving 0.73 μs after passing the mountaintop altitude.

Why the value 0.73 μs? Because the muons are moving at 0.994c, and the 6.7-μs interval from mountaintop to sea level is measured in Earth's frame of reference—not in the muons' frame. In the muons' frame, time dilation should reduce that interval to

$$\Delta t' = \Delta t \sqrt{1 - v^2/c^2} = (6.7 \ \mu s)(\sqrt{1 - 0.994^2})$$

$$= 0.73 \ \mu s.$$

Since the muons' decay is determined by their measure of time, we should therefore expect 414 muons surviving at sea level.

So what happens? When the experiment is done, muon counts of just over 400 per hour are observed at sea level. This is no subtle effect! The difference between 25 and 414 is dramatic. At 0.994c, the nonrelativistic description is hopelessly inadequate. At this speed, the factor $\sqrt{1 - v^2/c^2}$ is one-ninth, and time dilation is obvious.

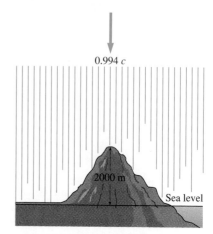

FIGURE 38-10 The rate at which high-speed muons are incident is measured on a mountaintop and again at sea level. The muons decay at such a rate that only a small fraction should survive to reach sea level. But the number measured at sea level is much larger, showing the effect of time dilation on muon lifetimes.

FIGURE 38-11 At departure, the twins are the same age.

The Twin Paradox

The phenomenon of time dilation has startling consequences, for it allows us to travel into the future! The famous "twin paradox" illustrates this possibility. One of two twins boards a fast spaceship for a journey to a distant star (Fig. 38-11). The other stays behind on Earth. Imagine that there are clocks at Earth and star, like the two clocks in frame S of our light-box experiment (Fig. 38-12a). There is a clock on the spaceship, like the one clock in our light-box frame S'. When the ship arrives at the distant star, less time will have elapsed on the ship clock than on the Earth and star clocks (Fig. 38-12b). Now the ship turns around and comes home. Again, the situation is identical to our light-box example, so again less time elapses on the ship clock, and the traveling twin arrives home younger than the earthbound twin (Fig. 38-13)! Depending on how far and how fast the traveling twin goes, the difference in ages could be arbitrarily large. The traveling twin could even return to Earth millions of years in the future, even though only hours had elapsed on the ship. But this is a one-way trip to the future! If the traveler doesn't like what he or she finds in the future, there is no going back!

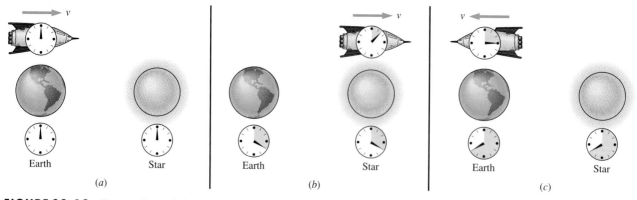

FIGURE 38-12 The traveling twin journeys to a distant star, then returns. Earth and star are at rest with respect to each other, and their clocks are synchronized. The figure is drawn from the Earth-star frame of reference. (a) Ship and Earth clocks agree as the ship leaves Earth. (b) When the ship reaches the star, less time has elapsed on its clock. (c) When the ship returns to Earth, its clock and Earth's no longer agree.

● **E X A M P L E   3 8 - 2**   THE TWIN PARADOX

Earth and a star are 10 light-years (ly) apart, measured in a frame at rest with respect to Earth and star. Twin A boards a space ship and travels at $0.80c$ to the star, then immediately turns around and returns to Earth at $0.80c$. Twin B remains behind. Determine the round-trip travel time in the Earth-star frame of reference and in the ship frame. By how much will the twins' ages differ when they get back together?

Solution
At $0.80c$, the time to go 10 ly in the Earth-star frame is just

$$\Delta t = \frac{10 \text{ ly}}{0.80 \text{ ly/y}} = 12.5 \text{ y}.$$

The round-trip time is then 25 y. Equation 38-4 for $\Delta t'$ then gives the one-way travel time in the ship frame:

$$\Delta t' = \Delta t \sqrt{1 - v^2/c^2} = (12.5 \text{ y})(\sqrt{1 - 0.80^2}) = 7.5 \text{ y}.$$

Then the round-trip time in the ship frame is 15 y, so the twins' ages differ by 10 y when twin A returns.

TIP   Years, Light-Years, and the Speed of Light   A light-year (ly) is the distance light travels in one year. By definition, therefore, the speed of light is 1 ly/y. It's often easiest in relativity to work in units where the speed of light is 1, whether those units be light-years and years, light-seconds and seconds, or whatever.

E X E R C I S E   A spacecraft makes a round trip to a point a distance ℓ from Earth, as measured in Earth's frame of reference. The ship travels at 60% of the speed of light. If the round trip takes 1 hour by the ship's clock, (a) what is the distance ℓ and (b) how much time elapses on Earth between the ship's departure and its return?

Answers: (a) 0.375 light-hours (4×10^{11} m); (b) 1.25 h

Some problems similar to Example 38-2: *7, 9, 14, 15* ●

The paradox in the twin example is not just that something strange happens to time—we already expect that of special relativity. But now look at things from the spaceship's frame of reference. Doesn't the spaceship see Earth recede into the distance, turn around, and come back? And then shouldn't the earthbound twin be younger? This is the paradox. It is resolved by considering what is *special* about the special theory of relativity. The special theory applies only to inertial—i.e., unaccelerated—frames of reference. The traveling twin must accelerate in order to return to Earth, and is therefore briefly in a noninertial frame. Absolute motion has no meaning in special relativity, but absolute acceleration does. The traveling twin feels inertial forces when the ship turns around, but the earthbound twin does not. Although we cannot say that one twin is moving and the other is not, we can say that one twin's motion changes and the other's does not. The situation is not symmetric, and that is why the traveling twin really does return younger.

Could it really happen? It could, and it has. Atomic clocks are now so accurate that experiments have been done to detect the minuscule time difference between a clock flown around the Earth and one left behind.

What if the traveling twin did not turn around? We could still argue that the ship clock runs slower than clocks on Earth and the star. But then isn't the situation symmetric? Couldn't the traveling twin argue that a clock on Earth should run slower than clocks on the ship? Yes—and we would find this to be true if we set up a series of clocks in the ship frame and measured time intervals on the Earth clock as Earth passed first one, then another of the ship clocks. But there is no contradiction. Unless the twins get together again, there is no way they can directly compare their clocks or their ages at the same place. Instead, they must compare one clock with a sequence of clocks that are all synchronized. And as we will soon see, clocks that are synchronized in one frame of reference are not synchronized in another frame! Only if the twins get back together can they compare just two clocks without having to worry about syn-

FIGURE 38-13 It is not only clock readings, but all manifestations of time that differ. The traveling twin really has aged less!

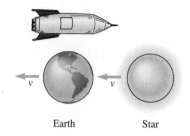

FIGURE 38-14 Earth-star trip viewed from the spaceship's frame of reference. Note that Earth-star distance is contracted, and the ship lengthened, relative to Fig. 38-12, which is drawn from the Earth-star frame.

chronization of distant clocks. And they can get back together only if at least one of them accelerates.

The Lorentz Contraction

In Example 38-2 the spaceship moved 10 ly in 12.5 y at speed $0.80c$. These quantities are related by the simple expression $\Delta x = v\,\Delta t$, where Δx is the distance between Earth and star *measured in the Earth-star frame of reference S*. Now an observer in the ship frame of reference S' sees Earth and star moving past at speed v. First Earth passes the ship, then the star passes (Fig. 38-14). We found that the time interval between these two events, measured in the ship frame, is $\Delta t' = \Delta t \sqrt{1 - v^2/c^2} = 7.5$ y. Since Earth and star are moving past at $v = 0.80\,c$, the distance between Earth and star as measured in the ship frame must be

$$\Delta x' = v\,\Delta t' = v\,\Delta t \sqrt{1 - v^2/c^2} = \Delta x \sqrt{1 - v^2/c^2}, \qquad (38\text{-}5)$$

or 6.0 ly in our example. This equation shows that the distance between two points is always greatest in a frame (the Earth-star frame S, in this example) fixed with respect to those points. In any other frame the distance is smaller. This phenomenon is called the **Lorentz contraction,** or the Lorentz-Fitzgerald contraction, after the Dutch physicist H. A. Lorentz and the Irish physicist George F. Fitzgerald, who, in the 1890s, independently proposed it as an ad hoc way of explaining the Michelson-Morley experiment. Only through Einstein's theory did the contraction acquire a solid conceptual basis.

Although we developed the Lorentz contraction using the distance between separate objects—Earth and a distant star—the effect occurs for any observer moving with respect to two points that are fixed with respect to each other. In particular, a rigid object like a meter stick or spaceship is shorter when measured by an observer with respect to whom it is moving. (An object's length in a frame in which it is at rest is called its **proper length,** although again relativity precludes our thinking of that frame as being in any way special.)

As with time dilation, do not look for some physical mechanism that squashes moving objects. Rather, it is space itself that is different for different observers. In order to accept the simple fact that absolute motion is meaningless, we must alter our common-sense notions of time and space. Lorentz contraction and time dilation are manifestations of that alteration.

● EXAMPLE 38-3   A STANFORD ELECTRON

At the Stanford Linear Accelerator Center (SLAC) (Fig. 38-15), subatomic particles are accelerated to high energies over a straight path whose proper length is 3.2 km. During a particular experiment, electrons are accelerated to 0.9999995 of the speed of light. In the SLAC frame, how long would it take electrons with this speed to travel the full length of the device?

How long would the trip take in the rest frame of the electrons? How long would the accelerator be in the rest frame of the electrons?

Solution

The electron speed is so close to that of light that the travel time is, to a very good approximation,

$$\Delta t = \frac{\Delta x}{c} = \frac{3.2 \times 10^3 \text{ m}}{3.0 \times 10^8 \text{ m/s}} = 1.1 \times 10^{-5} \text{ s}.$$

In the rest frame of the electrons, the time to traverse the accelerator is given by Equation 38-4, or

$$\Delta t' = \Delta t \sqrt{1 - v^2/c^2} = (1.1 \times 10^{-5} \text{ s}) \sqrt{1 - 0.9999995^2}$$

$$= (1.1 \times 10^{-5} \text{ s})(1.0 \times 10^{-3})$$

$$= 1.1 \times 10^{-8} \text{ s}.$$

Our time calculation shows that the relativistic factor $\sqrt{1 - v^2/c^2}$ is 10^{-3}, so in the electron frame of reference the accelerator length is

$$\Delta x' = \Delta x \sqrt{1 - v^2/c^2} = 3.2 \times 10^{-3} \text{ km} = 3.2 \text{ m}.$$

FIGURE 38-15 The Stanford Linear Accelerator is 3.2 km (2 miles) long. But to electrons moving through it at 0.999 999 5c, the accelerator is only 3.2 m long.

● **EXAMPLE 38-4** A STANFORD STUDENT

A physics student from New York flies to San Francisco to do an experiment at SLAC. She travels a distance of 4800 km on a plane going at 1000 km/h. How long does the trip take in a frame at rest with respect to Earth? How long does it take according to the student's watch? How far is it from New York to San Francisco in the airplane's frame of reference?

Solution
At 1000 km/h, the 4800-km trip takes

$$\Delta t = \frac{\Delta x}{v} = \frac{4800 \text{ km}}{1000 \text{ km/h}} = 4.8 \text{ hours}.$$

In the frame of the moving airplane, time and distance are altered by the relativistic factor $\sqrt{1 - v^2/c^2}$. The speed of the plane is 1000 km/h, or 278 m/s, so

$$\sqrt{1 - v^2/c^2} = \left[1 - \frac{(278 \text{ m/s})^2}{(3.0 \times 10^8 \text{ m/s})^2} \right]^{1/2}$$

$$\simeq 1 - (\tfrac{1}{2})(8.6 \times 10^{-13}) = 0.99999999999957.$$

Here we used the binomial theorem, $(1 + x)^n \simeq 1 + nx$ for $|nx| \ll 1$ because most calculators do not carry enough significant figures to distinguish the result from 1 (see Appendix A). We really need not carry the calculation further. The time on the student's watch is the same as the time on the ground to about five parts in 10^{13}, and the distance in the plane's frame is the same as the ground distance to within this same factor. The student need take no account of time dilation and Lorentz contraction, except in a physics exam or high-energy experiment!

EXERCISE Our Milky Way galaxy is about 100,000 light-years in diameter. What is its diameter as measured by an intergalactic spacecraft traveling at 0.96c?

Answer: 28,000 ly

Some problems similar to Examples 38-3 and 38-4: 6, 11, 12

Examples 38-3 and 38-4 show that relativistic effects are significant only at high relative velocities—so high that the quantity v^2/c^2 is comparable with 1. In our daily lives we have no experience with such velocities. It is for this reason that relativity so offends our common-sense notions of space and time. Those notions are built on a groundwork of limited experience that does not include high velocities. If we did move regularly with respect to our surroundings at

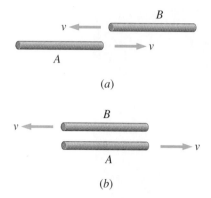

FIGURE 38-16 (*a*) In frame *S*, both sticks have the same speed v and both are contracted by the same amount. (*b*) Therefore their ends coincide at the same time.

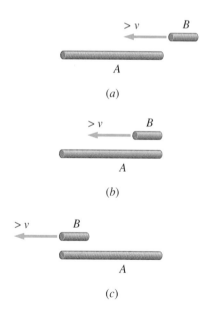

FIGURE 38-17 The passing sticks viewed in a frame *S'* at rest with respect to stick *A*. Relative speed is greater than the speed of either stick with respect to frame *S*. (*a*) First the left end of stick *B* passes the right end of *A*. (*b*) A while later, the right end of *B* passes the right end of *A*. This is event E_1. (*c*) Still later, the left end of *B* passes the left end of *A*. This is event E_2. Events E_1 and E_2 are not simultaneous in frame *S'*.

speeds near that of light, the relativity of space and time would be as obvious as our common-sense notions now seem.

For physicists working with high-energy elementary particles, relativistic effects *are* obvious. Unstable particles moving through the laboratory at high speeds live longer than they would at rest in the lab. And high-energy particle accelerators would not work if their design did not take relativity into account.

Events and Simultaneity

Consider two identical sticks *A* and *B*, each of proper length *L*. Suppose these sticks are moving toward each other. For a frame of reference *S* in which both sticks are moving at the same speed v, the situation is shown in Fig. 38-16. Both sticks are Lorentz contracted, but since both are moving at the same speed v relative to the frame *S*, both are contracted by the same amount and therefore have the same length. What happens as the sticks pass each other? First, the right end of stick *A* and the left end of stick *B* pass (Fig. 38-16*a*). A little while later, the right end of *A* passes the right end of *B*. At the same time, *because the sticks have the same length in S*, the left end of *A* passes the left end of *B* (Fig. 38-16*b*). The passing of the two right ends of the sticks is an event that we designate E_1. Similarly, the passing of the two left ends we designate E_2. We have shown that, in the frame *S*, the two events E_1 and E_2 are **simultaneous**—they occur at the same time.

Now look at the situation from a frame of reference *S'* in which stick *A* is at rest. In this frame, stick *B* moves toward stick *A*. Since we are in *S'*, we are at rest relative to stick *A*, and it has its proper length *L*. But stick *B* is contracted more than in frame *S* because of its higher relative velocity. The situation is shown in Fig. 38-17. As the figure indicates, the event E_1 occurs before E_2; the two events are not simultaneous in the frame *S'*. What happens in a frame *S''* at rest with stick *B*? As Fig. 38-18 shows, the events E_1 and E_2 are again not simultaneous, and this time event E_2 occurs first.

Isn't this all just an illusion arising from the apparent length differences due to motion of the sticks? Isn't the picture in frame *S* (Fig. 38-16) "really" the right one? No! Relativity theory assures us that all uniformly moving frames— including the frames *S*, *S'*, and *S''* of Figs. 38-16 through 38-18—are equally valid for describing physical reality. The length differences and the changes in time ordering of events E_1 and E_2 are not "apparent" and they are not "illusions." They arise from valid descriptions in different frames of reference, and each has equal claim to "reality." If you insist that one of the frames—say *S*—somehow has more validity, then you are reasserting the nineteenth-century notion that there is one favored reference frame in which alone the laws of physics are valid.

But how can observers disagree on the time order of events? Doesn't that violate causality? After all, if one event is a cause of another, we certainly expect the cause to precede the effect. It would be disturbing if some observer, with valid claim on "reality," found that cause and effect occurred in the reverse order. But there is no violation of causality. As we will soon show, the only events that can have their time order reversed are those that are so far apart in space, and so close in time, that not even light can travel fast enough to be at both events. There is no way that such events can influence each other, and therefore they cannot be causally related. In a very real sense it does not matter which event occurs first, and indeed different observers will disagree on their relative time order. For example, an event on Earth now and another occurring five minutes from now on the Sun cannot be causally related, for it takes light from the Sun eight minutes to reach Earth. For observers moving at high enough speeds relative to Earth and Sun, the solar event occurs first.

Only when events are close enough in space and separated enough in time so that light can travel from one to the other can the two be causally related. In that case, all observers will agree about their time order, although they may disagree about the actual time interval between the events. For example, an event on Earth now and another occurring fifteen minutes from now on the Sun could be causally related, and therefore all observers will agree that the terrestrial event occurs first. We will explore these notions more quantitatively in the next section.

The Lorentz Transformations

Our demonstration that the time order of events may be relative deals implicitly with the coordinates—position and time—of specific events, and suggests that those coordinates may differ for different observers. Similarly, time dilation and Lorentz contraction arise as specific instances of the way positions and times in one frame of reference are related to their values in another frame. We now seek more general expressions—called **Lorentz transformations**—relating the space and time coordinates of an event in two frames of reference in relative motion. Consider coordinate axes in a frame of reference S and in another frame S' moving in the x direction with speed v relative to S. Suppose that the origins of the two coordinate systems coincide at time $t = t' = 0$. If an event has coordinates x, y, z, and t in S, what are its coordinates x', y', z', and t' in S'? Were it not for relativity, we would expect the coordinates y, z, and the time t to remain unchanged from one coordinate system to the other. For an event occurring at time $t = 0$, when the two origins coincide, we would expect the x coordinates to be the same too. But as S' moves in the positive x direction relative to S, a given x value in S would correspond to a value $x' = x - vt$ in S' (Fig. 38-19).

How does relativity alter this coordinate transformation? First, there can be no change in the coordinates y and z at right angles to the relative motion, for if there were then in one frame distances along the y and z axes would be unambiguously shorter, and observers in both frames would agree about this (Fig. 38-20). But then there must be something special about the frame with the shorter y or z distances, and it is just such specialness that relativity prohibits.

Because of the Lorentz contraction of distances along the direction of relative motion, we expect that the simple expression $x' = x - vt$ will need

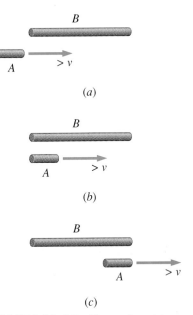

(a)

(b)

(c)

FIGURE 38-18 The passing sticks viewed in a frame of reference S'' at rest with respect to stick B. (a) First the right end of stick A passes the left end of B. (b) A while later, the left end of A passes the left end of B. This is event E_2. (c) Still later, the right end of A passes the right end of B. This is event E_1. The events E_1 and E_2 are not simultaneous in S'', and their time order is opposite what it was in S'.

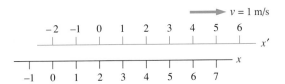

FIGURE 38-19 Nonrelativistic picture of two coordinate axes in relative motion. The x axis of frame S' moves to the right with speed v relative to S. At time $t = 0$ the two axes coincided. At a later time the x coordinates are related by $x' = x - vt$. Here $v = 1$ m/s, and the figure shows the two axes at $t = 2$ s.

FIGURE 38-20 A person holding a piece of chalk at $y' = 2$ in frame S' marks the y axis in S as the two axes pass. Observers in both frames agree unambiguously about the location of the mark. Unless it is at $y = 2$, there must be something special about one of the frames. Relativity precludes such "specialness." Thus coordinates perpendicular to the direction of relative motion are unchanged.

modification to be consistent with relativity. Any new expression we develop, however, must reduce to the nonrelativistic expression $x' = x - vt$ in the limit when $v \ll c$. A simple form that has this property is

$$x' = \gamma(x - vt), \tag{38-6}$$

where γ depends on the relative speed v. We could also transform the other way. The only difference is the direction of relative motion; frame S' is moving in the positive x direction relative to S, while S is moving in the negative x direction relative to S'. Therefore, the transformation from x' and t' to x should look like Equation 38-6 except with v replaced by $-v$:

$$x = \gamma(x' + vt'). \tag{38-7}$$

To see if we can make Equations 38-6 and 38-7 consistent with relativity, we impose the requirement that the speed of light be the same in both frames of reference. Suppose that a light flash goes off at the origin $x = 0$ at time $t = 0$. Since the origins of our frames coincide at this time, and since clocks at the origin in S and S' both read zero when the origins coincide, the light flash also occurs at $x' = 0$ and $t' = 0$ in frame S'. Let us call this event—the emission of the light flash—event E_1. Some time later, an observer at some position x in S observes the light flash. Let us call this event E_2. When does E_2 occur? Since light travels at speed c, we must have $x = ct$. Now in frame S' event E_2 has some coordinates x' and t'. But relativity requires that the speed of light in S' also be c, so we must have $x' = ct'$. Substituting $x = ct$ and $x' = ct'$ into Equations 38-6 and 38-7 gives

$$ct' = \gamma(ct - vt) = \gamma t(c - v)$$

and

$$ct = \gamma(ct' + vt') = \gamma t'(c + v).$$

Multiplying together the left-hand sides of these equations, and then the right-hand sides, and equating the results gives

$$c^2 t' t = \gamma^2 tt'(c - v)(c + v),$$

so

$$c^2 = \gamma^2(c^2 - v^2),$$

or

$$\gamma = \frac{1}{\sqrt{1 - v^2/c^2}}. \tag{38-8}$$

That we could find a value of γ depending on the relative velocity but not on the coordinates shows that our guess for the form of the transformation equations was correct. Equations 38-6 and 38-7, with γ given by Equation 38-8, are the relativistically correct transformations for the coordinates x and x'. Taking $v \ll c$ in Equation 38-8 shows that $\gamma \to 1$ in this limit, so our transformation equations correctly reduce to the nonrelativistic result at low relative velocities.

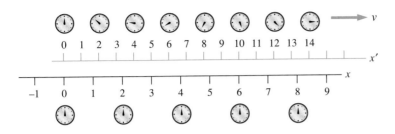

Solving Equations 38-6 and 38-7 simultaneously for t' and t (see Problem 25) gives the transformation equations for time:

$$t' = \gamma\left(t - \frac{vx}{c^2}\right) \tag{38-9}$$

and

$$t = \gamma\left(t' + \frac{vx'}{c^2}\right). \tag{38-10}$$

Because we know about time dilation, we should not be too surprised to find that measures of time differ between the two frames of reference. But why should the time in one frame depend not only on the time in the other frame but also on location in space? Because, as we have found, events that are simultaneous in one frame of reference are not simultaneous in a frame of reference moving relative to the first. In this case, our simultaneous events are the pointing of all clock hands to the same time. For the clocks in S', these events are simultaneous in S'. But they are not simultaneous in S. In fact, as Equation 38-9 shows, clocks that are farther to the right in S' read successively earlier times (Fig. 38-21). The term vx/c^2 in Equation 38-9 and its analog in Equation 38-10 account for this nonsynchronism of clocks in one frame as measured from the other frame.

Our earlier qualitative discussion of simultaneity can be made quantitative using the Lorentz transformations, as Example 38-5 and Problems 22 to 24 illustrate. Similarly, applying the Lorentz transformations to the coordinates describing the emission and return of the light flash in our light-box example results in a derivation of time dilation (see Problem 26). Table 38-1 summarizes the Lorentz transformations between coordinates in frames S and S', where S' is moving at speed v in the positive x direction relative to S.

▲ **TABLE 38-1**   THE LORENTZ TRANSFORMATIONS

S TO S'	S' TO S
$y' = y$	$y = y'$
$z' = z$	$z = z'$
$x' = \gamma(x - vt)$	$x = \gamma(x' + vt')$
$t' = \gamma(t - vx/c^2)$	$t = \gamma(t' + vx'/c^2)$

where $\gamma = \dfrac{1}{\sqrt{1 - v^2/c^2}}$

● **E X A M P L E   3 8 - 5**   GALACTIC FIREWORKS

Our Milky Way galaxy and the Andromeda galaxy are approximately at rest with respect to each other and are 2.0×10^6 light-years apart. At time $t = 0$ in the reference frame of these two galaxies, supernova explosions occur in both galaxies. Are these explosions simultaneous to the pilot of a spacecraft traveling at $0.80c$ from the Milky Way toward Andromeda? If not, what is the time interval between them, as measured in the spacecraft frame? Find also the spatial interval between the two explosions in the spacecraft frame.

Solution

The supernova explosions constitute two distinct events, and we're interested in their coordinates in the spaceship's frame of reference. Let the origin of the galaxy frame S be at the supernova in the Milky Way, and let the x axes of the galaxy frame S and spacecraft frame S' lie on the line connecting the two supernovae. Let the two frames coincide at time $t = t' = 0$. Then the space and time coordinates in S of the two supernova explosions are $x_1 = 0$, $t_1 = 0$, $x_2 = 2.0$ Mly, $t_2 = 0$. Similarly, the coordinates of the Milky Way explosion in the spacecraft frame S' are $x_1' = 0$, $t_1' = 0$. We seek the coordinates x_2' and t_2' of the Andromeda explosion in S'.

Referring to Table 38-1, we first calculate the relativistic factor γ:

$$\gamma = \frac{1}{\sqrt{1 - v^2/c^2}} = \frac{1}{\sqrt{1 - 0.80^2}} = 1.67.$$

Then using the Lorentz transformations, we have

$$t_2' = \gamma(t_2 - vx_2/c^2)$$

$$= (1.67)\left[0 \text{ y} - \frac{(0.80 \text{ ly/y})(2.0 \text{ Mly})}{(1 \text{ ly/y})^2}\right] = -2.7 \text{ My}.$$

and $x_2' = \gamma(x_2 - vt_2) = (1.67)[2.0 \text{ Mly} - (0.80 \text{ ly/y})(0 \text{ ly})]$

$$= 3.3 \text{ Mly}.$$

Do these results make sense? In the spacecraft frame, the Andromeda supernova occurs nearly three million years before the Milky Way supernova! (The minus sign tells us that t_2' is earlier than the time $t_1' = 0$ of the Milky Way supernova.) Here is an example of events that are simultaneous in one frame but not in another. To make matters worse, consider an observer moving at $0.80c$ from Andromeda toward the Milky Way. For an observer in this frame, we reverse the sign of v in the transformation equations, obtaining a time of $+2.7$ million years. For this observer, the Milky Way supernova occurs first! How can this be? This is no contradiction, and no violation of cause and effect. In the spaceship's frame S' the two supernova events occur 3.3 Mly apart in space but only 2.7 My apart in time. Light from the "earlier" event cannot travel to the "later" event, so there can be no causal influence between the two. It really doesn't matter which occurs first, and indeed which does depends on the observer.

Had we considered supernova events occurring not simultaneously but a long time apart— longer than 2 million years— in the galaxy frame of reference, we would find that all observers would agree on the time order of the two events, although not necessarily on the actual value of the time interval (see Problems 22 to 24).

E X E R C I S E   Coordinate system S' is moving along the x axis of system S at $0.90c$. The two origins coincide when clocks in both systems read noon. An event occurs at $x = 5.0$ light-hours, $t = 3$ P.M. in frame S. Find the position and time of this event in S'.

Answers: 5.28 lh, 8:34 A.M.

Some problems similar to Example 38-5:   21–23, 53, 54   ●

Relativistic Velocity Addition

If you're in an airplane moving at 1000 km/h relative to the ground and you walk toward the front of the plane at 5 km/h, common sense suggests that you move at 1005 km/h relative to the ground. But relativity implies that measures of time and distance vary among frames of reference in relative motion. For this reason the velocity of an object with respect to one frame does not simply add to the relative velocity between frames to give the object's velocity with respect to another frame. In the airplane your speed with respect to the ground is actually a little less than 1005 km/h as you stroll down the aisle of the plane, though the difference is insignificant at such a low speed.

The correct expression for **relativistic velocity addition** follows from the Lorentz transformations. Consider a frame of reference S and another frame S' moving in the positive x direction with speed v relative to S. Let their origins coincide at time $t = t' = 0$, so the Lorentz transformations of Table 38-1 apply. Suppose an object moves with velocity u' along the x' axis in S'. In our airplane example, S' would be the airplane frame of reference, u' the velocity at which you walk through the plane, and v the velocity of the plane relative to the ground, or S frame. We seek the velocity u of the object relative to the frame S (that is, your velocity relative to the ground as you walk in the plane).

In either frame, velocity is the ratio of change in position to change in time, or

$$u = \frac{\Delta x}{\Delta t}.$$

Designating the beginning of the interval Δt by the subscript 1 and the end by 2, we can use Equations 38-7 and 38-10 to write

$$\Delta x = x_2 - x_1 = \gamma[(x_2' - x_1') + v(t_2' - t_1')] = \gamma(\Delta x' + v\,\Delta t')$$

and

$$\Delta t = t_2 - t_1 = \gamma[(t_2' - t_1') + v(x_2' - x_1')/c^2] = \gamma(\Delta t' + v\,\Delta x'/c^2).$$

Forming the ratio of these quantities, we have

$$\frac{\Delta x}{\Delta t} = \frac{\Delta x' + v\,\Delta t'}{\Delta t' + v\,\Delta x'/c^2} = \frac{(\Delta x'/\Delta t') + v}{1 + v(\Delta x'/\Delta t')/c^2}.$$

But $\Delta x'/\Delta t'$ is the velocity u' of the object in frame S', and $\Delta x/\Delta t$ is the velocity u, so

$$u = \frac{u' + v}{1 + u'v/c^2}. \tag{38-11}$$

The numerator of this expression is just what we would expect from common sense. But this simple sum of two velocities is altered by the second term in the denominator, which is significant only when both the object's velocity u' and the relative velocity v between frames are comparable with c. Solving Equation 38-11 for u' in terms of u, v, and c gives the inverse transformation:

$$u' = \frac{u - v}{1 - uv/c^2}. \tag{38-12}$$

● **E X A M P L E   3 8 - 6**   COLLISION COURSE

Two spacecraft approach Earth from opposite directions, each moving at $0.80c$ relative to Earth, as shown in Fig. 38-22. How fast do the spacecraft move relative to each other?

Solution
Call the Earth frame of reference S', and let S be the frame of spacecraft A. Then S' is moving at speed $v = 0.80c$ relative to S, while spacecraft B is moving at $u' = 0.80c$ relative to S'. Then the velocity of B relative to A is given by Equation 38-11:

$$u = \frac{u' + v}{1 + u'v/c^2} = \frac{0.80c + 0.80c}{1 + (0.80c)(0.80c)/c^2} = \frac{1.6c}{1.64} = 0.98c.$$

The relative speed remains less than the speed of light. This result is quite general: Equations 38-11 and 38-12 imply that as long as an object moves at a speed $v < c$ relative to some frame of reference, its speed relative to any other frame of reference will also be less than c (see Problem 20).

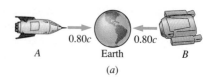

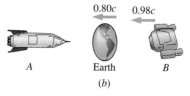

FIGURE 38-22   (a) Two spaceships approaching Earth at $0.80c$. (b) The situation in the frame of reference of the left-hand spaceship. Note changes in lengths of the ships and Earth. ●

● **E X A M P L E   3 8 - 7**   CATCH THAT LIGHT

A light wave moves past Earth at the speed of light c. You try to chase the light wave by hopping a fast spacecraft, moving at $0.95c$ relative to Earth. What is the speed of the light relative to the spacecraft?

Solution
Call the Earth frame S, so $u = c$, and the spacecraft frame S', so $v = 0.95c$. Then u', the speed of light relative to the spacecraft, is given by Equation 38-12:

$$u' = \frac{u - v}{1 - uv/c^2} = \frac{c - 0.95c}{1 - 0.95c^2/c^2} = \frac{0.05c}{0.05} = c.$$

We really didn't need to calculate this result, since a fundamental premise of relativity is that the speed of light is the same for all observers. The equations of relativistic velocity addition reflect this basic fact. No matter what the relative velocity v between two frames, light moving at c in one frame moves at c in any other frame. You cannot even begin to catch up with light!

E X E R C I S E   A spacecraft whizzes by Planet X at $0.75c$, in excess of the Galactic Federation speed limit. A space cop takes off from Planet X at $0.90c$ relative to the planet. Find the relative speed of cop and speeder.

Answer: $0.46c$

Some problems similar to Examples 38-6 and 38-7:   17–19 ●

38 - 5 ENERGY AND MOMENTUM IN RELATIVITY

Conservation of momentum and conservation of energy are cornerstones of Newtonian mechanics, where they hold in any inertial—i.e., uniformly moving—frame of reference. But both the momentum and energy of a particle are functions of its velocity, and we've just seen that relativity alters the Newtonian picture of how velocities transform from one reference frame to another. How, then, can momentum and energy be conserved in all frames of reference?

Momentum

In Newtonian mechanics the momentum of a particle with mass m and velocity $\mathbf{u}$ is $m\mathbf{u}$. (Here we use $\mathbf{u}$ for particle velocities, reserving $\mathbf{v}$ for the relative velocity between two reference frames.) But if this quantity is conserved in one frame of reference, then relativistic velocity addition suggests that it won't be conserved in another. The problem, however, lies not with momentum conservation but with our definition of momentum. The expression $m\mathbf{u}$ is an approximation valid only for speeds u much less than c. The measure of momentum valid at any speed is

$$\mathbf{p} = \frac{m\mathbf{u}}{\sqrt{1 - u^2/c^2}} = \gamma m\mathbf{u}, \qquad (38\text{-}13)$$

where γ is the familiar relativistic factor. The momentum given in expression 38-13 is conserved in all reference frames, and at low velocities it reduces to the Newtonian expression $\mathbf{p} = m\mathbf{u}$.

As $u \to c$ the factor γ grows arbitrarily large, and so does the relativistic momentum (Fig. 38-23). Since force is the rate of change of momentum, that means a very large force is required to produce even the slightest change in the velocity of a rapidly moving particle. This effect shows one answer to a common question about relativity: Why is it impossible to accelerate an object to the speed of light? The answer is that the object's momentum would approach infinity, and no matter how close to c it was moving it would still require infinite force to give it the last bit of speed needed to reach c.

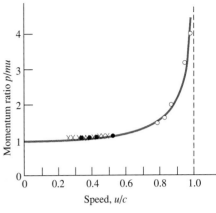

FIGURE 38-23 The ratio of the relativistic momentum to the Newtonian expression mu. Curve is a plot based on Equation 38-13, while data points correspond to actual measurements done on electrons. The relativistic momentum goes asymptotically to infinity as $u \to c$.

● EXAMPLE 38-8   PUSHING TOWARD c

Compare the momentum changes required to accelerate an electron (a) from rest to $0.01c$ and (b) from $0.98c$ to $0.99c$.

Solution

The momentum at each speed is given by Equation 38-13:

$$p(0) = \frac{mu}{\sqrt{1 - u^2/c^2}} = 0,$$

$$p(0.01c) = \frac{m(0.01c)}{\sqrt{1 - 0.01^2}} = 0.010\,mc,$$

$$p(0.98c) = \frac{m(0.98c)}{\sqrt{1 - 0.98^2}} = 4.92\,mc,$$

and $\quad p(0.99c) = \dfrac{m(0.99c)}{\sqrt{1 - 0.99^2}} = 7.02\,mc.$

In both cases the speed has increased by the same $0.01c$, but the momentum changes are very different: Going from rest to $0.01c$ gives $\Delta p = 0.010mc$, while from $0.98c$ to $0.99c$ the momentum change is $\Delta p = (7.02mc - 4.92mc) = 2.1mc$—a factor of 210 greater than the slow-speed case. To accomplish the second velocity change in the same time would therefore require a force 210 times greater. Note, incidentally, that we would have had sufficient accuracy using the Newtonian expression $p = mu$ in the low-speed case.

EXERCISE An electron is moving at $0.35c$. If its speed is doubled, by what factor does its momentum increase?

Answer: 2.62

Some problems similar to Example 38-8: 28, 30–32   ●

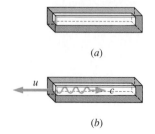

(a)

(b)

FIGURE 38-24 A light pulse is emitted at one end of a massive box. Since light carries momentum, the box recoils in the opposite direction.

Energy

The most widely known result of relativity is the famous equation $E = mc^2$. Here we show how that equation arises and what it means, then develop a general expression for relativistic energy and a relation between energy and momentum.

Einstein arrived at his equation through a simple "thought experiment." He imagined a closed, massive box of mass M and length L, initially at rest. A light flash is emitted from one end of the box. We found in Chapter 34 that light with energy E also carries momentum E/c; therefore, the box must recoil to conserve momentum (see Fig. 38-24). If the box is very massive its recoil speed u will be small compared with c, so we can express momentum conservation in the form $Mu = E/c$, or

$$u = \frac{E}{Mc}.$$

The light then moves down the box, taking a time

$$\Delta t = \frac{L}{c},$$

where again we assume that the box speed u is much less than c, so the distance traveled by the light is approximately L. In this time the box moves a very small distance Δx, given by

$$\Delta x = u\,\Delta t = \frac{EL}{Mc^2}.$$

Then the light flash hits the end of the box, transferring its momentum and bringing the box to a stop.

But now the box is in a new position. It looks as if its center of mass has moved, and yet the box is an isolated system whose center of mass cannot move! To escape this dilemma, Einstein assumed that light carries not only energy and momentum, but mass as well. If m is the mass carried by the light, we must have

$$mL = M\Delta x,$$

in order that center of mass of the system (box + light) will not move. Using our expression for Δx and solving for m gives

$$m = \frac{M\Delta x}{L} = \frac{M}{L}\frac{EL}{Mc^2} = \frac{E}{c^2},$$

or

$$E = mc^2,$$

where E is the energy of the light and m its equivalent mass.

Although we derived this expression for light energy, it is in fact a universal statement of the equivalence of mass and energy. Energy, like mass, exhibits

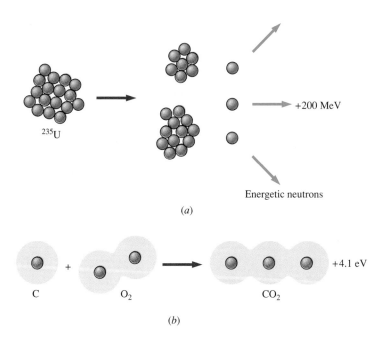

FIGURE 38-25 Energy release in (*a*) nuclear and (*b*) chemical reactions. A mass change occurs in both cases, but is approximately 10^7 times greater in the nuclear reaction.

inertia. A hot object is slightly harder to accelerate than a cold one because of the inertia of its thermal energy. A stretched spring is more massive than an unstretched one, because of its extra potential energy. When a system loses energy, it loses mass as well.

To the general public, $E = mc^2$ is synonymous with nuclear energy. This equation does describe mass changes that occur in nuclear reactions, but it applies equally well to chemical reactions and all other occurrences in which energy enters or leaves a system. If you weigh a nuclear power plant just after it has been refueled, then weigh it again a month later, you will find it weighs slightly less. If you weigh a coal-burning power plant and all the coal and oxygen that go into it for a month, and then weigh all the carbon dioxide and other combustion products that come out, you will find a discrepancy between the mass of what goes in and what comes out. If both plants produce the same amount of energy, the mass discrepancy will be the same for both. The only difference lies in the amount of mass released as energy in each individual reaction. The fissioning of a single uranium nucleus involves about 10 million times as much energy, and therefore mass, as the reaction of a single carbon atom with oxygen to make carbon dioxide (Fig. 38-25). That's why a coal-burning power plant consumes many hundred-car trainloads of coal each week, while a nuclear plant is refueled every 18 months with a few truckloads of uranium. Incidentally, neither process converts very much of the fuel mass to energy; if we could convert *all* the mass in a given object to energy, ordinary matter would be an almost limitless source of energy. Such conversion is in fact possible, but only in the annihilation of matter and antimatter. The opposite conversion also occurs, with a particle-antiparticle pair appearing where before there had been only energy (Fig. 38-26).

FIGURE 38-26 Pair creation events observed in a bubble chamber at Brookhaven National Laboratory. Each pair of curved paths represent the trajectories of a particle-antiparticle pair, bending in opposite directions in a magnetic field. The energy required to create the particles—mc^2 per particle—ultimately came from a single high-energy proton that collided with a stationary proton.

● EXAMPLE 38-9 ANNIHILATION

A positron is an antimatter particle with the same mass as the electron but the opposite electric charge. When an electron and a positron meet, they annihilate and produce a pair of gamma rays (bundles of electromagnetic wave energy) of equal energy. Find the energy of each gamma ray.

Solution
Here two electron masses annihilate to give two gamma rays. So the energy of each gamma ray is the energy equivalent of one electron mass, or

$$E = mc^2 = (9.11 \times 10^{-31} \text{ kg})(3.00 \times 10^8 \text{ m/s})^2$$

$$= 8.20 \times 10^{-14} \text{ J},$$

or 511 keV. The detection of 511-keV gamma rays from laboratory or astrophysical sources is a sure indication that electron-positron annihilation is occurring.

EXERCISE The Sun radiates energy at the rate of 3.85×10^{26} W. Find the rate at which it loses mass.

Answer: 4.28×10^9 kg/s

Some problems similar to Example 38-9: 38, 39 ●

Rest Energy and Kinetic Energy

Einstein's result shows that a mass m is equivalent to an energy $E = mc^2$. This quantity is called the **rest energy** because it is associated with the mass itself and not with any bulk motion the object may have. As usual, the energy of bulk motion is **kinetic energy.** We can find an expression for relativistic kinetic energy much as we did in Chapter 7 for the Newtonian case, by considering the work done on an object as it is accelerated. Problem 63 covers the details; the result is

$$K = \frac{mc^2}{\sqrt{1 - u^2/c^2}} - mc^2 = (\gamma - 1)mc^2, \tag{38-14}$$

where as usual $\gamma = 1/\sqrt{1 - u^2/c^2}$. This equation bears little resemblance to the Newtonian expression $K = \frac{1}{2}mu^2$. But Problem 43 uses the binomial approximation to show that the relativistic expression for kinetic energy really does reduce to $K = \frac{1}{2}mu^2$ for speeds u much less than c. Using Equation 38-14, we can now write an object's total energy as the sum of its kinetic energy K and rest energy mc^2:

$$E = \frac{mc^2}{\sqrt{1 - u^2/c^2}} = \gamma mc^2. \quad \text{(total energy)} \tag{38-15}$$

● EXAMPLE 38-10 A RELATIVISTIC ELECTRON

An electron has a total energy of 2.5 Mev. Find (a) its kinetic energy and (b) its speed.

Solution
In Example 38-9 we found that the rest energy of an electron is 511 keV, or 0.511 MeV. The electron's kinetic energy is the difference between its total energy and its rest energy, or

$$K = E - mc^2 = 2.5 \text{ MeV} - 0.511 \text{ MeV} = 1.99 \text{ MeV}.$$

Equation 13-15 shows that the total energy is just γ times the rest energy. Thus we have

$$\gamma = \frac{1}{\sqrt{1 - u^2/c^2}} = \frac{E}{mc^2} = \frac{2.5 \text{ MeV}}{0.511 \text{ MeV}} = 4.89.$$

Solving for u then gives

$$u = c\sqrt{1 - \frac{1}{4.89^2}} = 0.979c = 2.94 \times 10^8 \text{ m/s}.$$

E X E R C I S E   Find (a) the rest energy of a proton, and (b) its kinetic energy when moving at $0.98c$.

Answers: (a) 938 Mev; (b) 3.78 GeV

Some problems similar to Example 39-10:   40–42

●

The Energy-Momentum Relation

In Newtonian physics the equations $p = mu$ and $K = \frac{1}{2}mu^2$ yield the relation $p^2 = 2K/m$. Similarly, Problem 44 shows that in relativity we can combine the equations $p = \gamma mu$ and $E = \gamma mc^2$ to get

$$E^2 = p^2c^2 + (mc^2)^2, \tag{38-16}$$

where now the energy-momentum relation involves E rather than K because in relativity the energy includes both kinetic and rest energies. For a particle at rest, $p = 0$ and Equation 38-16 shows that the total energy is just the rest energy. For highly relativistic particles—those with speeds very near that of light—the rest energy is negligible and the total energy becomes very nearly $E = pc$. Some "particles"—like the photons that, in quantum physics, are "bundles" of electromagnetic energy, and possibly the neutral particles called neutrinos—have no rest mass. These particles exist only in motion at the speed of light, and for them Equation 38-16 gives the exact relation $E = pc$.

38-6 WHAT IS NOT RELATIVE

In relativity, space and time are not absolute but depend on the reference frame of the observer. But the speed of light c remains the same for all observers. Are there other such **relativistic invariants?** Yes—and these invariants, not the shifting measures of space and time, are at the basis of relativity's objective description of physical reality.

One invariant is electric charge. No matter what an electron's speed, all observers measure the same value for its charge. Other invariants may be formed from combinations of quantities that themselves are not invariant; for example, Equation 38-16 shows that the quantity $E^2 - p^2c^2$ for a particle is the same in all frames of reference and is in fact equal to the particle's rest energy.

In Newtonian physics the distance between two points is the same no matter who measures it. In relativity, distance depends on the observer. So does time. But there is a quantity, analogous to distance but incorporating time as well, that remains invariant. Called the **spacetime interval,** it is a kind of "four-dimensional distance," not between two points in space but between two events in space and time. The spacetime interval Δs is given by an expression that looks like a modified Pythagorean theorem:

$$(\Delta s)^2 = c^2(\Delta t)^2 - [(\Delta x)^2 + (\Delta y)^2 + (\Delta z)^2], \tag{38-17}$$

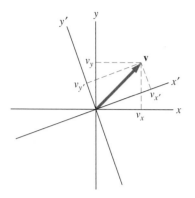

FIGURE 38-27 Although the x and y components of an ordinary vector depend on the choice of coordinate system, the length of the vector does not.

where the Δ quantities describe the differences between the space and time coordinates of two events. The invariance of Equation 38-17 follows directly from the Lorentz transformations (see Problem 45).

The invariance of the spacetime interval suggests that something absolute underlies the shifting sands of relativistic space and time. That absolute is **spacetime**—a four-dimensional framework linking space and time. The points in spacetime are events, specified by four coordinates. The time interval or space interval between two events depends on the particular frame of reference of the observer, but the spacetime interval—a four-dimensional "distance" that takes all four coordinates into account—is the same for all observers.

In more advanced treatments of relativity, it is convenient to consider four-dimensional vectors called **four-vectors.** The displacement between two events in spacetime—specified by the four quantities Δx, Δy, Δz, and Δt—is a four-vector, with "length" given by Equation 38-17. With two- and three-dimensional vectors in nonrelativistic physics, it is possible to break a vector into components in many different ways. Although the values of the individual components depend on your choice of coordinate system, the length of the vector does not (Fig. 38-27). Similarly, the individual space and time components of a four-vector depend on your choice of reference frame—that is, on your velocity. But the spacetime interval does not.

38-7 ELECTROMAGNETISM AND RELATIVITY

Historically, relativity arose from deep questions presented by Maxwell's equations with regard to electromagnetic waves. We have seen that relativity profoundly alters the basic concepts of space and time that stand at the foundation of Newtonian mechanics. As a result, fundamental ideas like momentum and energy must be altered for relativistic consistency. What analogous changes does relativity require of Maxwell's electromagnetic theory? The answer is simple: none. Maxwell's theory culminated in the prediction of light waves traveling through empty space at speed c. Relativity requires that the laws of physics be the same in all frames of reference in uniform motion. But that is exactly what Maxwell's equations suggest—that a light wave in one frame should be a light wave in any other frame, and that such a wave should have speed c with respect to any observer. Even the simple fact that electromagnetic induction occurs equally well when you move a magnet near a conductor, or a conductor near a magnet, suggests that only relative motion should be important in electromagnetism. Einstein thought a great deal about induction, and mentioned it at the beginning of his 1905 paper introducing special relativity. Even the title of that famous paper—"On the Electrodynamics of Moving Bodies"—shows how intimately related are electromagnetism and relativity. Maxwell's equations are relativistically correct, and require no modification.

Although electric and magnetic fields in any frame of reference obey the same Maxwell equations, this does not require the fields themselves to be independent of frame. If, for example, you sit in the rest frame of a point charge, you see a spherically symmetric point-charge field. If you move relative to the charge, you see a magnetic field as well, associated with the moving charge. Relativity accounts naturally for such a field transformation. In relativity, the electric and magnetic fields are not absolutes, but are manifestations of

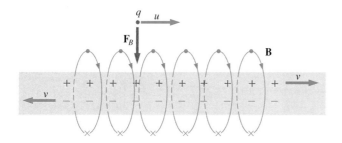

FIGURE 38-28 A current-carrying wire with equal densities of positive and negative charge moving in opposite directions. A magnetic field surrounds the wire, so a charge q moving parallel to the wire experiences a magnetic force toward the wire.

a more fundamental electromagnetic field. To one observer, this electromagnetic field breaks up in a certain way into electric and magnetic parts, while to another observer the individual electric and magnetic fields are different (see Problem 66).

We can illustrate the deep relationship that relativity imposes between electricity and magnetism by considering the force on a charged particle near a current-carrying wire. For simplicity, we consider a wire containing equal line charge densities of positive and negative charge, moving in opposite directions at the same speed v relative to the wire (Fig. 38-28).

Consider a particle of positive charge q a distance r from the wire. Since the densities of positive and negative charge are equal, the wire is neutral and has no electric field. If the particle is at rest with respect to the wire, it experiences no force. But there is a magnetic field associated with the current in the wire. The magnetic field lines encircle the wire, and the right-hand rule shows that they point out of the page at the location of the charged particle. Now suppose the particle is moving to the right with velocity $\mathbf{u}$ relative to the wire. It experiences a magnetic force $\mathbf{F} = q\mathbf{u} \times \mathbf{B}$; the right-hand rule shows that this force is toward the wire.

So the situation from the frame of reference of the wire is as follows: the wire is electrically neutral and therefore produces no electric field. But it does carry a current, and therefore produces a magnetic field. If a positively charged particle moves to the right, it experiences a magnetic force directed toward the wire. To describe the situation we needed to know about electric charges, about Ampère's law for magnetism, and about the magnetic force $q\mathbf{u} \times \mathbf{B}$—in short, about a variety of phenomena that were discovered independently during the nineteenth century.

Now let's look at the situation in the reference frame of the charged particle. Since the particle is moving to the right, the positive charges in the wire have a lower speed relative to the particle than do the negative charges. As measured by the particle, distances between the negative charges in the wire are therefore Lorentz contracted by *more* than the distances between the positive charges. But charge is invariant, so the charge density—the charge per unit length—is *greater* for the negative charges. So in the frame of the charged particle, *there is a net negative charge on the wire!* The negatively charged wire produces an electric field pointing toward the wire. As a result, our positively charged particle experiences an electric force toward the wire (Fig. 38-29). Of course there is still a magnetic field as well, but since the particle is at rest in its own frame of reference, it experiences no magnetic force. The force it does experience is entirely electric. What appeared as a magnetic phenomenon in the wire frame of reference—the existence of a force directed toward the wire—is explained entirely as an electric phenomenon in the particle's frame of reference.

FIGURE 38-29 The situation of Fig. 38-28 in the rest frame of the charged particle. The line of negative charges is Lorentz contracted more, and therefore the positively charged particle "sees" a net negative charge on the wire. Thus the particle experiences an electric force toward the wire.

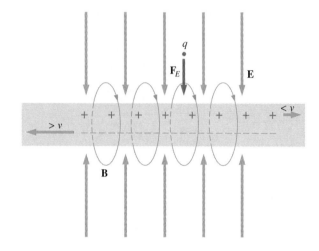

We have given two quite different descriptions of the force on a charged particle moving near a current-carrying wire. Our second description—from the frame of reference of the charged particle—required no knowledge of magnetism whatsoever. We only needed Coulomb's law for electric charge and the principle of relativity. And this illustrates a profound point: electricity and magnetism are not separate phenomena that happen to be related through Ampère's and Faraday's laws. Rather, they are two aspects of a single phenomenon—electromagnetism. In a universe obeying the principle of relativity, it is not even logically possible to have electricity without magnetism, or vice versa. Given Coulomb's law and the principle of relativity, all the rest of Maxwell's equations follow not as independent experimental results but as logically necessary consequences. Relativity provides for us the total unification of electricity and magnetism that we have hinted at since we began studying these phenomena.

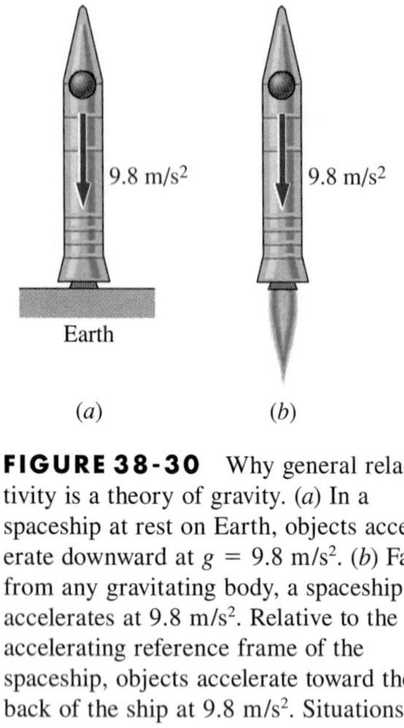

38-8 GENERAL RELATIVITY

The special theory of relativity is special because it is restricted to uniform motion. Following the development of special relativity, Einstein attempted to formulate a theory that would express the laws of physics in the same form in all frames of reference, including those in accelerated motion. But Einstein recognized that it is impossible to distinguish the effects of uniform acceleration from those of a uniform gravitational field (Fig. 38-30). Consequently, Einstein's general theory became a theory of gravity. Building on the notion of four-dimensional spacetime, Einstein introduced geometrical curvature of spacetime to account for gravity, and the result, published in 1916, is his **general theory of relativity.** The theory's predictions differ significantly from those of Newton's theory of gravity only in regions of very strong gravitational fields or when the overall structure of the universe is considered. By very strong fields we mean those of objects whose escape speed is comparable to that of light (see Problem 65). Because we have no direct laboratory experience of such fields, the general theory of relativity is not as solidly established as is the special theory. Nevertheless, general relativity is a cornerstone of modern astrophysics, playing

FIGURE 38-30 Why general relativity is a theory of gravity. (a) In a spaceship at rest on Earth, objects accelerate downward at $g = 9.8$ m/s². (b) Far from any gravitating body, a spaceship accelerates at 9.8 m/s². Relative to the accelerating reference frame of the spaceship, objects accelerate toward the back of the ship at 9.8 m/s². Situations (a) and (b) are impossible to distinguish, so a theory dealing with nonuniform motion must also consider gravity.

a crucial role in the physics of such bizarre objects as neutron stars and black holes. General relativity also addresses cosmological questions of the origin and ultimate fate of the universe. Research in astrophysics and cosmology, in turn, is increasingly confirming the predictions of general relativity.

CHAPTER SYNOPSIS

Summary

1. The **ether** was a hypothetical medium whose properties were supposed to explain the propagation of electromagnetic waves. In particular, such waves were supposed to have speed c relative to the ether.
2. The **Michelson-Morley experiment** and the observation of the aberration of starlight led to a contradiction in physics: Earth's motion through the ether could not be detected, yet Earth did not drag ether with it.
3. Einstein's **special theory of relativity** (1905) resolved the contradiction by asserting that uniform motion is undetectable by any experiment, mechanical or electromagnetic. Einstein did away with the ether, declaring simply that **the laws of physics are the same in all inertial frames of reference.** Mechanics and electromagnetism alike are included in Einstein's theory, so Maxwell's prediction of electromagnetic waves moving at speed c is correct in all frames of reference.
4. The simple statement that the laws of physics are the same in all frames of reference requires profound changes in our common-sense notions of space and time. These changes are described by the **Lorentz transformations,** which relate space and time measurements made in different frames of reference:

$$y' = y$$
$$z' = z$$
$$x' = \gamma(x - vt)$$
$$t' = \gamma(t - vx/c^2),$$

where

$$\gamma = \frac{1}{\sqrt{1 - v^2/c^2}}.$$

Particular manifestations of these transformations include **time dilation, Lorentz contraction,** and the relativistic velocity addition formulas.
5. Relativistic transformations result in new expressions for momentum:

$$\mathbf{p} = \frac{m\mathbf{u}}{\sqrt{1 - u^2/c^2}} = \gamma m\mathbf{u}$$

and total energy:

$$E = \frac{mc^2}{\sqrt{1 - u^2/c^2}} = \gamma mc^2,$$

for a particle of mass m moving with velocity $\mathbf{u}$. Kinetic energy is the difference between the total energy E and the **rest energy** mc^2. The existence of rest energy shows that matter and energy may be converted into each other.
6. Relativity links space and time into a four-dimensional framework called **spacetime.** Although individual space and time measurements depend on one's frame of reference, the **spacetime interval** between two events does not.
7. Maxwell's equations of electromagnetism are fully consistent with relativity. But the fields themselves are different in different frames of reference; what appears as a magnetic field in one frame may be partially electric in another, and vice versa. Relativity imposes a logical relationship between electricity and magnetism, in that neither phenomenon is possible without the other.
8. The **general theory of relativity** is Einstein's generalization of relativity to include accelerated reference frames. Because the effects of acceleration mimic those of gravity, general relativity is a theory of gravity.

Terms You Should Understand

(Pairs are closely related terms whose distinction is important; number in parentheses is chapter section where term first appears.)

ether (38-1)
Michelson-Morley experiment (38-2)
special relativity, general relativity (38-2, 38-8)
time dilation, Lorentz contraction (38-4)
proper time, proper length (38-4)
Lorentz transformations (38-4)
rest energy (38-5)
spacetime (38-6)
spacetime interval (38-6)

Symbols You Should Recognize

γ (38-4)

Problems You Should Be Able to Solve

finding times and distances for different observers in relative motion (38-4)

transforming space and time coordinates among coordinate systems in relative motion (38-4)

transforming velocities (38-4)

applying mass-energy equivalence (38-5)

evaluating relativistic energy and momentum (38-5)

Limitations to Keep in Mind

The special theory of relativity applies only to inertial frames of reference—i.e., those that are not accelerated.

QUESTIONS

1. Why was the Michelson-Morley experiment a more sensitive test of motion through the ether than independent measurements of the speed of light in two perpendicular directions?

2. Why was it necessary to repeat the Michelson-Morley experiment at different times throughout the year?

3. Why do we reject the idea that the ether frame of reference is the Earth frame?

4. What is special about the special theory of relativity?

5. Does relativity require that the speed of sound be the same for all observers? Why or why not?

6. How would the world be different if the speed of light were 160 km/h (100 miles per hour)? Would our "common-sense" notions change?

7. A friend argues that the speed of light cannot be the same for all observers, for if one of them is moving toward a light source then that one will clearly measure a higher speed. How would you refute this argument?

8. Time dilation is sometimes described by saying that "moving clocks run slow." In what sense is this true? In what sense does the statement violate the spirit of relativity?

9. If you are in a spaceship moving at $0.95c$ relative to Earth, do you perceive time to be passing more slowly than it would on Earth? Think! Is your answer consistent with the theory of relativity?

10. In our light-box example for time dilation, we found that a time interval between two events measured in frame S' was shorter than in frame S. But you could equally well say that frame S is moving relative to frame S', so clocks in S should "run slow" compared with those in S'. An observer in each frame should judge the clocks in the other frame to "run slow." Is this a contradiction?

11. To try to circumvent the difficulty of accelerating an object to the speed of light, you build a series of conveyor belts, all running in the same direction, and each moving 10 m/s relative to the one next to it (Fig. 38-31). You step from the ground onto the first conveyor belt, then to the next, and so forth. By the time you reach the 3×10^7th conveyor belt, you should be moving at c relative to the ground. Why doesn't this scheme work?

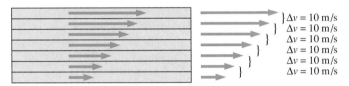

FIGURE 38-31 A series of conveyer belts, each moving at 10 m/s relative to its neighbors (Question 11).

12. If you took your pulse while traveling in a high-speed spacecraft, would it be faster, slower, or the same as on Earth?

13. The Andromeda galaxy is 2 million light-years from our Milky Way. Although nothing can go faster than light, it would still be possible to travel to Andromeda in much less than 2 million years. How is this possible?

14. Is matter converted to energy in a nuclear reactor? In a burning candle? In your body?

15. An unstretched rubber band is weighed on an extraordinarily sensitive scale. It is then stretched and weighed again. Is there a difference in the weight? Why or why not?

16. The rest energy of an electron is 511 keV. What is the approximate speed of an electron whose total energy is 1 GeV ($= 10^9$ eV)? You need not do any calculations!

17. An atom in an excited state emits a burst of light. What happens to the mass of the atom?

18. In some of the hottest parts of the universe, the thermal energy of particles may be many millions of electron volts. At such temperatures, the number of particles within a closed volume may vary. How is this possible? *Hint:* The mass of the electron is equivalent to about 0.5 MeV.

19. The electric field is not invariant, but changes from one frame to another. Is this a violation of relativity? Relativity requires that the laws of physics be the same in all frames of reference. Does this mean that all physical quantities must be the same?

20. The quantity $\mathbf{E} \cdot \mathbf{B}$ is invariant. What does this say about how different observers will measure the angle between $\mathbf{E}$ and $\mathbf{B}$ in a light wave?

PROBLEMS

Section 38-2 Matter, Motion, and the Ether

1. Consider an airplane flying at 800 km/h airspeed between two points 1800 km apart. What is the round-trip travel time for the plane (a) if there is no wind? (b) if there is a wind blowing at 130 km/h perpendicular to a line joining the two points? (c) if there is a wind blowing at 130 km/h along a line joining the two points? Ignore relativistic effects. (Why are you justified in doing so?)

2. What would be the difference in light travel times on the two legs of the Michelson-Morley experiment if the ether existed and if Earth moved relative to it at (a) its orbital speed relative to the Sun (Appendix E)? (b) $10^{-2}c$? (c) $0.5c$? (d) $0.99c$? Assume each light path is exactly 11 m in length, and that the paths are oriented parallel and perpendicular to the ether wind.

3. Figure 38-32 shows a plot of James Bradley's data on the aberration of light from the star γ Draconis, taken in 1727–1728. (a) From the data, determine the magnitude of Earth's orbital velocity. (b) The data very nearly fit a perfect sine curve. What does this say about the shape of Earth's orbit?

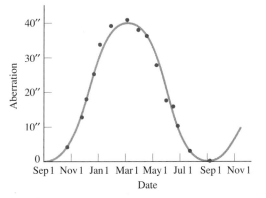

FIGURE 38-32 Bradley's data on the aberration of starlight (Problem 3). Aberration is measured in seconds of arc ($1'' = 1/3600°$).

4. Show that the time of Equation 38-2 is larger than that of Equation 38-1 as long as $0 < v < c$.

5. Suppose the speed of light differed by 100 m/s in two perpendicular directions. How long should the arms be in a Michelson interferometer if this difference is to cause the interference pattern to shift one-half cycle (i.e., a light fringe shifts to where a dark one was) relative to the pattern if there were no speed difference? Assume 550-nm light.

Section 38-4 Space and Time in Relativity

6. Two stars are 50 ly apart, measured in their common rest frame. How far apart are they to a spaceship moving between them at $0.75c$?

7. How long would it take a spacecraft traveling at 65% of the speed of light to make the 5.8×10^9 km journey from Earth to Pluto according to clocks (a) on Earth and (b) on the spacecraft?

8. A spaceship goes by at half the speed of light, and you determine that it is 35 m long. What is its length as measured in its rest frame?

9. Earth and Sun are 8.3 light-minutes apart, as measured in their rest frame. (a) What is the speed of a spacecraft that makes the trip in 5.0 min according to its on-board clocks? (b) What is the trip time as measured by clocks in the Earth-Sun frame?

10. The Andromeda galaxy is two million light-years from Earth, measured in the common rest frame of Earth and Andromeda (Fig. 38-33). Suppose you took a fast spaceship to Andromeda, so it got you there in 50 years as measured on the ship. If you sent a radio message home as soon as you reached Andromeda, how long after you left Earth would it arrive, according to timekeepers on Earth?

FIGURE 38-33 The Andromeda galaxy (Problem 10).

11. How fast would you have to move relative to a meter stick for its length to measure 99 cm in your frame of reference?

12. Electrons in a TV tube move through the tube at 30% of the speed of light. By what factor is the tube foreshortened in the electrons' frame of reference, relative to its length at rest?

13. You wish to travel to a star N light-years from Earth. How fast must you go if the one-way journey is to occupy N years of your life?

14. The nearest star beyond our solar system is about 4 light-years away. If a spaceship can get to the star in 5 years, as measured on Earth, (a) how long would the ship's pilot judge the journey to take? (b) How far from Earth would the pilot find the star to be?

15. Twins A and B live on Earth. On their 20th birthday, Twin B climbs into a spaceship and makes a round-trip journey

at 0.95c to a star 30 light-years distant, as measured in the Earth-star frame of reference. What are their ages when the twins are reunited?

16. Radioactive oxygen-15 decays at such a rate that half the atoms in a given sample decay every two minutes. If a tube containing 1000 O-15 atoms is moved at 0.80c relative to Earth for 6.67 minutes according to clocks at rest with respect to Earth, how many atoms will be left at the end of that time?

17. Two distant galaxies are receding from Earth at 0.75c, in opposite directions. How fast does an observer in one galaxy measure the other to be moving?

18. Two spaceships are having a race. The "slower" one moves past Earth at 0.70c, and the "faster" one moves at 0.40c relative to the slower one. How fast does the faster ship move relative to Earth?

19. Muons traveling vertically downward at 0.994c relative to Earth are observed from a rocket traveling upward at 0.25c. What speed does the rocket's crew measure for the muons?

20. Use relativistic velocity addition to show that if an object moves at speed $v < c$ relative to some uniformly moving frame of reference, then its speed relative to any other uniformly moving frame must also be less than c.

21. Earth and Sun are 8.33 light-minutes apart. Event A occurs on Earth at time $t = 0$, and event B on the Sun at time $t = 2.45$ min, as measured in the Earth-Sun frame. Find the time order and time difference between A and B for observers (a) moving on a line from Earth to Sun at 0.750c, (b) moving on a line from Sun to Earth at 0.750c, and (c) moving on a line from Earth to Sun at 0.294c.

22. Two civilizations are evolving on opposite sides of a galaxy, whose diameter is 10^5 light-years (Fig. 38-34). At time $t = 0$ in the galaxy frame of reference, civilization A launches its first interstellar spacecraft. Civilization B launches its first spacecraft 50,000 years later. A being from a more advanced civilization C is traveling through the galaxy at 0.99c, on a line from A to B. Which civilization does C judge to have first achieved interstellar travel, and how much in advance of the other?

23. Repeat the preceding problem, now assuming that civilization B lags A by 1 million years in the galaxy frame of reference.

24. Could there be observers who would judge the events in the two preceding problems to be simultaneous? If so, how must each be moving relative to the galaxy?

25. Derive the Lorentz transformations for time, Equations 38-9 and 38-10, from the transformations for space.

26. In the light box of Fig. 38-8, let event A be the emission of the light flash and event B its return to the source. Assign suitable space and time coordinates to these events in the frame in which the box moves with speed v. Apply the Lorentz transformations to show that the time $\Delta t'$ between the two events in the box frame is given by Equation 38-4.

27. Two spaceships are each 25 m long, as measured in their rest frames (Fig. 38-35). Ship A is approaching Earth at 0.65c. Ship B is approaching Earth from the opposite direction at 0.50c. Find the length of ship B as measured (a) in Earth's frame of reference and (b) in ship A's frame of reference.

$v = 0.65c$ $v = 0.50c$

FIGURE 38-35   Problem 27 (the drawing is in Earth's frame of reference).

Section 38-5 Energy and Momentum in Relativity

28. By what factor does the momentum of an object change if you double its speed when its original speed is (a) 25 m/s and (b) 1.0×10^8 m/s?

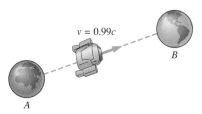

$v = 0.99c$

FIGURE 38-34   Civilizations A and B are on opposite sides of a galaxy, a distance of 10^5 light-years in the galaxy frame of reference (Problems 22, 23, 24).

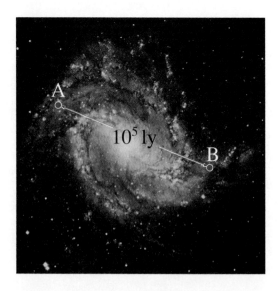

10^5 ly

29. At what speed will the momentum of a proton (mass 1 u) equal that of an alpha particle (mass 4 u) moving at $0.5c$?

30. At what speed will the Newtonian expression for momentum be in error by 1%?

31. A particle is moving at $0.90c$. If its speed increases by 10%, by what factor does its momentum increase?

32. Compare the momentum changes needed to boost a spacecraft (a) from $0.1c$ to $0.2c$ and (b) from $0.8c$ to $0.9c$.

33. Find (a) the total energy and (b) the kinetic energy of an electron moving at $0.97c$.

34. A TV tube accelerates electrons from rest through a potential difference of 30 kV. Find their final speed.

35. At what speed will the relativistic and Newtonian expressions for kinetic energy differ by 10%?

36. Find (a) the speed and (b) the momentum of a proton whose kinetic energy is 500 MeV.

37. Among the most energetic cosmic rays ever detected are protons with energies around 10^{20} eV. Find the momentum of such a proton, and compare with that of a 25-mg insect crawling at 2 mm/s (Fig. 38-36).

(a) (b)

FIGURE 38-36 (a) A cosmic ray proton slams into the upper atmosphere, creating a shower of elementary particles. How does the proton's energy compare with that of a crawling bug (b)? (Problem 37)

38. A large city consumes electrical energy at the rate of 10^9 W. If you converted all the rest mass in a 1-g raisin to electrical energy, for how long could it power the city?

39. In a nuclear fusion reaction, two deuterium nuclei (^{2}H) combine to give a helium nucleus (^{3}He) plus a neutron. The energy released in the process is 3.3 MeV. By how much do the combined masses of the helium nucleus and neutron differ from the combined masses of the original deuterium nuclei?

40. Find the kinetic energy of an electron moving at (a) $0.0010c$, (b) $0.60c$, and (c) $0.99c$. Use suitable approximations where possible.

41. Find the speed of an electron with kinetic energy (a) 100 eV, (b) 100 keV, (c) 1 MeV, (d) 1 GeV. Use suitable approximations where possible.

42. How much energy would it take to accelerate a proton (a) from rest to $0.10c$, (b) from $0.50c$ to $0.55c$, (c) from $0.95c$ to $0.99c$, and (d) from $0.99c$ to c.

43. Use the binomial approximation (Appendix A) to show that Equation 38-14 reduces to the Newtonian expression for kinetic energy in the limit $u \ll c$.

44. Show that Equation 38-16 follows from the expressions for relativistic momentum and total energy.

Section 38-6 What Is Not Relative

45. Show from the Lorentz transformations that the spacetime interval of Equation 38-17 has the same value in all frames of reference.

46. A spaceship travels at $0.80c$ from Earth to a star 10 light-years distant, as measured in the Earth-star frame (Fig. 38-37). Let event A be the ship's departure from Earth and event B its arrival at the star. (a) Find the distance and time between the two events in the Earth-star frame. (b) Find the distance and time between the two events in the ship frame. *Hint:* The distance in the ship frame is the distance an observer has to move *with respect to that frame* to be at both events—not the same as the Lorentz-contracted distance between Earth and star. (c) Compute the square of the spacetime interval in both frames to show explicitly that it is invariant.

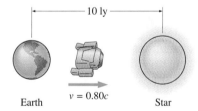

$\vdash$ 10 ly $\dashv$

Earth $v = 0.80c$ Star

FIGURE 38-37 Problem 46.

47. Use Equation 38-17 to calculate the square of the spacetime interval between the events (a) of Problem 22 and (b) of Problem 23. Comment on the signs of your answers in relation to the possibility of a causal relation between the events.

48. A light beam is emitted at event A and arrives at event B. Show that the spacetime interval between the two events is zero.

Paired Problems

(Both problems in a pair involve the same principles and techniques. If you can get the first problem, you should be able to solve the second one.)

49. An extraterrestrial spacecraft passes Earth and 4.5 s later, according to its clocks, it passes the moon. Find its speed.

50. How fast would you have to go to reach a star 200 light-years distant in a 75-year human lifetime?

51. An electron moves down a 1.2-km-long particle accelerator at $0.999992c$. In the electron's frame, (a) how much time does the trip take and (b) how long is the accelerator?

52. An advanced civilization has developed a spaceship that goes, with respect to the galaxy, only 50 km/s slower than light. (a) How long, according to the ship's crew, does it take to cross the 100,000-ly diameter of the galaxy? (b) What is the galactic diameter as measured in the ship's frame of reference?

53. Event A occurs at $x = 0$ and $t = 0$ in a frame of reference S. Event B occurs at $x = 3.8$ light-years, $t = 1.6$ years in S. Find (a) the distance and (b) the time between A and B in a frame S' moving at $0.80c$ along the x axis of S.

54. Two nuclear reactions occur in a particle accelerator, at points 45 cm apart and separated by 6.8 ns in time, as measured in the accelerator's frame of reference. Find (a) the distance and (b) the time between the events as measured in the reference frame of a proton moving at $0.92c$ on a line between the two.

55. When a particle's speed doubles, its momentum increases by a factor of 3. What was the original speed?

56. Find the speed of an electron whose momentum is 50% greater than the value given by the Newtonian expression $p = mu$.

Supplementary Problems

57. How fast would you have to travel to reach the Crab Nebula, 6500 light-years from Earth, in 20 years? Give your answer to 7 significant figures.

58. At what speed are a particle's kinetic and rest energies equal?

59. A cosmic ray proton with energy 20 TeV is heading toward Earth. What is Earth's diameter measured in the proton's frame of reference?

60. Plot the ratio of relativistic to Newtonian kinetic energy as a function of speed, and from your graph find the speed where the relativistic kinetic energy is 50% greater than the Newtonian value.

61. When the speed of an object increases by 5%, its momentum goes up by a factor of 5. What was the original speed?

62. Use the Lorentz transformations to show that if two events are separated in space and time so that a light signal leaving one event could not reach the other, then there is an observer for whom the two events are simultaneous. Show that the converse is true as well: that if a light signal could get from one event to the other, then no observer will find them simultaneous.

63. By writing force as the time derivative of relativistic momentum, formulate a relativistic version of Section 7-5's derivation of the expression for kinetic energy, and show that the result is Equation 38-14.

64. A source emitting light with frequency f moves toward you at speed u. By considering both time dilation and the effect of wavefronts "piling up" as shown in Fig. 17-20, show that you measure a Doppler shifted frequency given by

$$f' = f\sqrt{\frac{c + u}{c - u}}.$$

Use the binomial approximation to show that this result can be written in the form of Equation 17-12 for $u \ll c$.

65. Use Equation 9-7 to estimate the size to which you would have to squeeze each of the following before escape speed at its surface approximated the speed of light: (a) Earth; (b) the Sun; (c) the Milky Way galaxy, containing about 10^{11} solar masses. Your answers show why general relativity is not needed for most astronomical calculations.

66. Consider a line of positive charge with line charge density λ, as measured in a frame S at rest with respect to the charges. (a) Show that the electric field a distance r from this charged line has magnitude $E = \lambda/2\pi\varepsilon_0 r$, and that there is no magnetic field (no relativity needed here). Now consider the situation in a frame S' moving at speed v parallel to the line of charge. (b) Show that the line charge density as measured in S' is given by $\lambda' = \gamma\lambda$, with $\gamma = 1/\sqrt{1 - v^2/c^2}$. (c) Use the result of (b) to find the electric field in S'. Since the line of charge is moving with respect to S', there is a current in S'. (d) Find an expression for this current and (e) for the magnetic field it produces. Show that the quantities (f) $\mathbf{E} \cdot \mathbf{B}$ and (g) $E^2 - B^2$ are the same in both frames of reference. (In fact, these quantities are always invariant.)

CHAPTER 39

LIGHT AND MATTER: WAVES OR PARTICLES?

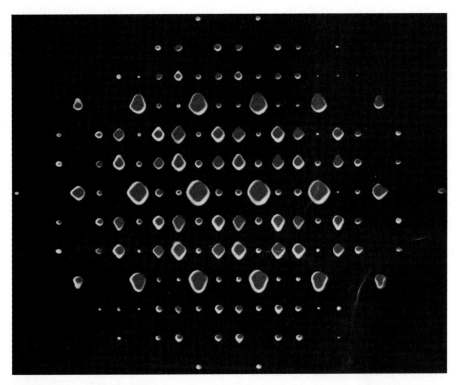

This pattern, produced by diffraction of an electron beam from a titanium-nickel alloy, shows that electrons exhibit a wave-like character. Quantum theory has blurred the distinction between particles and waves, showing that particles have a wave-like aspect and that light has a particle-like aspect.

Newtonian mechanics and Maxwell's equations of electromagnetism constitute the core of **classical physics,** providing a rich and deep understanding of physical reality. Application of classical physics has led to many of the technological developments essential to modern civilization. Although these theories were firmly established by the middle of the nineteenth century, they remain central to the work of many scientists and engineers. They may be classical, but they are also vitally contemporary.

Nevertheless, at the end of the nineteenth century a few seemingly minor phenomena defied classical explanation. Most physicists felt that it was only a matter of time before these, too, came under the classical umbrella. But that was not to be. We've already seen how questions about the speed of light led to a radical restructuring of our fundamental notions of space and time. Other questions, especially those concerning the behavior of matter at the atomic scale, were to bring an even more radical transformation of physical thought.

The essence of the new physics lies in the ideas of **quantum physics,** developed first to explain the workings of the atom and subsequently pushed into the nuclear realm and then to the subnuclear world of elementary particles. In this chapter we introduce some phenomena that led to the ideas of quantum physics and explore early attempts to explain them. The next chapter gives a fuller account of the complete theory called quantum mechanics, and subsequent chapters explore its application to atoms, molecules, nuclei, and other systems.

39-1 TOWARD THE QUANTUM THEORY

Are matter and energy continuously divisible, or are there some minimum quantities of each from which all else is built? The essential difference between classical and quantum physics is that the former answers this question "in principle, yes," while the latter says definitively "no." Instead, most physical quantities in quantum physics are **quantized,** coming only in certain discrete values.

The idea that physical quantities might come in discrete "chunks" is not new to modern quantum theory. Some 2400 years ago the Greek philosopher Democritus proposed that all matter consists of indivisible atoms of different types. By the start of the twentieth century a more scientifically grounded atomic theory was widely accepted. J. J. Thomson's discovery of the electron in 1897 showed that atoms might be divisible after all, but at the same time it revealed a finer division of matter into discrete "chunks." Millikan's 1909 oil-drop experiment (Section 23-2) showed that electric charge is similarly quantized, seemingly available only in integer multiples of the elementary charge e. Discovery of the proton and later the neutron further solidified the notion that matter is comprised of fundamental building blocks with definite and unchanging values for their various physical properties.

Quantization of matter into particles with discrete properties is not incompatible with classical physics as long as those particles behave according to classical laws—in particular, that they move continuously through space and can have *any* amount of energy. Add electromagnetism to the picture and the classical viewpoint requires that the fields be truly continuous, exerting forces on charged particles and perhaps changing, in a gradual and continuous way, the particles' energies.

The startling fact of quantum physics is that this classical behavior does not occur at the atomic scale; instead, energy itself is in many instances quantized. Reconciling the implications of that fact with our commonsense notions of matter and motion has proved impossible; instead, the quantum world speaks a different language, one in which deeply ingrained ideas about causality and the solid reality of matter seem no longer to apply. Here we look at three distinct phenomena that force on us the idea of energy quantization.

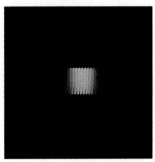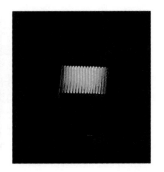

FIGURE 39-1 As the temperature of an electrically heated coil increases, it emits increasing amounts of radiation that peaks at shorter and shorter wavelengths. The color, correspondingly, changes from a dull red to a bright yellow-white.

39-2 BLACKBODY RADIATION

Heat a solid object hot enough and it begins to glow, emitting electromagnetic radiation in the form of light. As we saw in Section 19-6, the total power radiated is proportional to the fourth power of the temperature. There's also a change in wavelength with increasing temperature: The first visible glow is a dull red, changing with higher temperatures to orange and then yellow colors corresponding to ever shorter wavelengths (Fig. 39-1).

The radiation emitted by a dense, glowing substance that is a perfect absorber of radiation is **blackbody radiation,** so called because the substance absorbs all radiation incident on it and thus appears black when viewed in reflected light only. Although many materials behave approximately as blackbodies, an excellent approximation to a blackbody can be achieved using a hollow piece of *any* material with a small hole in it. As Fig. 39-2 shows, any radiation entering the hole undergoes multiple reflections from the interior walls and is eventually absorbed. The hole, therefore, is a nearly perfect absorber, so when the material is heated the radiation emerging from it will be blackbody radiation.

Experimental study of blackbody radiation shows several characteristic features:

1. The radiation covers a continuous range of wavelengths, with the total power radiated at all wavelengths combined being given by the Stefan-Boltzmann law that we introduced in Chapter 19:

$$P_{\text{blackbody}} = \sigma A T^4, \qquad (39\text{-}1)$$

where A is the area of the radiating surface, T is its temperature, and $\sigma = 5.67 \times 10^{-8}$ W/m$^2 \cdot$K^4 is the Stefan-Boltzmann constant.

2. The wavelength at which the most radiation occurs is inversely proportional to the temperature; specifically

$$\lambda_{\text{max}} T = 2.898 \times 10^{-3} \text{ m} \cdot \text{K}. \qquad (39\text{-}2)$$

This relation is called **Wien's displacement law.**

3. The distribution of wavelengths depends only on the temperature, not on the material of which the blackbody is made.

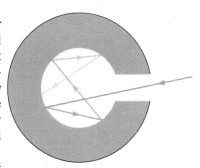

FIGURE 39-2 A piece of material containing a cavity and a small hole. Any radiation incident on the hole undergoes multiple reflections inside the cavity, and its energy is eventually absorbed. Very little radiation escapes the hole, which therefore appears black. The hole is a nearly perfect blackbody.

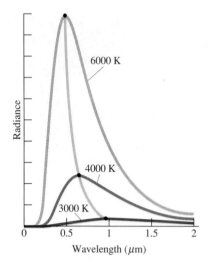

FIGURE 39-3 Radiance of a blackbody as a function of wavelength. Gray curve shows the shift in peak radiance toward shorter wavelengths. Note also the rapid increase in total radiation with increasing temperature.

Figure 39-3 shows the intensity of blackbody radiation as a function of wavelength for several temperatures.

Microscopically, blackbody radiation is associated with the thermal motions of atoms and molecules, so it's not surprising that the radiation increases with temperature. In the late 1800s physicists tried to apply the laws of electromagnetism and statistical mechanics to explain the experimental observations of blackbody radiation. They met with some success in describing such aspects as the T^4 dependence of the total energy radiated, and the shifting of the radiation distribution toward shorter wavelengths with increasing temperature, but they could not reproduce the actual observed distribution at all wavelengths.

In 1900 the German physicist Max Planck formulated an equation that fit the observed radiance-versus-wavelength curves of blackbody radiation at all wavelengths. Planck's equation describes the radiation arising from the thermal vibration of molecules, and in terms of the emitted radiation it reads

$$R(\lambda, T) = \frac{2\pi hc^2}{\lambda^5(e^{hc/\lambda kT} - 1)}. \tag{39-3}$$

Here R—called the radiance—is the power radiated per unit area per unit wavelength; multiplying by a small wavelength band $\Delta\lambda$ would then give the intensity in that band. Two familiar quantities in Equation 39-3 are Boltzmann's constant $k = 1.38\times10^{-23}$ J/K, introduced in Chapter 20, and the speed of light c. A new quantity is the constant h, whose value is chosen to make the equation fit experimental data.

In the weeks following its presentation, Planck searched for a physical interpretation of his new law. Through arguments involving entropy and statistical mechanics, he showed that the equation has a remarkable implication:

The energy of a vibrating molecule is quantized, meaning it can have only certain discrete values. Specifically, if f is the vibration frequency, then the energy must be an integer multiple of the quantity hf:

$$E = nhf, \qquad n = 0, 1, 2, 3, \ldots \tag{39-4}$$

where h is the constant Planck introduced in Equation 39-3. Today we know h as one of the fundamental constants of nature and call it **Planck's constant.** Its value is approximately 6.63×10^{-34} J·s, and it is precisely because h is so small that quantum phenomena usually become obvious only when we delve into the atomic and molecular realm. Planck's quantization of the energy of vibrating molecules implies further that a molecule can absorb or emit energy only in discrete "bundles" of size hf, and that in doing so it jumps abruptly from one of its allowed energy levels to another (Fig. 39-4). (Later developments showed that Planck was correct about the size of the energy jumps, but that the factor n in Equation 39-4 should actually be $n + \frac{1}{2}$.)

Planck himself was a very conservative scientist and was reluctant to accept or elaborate on his theory's evident disagreement with classical physics; nevertheless, his revolutionary work won Planck the 1918 Nobel Prize. Other physicists subsequently emphasized the contrast between Planck's work and an ear-

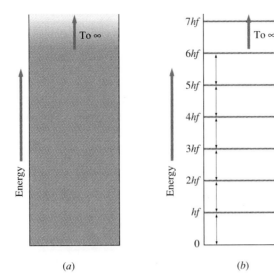

(a)   (b)

FIGURE 39-4 (a) In classical physics, a vibrating molecule can have any energy, and its energy can change through absorption or emission of any amount of energy. (b) Allowed energies in Planck's theory are integer multiples of hf, and can change only through absorption or emission of discrete energy "bundles" of size hf. Two-headed arrows indicate these allowed transitions. Energy-level diagrams like this are used frequently in quantum physics, and usually, as here, the horizontal axis has no physical significance.

lier classical treatment of blackbody radiation. That treatment, based on the assumption that energy is shared equally among all possible vibrational modes, had led to the **Rayleigh-Jeans law** for the radiance of a blackbody:

$$R(\lambda, T) = \frac{2\pi ckT}{\lambda^4}. \tag{39-5}$$

Not only did the Rayleigh-Jeans prediction contradict experimental measurements, but it also led to the absurd conclusion that every glowing object should emit electromagnetic energy at an infinite rate, with that energy concentrated at the shortest wavelengths (Fig. 39-5). Since the shortest wavelength radiation known at the time was ultraviolet, this phenomenon was called the **ultraviolet catastrophe.** In Planck's equation the exponential term in the denominator grows rapidly with decreasing wavelength, diminishing the radiance and averting the ultraviolet catastrophe. Problems 13, 86, and 87 show that Planck's law reduces to the Rayleigh-Jeans law for long wavelengths, and that it also leads to Wien's displacement law and the Stefan-Boltzmann law.

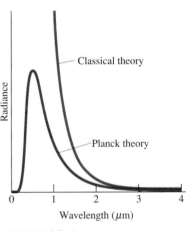

FIGURE 39-5 Radiance versus wavelength for blackbody radiation from an object at 6000 K. Classical theory agrees with experiment and with Planck's theory at long wavelengths, but predicts an excess of radiation at short wavelengths—the "ultraviolet catastrophe."

● **EXAMPLE 39-1**   LIGHT–BULB EFFICIENCY, BLACKBODY RADIATION, AND THE UNIVERSE

A standard incandescent light bulb's filament temperature is about 3000 K. (a) Find the peak wavelength emitted by the bulb. (b) How does the radiance at 550 nm—the approximate center of the visible spectrum—compare with that at the peak wavelength?

Solution

Equation 39-2 relates the peak wavelength and temperature, so for the light bulb we have

$$\lambda_{max} = \frac{2.898 \times 10^{-3}\ \text{m·K}}{T} = \frac{2.898 \times 10^{-3}\ \text{m·K}}{3000\ \text{K}} = 966\ \text{nm}.$$

That this wavelength lies in the infrared shows why incandescent lamps are not particularly efficient—they emit more heat (infrared radiation) than light. To compare the radiance at 550 nm with that at the 966-nm peak, we take the ratio of the radiances given by Planck's law, Equation 39-3:

$$\frac{R(\lambda_2, T)}{R(\lambda_1, T)} = \frac{2\pi hc^2/[\lambda_2^5(e^{hc/\lambda_2 kT} - 1)]}{2\pi hc^2/[\lambda_1^5(e^{hc/\lambda_1 kT} - 1)]} = \frac{\lambda_1^5(e^{hc/\lambda_1 kT} - 1)}{\lambda_2^5(e^{hc/\lambda_2 kT} - 1)}$$

$$= 0.38,$$

where we used $\lambda_2 = 550$ nm, $\lambda_1 = 966$ nm, $T = 3000$ K, and the values for the constants c, h, and k. Thus the light bulb's brightness in the center of the visible spectrum is only about 40% of its peak brightness. That is, incidentally, why color film intended for daylight applications gives incorrect color rendition when used indoors under ordinary light bulbs.

EXERCISE The universe is bathed in "cosmic microwave background radiation," which originated about 10^5 years after the "big bang" event that began the universe (more on this in Chapter 45). Intensity of the cosmic background radiation peaks at a wavelength of about 1 mm. To what temperature does this blackbody radiation correspond?

Answer: 3 K

Some problems similar to Example 39-1: 2, 4, 7, 10, 11 ●

39-3 PHOTONS

Planck's hypothesis quantized the energies of vibrating molecules and showed that they could exchange energy with electromagnetic radiation only in discrete "bundles" of size hf. But what about the radiation itself? Is its energy also quantized?

The Photoelectric Effect

Experiments begun by Hertz in 1887 and continued by other physicists in the early 1900s showed that some metals emit electrons when illuminated with light. This **photoelectric effect** is studied by putting a metal surface in an evacuated glass container that also contains a second conducting electrode (Fig. 39-6). Making the second electrode positive with respect to the emitting surface attracts the electrons, and the resulting current provides a measure of the number of electrons ejected. If the second electrode is sufficiently negative with respect to the emitting surface, on the other hand, then the electron energy is not enough to overcome the potential difference and the current ceases. The value of this so-called stopping potential thus provides a measure of the ejected electrons' maximum kinetic energy.

FIGURE 39-6 Apparatus for studying the photoelectric effect. An adjustable filter determines the wavelength of light incident on the apparatus. Light passes through a hole in a metal electrode and strikes a metal surface, where it ejects electrons. Varying the magnitude and polarity of the voltage between the electrode and the metal surface allows measurement of the number and energy of the ejected electrons.

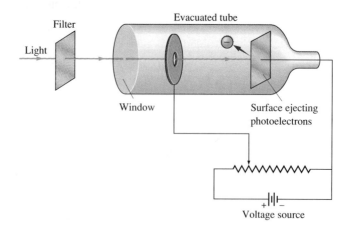

▲ **T A B L E 3 9 - 1** WORK FUNCTIONS ϕ FOR SELECTED ELEMENTS

SYMBOL	NAME	ϕ, eV	ϕ, 10^{-19} J
Ag	Silver	4.26	6.82
Al	Aluminum	4.28	6.86
Cs	Cesium	2.14	3.43
Cu	Copper	4.65	7.45
K	Potassium	2.30	3.68
Na	Sodium	2.75	4.41
Ni	Nickel	5.15	8.25
Si	Silicon	4.85	7.77

Classical physics suggests that the photoelectric effect should occur because an electron in a metal experiences a force in the oscillating electric field of a light wave. As the electron continuously absorbs energy from the wave, the amplitude of its oscillatory motion should grow until eventually it has enough energy to escape from the metal. Because the energy in a wave is spread throughout the entire wave, it might take a while for a single tiny electron to absorb enough energy to escape. Increasing the light intensity should increase the electric field in the light wave, so should result in the electron's being ejected sooner and with more energy. Changing the wave frequency should have little effect on the behavior of the free electrons in the metal.

The photoelectric effect does occur, but not in the way classical physics suggests. The major disagreements are the following:

1. Electrons are ejected almost immediately, even in dim light.
2. The maximum electron energy is independent of the light intensity.
3. Below a certain cutoff frequency *no* electrons are emitted, no matter how strong the light. Above the cutoff frequency electrons are emitted with a maximum energy that increases in proportion to the light-wave frequency.

In 1905, the same year in which he formulated the special theory of relativity, Albert Einstein offered an explanation for the photoelectric effect. Einstein suggested that the energy of an electromagnetic wave is not spread throughout the wave, but is instead concentrated in "bundles" called **quanta** or **photons.** Einstein applied to these photons the same energy-quantization condition Planck had already proposed for molecular vibrations: that photons in light with frequency f have energy hf, where again h is Planck's constant:

$$E = hf. \qquad \text{(energy of a photon)} \qquad (39\text{-}6)$$

The more intense the light, the more photons—but the energy of each photon is unrelated to the light intensity.

Einstein's idea readily explains the nonclassical aspects of the photoelectric effect. He held that for each material there is a minimum energy—called the work function, ϕ, required to eject an electron. (Table 39-1 lists work functions for selected elements.) Since the energy in a photon of light with frequency f is hf, the photons in low-frequency light have less energy than the work function and are therefore unable to eject electrons—no matter how many photons there are. At the cutoff frequency the photon energy equals the work function, and the

FIGURE 39-7 Results of a photoelectric experiment. White line shows stopping potential as a function of light frequency and wave length. The stopping potential in volts is a direct measure of the electron energy in eV. Slope of the straight line is numerically equal to Planck's constant, which confirms Equation 39-7. Can you tell from the graph which material in Table 39-1 is the emitting surface?

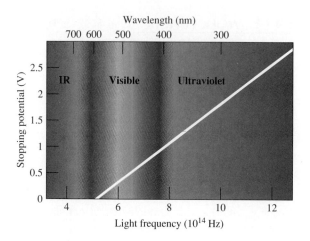

photons have just enough energy to eject electrons. As the frequency increases still further the electrons emerge with maximum kinetic energy K equal to the difference between the photon energy and the work function:

$$K_{max} = hf - \phi. \tag{39-7}$$

Thus, the electrons' maximum kinetic energy depends only on the photon energy—that is, on the light frequency but not on its intensity (Fig. 39-7). Finally, the immediate ejection of electrons follows because an individual photon delivers its entire bundle of energy to an electron all at once. Einstein received the 1921 Nobel Prize for his explanation of the photoelectric effect; his special and general theories of relativity were at that time sufficiently controversial that they bore no mention in the Nobel citation. In 1914 Millikan, who had earlier demonstrated the quantization of electric charge, carried out meticulous photoelectric experiments that confirmed Einstein's hypothesis and helped earn Millikan the 1923 Nobel prize.

◼ APPLICATION THE PHOTOMULTIPLIER

The photoelectric effect is put to good use in the photomultiplier, an extremely sensitive device for detecting light. Figure 39-8 shows a typical photomultiplier, consisting of an evacuated tube containing a number of electrodes held at increasing electric potentials. Light is incident at one end on the lowest voltage electrode, called the photocathode. Photoelectrons ejected by the photoelectric effect are attracted to the higher potential of the next electrode, called a dynode. Each electron is accelerated with enough energy that it releases several additional electrons when it hits the dynode. These accelerate to the next dynode, again releasing several electrons. This process continues through a chain of typically 10 dynodes, resulting in "electron multiplication" factors as high as 10^9. For each electron ejected at the photocathode, that means there may be as many as a billion electrons striking the final electrode. These produce a measurable pulse of electric current.

Typical photomultipliers have so-called quantum efficiencies around 20%, meaning that only 20% of the incident photons actually dislodge photoelectrons. But in most applications that inefficiency is more than compensated for by the enormous gain due to electron multiplication.

Photomultipliers are widely used in research, industry, environmental monitoring, laser-based distance measurements, radiation detectors, and medical instrumentation. In some application they have been superseded by solid-state devices based on semiconductor technology, but for very sensitive and high-speed measurements, the photomultiplier remains at present the detector of choice.

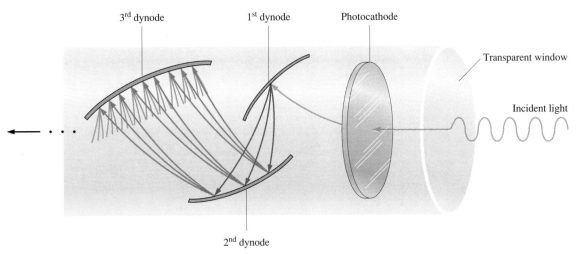

FIGURE 39-8 Part of a photomultiplier tube, showing the multiplying effect as electrons hit each dynode.

● EXAMPLE 39-2 DESIGNING A PHOTOMULTIPLIER

A photomultiplier is to respond to visible light with wavelengths shorter than 575 nm. Choose a suitable photocathode material from Table 39-1, and find the maximum kinetic energy of an electron that can be ejected by visible light.

Solution
Equation 39-7 shows that a photosensitive material will just barely eject electrons when the photon energy hf is equal to the work function ϕ. With $\lambda = c/f$, this gives a maximum work function

$$\phi = \frac{hc}{\lambda} = \frac{(6.63 \times 10^{-34} \text{ J·s})(3.00 \times 10^8 \text{ m/s})}{(575 \times 10^{-9} \text{ m})}$$

$$= 3.46 \times 10^{-19} \text{ J}.$$

The only material in Table 39-1 with a work function below this value is cesium, with $\phi = 3.43 \times 10^{-19}$ J.

The minimum wavelength—and therefore maximum photon energy—in visible light is about 400 nm. Using this value in Equation 39-7 with $f = c/\lambda$ gives

$$K = \frac{hc}{\lambda} - \phi$$

$$= \frac{(6.63 \times 10^{-34} \text{ J·s})(3.00 \times 10^8 \text{ m/s})}{400 \times 10^{-9} \text{ m}} - 3.43 \times 10^{-19} \text{ J}$$

$$= 1.54 \times 10^{-19} \text{ J}$$

for the maximum photoelectron energy.

Although cesium has the lowest elemental work function, photocathodes made with appropriate compounds operate throughout the visible and even into the infrared regions of the spectrum.

EXERCISE Which materials in Table 39-1 exhibit the photoelectric effect for at least some wavelengths of visible light? Comment on where they occur in the periodic table (see inside back cover).

Answer: cesium, potassium, sodium—all alkali metals from column 1 of the periodic table

Some problems similar to Example 37-2: 20–22, 25

Waves or Particles?

The work of Planck and Einstein applies quantization of energy to two quite different systems, vibrating molecules in Planck's case and electromagnetic waves in Einstein's. In positing the existence of photons, Einstein gave the first

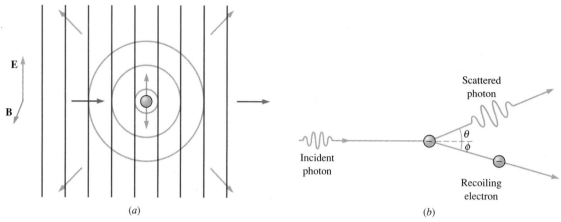

FIGURE 39-9   Scattering of electromagnetic radiation by an electron. (*a*) In the classical picture, the electron oscillates at the wave frequency and so reradiates at this frequency. Vertical lines are the incident wavefronts; circles are reradiated wavefronts. (*b*) In the quantum picture, the electron undergoes a particle-like collision with a quantum of electromagnetic radiation. As a result the photon energy decreases and its wavelength increases.

inklings of the **wave-particle duality**—the seemingly dual nature of light, which acts in some situations like a wave and in others, as in the photoelectric effect, more like a localized particle. We now turn to another phenomenon that demonstrates light's particle-like aspect. Later we will see how the wave-particle duality encompasses not only light but matter as well.

The Compton Effect

In 1923 the American physicist Arthur Holly Compton, at Washington University in St. Louis, did an experiment that dramatically confirmed the particle-like aspect of electromagnetic radiation. Although Compton's work came much later in the history of quantum theory than did Einstein's 1905 interpretation of the photoelectric effect, we include it here because it so strongly corroborates Einstein's photon hypothesis.

Compton was studying the interaction of electromagnetic radiation—specifically, x rays—with electrons. Classically, an electron subject to an electromagnetic wave should undergo oscillatory motion, driven by the wave's oscillating electric field. Since accelerated charge is the source of electromagnetic waves, the electron should itself produce electromagnetic waves *of the same frequency as the incident waves* (Fig. 39-9*a*). From the considerations of Section 34-9, the electron should radiate in essentially all directions, with maximum radiation perpendicular to its oscillatory motion.

Compton and his co-workers subjected a graphite target to an x-ray beam, and measured the intensity of scattered x rays as a function of wavelength for different scattering angles. Remarkably, they found the greatest concentration of scattered x rays at a wavelength *longer* than that of the incident radiation (Fig. 39-10). They interpreted their results as implying that particle-like photons had collided with free electrons in the graphite, losing energy to the electrons and therefore, since $E = hf$, emerging with lower frequency and correspondingly longer wavelength (Fig. 39-9*b*). (Another peak, which appears at the incident wavelength, is due to scattering from the atoms' tightly bound inner electrons. This scattering occurs without significant energy loss.)

We saw in Chapter 34 that light with energy E has momentum $p = E/c$. Therefore a photon of frequency f should have momentum $p = hf/c$. Treating the Compton scattering process as a relativistic collision between a particle with this momentum and an initially stationary electron leads to the conclusion that the scattered photon should have its wavelength shifted by an amount given by

$$\Delta\lambda = \frac{h}{mc}(1 - \cos\theta), \qquad (39\text{-}8)$$

where m is the electron mass and θ is the angle through which the photon's direction is changed (see Fig. 39-9b). This equation is in excellent agreement with data from scattering experiments (Fig. 39-10).

The term h/mc in Equation 39-8 is called the **Compton wavelength** of the electron and gives the wavelength shift for a photon scattering at $\theta = 90°$. Its value is readily calculated:

$$\lambda_c = \frac{h}{mc} = 0.00243 \text{ nm}.$$

Equation 39-8 shows that the largest wavelength shift will be $2\lambda_C$, occurring at $\theta = 180°$. For the shift to be noticeable it should be a significant fraction of the incident wavelength, which therefore cannot be too many times the Compton wavelength. For x rays, λ is in the range from approximately 0.01 nm to 10 nm, and therefore, detection of the Compton shift in x-rays is already a difficult experiment. It would be totally impossible with visible light. Scattering from more massive particles than electrons only makes matter worse, since a particle's Compton wavelength is inversely proportional to its mass.

Today, Compton scattering with gamma rays is a widely used technique for studying the structure of matter. For example, abnormalities in human bone can be detected through Compton scattering of gamma rays emitted by a small radioactive source embedded in bone. And the inverse Compton effect—the scattering of a rapidly moving electron off a photon—is a common process in high-energy astrophysical systems and is used in the laboratory to produce beams of gamma radiation.

The wavelength shift in Compton scattering simply admits no classical explanation. Coming after a decade of experimental and theoretical work that pointed increasingly to quantization as the essence of the atomic world, Compton's experimental results were for many physicists the convincing evidence for the reality of quanta.

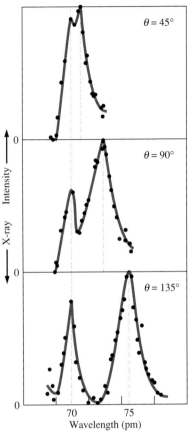

FIGURE 39-10 Compton's results for scattering of x rays with average wavelength about 71 pm, for scattering angles of 45°, 90°, and 135°. The right-hand peak is from photons scattering off free electrons, and shows clearly the wavelength shift that is the hallmark of the Compton effect. The left-hand peak, at the incident wavelength, is from photons scattering off tightly bound atomic electrons, which cannot absorb significant energy from the photons.

● **EXAMPLE 39-3** COMPTON SCATTERING

A 0.0650-nm x-ray photon scatters off an electron at 75° to its initial direction. Find (a) the wavelength of the scattered photon and (b) the energy gained by the electron.

Solution

The wavelength shift is given by Equation 39-8:

$$\Delta\lambda = \lambda_c(1 - \cos\theta) = (0.00243 \text{ nm})(1 - \cos 75°)$$

$$= 0.00180 \text{ nm},$$

where we used $h/m_e c = \lambda_c$. The wavelength of the scattered photon is, therefore,

$\lambda = \lambda_0 + \Delta\lambda = 0.0650 \text{ nm} + 0.0018 \text{ nm} = 0.0668 \text{ nm}.$

The photon's energy loss is the electron's gain; with $E = hf = hc/\lambda$ for the photon, we have

$$\Delta E = \frac{hc}{\lambda_0} - \frac{hc}{\lambda} = hc\left(\frac{1}{\lambda_0} - \frac{1}{\lambda}\right)$$

$$= (6.63\times10^{-34} \text{ J·s})(3.00\times10^8 \text{ m/s})$$

$$\times \left(\frac{1}{6.5\times10^{-11} \text{ m}} - \frac{1}{6.68\times10^{-11} \text{ m}}\right)$$

$$= 8.25\times1\times10^{-17} \text{ J} = 515 \text{ eV}.$$

EXERCISE   A gamma ray undergoes a 45° Compton scattering. If its wavelength increases by 25%, find (a) its initial wavelength and (b) its initial energy in electron-volts.

Answers: (a) 2.85 pm; (b) 436 keV

Some problems similar to Example 39-3:   27, 30, 31

39-4 ATOMIC SPECTRA AND THE BOHR ATOM

In Chapter 34 we found that accelerated charges are the source of electromagnetic radiation. By 1900 it was known that atoms contain negative electrons as well as regions of positive charge; by 1911 experiments by Ernest Rutherford and his colleague Hans Geiger and student Ernest Marsden had localized the positive charge in the tiny but massive nucleus. Electrons should orbit the nucleus under the influence of the electric force, radiating electromagnetic wave energy as they accelerated in their orbits. In fact, a classical calculation shows atomic electrons should quickly radiate away all their energy and spiral into the nucleus. Thus, the very existence of atoms seems at odds with classical physics.

The Hydrogen Spectrum

A more subtle problem involving radiation from atoms dates to 1804, when William Wollaston noticed apparent lines of demarcation between some of the colors dispersed by a prism. Ten years later the German optician Josef von Fraunhofer dispersed the solar spectrum sufficiently that he could see hundreds of narrow, dark lines against the otherwise continuous spectrum of colors (Fig. 39-11a). Studies of light emitted by diffuse gases excited by electric discharges

FIGURE 39-11   (a) The solar spectrum is a continuous band of colors overlain by dark lines where specific colors have been absorbed in the solar atmosphere. Prominent line in the red is hydrogen-alpha, and the pair in the yellow are the 589-nm sodium doublet. (b) Spectral lines emitted by a diffuse sample of hydrogen gas. This is the Balmer series of lines, with Hα the bright line in the red.

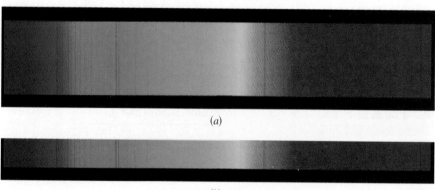

(a)

(b)

show similar **spectral lines,** these bright against an otherwise dark background (Fig. 39-11*b*). We know that bright-line spectra—called **emission spectra**—are produced when atoms emit light of discrete frequencies. **Absorption spectra,** in contrast, arise when atoms in a diffuse gas absorb certain discrete frequencies of light from a continuous source. What Fraunhofer saw in sunlight were absorption lines arising from the diffuse gas overlying the bright visible surface of the Sun. We emphasize the word *diffuse:* Discrete spectra generally arise only when a gas is sufficiently diffuse that light from one atom stands a strong chance of escaping the gas before it interacts with other atoms. In dense gases, or in solids and liquids, multiple interactions result in the continuous spectrum characteristic of blackbody radiation.

But why should atoms emit and absorb light of discrete frequencies? In 1884, several decades before a satisfactory answer was given, a Swiss schoolteacher named Johann Balmer realized that the wavelengths of the first four lines in the visible spectrum of hydrogen (Fig. 39-11*b*) were related by the equation

$$\frac{1}{\lambda} = R_H\left(\frac{1}{2^2} - \frac{1}{n^2}\right),$$

where $n = 3, 4, 5, 6 \ldots$ and R_H is a constant called the **Rydberg constant** for hydrogen, with value approximately 1.097×10^7 m^{-1}. Other series of lines in the hydrogen spectrum were soon found, and Balmer's equation was generalized to read

$$\frac{1}{\lambda} = R_H\left(\frac{1}{n_2^2} - \frac{1}{n_1^2}\right), \tag{39-9}$$

where $n_1 = n_2 + 1, n_2 + 2, \ldots$. The Balmer series of lines has $n_2 = 2$; the Lyman series, in the ultraviolet, has $n_2 = 1$, and the infrared Paschen series has $n_2 = 3$. There are in fact infinitely many such series, corresponding to $n_2 = 1, 2, 3, \ldots$.

But why should atoms emit light in discrete spectral lines? And why should the hydrogen lines form patterns with the simple regularity described by Equation 39-9?

The Bohr Atom

In 1913 the great Danish physicist Niels Bohr (Fig. 39-12) proposed an atomic theory that accounted for the spectral lines of hydrogen. In the **Bohr atom** the electron moves in a circular orbit about the nucleus, held in orbit by the electric force. Classically, any orbital radius and correspondingly any energy and any angular momentum should be possible. But Bohr introduced quantization to the atom, stating that only certain orbits were possible—namely, those whose angular momentum is an integer multiple of Planck's constant divided by 2π. Angular momentum quantization implies energy quantization, which, as we soon show, leads to Equation 39-9 for the hydrogen spectral lines.

Bohr's theory asserted that an electron in any of the allowed orbits does not radiate energy, in direct contradiction to the predictions of classical electromagnetism. But an electron can jump from one orbit to another, emitting or absorbing a photon whose energy is equal to the energy difference between the two

FIGURE 39-12 Danish physicist Niels Bohr (left) was a powerful influence on many of the twentieth century's greatest physicists. In his later years he was also active in humanitarian causes, especially the effort to control nuclear weapons. Here Bohr is discussing quantum mechanics with Max Planck.

orbital levels. We can therefore find the expected photon energies—and thus the corresponding wavelengths—if we know the allowed energy levels.

To find the quantized atomic energy levels in Bohr's model, consider a hydrogen atom consisting of a fixed proton and an electron in circular orbit about the proton, held in its orbit by the electric force. Treating the proton as fixed is a good approximation because its mass is nearly 2000 times the electron's. We consider only electron speeds much less than that of the light, which is an excellent approximation in hydrogen and hydrogen-like atoms.

In Example 13-2 we found that the angular momentum of a particle with mass m and speed v, moving in a circular path of radius r, is mvr. Thus, Bohr's quantization condition reads

$$mvr = n\hbar, \tag{39-10}$$

where $n = 1, 2, 3, \ldots$ and where we define $\hbar \equiv h/2\pi$. (This quantity, called "h bar," is useful because the combination $h/2\pi$ occurs frequently in quantum mechanics.) We need to relate the electron's angular momentum to its energy so we can find out what Equation 39-10 implies about energy quantization.

We studied circular orbits for the inverse-square force of gravity in Chapter 9, where we found that the kinetic and potential energy in a circular orbit are related by $K = -\frac{1}{2}U$, where the zero of potential energy is taken at infinity. The total energy $K + U$ is therefore $\frac{1}{2}U$. These results hold for any $1/r^2$ force, including the electric force. In the electric case the potential energy U is the point-charge potential of the proton, ke/r, multiplied by the electron charge, $-e$. Then the total energy is

$$E = \frac{1}{2}U = -\frac{ke^2}{2r}.$$

The minus sign has a simple interpretation: it means the electron is *bound* to the proton, in that it would take energy to separate them. Solving this equation for r then gives

$$r = -\frac{ke^2}{2E}. \tag{39-11}$$

Since the kinetic energy is $K = -\frac{1}{2}U = -E$, we also have $\frac{1}{2}mv^2 = -E$ or

$$v = \sqrt{-2E/m}.$$

(Remember that E is negative, so the quantity under the square root is positive.) Using our expressions for r and v in the quantization condition of Equation 39-10 gives

$$m\sqrt{-2E/m}\left(-\frac{ke^2}{2E}\right) = n\hbar.$$

Solving for the energy E, we find

$$E = -\frac{k^2e^4m}{2\hbar^2n^2}.$$

It is convenient to define the **Bohr radius,** a_0, given by

$$a_0 = \frac{\hbar^2}{mke^2} = 0.0529 \text{ nm};$$

with this definition our expression for the energy becomes

$$E = -\frac{ke^2}{2a_0}\left(\frac{1}{n^2}\right). \tag{39-12a}$$

Equation 39-12a gives us the allowed energy levels under Bohr's quantization condition. Evaluating this expression for the case $n = 1$ gives $E_1 = -2.18 \times 10^{-19}$ J $= -13.6$ eV; it's then convenient to write Equation 39-12a numerically in the form

$$E = -\frac{13.6 \text{ eV}}{n^2}, \tag{39-12b}$$

where in both forms $n = 1, 2, 3, \ldots$. The lowest energy state, $n = 1$, is called the **ground state,** while the others are **excited states.**

Now we have the allowed energy levels. What about spectra? Recall that Bohr's theory allows the electron to jump between energy levels, emitting or absorbing a photon whose energy hf is equal to the energy difference between the levels. So imagine an electron going from a higher level n_1 to a lower level n_2. The energy difference, according to Equation 39-12a, is

$$\Delta E = -\frac{ke^2}{2a_0}\left(\frac{1}{n_1^2} - \frac{1}{n_2^2}\right) = \frac{ke^2}{2a_0}\left(\frac{1}{n_2^2} - \frac{1}{n_1^2}\right),$$

and this is equal to the energy of the emitted photon. But the photon energy is $\Delta E = hf = hc/\lambda$, and therefore $1/\lambda = \Delta E/hc$ or, using our expression for ΔE,

$$\frac{1}{\lambda} = \frac{ke^2}{2a_0 hc}\left(\frac{1}{n_2^2} - \frac{1}{n_1^2}\right).$$

This looks just like Equation 39-9 for the hydrogen spectral lines, except that $ke^2/2a_0hc$ replaces the Rydberg constant R_H. Evaluating this quantity gives

$$\frac{ke^2}{2a_0 hc} = \frac{(9.0 \times 10^9 \text{ N·m}^2/\text{C}^2)(1.6 \times 10^{-19} \text{ C})^2}{(2)(5.29 \times 10^{-11} \text{ m})(6.63 \times 10^{-34} \text{ J·s})(3.00 \times 10^8 \text{ m/s})}$$
$$= 1.09 \times 10^7 \text{ m}^{-1},$$

which is very close to the experimentally observed Rydberg constant. Using more accurate values for the fundamental constants would show that the two are the same to within about 1 part in 2000, with nearly all of the discrepancy arising from our approximation that the proton is stationary.

Bohr's theory of quantized angular momentum thus accounts brilliantly for the observed spectrum of hydrogen. We can understand the origin of the various

FIGURE 39-13 Energy-level diagram for the Bohr model of the hydrogen atom, showing transitions responsible for the first three series of spectral lines. Each series arises in jumps to a common final state; $n = 2$ gives the visible and ultraviolet lines of the Balmer series.

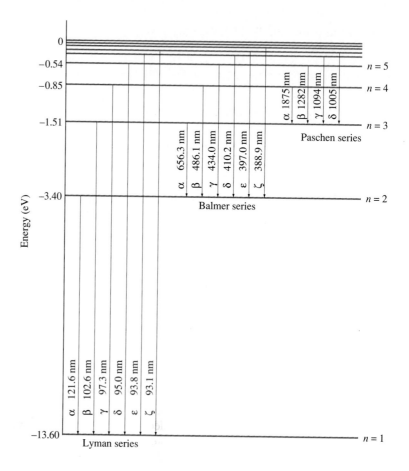

spectral line series using Fig. 39-13, which is an **energy level diagram** for the Bohr model of hydrogen. The diagram shows the energy levels associated with the different values of n in Equation 39-12 as "shelves" on which electrons can be placed. It also shows the various possible transitions among states, grouped in sets with a common final state. These groupings account for the different series of spectral lines.

Knowing the energy levels of Equation 39-12, we can also find the radii of the allowed electron orbits, as given by Equation 39-11:

$$r = -\frac{ke^2}{2E} = \left(\frac{ke^2}{2}\right)\left(\frac{2a_0 n^2}{ke^2}\right) = n^2 a_0. \qquad (39\text{-}13)$$

Thus, the lowest energy orbit has a radius of one Bohr radius, with higher energy orbits growing rapidly with increasing n. A hydrogen atom in its ground state— $n = 1$—has a diameter of two Bohr radii, or about 0.1 nm. As we will see in the next chapter, the Bohr model's precise electron orbits are not compatible with the fully developed theory of quantum mechanics; nevertheless, Equation 39-13 gives correctly the approximate size of atoms.

● **EXAMPLE 39-4** BIG ATOMS

The tendency of physical systems toward states of lowest energy holds for atoms, and therefore, hydrogen atoms are normally found in their ground state, with diameter approximately 0.1 nm. But in the diffuse hydrogen gas of interstellar space, atoms can be found in highly excited states with sizes approaching a fraction of a millimeter. Transitions between adjacent energy levels of these so-called Rydberg atoms result in emission of photons at radio frequencies, and therefore, radio astronomers can study the enormous atoms. One of the longest wavelengths observed is the so-called 272α, corresponding to a transition from $n = 273$ to $n = 272$. (a) What is the diameter of a hydrogen atom in the $n = 273$ state? (b) At what wavelength and frequency should a radio telescope be set to observe this transition?

Solution

(a) Equation 39-13 gives the atomic diameter:

$$d = 2r = 2n^2 a_0 = (2)(273^2)(0.0529 \times 10^{-9} \text{ m}) = 7.9 \text{ } \mu\text{m}.$$

This is just about the size of a red blood cell!

(b) To find the transition wavelength we use Equation 39-9 with $n_1 = 273$ and $n_2 = 272$;

$$\lambda = \left[R\left(\frac{1}{n_2^2} - \frac{1}{n_1^2}\right) \right]^{-1}$$

$$= \left[(1.097 \times 10^7 \text{ m}^{-1})\left(\frac{1}{272^2} - \frac{1}{273^2}\right) \right]^{-1} = 92 \text{ cm}.$$

Using $f = c/\lambda$ gives the corresponding frequency of 325 MHz, which happens to lie in a gap between the VHF and UHF bands used for television broadcasting.

EXERCISE Find the wavelength of the spectral line Hα, corresponding to the transition from $n = 3$ to $n = 2$.

Answer: 656.3 nm

Some problems similar to Example 39-4: 38, 45, 74 ●

Equation 39-12 shows, and Fig. 39-13 suggests, that there are infinitely many electron energy levels between the ground state at -13.6 eV and the zero of energy. It's possible to give an atomic electron enough energy to bring it above the $E = 0$ level, but then it is no longer bound to its proton and is therefore not part of the atom. The process of removing an electron is called **ionization,** and Equation 39-12b and Fig. 39-13 show that it takes 13.6 eV to ionize a hydrogen atom in its ground state. This quantity is called the **ionization energy.**

The Franck-Hertz Experiment

The Bohr model gained credibility with a 1914 experiment by James Franck and Gustav Hertz (nephew of Heinrich Hertz). Franck and Hertz used a tube filled with mercury vapor at low pressure, containing three metal electrodes (Fig. 39-14a). At one end, a hot cathode emitted electrons. The electrons were accelerated toward a metal grid at a higher potential than the cathode, and many passed through the grid to reach the anode and thus produce a current measured by the ammeter.

Franck and Hertz recorded a rapid increase in current as the grid-cathode potential was increased from zero. But then, at approximately 4.9 V, the current dropped abruptly (Fig. 39-14b). They interpreted this as meaning that a mercury atom could absorb 4.9 eV of an electron's kinetic energy, but not less. Increasing the voltage still further led to an increase in current, as presumably electrons that had given up their energy to mercury atoms were again accelerated. But at about 9.8 V the current dropped again, indicating that the accelerated electrons could again give up a second 4.9 eV of energy to the mercury atoms. Thus, the Franck-Hertz experiment gave strong evidence that mercury has discrete energy levels, with the first excited level 4.9 eV above the ground

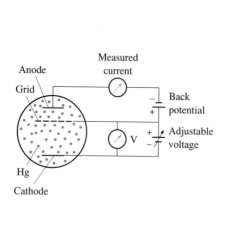

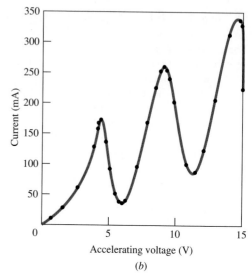

(a)

(b)

FIGURE 39-14 (a) Schematic diagram of the Franck-Hertz apparatus. (b) A plot of current versus voltage as measured by the meters shown in (a). Each time the voltage increases by 4.9 V, the current drops. Franck and Hertz interpreted these results as indicating an excitation of mercury atoms to higher energy states each time 4.9 eV of energy became available.

state.* If that interpretation was correct, then mercury atoms dropping back to the ground state should emit light with wavelength corresponding to a photon energy of 4.9 eV. Spectroscopic observations indeed showed the presence of a spectral line at the appropriate wavelength (see Problem 39).

Limitations of the Bohr Model

Bohr's theory proved astoundingly successful in explaining the hydrogen spectrum. It also explains the spectra of hydrogen-like atoms—those with all but one of their electrons removed—with the appropriate change in the value of the nuclear charge. And it has some success in predicting the spectra of atoms like lithium and sodium that have a single valence electron beyond a group of more tightly bound electrons. But it cannot account for the spectrum of more complicated atoms, even the simple two-electron helium. And even with hydrogen, there are subtle spectral details that the Bohr model does not address. Furthermore, like Planck's original quantum hypothesis, Bohr's quantization of atomic energy levels lacked a convincing theoretical basis. We will see in the next chapter how the much more comprehensive theory of quantum mechanics overcomes these limitations.

39-5 MATTER WAVES

In classical physics, light is purely a wave phenomenon. Einstein's introduction of energy "bundles"—photons—gave light a particle-like quality as well. In 1923, ten years after Bohr published his atomic theory, a French prince named

---

* The results of the Franck-Hertz experiment turned out as clearcut as they did because of special circumstances; see G. F. Hanne, "What really happens in the Franck-Hertz experiment with mercury?" *American Journal of Physics,* 1988, Vol. 56, pp. 696–700.

Louis de Broglie (pronounced "de Broy") set forth a remarkable hypothesis in his doctoral thesis. If light has both wave-like and particle-like properties, he reasoned, why shouldn't particles of matter also exhibit both properties?

We saw in Chapter 34 that light with energy E also carries momentum $p = E/c$. Combined with Equation 39-6, that means a photon of light with frequency f has momentum $p = hf/c$. Since $f\lambda = c$, the photon's momentum and wavelength are therefore related by

$$\lambda = \frac{h}{p}. \qquad \text{(de Broglie wavelength)} \qquad (39\text{-}14)$$

FIGURE 39-15 The allowed electron orbits in the Bohr atom are those that can fit an integral number of de Broglie wavelengths. Blue curve represents the de Broglie standing wave for the $n = 7$ level; red curve does not satisfy the Bohr condition.

De Broglie proposed that this same relation should hold for particles of matter; at nonrelativistic speed, for example, an electron should have associated with it a "wavelength" given by h/mv.

De Broglie used his matter-wave hypothesis to explain why atomic electron orbits should be quantized. He proposed that Bohr's allowed orbits were those in which standing waves could exist (Fig. 39-15), in much the same way that a violin string can support only certain frequencies of standing waves. Suppose that n full wavelengths of a de Broglie electron wave fit around the circumference of the electron's circular orbit. Then we must have

$$n\lambda = n\frac{h}{p} = n\frac{h}{mv} = 2\pi r,$$

with r the orbit radius. Multiplying both sides by $mv/2\pi$ then gives $mvr = nh/2\pi = n\hbar$, which is Bohr's quantization condition. Thus, de Broglie's hypothesis provides a natural explanation for the quantization of atomic energy levels.

If matter has wave-like properties, why don't we notice them all the time? As in optics, wave effects become important only in systems whose size is comparable to or smaller than a wavelength. For macroscopic objects, and even for electrons moving at substantial speed in, say, a TV tube, the electron wavelength is so small that wave properties are unnoticeable. But when an electron is confined to small spaces—like the lower energy orbits of the Bohr atom—its wave properties are dominant.

● **EXAMPLE 39-5** BASEBALLS AND ELECTRONS

Find the de Broglie wavelength of (a) a 150-g baseball pitched at 45 m/s and (b) an electron moving at 1.0×10^6 m/s. Compare with the sizes of home plate and an atom, respectively.

Solution
Equation 39-14 gives the wavelengths:

$$\lambda_{\text{baseball}} = \frac{h}{p} = \frac{h}{mv} = \frac{6.63 \times 10^{-34} \text{ J·s}}{(0.15 \text{ kg})(45 \text{ m/s})} = 9.8 \times 10^{-35} \text{ m}.$$

This is about 34 orders of magnitude smaller than home plate, showing that the baseball's wave properties are totally negligible in the systems with which it interacts. For the electron, on the other hand,

$$\lambda_{\text{electron}} = \frac{h}{mv} = \frac{6.63 \times 10^{-34} \text{ J·s}}{(9.11 \times 10^{-31} \text{ kg})1(1.0 \times 10^6 \text{ m/s})} = 0.73 \text{ nm},$$

several times the size of an atom. Thus we should expect wave effects to be prominent when electrons with this speed interact with atoms.

EXERCISE At what speed would a proton's de Broglie wavelength equal the approximately 0.1-nm diameter of the hydrogen atom?

Answer: 4 km/s

Some problems similar to Example 39-5: 47, 48, 50, 52 ●

■ APPLICATION THE ELECTRON MICROSCOPE

In Chapter 37 we saw that the resolution of an optical system—the minimum separation between two distinguishable points—is on the order of the wavelength of the illuminating light. That limits the resolution of conventional optical microscopes to some hundreds of nanometers. But the wavelengths of electrons can be made much smaller than that of visible light, and therefore **electron microscopes** can achieve resolutions on the order of 1 nm or less. The associated magnifications approach 10^6.

Figure 39-16 shows a simplified diagram of an electron microscope used for high magnification of small objects. An electric field accelerates electrons to energies of $50 - 100$ keV, with corresponding wavelengths on the order of 0.005 nm. The electron beam then enters the microscope's "optical" system, consisting of magnetic fields that focus the electron beam in the way glass lenses focus light in a conventional microscope. Aberration in these lenses, rather than diffraction effects, ultimately limits the resolution of the electron microscope at electron energies above about 100 keV. An object placed in the beam scatters electrons, resulting in darker regions on a screen, film, or other detector (Fig. 39-17a).

Because the electron beam would be scattered by air molecules, the electron microscope must be evacuated. This complicates the preparation of biological samples, which must be desiccated and mounted on substrates that themselves do not significantly scatter the beam. Nevertheless, the electron microscope has become an indispensable tool not only in biology, but also in chemistry and metallurgy.

A related instrument, the scanning electron microscope, affords dramatic three-dimensional imaging at magnifications in the range from about 10 to 10^5 (Fig. 39-17b).

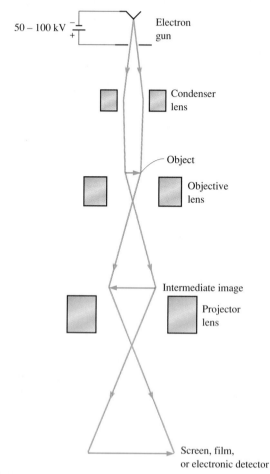

FIGURE 39-16 An electron microscope. The electron gun accelerates electrons to energies of 50–100 keV. The optical system then uses magnetic fields as lenses to focus the beam and form an image.

FIGURE 39-17 (a) Conventional electron micrograph of a tuberculosis bacterium. (b) Scanning electron micrograph showing blood cells in an arteriole.

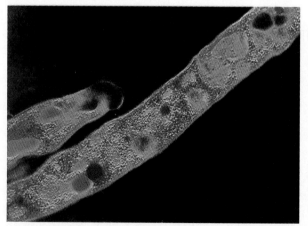

(a)

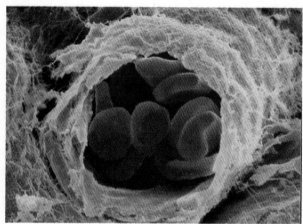

(b)

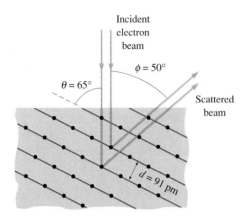

Incident electron beam

$\phi = 50°$

$\theta = 65°$

Scattered beam

$d = 91$ pm

FIGURE 39-18 In reflecting an electron beam off a nickel crystal, Davisson and Germer found maximum reflection at $\theta = 50°$—the angle one would expect for diffraction of waves with the de Broglie wavelength of the incident electrons.

Electron Diffraction

In 1927 the American physicists Clinton J. Davisson and Lester H. Germer gave a convincing experimental verification of de Broglie's matter-wave hypothesis. Davisson and Germer were not actually searching for de Broglie's waves, but were studying the interaction of an electron beam with a nickel crystal. They found that the intensity of the scattered beam peaked when the angle between the incident and scattered beams was 50° (Fig. 39-18). Davisson and Germer noted the similarity between this phenomenon and the peaks observed in x-ray diffraction from the planes of atoms in a crystal, and concluded that they were seeing the results of interference among the electron waves scattered from the nickel atoms. As the example below shows, their measurement also confirmed quantitatively de Broglie's formula for the electron wavelength.

● **EXAMPLE 39-6** ELECTRON DIFFRACTION

The electron energy in Davisson and Germer's experiment was 54 eV, and the spacing between atomic planes in their nickel crystal was 0.091 nm. Show that their detection of a scattered peak at $\phi = 50°$ in Fig. 39-18 is consistent with de Broglie's expression for the electron wavelength (Equation 39-14).

Solution

We need the electron momentum to find the wavelength. Writing $k = \frac{1}{2}mv^2$ and then solving for the momentum gives $p = mv = \sqrt{2mK}$. Then the de Broglie wavelength is

$$\lambda = \frac{h}{p} = \frac{h}{\sqrt{2mK}}$$

$$= \frac{6.63 \times 10^{-34} \text{ J·s}}{\sqrt{(2)(54 \text{ eV})(1.6 \times 10^{-19} \text{ J/eV})(9.11 \times 10^{-31} \text{ kg})}}$$

$$= 1.67 \times 10^{-10} \text{ m} = 0.167 \text{ nm}.$$

If we treat the scattering of the electron waves like the x-ray scattering of Fig. 37-27, we can apply the Bragg condition (Equation 37-8): $2d \sin\theta = m\lambda$. Comparison of Figs. 37-27 and 39-18 shows that the angle θ in the Bragg condition is 65° for Davisson and Germer's experiment. Using the calculated de Broglie wavelength and the 0.091-nm atomic spacing, we expect a first-order diffraction peak at

$$\theta = \sin^{-1}\left(\frac{\lambda}{2d}\right) = \sin^{-1}\left(\frac{0.167 \text{ nm}}{(2)(0.091 \text{ nm})}\right) = 67°,$$

in close agreement with the experimental result that $\theta = 65°$.

EXERCISE Find the angle ϕ in Fig. 39-18 if 100-eV electrons are incident on the same crystal.

Answer: 95°

Some problems similar to Example 39-6: 53, 54 ●

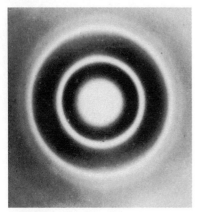

FIGURE 39-19 Diffraction pattern produced by passing an electron beam through a circular aperture shows that electrons have a wave-like character.

Shortly after Davisson and Germer's work, the Scottish physicist George Thomson passed an electron beam through a small aperture and observed electron diffraction directly (Fig. 39-19). Thomson was the son of J. J. Thomson, who had discovered the electron in 1897. In a familial manifestation of the wave-particle duality, father Thomson identified the electron as a particle while his son found it to be a wave. Since Thomson's work, similar diffraction experiments have shown the wave nature of neutrons and other particles.

39-6 THE UNCERTAINTY PRINCIPLE

In classical physics it is, in principle, possible to know the exact position and velocity of a particle and, therefore, to predict with certainty its future behavior. But not so in quantum mechanics! In 1927 the German physicist Werner Heisenberg presented his famous **uncertainty principle,** which states that certain pairs of quantities cannot be measured simultaneously with arbitrary precision. Position and momentum constitute one such pair; if we measure a particle's position to within an uncertainty Δx, then we cannot simultaneously determine its momentum to an accuracy better than Δp, where

$$\Delta x \, \Delta p \geq \hbar.^* \qquad \text{(uncertainty principle)} \qquad (39\text{-}15)$$

Why this limitation? The fundamental reason is quantization. To measure some property of a system requires interacting with that system—for example, shining light on it. That interaction invariably involves energy, and the interaction energy disturbs the system slightly. As a result of this disturbance, values inferred from the measurement are no longer quite the right ones. In classical physics the energy can be made as small as possible, resulting in an arbitrarily small disturbance. But in quantum theory the minimum energy is a single quantum, like a photon of light, and thus, the disturbance cannot be made arbitrarily small.

So why not use lower frequency light, whose photon energy hf is smaller? Because lower frequency means longer wavelength and, as we found in Chapter 37, precise imaging requires light whose wavelength is much shorter than the system being imaged. Heisenberg summarized this dilemma with the "thought experiment" shown in Fig. 39-20. He imagined a "quantum microscope" observing a single photon interacting with an electron. Using a short-wavelength photon allows precise localization of the electron (Fig. 39-20a). But short wavelength means high frequency and thus, through Equation 39-6, high photon energy. The high-energy photon imparts considerable momentum to the electron, and thus, any knowledge we had of the electron's momentum is seriously degraded by the very act that determines its position. We might instead choose a long-wavelength, low-energy photon, as in Fig. 39-20b. Then there is little disturbance, and we can get an accurate assessment of the electron's momentum. But now the photon wavelength is so long that diffraction effects severely limit our ability to tell just where the electron was. So we can measure the

---

* The right-hand side of this inequality is approximate; the important point is that the product cannot exceed a quantity on the order of $\hbar$. The absolute minimum value for the uncertainty is $\frac{1}{2}\hbar$ and occurs only when the waveform of the particle's de Broglie wave has a certain shape.

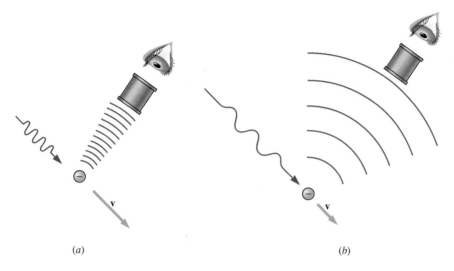

FIGURE 39-20 Heisenberg's "quantum microscope." (a) A short-wavelength, high-energy photon interacts with an electron. Diffraction is minimal, and the electron's position is accurately determined. But the high-energy photon disturbs the electron's trajectory, leaving its momentum very uncertain. (b) Using a long-wavelength, low-energy photon minimizes the disturbance and allows accurate determination of the electron's momentum. But now diffraction is significant, and the measurement does not give an accurate position.

(a)   (b)

● **EXAMPLE 39-7** MICROELECTRONICS

A beam of aluminum atoms is used to "dope" part of a semiconductor chip to give it the proper electrical characteristics. If the electron velocity is required to be $(4.5000 \pm 0.00001) \times 10^4$ m/s, how accurately can the electrons be localized?

Solution
With a velocity spread of $\pm 0.00001 \times 10^4$ m/s, the maximum tolerable uncertainty in the atoms' velocity is $\Delta v = 0.00002 \times 10^4$ m/s, or 0.20 m/s. Then corresponding momentum uncertainty is

$$\Delta p = m\Delta v = (26.98 \text{ u})(1.66 \times 10^{-27} \text{ kg/u})(0.20 \text{ m/s})$$

$$= 8.96 \times 10^{-27} \text{ kg·m/s},$$

where we found the atomic weight of aluminum in Appendix D. Equation 39-15 then gives the uncertainty in position:

$$\Delta x \geq \frac{\hbar}{\Delta p} = \left(\frac{1.055 \times 10^{-34} \text{ J·s}}{8.96 \times 10^{-27} \text{ kg·m/s}} \right) = 12 \text{ nm}.$$

This is about 100 atomic diameters and shows that the uncertainty principle will ultimately limit our ability to manufacture very small microelectronic structures by the means now used.

EXERCISE You measure the velocity of a 3.0-g ping-pong ball to an accuracy of ± 0.01 mm/s. Find the minimum uncertainty in its position, and compare with the size of the ball.

Answer: 1.8×10^{-27} m, totally negligible compared with the ball's size

Some problems similar to Example 39-7: 55, 57, 58 ●

electron's position accurately, but only at the expense of very imprecise knowledge of its momentum. Or we can measure its momentum accurately, but then we'll know very little about its position. With a photon of intermediate wavelength we could measure both quantities, but neither with perfect accuracy. Equation 39-15 establishes the minimum uncertainties possible.

The uncertainty principle is intimately connected with de Broglie's wave hypothesis. Suppose we pass an electron beam through a slit, as shown in Fig. 39-21. Then we know the vertical component of the electrons' position to within the slit width. But the electrons have a wave-like aspect, and they diffract in passing through the slit. If the slit width is much greater than the wavelength, then, as we found in Chapter 37, the diffracted beam does not spread much (Fig. 39-21a). We can therefore be quite sure of the electrons' vertical momentum; in this case it is very close to zero. But the slit is wide, so there is considerable

FIGURE 39-21 (*a*) An electron beam passing through a wide slit shows little diffraction, and therefore the vertical component of an electron's momentum is known to be very nearly zero. With a wide slit, though, an electron's position is not well known. (*b*) A narrow slit localizes the electron, increasing the accuracy with which its vertical position is known. But diffraction increases the uncertainty in its vertical momentum.

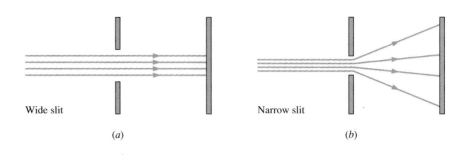

Wide slit   (*a*)   Narrow slit   (*b*)

uncertainty in the vertical position. We can find the position more accurately by making the slit smaller, but then the diffracted beam widens and the uncertainty in the vertical momentum increases. So the wave nature of matter imposes a tradeoff: the more accurately we know the position of a particle, the less accurately can we know its momentum, and vice versa.

It sounds like the uncertainty principle only diminishes our knowledge. But actually, the principle provides useful estimates of the size and energy scales in atomic and subatomic systems, as the example below illustrates.

● EXAMPLE 39-8 ATOMIC AND NUCLEAR ENERGIES

Use the uncertainty principle to estimate the minimum energy possible for (a) an electron confined to a one-dimensional region of atomic dimensions (about 0.1 nm) and (b) a proton confined to a region of nuclear dimensions (about 1 fm).

Solution
The key to making these estimates is to understand that momentum is a vector quantity. We could know its exact magnitude and still be uncertain about it if we didn't know its direction. In this example we have particles confined in one-dimensional regions. The uncertainty in position is at most the width Δx of the region. The particles can't be at rest because then we would know their exact momentum—zero—in violation of the uncertainty principle. The minimum momentum p they could have, in fact, is such that the uncertainty associated with not knowing which way the particles are going satisfies the uncertainty principle. Not knowing the direction amounts to being uncertain by the difference between the momentum in one direction and the momentum in the other, or $\Delta p = p - (-p) = 2p$. Then the uncertainty principle (Equation 39-15) gives $2p\Delta x \geq \hbar$, or

$$p \geq \frac{\hbar}{2\Delta x}.$$

Using $p = mv$ and $K = \frac{1}{2}mv^2$, we can write the kinetic energy as $K = p^2/2m$, and therefore, we must have

$$K \geq \frac{1}{2m}\left(\frac{\hbar}{2\Delta x}\right)^2.$$

Using the electron mass for m and $\Delta x = 0.1$ nm then gives a minimum energy of about 1 eV, roughly the energy scale for atomic energy levels. Using the proton mass and the nuclear size gives a minimum energy of 5 MeV. The ratio of these energies, a factor of about 10^7, indicates the dramatic difference between chemical energy sources—that ultimately involve rearrangement of atomic electrons—and nuclear sources that disrupt the nucleus itself.

EXERCISE An electron is somewhere in a molecule 15 nm long. What is its minimum kinetic energy?

Answer: 42 μeV

Some problems similar to Example 39-8: 60, 84 ●

Energy-Time Uncertainty

A second pair of variables that defy simultaneous accurate measurement are the energy of a system and the time it takes to measure that energy. The uncertainty ΔE in the energy is related to the measurement time Δt through the inequality

$$\Delta E \, \Delta t \geq \hbar. \qquad (39\text{-}16)$$

One effect of energy-time uncertainty is to render atomic and nuclear energy levels inexact and, therefore, to broaden spectral lines. If an atom were forever in a fixed energy state, we could take infinitely long to measure its energy and therefore make ΔE arbitrarily small. But excited states of atoms have characteristic lifetimes (typically $\sim 10^{-8}$ s), which limits the measurement time and therefore sets a minimum uncertainty in the energy level. Problems 64 and 65 explore energy-time uncertainty.

■ APPLICATION THE MÖSSBAUER EFFECT

In most atomic systems, the Doppler shift and other effects of the atoms' thermal motion broaden spectral lines far beyond the minimum width set by the uncertainty principle. But in 1957 Rudolf L. Mössbauer discovered that gamma rays emitted by nuclei tightly bound in solids could produce line widths limited in sharpness only by the uncertainty principle. The reason is that the entire solid—some 10^{20} atoms or so—absorbs the recoil momentum associated with the gamma-ray emission, and therefore, any change in gamma-ray energy due to the recoiling nucleus is negligible. One of the sharpest energy transitions achieved in the Mössbauer effect is with the isotope tantalum-181, which emits a 6.2-keV gamma ray with a lifetime of 6.8 μs. Applying Equation 39-16 with $\Delta t = 6.8$ μs then gives $\Delta E = 10^{-10}$ eV, or about 1 part in 10^{14} of the 6.2-keV gamma-ray energy.

In a typical Mössbauer experiment, a gamma-ray source is set into slow oscillatory motion relative to a sample under study. Because of the sharpness of the gamma-ray spectrum, the sample absorbs gamma rays only when the Doppler shift arising from the source motion puts the gamma-ray wavelength, and therefore, the corresponding energy, at a level appropriate to a transition between energy levels in the sample material. The sharpness of the gamma-ray spectra allows very accurate discrimination of different energy levels.

Mössbauer spectroscopy is widely used throughout the sciences. Nuclear physicists measure properties of atomic nuclei, while solid-state physicists study magnetic interactions in alloys. Chemists use Mössbauer spectroscopy to discern the arrangements of electrons in atoms, and biologists study how iron is incorporated into hemoglobin molecules. Archaeologists have even used the Mössbauer effect to trace the geographical distribution of ancient pottery.

Mössbauer spectroscopy is so sensitive that physicists R. V. Pound and G. Rebka of Harvard University used it in 1960 to confirm one aspect of Einstein's general theory of relativity, by measuring the so-called gravitational redshift of gamma-ray photons as they traversed a 23-m-high vertical shaft. By varying the source speed until a fixed target of the same material absorbed the redshifted gamma rays, they were able to measure the value of the redshift and thus confirmed Einstein's prediction.

Observers, Uncertainty, and Causality

The uncertainty principle moves the observer from a passive onlooker to an active participant in physical events. To observe is necessarily to disturb, and quantum theory is therefore very much concerned with the role of the observer and the process of measurement. The uncertainty principle is fundamentally a statement about what can and cannot be learned through measurement.

Position and momentum cannot be measured simultaneously with perfect accuracy. Surely, though, a particle has well-defined values of both, even though we cannot know them? The answer is apparently no. The standard interpretation of quantum mechanics suggests that it makes no sense to talk about what cannot be measured, and recent experiments have ruled out "hidden variables" that might be active at a lower level to guide the particle in a deterministic path. Its wave aspect makes a particle a "fuzzy" thing, and it really makes no sense to think of the particle as a tiny ball with definite momentum and position. For that reason it also makes no sense to think of the particle's future as being fully determined in the sense that Newton's laws determine the future path of, say, a baseball. We are left with uncertainty—or indeterminacy, as Heisenberg's word also translates—as a fundamental fact of our universe.

39-7 THE WAVE-PARTICLE DUALITY, COMPLEMENTARITY, AND THE CORRESPONDENCE PRINCIPLE

One of the most disturbing aspects of quantum theory is the wave-particle duality—the seemingly contradictory fact that matter and light have both wave-like and particle-like properties. If this bothers you, you're in good company: it bothered the early quantum physicists, too (Fig. 39-22). Werner Heisenberg, who first stated the uncertainty principle, described his frustration in trying to understand the quantum world:

> I remember discussions with Bohr which went through many hours till very late at night and ended almost in despair; and when at the end of the discussion I went alone for a walk in the neighboring park I repeated to myself again and again the question: Can nature possibly be as absurd as it seems to us in these atomic experiments?*

FIGURE 39-22 Niels Bohr, Werner Heisenberg, and Wolfgang Pauli in an intense discussion of quantum mechanics.

Bohr dealt with the wave-particle duality through his **principle of complementarity.** The wave and particle pictures, he said, are complementary aspects of the same reality. If we do an experiment to measure a wave-like property—for example, the diffraction of electrons—then we find wave properties but not particle properties. If we do an experiment to measure a particle-like property—for example, localizing an electron—then we will not find wave properties. The two measurements require different experiments, and we can't perform both simultaneously on the same entity. So we will never catch wave and particle in an outright contradiction, and the answer to the question "Which is it, wave or particle?" has to be that it's both, and that which you find depends on what experiment you choose to perform.

Bohr articulated a second principle that helps reconcile the seeming contradiction between classical and quantum physics. His **correspondence principle** states that the predictions of classical and quantum physics should agree in situations where the size of individual quanta is negligible. Taking $h \rightarrow 0$ in Planck's law, for example, gives the classical Rayleigh-Jeans law (see Problem 13). Or, for large values of n, the energies of adjacent atomic states in the Bohr model become so close that the levels appear essentially as a continuum of allowed energies—as expected in classical physics. Or consider a 1000-W radio beam; the photon energy hf is so small that the beam contains an enormous number of photons, and we can consider the energy distributed essentially continuously over the beam. But in a 1000-W x-ray beam, the photon energy is much higher and the number of photons correspondingly fewer; it's therefore difficult to avoid the fundamental fact of energy quantization in this beam. Visible light lies somewhere in between; we can often treat its energy as being continuously distributed, except when it interacts with systems as small as individual atoms.

---

* Heisenberg, Werner. *Physics and Philosophy: The Revolution in Modern Science* (New York: Harper & Brothers, 1962).

CHAPTER SYNOPSIS

Summary

1. **Quantum physics** describes matter and energy on the small scale. The essential difference between quantum and classical physics is that some physical quantities in quantum physics come only in certain discrete values, or quanta.

2. The wavelength distribution of **blackbody radiation** could not be explained using classical physics. Only by introducing the notion that the energies of vibrating molecules are quantized in units of hf could Planck give a mathematical description of the blackbody spectrum. The constant h, called **Planck's constant**, sets the fundamental scale for quantum phenomena.

3. Einstein's explanation of the **photoelectric effect** introduced quantization of electromagnetic wave energy in the form of **photons** of energy hf, where f is the wave frequency. Einstein's proposal gave the first hint of the **wave-particle duality.**

4. Experiments with the **Compton effect** reinforced the view that light has a particle-like aspect, by demonstrating that a photon and electron undergo particle-like collisions from which the photon emerges with lower energy and longer wavelength.

5. Light from glowing gases shows discrete **spectral lines** whose wavelength, for hydrogen, follows the simple formula

$$\frac{1}{\lambda} = R_H\left(\frac{1}{n_2^2} - \frac{1}{n_1^2}\right),$$

where n_1 and n_2 are any integers with $n_1 > n_2$. Bohr was able to explain the hydrogen spectrum by postulating that electrons circle the nucleus in only certain allowed orbits whose angular momentum mvr is quantized according to the relation $mvr = nh/2\pi$, where n is an integer. Bohr's theory shows that the resulting electron energies are given by

$$E = -\frac{13.6 \text{ eV}}{n^2},$$

and that the radius of the electron orbit is

$$r = a_0 n^2,$$

where a_0 is the **Bohr radius** given by $a_0 = \hbar^2/mke^2 = 0.0529$ nm. The **Franck-Hertz experiment** helped confirm Bohr's theory by showing that atoms can absorb only certain discrete amounts of energy.

6. De Broglie proposed that particles of matter have a wave-like aspect, with their wavelength given by

$$\lambda = \frac{h}{p},$$

where p is the particle's momentum. Bohr's quantization condition for atomic electron orbits follows under the assumption that the allowed orbits are those that can accommodate an integer number of de Broglie electron wavelengths. Experiments by Davisson and Germer and by G. Thomson demonstrated wave interference and diffraction effects in electron beams.

7. Heisenberg's **uncertainty principle** states that it is impossible to measure simultaneously and with perfect accuracy both the position and momentum of a particle; in particular,

$$\Delta p \, \Delta x \geq \hbar,$$

where Δp and Δx are the uncertainties in position and momentum, respectively, and where $\hbar = h/2\pi$. Similarly, it is impossible to measure energy with perfect accuracy in a finite time.

8. Bohr's **principle of complementarity** reconciles the seeming contradiction inherent in the wave-particle duality. The principle states that wave and particle are complementary aspects of reality, and that any given experiment detects one aspect or the other, but never both simultaneously. Bohr's **correspondence principle** states that quantum and classical physics must agree in situations where individual quanta are negligibly small.

Terms You Should Understand

(Pairs are closely related terms whose distinction is important; number in parentheses is chapter section where term first appears.)

classical physics, quantum physics (introduction)
blackbody radiation (39-2)
Planck's constant (39-2)
ultraviolet catastrophe (39-2)
photoelectric effect (39-3)
wave-particle duality (39-3, 39-7)
spectral line (39-4)
Bohr atom (39-4)
ground state, excited state (39-4)
de Broglie wavelength (39-5)
uncertainty principle (39-6)
principle of complementarity (39-7)
correspondence principle (39-7)

Symbols You Should Recognize

h, $\hbar$ (39-2, 39-4)
λ_c (39-3)
a_0 (39-4)
Δp, Δx, ΔE, Δt (39-6)

Problems You Should Be Able to Solve

evaluating radiance of blackbodies (39-2)

solving for electron energies and other quantities in the photoelectric effect (39-3)

analyzing Compton scattering (39-3)

evaluating energy levels and orbital radii in the Bohr atom (39-4)

finding spectral lines for hydrogen and hydrogen-like atoms (39-4)

evaluating de Broglie wavelengths (39-5)

applying the uncertainty principle (39-6)

Limitations to Keep in Mind

The theories outlined in this chapter represent first attempts at explaining quantum phenomena, and are neither fully accurate nor universally applicable.

QUESTIONS

1. Can you explain the workings of the telephone using classical physics? What about the electron microscope? Why or why not?
2. Why does classical physics predict that atoms should collapse?
3. Looking at the night sky, you see one star that appears reddish, another yellowish, and another bluish. Compare their temperatures.
4. Any object warmer than absolute zero radiates energy. So why can't you see warm objects in a dark room?
5. Imagine an atom that, unlike hydrogen, had only three energy levels. If these levels were evenly spaced, how many spectral lines would result? How would their wavelengths compare?
6. What colors of visible light have the highest energy photons?
7. Why is the immediate ejection of electrons in the photoelectric effect surprising from a classical viewpoint?
8. Suppose the Compton effect were significant at radio wavelengths. What problems might this present for radio and TV broadcasting?
9. How are the wave-particle duality and the uncertainty principle related?
10. Why is the Compton effect seen only with x rays and gamma rays?
11. For what photon scattering angle θ will the electron in a Compton scattering event gain the most energy?
12. How many spectral lines are in the entire Balmer series?
13. Why are the lines of the Lyman series in the ultraviolet while some Balmer lines are visible?
14. Why does the photoelectric effect suggest that light has particle-like properties?
15. Energy-time uncertainty limits the precision with which we can know the mass of unstable particles (those that decay after a finite time). Why?
16. Why is the Compton effect a particularly convincing demonstration that electromagnetic radiation has particle-like properties?
17. Equation 39-14 suggests that a particle that's not moving should have infinite wavelength. Speculate on what this might mean.
18. If you measure a particle's position with perfect accuracy, what do you know about its momentum?
19. How might our everyday experience be different if Planck's constant had the value 1 J·s?
20. Why is the energy given by Equation 39-12 negative?

PROBLEMS

Section 39-2 Blackbody Radiation

Note: Most answers in this section are very sensitive to the exact values used for the constants h and c.

1. If you increase the temperature of a blackbody by a factor of 2, by what factor does its radiated power increase?
2. The surface temperature of the star Rigel is 10^4 K. (a) What is the power per square meter radiated by Rigel? (b) At what wavelength is the maximum power radiated? In what spectral region is this?
3. At what wavelength does Earth, approximated as a 286-K blackbody, radiate the most energy?
4. Spacecraft instruments measure the radiation from a passing asteroid and find that it peaks at 40 μm wavelength. Assuming the asteroid is a blackbody, what is its surface temperature?
5. According to Planck's theory, what is the minimum nonzero energy of a molecule with vibration frequency 3.4×10^{14} Hz?
6. An oscillating molecule drops from one allowed energy level to the next lower one. If its energy changes by 2.2 eV, what is its oscillation frequency?
7. The Sun approximates a blackbody at 5800 K. Find the wavelength at which the Sun emits the most energy.

8. What is the power per unit area emitted by a 3000-K incandescent lamp filament in the wavelength interval from 500 nm to 502 nm? *Hint:* The radiance given by Equation 39-3 is the power per unit area *per unit wavelength interval.*

9. Find the temperature range over which a blackbody will radiate the most energy at a wavelength in the visible range (400 nm–700 nm).

10. By what factors does the radiance of a blackbody at 600 nm increase as the temperature is increased from 1000 K to (a) 1500 K and (b) 2000 K?

11. Treating the Sun as a 5.8-kK blackbody, how does its ultraviolet radiance at 200 nm compare with its visible radiance at its peak wavelength of 500 nm?

12. For a 2.0-kK blackbody, by what percentage is the Rayleigh-Jeans law in error at wavelengths of (a) 1.0 mm, (b) 10 μm, and (c) 1.0 μm?

13. Use the series expansion for e^x (Appendix A) to show that Planck's law (Equation 39-3) reduces to the Rayleigh-Jeans law (Equation 39-5) when $\lambda \gg hc/kT$.

Section 39-3 Photons

14. Find the energy in electron-volts of (a) a 1.0-MHz radio photon, (b) a 5.0×10^{14}-Hz optical photon, and (c) a 3.0×10^{18}-Hz x-ray photon.

15. What is the wavelength of a 6.5-eV photon? In what spectral region is this?

16. A microwave oven uses electromagnetic radiation at 2.4 GHz. (a) What is the energy of each microwave photon? (b) At what rate does a 625-W oven produce photons?

17. A red laser at 650 nm and a blue laser at 450 nm emit photons at the same rate. How do their total power outputs compare?

18. (a) Find the Compton wavelength for the proton. (b) Find the energy, in electron-volts, of a gamma ray whose wavelength equals the proton's Compton wavelength.

19. Find the rate of photon production by (a) a radio antenna broadcasting 1.0 kW at 89.5 MHz, (b) a laser producing 1.0 mW of 633-nm light, and (c) an x-ray machine producing 0.10-nm x rays with a total power of 2.5 kW.

20. What is the work function of a material whose photoelectric cutoff wavelength is 390 nm?

21. Electrons in a photoelectric experiment emerge from an aluminum surface with maximum kinetic energy of 1.3 eV. What is the wavelength of the illuminating radiation?

22. What is the maximum work function for a surface to emit electrons when illuminated with 900-nm infrared light?

23. (a) Find the cutoff frequency for the photoelectric effect in copper. (b) Find the maximum energy of the ejected electrons if the copper is illuminated with light of frequency 1.8×10^{15} Hz.

24. The stopping potential in a photoelectric experiment is 1.8 V when the illuminating radiation has wavelength 365 nm.

(a) What is the work function of the emitting surface? (b) What would be the stopping potential for 280-nm radiation?

25. Which materials in Table 39-1 exhibit the photoelectric effect *only* for wavelengths shorter than 275 nm?

26. What is the initial wavelength of a photon that loses half its energy when it Compton scatters from an electron and emerges at 90° to its initial direction of motion?

27. A photon with wavelength 15 pm Compton scatters off an electron, its direction of motion changing by 110°. What fraction of the photon's initial energy is lost to the electron?

28. When light shines on a potassium surface, the maximum speed with which electrons leave the surface is 4.2×10^5 m/s. Find the wavelength of the light.

29. The maximum electron energy in a photoelectric experiment is 2.8 eV. When the wavelength of the illuminating radiation is increased by 50%, the maximum electron energy drops to 1.1 eV. Find (a) the work function of the emitting surface and (b) the original wavelength.

30. A 150-pm x-ray photon undergoes Compton scattering with an electron, and emerges at 135° to its original direction of motion. Find (a) the wavelength of the scattered photon and (b) the kinetic energy of the electron.

31. An electron is initially at rest. What will be its kinetic energy after a 0.10-nm x-ray photon scatters from it at 90° to its original direction of motion?

32. Find the scattering angle for which a 7.5-MeV photon undergoing Compton scattering with an electron loses half its energy.

33. What is the minimum photon energy for which it is possible for the photon to lose half its energy undergoing Compton scattering with an electron?

Section 39-4 Atomic Spectra and the Bohr Atom

34. Calculate the wavelengths of the first three lines in the Lyman series for hydrogen.

35. Which spectral line of the hydrogen Paschen series ($n_2 = 3$) has wavelength 1282 nm?

36. The wavelengths of a spectral line series tend to a limit as $n_1 \to \infty$. Evaluate the series limit for (a) the Lyman series and (b) the Balmer series in hydrogen.

37. What is the maximum wavelength of light that can ionize hydrogen in its ground state? In what spectral region is this?

38. A Rydberg hydrogen atom makes a downward transition to the $n = 225$ state. If the photon emitted has energy 9.32 μeV, what was the original state?

39. What was the wavelength of light emitted as the electrons excited in the Franck-Hertz experiment dropped back to their ground states?

40. A hydrogen atom is in its ground state when its electron absorbs 48 eV in an interaction with a photon. What is the energy of the resulting free electron?

41. At what energy level does the Bohr hydrogen atom have diameter 5.18 nm?

42. How much energy does it take to ionize a hydrogen atom in its first excited state?

43. Ultraviolet light with wavelength 75 nm shines on a gas of hydrogen atoms in their ground states, ionizing some of the atoms. What is the energy of the electrons freed in this process?

44. A hydrogen atom in its ground state absorbs a 12.75-eV photon and ends up in an excited state. It then drops to the next lower state. What is the energy of the photon emitted in this downward transition?

45. Helium with one of its two electrons removed acts very much like hydrogen, and the Bohr model successfully describes it. Find (a) the radius of the ground-state electron orbit and (b) the photon energy emitted in a transition from the $n = 2$ to the $n = 1$ state in this singly ionized helium. *Hint:* The nuclear charge is $2e$.

46. What is the radius of the first Bohr orbit in O^{7+} (i.e., oxygen with 7 of its 8 electrons removed)?

Section 39-5 Matter Waves

47. Find the de Broglie wavelength of (a) Earth, in its 30-km/s orbital motion, and (b) an electron moving at 10 km/s.

48. How slowly must an electron be moving for its de Broglie wavelength to equal 1 mm?

49. A proton and an electron have the same de Broglie wavelength. How do their speeds compare, assuming both are much less than that of light?

50. Find the de Broglie wavelength of electrons with kinetic energies of (a) 10 eV, (b) 1.0 keV, and (c) 10 keV.

51. Through what potential difference should you accelerate an electron from rest so its de Broglie wavelength will be the size of a hydrogen atom, about 0.1 nm?

52. What is the minimum electron speed that would make an electron microscope superior to an optical microscope using 450-nm light?

53. A Davisson-Germer type experiment using nickel gives peak intensity in the reflected beam when the incident and reflected beams have an angular separation of 100°. What is the electron energy?

54. A beam of 75-eV electrons impinges on an unknown crystal, producing a first-order diffraction peak at 65° from the incident beam. What is the spacing of the atomic planes in the crystal?

Section 39-6 The Uncertainty Principle

55. A proton is confined to a space 1 fm wide (about the size of the atomic nucleus). What is the minimum uncertainty in its velocity?

56. Is it possible to measure an electron's velocity to an accuracy of ± 1.0 m/s while simultaneously finding its position to an accuracy of ± 1.0 μm? What about a proton?

57. A proton has velocity $\mathbf{v} = (1500 \pm 0.25)\hat{\mathbf{i}}$ m/s. What is the uncertainty in its position?

58. An electron is moving in the $+x$ direction with speed measured at 5.0×10^7 m/s, to an accuracy of $\pm 10\%$. What is the minimum uncertainty in its position?

59. Find the minimum energy for a neutron in a uranium nucleus whose diameter is 15 fm.

60. An electron is trapped in a "quantum well" 20 nm wide. What is the minimum speed it could have?

61. A proton is moving along the x axis with speed $v = (1500 \pm 0.25)$ m/s, but its direction ($+$ or $-x$) is unknown. Find the uncertainty in its position. *Hint:* What is the difference between the two extreme possibilities for the *velocity?*

62. An electron beam is accelerated at the back of a TV tube and then heads toward the center of the screen with horizontal velocity 2.2×10^7 m/s. As the electrons leave the acceleration region, their vertical position is known to within ± 45 nm. Find the minimum angular spread in the beam, as set by the uncertainty principle.

63. An electron is moving at 10^6 m/s, and you wish to measure its energy to an accuracy of $\pm 0.01\%$. What is the minimum time necessary for the measurement?

64. A cosmic-ray particle interacts with an energy-measuring device for a mere 12 fs. What is the minimum uncertainty in its energy?

65. The lifetimes of unstable particles set energy-time uncertainty limits on the accuracy with which the rest energies—and hence the masses—of those particles can be known. The particle known as the neutral pion has rest energy 135 MeV and lifetime 8×10^{-17} s. Find the uncertainty in its rest energy (a) in eV and (b) as a fraction of the rest energy.

66. An electron has momentum of magnitude p, but the direction is unknown. (a) Argue that the uncertainty in its momentum is therefore $2p$. (b) Compare the minimum uncertainty in its position with its de Broglie wavelength. Why might the two be related?

Paired Problems

(Both problems in a pair involve the same principles and techniques. If you can get the first problem, you should be able to solve the second one.)

67. (a) Find the wavelength at which a 2200-K blackbody has its peak radiance. (b) How does the radiance for 690-nm red light compare with the peak radiance?

68. The radiance of a blackbody peaks at 660 nm. (a) What is its temperature? (b) How does the radiance at 400 nm compare with that at 700 nm?

69. A photocathode ejects electrons with maximum energy 0.85 eV when illuminated with 430-nm blue light. Will it eject electrons when illuminated with 633-nm red light, and if so what will be the maximum electron energy?

70. A photocathode ejects electrons with maximum energy 2.8 eV when illuminated with 360-nm ultraviolet light.

Will it eject electrons when illuminated with 840-nm infrared light, and if so what will be the maximum electron energy?

71. Find the initial wavelength of a photon that loses 20% of its energy when it Compton scatters from an electron through an angle of 90°.

72. A photon Compton scatters from an electron, emerging at 140° to its initial direction and with an energy of 85 keV. Find the photon's initial energy.

73. (a) Find the energy of the highest energy photon that can be emitted as the electron jumps between two adjacent allowed energy levels in the Bohr hydrogen atom. (b) Which energy levels are involved?

74. Find (a) the wavelength and (b) the energy in eV of the photon emitted as a Rydberg hydrogen atom drops from the $n = 180$ level to the $n = 179$ level.

75. In which of the following will wave properties be more evident: an electron moving at 10 Mm/s or a proton moving at 1 km/s?

76. Give the ratio of de Broglie wavelength to diameter for (a) a 100-g ball, 10 cm in diameter, moving at 5 m/s; (b) a 4-u alpha particle, 3.4 fm in diameter, moving at 20 Mm/s; and (c) a 238-u uranium nucleus, 13 fm in diameter, moving at 1.4 km/s. (d) For which will wave properties be negligible, and for which will they be completely dominant?

77. An electron is known to be within about 0.1 nm of the nucleus of an atom. What is the uncertainty in its velocity?

78 .Typically, an atom remains in an excited state for about 10^{-8} s before it drops to a lower energy state, emitting a photon in the process. What is the uncertainty in the energy of this transition?

Supplementary Problems

79. An electron is accelerated from rest through a 4.5-kV potential difference. What is its de Broglie wavelength?

80. A photomultiplier consists of a photocathode, 8 dynodes, and an anode that collects the electrons from the final dynode. The photocathode's active surface is a circle 2.5 cm in diameter, and its quantum efficiency is 7.5% for light with wavelength 633 nm. Each dynode releases 3 electrons for each electron incident on it. Find the current at the anode if the cathode is illuminated with 633-nm light with intensity 6.5 mW/m².

81. A photon's wavelength is equal to the Compton wavelength of a particle with mass m. Show that the photon's energy is equal to the particle's rest energy.

82. Show that the frequency range occupied by the hydrogen spectral line series involving transitions ending at the nth level is given by $\Delta f = cR_H/(n + 1)^2$.

83. A photon with initial energy E_0 undergoes Compton scattering with an initially stationary electron, and the scattered photon emerges at an angle θ to its initial direction. Use the conservation of relativistic momentum and energy to show that the electron's recoil angle ϕ, shown in Fig. 39-9b, can be written

$$\tan \phi = \frac{\cot(\theta/2)}{1 + E_0/m_ec^2}.$$

Hint: A photon of energy E carries momentum E/c; don't forget that the photon's energy changes as a result of the scattering.

84. (a) Show that the energy required for an electron to escape from the surface of a nucleus of radius R and charge Ze is kZe^2/R, where k is the Coulomb constant. Evaluate for the oxygen nucleus, with $Z = 8$ and $R = 3$ fm. (b) Use the uncertainty principle to find the minimum energy for an electron confined to this nucleus. (You'll need to use the relativistic energy-momentum relation, which you can simplify to $E = pc$ because the energy is much higher than the rest energy.) (c) Compare your answers to (a) and (b) to show that your nucleus cannot contain an electron.

85. A photon undergoes a 90° Compton scattering off a stationary electron, and the electron emerges with *total* energy γm_ec^2, where γ is the relativistic factor introduced in Chapter 38. Find an expression for the initial photon energy.

86. Show that Wien's displacement law follows from Planck's law (Equation 39-3). *Hint:* Differentiate Planck's law with respect to wavelength.

87. The total power per unit area emitted by a blackbody is given by integrating the power per area per unit wavelength (Equation 39-3) over all wavelengths. Carry out this integration to show that the radiated power per unit area is proportional to the fourth power of the temperature. Show that your result also implies that the Stefan-Boltzmann constant is given by $\sigma = 2\pi^5k^4/15c^2h^3$. *Hint:* Use the quantity $u = hc/\lambda kT$ as your integration variable, and consult a table of definite integrals.

QUANTUM MECHANICS

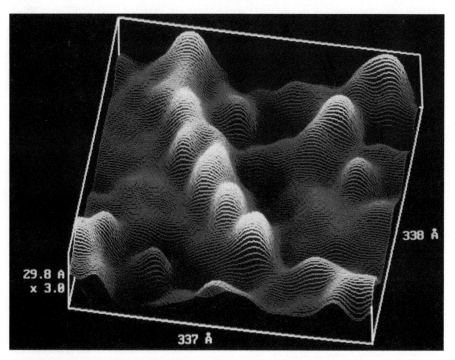

29.8 Å
x 3.0

338 Å

337 Å

Scanning tunneling microscope image of a portion of a DNA molecule shows a region only 34 nm on a side. This new type of microscope exploits the phenomenon of quantum tunneling, made possible by the wave-like aspect of matter.

The ideas developed in the preceding chapter comprise the core of what is called the **old quantum theory.** The old quantum theory introduced the basic concepts of quantum physics, and it was remarkably successful in explaining a number of quantum phenomena—for example, blackbody radiation, the photoelectric effect, and the hydrogen spectrum. On the other hand, it could not treat even the simplest of multielectron atoms, and it left some subtle spectral features unexplained. Furthermore, the old quantum theory was a hodge-podge of separate but loosely related propositions, each developed to explain a particular phenomenon; it lacked coherence and clear guiding principles.

Is there a more coherent theory that predicts the behavior of systems at the atomic and subatomic scales, and that offers a satisfying description of how such systems really work? The answer is at once an emphatic yes and a disappointing no. Yes, because the theory of **quantum mechanics** developed in the 1920s predicts with remarkable precision many of the observed properties of atomic systems, including their energies, the wavelengths of spectral lines they emit, and even the lifetimes of excited atoms. No, because quantum mechanics cannot give us a satisfying visual *picture* of the atomic and subatomic world. The uncertainty principle and wave-particle duality are essential aspects of quantum mechanics; indeed, they can be taken as postulates on which the theory is built. So quantum mechanics will never give a Newtonian description in which positions and momenta of particles are known precisely and their behavior therefore determined. Any picture we formulate in our minds of electrons or photons whizzing around like miniature baseballs is simply inappropriate at the atomic scale. But quantum mechanics does provide a self-consistent mathematical description that lets us explore and predict the behavior of atoms, the organization of chemical elements, the physics of semiconductors and superconductors, the extraordinary behavior of matter at low temperature, the formation of white dwarf stars, the operation of lasers, and a host of other phenomena that are crucial in our technological world and for which classical physics is at best inaccurate and at worst totally inappropriate. In this chapter we explore the mathematical structure and physical interpretation of quantum mechanics. In Chapters 41 and 42, we will apply quantum mechanics first to the atom and then to more complex systems like semiconductor materials that involve the quantum-mechanical interaction of many atoms.

40-1 PARTICLES, WAVES, AND PROBABILITY

Photons and Light Waves

In Maxwell's electromagnetic theory (Chapter 34) we had a seemingly complete description of light as an electromagnetic wave. Now we find, through the photoelectric and Compton effects, that light sometimes manifests itself as particles. What is the connection between the wave and particle descriptions?

In a photoelectric experiment, the *energy* of the ejected electrons depends on the wavelength of the light, not on its intensity. But the *number* of ejected electrons does depend on the intensity. Since an electron is ejected when it absorbs a photon, we conclude that the number of photons in the incident light is proportional to the light intensity. Now, the intensity of an electromagnetic wave depends on the square of the electric or magnetic field (recall Equations 34-20b, c). The fields, in turn, obey Maxwell's equations, so one aspect of a photoelectric experiment—namely, the number of electrons ejected—is related to Maxwell's description of light as an electromagnetic wave.

Can we quantify further the relation between wave intensity and number of photons? Yes, but only in a statistical sense. The ejection of individual electrons in a photoelectric experiment is quite random. The uncertainty principle prevents us from following a photon trajectory and therefore from predicting when and where an electron will be ejected. All we can say is that electrons are more likely to be ejected where the wave intensity is greater. Specifically, the probability that an electron will be ejected is directly proportional to the intensity of the incident electromagnetic waves—that is, to the square of the electric or magnetic field. More generally, the probability of finding a photon in a beam of electromagnetic waves is directly proportional to the wave intensity (Fig. 40-1).

In this quantum-mechanical description, the fields still evolve according to Maxwell's equations. For example, the fields of an electromagnetic wave undergo-

Light wave

Photon beam

FIGURE 40-1 The probability of finding a photon is proportional to the intensity of the electromagnetic wave. Here we show an electromagnetic wave whose intensity varies in space and the corresponding distribution of photons. Actually, the picture itself is only suggestive, for by localizing the individual photons we would have lost any knowledge of their wave properties. It is better to think of the wave intensity as determining the probability of finding a photon *when you look for it*. In the standard interpretation of quantum mechanics, it makes no sense to ask where the photons are, or even whether they are there at all, when you're not trying to detect them.

FIGURE 40-2 Computer simulation showing the development of a two-slit interference pattern from random photon events. The probability of a photon landing at a particular point is proportional to the electromagnetic wave intensity. The statistical pattern becomes evident only with a large number of photons. (*a*) 50 photons; (*b*) 250 photons; (*c*) 1000 photons; (*d*) 10,000 photons.

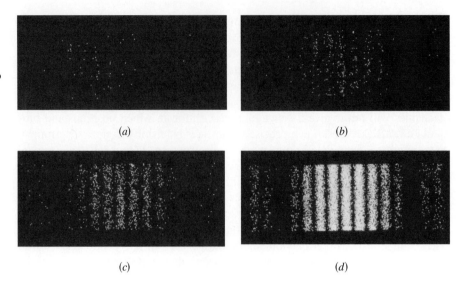

ing double-slit interference develop regions of maximum and minimum wave intensity—the bright and dark bands of the interference pattern. But the wave fields do not completely determine everything that happens; instead, they fix only the *probability* that individual photons will be detected in the interference pattern. That's why a very short exposure or low-intensity beam results not in a weak version of the interference pattern but in a seemingly arbitrary pattern. Only when large numbers of photons have been detected does the statistical pattern emerge (Fig. 40-2).

In quantum mechanics, then, the relation between the wave and particle aspects of light is this: As long as we don't try to detect the light, it propagates as a wave governed by Maxwell's equations. We can use those equations, and the theory of physical optics (Chapter 37) that is based on them, to predict the behavior of the waves. But when we detect the light, we do so through interactions involving individual photons. Those interactions are random events whose probability depends on the wave intensity—that is, on the square of the wave fields.

Electrons and Matter Waves

The quantum-mechanical description of matter is similar to that of light. Adopting de Broglie's hypothesis, we assume that the particles of matter are associated with some type of wave. Just as the probability of finding a photon is proportional to the intensity of light—that is, to the *square* of the electromagnetic fields—so the probability of finding a particle is proportional to the square of the quantity characterizing the matter wave (see Fig. 40-3). And as with light, the particle nature of matter manifests itself only through interactions with other matter or with light; when we aren't trying to detect a particle, its behavior is governed by the evolution of the associated matter wave.

A striking confirmation that a wave description applies to matter as well as light is the double-slit experiment with electrons (Fig. 40-4*a*). The resulting distribution of electrons at the detector shows the same characteristic double-slit interference pattern we've come to expect for light (Fig. 40-4*b*). It's impossible to explain this result if the electrons follow well-defined, classical trajectories.

Matter wave

Electron beam

FIGURE 40-3 A beam of electrons and its associated matter wave. Where the wave amplitude is larger there is a higher probability of finding an electron. The situation is exactly analogous to that of Fig. 40-1, and the same statistical interpretation and reservations apply; in particular, it makes no sense to talk about individual electrons unless we set out to detect them.

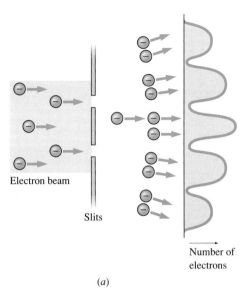

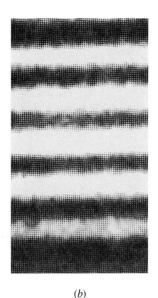

(a)   (b)

FIGURE 40-4 (a) A double-slit experiment with electrons. Picturing of the electrons is only suggestive; they cannot be considered as particles except when they interact with the detector. Curve at right shows the number of electrons as a function of position along the detector. (b) Experimental result for the electron distribution at the detector shows the characteristic fringes of double-slit interference.

Instead, the electrons' behavior in the double-slit system is determined by their wave nature; only when they interact with the detector do they behave as particles.

With light, we know the underlying laws that govern the evolution of the waves that, in turn, determine the probabilities of detecting photons. Those laws are Maxwell's equations and the wave equation that can be derived from them. But what equation describes the matter waves? Answering that question culminated the development of quantum mechanics during the first three decades of this century. Actually, two quite separate approaches were taken to the quantum description of matter. In 1925, at the age of 24, the German physicist Werner Heisenberg collaborated with Max Born and Pascual Jordan to develop **matrix mechanics**—a quantum theory formulated in the abstract mathematical language of matrices. At nearly the same time, the Austrian physicist Erwin Schrödinger (Fig. 40-5) formulated his theory of **wave mechanics.** Schrödinger soon showed that his theory and Heisenberg's were, in fact, equivalent. Heisenberg and Schrödinger received the 1932 and 1933 Nobel Prizes in Physics (the latter shared with Paul Dirac), respectively, for their contributions to the development of quantum mechanics. Because Schrödinger's work has a more obvious physical interpretation in terms of waves, we will continue our discussion of quantum mechanics entirely in terms of Schrödinger's formulation.

FIGURE 40-5 Erwin Schrödinger.

40-2 THE SCHRÖDINGER EQUATION

Schrödinger's challenge was to find a wave equation describing the behavior of matter waves. Guided by earlier quantum-mechanical thought and experiments, Schrödinger developed a quantum-mechanical wave equation consistent with the quantum relations $E = hf$ and $\lambda = h/p$, and with the Newtonian relation $E = p^2/2m$ between kinetic energy and momentum. For a particle of mass m and energy E moving in one dimension, with potential energy $U(x)$ a function of position only, Schrödinger postulated a spatial variation of the associated matter waves given by

$$-\frac{\hbar^2}{2m}\frac{d^2\psi(x)}{dx^2} + U(x)\psi(x) = E\psi(x), \qquad (40\text{-}1)$$

where the quantity ψ is a function of x that characterizes the matter waves. Since it gives information only about the spatial variation of the waves, Equation 40-1 is called the **time-independent Schrödinger equation.** The full **wave function** satisfying the time-dependent equation consists of a solution of Equation 40-1 multiplied by a harmonic oscillation at frequency $f = E/h$, with E the particle energy.

The Schrödinger equation invites two questions. First, how do we know it's right, and, second, what is the physical meaning of the wave function ψ? The answer to the first question is clear: the equation provides a description of physical reality that is in remarkable agreement with experiments. And not only a few arcane experiments; as we will see, Schrödinger's equation goes a long way toward explaining the structure of atoms, their chemical properties, and indeed the entire science of chemistry. Furthermore, as we will soon see, the Schrödinger description agrees with Newtonian mechanics for macroscopic systems so large that the "graininess" or quantization at the atomic level is not obvious.

The Meaning of ψ

Physicists and philosophers have been debating the meaning of ψ since Schrödinger first wrote his equation, and the debate continues today. In the standard interpretation, first articulated by Max Born in 1926, ψ itself is not an observable quantity. It manifests itself only in the statistical distributions of particle detections. More specifically, the probability per unit volume (also called the **probability density**) that we will find the particle described by the wave function ψ is given by ψ^2. For a particle confined to one dimension, the probability density becomes the probability per unit length, and we interpret ψ to mean that the probability of finding the particle in a small interval dx at position x is

$$\text{Probability} = \psi^2(x)\,dx. \qquad (40\text{-}2)$$

Here we write $\psi^2(x)$ to indicate that the probability may vary from place to place. We can interpret Equation 40-2 in two ways. At face value, it is the probability that a single experiment, with a detector set up to find particles in the interval dx at position x, will detect the particle. Or, if we do many such experiments, the equation gives the fraction of the time that we should expect to find a particle in our detector.

But *what is* ψ? How can it be totally unobservable yet govern the behavior of matter? There can't be a direct causal link between the wave function and individual particles, since ψ determines only the *probability* that a particle will behave in a certain way. Think about this! In quantum mechanics the outcome of an experiment is not fully determined. The Schrödinger equation describes only the probability of a given outcome. How can this be? The quantum world is so different, according to the standard interpretation, that our macroscopic language, concepts, and pictorial models are simply inadequate. In particular, macroscopic causality is replaced by a microscopic indeterminancy in which quantum events are truly random; only the overall statistical pattern of those events is governed by physical laws.

The philosophical implications of quantum mechanics have been a matter of debate since the theory was formulated (Fig. 40-6). A central theme in this debate is the possibility of "hidden variables," physical quantities that might be hidden from us by the uncertainty principle but that might nevertheless govern the microscopic world in a fully deterministic way. Experiments of the early 1980s have placed severe restrictions on such hidden-variable theories, but fascinating discussions on the interpretation of quantum mechanics continue.* Here, however, we turn to the Schrödinger equation to see how it is used in analyzing quantum-mechanical systems.

FIGURE 40-6 Niels Bohr and Albert Einstein discussing the implications of quantum mechanics. Einstein never accepted the probabilistic interpretation of quantum mechanics, as indicated in his paraphrased remark "God does not play dice."

Normalization and Other Constraints on the Wave Function

In one dimension, the quantity $\psi^2 \, dx$ represents the probability of finding the particle associated with the wave function $\psi(x)$ in the interval dx at some position x. But the particle *must* be *somewhere*. Therefore, if we sum the probabilities of finding the particle in all such intervals dx along the x axis, the result must be 1; there must be a 100% chance that we will find the particle somewhere. Since the probability density may vary with position, that sum becomes an integral, and we have

$$\int_{-\infty}^{+\infty} \psi^2 \, dx = 1. \qquad \text{(normalization condition)} \qquad (40\text{-}3)$$

Equation 40-3 is called the **normalization condition;** once we have found an appropriate solution $\psi(x)$ to the Schrödinger Equation 40-1, this condition can be used to set the overall amplitude of the function ψ.

The Schrödinger equation contains the second derivative of ψ. In order that this term be well-defined, both ψ itself and its first derivative must be continuous. (An exception to the continuity condition on $d\psi/dx$—possible only in unrealistic example situations—occurs if the potential energy U becomes infinite.)

40-3 THE INFINITE SQUARE WELL

We now solve the Schrödinger equation for a particularly simple system—a particle trapped in one dimension between two perfectly rigid walls. Although unrealistic in some respects, this system nevertheless is a surprisingly good approximation to some real quantum systems, including some new electronic devices and simple nuclei like deuterium. More importantly, its analysis illustrates the general procedure for applying the Schrödinger equation and shows how energy quantization emerges from Schrödinger's theory.

*See Abner Shimony, "The Reality of the Quantum World," *Scientific American*, January 1988, Vol. 258.

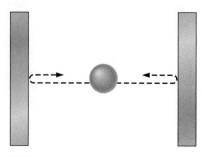

FIGURE 40-7 A particle trapped between rigid walls. Classically, it moves back and forth with constant total energy. That energy can have any value.

In classical physics, a particle trapped between rigid walls moves back and forth with constant speed (Fig. 40-7). In the absence of friction or other losses, the particle's energy remains constant at its initial value. And in classical physics, that value can be anything.

We can describe the particle's situation in terms of its potential-energy curve. (We will be talking a great deal about potential-energy curves for a while, and you might find it helpful to review Section 8-4 at this point.) Since the particle is subject to no forces while it's between the walls, its potential energy is constant in this region, and we can fix the arbitrary zero of potential energy by setting that constant to zero. If the walls are perfectly rigid then the particle should not be able to penetrate them at all, no matter what its energy. This means that the potential energy becomes abruptly infinite at the walls. Then the potential-energy curve for our particle looks like Fig. 40-8; you can see from the figure why this curve is called an **infinite square well.**

We now consider the quantum-mechanical description of the particle in a square well. The particle has a wave function whose time-independent part is given by the Schrödinger equation (Equation 40-1):

$$-\frac{\hbar^2}{2m}\frac{d^2\psi}{dx^2} + U(x)\psi = E\psi,$$

where the potential energy $U(x)$ is that of the square well in Fig. 40-8:

$$U = 0 \text{ for } 0 < x < L$$

$$U = \infty \text{ for } x < 0 \text{ or } x > L.$$

Physically, we set $U = \infty$ so there is no chance that the particle can penetrate the rigid walls. Quantum-mechanically, this means that the function ψ must be exactly zero in the region where $U = \infty$, giving zero probability of finding the particle beyond the walls. Whatever the wave function ψ looks like, then, it must go to zero at $x = 0$ and at $x = L$, and must remain zero beyond those points. All we need to calculate, then, is ψ inside the well, where $0 < x < L$.

Within the well, $U = 0$ and the Schrödinger equation becomes

$$-\frac{\hbar^2}{2m}\frac{d^2\psi}{dx^2} = E\psi. \tag{40-4}$$

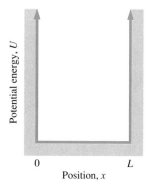

FIGURE 40-8 Infinite square well potential-energy curve describes a particle constrained to move in one dimension between rigid walls a distance L apart.

Following de Broglie, we expect that the solutions inside the well should take the form of standing waves with nodes at $x = 0$ and $x = L$. One possible form for such waves is sinusoidal:

$$\psi = A\sin kx,$$

where A and k are constants. To meet the condition $\psi = 0$ at $x = L$, the argument of the sine function must be a multiple of π when $x = L$; that is, $kL = n\pi$, where n is any integer. Then $k = n\pi/L$, and our waves have the form

$$\psi = A\sin\left(\frac{n\pi x}{L}\right), \tag{40-5}$$

with the constant A still undetermined. Equation 40-5 represents standing waves with nodes at the ends of our square well, but does it satisfy the Schrödinger equation? We can find out by substituting Equation 40-5 into Equation 40-4. We need not only ψ but also its second derivative; twice differentiating Equation 40-5 gives

$$\frac{d^2\psi}{dx^2} = -A\frac{n^2\pi^2}{L^2}\sin\left(\frac{n\pi x}{L}\right).$$

Substituting ψ and $d^2\psi/dx^2$ into Equation 40-4 gives

$$\left(-\frac{\hbar^2}{2m}\right)\left[-A\frac{n^2\pi^2}{L^2}\sin\left(\frac{n\pi x}{L}\right)\right] = EA\sin\left(\frac{n\pi x}{L}\right),$$

which reduces to

$$E = \frac{n^2\pi^2\hbar^2}{2mL^2} = \frac{n^2h^2}{8mL^2}, \qquad (40\text{-}6)$$

where we used $\hbar = h/2\pi$. What is Equation 40-6 telling us? If it is an identity, it says that our proposed solution 40-5 does indeed satisfy the Schrödinger equation. And Equation 40-6 can be an identity—provided the particle energy E has a value given by the right-hand side of the equation, with n an integer.

Our standing-wave solutions and the associated condition 40-6 show how the quantization of energy arises naturally from the mathematics of the Schrödinger equation. Physically, of course, the reason for quantization remains as de Broglie had postulated: matter waves in a confined system must be standing waves with an integer number of half wavelengths. Although de Broglie's hypothesis and the Schrödinger equation lead to exactly the same conclusion in this simple case of an infinite square well, we will see that with more complicated potential-energy functions only the Schrödinger equation can give us the full story.

The integer n that appears in the wave function 40-5 and in the energy quantization condition of Equation 40-6 is called the **quantum number** for the particle in the square well. The physical state of a quantum-mechanical system is called a **quantum state.** In this simple case, one quantum number suffices to specify the quantum state, which then tells us everything quantum mechanics has to say about the situation. As far as the Schrödinger equation is concerned, it looks like all integer values of n are allowed. The choice of negative or positive n has no physical significance since ψ^2 has the same value with either sign of ψ; for this reason, negative n's are redundant. But $n = 0$ implies $\psi = 0$ everywhere, giving no chance of finding the particle anywhere. This case cannot satisfy the normalization condition of Equation 40-3, leaving only positive integer values of n.

With only nonzero n's allowed, Equation 40-6 shows that the particle's energy is always positive; zero energy is not allowed. The lowest possible energy is $h^2/8mL^2$, obtained with $n = 1$. This lowest energy is called the **ground-state energy;** the corresponding wave function is the **ground-state wave function.** A nonzero ground-state energy is a common feature of quantum systems, and one with no classical counterpart. This nonzero minimum energy has profound implications, especially in systems where typical energies are small. At very low

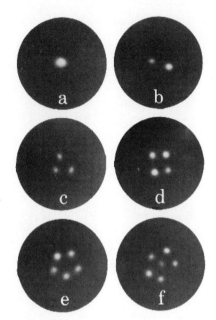

FIGURE 40-9 Quantization in a macroscopic system. Photo is a top view of a rotating container of liquid helium at 0.1 K. White dots are vortices (whirlpools). The angular momentum of these vortices is quantized, so as the container rotates more rapidly, more vortices appear. In an ordinary fluid, there would be a single vortex whose strength increased continuously with increasing rotation rate.

temperatures, for example, classical physics suggests that all motion should stop as particle energies approach zero. But quantum mechanics tells us that the lowest possible energy for a particle is not zero. As a system—even a macroscopic one—cools to very low temperatures, this difference between classical and quantum descriptions can become very apparent in the behavior of the system (Fig. 40-9).

Why is it not possible to have a zero-energy particle in our square well? The answer lies in the uncertainty principle. If the energy were exactly zero, then the particle's momentum would also be exactly zero—and we would know its momentum precisely. In other words, the uncertainty Δp in the momentum would be zero. However, we know that the particle is confined to the well; therefore, the uncertainty Δx in its position is L. So $\Delta p \Delta x$ would be zero, in violation of the uncertainty principle. Further, we can show explicitly that the ground-state energy $E_1 = h^2/8mL^2$ is consistent with the uncertainty principle. Within the well the energy is all kinetic, and we can use the relation $E = p^2/2m$ (that itself comes from $E = \frac{1}{2}mv^2$) to estimate the square of the corresponding ground-state momentum:

$$p_1^2 = 2mE_1 = 2m\frac{h^2}{8mL^2} = \frac{h^2}{4L^2}.$$

But in a one-dimensional square well, the particle can be moving in either of two directions. If we think it is moving one way with momentum p, it might really be moving the other way—that is, with momentum $-p$. The uncertainty in its momentum would then be $p - (-p) = 2p$. Our result above for the likely value of p^2 then suggests that the uncertainty in the particle's momentum is $2\sqrt{p^2}$, or

$$\Delta p = 2\sqrt{\frac{h^2}{4L^2}} = \frac{h}{L}.$$

Since the particle is in the well, $\Delta x = L$. Then

$$\Delta p \Delta x = \left(\frac{h}{L}\right)L = h = 2\pi\hbar.$$

Since this product of uncertainties is greater than $\hbar$, the uncertainty principle is satisfied.

● EXAMPLE 40-1 HINTS OF ATOMIC STRUCTURE

What is the ground-state energy of an electron in an infinite square well whose width L is the approximate diameter of a hydrogen atom, about 10^{-10} m? Give the answer in both joules and electron volts.

Solution

With $n = 1$ for the ground state, and with $L = 1 \times 10^{-10}$ m, Equation 40-6 gives:

$$E = \frac{h^2}{8mL^2}$$

$$= \frac{(6.63 \times 10^{-34} \text{ J·s})^2}{(8)(9.11 \times 10^{-31} \text{ kg})(1 \times 10^{-10} \text{ m})^2} = 6 \times 10^{-18} \text{ J}.$$

Since 1 eV = 1.6×10^{-19} J, this is

$$E = \frac{6 \times 10^{-18} \text{ J}}{1.6 \times 10^{-19} \text{ J/eV}} = 40 \text{ eV},$$

where in this approximate calculation we've rounded to one significant figure. Although a square-well potential is very different from the electrical potential energy associated with electrical forces in an atom, this example nevertheless provides a rough estimate of the energies we should expect to find for electrons in atoms—typically in the tens-of-eV range.

EXERCISE Find the width of an infinite square well containing a proton whose ground-state energy is 10 MeV.

Answer: 4.5 fm, about the size of a nucleus

Some problems similar to Example 40-1: 8, 9, 11, 13

The ground-state energy is the minimum possible for a particle in a square well. But it can have any of the other energies given by Equation 40-6 with $n > 0$. Figure 40-10 shows an **energy-level diagram** for a particle in an infinite square well. In Chapter 39 we used a similar diagram (Fig. 39-13) to represent atomic energy levels in the Bohr atom.

Normalization, Probabilities, and the Correspondence Principle

In our solution for the infinite square well, we still don't know the value of the constant A in the wave function. This constant can be determined from the normalization condition 40-3:

$$\int_{-\infty}^{\infty} \psi^2 \, dx = 1.$$

Inside the well, ψ is given by Equation 40-5. Outside the well, ψ is exactly zero. So we can write the normalization condition as an integral involving only Equation 40-5 and the interval $0 < x < L$:

$$\int_{0}^{L} A^2 \sin^2\left(\frac{n\pi x}{L}\right) dx = 1.$$

If we divided $\int_{0}^{L} \sin^2(n\pi x/L)\, dx$ by the well width L, we would have the average of $\sin^2$ over an integer number of half cycles—or just $\frac{1}{2}$. So the integral of $\sin^2(n\pi x/L)$ from 0 to L is L times $\frac{1}{2}$, and therefore

$$A^2 \frac{L}{2} = 1,$$

or

$$A = \sqrt{\frac{2}{L}}.$$

The complete wave function is then given by

$$\psi_n = \sqrt{\frac{2}{L}} \sin\left(\frac{n\pi x}{L}\right), \tag{40-7}$$

where the subscript n refers to the function associated with the nth quantum state. Figure 40-11 shows ψ functions for the ground state and several excited states.

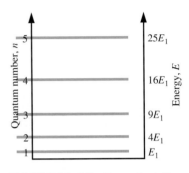

FIGURE 40-10 Energy-level diagram for a particle in an infinite square well. The levels are not evenly spaced because the energy (Equation 40-6) is proportional to n^2.

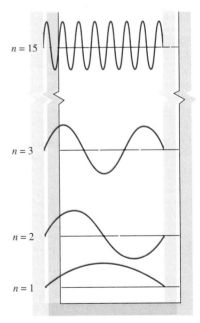

FIGURE 40-11 Wave functions for a particle in an infinite square well. Horizontal axis for each function is at the appropriate energy level, with break indicating that the $n = 15$ state has a much higher energy.

Where are we likely to find the particle? Classically, it would be moving back and forth at constant speed, and therefore would be equally likely to be anywhere in the well. Quantum-mechanically, the probability of finding it at some position x is proportional to the probability density ψ^2 at that point. Figure 40-12 shows the probability densities given by squaring the wave functions of Fig. 40-11. For $n = 1$ we're clearly more likely to find the particle near the middle of the well—in marked contrast to the classical prediction of equal probability everywhere. For other low-n states there are obvious regions of high and low probability, again in contrast to the classical prediction. But as the quantum number increases the maxima and minima of the probability density get closer together. Any instrument we use to find the electron has a finite resolution, and once the periodicity of the wave function drops below that resolution, then we measure an average probability, which is essentially constant over the interval (see Fig. 40-12).

We have here a manifestation of Bohr's correspondence principle: At large values of the quantum number n the interval between adjacent energy levels becomes small compared with the energy itself, and a measurement—in this case of the electron's position—gives results in agreement with classical physics. But classical physics is totally inadequate at low n, where the nonclassical zero-point energy and energy quantization are most evident.

● **EXAMPLE 40-2**   QUANTUM PROBABILITY

A particle is in the ground state of an infinite square well that extends from $x = 0$ to $x = L$. What is the probability that it will be found in the leftmost quarter of the well?

Solution
In deriving the normalization condition 40-3, which expresses mathematically that the particle has probability 1 of being found *somewhere*, we integrated the probability density ψ^2 over all values of x. Similarly, to determine the probability of finding the particle in a particular region, we integrate ψ^2 over that region. Here the region is the left-hand fourth of the well, given by $0 < x < \frac{1}{4}L$. Then the probability is

$$P = \int_0^{L/4} \psi^2 \, dx.$$

The ground-state wave function, given by Equation 40-7 with $n = 1$, is $\psi_1 = \sqrt{2/L}\, \sin(\pi x/L)$, so we have

$$P = \frac{2}{L} \int_0^{L/4} \sin^2\left(\frac{\pi x}{L}\right) dx.$$

We can integrate using the table at the end of Appendix A; the result is

$$P = \frac{2}{L}\left(\frac{x}{2} - \frac{\sin(2\pi x/L)}{4\pi/L}\right)\Bigg|_0^{L/4} = \frac{2}{L}\left(\frac{L}{8} - \frac{L}{4\pi}\right) = \frac{1}{4} - \frac{1}{2\pi}$$

$$= 0.091 .$$

This is considerably lower than the probability $P = 0.25$ we would expect classically for finding the particle in any quarter of the well, and reflects the lower value of ψ^2 nearer the well ends. Problem 54 repeats the calculation of this example for arbitrary quantum numbers, and shows that classical and quantum probabilities agree at large n.

EXERCISE   Determine the probability of finding a ground-state particle in the central quarter of an infinite square well (i.e., in the range $\frac{3}{8}L < x < \frac{5}{8}L$), and compare with the classical prediction.

Answer: 0.475, nearly twice the classical value

Some problems similar to Example 40-2: 23, 47, 48, 55   ●

The infinite square well we have analyzed is not a very realistic model for many quantum-mechanical systems (although it is a good approximation in multilayer technologies now being explored by solid-state physicists). But it has given us insights into a number of important quantum phenomena that are shared by atoms and other real systems. These include the quantization of energy levels, the existence of a nonzero ground-state energy, probability distributions that differ from classical predictions, and the correspondence of quantum and classical physics at high quantum numbers. In the next chapter we will apply the Schrödinger equation to atoms, using the same techniques and finding many of the same phenomena we did here. First, however, we examine other simple quantum systems that give rise to new and surprising behavior quite at odds with classical physics.

40-4 THE HARMONIC OSCILLATOR

In Chapter 15 we studied the motion of a particle subject to a restoring force directly proportional to the particle's displacement from its equilibrium position. Using classical physics, we found that the particle undergoes simple harmonic motion, oscillating sinusoidally at an angular frequency ω determined by its mass and the nature of the restoring force. Although our paradigm system was a mass on a spring, we pointed out (see Fig. 15-24) that any other system whose potential energy is a quadratic function of displacement also constitutes a simple harmonic oscillator. That includes many systems at the atomic and molecular level. And our most fundamental theories of physics—the so-called quantum field theories—treat the entire universe of particles and fields as essentially a huge ensemble of harmonic oscillators. Understanding the quantum-mechanical harmonic oscillator is therefore crucial in describing the behavior of matter at the smallest scales.

We now use the Schrödinger equation to investigate the behavior of a quantum-mechanical harmonic oscillator. If the system were a mass on a spring, it would have potential energy $U = \frac{1}{2}kx^2$ (Equation 8-4) and frequency $\omega = \sqrt{k/m}$, with k the spring constant and m the particle mass. Combining these equations, we can write the potential energy in terms of the frequency and mass:

$$U = \frac{1}{2}m\omega^2 x^2.$$ (40-8)

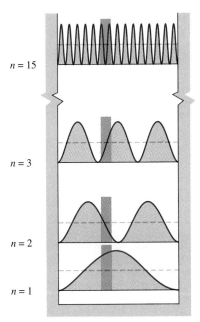

FIGURE 40-12 Classical (dashed) and quantum (solid) probability densities for a particle in an infinite square well. Yellow bands represent the resolution of an instrument used to detect the particle; for large n the instrument averages over one or more cycles and therefore detects the particle with an average probability everywhere equal to the classical value.

A quantum-mechanical harmonic oscillator is more likely to be an electron in some electric field configuration that gives rise to a restoring force, or an atom vibrating at the end of a molecular bond; for this reason Equation 40-8, which makes no explicit reference to springs and spring constants, is an appropriate form for the potential energy of a general harmonic oscillator.

Figure 40-13 shows a plot of the parabola described by Equation 40-8. This curve is analogous to Fig. 40-8 for the infinite square well. In contrast to the discontinuous potential-energy curve of the square well, we have in Equation 40-8 a single continuous function giving the harmonic oscillator's potential energy everywhere. Using this function for $U(x)$ in the Schrödinger equation (Equation 40-1) gives

$$-\frac{\hbar^2}{2m}\frac{d^2\psi}{dx^2} + \frac{1}{2}m\omega^2 x^2\psi = E\psi.$$ (40-9)

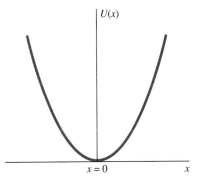

FIGURE 40-13 The potential-energy curve for a simple harmonic oscillator is a parabola.

Solving this differential equation involves advanced mathematical techniques; here we give only the ground-state solution:

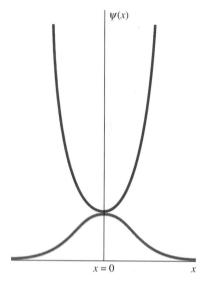

FIGURE 40-14 The functions $e^{+m\omega x^2/2\hbar}$ (blue) and $e^{-m\omega x^2/2\hbar}$ (red) both satisfy the Schrödinger equation 40-9 for the harmonic oscillator, but only the function with the negative exponent can be normalized. Aside from the normalization constant $\pi^{-1/4}$, this latter function is ground-state wave function.

$$\psi_0 = \left(\frac{m\omega}{\pi\hbar}\right)^{1/4} e^{-m\omega x^2/2\hbar} \tag{40-10}$$

and let you show (Problem 28) that it satisfies Equation 40-9. In so doing you will also find that the ground-state energy is $\frac{1}{2}\hbar\omega$.

Many other functions also satisfy Equation 40-9, but most of them approach infinity at large values of x and, therefore, cannot be normalized (Fig. 40-14). The normalizable solutions—those that represent particle wave functions—occur only for certain discrete values of the energy E, given by

$$E_n = \left(n + \frac{1}{2}\right)\hbar\omega, \tag{40-11}$$

where $n = 0$ now corresponds to the ground state. Figure 40-15 shows an energy-level diagram for the harmonic oscillator; note the even spacing implied by Equation 40-11. Incidentally, the additive factor $\frac{1}{2}$ in Equation 40-11 shows that Planck was wrong in his hypothesis that the allowed harmonic oscillator energies should be multiples of $hf(= \hbar\omega)$. Planck's spectral distribution (Equation 39-3) is nevertheless correct, but he did not foresee the existence of nonzero ground-state energy.

The even spacing between energy levels of the harmonic oscillator as shown in Fig. 40-15 is in marked contrast to the situation in atoms (see Fig. 39-13) or in the infinite square well (see Fig. 40-10). A quantum harmonic oscillator emits or absorbs photons as it makes transitions among adjacent levels, and the even level spacing implies that all transitions between adjacent levels of a pure harmonic oscillator involve photons of the same energy.

● EXAMPLE 40-3   A HIGH-POWER LASER

Among the most intense lasers are those powered by chemical reactions that produce harmonic-oscillator-like molecules in excited states. In a laser using hydrogen fluoride molecules, developed under the "Star Wars" antimissile program, much of the emission occurs as a result of transitions between the $n = 1$ harmonic oscillator level and the ground state. The infrared light emitted in these transitions has wavelength 2.7 μm. What is the ground-state energy, in electron-volts, of the HF molecules?

Solution
Equation 40-11 shows that the spacing between adjacent energy levels is $\hbar\omega$ and that the ground state energy is $E_0 = \frac{1}{2}\hbar\omega$. The energy difference $\hbar\omega$ between the $n = 1$ state and the ground state ($n = 0$) is the energy of a 2.7-μm photon. In Chapter 39 we saw that $E = hc/\lambda$ for a photon of wavelength λ. Thus, we have $\hbar\omega = hc/\lambda$, or, for the ground-state energy,

$$E_0 = \frac{1}{2}\hbar\omega = \frac{hc}{2\lambda} = \frac{(6.63\times10^{-34}\ \text{J·s})(3.0\times10^8\ \text{m/s})}{(2)(2.7\times10^{-6}\ \text{m})(1.6\times10^{-19}\ \text{J/eV})}$$
$$= 0.23\ \text{eV}.$$

This is the minimum possible energy for the hydrogen fluoride molecule. We will explore chemical lasers further in Chapter 42.

EXERCISE An electron is trapped in a harmonic oscillator potential. Find the classical angular frequency of oscillation if the spacing between adjacent energy levels is 2.8 eV.

Answer: $4.25\times10^{15}\ \text{s}^{-1}$

Some problems similar to Example 40-3: 24, 25, 27   ●

With the infinite square well, we found that the probability density for the low-energy states bore no resemblance to what we would expect from classical physics. The same is true for the harmonic oscillator. If we consider a classical harmonic oscillator—for example, a mass moving back and forth on a spring—we find the mass moving slowly near the extremes of its displacement, but rapidly through its equilibrium position. We are most likely to find the particle where it spends the most time—namely, near the turning points—and least likely to find it where it's moving fastest—namely, near its equilibrium position. Figure 40-16a shows this classical probability density, along with the probability density $\psi_0^2(x)$ associated with the quantum-mechanical ground state. As with the infinite square well, the classical and quantum probability densities for the ground state are completely at odds. Quantum-mechanically, we're most likely to find the particle in its equilibrium position, where classical physics says the probability is lowest. As the energy level rises, however, the quantum description becomes more like the classical one, with the probability greatest near the classical turning points (Fig. 40-16, parts b–e). We found the same result for the infinite square well. Again, we see Bohr's correspondence principle at work: In the limit of large quantum numbers, quantum mechanics gives the same predictions as classical physics. Only when the energy is reasonably close to the ground-state energy do the new effects of quantum mechanics become dramatically obvious.

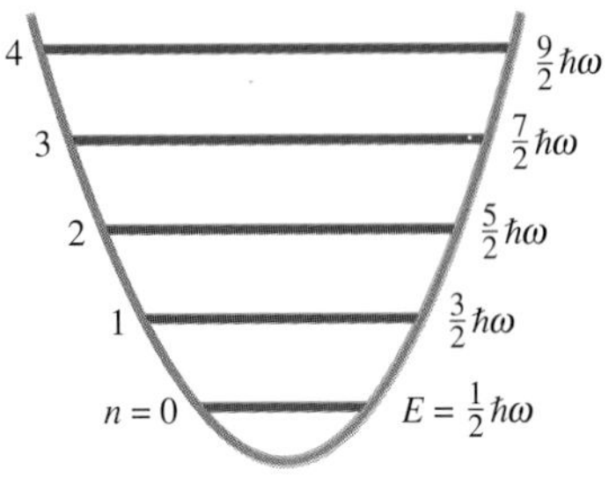

FIGURE 40-15 Energy-level diagram for a quantum-mechanical harmonic oscillator, superposed on the potential-energy curve. Energy levels are evenly spaced at intervals of $\hbar\omega$ ($= hf$), starting with the ground state at energy $\frac{1}{2}\hbar\omega$.

40-5 QUANTUM TUNNELING

One of the most remarkable features of Fig. 40-16 is the nonzero probability of finding a quantum-mechanical harmonic oscillator beyond its classical turning points—the points at which its kinetic energy has been converted entirely to potential energy. This unusual situation, which seems at first to violate energy conservation, has no counterpart in the classical description of matter.

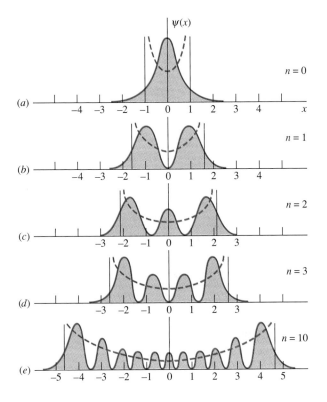

FIGURE 40-16 Probability densities for the $n = 0$, 1, 2, 3, and 10 states of the harmonic oscillator. Dashed curve is the classical prediction, in which the probability density is highest where the particle is moving most slowly—namely, near its turning points (marked with vertical lines). Quantum mechanically, the probability density in the ground state (a) is completely at odds with the classical prediction. But at higher energy levels (e) the probability density (averaged over the rapid fluctuations in ψ) approaches the classical prediction. Note the increase in amplitude of the motion with increasing energy, a feature common to both classical and quantum analysis. The classical probability drops abruptly to zero beyond the turning points, but in quantum mechanics there remains a nonzero chance of finding the particle beyond the classical turning points. The horizontal axis is position, in units of the classical turning point at the ground-state energy.

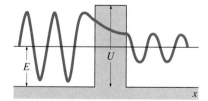

FIGURE 40-17 A potential barrier of height U, showing the wave function for a particle incident from the left with energy E less than the barrier energy U. Classically, the particle should rebound and remain to the left of the barrier. But quantum-mechanically, there is a nonzero probability—reflected in the nonzero wave function on the right—that it will penetrate the barrier and emerge on the other side.

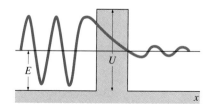

FIGURE 40-18 For a massive particle, the wave function drops rapidly in the barrier, giving a negligible probability of penetration. Compare with the case of a low-mass particle shown in Fig. 40-17.

Another example of penetration into a classically forbidden region is provided by a particle encountering a potential barrier (Fig. 40-17). Classically, a particle whose total energy lies below the barrier energy is confined to one side of the barrier. (It can't be inside the barrier because its kinetic energy $\frac{1}{2}mv^2$ would have to be negative—a clear impossibility.) If we solve the Schrödinger equation for this potential-energy curve, however, we find oscillatory solutions on either side of the barrier, joined according to the continuity conditions on ψ and $d\psi/dx$ by exponential functions within the barrier (see Problems 30, 31 and 56). Such a solution is shown superimposed on the barrier in Fig. 40-17. The probability density ψ^2 associated with this solution remains nonzero through the barrier, and continues to give a nonzero probability of finding the particle on the far side— implying that a particle initially on one side of the barrier may later be found on the other side.

How likely is this phenomenon, called **quantum tunneling?** That depends on the relation of the particle energy to the barrier energy, and also on the width of the barrier. As you can show in Problem 30, the ψ function inside the barrier involves exponential functions of the form $e^{\pm\sqrt{2m(U_0-E)}x/\hbar}$, where U_0 is the barrier height. In general, these exponentials drop very rapidly across the barrier width unless the particle energy E is very close to the barrier energy or the particle mass m is very small. The probability that a particle will be found on the far side of the barrier is therefore very small when the mass m is large, so quantum tunneling is essentially a microscopic phenomenon (Fig. 40-18).

It looks as if tunneling violates the conservation-of-energy principle. But conservation of energy is saved by the uncertainty principle. Suppose we try to catch a particle in the act of tunneling. If we know the particle is within the barrier, the uncertainty in its position is no greater than the barrier width. Then the uncertainty principle sets a minimum value for the uncertainty in the particle's momentum, and this in turn implies a minimum energy. A quantitative analysis shows that minimum to be such that we can no longer be sure that the particle energy is below the barrier energy. But quantum tunneling really does occur, as we can confirm experimentally by measuring particle energies on both sides of the barrier. They can both be below the barrier height, and yet somehow particles have gone from one side to the other.

If we don't try to detect a particle within the barrier, its penetration is a purely wave phenomenon to which our particulate energy considerations don't apply. Again we see the wave-particle duality at work: If we don't observe the particle, its behavior is governed by the associated waves and may result in most unparticle-like phenomena such as tunneling. If we do try to catch it in the act of such behavior, it ceases to be wavelike and the surprising phenomena are not there.

Tunneling is important in a number of quantum-mechanical phenomena and is exploited in several technological devices. From our human perspective, perhaps the most significant result of quantum tunneling is that the Sun shines. The Sun's energy source is the fusion of hydrogen nuclei into helium that takes place in the solar core. But even at a temperature of 15 MK, the average thermal energy is far less than necessary classically to overcome the electrical repulsion of the nuclei. Quantum-mechanically, though, nuclei can tunnel through this "coulomb barrier" and "fall" into the deep potential well associated with the nuclear force (Fig. 40-19). The subsequent release of energy keeps the Sun shining and makes life as we know it possible on our planet.

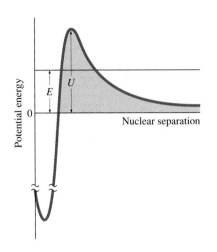

FIGURE 40-19 Potential-energy curve for a pair of deuterium nuclei, showing the barrier associated with the repulsive electrical force and the deep well arising from the strongly attractive nuclear force. Horizontal line is the energy E of a particle that, according to classical physics, cannot penetrate the Coulomb barrier. Quantum mechanically, though, it can tunnel through the barrier and fall into the potential well, releasing energy. Such tunneling is ultimately responsible for the fusion reactions that make the Sun shine.

Convincing evidence for quantum tunneling is also provided by the process of alpha decay, whereby a heavy nucleus like uranium-238 emits a helium nucleus (also called an alpha particle). The two protons and two neutrons making up the alpha particle are trapped in a nuclear potential well qualitatively like that of Fig. 40-19. Just as a particle can tunnel into the well from outside, so can the alpha particle tunnel out. Measurement of the alpha particle energy far from the nucleus shows that its total energy is much less than needed to overcome the barrier that held it in the nucleus. Observation of alpha decay thus confirms that tunneling occurs. A quantum-mechanical calculation also gives the probability of alpha emission, and from it the decay rate of a given radioactive sample. First carried out in 1928 by George Gamow, and independently by E. U. Condon and R. W. Gurney, such quantum-tunneling analysis provided one of the major verifications of the new quantum mechanics.

■ APPLICATION THE SCANNING-TUNNELING MICROSCOPE

Conventional optical microscopes cannot resolve objects smaller than the wavelength of visible light—several hundred nanometers, or several thousand times the size of an atom. Electron microscopes achieve higher resolution using a beam of electrons in place of light. But quantum mechanics shows that even electrons have wavelike properties and an associated wavelength. For high resolution, the electron wavelength must be small; since $\lambda = h/p$, small electron wavelength implies a large momentum and correspondingly large energy. The high-energy electrons in an electron microscope beam penetrate deeply into the material under observation. The electron microscope is therefore good at resolving internal structure, but it reveals little about the surface of a material. Yet surface structure has become increasingly important to semiconductor engineers, as their microelectronic devices continue to shrink in size, and the ability to picture surface details is also necessary for biologists studying the work-

ings of viruses, and for chemists concerned with reactions that take place on the surfaces of materials.

The **scanning-tunneling microscope,** developed in the 1980s by Heinrich Rohrer and Gerd Binnig of IBM Zurich Research Laboratory in Switzerland, makes explicit use of quantum tunneling to resolve surface details down to the size of a single atom (Fig. 40-20). Their work brought Rohrer and Binnig the 1986 Nobel Prize in Physics, shared also with Ernst Ruska for his 1931 invention of the conventional electron microscope.

To see how the tunneling microscope works, consider the electrons in a material whose surface is to be studied. Classically, those electrons are bound inside the material by a steep potential-energy jump at the surface; this jump is associated with the electrostatic attraction between the electrons and the rest of the material. The material surface therefore marks the edge of a finite potential well in which classical physics suggests the elec-

FIGURE 40-20   Heinrich Roher (left) and Gerd Binning with an early model of their scanning-tunneling microscope.

trons should be bound. But quantum-mechanically, electron wave functions extend beyond the surface, into the classically forbidden region (Fig. 40-21a). Since there are many electrons in the material, we can picture the material as surrounded by a "cloud" of electrons that have quantum-mechanically tunneled

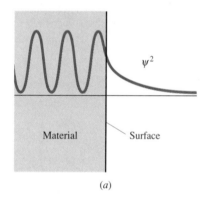

(a)

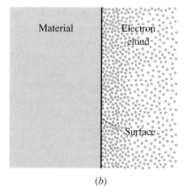

(b)

FIGURE 40-21   (a) The probability density ψ^2 extends beyond a material surface, giving a finite chance of finding electrons outside the material (b) Statistically, an electron "cloud" surrounds the material. The cloud density decreases exponentially with distance from the surface.

FIGURE 40-22   Conventional scanning electron microscope image of a tunneling microscope needle. Such a needle may be only one atom wide at its tip.

beyond the material surface (Fig. 40-21b). As Problem 30 shows, the wave function decreases exponentially with distance into the barrier—that is, with distance from the material surface. Statistically, this means that the density of the electron cloud also decreases exponentially.

In the scanning-tunneling microscope, a very sharp needle (Fig. 40-22; sometimes so sharp that a single atom sits at its tip!) is brought so close—typically less than 1 nm—to the surface under study that the electron clouds of needle and surface overlap. Application of a small potential difference between needle and surface then results in an electric current. In potential-energy terms, electrons carrying this current are tunneling through a potential barrier between surface and needle (Fig. 40-23); the current is therefore called a **tunneling current.** Because the wave function and therefore the electron cloud density decrease exponentially with distance from the surface, the tunneling current is very sensitive to the surface-needle separation. By scanning the needle parallel to the surface and recording the current, an image of the surface can be constructed. (In practice, the image construction method is more complicated. Feedback mechanisms are used to move the needle vertically to keep the current and therefore the surface-needle separation constant. Then the needle itself traces out the contours of the surface.)

Successful operation of the scanning-tunneling microscope requires precise control of the needle position to within 0.01 nm, or less than one-tenth of a typical atomic diameter. Needle motion is controlled with so-called piezoelectric materials, whose size changes slightly in response to an applied electric field. External disturbances are isolated and damped through an elaborate combination of springs and eddy-current damping. The result is a device that has already made significant contributions in physics, biology, chemistry, and semiconductor engineering. Figure 40-24 shows some images produced with scanning-tunneling microscopes.

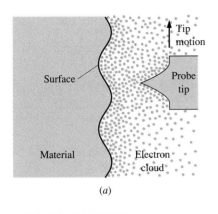

(a)

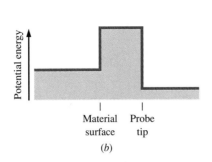

(b)

FIGURE 40-23 (a) The scanning-tunneling microscope detects currents flowing between the tip of a needle-like probe and the surface under study. Because the electron cloud density falls exponentially with distance from the surface, the current is a sensitive measure of that distance. By scanning the needle parallel to the surface, an image of the irregular surface can be constructed. (b) Potential energy curve for the region from material to needle tip. The electron current flows by tunneling through the potential barrier.

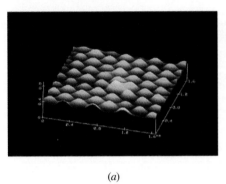

(a)

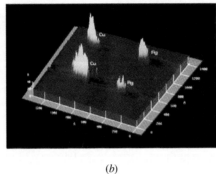

(b)

FIGURE 40-24 (a) An STM image of a graphite surface. The bumps are individual atoms in their regular crystal lattice, and a 1-atom defect is visible near the center. The region shown measures only 1.6 nm on a side. (b) In 1992 scientists at the University of California developed the world's smallest batteries. Two of the batteries appear in this STM image; each consists of a pile of copper (Cu) and gold (Ag) atoms only about 80 nm apart. The region shown measures 140 nm on a side, with axes labeled in angstrom units (1Å = 0.1 nm).

40-6 FINITE POTENTIAL WELLS

Both the infinite square well and the ideal harmonic oscillator involve potential wells of infinite depth. A particle is necessarily bound to such a well; it cannot escape to arbitrarily large distances. Its quantized energy states are therefore called **bound states.** But what if we have a well of finite depth, as shown in Fig. 40-25? We will state without mathematical analysis some properties of the resulting quantum states; aspects of the mathematics are covered in Problems 30 and 55–57. Provided the well is not too shallow, there are, in general, bound states with quantized energy levels whose wave functions resemble those of the infinite square well (Fig. 40-26). The main difference between these and the states of the infinite square well lies in the nonzero probability of quantum tunneling into the classically forbidden region outside the well.

The bound states, with their quantized energy levels, describe particles whose energies lie below the well height. Although those particles can penetrate somewhat into the classically forbidden region beyond the edges of the well, the exponential decrease in ψ ensures that they are very unlikely to be found far from the well. Particles whose energies exceed the well height, however, should be free to move anywhere. Quantum-mechanically, this freedom is reflected in a ψ function that is everywhere oscillatory, without exponential decay (see Problem 30). Furthermore, particles in these **unbound states** can have any energy whatsoever

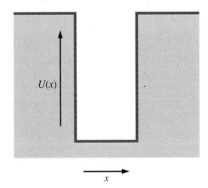

FIGURE 40-25 Potential-energy curve for a finite square well.

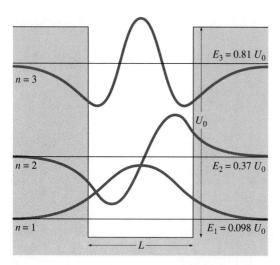

FIGURE 40-26 Bound-state wave functions for a finite square well, superimposed on the associated energy levels. Within the well the solutions are oscillatory, while outside they decay exponentially. For this particular combination of well depth and width (given by $\sqrt{mU_0L^2/2\hbar^2} = 4$), there are only three bound states.

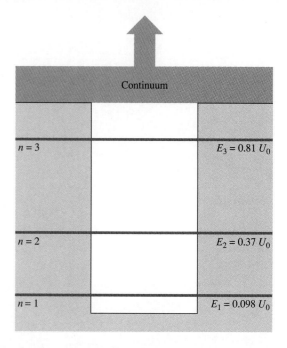

FIGURE 40-27 A complete energy-level diagram for a finite square well shows discrete bound states as well as a continuum of unbound states extending to arbitrarily high energy.

as long as it exceeds the well height; unbound energy values are not quantized. Rather, there is a **continuum** of allowed energies above the well height, in contrast to the discrete, quantized levels below. Figure 40-27 is an energy-level diagram reflecting the presence of quantized, bound states as well as the continuum of unbound states. We will again find the presence of both bound and continuum states when we study the atom in the next chapter.

40-7 QUANTUM MECHANICS IN THREE DIMENSIONS

Our study of one-dimensional quantum systems has revealed important features of the quantum world, like energy quantization. But atoms and most other quantum systems are inherently three dimensional. The wave function then depends on all three spatial variables, and the Schrödinger equation must be generalized to involve these variables. We will not use the full three-dimensional Schrödinger equation here but will simply point out some new features that arise in three-dimensional quantum systems.

We characterized the quantum states of a one-dimensional square well by a single integer quantum number n. That number arose mathematically from the condition that the wave function within the well include an integer number of half wavelengths. Solving the Schrödinger equation under these conditions gave a unique energy level associated with each value of n. In a similar way, the potential-energy function for a one-dimensional harmonic oscillator led to allowed energy levels described in terms of a single quantum number.

In three dimensions, constraints analogous to those imposed by the boundaries of a one-dimensional well or the potential-energy function of a one-dimensional harmonic oscillator lead to independent quantum numbers for each dimension (Fig. 40-28). For each set of three quantum numbers there is an associated energy

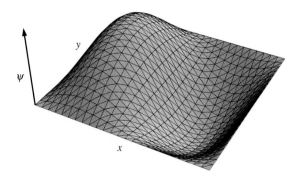

FIGURE 40-28 The wave function for a particle confined to a square region in two dimensions. The function (which satisfies the two-dimensional Schrödinger equation) is $\sin(n_x \pi x/L) \sin(n_y \pi y/L)$, with $n_x = 2$ and $n_y = 1$. Note that the two different quantum numbers correspond to a full wavelength in the x direction and a half wavelength in the y direction. An analogous expression would hold for a particle confined in three dimensions.

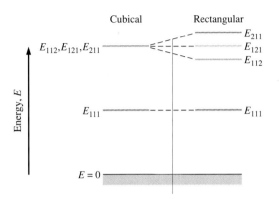

FIGURE 40-29 Energy-level diagrams showing the ground state and first excited state for a particle in a three-dimensional box. At left the box is cubical and the excited level is threefold degenerate. At right, the sides L_x, L_y, and L_z are slightly unequal, and chosen so that $L_x < L_y < L_z$ in such a way that the ground-state energy is unchanged. Introducing this asymmetry to the box removes the degeneracy, splitting the first excited state into three closely spaced but distinct levels.

level. For a particle of mass m confined to a cubical box of side L, for example, a generalization of the one-dimensional square well leads to the energy levels

$$E = \frac{h^2}{8mL^2}(n_x^2 + n_y^2 + n_z^2), \qquad (40\text{-}12)$$

where the n's are the quantum numbers associated with each spatial dimension. As in one dimension, the allowed values for the n's are positive integers. Thus, the ground state has $n_x = n_y = n_z = 1$. But what's the first excited state? It could be $n_x = 2$, $n_y = n_z = 1$. But it could equally well be $n_x = n_y = 1$, $n_z = 2$, or $n_x = n_z = 1$, $n_y = 2$, since all three of these combinations give the same energy.

When two or more quantum states have the same energy, those states are said to be **degenerate.** The first excited state of a particle confined to a cubical box, for example, is threefold degenerate, meaning there are three distinct states with the same energy. Degeneracy is a new feature of quantum mechanics in two and three dimensions. It is often associated with symmetry of the quantum-mechanical system. In the cubical box, for example, it is the fact that the sides have equal lengths that allows different combinations of quantum numbers to give exactly the same energy. Making all three sides different would remove the degeneracy, splitting what was a single energy level into three levels (Fig. 40-29). The same thing happens in more realistic quantum systems. For example, imposing a magnetic field on an otherwise spherically symmetric atom breaks the symmetry and may split energy levels that were previously degenerate (Fig. 40-30).

40-8 RELATIVISTIC QUANTUM MECHANICS

Like Newtonian physics, quantum mechanics based on the Schrödinger equation is a nonrelativistic theory—that is, it is not consistent with the requirement of special relativity that the laws of physics be the same in all inertial frames. It is therefore an approximation valid only at speeds much less than that of light. For

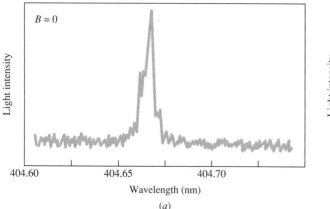 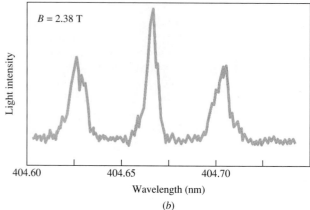

FIGURE 40-30 (*a*) Spectral line at 404.66 nm produced by mercury atoms undergoing transitions from the $n = 7$ level to the $n = 6$ level. The upper level is actually threefold degenerate. (*b*) Application of a 2.38-T magnetic field breaks the symmetry and removes the degeneracy, giving three lines where before there was one. Plots were taken with a grating spectrometer.

many applications, including most in atomic, molecular, solid-state, and low-temperature physics, the nonrelativistic approximation $v \ll c$ is valid and the Schrödinger equation applies. But when particle speeds become significant compared with the speed of light, the Schrödinger equation becomes inadequate and must be replaced with a truly relativistic wave equation. And even for slowly moving particles, the requirement of relativistic invariance leads to some remarkable new phenomena.

The Dirac Equation and Antiparticles

In 1928, the English physicist Paul Dirac formulated a relativistic wave equation applicable to electrons. In the process, Dirac encountered several unexpected mathematical requirements with deep physical implications.

In Schrödinger's quantum mechanics, as well as in Newtonian physics, the nonrelativistic assumption $v \ll c$ is inherent in the expression $E = \frac{1}{2}mv^2$ for the energy of a free particle. The corresponding relation between energy and momentum, $E = p^2/2m$, is at the heart of the Schrödinger equation. Actually this expression is an approximation to the more general form we developed in Chapter 38:

$$E^2 = m^2c^4 + p^2c^2.$$

Dirac's relativistic wave equation incorporated this relativistic energy-momentum relation in place of the Newtonian expression used in the Schrödinger equation. But Dirac faced a problem: the relativistic expression implies two possible values for the energy E, depending on which sign is used when one takes the square root. Dirac found that it was not possible to dismiss the negative root, and that therefore quantum states with negative energy must be possible. Through a rather involved argument, Dirac showed that the existence of these negative-energy states implied the possibility of a particle identical in mass to the electron but carrying a positive charge. The 1932 discovery of this particle, called the

positron, was a vindication of Dirac's brilliant theorizing. Although Dirac's theory applies specifically to the electron, we now know that every particle has an associated **antiparticle,** identical in mass but opposite in electrical, magnetic, and related properties.

Since a particle-antiparticle pair has no net electric charge, the creation of such a pair does not violate charge conservation. Einstein's mass-energy equivalence therefore implies that such **pair creation** is possible given an energy of at least $2mc^2$, where m is the mass of each particle. In a typical pair-creation event, a photon of energy at least twice the rest energy of a given particle disappears, and in its place appear the particle and its antiparticle (Fig. 40-31). The opposite process, annihilation, also occurs as a particle and its antiparticle disappear to form a pair of photons.

The concept of antiparticles has proved essential in high-energy physics and in studies of the very early universe. Essentially, particle-antiparticle pairs form when available energy—in a particle accelerator beam, or in the thermal energy just following the big bang—exceeds the threshold $2mc^2$ for formation of a pair of particles of mass m. In such a high-energy environment the equivalence of mass and energy is far more obvious than in the cool, low-energy world of our everyday experience.

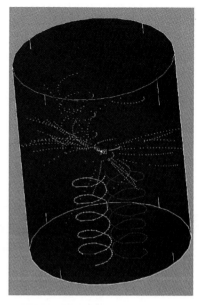

FIGURE 40-31 Pair creation. A high-energy photon produced an electron-positron pair whose trajectories spiral in opposite directions about a magnetic field. This image was made with the ALEPH detector at the Large Electron–Positron Collider at CERN.

Electron Spin

Another unexpected mathematical result of Dirac's work was his discovery that the relativistic wave function could not be a scalar; instead, it proved necessary to work with matrices. Dirac showed that this mathematical situation implied physically that the electron must possess an intrinsic angular momentum of magnitude $(\sqrt{3}/2)\hbar$—something that other quantum physicists had already inferred from experiments, but without any theoretical grounding. The existence of this angular momentum, called **spin,** has enormous significance in quantum mechanics, as we will examine extensively in the next chapter.

For now, we note simply that electron spin is grounded ultimately in the requirement of relativistic invariance, and is a relativistic effect that applies to *all* electrons, regardless of speed. In this sense, the intrinsic angular momentum of an electron is like its rest energy: both are consequences of relativistic invariance and both exist even for electrons at rest.

CHAPTER SYNOPSIS

Summary

1. Experimentally, matter and light are observed to manifest both particlelike and wavelike behavior. Quantum mechanics describes this situation through a statistical linkage between the **wave function** and the probability of finding the corresponding particle. The probability of finding a particle is proportional to the square of the associated wave amplitude.

    a. For light, the appropriate wave equation derives from Maxwell's equations for the electric and magnetic fields. The corresponding particle is the photon, and the probability of finding a photon is proportional to the light intensity—that is, to the square of the electric or magnetic field strength.

    b. For matter moving nonrelativistically, the appropriate wave equation is the **Schrödinger equation.** The associated particles are most likely to be found where the square of the wave function is greatest.

2. The **time-independent Schrödinger equation,**

$$-\frac{\hbar^2}{2m}\frac{d^2\psi}{dx^2} + U(x)\psi = E\psi,$$

governs the spatial behavior of matter waves associated with particles moving in one dimension.

a. The wave function ψ is itself an unobservable quantity. Its square, however, gives the **probability density,** or probability per unit volume (unit length in one dimension) of finding the particle. That is, in one dimension the probability of finding the particle in some small interval dx at position x is given by

$$\text{Probability} = \psi^2(x)\,dx.$$

b. In the standard interpretation of quantum mechanics, the link between ψ and the finding of an actual particle is always statistical; the strict causality and determinism of classical physics are replaced by a fundamental indeterminism in the microscopic world. Only the probability that a particular outcome will occur is governed by exact laws.

c. A particle must be found *somewhere*. Mathematically, this is expressed through the **normalization condition:**

$$\int_{-\infty}^{\infty} \psi^2(x)\,dx = 1.$$

d. The wave function ψ is further constrained to be a continuous function, and its derivative $d\psi/dx$ must also be continuous except where the potential energy $U(x)$ goes discontinuously to infinity.

3. Solving the Schrödinger equation involves finding functions that satisfy the equation for a given potential energy $U(x)$. Of these functions, only those that are normalizable and that meet the continuity conditions are acceptable solutions. In general, potential-energy functions that would confine a classical particle to a finite region result in **energy quantization**—that is, in solutions to the Schrödinger equation for only certain discrete values of the total energy E.

4. A particle in an infinite square well provides a simple application of the Schrödinger equation. Here the discrete energy levels are given by

$$E_n = \frac{n^2 h^2}{8mL^2},$$

where n, a positive integer, is the **quantum number.** The lowest possible energy, in this case $h^2/8mL^2$, is the **ground-state energy.** In general, zero energy is not permitted for a quantum-mechanical particle in a confined space—a manifestation of the uncertainty principle.

5. For the quantum-mechanical harmonic oscillator, a good approximation to many atomic and molecular systems, the energy levels are given by

$$E_n = \left(n + \frac{1}{2}\right)\hbar\omega,$$

where the ground-state energy $\frac{1}{2}\hbar\omega$ corresponds to $n = 0$.

6. Solutions $\psi(x)$ to the Schrödinger equation can have nonzero values in regions where particle motion is not possible in classical physics. As a result, a particle may tunnel through a finite-width barrier even though classical physics suggests it has insufficient energy. Such **quantum tunneling** is important in many nuclear reactions including fusion and alpha decay, and is exploited in technologies including some microelectronic devices and the scanning-tunneling microscope.

7. For potential wells of finite depth, particle energies are quantized only for levels below the well top. Any energy above that level is possible, giving rise to a **continuum** of states.

8. In three dimensions, three quantum numbers are needed to specify the quantum state of a particle. Different combinations of quantum numbers, representing distinct quantum states, may give the same quantized energy. When this happens, the associated energy level is said to be **degenerate.**

9. For electrons moving at relativistic speeds, the Schrödinger equation must be replaced with Dirac's relativistic wave equation. In deriving that equation, Dirac predicted the existence of **antiparticles,** a prediction soon confirmed experimentally with the discovery of the positron in 1932. Dirac's work also shows that relativistic invariance requires that all electrons have an intrinsic angular momentum called **spin.**

Terms You Should Understand

(Pairs are closely related terms whose distinction is important; number in parentheses is chapter section where term first appears.)

old quantum theory, quantum mechanics (introduction)
Schrödinger equation (40-2)
wave function (40-2)
probability density (40-2)
normalization (40-2)
quantum number (40-3)
square-well potential (40-3, 40-6)
quantum tunneling (40-5)
bound state, unbound state (40-6)
continuum (40-6)
degeneracy (40-7)
antimatter (40-8)
spin (40-8)

Symbols You Should Recognize

ψ (40-2)

Problems You Should Be Able to Solve

solving the Schrödinger equation for simple potentials (40-3)
evaluating energy levels for the infinite square well (40-3)
evaluating energy levels for the quantum harmonic oscillator (40-4)
finding energy levels and degeneracy in three-dimensional square wells (40-7)

Limitations to Keep in Mind

The Schrödinger equation is nonrelativistic, and therefore only applies when particle velocities are much less than the speed of light.

Quantum mechanics defies explanation in terms of Newtonian particles with well-defined trajectories. This is especially true at low quantum numbers; at higher quantum numbers, the correspondence principle shows how the quantum and classical views converge.

QUESTIONS

1. Explain qualitatively why a particle confined to a finite region cannot have zero energy.
2. Does quantum tunneling violate energy conservation? Explain.
3. In describing the quantum-mechanical behavior of matter, we have settled on a wave equation (the Schrödinger equation) to express the underlying physical law. Does this mean that the wave properties of matter are in some way more important than the particle properties? Explain.
4. Bohr's correspondence principle states that quantum and classical mechanics must agree in a certain limit. What limit? Give an example.
5. Our analysis of the infinite square well shows that it is possible to know exactly the energy of a quantum-mechanical particle confined to a finite region. How is this not a violation of the uncertainty principle? After all, knowing the energy amounts to knowing the speed and therefore the magnitude of the momentum.
6. Why is the blue curve of Fig. 40-14 not a suitable ψ function? After all, it satisfies the Schrödinger equation.
7. Consider two states of a particle in a cubical box that correspond to the same energy (for example, the states n_x, n_y,

$n_z = 1, 1, 2$ and $n_x, n_y, n_z = 2, 1, 1$). Is there any physically measurable property that could be used to distinguish these states? Explain.
8. In terms of de Broglie's matter wave hypothesis, how does making the sides of a box different lengths remove some of the degeneracy associated with a particle confined to that box?
9. A particle is confined to a two-dimensional box whose sides are in the ratio $1:2$. Are any of its energy levels degenerate? If so, give an example. If not, why not?
10. Is there any ratio of sides of a two-dimensional box for which a particle confined to the box has no degenerate energy levels? If so, give a ratio. If not, why not?
11. What did Einstein mean by his remark, loosely paraphrased, that "God does not play dice"?
12. Some philosophers argue that the strict determinism of classical physics is inconsistent with human free will, but that the indeterminacy of quantum mechanics does leave room for free will. Others claim that physics has no bearing on the question of free will. What do you think?

PROBLEMS

Section 40-2 The Schrödinger Equation

1. What are the units of the wave function $\psi(x)$ in a one-dimensional situation?
2. The wave function for a particle is given by $\psi = Ae^{-x^2/a^2}$, where A and a are constants. (a) Where is the particle most likely to be found? (b) Where is the probability per unit length half its maximum value?
3. The solution to the Schrödinger equation for a particular potential is $\psi = 0$ for $|x| > a$, and $\psi = A\sin(\pi x/a)$ for $-a \le x \le a$, where A and a are constants. In terms of a, what value of A is required to normalize ψ?
4. Find an expression for the normalization constant A for the wave function given by $\psi = 0$ for the case $|x| > b$ and by $\psi = A(b^2 - x^2)$ for $-b \le x \le b$.
5. Use a table of definite integrals or symbolic math software to help evaluate the normalization constant A in the wave function of Problem 2.

6. Suppose ψ_1 and ψ_2 are both solutions of the Schrödinger equation for the same energy E. Show that $a\psi_1 + b\psi_2$ is also a solution, where a and b are arbitrary constants.
7. Describe the potential, in terms of x and the particle energy E, that would result in the wave function of Problem 4.

Section 40-3 The Infinite Square Well

8. Determine the ground-state energy for an electron in an infinite square well of width 10.0 nm.
9. What is the width of an infinite square well in which a proton cannot have an energy less than 100 eV?
10. One reason we don't notice the effects of quantum mechanics in everyday life is that Planck's constant h is too small. Treating yourself as a particle in a room-sized one-dimensional infinite square well, how big would h have to be if your minimum possible energy corresponded to a speed of

1.0 m/s? Take the width of the room to be 2.6 m and your mass to be 60 kg.

11. A particle is confined to an infinite square well of width 1.0 nm. If the energy difference between the ground state and the first excited state is 1.13 eV, is the particle an electron or a proton?

12. An electron is trapped in an infinite square well 25 nm wide. Find the wavelengths of the photons emitted in the following transitions: (a) $n = 2$ to $n = 1$; (b) $n = 20$ to $n = 19$; (c) $n = 100$ to $n = 1$.

13. An electron drops from the $n = 7$ to the $n = 6$ level of an infinite square well 1.5 nm wide. Find (a) the energy and (b) the wavelength of the photon emitted.

14. Show explicitly that the difference between adjacent energy levels in an infinite square well becomes arbitrarily small compared with the energy of the upper level, in the limit of large quantum number n.

15. An electron is in a narrow molecule 4.4 nm long, a situation that approximates a one-dimensional infinite square well. If the electron is in its ground state, what is the maximum wavelength of electromagnetic radiation that can cause a transition to an excited state?

16. A snail of mass 3 g crawls between two rocks 15 cm apart at a speed of 0.5 mm/s. Treating this system like a particle in an infinite square well, determine the approximate quantum number. Does the correspondence principle permit the use of the classical approximation in this case?

17. Repeat Example 40-1 for a proton trapped in a nuclear-size square well of width 1 fm. Comparison with the result of Example 40-1 gives a rough estimate of the energy difference between nuclear and chemical reactions.

18. The ground-state energy for an electron in square well A is equal to the energy of the first excited state for an electron in square well B. How do the widths of the two wells compare? Both are infinitely deep.

19. An electron drops from the $n = 5$ level to the $n = 2$ level of an infinite square well, emitting a 1.4-eV photon in the process. Find the width of the square well.

20. Electrons in an ensemble of many square-well systems are initially in the $n = 4$ state. The wells all have the same width, 0.10 nm. Find the wavelengths of all spectral lines emitted as the electrons cascade downward to the ground state through all possible transitions.

21. Sketch the probability density for the $n = 2$ state of an infinite square well extending from $x = 0$ to $x = L$, and determine where the particle is most likely to be found.

22. An infinite square well extends from $-L/2$ to $L/2$. (a) Find expressions for the normalized wave functions for a particle of mass m in this well. You will need separate expressions for even and odd quantum numbers. (b) Find the corresponding energy levels.

23. A particle is in the ground state of an infinite square well. What is the probability of finding the particle in the left-hand third of the well?

Sections 40-4, 40-5, and 40-6 The Harmonic Oscillator, Quantum Tunneling, and Finite Potential Wells

24. A particle in a harmonic oscillator potential has a ground-state energy of 0.14 eV. What would be the classical frequency f of the oscillator?

25. What is the ground-state energy for a particle in a harmonic oscillator potential whose classical angular frequency is $\omega = 1.0 \times 10^{17}$ s^{-1}?

26. A 10-g mass is attached to an ideal spring of spring constant $k = 150$ N/m, and set oscillating with amplitude 12 cm. To what quantum number does this oscillation correspond?

27. A hydrogen chloride molecule may be modeled as a hydrogen atom (mass 1.67×10^{-27} kg) on a spring; the other end of the spring is attached to a rigid wall (the massive chlorine atom). If the minimum photon energy that will promote this molecule to its first excited state is 0.358 eV, find the "spring constant."

28. Show that the two functions $Ae^{\pm m\omega x^2/2\hbar}$ both satisfy the Schrödinger equation for a one-dimensional harmonic oscillator (Equation 40-9), and determine the corresponding energies.

29. For what quantum numbers is the spacing between adjacent energy levels in a harmonic oscillator less than 1% of the actual energy?

30. Figure 40-32 shows a finite square well. Consider a particle of mass m whose energy E lies between 0 and U_0. (a) Write two forms of the Schrödinger equation for this particle, corresponding to the interior ($U = 0$) and exterior ($U = U_0$) of the well. (b) Show that the equation for the well interior is satisfied by functions of the form $A \sin(\sqrt{2mE}\, x/\hbar)$ and $A \cos(\sqrt{2mE}\, x/\hbar)$, where A is an arbitrary constant. (c) Show that the equation for the well exterior is satisfied by functions of the form $Ae^{\pm\sqrt{2m(U_0-E)}x/\hbar}$. *Note:* You are not solving the whole problem because your solutions aren't necessarily consistent at the well boundaries. Problems 55–57 treat the full solution.

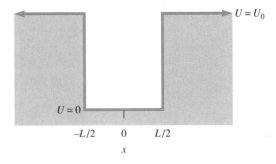

FIGURE 40-32 Problems 30, 31.

31. When the square well width L of Fig. 40-32, the well depth U_0, and the particle mass m are related by

$\sqrt{mU_0L^2/2\hbar^2} = 4$, a detailed analysis shows that the ground-state energy is given by $E_1 = 0.098U_0$ and the corresponding wave function is approximately

$$\psi_1 = 17.9\frac{1}{\sqrt{L}}e^{7.60x/L} \qquad (x \le -\tfrac{1}{2}L)$$

$$\psi_1 = 1.26\frac{1}{\sqrt{L}}\cos(2.50x/L) \qquad (-\tfrac{1}{2}L \le x \le \tfrac{1}{2}L)$$

$$\psi_1 = 17.9\frac{1}{\sqrt{L}}e^{-7.60x/L} \qquad (x \ge \tfrac{1}{2}L).$$

Show that both ψ_1 and its first derivative are continuous at the well edge. (That is, show that the second and third expressions above have the same value and the same derivative at $x = \tfrac{1}{2}L$.) Because the functions given are approximate, your results won't be exactly the same, but should agree to two significant figures. *Hint:* Be sure to set your calculator to radians before computing the trig functions! (b) Plot the function $\sqrt{L}\psi_1$ as a function of x/L, indicating the well edges on your plot.

32. For the state described in the preceding problem, find the probability that the particle will be found somewhere in the classically forbidden region outside the well. *Hint:* Symmetry allows you to calculate the probability for $x \ge \tfrac{1}{2}L$, then double the result. Also see Example 40-2, and consult the integral table in Appendix A if necessary.

33. The probability that a particle of mass m and energy $E < U$ will tunnel through the potential barrier of Fig. 40-17 is approximately

$$P = e^{-2\sqrt{2m(U-E)}L/\hbar},$$

where L is the barrier width. Evaluate this probability (a) for a 2.8-eV electron incident on a 1.0-nm wide barrier 4.0 eV high and (b) for a 1200-kg car moving at 15 m/s striking a 1.0-m-thick stone wall requiring, classically, 150 kJ to breach it.

Section 40-7 The Schrödinger Equation in Three Dimensions

34. If all sides of a cubical box are doubled, what happens to the ground-state energy of a particle in that box?

35. A very crude model for an atomic nucleus is a cubical box about 1 fm on a side. What would be the energy of a gamma ray emitted if a proton in such a nucleus made a transition from its first excited state to the ground state?

36. (a) Use Equation 40-12 to draw an energy-level diagram for the first six energy levels of a particle in a cubical box, in terms of $h^2/8mL^2$, and (b) give the degeneracy of each.

37. The generalization of the Schrödinger equation to three dimensions is

$$-\frac{\hbar^2}{2m}\left(\frac{\partial^2\psi}{\partial x^2} + \frac{\partial^2\psi}{\partial y^2} + \frac{\partial^2\psi}{\partial z^2}\right) + U(x,y,z)\psi = E\psi.$$

(a) For a particle confined to the cubical region $0 \le x \le L$, $0 \le y \le L$, $0 \le z \le L$, show by direct substitution that the equation is satisfied by wave functions of the form $\psi(x,y,z) = A\sin(n_x\pi x/L)\sin(n_y\pi y/L)\sin(n_z\pi z/L)$, where the n's are integers and A is a constant. (b) In the process of working part (a), verify that the energies E are given by Equation 40-12.

Section 40-8 Relativistic Quantum Mechanics

38. (a) Find the minimum energy of a photon needed to create an electron-positron pair. (b) Find the corresponding photon frequency and wavelength, and use Fig. 34-11 to determine in what part of the electromagnetic spectrum it lies.

39. Early in the history of the universe, the temperature was so high that thermal energy resulted in frequent pair creation. Find the approximate temperature such that the thermal energy kT would be enough to create (a) an electron-positron pair and (b) a proton-antiproton pair.

40. A muon and its antiparticle annihilate, producing a pair of 106-MeV gamma-ray photons. Find the mass of the muon.

Paired Problems

(Both problems in a pair involve the same principles and techniques. If you can get the first problem, you should be able to solve the second one.)

41. An alpha particle (mass 4 u) is trapped in a uranium nucleus of diameter 15 fm. Treating the system as a one-dimensional square well, what would be the minimum energy for the alpha particle?

42. A laser emits 1.96-eV photons. If this emission is due to electron transitions from the $n = 2$ to $n = 1$ states of an infinite square well, what is the width of the well?

43. What is the probability of finding a particle in the central 80% of an infinite square well, assuming it's in the ground state?

44. A particle detector has a resolution 15% of the width of an infinite square well. What is the chance that the detector will find a particle in the ground state of the square well if the detector is centered on (a) the midpoint of the well or (b) a point one-fourth of the way across the well?

45. A harmonic oscillator emits a 1.1-eV photon as it undergoes a transition between adjacent states. What is its classical oscillation frequency?

46. An ensemble of harmonic oscillators undergoes transitions from the first excited state to the ground state, and the spectrum from the ensemble shows a line at 5.6 μm. Find the frequency of the oscillators.

47. An electron is confined to a cubical box. For what box width will a transition from the first excited state to the ground state result in emission of a 950-nm infrared photon?

48. A proton is confined to a cubical box 2.5 fm on a side. Find the minimum energy of a gamma ray that could be emitted in transitions among any of the first 6 energy levels.

Supplementary Problems

49. For what quantum state is the probability of finding a particle in the left-hand quarter of an infinite square well equal to 0.303?

50. A 9-W laser beam shines on an ensemble of 10^{24} electrons, each in the ground state of a one-dimensional infinite square well 0.72 nm wide. The photon energy is just enough to raise an electron to its first excited state. How many electrons can be so excited if the beam shines for 10 ms?

51. A large number of electrons are all confined to infinite square wells 1.2 nm wide. They are undergoing transitions among all possible states. How many (a) visible (400 nm to 700 nm) and (b) infrared lines will there be in the spectrum emitted by this ensemble of square-well systems?

52. Show that the turning points for a classical harmonic oscillator of mass m with energy given by Equation 40-11 are located at $x = \pm \sqrt{\dfrac{\hbar}{m\omega}(2n + 1)}$.

53. Consider an infinite square well with a steplike potential in the bottom, as shown in Fig. 40-33. Without solving any equations, sketch what you think the wave function should look like for a particle whose energy is (a) less than the step height and (b) greater than the step height.

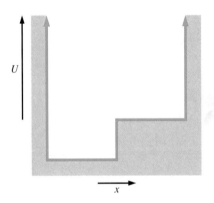

FIGURE 40-33 Problem 53.

54. A particle is in the nth quantum state of an infinite square well. (a) Show that the probability of finding it in the left-hand quarter of the well is

$$P = \frac{1}{4} - \frac{\sin(n\pi/2)}{2n\pi}.$$

(b) Show that for even n the probability is exactly $\frac{1}{4}$. (c) Show that for odd n the probability approaches the classical value $\frac{1}{4}$ as $n \to \infty$.

55. The next three problems illustrate the matching procedure necessary in solving the Schrödinger equation for potential wells of finite depth. Consider a particle of mass m in the semi-infinite well shown in Fig. 40-34. It will be convenient to work with dimensionless forms of the particle energy E and well depth U_0, in units of $\hbar^2/2mL^2$: $\varepsilon \equiv 2mL^2E/\hbar^2$

and $\mu \equiv 2mL^2U_0/\hbar^2$. Assume that $E < U_0$ or, equivalently, $\varepsilon < \mu$. Show by substitution that the following wave functions satisfy the Schrödinger equation in the regions indicated:

$$\psi_1 = A \sin(\sqrt{\varepsilon}\,x/L) \qquad (0 \leq x \leq L)$$

and

$$\psi_2 = Be^{-\sqrt{\mu - \varepsilon}\,x/L} \qquad (x \geq L),$$

where A and B are arbitrary constants.

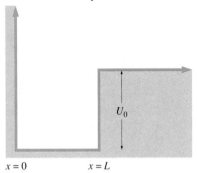

FIGURE 40-34 Problems 55–57.

56. The wave functions of the preceding problem satisfy the Schrödinger equation, but they don't necessarily meet the conditions that ψ and its derivative be continuous; in particular, both quantities could change abruptly at the well edge $x = L$. To achieve continuity of ψ, evaluate ψ_1 and ψ_2 at $x = L$ and set them equal. To achieve continuity of $d\psi/dx$, first differentiate ψ_1 and ψ_2, then evaluate the derivatives at $x = L$, and then set them equal. The two continuity conditions should result in these equations:

$$A \sin(\sqrt{\varepsilon}) = Be^{-\sqrt{\mu - \varepsilon}}$$

and

$$\sqrt{\varepsilon}A \cos(\sqrt{\varepsilon}) = -\sqrt{\mu - \varepsilon}\,Be^{-\sqrt{\mu - \varepsilon}}$$

Use the first of these equations to eliminate the quantity $Be^{-\sqrt{\mu - \varepsilon}}$ from the second equation, and show that the result is

$$\tan(\sqrt{\varepsilon}) = -\sqrt{\frac{\varepsilon}{\mu - \varepsilon}}.$$

This equation determines the values of ε that give the quantized energy levels.

57. The equation derived in the preceding problem cannot be solved algebraically since the unknown ε appears in a trig function and under the square root. It can be solved graphically, by plotting both sides on the same graph and determining where they intersect. It can also be solved by trial and error on a calculator, by using a calculator with a root-finding routine, or by computer. Use one of these methods to find all possible values of ε for (a) $\mu = 2$, (b) $\mu = 20$, and (c) $\mu = 50$. *Note:* The number of solutions varies with μ; there may be no solutions, meaning no bound states are possible, or there may be one or more bound states.

CHAPTER 41

ATOMIC PHYSICS

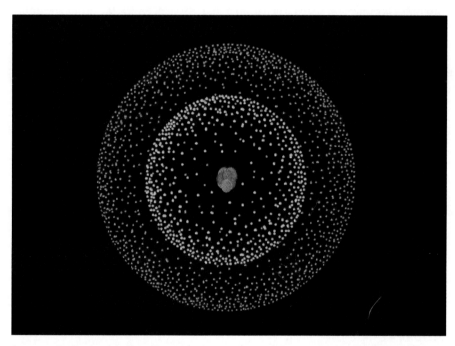

Quantum mechanics reveals the atom as a nucleus surrounded by a "cloud" of probability indicating the likelihood of finding the atomic electrons at particular locations. This computer-generated model of a beryllium atom shows that the electron "cloud" consists of two distinct shells.

In the preceding chapter we developed some key ideas of quantum mechanics, largely by exploring the Schrödinger equation in application to simple systems. Now we turn to the atom and show how the Schrödinger equation, along with two important results from relativistic quantum mechanics, leads us to a description of atomic spectra and to an explanation of the periodic table and the chemical properties of the elements. We will deal most thoroughly with the simplest atom, hydrogen. Even with this case, we must skip many mathematical details that are beyond the scope of this book. Nevertheless, we will make plausible a number of important atomic properties, many of which have analogs in the simpler systems of the preceding chapter.

41-1 THE HYDROGEN ATOM

Like the particle in a box that we considered in the preceding chapter, the electron in a hydrogen atom is also a particle confined by a force associated with a three-dimensional potential-energy function. For the electron, the potential energy is that of the electrostatic attraction between the electron and proton. In Chapter 25 we found that the electric potential associated with a point charge q is $V(r) = kq/r$, where r is the distance from the point charge and where the zero of potential is at infinity. In treating the hydrogen atom, we consider the more massive proton (1836 times the electron mass) to be fixed at the origin. The proton's charge is $+e$, so the electric potential associated with the proton becomes $V(r) = ke/r$. Recall that electric potential is the potential energy *per unit charge;* to get the potential energy of an electron in the field of the proton, we multiply the potential by the electron charge, $-e$, to get

$$U(r) = \frac{-ke^2}{r}. \tag{41-1}$$

This is the potential-energy function to be used in the Schrödinger equation for the electron in a hydrogen atom. Figure 41-1 shows a plot of Equation 41-1.

The Schrödinger Equation in Spherical Coordinates

Substituting Equation 41-1 for $U(x, y, z)$ in the three-dimensional Schrödinger equation (see Problem 40-37) gives the equation we must solve to find the structure of the hydrogen atom. But our potential energy is expressed in terms of the radial distance r, while the Schrödinger equation is in rectangular coordinates x, y, z. We could write r in terms of x, y, z, but the resulting equation would be unenlightening because rectangular coordinates do not reflect the natural spheri-

FIGURE 41-1 Potential energy of the electron in a hydrogen atom (in eV), as a function of position in a plane containing the proton. Horizontal axes are distances from the proton in units of the Bohr radius $a_0 = 0.0529$ nm.

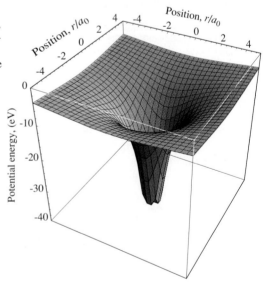

cal symmetry of the potential-energy function. Instead, it is easiest to work in **spherical coordinates,** where the position of a point is specified by giving its distance r from the origin along with two angles θ and ϕ that specify its orientation (Fig. 41-2). Converting to spherical coordinates is straightforward but tedious; the result is the equation

$$-\frac{\hbar^2}{2mr^2}\left[\frac{\partial}{\partial r}\left(r^2\frac{\partial\psi}{\partial r}\right) + \frac{1}{\sin\theta}\frac{\partial}{\partial\theta}\left(\sin\theta\frac{\partial\psi}{\partial\theta}\right) + \frac{1}{\sin^2\theta}\frac{\partial^2\psi}{\partial\phi^2}\right] - \frac{ke^2}{r}\psi \tag{41-2}$$
$$= E\psi,$$

where we have used Equation 41-1 for the potential energy function, and where ψ is now a function of the spherical coordinates r, θ, ϕ of Fig. 41-2.

Although Equation 41-2 looks forbidding, it can be solved readily using more advanced techniques than you have probably encountered. For total energy E less than zero, corresponding to bound states in the potential well of Fig. 41-1, it turns out that most solutions become infinite at large values of r and are therefore not normalizable. As with simpler one-dimensional systems, only certain values of the energy E result in acceptable bound-state solutions. For total energy greater than zero, the electron is unbound and any energy proves possible, as with the finite square well.

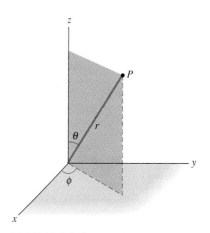

FIGURE 41-2 Spherical coordinates locate a point P in terms of its distance r from the origin, the angle θ from the z axis, and the angle ϕ that measures position relative to the x axis.

The Hydrogen Ground State

In general, solutions to Equation 41-2 depend on all three variables r, θ, ϕ. But some solutions, including the ground state, are spherically symmetric—that is, they depend only on r. Here we show that the ground state has the form of a simple exponential, and in the process derive the ground-state energy. Consider the function

$$\psi = Ae^{-r/a_0}, \tag{41-3}$$

where A and a_0 are as yet undetermined constants, the latter with the units of length. For this spherically symmetric function, nothing depends on the angular variables θ and ϕ, so derivatives with respect to those variables are strictly zero. Since we are then dealing with a function of only one variable, we can write total instead of partial derivatives. Equation 41-2 then becomes

$$-\frac{\hbar^2}{2mr^2}\frac{d}{dr}\left(r^2\frac{d\psi}{dr}\right) - \frac{ke^2}{r}\psi = E\psi, \tag{41-4}$$

where ψ is now a function of r only. Substituting the proposed solution 41-3 into this equation gives good practice in differentiating (see Problem 19); the result, after some algebra, is

$$-\frac{\hbar^2}{2ma_0^2} + \frac{\hbar^2}{mra_0} - \frac{ke^2}{r} = E_1, \tag{41-5}$$

where we write E_1 to designate the ground-state energy. Equation 41-5 must be satisfied for all values of r. But this can be the case only if the two r-dependent terms cancel; that is, if

$$\frac{\hbar^2}{mra_0} = \frac{ke^2}{r}$$

or

$$a_0 = \frac{\hbar^2}{mke^2} = 5.29 \times 10^{-11} \text{ m} = 0.0529 \text{ nm}.$$

Substituting this result into Equation 41-5 then gives the ground-state energy:

$$E_1 = -\frac{\hbar^2}{2ma_0^2} = -2.18 \times 10^{-18} \text{ J} = -13.6 \text{ eV},$$

where the minus sign shows that the atom is a bound system. Equivalently, it would take 13.6 eV or more to bring the electron to an unbound state with $E > 0$—that is, to ionize the atom. Our results show that Equation 41-3 is indeed a solution to the Schrödinger equation for the hydrogen atom, corresponding to energy $E_1 = -13.6$ eV.

In deriving expressions for a_0 and E_1, we've shown how Schrödinger's theory gives rise to two of the fundamental parameters of atomic physics. We've actually seen these parameters before: a_0 is the **Bohr radius,** defined in Chapter 39 as the orbital radius of the smallest allowed electron orbit in Bohr's atomic theory, while our E_1 is the same as the electron energy in the lowest Bohr orbit. But while the ground-state energies of the Bohr and Schrödinger theories agree, the interpretation of these two theories differs dramatically, and, as we will see later, only the Schrödinger theory describes correctly the finer details of atomic structure. In particular, the Bohr theory still clings to the classical notion of electrons in circular orbits about the nucleus. But the Schrödinger theory is truly quantum-mechanical, with the electron represented by the wave function ψ and associated probability distribution. The Bohr radius a_0 no longer represents an actual orbital radius but instead determines the atomic size only in a statistical sense. Since the wave function falls off exponentially as e^{-r/a_0}, the probability of finding the electron becomes negligible at distances from the nucleus much greater than a_0.

The Radial Probability Distribution

Where are we most likely to find the electron in a hydrogen atom that is in its ground state? Figure 41-3 shows the probability density ψ^2 associated with Equation 41-3. At first glance it looks as if the electron is most likely to be at $r = 0$, right inside the nucleus! With the one-dimensional systems of the preceding chapter, that interpretation would have been correct. But here we have to be more careful. What we really want to know is this: at what radial distance from the nucleus are we most likely to find the electron? The probability density ψ^2 is the probability *per unit volume* of finding the electron at some point. What we want is the probability *per unit radial distance*. In a one-dimensional system, probability per unit length and per unit volume are proportional (see Equation 40-2 and

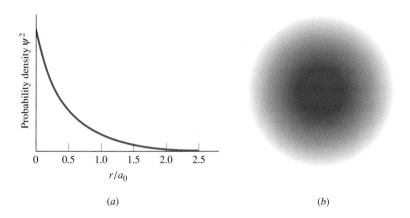

(a)   (b)

FIGURE 41-3 Probability density ψ^2
for the hydrogen ground state of Equation
41-3. (a) is a plot of ψ^2 as a function of
distance from the proton, while (b) shows
the "electron cloud," whose intensity is
proportional to the probability density.

preceding discussion). But in three dimensions they need not be. Figure 41-4
shows why.

Consider a thin spherical shell of radius r and thickness dr centered on the
nucleus. The area of the shell is $4\pi r^2$, so its volume is $dV = 4\pi r^2 dr$. Since the
ground-state wave function depends only on radius, so does the probability density, and the probability of finding the electron in the shell is given by

$$\text{Probability} = \psi^2 dV = 4\pi r^2 \psi^2 dr.$$

We define the **radial probability density,** $P(r)$, as the probability per unit radius
of finding the electron at radial distance r. Then the probability itself is $P(r)dr$,
and we see from the equation above

$$P(r) = 4\pi r^2 \psi^2, \tag{41-6}$$

or $4\pi r^2$ times the probability per unit volume. Physically, the two probability
densities differ because spherical shells of fixed thickness dr increase in volume
as their radii increase.

When we ask at what radius are we most likely to find the electron, we are
asking where the radial probability density of Equation 41-6 is a maximum.
Multiplying ψ^2 from Equation 41-3 by $4\pi r^2$ to get the radial probability density
for the hydrogen ground state gives

$$P_1(r) = 4\pi r^2 A^2 e^{-2r/a_0}.$$

Figure 41-5 is a plot of this radial probability density. As the figure shows, and as
you can prove in Problem 20, the maximum occurs at $r = a_0$, showing that the
electron is most likely to be found at the distance suggested by Bohr for the
ground-state orbit. But in contrast to Bohr's theory, the electron is *not* in orbit,
and it need not be found at a distance of one Bohr radius. In quantum mechanics,
the atom is a fuzzy thing. A hydrogen atom is roughly two Bohr radii in diameter,
but that statement is necessarily vague because of the nonzero probability of
finding the electron either closer to or farther from the nucleus.

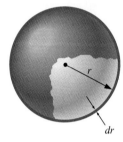

FIGURE 41-4 The volume of a
spherical shell of radius r and thickness
dr is $dV = 4\pi r^2 dr$. The probability per
unit radial distance is thus $4\pi r^2$ times the
probability per unit volume.

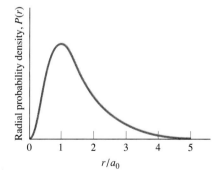

FIGURE 41-5 Radial probability
density for the hydrogen ground state.
The electron is most likely to be found
one Bohr radius from the nucleus.

● EXAMPLE 41-1 NORMALIZATION AND THE PROBABILITY DISTRIBUTION

Determine the normalization constant A of Equation 41-3 and, using the result, calculate the probability that the electron in the hydrogen ground state will be found beyond the Bohr radius.

Solution

The electron must be *somewhere,* so integrating the probability density ψ^2 over all space must give 1; mathematically this gives the normalization condition

$$\int_{\text{all space}} \psi^2 dV = 1.$$

With the spherically symmetric ground-state function, we found that $\psi^2 dV$ can be written $4\pi r^2 \psi^2 dr$. Using Equation 41-3 for ψ, the normalization condition then becomes

$$\int_{r=0}^{r=\infty} 4\pi A^2 r^2 e^{-2r/a_0} dr = 1.$$

We could evaluate the integral using integration by parts; however, the result is given in the integral table at the end of Appendix A. Replacing x by r and a by $-2/a_0$ in the table's expression for $\int x^2 e^{ax} dx$, we have

$$\int_{r=0}^{r=\infty} 4\pi A^2 r^2 e^{-2r/a_0} dr$$

$$= 4\pi A^2 \left\{ \frac{r^2 e^{-2r/a_0}}{(-2/a_0)} - \frac{2}{(-2/a_0)} \left[\frac{e^{-2r/a_0}}{(-2/a_0)^2} \left(-\frac{2}{a_0} r - 1 \right) \right] \right\} \Bigg|_{r=0}^{r=\infty}$$

$$= 1.$$

The expression in curly brackets vanishes at $r = \infty$, and at $r = 0$ the exponentials are just 1, so we have

$$4\pi A^2 \left[0 - \left(-\frac{1}{4} a_0^3 \right) \right] = 1,$$

or

$$A = \frac{1}{\sqrt{\pi a_0^3}}.$$

The probability of finding the electron beyond the Bohr radius is given by integrating the probability density over the region of interest—from $r = a_0$ to $r = \infty$:

$$P(r > a_0) = \int_{a_0}^{\infty} 4\pi r^2 \psi^2 dr.$$

This is the same integral that arose in evaluating the normalization constant A, except that now we evaluate between a_0 and infinity. Making this change in the limits, we are led to

$$P(r > a_0) = 4\pi A^2 a_0^3 \left(\frac{1}{2} e^{-2} + \frac{3}{4} e^{-2} \right) = 5\pi A^2 a_0^3 e^{-2}.$$

With $A^2 = 1/\pi a_0^3$, this becomes

$$P(r > a_0) = 5e^{-2} \approx 0.677.$$

Thus in about two of three cases, the electron will be found beyond the Bohr radius. In Problem 4, you can do a similar calculation to show that there is about one chance in four of finding the electron beyond two Bohr radii.

EXERCISE Find the probability that the electron in the hydrogen ground state will be found within $\frac{1}{2} a_0$ of the nucleus.

Answer: 0.080

Some problems similar to Example 41-1: 4, 51, 52
●

Excited States of Hydrogen

So far we have examined only the ground state of the hydrogen atom. It turns out that this state—described by the wave function 41-3—is the only state associated with the -13.6-eV ground-state energy. But Equation 41-2 admits many more normalizable solutions, corresponding to the excited states of hydrogen.

In general, each energy level is associated with one spherically symmetric wave function and a number of nonsymmetric ones. For historical reasons, the spherically symmetric states are called **s-states.** The distinct energy levels are labeled by the quantum number n, called the **principal quantum number.** The ground state, for example, is the 1s state. The energy of the nth level, derivable from the Schrödinger equation, turns out to agree exactly with the earlier Bohr theory:

$$E_n = -\frac{1}{n^2} \frac{\hbar^2}{2ma_0^2} = \frac{E_1}{n^2} = \frac{-13.6 \text{ eV}}{n^2}. \tag{41-7}$$

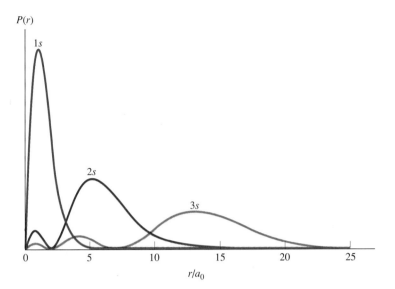

P(r)

The spherically symmetric state with energy E_2—that is, the $2s$ state—has wave function given by

$$\psi_{2s} = \frac{1}{4\sqrt{2\pi a_0^3}}\left(2 - \frac{r}{a_0}\right)e^{-r/2a_0}. \qquad (41\text{-}8)$$

By substituting this function into Equation 41-4, you can verify that the energy E_2 is given by Equation 41-7 (see Problem 61). The radial probability densities for the first three spherically symmetric states are plotted in Fig. 41-6; notice that the excited states correspond to larger, more "smeared-out" atoms.

Although we are discussing hydrogen specifically, our results are easily generalized to any single-electron atom—that is, to an atom of atomic number Z ionized $Z - 1$ times. For such an atom the potential energy function becomes $-kZe^2/r$, and our calculations go through as before except that the factor e^2 is replaced by Ze^2. Then the energy levels become

$$E_n = -\frac{Z^2}{n^2}\frac{\hbar^2}{2ma_0^2} = \frac{Z^2 E_1}{n^2} = -\frac{(13.6 \text{ eV})Z^2}{n^2}, \qquad (41\text{-}9)$$

reflecting the tighter binding of the more highly charged nucleus (see Problem 63).

Orbital Quantum Numbers and Angular Momentum

So far we have discussed only spherically symmetric states and the principal quantum number n that determines the energy. In these states, it turns out that the angular momentum associated with the electron's motion about the nucleus—also called the **orbital angular momentum**—is exactly zero. (This makes sense because the electron probability distribution is spherically symmetric, so there is no preferred axis about which angular momentum might be defined. But it contrasts with the Bohr theory prediction—which is incorrect—that the ground-state orbital angular momentum should be $\hbar$.) But there are other solutions to the

Schrödinger equation for the hydrogen atom, solutions that involve all three variables and are therefore not spherically symmetric. These solutions correspond to nonzero values of the orbital angular momentum.

For a given principal quantum number n, there are in fact n distinct solutions with different angular momenta that share the nth energy level given by Equation 41-7. (The fact that these different states have essentially the same energy is related to the $1/r$ nature of the Coulomb potential; so-called fine structure effects that we will discuss shortly can actually lead to small differences in energy among states with the same n.) The different angular momentum states are labeled with the quantum number ℓ, called the **orbital quantum number.** The range for ℓ is from 0 to $n - 1$, so the ground state ($n = 1$) corresponds to the single value $\ell = 0$. Higher energy levels, however, are degenerate. The orbital quantum number ℓ determines the magnitude, L, of the angular momentum associated with the electron's motion:

$$L = \sqrt{\ell(\ell + 1)}\hbar. \qquad (41\text{-}10)$$

● **EXAMPLE 41-2** ORBITAL ANGULAR MOMENTUM

What are the possible values for the magnitude of the orbital angular momentum of an electron in the $n = 3$ energy level of hydrogen?

Solution

At the nth energy level, there are n different angular momentum values associated with the ℓ values $\ell = 0, 1, \ldots n - 1$. Here $n = 3$ so $\ell = 0, 1,$ or 2. The corresponding magnitudes of the orbital angular momentum are, from Equation 41-10,

$$\ell = 0: L = \sqrt{0(0 + 1)}\hbar = 0,$$
$$\ell = 1: L = \sqrt{1(1 + 1)}\hbar = \sqrt{2}\hbar,$$
$$\ell = 2: L = \sqrt{2(2 + 1)}\hbar = \sqrt{6}\hbar.$$

EXERCISE Find the maximum value for the orbital angular momentum of a hydrogen atom in the $n = 5$ state.

Answer: $2\sqrt{5}\hbar$

Some problems similar to Example 41-2: 5, 6, 8, 9 ●

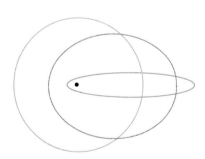

FIGURE 41-7 Classical electron orbits with the same energy but different angular momenta. Any value of L from zero to the maximum L of the circular orbit is allowed in classical physics.

For historical reasons* states with the ℓ values 0, 1, 2, 3, 4, 5, . . . are given the letter labels s, p, d, f, g, h, ... These are combined with the principal quantum number n to specify both the energy and angular momentum of a state. Thus, the ground state is 1s, while the $n = 2$ state with $\ell = 1$ is the 2p state. (The lowercase letters s, p, d, . . . are used in discussing individual electrons, while the corresponding capital letters denote orbital angular momentum states of an entire atom. For one-electron hydrogen, the two are the same.)

In the quantization of orbital angular momentum, we have another nonclassical result of quantum mechanics, for in classical physics an electron of a given energy should be able to have any angular momentum, up to a maximum value that corresponds to a circular orbit (Fig. 41-7). At large n, however, the number of allowed angular momentum values (namely, n) is so large that there appears to be a nearly continuous range of values. Again, we see Bohr's correspondence principle at work: for large quantum numbers, the quantum description ap-

*The first four letters stand for the spectroscopists' description of spectral lines as "sharp," "principal," "diffuse," and "fundamental." The remaining letters are alphabetical.

proaches the classical limit. But for small n, the quantum-mechanical discreteness of both energy and angular momentum is clearly evident, and stands in marked contrast to the classical prediction.

● EXAMPLE 41-3   QUANTIZING THE MOON

Estimate the orbital angular momentum of the moon, and find the corresponding ℓ value. Then find the difference in angular momenta between this ℓ value and the next lower value, and compare with the angular momentum itself.

Solution

Although the Earth-moon system is not an electrostatically bound hydrogen atom, we can still use Equation 41-10 to estimate the quantized angular momentum values. From Chapter 12, we have $L = mvr$ for an object of mass m moving at speed v in a circle of radius r. For the moon, these three quantities are given in Appendix E, and we have

$$L = mvr = (0.0735 \times 10^{24} \text{ kg})(1.0 \times 10^3 \text{ m/s})(0.385 \times 10^9 \text{ m})$$

$$= 2.8 \times 10^{34} \text{ J} \cdot \text{s}.$$

(Check that the units can be written as J·s.) Since $\hbar$ is so small, ℓ in Equation 41-10 will clearly be very large, so we can approx-

imate $\ell + 1 \simeq \ell$ to avoid solving a quadratic equation. With this approximation, Equation 41-10 becomes $L \simeq \ell\hbar$, or

$$\ell = \frac{L}{\hbar} = \frac{2.8 \times 10^{34} \text{ J} \cdot \text{s}}{1.06 \times 10^{-34} \text{ J} \cdot \text{s}} = 3 \times 10^{68}.$$

This huge ℓ value indicates that the system is essentially at the classical limit. With $L \simeq \ell\hbar$, the difference between adjacent angular momentum values is just $\hbar$. This is such a tiny fraction ($\simeq 10^{-68}$) of the angular momentum that quantization is totally unobservable.

EXERCISE   A hydrogen atom in an interstellar gas cloud is in the $n = 293$ state. What is the difference between the maximum possible value for its angular momentum and the next lower value, as a fraction of the maximum value?

Answer: 0.34%

Some problems similar to Example 41-3:   11, 14 ●

Space Quantization

Angular momentum is a vector, and in quantum mechanics it turns out that the angular momentum vector is quantized not only in magnitude but also in direction—a phenomenon called **space quantization.** Space quantization of orbital angular momentum gives rise to a third quantum number, designated m_ℓ. Space quantization becomes evident when an atom is in a magnetic field that establishes a preferred axis along which the angular momentum component can be measured; for this reason m_ℓ is called the **orbital magnetic quantum number.**

Space quantization requires, in particular, that the component L_z of orbital angular momentum along any chosen axis can have only values given by

$$L_z = m_\ell \hbar, \tag{41-11}$$

where m_ℓ takes on integer values from $-\ell$ to ℓ, with ℓ the orbital quantum number. Thus an $\ell = 1$ state can have one of the three possible m_ℓ values $-1, 0$, or $+1$, corresponding to angular momentum components $-\hbar$, 0, or $+\hbar$ along some axis. Since the magnitude of the angular momentum in an $\ell = 1$ state is $\sqrt{2}\hbar$ (see Example 41-2), none of these values corresponds to full alignment with the given axis. Instead, we can think geometrically of the angular momentum vectors as being constrained to lie at angles $\cos^{-1}(L_z/L)$ to the axis; for $\ell = 1$ these angles are $\pm 45°$ and $0°$ (see Fig. 41-8a). Although knowing the angle is useful for diagramming the angular momentum vector, we emphasize that the

FIGURE 41-8 (*a*) Possible orientations for the orbital angular momentum vector in the $\ell = 1$ state, where $L = \sqrt{2}\hbar$. More generally, there are $2\ell + 1$ possible orientations; (*b*) shows the five possibilities for $\ell = 2$, where $L = \sqrt{6}\hbar$. Only the z component is fixed for a given state; x and y components are uncertain.

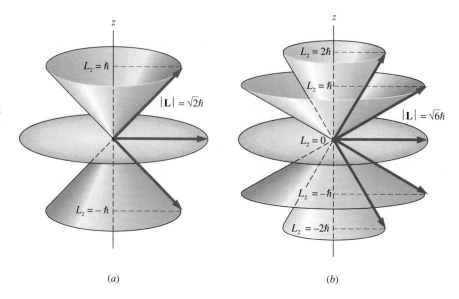

(*a*)　　　　　　　(*b*)

quantum numbers ℓ and m_ℓ tell everything there is to know about quantized orbital angular momentum. Quantum physicists, therefore, are usually not concerned with the orientation of angular momentum vectors.

● **EXAMPLE 41-4** SPACE QUANTIZATION

Derive an expression for the minimum angle between the oribital angular momentum vector and a chosen axis for arbitrary quantum numbers ℓ, and show that this angle approaches zero as $\ell \to \infty$.

Solution
The component of the orbital angular momentum on the axis is $L_z = m_\ell \hbar$. The magnitude of the angular momentum, given by Equation 41-10, is $L = \sqrt{\ell(\ell + 1)}\hbar$. Figure 41-9 shows that the angle θ between the angular momentum vector and the axis is given by

$$\cos \theta = \frac{L_z}{L} = \frac{m_\ell \hbar}{\sqrt{\ell(\ell + 1)}\hbar}.$$

The minimum θ occurs when the cosine has the largest possible magnitude. Since the maximum possible m_ℓ is ℓ, we have

$$\cos \theta_{\min} = \frac{\ell}{\sqrt{\ell(\ell + 1)}}.$$

As $\ell \to \infty$, we can neglect the 1 in the term $\ell + 1$, so $\sqrt{\ell(\ell + 1)} \to \sqrt{\ell^2} = \ell$ as $\ell \to \infty$. Then $\cos \theta \to 1$, so $\theta \to 0$, and we have shown that the angle becomes arbitrarily small for large ℓ. Again, this is a manifestation of the correspondence principle. At large quantum numbers the classically allowed sit-

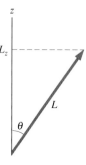

FIGURE 41-9 With magnitude L and component L_z on the axis, the angular momentum vector makes an angle θ with the z axis, where $\cos \theta = L_z/L$.

uation $\theta = 0$ is approached by quantum mechanics as well. Indeed, at large ℓ the number of allowed m_ℓ values is also large, so space quantization becomes unnoticeable.

EXERCISE What are the possible angles between the angular momentum vector and the z axis for an electron in the $\ell = 3$ state?

Answer: 30°, 54.7°, 73.2°, 90°, 107°, 125°, 150°

Some problems similar to Example 41-4: 16, 24 ●

41-2 ELECTRON SPIN

Detailed observation of the hydrogen spectrum shows that spectral lines exhibit a fine splitting; where a lower resolution spectrum shows one spectral line, at higher resolution there appears a closely spaced pair of lines. This splitting could not be explained using the three quantum numbers n, ℓ, and m_ℓ. In 1925 the Austrian physicist Wolfgang Pauli suggested that a fourth quantum number, capable of taking only two values, might be needed. Soon Samuel Goudsmit and George Uhlenbeck (Fig. 41-10), graduate students at the University at Leiden, realized that the spectral splitting could be explained if this fourth quantum number were associated with an intrinsic angular momentum, or **spin,** carried by the electron. Their proposal was without firm theoretical grounding; later, as we indicated in Chapter 40, Paul Dirac showed that electron spin follows from the requirement of relativistic invariance. Spin is an inherently quantum-mechanical property with no classical analog. Although spin can be visualized crudely by imagining the electron to be a small sphere spinning about an axis, this classical picture of a spinning sphere is really inappropriate.

Spin angular momentum is quantized in a manner similar to that of orbital angular momentum. But unlike the orbital quantum number ℓ that takes a range of integer values, the electron spin quantum number s has only the single value $s = \frac{1}{2}$. The electron is therefore called a **spin-$\frac{1}{2}$** particle. The magnitude of the spin angular momentum is related to the spin quantum number in the same way that the magnitude of orbital angular momentum is related to the orbital quantum number ℓ; that is, by an equation like Equation 41-10:

$$S = \sqrt{s(s + 1)}\hbar. \tag{41-12}$$

Since s takes only the value $\frac{1}{2}$, Equation 41-12 shows that the magnitude of the electron spin angular momentum is $S = \frac{\sqrt{3}}{2}\hbar$.

Spin angular momentum also exhibits space quantization. That is, the component of spin along a chosen axis can take only values given by an equation analogous to Equation 41-11:

$$S_z = m_s\hbar, \tag{41-13}$$

where the quantum number m_s has only the two possible values $-\frac{1}{2}$ and $+\frac{1}{2}$ (m_s is thus analogous to the orbital magnetic quantum number m_ℓ, whose values range from $-\ell$ to $+\ell$ in increments of 1). Figure 41-11 shows space quantization of electron spin.

Magnetic Moment of the Electron

Together, the electron's spin and electric charge mean the electron behaves like a miniature current loop, with an intrinsic magnetic dipole moment. The dipole moment vector **M** associated with the spin angular momentum vector **S** is given by

$$\mathbf{M} = -\frac{e}{m}\mathbf{S}, \tag{41-14}$$

FIGURE 41-10 George Uhlenbeck (left) and Samuel Goudsmit (right) who, as graduate students, first proposed electron spin.

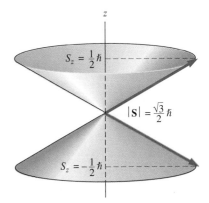

FIGURE 41-11 Space quantization of electron spin. The component of spin angular momentum on a given axis can take only the value $\pm\frac{1}{2}\hbar$.

with e/m the charge-to-mass ratio of the electron. (The minus sign reflects the electron's negative charge, and we consider here *only* the magnetic moment arising from electron spin. An additional magnetic moment results from orbital motion in states with nonzero ℓ.) Since the component of **S** on any axis can take only the values $\pm\frac{1}{2}\hbar$, the components of the magnetic moment can only be

$$M_z = \pm\frac{e\hbar}{2m}. \tag{41-15}$$

The quantity $e\hbar/2m$ is a fundamental unit for measuring magnetic moments called the **Bohr magneton;** its value is 9.27×10^{-24} A·m^2.

The ratio of magnetic moment to spin angular momentum is twice what we would expect classically for a charged particle in circular motion. Like spin itself, the factor of 2 is a relativistic effect first explained by Dirac. Actually, the factor is not quite 2 but approximately 2.00232, a result first established by Lamb and Retherford in 1947, and justified theoretically by Schwinger using the theory of quantum electrodynamics.

The Stern-Gerlach Experiment

In 1922, Otto Stern and Walther Gerlach at the University of Hamburg performed an experiment that demonstrated the quantization of atomic angular momentum vectors. The **Stern-Gerlach experiment** used a nonuniform magnetic field to separate a beam of silver atoms according to the component of their angular momentum along the field direction. (Recall from our study of magnetism that a magnetic dipole in a nonuniform magnetic field experiences a net force. Atoms with different angular momentum components and therefore differently oriented magnetic moments are thus deflected by different amounts.) The experiment was repeated in 1927 by Phipps and Taylor in a way that gives an unambiguous verification of quantized electron spin. Phipps and Taylor used hydrogen atoms in the ground state; as we have seen, this state has zero orbital angular momentum, so the only angular momentum effects are due to electron spin. Classically, a beam of hydrogen should be split into a continuous band corresponding to angular momentum components from $\frac{-\sqrt{3}}{2}\hbar$ to $\frac{+\sqrt{3}}{2}\hbar$. But when the experiment is done, the beam always splits into two beams, corresponding to the two angular momentum components $\pm\frac{1}{2}\hbar$. Figure 41-12 shows the experiment.

FIGURE 41-12 In the Stern-Gerlach experiment, as repeated by Phipps and Taylor, a beam of ground-state hydrogen atoms passes through a nonuniform magnetic field, established here by concave and convex magnet pole pieces. Atoms deflect according to the orientations of their magnetic moments, which in turn depend on electron spin orientation. (*a*) Classically, any spin is possible, and the beam should spread into a band. (*b*) In reality, the beam splits in two, corresponding to the two possible spin angular momentum components $+\frac{1}{2}\hbar$ and $-\frac{1}{2}\hbar$.

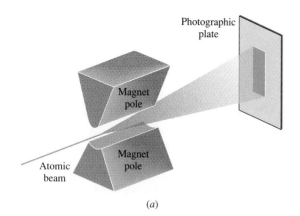

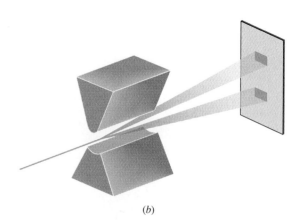

(*a*)

(*b*)

Spins of Other Particles

Spin angular momentum as proposed by Goudsmit and Uhlenbeck and confirmed theoretically by Dirac applies specifically to the electron. We now know that particles in general carry quantized intrinsic angular momentum or spin. Some, like electrons, protons, and neutrons, have a spin quantum number $s = \frac{1}{2}$. According to Equation 41-12, these spin-$\frac{1}{2}$ particles have spin angular momentum of $\frac{\sqrt{3}}{2}\hbar$ and, according to Equation 41-13, the spin component on a given axis takes only the values $\pm\frac{1}{2}\hbar$. Other particles—including photons and some composite nuclei—are spin-1 particles, for which $s = 1$. There are profound and fascinating differences in behavior between particles with half-integer and integer spins, as we describe shortly. Whatever the spin quantum number s, particle spins obey the quantization relations of Equations 41-12 and 41-13, with values of m_s ranging from $-s$ to s in increments of $\hbar$.*

● EXAMPLE 41-5 SPIN OF THE DEUTERON

The spins of its constituent proton and neutron combine to make the deuteron a spin-1 particle. Determine the magnitude of its spin angular momentum and the possible values of its spin component on a given axis. Construct a vector diagram analogous to Fig. 41-11.

Solution
Equation 41-12 gives the magnitude of the spin angular momentum; with $s = 1$ for the spin-1 deuteron, we have

$$S = \sqrt{s(s + 1)}\hbar = \sqrt{2}\hbar.$$

The possible values of m_s range from $-s$ to s in increments of 1; thus, $m_s = -1, 0,$ or 1. The corresponding angular momentum components, given by Equation 41-13, are $-\hbar, 0,$ and $\hbar$. Figure 41-13 shows the resulting vector diagram.

EXERCISE Find a minimum angle between the spin vector and the z axis for a spin-$\frac{1}{2}$ particle.

Answer: 55°

Some problems similar to Example 41-5: 16, 24

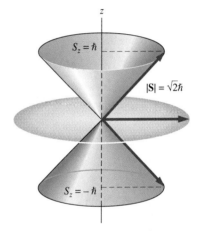

FIGURE 41-13 Space quantization for the deuteron, a spin-1 particle (Example 41-5). Compare with Fig. 41-8a for orbital angular momentum with $\ell = 1$.

Total Angular Momentum and Spin-Orbit Coupling

The total angular momentum of the atom, designated **J**, is the vector sum of the orbital and spin angular momenta; that is,

$$\mathbf{J} = \mathbf{L} + \mathbf{S}. \qquad (41\text{-}16)$$

A detailed quantum-mechanical analysis shows that the magnitude of **J** is quantized according to a rule similar to those for orbital and spin angular momenta:

---

*Photons are an exception to this rule; the fact that they are massless turns out to eliminate the state $m_s = 0$ that would otherwise be possible for a spin-1 particle.

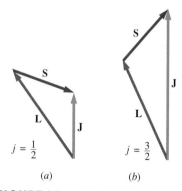

FIGURE 41-14 Two possibilities for coupling of spin and orbital angular momentum. Case shown is for $\ell = 1$, so Equation 41-18a gives $j = \frac{1}{2}$ or $j = \frac{3}{2}$.

$$J = \sqrt{j(j+1)}\hbar. \tag{41-17}$$

For an atom with a single electron, the quantum number j takes the values

$$j = \ell \pm \frac{1}{2} \qquad \text{for } \ell \neq 0 \tag{41-18a}$$

$$j = \frac{1}{2} \qquad \text{for } \ell = 0. \tag{41-18b}$$

The state of an atom with total angular momentum of magnitude J is often given by listing the principal quantum number, the capital letter designating the orbital angular momentum (S, P, D, F, G, . . .), and, as a subscript, the j value. Thus a hydrogen atom with $n = 3$, $\ell = 2$, and $j = \frac{3}{2}$ would be designated $3D_{3/2}$.

Total angular momentum also exhibits space quantization, with the possible values for the component of **J** on some axis given by

$$J_z = m_j \hbar, \tag{41-19}$$

where the quantum number m_j takes values from $-j$ to j, with a difference of 1 between successive m_j values (i.e., $m_j = -j, -j+1, -j+2, \ldots, j$).

Although derivation of these so-called **angular momentum coupling rules** is not at all straightforward, we can understand them in terms of simple vector diagrams. Recall that the magnitude of the electron's spin angular momentum is $\frac{\sqrt{3}}{2}\hbar$ and its component on a given axis takes only the two values $\pm \frac{1}{2}\hbar$. When $\ell = 0$, there is no orbital angular momentum, and Equations 41-17 and 41-18b then show that the magnitude of the total angular momentum is just that of the spin. When $\ell \neq 0$, spin and orbital angular momentum add in accordance with Equation 41-18a to produce a total angular momentum vector **J** whose magnitude is given by Equation 41-17 (Fig. 41-14).

Space quantization with respect to any axis applies to the total angular momentum vector; Fig. 41-15 shows possible orientations of **J** and the corresponding J_z components for the cases $j = \frac{1}{2}$ and $j = \frac{3}{2}$ of Fig. 41-14.

FIGURE 41-15 Space quantization of the total angular momentum for the case (a) $j = \frac{1}{2}$ and (b) $j = \frac{3}{2}$, showing the allowed orientations of the total angular momentum vector **J**.

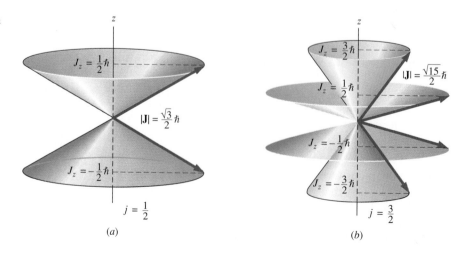

● **EXAMPLE 41-6**   TOTAL ANGULAR MOMENTUM

Find the possible magnitudes of the total angular momenta for a hydrogen atom in an $\ell = 2$ state. For each possible J value, how many possible values are there for the component of **J** on a given axis?

Solution

For $\ell = 2$, Equation 41-18a gives $j = \frac{3}{2}$ or $j = \frac{5}{2}$. Then, from Equation 41-17, we have

$$J = \sqrt{j(j+1)}\hbar = \sqrt{\tfrac{3}{2}(\tfrac{3}{2}+1)}\hbar = \frac{\sqrt{15}}{2}\hbar$$

or

$$J = \sqrt{j(j+1)}\hbar = \sqrt{\tfrac{5}{2}(\tfrac{5}{2}+1)}\hbar = \frac{\sqrt{35}}{2}\hbar.$$

Figure 41-16 shows vector diagrams for the angular momentum coupling in these two cases.

For the $j = \frac{3}{2}$ state, there are four possible values of m_j, ranging from $-j$ to j; they are $-\frac{3}{2}, -\frac{1}{2}, \frac{1}{2},$ and $\frac{3}{2}$, corresponding to four values of the total angular momentum component given by Equation 41-19. For the $j = \frac{5}{2}$ state, there are six possible m_j's and correspondingly six values for the angular momentum component.

EXERCISE   Find the maximum possible angular momentum for a hydrogen atom in an $\ell = 3$ state.

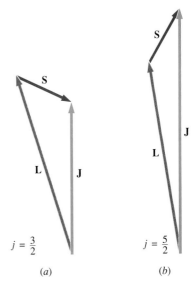

$j = \frac{3}{2}$ (a) $j = \frac{5}{2}$ (b)

FIGURE 41-16   Vector diagrams for spin-orbit coupling with $\ell = 2$ (Example 41-6).

Answer: $\dfrac{3\sqrt{7}}{2}\hbar$

Some problems similar to Example 41-6: 25–27 ●

The two possible values of j for a given ℓ correspond to two distinct quantum states, and these states actually have slightly different energies. This energy difference can be understood using the following semiclassical argument. In the reference frame of the electron, the positively charged proton appears to be moving, giving rise to a magnetic field experienced by the electron; the direction of this field turns out to be parallel to the electron's orbital angular momentum **L**. As Equation 41-14 shows, the negative charge of the electron means that its magnetic moment **M** is *antiparallel* to the spin vector **S.** The two possible spin orientations shown in Fig. 41-11 correspond to two different orientations of the electron's magnetic moment in the magnetic field of the nucleus. In Section 29-5, we found that the potential energy of a magnetic dipole in a magnetic field depends on the orientation of dipole moment and field, with maximum potential energy when the magnetic moment and field are antiparallel (see Equation 29-12). This magnetic potential energy contributes to the total energy of the atom, and therefore the two possible spin orientations correspond to slightly different energies. Because the spin vector **S** and magnetic moment **M** are antiparallel, the more-nearly parallel alignment of **S** and **L**—corresponding to $j = \ell + \frac{1}{2}$— has **M** turned farther away from **L** and therefore from the magnetic field **B,** so this state has higher energy.

In hydrogen, the magnitude of the energy difference between the $j = \frac{1}{2}$ and $j = \frac{3}{2}$ states of the first excited level is only 5×10^{-5} eV, far smaller than the separation between this level and the ground state. Because the $n = 2$, $\ell = 1$

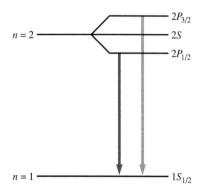

FIGURE 41-17 Energy-level diagram showing spin-orbit splitting in hydrogen. Left side shows $n = 1$ and $n = 2$ levels if there were not splitting; at right the $2P$ state splits because of the spin-orbit interaction, while the zero-angular-momentum S states are unaffected. Allowed transitions to the ground state are indicated by arrows; they result in spectral lines of slightly different wavelengths. In hydrogen the $2P$ levels actually differ by only about 10^{-5} of the separation between the $n = 1$ and $n = 2$ levels.

state is actually two states of slightly different energy, hydrogen atoms undergoing transitions from these states to the ground state emit two spectral lines slightly separated in wavelength. The term **fine structure** describes the splitting associated with spin-orbit and related effects. In the present example, the split spectral line is called a **doublet.** (Transitions from the $n = 2$, $\ell = 0$ state to the ground state are forbidden; otherwise the line would be a triplet.) Figure 41-17 is an energy-level diagram showing the effect of spin-orbit splitting in hydrogen, while Fig. 41-18 shows the doublet spectral line resulting from the analogous transition in sodium. (The effect is difficult to observe in hydrogen.)

The spin-orbit effect results from a magnetic field internal to the atom itself. But splitting of energy levels also occurs in an external magnetic field, and is called the **Zeeman effect.** We showed an example of Zeeman splitting in Fig. 37-23.

Since it has zero orbital angular momentum, the ground state does not exhibit spin-orbit splitting. But interaction of the electron spin with the magnetic dipole moment of the nucleus does result in an even finer splitting known as **hyperfine structure.** The transition between the two hyperfine levels of the hydrogen ground state—corresponding physically to a change in the orientation of the electron spin vector—involves a photon of 21-cm wavelength. Radio astronomers use the 21-cm hydrogen radiation to map regions of interstellar hydrogen in the cosmos (Fig. 41-19).

41-3 THE PAULI EXCLUSION PRINCIPLE

In trying to understand why atomic electrons distributed themselves as they did, Pauli was led in 1924 to his famous **exclusion principle** which, loosely, states the following:

> **Two electrons cannot be in the same quantum state.**

Since the quantum state of an electron includes its spin orientation specified by the spin quantum number m_s, the exclusion principle means that at most two electrons can occupy a state whose other quantum numbers are the same.

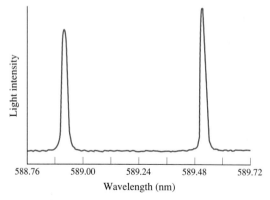

FIGURE 41-18 A high-resolution spectrometer provides a close look at the doublet spectral lines resulting from the $3P_{1/2} \rightarrow 3S_{1/2}$ and $3P_{3/2} \rightarrow 3S_{1/2}$ transitions in sodium. The sodium doublet also shows prominently near the center of the solar spectrum in Fig. 39-11a.

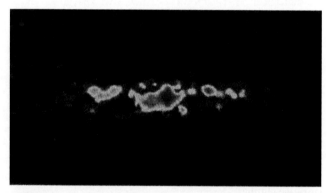

FIGURE 41-19 This image of the galaxy NGC 784 was made with the Very Large Array radio telescope tuned to the 21-cm line of hydrogen. The image shows the distribution of neutral hydrogen in the galaxy.

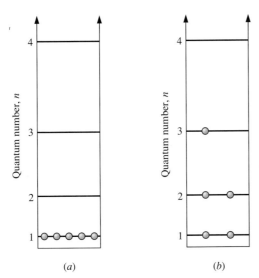

FIGURE 41-20 (*a*) If the exclusion principle did not hold, the lowest energy configuration of a square well containing five electrons would have them all in the ground state. (*b*) The exclusion principle permits only two electrons, their spin components opposite, to occupy a single energy level.

The Pauli exclusion principle has profound implications of the behavior of multielectron systems, for it requires that most electrons in such a system remain in high-energy states (Fig. 41-20). If the exclusion principle did not hold, atomic electrons would collapse to the ground state and there would be no such thing as chemistry or life! And, as the application below indicates, the exclusion principle even manifests itself in objects at the astrophysical scale.

■ APPLICATION WHITE DWARFS AND NEUTRON STARS

The endpoint of a star's life comes when nuclear fusion ceases in the stellar interior. Then there is no longer sufficient pressure to maintain the star against its own gravity, and collapse ensues. For stars with more than about five times the Sun's mass, no known force can halt the collapse, and the star beomes a black hole—a singular configuration governed by the general theory of relativity. But for less massive stars, including our Sun, collapse is ultimately halted by a quantum-mechanical pressure arising from the exclusion principle.

In the eventual collapse of the Sun, some five billion years hence, all the electrons will drop to the lowest possible energy states. But the exclusion principle asserts that only two electrons, with their spin components opposite, can occupy each energy level. So most of the electrons end up in high-energy states (imagine putting 10^{57} electrons in the square well of Fig. 41-20). Consequently they have substantial momentum and therefore

exert considerable pressure. It is this **degenerate electron pressure**—quite independent of temperature, unlike the pressure of an ordinary gas—that sustains the star against further collapse. In its final configuration, the resulting **white dwarf** star is only about the size of the Earth (Fig. 41-21).

In stars more massive than 1.4 times the Sun's mass, degenerate electron pressure is insufficient to counter the gravitational force, and the star collapses further until electrons and protons join to form neutrons. The neutrons, too, develop a degenerate pressure that halts the collapse. At this point the former star is only a few kilometers in diameter, and is called a **neutron star** (Fig. 41-21). Neutron stars were first detected in 1968 by British astronomer Jocelyn Bell, who noted regular pulses of radio emission from cosmic sources. These sources, first called pulsars, were later identified as rapidly rotating neutron stars.

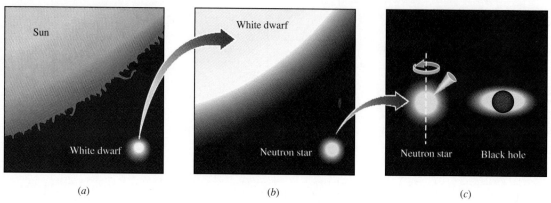

FIGURE 41-21 Comparative sizes of the Sun, a white dwarf, a neutron star, and a black hole. The white dwarf and neutron star are supported against gravitational collapse by pressure that arises as a consequence of the exclusion principle. Neutron star is shown in rapid rotation, emitting a searchlight-like beam of radiation; black hole is surrounded by a disk of accreting gas.

The Pauli exclusion principle quickly became a fundamental rule in the developing quantum mechanics of the late 1920s. But physicists remained dissatisfied by the need to invoke this seemingly ad hoc rule with no theoretical basis. Late in the 1930s, following detailed analysis of relativistic quantum theories, Pauli finally showed that the exclusion principle, like the existence of spin, is ultimately grounded in the requirement of relativistic invariance. More specifically, Pauli found that particles whose spin quantum number s is a half-integer (collectively called **fermions**) must necessarily obey the exclusion principle. On the other hand, particles with integer spin (called **bosons**) *do not* obey the exclusion principle. Photons, for example, are spin-1 particles and therefore an arbitrarily large number of them can occupy exactly the same quantum state. The laser, with its intense, coherent beam of light, is possible because the many photons comprising the beam are essentially all in the same state (Fig. 41-22).

41-4 MULTIELECTRON ATOMS AND THE PERIODIC TABLE

The Chemical Elements

FIGURE 41-22 The intensity and coherence of laser beams are possible only because photons, being spin-1 particles, do not obey the exclusion principle; thus, many photons can occupy the same quantum state. Here a high-intensity beam from an argon laser (blue) is used to crystallize silicon in semiconductor manufacturing. A less intense beam from a helium-neon laser (red) guides the argon laser beam.

Our modern understanding of the chemical elements developed in the late eighteenth century. At that time, the French chemist Antoine Lavoisier found that burning resulted from a combination of substances with the newly discovered gas oxygen. Lavoisier made the first reasonable list of elements, and recognized which were combined into compounds. Since that time, elements have been defined as substances that cannot be decomposed by chemical means (Fig. 41-23).

Chemists proceeded to discover formulas for various compounds, such as H_2O for water. From the formulas, they could determine the relative atomic masses of various elements. In 1864, the British chemist John Newlands discovered that similar chemical properties occurred periodically when elements were arranged in order of atomic mass. The periodic arrangement was clarified further by the German chemist Lothar Meyer and the Russian chemist Dmitri Mendeleev. In

1869 Mendeleev set up a periodic table based on chemical properties of the approximately 60 elements then known. He left blanks where necessary to maintain the alignment of chemical properties. Mendeleev is credited with the leap of genius it took to predict the properties of the elements that would fill the blanks in his table. Three such elements—gallium, scandium, and germanium—were soon found. Though there were still uncertainties in Mendeleev's table, the existence of periodicity suggested an underlying ordered structure in the composition of atoms. The periodic table was further refined early in the twentieth century through the study of x-ray spectra produced by bombardment of materials with high-energy electrons. Through these studies, the table could be organized by atomic number Z (the number of protons in the atomic nucleus), rather than by atomic weight. When this was done, a number of elements missing from earlier periodic tables were identified. The modern periodic table is shown in Fig. 41-24 and is printed with atomic weights inside the back cover.

Explaining the Periodic Table

The orderly arrangement of elements in the periodic table has greatly enhanced our understanding of chemistry and our ability to formulate new and useful compounds. But why does nature exhibit this regularity? The answer lies in the Schrödinger equation and the exclusion principle.

Solution of the Schrödinger equation for multielectron atoms is complicated by the interactions among the electrons; simple, analytic solutions like those for hydrogen are generally not available. But qualitatively, we still find energy levels characterized by the principal quantum number n. Each such level is called a **shell**; for historical reasons, the shells $n = 1, 2, 3, \ldots$ are also labeled with the letters $K, L, M, \ldots$. As with hydrogen, an electron at the nth energy level can take on any of the n values $\ell = 0, 1, 2, \ldots, n - 1$ for the orbital quantum number ℓ that determines its orbital angular momentum. The different angular momentum states within a shell are termed **subshells**; subshells with the values

FIGURE 41-23 The chemical elements exhibit an amazing variety of forms, colors, and other physical and chemical properties. All these are ultimately traceable to atomic structure, which in turn follows from the principles of quantum physics. On glass dishes, clockwise from top left, are sodium, chromium, sulfur, iodine, and iron. Also shown are zinc and copper strips and a magnesium ribbon. Invisible gaseous elements—primarily oxygen and nitrogen—comprise the surrounding air.

FIGURE 41-24 The periodic table. A larger version, with atomic weights, is printed inside the back cover. Blue represents nonmetals, pink represents metals, and yellow represents semimetals.

1 H																	2 He
3 Li	4 Be											5 B	6 C	7 N	8 O	9 F	10 Ne
11 Na	12 Mg											13 Al	14 Si	15 P	16 S	17 Cl	18 Ar
19 K	20 Ca	21 Sc	22 Ti	23 V	24 Cr	25 Mn	26 Fe	27 Co	28 Ni	29 Cu	30 Zn	31 Ga	32 Ge	33 As	34 Se	35 Br	36 Kr
37 Rb	38 Sr	39 Y	40 Zr	41 Nb	42 Mo	43 Tc	44 Ru	45 Rh	46 Pd	47 Ag	48 Cd	49 In	50 Sn	51 Sb	52 Te	53 I	54 Xe
55 Cs	56 Ba	57-71 Lanthanide series	72 Hf	73 Ta	74 W	75 Re	76 Os	77 Ir	78 Pt	79 Au	80 Hg	81 Tl	82 Pb	83 Bi	84 Po	85 At	86 Rn
87 Fr	88 Ra	89-103 Actinide series	104 Rf	105 Ha	106 Sg	107 Ns	108 Hs	109 Mt									

Atomic number — 2 He — Symbol

Lanthanide series	57 La	58 Ce	59 Pr	60 Nd	61 Pm	62 Sm	63 Eu	64 Gd	65 Tb	66 Dy	67 Ho	68 Er	69 Tm	70 Yb	71 Lu
Actinide series	89 Ac	90 Th	91 Pa	92 U	93 Np	94 Pu	95 Am	96 Cm	97 Bk	98 Cf	99 Es	100 Fm	101 Md	102 No	103 Lr

$\ell = 0, 1, 2, 3, \ldots$ are labeled with the letters $s, p, d, f, \ldots$. Finally, for each subshell there are $2\ell + 1$ possible values of the magnetic orbital quantum number m_ℓ, ranging from $-\ell$ to ℓ. A state characterized by all three quantum numbers n, ℓ, and m_ℓ is called an **orbital.** Table 41-1 summarizes shell-structure notation in relation to quantum numbers. For completeness, the table lists also the spin quantum number m_s.

The structure of a multielectron atom is determined by the quantum states of its constituent electrons—that is, by their distribution among the shells, subshells, and orbitals. That distribution, in turn, is determined by the exclusion principle. According to the exclusion principle, no two electrons can be in exactly the same quantum state—that is, they cannot have the same values for all four quantum numbers n, ℓ, m_ℓ, and m_s. Since an atomic orbital is characterized by the three quantum numbers n, ℓ, and m_ℓ, the exclusion principle implies that at most two electrons can occupy a single orbital.

We're now ready to understand the ground-state electronic structure of multi-electron atoms. The simplest is helium, with two electrons. The K shell ($n = 1$) is the lowest possible energy level. As Table 41-1 shows, only the zero-angular-momentum s subshell is permitted within the K shell, and within that subshell only the single orbital corresponding to $m_\ell = 0$ is allowed. But that orbital can accommodate two electrons. So in the ground state of helium, both electrons occupy the s subshell of the K shell. Atomic physicists describe this situation with the shorthand notation $1s^2$, where 1 stands for the principal quantum number n (i.e., for the K shell), s for the subshell, and the superscript 2 for the number of electrons in that subshell. The corresponding notation for hydrogen would be $1s^1$.

After helium comes lithium, with three electrons. From our analysis of helium, we know that the K shell is full with two electrons. So the third electron goes into the L shell, or $n = 2$ energy level. Of the subshells in the L shell, the s subshell turns out to have slightly lower energy than the others, so the third electron occupies the s subshell. Then the electronic configuration of lithium is $1s^22s^1$; i.e., a helium-like core with a single outer electron in the s subshell of the $n = 2$ level.

Beryllium, with four electrons, fills the $1s$ and $2s$ subshells; its designation is $1s^22s^2$. The fifth electron of boron then goes into the $2p$ subshell, giving the structure $1s^22s^22p^1$. Table 41-1 shows that a p subshell ($\ell = 1$) allows three m_ℓ values—i.e., three orbitals, capable of holding a total of six electrons. As we advance in atomic number, electrons continue to fill the p subshell. Finally, at neon ($Z = 10$), the $2p$ subshell is full. Only with the next element, sodium ($Z = 11$), does the $n = 3$ shell begin to fill. Table 41-2 lists electronic configurations for the elements hydrogen ($Z = 1$) through argon ($Z = 18$), along with their ionization energies (that is, the energy required to remove the outermost electron).

▲ **TABLE 41-1** ATOMIC SHELL STRUCTURE NOTATION

QUANTUM NUMBER	SHELL-STRUCTURE NOTATION	ALLOWED VALUES	LETTER LABELS	NUMBER OF STATES
n	Shell	$1, 2, 3, \ldots$	$K, L, M, \ldots$	Infinite
ℓ	Subshell	$0, 1, 2, \ldots, n-1$	$s, p, d, f, \ldots$	n
m_ℓ	Orbital	$-\ell, -\ell+1, \ldots, \ell-1, \ell$	–	$2\ell + 1$
m_s	—	$-\frac{1}{2}, +\frac{1}{2}$		2

▲ **TABLE 41-2** ELECTRONIC CONFIGURATIONS AND IONIZATION
ENERGIES OF ELEMENTS 1–18

ATOMIC NUMBER, Z	ELEMENT	ELECTRONIC CONFIGURATION	IONIZATION ENERGY, eV
1	H	$1s^1$	13.60
2	He	$1s^2$	24.60
3	Li	$1s^22s^1$	5.390
4	Be	$1s^22s^2$	9.320
5	B	$1s^22s^22p^1$	8.296
6	C	$1s^22s^22p^2$	11.26
7	N	$1s^22s^22p^3$	14.55
8	O	$1s^22s^22p^4$	13.61
9	F	$1s^22s^22p^5$	17.42
10	Ne	$1s^22s^22p^6$	21.56
11	Na	$1s^22s^22p^63s^1$	5.138
12	Mg	$1s^22s^22p^63s^2$	7.644
13	Al	$1s^22s^22p^63s^23p^1$	5.984
14	Si	$1s^22s^22p^63s^23p^2$	8.149
15	P	$1s^22s^22p^63s^23p^3$	10.48
16	S	$1s^22s^22p^63s^23p^4$	10.36
17	Cl	$1s^22s^22p^63s^23p^5$	13.01
18	Ar	$1s^22s^22p^63s^23p^6$	15.76

Gaps mark ends of periodic-table rows.

The chemical behavior of an atom is determined primarily by its outermost electrons, both because these electrons interact most directly with nearby atoms and because they are most weakly bound to their nuclei. A look at Table 41-2 shows that the configurations of outer electrons for the elements lithium through neon are the same as the corresponding configurations for sodium through argon. The chemical properties of the corresponding atoms are therefore similar. Both lithium and sodium, for example, have a single electron in their outermost shell. As the relatively low ionization energy suggests, this electron is loosely bound, so it readily interacts with other atoms, giving these elements their extreme reactivity. Neon and argon, in contrast, both have completely filled outermost shells. All the outer-shell electrons have essentially the same energy; the corresponding ionization energy is high; and there is little tendency for these electrons to interact with those of other atoms. As a result, argon and neon are **inert**—that is, they do not readily form chemical compounds, and at normal temperatures their lack of interaction makes them gases. Other element pairs also share similar properties. Fluorine and chlorine, for instance, each need only one more electron to achieve the energetically favorable situation of an inert gas with its filled outer shell. Consequently these elements readily accept electrons. Materials like common salt, NaCl, owe their high melting points to the strong bond that results when reactive sodium gives up its outer electron to electron-accepting chlorine and the resulting positive and negative ions bind strongly by the electrostatic force. We will consider such molecular bonding in the next chapter.

As we move beyond argon ($Z = 18$), things become slightly more complicated. In these multielectron atoms, shielding effects of the inner electrons result in the $4s$ state having lower energy than the $3d$ states. Potassium ($Z = 19$) thus has the electronic configuration $1s^22s^22p^63s^23p^64s^1$ rather than $1s^22s^22p^63s^23p^63d^1$ as would occur if the $n = 4$ state had a higher energy than any $n = 3$ state. After potassium comes calcium, with two electrons in its single

4s orbital. But the $4p$ orbitals do have higher energy than the $3d$ orbitals, so elements beyond calcium begin filling the $3d$ orbitals. The next ten elements, scandium through zinc, have chemical properties that vary only slightly because their outermost electrons remain $4s$ electrons; collectively, they are called **transition elements.*** Finally, elements 31 (gallium) through 36 (krypton) repeat the pattern of aluminum through argon shown in Table 41-2, as their $4p$ orbitals fill with electrons. Krypton, with its outer p subshell full, is again an inert gas.

● **EXAMPLE 41-7**   ELECTRONIC CONFIGURATION OF IRON

Given that the $4s$ states fill before the $3d$ states in the transition element iron (Fe), but with no other anomalies, determine the electronic configuration of iron.

Solution

Consulting the periodic table (Fig. 41-24 or inside back cover), we find that iron has atomic number $Z = 26$—that is, it has eight more electrons than the last entry, argon, in Table 41-2. So its inner structure is like that of argon, and we need determine only the configuration of the remaining eight electrons. In argon, the $3p$ orbitals are full; the next $n = 3$ orbitals are the d orbitals, but we are told the $4s$ orbital fills first. So two electrons fill the

single $4s$ orbital, leaving six in the $3d$ orbitals. (Since the d subshell has five orbitals, accommodating up to ten electrons, the iron d subshell can handle all the remaining electrons.) So the electronic structure of iron is $1s^2 2s^2 2p^6 3s^2 3p^6 3d^6 4s^2$. As a check, note that the superscripts add to give a total of 26 electrons.

EXERCISE   What is the electronic configuration of the transition element titanium ($Z = 22$)?

Answer: $1s^2 2s^2 2p^6 3s^2 3p^6 3d^2 4s^2$

Some problems similar to Example 41-7:   33–35 ●

We can now understand the organization of the periodic table, shown in Fig. 41-24. The vertical columns contain elements with the same outer-electron configurations, resulting in similar chemical properties. Each row starts with an element whose outermost shell contains a single s electron; these are hydrogen and the highly reactive alkali metals. Each row ends with an inert gas, its outermost p subshell full. Traversing the first row involves filling the $1s$ orbital only; since this orbital holds at most two electrons, there are only two elements in the first row. The second row has eight elements, associated with the filling of the $2s$ and $2p$ orbitals as shown in Table 41-2. The third row is like the second, but with the $3s$ and $3p$ orbitals filling. Because the $4s$ orbitals have lower energy than the $3d$ orbitals, the third row ends with an inert gas whose $3p$ orbitals are full, and the fourth row begins as the $4s$ orbital begins to fill. But then come the elements $Z = 21$ through $Z = 30$, in which the $3d$ orbitals are filling; these make for ten additional elements in the fourth row. The fifth row is a repeat of the fourth, as first the $5s$ orbitals fill, then the higher energy $4d$ orbitals, then the remaining $5p$ orbitals. The sixth row begins with element 55, cesium, as the $6s$ orbitals fill before either the $5d$ or $4f$ orbitals. Two elements later, at lanthanum, the $4f$ orbitals begin to fill. Since their outer electrons remain $6s$ electrons, these elements have very similar properties. Together, they constitute the **lanthanide series.** Because its elements are so similar chemically, and to avoid stretching the sixth row of the periodic table, the lanthanide series is usually printed separately below the main table. Row seven repeats row six, with the **actinide series** again printed below. The actinides include elements 89 (actinium) through 103 (lawren-

---

*Chromium ($Z = 24$) and copper ($Z = 29$) are minor exceptions; in these an extra electron goes into the $3d$ orbitals, leaving only a single $4s$ electron.

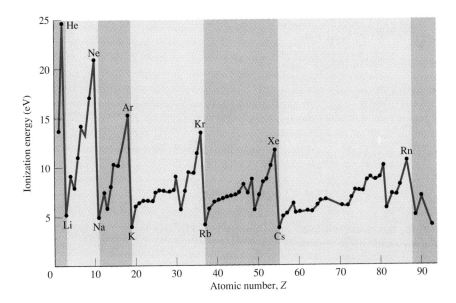

FIGURE 41-25 Ionization energy as a function of atomic number Z. Shaded bands mark the rows of the periodic table.

cium, named for the inventor of the cyclotron). Elements beyond 92 (uranium) are radioactive with half-lives short compared with the age of the Earth. They are not found naturally but are produced in particle accelerators, fission reactors, and nuclear weapons. Elements 104 through 109 follow the actinide series and partially complete the seventh row of the periodic table. This group, some of which were first synthesized only in the 1980s, are generally very short lived. However, theories of nuclear physics predict that longer lived nuclei should occur among the even-numbered elements from 110 to 116.

The periodic ordering of the chemical elements is reflected not only in the periodic table, but also in their ionization energies. Figure 41-25 is a plot of ionization energy versus atomic number. A peak at each inert gas is followed by an abrupt drop to the easily ionized alkali metal that starts the next row of the periodic table. Then the ionization energies generally increase across each row, showing that the atoms become more reluctant to give up their electrons and more willing to accept electrons as they combine chemically with other atoms.

Note the crucial role the exclusion principle plays in this discussion of the chemical elements. Without that principle, every atom in its ground state would have all its electrons in the 1s orbital. There would be no qualitative distinction among the elements, and the science of chemistry would not exist. Nor would there be any chemists or physicists; life itself would be impossible without the rich diversity of chemical compounds formed from the different elements.

41-5 TRANSITIONS AND ATOMIC SPECTRA

Emission and absorption of photons with specific energies provide the most direct experimental manifestation of quantized atomic energy levels, and give rise to the spectral lines that permit precise analysis of atomic systems from the laboratory to distant astrophysical objects (Fig. 41-26). Even the simple hydrogen atom exhibits a myriad of quantum states. In multielectron atoms, the possibilities for

FIGURE 41-26 Spectrum from the supernova remnant Puppis A shows strong lines of oxygen and neon. The absence of other elements suggests that there was a pure oxygen layer within the star that exploded, and thus confirms theories of stellar evolution. Lines are labeled with element symbol, a Roman numeral indicating the degree of ionization, and the wavelength in nm.

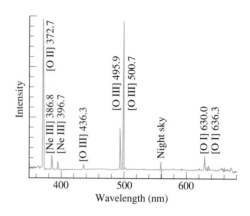

electronic excitation are even more numerous. The spectra of atoms reflect this rich array of available quantum states.

Selection Rules

Of the conceivable transitions between energy levels, not all are equally likely. So-called **selection rules** determine which are **allowed transitions**—those most likely to occur. One selection rule limits allowed transitions to those for which the orbital quantum number ℓ changes by $\Delta\ell = \pm 1$; this and other selection rules are related to conservation of angular momentum. Quantum mechanics also provides a way of calculating transition probabilities or, equivalently, the mean lifetimes of excited states. For outer electrons, excited states that de-excite by allowed transitions have typical lifetimes on the order of 10^{-9} s.

Transitions that are not allowed by selection rules are called **forbidden transitions;** most are not strictly impossible but just extremely unlikely. States that can lose energy only by forbidden transitions are **metastable states;** their lifetimes are many orders of magnitude longer than the nanosecond timescale for allowed transitions (Fig. 41-27). Forbidden spectral lines are especially valuable probes of the solar corona and other low-density astrophysical gases in which collisions are rare, and atoms can, therefore, remain in metastable states.

Optical Spectra

Spectral lines in or near the visible region result from transitions involving electrons in the incompletely filled outer shells of atoms. The alkali metals, with a single s electron in their outer shells, therefore produce spectra qualitatively similar to that of hydrogen. However, the more complicated structure of a multi-electron atom causes shifts in energy levels not observed in hydrogen. Figure 41-28 shows energy-level diagrams and allowed transitions for hydrogen and sodium. The diagram for an atom with more than one electron in its outer shell is even more complicated.

Each of the transitions shown in Fig. 41-28 corresponds to a spectral line of specific wavelength. At higher resolution, many spectral lines are observed to be doublets or triplets of closely spaced lines. The p levels, for example, are split into two by the spin-orbit splitting; therefore, transitions between s and p levels result in doublet lines. Figure 41-29 shows part of the energy-level diagram for sodium in more detail, with doublet lines resulting from spin-orbit splitting now evident. (The sodium lines themselves were shown in Fig. 41-18.)

FIGURE 41-27 Phosphorescent materials emit light after their atoms are excited to long-lived metastable states. Such materials are used in a variety of applications from TV picture tubes to day-glo paints.

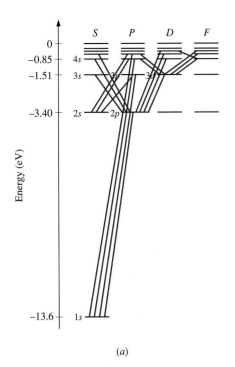

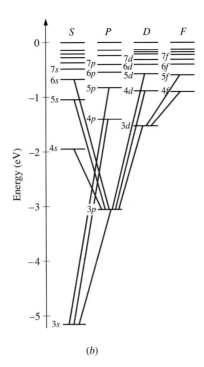

FIGURE 41-28 Simplified energy-level diagrams for (*a*) hydrogen and (*b*) sodium, neglecting spin-orbit splitting. The letters *S*, *P*, *D*, . . . correspond to the $\ell = 0, 1, 2, \ldots$ states. Only the outermost electron of sodium participates in optical transitions, so the lowest level shown is the 3*s* level of this electron in the sodium ground state. Note that the 4*s* and 4*p* levels are widely separated in sodium, with the 3*d* level between them; this explains why the 4*s* orbital fills before the 3*d*. Allowed transitions are shown in purple.

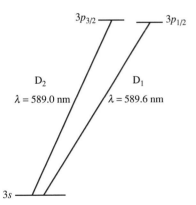

FIGURE 41-29 A portion of the energy-level diagram for sodium, showing that the spin-orbit effect splits the 3*p* level into two closely spaced states. Energy difference (vertical separation) here is exaggerated; it would be imperceptible on the scale of this diagram. Two slightly different wavelengths result from the $3p \rightarrow 3s$ transitions; these constitute the sodium D doublet shown in Fig. 41-18. Transitions to 3*p* from the higher *s* states (not shown) also result in doublet lines.

● **EXAMPLE 41-8** THE SODIUM DOUBLET

Use the sodium-*D* wavelengths shown in Fig. 41-29 to calculate the energy difference between the 3*p* states of sodium.

Solution
Using $E = hf$ and $f\lambda = c$, we can write $E = hc/\lambda$ for the energy of each transition. Then the energy splitting is

$$\Delta E = \frac{hc}{\lambda_1} - \frac{hc}{\lambda_2} = hc\left(\frac{1}{\lambda_1} - \frac{1}{\lambda_2}\right)$$

$$= (6.63 \times 10^{-34} \text{ J·s})(3.00 \times 10^8 \text{ m/s})$$

$$\times \left(\frac{1}{589.0 \times 10^{-9} \text{ m}} - \frac{1}{589.6 \times 10^{-9} \text{ m}}\right)$$

$$= 3.44 \times 10^{-22} \text{ J} = 2.15 \times 10^{-3} \text{ eV}.$$

This fine-structure splitting is only about 0.1% of the sodium-*D* photon energies but is readily measured with spectroscopic techniques.

EXERCISE The transition $6s \rightarrow 4p$ in sodium results in doublet spectral lines at 1.6374 μm and 1.6388 μm. What is the splitting between the two energy levels of the 4*p* state?

Answer: 1.038×10^{-22} J = 648 μeV

Some problems similar to Example 41-8: 39–41 ●

X-ray Spectra

With sufficient energy, it is possible to eject an electron from an inner atomic orbital. Then an electron in a higher-energy orbital drops into the unoccupied state, emitting a photon whose energy is equal to the difference between the two atomic energy levels. For inner-shell electrons, those energies are typically measured in keV, corresponding to wavelength in the range 0.01 to 1 nm—that is, in the x-ray region of the spectrum.

A simple way to produce x-ray photons is to bombard a metal target with a beam of high-energy electrons (Fig. 41-30). Most of the electrons decelerate rapidly on hitting the target, resulting in a broad band of x-ray radiation (recall from Chapter 34 that accelerated charges produce electromagnetic radiation). But some electrons undergo collisions with inner-shell electrons in the target atoms, knocking them out and initiating the chain of transitions that produce x-ray photons of specific energies—the **characteristic x rays** whose spectrum is characteristic of the target material. In analogy with the hydrogen spectrum, x-ray spectral lines are labeled $K\alpha$, $K\beta$, $K\gamma$, . . . for transitions from the $L, M, N, \ldots$ shells to the K shell. Transitions from the $M, N, O, \ldots$ shells to the L shell, respectively, are labelled $L\alpha$, $L\beta$, $L\gamma$, . . . , and so forth. Figure 41-31 shows x-ray spectra obtained by bombarding molybdenum with high-energy electrons.

We can estimate the energy of an x-ray transition by considering that the electron undergoing the transition experiences not the full electric field of the nucleus but a field reduced by the effect of any electrons closer to the nucleus. In the $L \rightarrow K$ transition that gives rise to the $K\alpha$ line, for instance, the electron sees an effective electric charge of $(Z - 1)e$ from the Z protons in the nucleus (charge Ze) and the remaining K-shell electron (charge $-e$). So the electron making the transition is in roughly the same circumstances as the electron in a single-electron atom of atomic number $Z - 1$. More generally, an electron in the nth energy level sees an effective nuclear charge Z_{eff} that is reduced approximately by the number of electrons in the shells below the upper level. For transitions across several

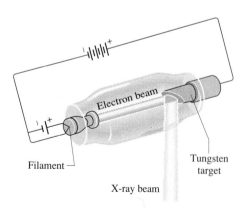

FIGURE 41-30 An x-ray tube. Current from the small battery heats the filament, which boils off electrons. Potential difference between filament and target—tens of kilovolts—accelerates electrons toward the target. X-ray production occurs when the electrons decelerate rapidly on hitting the target.

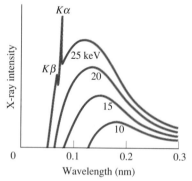

FIGURE 41-31 X-ray spectra obtained by bombarding molybdenum with electrons. Curves are labeled with the electron energy in keV. The highest energies are sufficient to eject inner-shell electrons from molybdenum atoms. As other electrons drop into the vacant states, x-ray photons corresponding to the $K\alpha$ and $K\beta$ spectral lines are emitted. Also shown is the broad band of x radiation associated with the electrons' rapid deceleration as they hit the target.

levels, Z_{eff} is more difficult to calculate, since the shielding of the nucleus changes from shell to shell. Once Z_{eff} is known, the approximate transition energy can be obtained by subtracting the energy levels given by Equation 41-9:

$$\Delta E = E_{n_1} - E_{n_2} \simeq -\frac{(13.6 \text{ eV})Z_{eff}^2}{n_1^2} - \left[-\frac{(13.6 \text{ eV})Z_{eff}^2}{n_2^2} \right]$$

$$= (13.6 \text{ eV})Z_{eff}^2 \left(\frac{1}{n_2^2} - \frac{1}{n_1^2} \right). \tag{41-20}$$

● **EXAMPLE 41-9** AN X-RAY TRANSITION

Estimate the energy and wavelength of the $L\alpha$ line in molybdenum ($Z = 42$).

Solution
The transition $L\alpha$ is from the M shell ($n = 3$) to the L shell ($n = 2$). Interior to the M shell are the K shell, holding two electrons, and the L shell, holding at most two s electrons and six p electrons. Since the M-shell electron is making a transition to the L shell, one of the L-shell electrons is missing. So there are nine electrons interior to the M shell, and $Z_{eff} = 42 - 9 = 33$. Applying Equation 41-20 then gives

$$\Delta E = (13.6 \text{ eV})Z_{eff}^2 \left(\frac{1}{n_2^2} - \frac{1}{n_1^2} \right)$$

$$= (13.6 \text{ eV})(33^2) \left(\frac{1}{2^2} - \frac{1}{3^2} \right) = 2057 \text{ eV}.$$

The corresponding wavelength is then

$$\lambda = \frac{hc}{\Delta E} = \frac{(6.63 \times 10^{-34} \text{ J·s})(3.00 \times 10^8 \text{ m/s})}{(2057 \text{ eV})(1.6 \times 10^{-19} \text{ J/eV})} = 0.60 \text{ nm}.$$

EXERCISE Estimate the energy of the $K\alpha$ transition in tungsten ($Z = 74$).

Answer: 54 keV

Some problems similar to Example 41-9: 43, 46, 57, 58

We emphasize that the use of an effective Z in Equation 41-20 is only a crude approximation; detailed analysis of inner-shell electron energies requires more realistic approximations to the Schrödinger problem for the multielectron atom. Nevertheless, Equation 41-20 and related forms do provide reasonable fits to experimental data, especially when Z_{eff} is adjusted to help the fit. For example, the difference between the actual atomic number Z and Z_{eff} for $L\alpha$ transitions turns out to be about 7.4 instead of the estimate of 9 we made in Example 41-9. The use of empirical equations relating the wavelengths of x-ray spectral lines to atomic number was historically important in establishing the periodic table. In 1914, Henry G. J. Moseley in Great Britain made a plot of the square root of x-ray frequencies ($f = c/\lambda$) versus atomic number Z. Equation 41-20 shows that such a plot should be a straight line, and that is what Moseley found experimentally (Fig. 41-32). He was then able to determine the atomic numbers of many elements where before only the atomic weights were known. Recasting the periodic table in terms of atomic number rather than atomic weight made possible our modern understanding of the chemical elements.

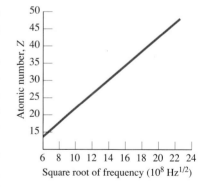

FIGURE 41-32 Moseley's plot showing square root of $K\alpha$-line frequencies as a function of atomic number Z.

Atomic x-ray transitions find wide practical use (Fig. 41-33). Astrophysicists studying high-energy processes in the cosmos use x-ray spectrometers on satellites to identify x-ray emission lines produced by highly excited atoms. Geologists bombard rocks with high energy electrons, knocking out inner-shell electrons and

FIGURE 41-33 X-ray applications. (*a*) X-ray spectrum produced by excitation of air pollutants trapped on a filter. The labeled spectral-line peaks show the presence of lead (Pb) and arsenic (As) in the samples. (*b*) X-ray image shows metal pins used to repair a human hip joint.

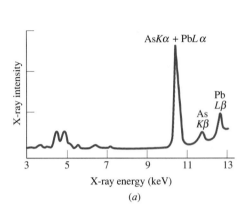

(a)

(b)

causing x-ray emission. Analysis of the resulting x-ray spectra helps identify abundances of elements within the rock. Airborne pollutants are readily measured by passing air through a filter, exciting the atoms caught on the filter, and analyzing the resulting x-ray spectrum. Absorption of x-rays is also significant. The element barium exhibits particularly strong x-ray absorption; in medical applications, barium is used to coat the intestinal tract, making it opaque to x-rays.

Spontaneous and Stimulated Transitions

What causes an electron to jump from one energy level to another? In the case of an upward transition, the electron must absorb the appropriate amount of energy. Generally, that energy is supplied by a photon whose energy is equal to the energy difference between the two levels; the process is then called **stimulated absorption** (Fig. 41-34*a*). (Upward transitions can result from other processes as well,

FIGURE 41-34 Interaction of photons with atomic electrons. Horizontal lines denote two atomic energy levels and the wave a photon with energy equal to the energy difference between the two levels. (*a*) In stimulated absorption, one photon is absorbed and the electron jumps to the higher level. (*b*) In spontaneous emission, a photon is emitted as the electron jumps spontaneously to the lower level. (*c*) In stimulated emission, the nearby passage of a photon stimulates a drop from the higher level. A second photon is emitted in the process. The two photons have the same energy and phase, and thus interfere constructively.

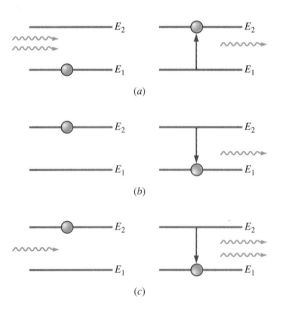

as, for example, in an energetic collision between two atoms or the interaction of a free electron with atomic electrons.)

For most downward transitions, however, we can identify no specific cause. An electron spontaneously jumps from a higher to a lower energy level and a photon is emitted; the process is termed **spontaneous emission** (Fig. 41-34b). Although an individual spontaneous emission is a random event, quantum mechanics permits us to calculate the probability per unit time for that event to occur; the inverse of that probability is the mean lifetime of the excited state.

In 1917 Einstein recognized a third possibility: excited atoms can be stimulated to drop into lower energy states by the nearby passage of a photon, again of energy appropriate to the transition. A photon is emitted in the downward transition, and the stimulating and emitted photon have the same energy and phase. This process, called **stimulated emission,** is the reverse of stimulated absorption (Fig. 41-34c).

Spontaneous emission, stimulated absorption, and stimulated emission play an important role in the transfer of radiation through gases. And stimulated emission is responsible for an important technological development of the late twentieth century: the laser.

■ APPLICATION THE LASER

As Fig. 41-34c suggests, stimulated emission provides a way of increasing the number of photons of a given energy and phase. In the 1940s Charles Townes in the United States and Nikolai Basov and Alexandr Prokhorov in Russia recognized the practical importance of this increase. They developed a device called the **maser,** whose acronym stands for **m**icrowave **a**mplification by **s**timulated **e**mission of **r**adiation. Today masers find use as low-noise amplifiers in microwave applications, as highly accurate clocks based on the sharply defined frequency of the atomic transition stimulated in the maser, and as systems for probing fundamental properties of atoms.

In the 1950s, scientists worked to extend the maser principle of stimulated emission into the visible part of the spectrum. The result was the **laser,** for **l**ight **a**mplification by **s**timulated **e**mission of **r**adiation. The key to laser action is to establish a large population of atoms in an excited state; since most atoms are normally in the ground state, this excited situation is termed a **population inversion** (Fig. 41-35). For most excited states, the atoms would normally decay very rapidly by spontaneous emission, and the population inversion could not be maintained. For this reason, the excited level is usually chosen to be a metastable one, whose lifetime is relatively long. Excitation is then a two-step process: first, atoms are excited to an energy state higher than the desired metastable state; then they quickly decay by spontaneous emission to the metastable state where they are "stuck" for a relatively long time by the lack of an allowed transition back to the ground state (Fig. 41-36). The excitation process is called **pumping** and the energy source for the excitation is the **pump.** A wide variety of pumps are used, including flash lamps, sunlight, other lasers, electric discharges, chemical reactions, and, in the x-ray laser once proposed for ballistic missile defense, nuclear explosions.

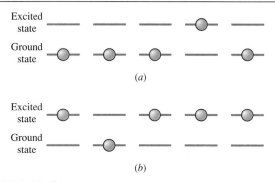

FIGURE 41-35 A group of atoms, represented as pairs of energy levels. (a) Normally, most atoms are in the ground state, although random thermal energy may occasionally excite some to higher levels. (b) In a population inversion, most atoms are in excited states.

If they were isolated, the excited atoms would remain in the metastable state for a long time, occasionally and randomly dropping to the ground state. The result would be a feeble, incoherent emission of photons. But when one atom drops to the ground state, it emits a photon that passes by many other excited atoms. The passing photon stimulates the others to emit as well. Now more photons are available, and more stimulated emission takes place. The process snowballs, resulting in a burst of photons of the same energy and phase. In a so-called **pulsed laser,** the pump is again applied to restore the population inversion, and another photon pulse is obtained. In a **cw laser** (for continuous wave), pumping and lasing occur continuously, resulting in continuous light output.

Working at Columbia University in the 1950s, Townes and Arthur Schawlow realized that stimulated emission could be

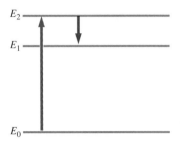

FIGURE 41-36 Pumping a laser usually involves a two-step process. First, atoms are excited from the ground state (E_0) to higher energy level (E_2). Then they drop spontaneously to the metastable level E_1, where they remain until passing photons stimulate transitions back to the ground state.

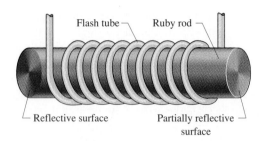

FIGURE 41-38 A ruby laser. The flash tube is the pump that excites chromium atoms in the ruby rod. Mirrors at either end form an optical cavity; one mirror is partially silvered to let the beam emerge.

enhanced by enclosing the radiating medium in a cavity with reflective ends. Photons bouncing back and forth in the cavity would then cause a high level of stimulated emission. One end of the cavity could be made only partially reflective, so light could emerge to form the laser beam. The first working laser was built in 1960 by Theodore Maiman (Fig. 41-37). In Maiman's laser, a flash tube surrounded a ruby rod; light from the flash excited chromium ions in the ruby to a metastable state; stimulated emission at a wavelength of 694 nm followed. Mirrors at either end of the rod provided the cavity, with one mirror partially silvered to allow the beam to emerge. Figure 41-38 is a schematic diagram of this ruby laser.

Since 1960, a myriad of laser types has been developed. Virtually anything can be used as the lasing medium, as long as a population inversion can be established. Media now in use

include gases, solids, liquids, semiconductor materials, and ionized plasmas. Natural laser action has even been detected in interstellar gas clouds. Some lasers, particularly those using chemical dyes as the lasing medium, are tunable over a broad range of wavelengths.

A laser commonly used in elementary physics experiments is the **helium-neon laser,** containing a mixture of 80% helium and 20% neon. The operation of this device is slightly more complicated than the two-step excitation process of Fig. 41-36. Electric current passes through the gas mixture, exciting helium by electron-ion collisions. The metastable excited state, 20.61 eV above the ground state, is very close to a 20.66-eV excited state of neon. Collisions between helium and neon populate this 20.66-eV neon level. The neon atoms drop to a lower energy level by stimulated emission, resulting in red laser light at

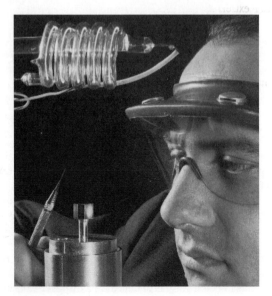

FIGURE 41-37 Theodore Maiman, who made the first laser in 1960, with his device. Coiled structure is the flash tube surrounding the ruby rod and providing the pump energy.

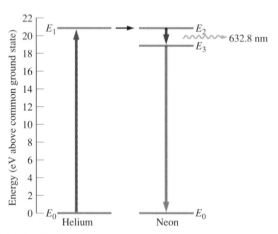

FIGURE 41-39 Energy-level diagram for the helium-neon laser. Electron-ion collisions excite helium to the metastable level E_1 at 20.61 eV. Collisions between helium and neon then transfer energy to the neon atoms, exciting them to the 20.66-eV level E_2. Laser emission occurs as the neon atoms drop by stimulated emission to the level E_3. Finally, they drop spontaneously to the ground state and are ready to begin the process again.

FIGURE 41-40 Coherence of laser light allows laser beams to travel considerable distances with minimal spreading. Here a laser is aimed at the moon in a lunar ranging experiment.

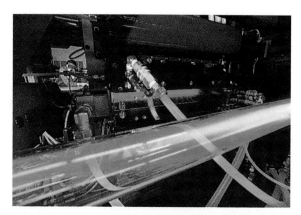

FIGURE 41-41 This high-power laser beam is used to separate uranium isotopes.

632.8 nm. Finally, the neon atoms fall to the ground state by spontaneous emission, and the process continues. Figure 41-39 shows an energy-level diagram for the He-Ne laser. This laser, like most, is not very efficient. Note from the energy-level diagram how little of the neon excitation energy is emitted in the laser beam. Problem 47 explores the energetics of the He-Ne laser.

The special characteristics of a laser beam are as follows: (1) The light is extremely monochromatic. Because the beam results entirely from transitions between a pair of discrete energy levels, all photons have essentially the same energy. (2) The beam is coherent; in wave terms, waves emitted by all the atoms are in phase, and interfere constructively to form a single wave whose wavefronts are well-defined surfaces that advance in step. This coherence allows the beam to travel long distances with minimal spreading (Fig. 41-40), and also makes possible very precise focusing. The fundamental limitation of optical diffraction is often the only barrier to a perfectly focused laser spot. (3) The beam can be made very powerful since the stimulated-emission process extracts energy from a great many atoms simultaneously. Since photons are spin-1 particles that do not obey the exclusion principle, there is no limit to the number of photons with the same energy and phase that can participate in a laser beam. Although typical lasers used in introductory physics laboratories have power outputs in the milliwatt range,

large continuous-wave lasers with power outputs in excess of 1 megawatt are now available, and pulsed outputs of 100 MW and higher have been achieved (Fig. 41-41). Multiple laser systems, such as those used for laser fusion experiments, reach peak power outputs of 10^{14} W—about 100 times the output of all the world's electric generating plants.

In the several decades since the first laser, applications of lasers have become ubiquitous. Small lasers scan price codes at supermarket checkouts and faithfully extract audio signals and computer data from compact discs. A laser beam makes possible the versatility and quality of laser-printed computer output. Medical uses increase each year; eye surgery in particular has been revolutionized by the laser. Lasers help surveyors establish straight lines over long distances. On a grander scale, they provide extremely accurate measurement of the distance to the moon. Small solid-state lasers provide drivers for fiber-optic systems, encoding telephone signals and computer data as pulses of light. Lasers entertain us with dazzling light shows. In science and engineering, they are used to provide precise excitation of atomic and molecular states, allowing analysis of materials and elucidation of fundamental physical principles. Experiments that were once difficult to perform—like Young's double-slit interference experiment—become trivial with lasers. Through the process of **holography**, laser wavefronts are captured on photographic emulsions from which three-dimensional images are reconstructed. Applications of holography include measurement of very fine motions—on the order of a wavelength of light—in connection with measurements of stress in materials. At higher powers, laser beams serve as precision cutting tools, eating their way through metals. Ultrafast lasers provide picosecond pulses of light, probing phenomena that occur on extremely short timescales. And as laser applications proliferate, so do the wavelengths available from lasers. Currently, lasers are available with outputs in the radio, the infrared, the visible, the ultraviolet, and the x-ray regions of the spectrum. A gamma-ray laser, using population inversions of excited nuclear states, is under development. Figure 41-42 samples the diversity of laser applications.

(a)

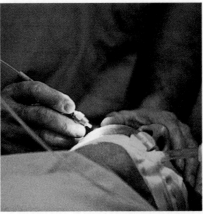

(b)

(c)

FIGURE 41-42 Laser applications include (a) cutting steel, (b) treating skin cancer, (c) reading bar codes at a supermarket checkout, and (d) measuring fluid flow velocity.

(d)

CHAPTER SYNOPSIS

Summary

1. The hydrogen atom is described by the Schrödinger equation in spherical coordinates. Especially simple states are those that depend only on the radial coordinate r, among them the ground state. The probability of finding an electron at a given position is best described by the **radial probability density,** given by $P(r) = 4\pi r^2 \psi^2$. In general, solutions to the Schrödinger equation involve all three coordinates, and are described by three quantum numbers:

   a. The **principal quantum number** n determines the energy:

$$E_n = -\frac{1}{n^2}\frac{\hbar^2}{2ma_0^2} = \frac{E_1}{n^2} = \frac{-13.6 \text{ eV}}{n^2}.$$

   b. The **orbital quantum number,** ℓ, determines the magnitude of the angular momentum:

$$L = \sqrt{\ell(\ell+1)}\,\hbar,$$

   where allowed values of ℓ are integers ranging from 0 to $n-1$.

   c. The **orbital magnetic quantum number** m_ℓ determines the component of orbital angular momentum along a given axis:

$$L_z = m_\ell \hbar,$$

   where m_ℓ takes integer values between $-\ell$ and $+\ell$. This quantization of the direction of the angular momentum vector **L** is called **space quantization.**

2. The electron carries an intrinsic angular momentum called **spin.** In general, the spin of a particle is described by a quantum number s that takes a single value; for electrons and other **spin-$\frac{1}{2}$ particles,** including protons and neutrons,

$s = \frac{1}{2}$. The magnitude of the spin angular momentum is given by

$$S = \sqrt{s(s + 1)}\hbar;$$

for electrons $S = (\sqrt{3}/2)\hbar$. Like orbital angular momentum, spin exhibits space quantization. The component S_z of spin on a given axis can take on only discrete values, spaced at intervals of $\hbar$:

$$S_z = m_s\hbar,$$

where for a spin-$\frac{1}{2}$ particle like the electron, m_s takes only the values $\pm\frac{1}{2}$. Spin is ultimately a consequence of relativistic invariance but was discovered experimentally before its theoretical explanation was known. The **Stern-Gerlach experiment** and related experiments provide evidence for electron spin and its space quantization.

3. The total angular momentum of a hydrogen atom is the vector sum of the orbital and spin angular momenta of its single electron. Magnetic interaction between magnetic dipole moments associated with electron spin and the orbital angular momentum—called **spin-orbit coupling**—results in **fine-structure splitting** of atomic energy levels. Such splitting occurs only for states with nonzero angular momentum; that is, for $\ell \neq 0$ states. **Hyperfine structure,** which arises from the interaction of the spin magnetic moment with the nuclear magnetic moment, occurs in $\ell = 0$ states as well.

4. Pauli's **exclusion principle** states that no two electrons can be in the same state. Since an electron has two possible spin orientations, the exclusion principle means that at most two atomic electrons can be in a state with the same quantum numbers n, ℓ, and m_ℓ. The exclusion principle applies only to particles of half-integer spin; integer-spin particles like photons can have arbitrarily many particles in the same state.

5. In multielectron atoms, the exclusion principle governs the organization of electrons among the various quantum states, designated **shells, subshells,** and **orbitals.** Chemical properties of the elements are determined by the outermost atomic electrons. The regular arrangement of elements along each row in the **periodic table** reflects the filling of the outermost atomic shells; the inert gas at the end of each row has a completely full outer p subshell and does not readily participate in chemical reactions.

6. Transitions among atomic energy levels are governed by **selection rules,** many related to underlying conservation laws.
   a. The visible and near-visible spectra of atoms are the result of transitions involving the outermost electrons. At high resolution, spectral lines are often split by effects like the spin-orbit splitting.
   b. X-ray spectra result from transitions involving inner-shell electrons, typically when one such electron is ejected from the atom and another drops into its place.
   c. Atomic transitions include **stimulated absorption,** in which an atom absorbs a photon as an atomic electron jumps to a higher energy level; **spontaneous emission,** in which an electron jumps spontaneously to a lower level, resulting in emission of a photon; and **stimulated emission,** whereby a passing photon stimulates an excited atom to undergo a downward transition and emit a photon of the same energy and phase. Stimulated emission is the essential process in the **laser,** a source of intense and coherent electromagnetic radiation.

Terms You Should Understand

(Pairs are closely related terms whose distinction is important; number in parentheses is chapter section where term first appears.)

radial probability density (41-1)
orbital angular momentum (41-1)
principal quantum number, orbital quantum number, orbital magnetic quantum number (41-1)
space quantization (41-1)
Stern-Gerlach experiment (41-2)
fine structure, spin-orbit effect (41-2)
exclusion principle (41-3)
fermion, boson (41-3)
shell, subshell, orbital (41-4)
selection rules (41-5)
allowed transition, forbidden transition (41-5)
spontaneous emission, stimulated emission (41-5)
population inversion (41-5)

Symbols You Should Recognize

n, ℓ, m_ℓ (41-1)
s, m_s, j, m_j (41-1)
s, p, d, f, . . . (41-2)
S, P, D. F, . . . (41-2)
K, L, M, . . . (41-2)
symbols like $3D_{3/2}$ (41-2)
electronic configurations like $1s^2 2s^2 2p^6$ (41-4)

Problems You Should Be Able to Solve

evaluating energies and angular momenta associated with states of one-electron atoms (41-1)
finding probabilities for locating electrons in one-electron atoms (41-1)
evaluating spin angular momentum (41-2)
finding allowed orientations for spin and other angular momentum vectors (41-2)
finding allowed values for total angular momentum (41-2)
determining electronic structure of atoms (41-4)
relating energy-level diagrams and spectra (41-5)

Limitations to Keep in Mind

Closed-form solutions of the Schrödinger equation apply only to one-electron atoms.

Optical spectra of multielectron atoms are much more complex than those discussed in this chapter.

The analysis presented here for x-ray line spectra is only a crude approximation.

QUESTIONS

1. The electron in a hydrogen atom is very crudely like a particle confined to a three-dimensional box. In the atom, what plays the role of the confining box?

2. A friend who has not studied physics asks you the size of a hydrogen atom. How do you answer?

3. How many quantum numbers are required to specify fully the state of a hydrogen atom? Why? (Don't forget about spin.)

4. Both the Bohr and Schrödinger theories of the hydrogen atom predict the same ground-state energy. Do they agree about the angular momentum in the ground state? Explain.

5. Plotted as a function of radius r, the probability density ψ^2 for the hydrogen ground state shows a maximum right at the nucleus, $r = 0$ (see Fig. 41-3). Why is it not correct to infer that the electron is most likely to be found at $r = 0$?

6. Describe space quantization, and explain why it is most obviously at odds with classical physics for states with low orbital quantum number ℓ.

7. The potential-energy function for the electron in a hydrogen atom is spherically symmetric. Does this mean that all possible ψ functions for the electron depend only on radial distance from the proton? Explain.

8. Is it possible for a hydrogen atom to be in the $2d$ state? Explain.

9. Can the component of a quantized angular momentum measured on a given axis ever equal the magnitude of the angular momentum vector? Explain.

10. The electron is a spin-$\frac{1}{2}$ particle. Does this mean the intrinsic angular momentum of the electron is $\frac{1}{2}\hbar$? Explain.

11. How does the Stern-Gerlach experiment provide convincing evidence for space quantization?

12. Why is there no spin-orbit splitting in the ground state of hydrogen?

13. In what way is the exclusion principle responsible for the diversity of chemical elements?

14. Helium and lithium exhibit very different chemical behavior, yet they differ by only one unit of nuclear charge. Explain how the exclusion principle brings about the dramatic difference in chemical behavior.

15. Are forbidden transitions really forbidden? Explain and contrast with allowed transitions.

16. Why is an inner-shell vacancy generally required for an atom to emit an x-ray photon?

17. Optical spectra change qualitatively from one element to the next, while x-ray spectra exhibit only a gradual, continuous change from one element to the next. What might explain the difference?

18. Why is stimulated emission essential for laser action?

19. Why do most lasers use a metastable state of the lasing medium?

PROBLEMS

Section 41-1 The Hydrogen Atom

1. Using physical constants accurate to four significant figures (see the inside front cover), verify the numerical values of the Bohr radius a_0 and the hydrogen ground-state energy E_1.

2. A group of hydrogen atoms are in the same excited state. It is found that photons with at least 1.5 eV energy are required to ionize these atoms. What is the quantum number n for the initial excited state?

3. Find (a) the probability density and (b) the radial probability density at $r = \frac{1}{2}a_0$ for the hydrogen ground state. Comment on the relative sizes of these quantities.

4. Repeat the calculation in the second half of Example 41-1 to find the probability that an electron in the hydrogen ground state will be found beyond two Bohr radii.

5. What is the maximum possible magnitude for the orbital angular momentum of an electron in the $n = 7$ state of hydrogen?

6. Which of the following is not a possible value for the magnitude of the orbital angular momentum in hydrogen? (a) $\sqrt{12}\hbar$; (b) $\sqrt{20}\hbar$; (c) $\sqrt{30}\hbar$; (d) $\sqrt{40}\hbar$; (e) $\sqrt{56}\hbar$.

7. The orbital angular momentum of the electron in a hydrogen atom has magnitude 2.585×10^{-34} J·s. What is the minimum possible value for its energy?

8. What is the orbital quantum number for an electron whose orbital angular momentum has magnitude $L = \sqrt{30}\hbar$?

9. Determine the principal and orbital quantum numbers for a hydrogen atom whose electron has energy -0.850 eV and orbital angular momentum of magnitude $\sqrt{12}\hbar$.

10. Find (a) the energy and (b) the magnitude of the orbital angular momentum for an electron in the $5d$ state of hydrogen.

11. A 1200-kg car rounds a turn of radius 150 m at a speed of 10 m/s. Assuming its angular momentum is quantized according to Equation 41-10, find the approximate value for ℓ.

12. A hydrogen atom is in the $6f$ state. What are (a) its energy and (b) the magnitude of its orbital angular momentum?

13. Give a symbolic description for the state of the electron in a hydrogen atom when the total energy is -1.51 eV and the orbital angular momentum is $\sqrt{6}\hbar$.

14. The maximum possible angular momentum for a hydrogen atom in a certain state is $30\sqrt{11}\hbar$. Find (a) the principal quantum number and (b) the energy.

15. Which of the following pairs of energy and magnitude of orbital angular momentum are possible for a hydrogen atom, and to what n and ℓ values do they correspond? (a) -0.544 eV, 3.655×10^{-34} J·s; (b) -1.51 eV, 3.655×10^{-34} J·s; (c) -1.51 eV, 5.842×10^{-34} J·s; (d) -3.4 eV, 1.492×10^{-34} J·s

16. A hydrogen atom is in a state with $\ell = 2$. What are the possible angles its orbital angular momentum vector can make with an arbitrary axis?

17. A hydrogen atom has energy $E = -0.850$ eV. What are the maximum possible values for (a) the magnitude of its orbital angular momentum and (b) the component of that angular momentum on a chosen axis?

18. An electron in hydrogen is in the $5f$ state. What possible values, in units of $\hbar$, could a measurement of the orbital angular momentum component on a given axis yield?

19. Substitute Equation 41-3 for ψ in Equation 41-4, and carry out the indicated differentiation to show that the result is Equation 41-5.

20. Differentiate the radial probability density for the hydrogen ground state, and set the result to zero to show that the electron is most likely to be found at one Bohr radius.

Section 41-2 Electron Spin

21. Verify the value of the Bohr magneton in Equation 41-15.

22. Theories of quantum gravity predict the existence of a spin-2 particle called the graviton. What would be the magnitude of the graviton's spin angular momentum?

23. Some very short-lived particles known as delta resonances have spin $\frac{3}{2}$. Find (a) the magnitude of their spin angular momentum and (b) the number of possible spin states.

24. What is the minimum angle between the spin vector and the z axis for (a) a spin-$\frac{3}{2}$ particle and (b) a spin-2 particle?

25. What are the possible j values for a hydrogen atom in the $3D$ state?

26. Repeat the preceding problem for the case where you know only that the principle quantum number is 3; that is, ℓ might have any of its possible values.

27. Draw vector diagrams similar to Fig. 41-14 for spin-orbit coupling of an electron in the $\ell = 3$ state, and, for the lower

of the two j values, construct also a diagram similar to Fig. 41-15a.

28. A hydrogen atom is in the $4F_{5/2}$ state. What are (a) its energy in units of the ground-state energy, (b) the magnitude of its orbital angular momentum in units of $\hbar$, and (c) the magnitude of its total angular momentum in units of $\hbar$? (d) Which of (b) or (c) is greater? Why?

Section 41-3 The Pauli Exclusion Principle

29. Suppose you put five electrons into an infinite square well. (a) How do the electrons arrange themselves to achieve the lowest total energy? (b) Give an expression for this energy in terms of the electron mass m, the well width L, and Planck's constant.

30. A harmonic oscillator potential of natural frequency ω contains eight electrons and is in the state of lowest energy. (a) What is that energy? (b) What would be the energy if the electrons were replaced by spin-1 particles of the same mass?

31. A harmonic oscillator potential of natural frequency ω contains a number of electrons and is in the state of lowest energy. If that energy is $6.5\hbar\omega$, (a) how many electrons are in the potential well and (b) what is the energy of the highest energy electron?

Section 41-4 Multielectron Atoms and the Periodic Table

32. Use shell notation to characterize the outermost electron of rubidium.

33. Write the full electronic structure of scandium.

34. Write the full electronic structure of bromine.

35. Consult the footnote on page 1108 to determine the electronic configuration of copper.

36. Using the table below, make a plot of atomic volume versus atomic number, for the elements from $Z = 30$ to $Z = 59$ listed in the table. Comment on the structure of your graph in relation to the periodic table, the electronic structures of atoms, and their chemical properties. (Volumes are in units of 10^{-30} m³.)

Z	V	Z	V	Z	V
30	7.99	40	26.1	50	11.2
31	12.5	41	20.2	51	8.78
32	6.54	42	18.8	52	6.88
33	4.99	43	17.5	53	5.28
34	3.71	44	16.2	54	4.19
35	2.85	45	12.8	55	95.9
36	2.57	46	12.0	56	51.6
37	70.3	47	11.2	57	49.0
38	37.2	48	10.5	58	46.5
39	28.3	49	17.2	59	44.0

Section 41-5 Transitions and Atomic Spectra

37. An electron in a highly excited state of hydrogen ($n_1 \gg 1$) drops into the state $n = n_2$. What is the lowest value of n_2 for which the emitted photon will be in the infrared ($\lambda > 700$ nm)?

38. A solid-state laser made from lead-tin selenide has a lasing transition at a wavelength of 30.0-μm. If its power output is 2.0 mW, how many lasing transitions occur each second?

39. The $4f \rightarrow 3p$ transition in sodium produces a spectral line at 567.0 nm. What is the energy difference between these two levels?

40. The $4p \rightarrow 3s$ transition in sodium produces a doublet spectral line at 330.2 and 330.3 nm. What is the energy splitting of the $4p$ level?

41. The $4s \rightarrow 3p$ transition in sodium produces a doublet spectral line at 1138.1 nm and 1140.4 nm. Combine this fact with the discussion in Example 41-8 to find an accurate value for the energy difference between the $3s$ and $4s$ states in sodium.

42. Show that the wavelength λ in nm of a photon whose energy E is given in eV is

$$\lambda = \frac{1240}{E}.$$

43. Estimate the wavelength of the $K\alpha$ x-ray line in calcium.

44. Estimate the minimum electron energy that could excite the $K\alpha$ line in molybdenum, and compare with Fig. 41-31. *Hint:* You need to eject a K-shell electron entirely from the atom.

45. What is the approximate minimum accelerating voltage for an x-ray tube with an iron target to produce the $L\alpha$ line? *Hint:* See Problem 44.

46. Use the x-ray spectrum shown in Fig. 41-33 to estimate Z_{eff} for the $L\beta$ transition in lead.

47. Use information from Fig. 41-39 (a) to find the energy of the level E_3 and (b) to determine the maximum fraction of the energy delivered to excite neon atoms in a He-Ne laser that actually ends up as laser light.

48. A selection rule for the infinite square well allows only those transitions in which n changes by an odd number. Suppose an infinite square well of width 0.200 nm contains an electron in the $n = 4$ state. (a) Draw an energy-level diagram showing all allowed transitions that could occur as this electron drops toward the ground state (include transitions from lower levels that could be reached from $n = 4$). (b) What are all the possible photon energies emitted in these transitions?

49. An ensemble of square-well systems of width 1.17 nm all contain electrons in highly excited states. They undergo all possible transitions in dropping toward the ground state, obeying the selection rule that Δn must be odd. (a) What wavelengths of visible light are emitted? (b) Is there any infrared emission? If so, how many spectral lines fall in the infrared?

50. For hydrogen, fine-structure splitting of the $2p$ state is only about 5×10^{-5} eV. What percentage is this difference of the photon energy emitted in the $2p \rightarrow 1s$ transition? Your answer shows why it's hard to observe spin-orbit splitting in hydrogen.

Paired Problems

(Both problems in a pair involve the same principles and techniques. If you can get the first problem, you should be able to solve the second one.)

51. Find the probability that the electron in the hydrogen ground state will be found in the radial distance range $r = a_0 \pm 0.1a_0$.

52. A gas contains 1000 hydrogen atoms in their ground states. If you locate their electrons, in how many atoms, on average, will they be within 3 Bohr radii of the nucleus?

53. Find the spacing between adjacent orbital angular momentum values for hydrogen in the $n = 2$ state.

54. What is the most orbital angular momentum that could be added to an atomic electron initially in the $6d$ state, without changing its principal quantum number? What would be the new state?

55. (a) Write the alpha-numeric description of a hydrogen atom whose energy is one-ninth of the ground-state energy (recall that the latter is negative, so this is an excited state), whose orbital angular momentum has magnitude $\sqrt{6}\hbar$, and whose spin and orbital angular momentum are as nearly aligned as possible. (b) How many m_j values are there for this state?

56. A hydrogen atom is in an F state. (a) Find the possible values for the magnitude of its total angular momentum and, (b) for the state with the greatest angular momentum, the number of possible values for the component of $\mathbf{J}$ on a given axis.

57. Estimate the energy of the $L\alpha$ x-ray transition in arsenic.

58. Estimate the wavelength of the $L\alpha$ x-ray transition in lead.

Supplementary Problems

59. Verify that the normalization constant $1/4\sqrt{2\pi a_0^3}$ in Equation 41-8 is correct.

60. A hydrogen atom is in the $2s$ state. What is the probability that its electron will be found (a) beyond one Bohr radius and (b) beyond 10 Bohr radii?

61. Substitute the ψ_{2s} wave function (Equation 41-8) into Equation 41-4 to verify that the equation is satisfied and that the energy is given by Equation 41-7 with $n = 2$.

62. Show that the maximum number of electrons in the nth shell of an atom is $2n^2$.

63. (a) Verify Equation 41-9 by considering a single-electron atom with nuclear charge Ze instead of e. (b) Calculate the ionization energies for single-electron versions of helium, oxygen, iron, lead, and uranium.

64. Form the radial probability density $P_2(r)$ associated with the ψ_{2s} state of Equation 41-8, and find the most probable radial position for the electron.

CHAPTER 42

MOLECULAR AND SOLID-STATE PHYSICS

Atoms bind together to form molecules, shown here in a computer simulation.

Quantum mechanics is remarkably successful in explaining the properties of atoms. But what happens when we join atoms to form more complex structures like molecules and solids? The laws of quantum mechanics govern even these many-particle systems. In principle, we could treat a complicated molecule like DNA by solving the Schrödinger equation for all its constituent electrons and nuclei. Recently, the availability of supercomputers has made calculations of molecular structure increasingly successful, although we are still far from computing the structure of DNA. Nevertheless, in most applications it remains appropriate to think of multiparticle systems in more qualitative terms, categorizing them according to the mechanisms that bind their constituent particles.

42-1 MOLECULAR BONDING

The binding of atoms into molecules involves both electrical forces and peculiarly quantum-mechanical effects associated with the exclusion principle. Although individual atoms are electrically neutral, the distribution of positive and negative charge within them nevertheless results in attractive or repulsive forces. When atoms are squeezed closely together, interactions involving spins of their outermost electrons also result in attractive or repulsive interactions. For atoms with unfilled outer shells, it is energetically favorable for electrons to pair with opposite spins; this causes an attractive interaction. When the outer atomic shells are filled, two electrons originally in identical states in the separate atoms are forced into different states as the atoms are brought together. This effect occurs because the two atoms become effectively a single system to which the exclusion principle applies. One or more electrons are forced into higher-energy states, raising the overall energy and giving rise to a repulsive interaction. Finally, if atoms are pushed very closely together, the electrical repulsion of the nuclei becomes important. Ultimately, the balance of attractive and repulsive interactions determines the equilibrium configuration of a molecule. In energy terms, we can think of a stable molecule as a minimum-energy configuration of the electrons and nuclei making up two or more atoms (Fig. 42-1). Although such force and energy considerations ultimately govern all molecules, it is convenient to distinguish several molecular binding mechanisms, based on which of the attractive and repulsive interactions are most important.

Ionic Bonding

As we saw in the preceding chapter, elements near the left side of the periodic table have relatively few electrons in their outermost shells and correspondingly low ionization energies. In contrast, elements near the right side of the table exhibit nearly filled shells and consequently have strong affinities for electrons. When atoms from these different regions of the periodic table come together, relatively little energy is needed to transfer one or more electrons between them. Sodium, for example, has an ionization energy of 5.1 eV, meaning that this much energy must be supplied to make an Na^+ ion. Chlorine, at the opposite end of the periodic table, has such a strong electron affinity that the energy of a Cl^- ion is actually 3.8 eV below that of a neutral Cl atom. Thus an expenditure of only 1.3 eV (5.1 eV − 3.8 eV) is required to transfer the outermost electron from a sodium to a chlorine atom. The resulting ions are strongly attracted, and they come together until they reach equilibrium at an internuclear separation of about 0.24 nm. The total energy of the pair is then 4.2 eV below that of neutral chlorine and sodium atoms at large separation (Fig. 42-2). Since it would therefore take 4.2 eV to separate the atoms, this quantity is called the **dissociation energy.**

Because the minimum-energy sodium-chlorine structure consists of ions bound by the electrostatic force, the binding mechanism is termed **ionic bonding.** Ionic bonding generally occurs not between isolated pairs of atoms but in crystalline solids. Because the building blocks of an ionically bound substance are electrically charged, each can bind to several of the opposite charge, giving rise to the regular pattern of a crystalline solid (Fig. 42-3). Because the electro-

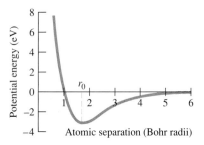

FIGURE 42-1 Potential energy of a pair of hydrogen atoms, as a function of the distance between their nuclei. The minimum of the curve marks the internuclear separation r_0 of an H_2 molecule at equilibrium.

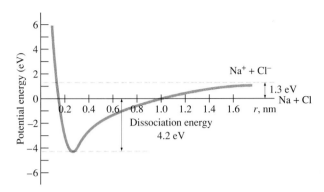

FIGURE 42-2 Potential-energy curve for a Na^+ and Cl^- ion. The zero of potential energy corresponds to neutral Na and Cl atoms at infinite separation. Here the curve goes toward $+1.3$ eV at large separation, reflecting the energy needed to transfer an electron from Na to Cl to form the ion pair.

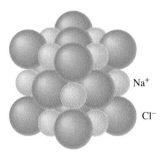

FIGURE 42-3 A sodium chloride crystal is a regular array of sodium and chlorine ions, bound by the electrostatic force. In the crystalline solid, the effect of multiple-ion interactions increases the internuclear spacing from 0.24 nm of the isolated ion pair (Fig. 42-2) to 0.28 nm.

static force between ions is relatively strong, ionic solids are tightly bound and therefore have high melting points (801°C for NaCl). And because all electrons are bound to individual nuclei, there are no free electrons and therefore ionic solids are excellent electrical insulators.

Covalent Bonding

In an ionic bond, each electron is associated with only one ion. In a **covalent bond,** on the other hand, electrons are shared among atoms. Covalent bonds can occur between atoms whose outermost shells are not full, and whose outer electrons can therefore pair with opposite spins. The simplest example of a covalent bond is the hydrogen molecule, H_2. Since each hydrogen atom has only a single $1s$ electron, each could accommodate in its $1s$ shell a second electron with opposite spin. When two hydrogen atoms come together, quantum mechanics predicts a molecular ground state in which both electrons share a single orbital, with the greatest probability of finding the electrons between the nuclei (Fig. 42-4). Dissociation energies for covalent bonds are, like those of ionic bonds, on the order of a few eV.

With their outermost molecular orbitals full (i.e., containing paired electrons of opposite spin), covalently bonded molecules often have no room for another electron in their structures. For example, the addition of a third hydrogen atom to an H_2 molecule is not possible because the ground-state orbital already contains two electrons with opposite spins, and the exclusion principle requires that a third electron go into a higher energy state. The energy of that state is, in fact, higher than that of an H_2 molecule and a distant H atom; for this reason H_3 is not a stable molecule. Because their outermost molecular orbitals are full, covalent molecules interact only weakly, and as a result many common covalent materials—for example, H_2, CO, N_2, and H_2O—are either gases or liquids at ordinary temperatures. In other cases covalent bonds can form crystalline structures; an example is diamond, formed when each carbon atom bonds covalently to its four nearest neighbors (Fig. 42-5).

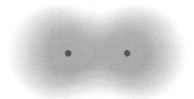

FIGURE 42-4 Probability density for the ground state of molecular hydrogen (H_2). Binding of the molecule results from sharing of the two electrons, here evident in the enhanced probability density between the nuclei.

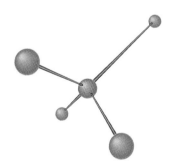

FIGURE 42-5 In the diamond crystal, each atom bonds covalently to its four nearest neighbors.

■ APPLICATION BUCKMINSTERFULLERINE

Molecular bonding has been with us for a long time, but nature continues to surprise us. As early as the 1960s, chemists suggested that carbon atoms might bond to produce large, hollow molecules. Such molecules were first identified in the 1980s, and in 1990 they were produced in abundance. They join diamond and graphite as a third form of pure carbon.

The paradigm of this new molecular family is C_{60}, named buckminsterfullerine after the American philosopher, designer, and engineer R. Buckminster Fuller, who is best known for the geodesic dome. With its 60 covalently bonded carbon atoms identically positioned at the corners of adjacent pentagons and hexagons, buckminsterfullerine resembles a soccer ball and is the roundest molecule possible (Fig. 42-6a). Other fullerines include asymmetric "buckybabies" such as C_{32} and C_{44}, the rugby-ball-shaped C_{70}, and giants such as C_{540} and C_{960}.

Buckminster Fuller argued that his geodesic domes are among the strongest of structures, and his namesake molecules

are similarly stable and resilient. Yet despite their stability, fullerines show remarkable chemical reactivity. Other atoms and molecular complexes attach readily to the outside of the fullerine "cage", and individual atoms can be "trapped" inside. In 1991 scientists at Bell Laboratories made the startling discovery that C_{60} compounds with potassium or rubidium are superconductors, and in the same year Japanese scientists synthesized "buckytubes"—tiny, tough, carbon cylinders (Fig. 42-6b). Finally, C_{60} molecules themselves can fit into other chemical structures, as shown in Fig. 42-6c.

Fullerines are easy to produce by vaporizing carbon; they form, for example, in sooty candle flames and in interstellar space. Chemists and physicists are only just beginning to understand these remarkable manifestations of molecular bonding, which will almost certainly see widespread technological applications in the coming decades.

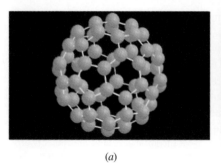 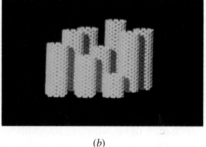 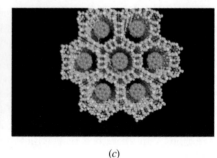

(a)   (b)   (c)

FIGURE 42-6 A few of the many fullerine forms. (a) Buckminsterfullerine, C_{60}, is a symmetric arrangement of 60 carbon atoms. (b) "Buckytubes" are tubelike structures made from carbon atoms. (c) Buckminsterfullerine molecules nestle inside a crystal of zeolite, an aluminum silicate mineral.

Hydrogen Bonding

If water consists of covalent molecules, why does it ever take on a solid form? The answer lies in **hydrogen bonds,** a type of weak bond that forms when the positively charged proton of a hydrogen nucleus attracts negative parts of other molecules. In ice, for example, hydrogen bonds link a proton in one H_2O molecule to the oxygen atom in another (Fig. 42-7a). This bonding is most accurately described as an attraction between electric dipoles composed of OH structures in each molecule. These dipoles occur because the electrons in H_2O are statistically more likely to be found in the vicinity of the oxygen, giving the oxygen a slight negative charge and leaving the hydrogens slightly positive. This statistical effect results, on the average, in the transfer of much less than one negative elementary charge to the oxygen. For this reason, and because dipole-dipole forces fall much more rapidly with distance than point-charge forces,

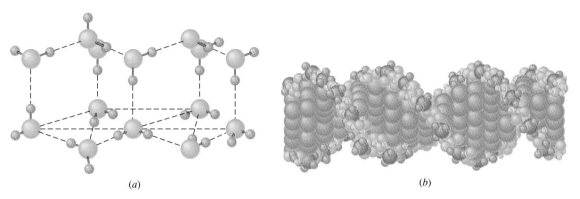

FIGURE 42-7 Hydrogen bonding. (*a*) Hydrogen bonds (shown as dashed lines) join water molecules in an ice crystal. (*b*) In DNA, hydrogen bonds hold two covalently bonded chains of atoms in the double-helix structure.

hydrogen bonds are much weaker than ionic or covalent bonds. A typical hydrogen-bond energy is 0.1 eV. Hydrogen bonds are often important in determining the overall physical configuration of complicated molecules. In DNA, for example, covalent bonds join individual atoms to form long chains; the chains are then linked into the famous double-helix structure by hydrogen bonds (Fig. 42-7*b*).

You might expect hydrogenlike bonds to occur any time part of a molecule is positively charged. But the bond is unique to hydrogen, for this reason: The core of a hydrogen atom consists only of the proton, about 10,000 times smaller than the atom. This tiny proton can easily nestle close to negatively charged structures, giving rise to a bond of nonnegligible strength. With any other atom, the core consists of a nucleus and the tightly bound inner electrons, giving a structure that is more atomic size than nuclear size. These larger cores cannot fit tightly between other atomic cores, and so cannot form hydrogenlike bonds of significant strength.

Van der Waals Bonding

In Section 23-6 we showed how the weak van der Waals force results from electrostatic interaction between induced dipole moments of otherwise nonpolar molecules. In gases of covalent molecules, the van der Waals force causes deviations from the perfect gas law that are most pronounced at high densities. But as the temperature drops, this weakly attractive force becomes effective in binding molecules into liquids and solids. Liquid and solid forms of oxygen (O_2) and nitrogen (N_2), for example, are held together by van der Waals bonding.

Metallic Bonding

In a metal, the outermost atomic electrons are quite free of the individual nuclei, and can move throughout the material. The metal forms a crystal lattice of positive ions, bound together by this "electron gas" spread through the metal. The relatively free electrons give a metal its high electrical and thermal conductivities.

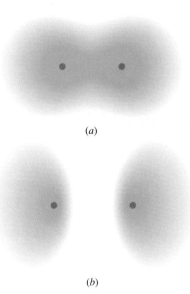

(a)

(b)

FIGURE 42-8 Electron probability densities for (a) the ground state and (b) the first excited state of the hydrogen molecule (H_2). Note that the excited state has a greater internuclear separation.

42-2 MOLECULAR ENERGY LEVELS

In a molecule, electrons and nuclei are bound into a single structure by electric forces. Like an atom or other quantum-mechanical bound system, the possible energy levels of a molecule are quantized. As in atoms, differences among molecular energy levels may be associated with different electronic configurations (Fig. 42-8). But molecules are more complex structures than atoms, and molecular energy can take additional forms.

In Chapter 21, we found that a complete description of the specific heats of gases required that we consider the rotational and vibrational motion of individual molecules. We even hinted at quantum mechanics, as we pointed out that each of these modes of molecular motion could absorb only certain discrete amounts of energy (see Fig. 21-24 and accompanying text). Here, in a quantum-mechanical treatment of molecular energetics, we consider rotational and vibrational energy states as well as electronic configuration.

Rotational Energy Levels

If a molecule is rotating, it has angular momentum whose magnitude, from Equation 13-4, is

$$J = I\omega,$$

where I is the rotational inertia and ω the angular speed, and where we use J rather than L for consistency with the molecular physics literature. In the preceding chapter, we found that atomic angular momentum is quantized, with magnitude given by

$$J = \sqrt{j(j + 1)}\hbar. \qquad (42\text{-}1)$$

Not surprisingly, the angular momentum of molecular rotation, $J = I\omega$, also obeys the quantization condition of Equation 42-1, where the quantum number j takes on integer values 0, 1, 2, 3, ... But then the rotational energy, which from Equation 12-30 is

$$E_{\text{rot}} = \tfrac{1}{2}I\omega^2,$$

must also be quantized. Solving the equation $J = I\omega$ for ω allows us to write the energy as

$$E_{\text{rot}} = \tfrac{1}{2}I\left(\frac{J}{I}\right)^2 = \frac{J^2}{2I}.$$

Applying the quantization condition 42-1 for J, we then have the quantized rotational energy levels:

$$E_{\text{rot}} = \frac{\hbar^2}{2I}j(j + 1), \qquad j = 0, 1, 2, 3, \ldots . \qquad (42\text{-}2)$$

Molecules, like atoms, can jump between energy levels by absorbing or emitting photons. Conservation of angular momentum and the fact that the photon is a spin-1 particle limit the allowed transitions to those for which $\Delta j = \pm 1$, so the only allowed jumps are between adjacent rotational levels. Measuring the photon energy associated with rotational transitions allows us to infer the rotational inertia through Equation 42-2, and thus to learn about the structures of molecules.

● EXAMPLE 42-1   HOW BIG IS A MOLECULE?

The spectrum of a gas of molecules undergoing transitions among rotational states shows a series of sharp lines at wavelengths corresponding to the energy differences between adjacent states (see Fig. 42-12 in the next section). The energy difference between adjacent spectral lines is constant; for hydrogen chloride (HCl), this difference is 2.63 meV (2.63×10^{-3} eV). Use this experimental result to calculate the rotational inertia of the HCl molecule. Then, making the approximation that the massive chlorine atom stays nearly fixed while the hydrogen rotates rigidly about it (Fig. 42-9), find an expression for the rotational inertia in terms of the hydrogen mass and internuclear separation. By equating with your spectroscopically measured rotational inertia, determine the approximate internuclear separation in HCl.

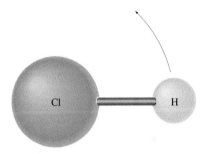

FIGURE 42-9 Dumbbell model for a rotating HCl molecule. Since the chlorine atom is much more massive than the hydrogen, it is a good approximation to consider the chlorine fixed. Then the rotational inertia is due entirely to the hydrogen.

Solution
Equation 42-2 shows that the difference between the jth and $(j - 1)$th energy levels is

$$\Delta E_{j \to (j-1)} = \frac{\hbar^2}{2I}[j(j+1) - (j-1)j] = \frac{\hbar^2 j}{I}.$$

This is the energy of a single spectral line corresponding to a transition between the jth and $(j - 1)$th levels. An adjacent spectral line, corresponding to a transition between the $(j - 1)$th and $(j - 2)$th levels, would similarly have $\Delta E_{(j-1) \to (j-2)} = \hbar^2(j - 1)/I$. So the difference $\Delta(\Delta E)$ between adjacent spectral lines is

$$\Delta(\Delta E) = \frac{\hbar^2 j}{I} - \frac{\hbar^2(j-1)}{I} = \frac{\hbar^2}{I},$$

a result that is independent of j. Solving for the rotational inertia I, and using our 2.63-meV value for $\Delta(\Delta E)$ gives

$$I = \frac{\hbar^2}{\Delta(\Delta E)} = \frac{(1.055 \times 10^{-34} \text{ J·s})^2}{(2.63 \times 10^{-3} \text{ eV})(1.6 \times 10^{-19} \text{ J/eV})}$$

$$= 2.65 \times 10^{-47} \text{ kg·m}^2.$$

For a point mass m moving in a circle of radius R, the rotational inertia given by Equation 12-15 is just $I = mR^2$, so

$$R = \sqrt{\frac{I}{M}} = \sqrt{\frac{2.65 \times 10^{-47} \text{ kg·m}^2}{1.67 \times 10^{-27} \text{ kg}}} = 1.26 \times 10^{-10} \text{ m}$$

$$= 0.126 \text{ nm},$$

slightly larger than an isolated hydrogen atom. (Here we used the proton mass as an excellent approximation to the hydrogen mass.) Physical chemists use the method of this example to determine molecular bond lengths.

In this example, we made the approximation that the more massive Cl atom remained essentially fixed, with the rotational inertia determined entirely by the hydrogen atom. A more accurate calculation—essential in a diatomic molecule whose constituent atoms had more comparable masses—would account for the motion of both atoms in determining the rotational inertia (see Problems 10 and 52). This calculation is also approximate in that the wavelength given is for a hypothetical molecule with no vibrational energy. The nonzero ground-state vibrational energy of a real molecule results in a slight stretching of the molecular bond and, therefore, in a shift in the rotational transitions.

EXERCISE The equilibrium bond length in hydrogen bromide (HBr) is 0.141 nm. Treating the bromine as essentially at rest, determine the energy of the first rotational excited state.

Answer: 2.09 meV

Some problems similar to Example 42-1: 6, 7, 9, 10, 51 ●

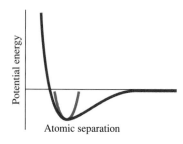

FIGURE 42-10 Near its minimum, the molecular potential-energy curve approximates a parabola. For small-amplitude displacements from equilibrium, the molecule is thus a quantum-mechanical harmonic oscillator.

As Example 42-1 indicates, transitions among rotational energy states generally involve fairly low-energy photons, typically in the microwave region of the spectrum.

Vibrational Energy Levels

The equilibrium configuration of a molecule corresponds to the minimum of the molecular potential-energy curve. In the vicinity of that minimum the curve is generally well approximated by a parabola (Fig. 42-10; see also Fig. 15-24). In Chapter 15, we saw that parabolic potential-energy curves result in simple harmonic motion, and in Chapter 40 we examined the quantum harmonic oscillator by using a parabolic potential-energy curve in the Schrödinger equation. There, we found that quantized vibrational energy levels are given by

$$E_{vib} = (n + \tfrac{1}{2})\hbar\omega, \tag{42-3}$$

where the quantum number n takes on integer values 0, 1, 2, 3, . . . , and where ω is the natural frequency for classical harmonic oscillations of the molecule. The selection rule for harmonic oscillators limits allowed transitions to those with $\Delta n = \pm 1$, so $\hbar\omega$ is the energy of photons emitted or absorbed in allowed transitions among vibrational energy levels. (Actually, the small-amplitude approximation is often justified only for the lower quantum states, so Equation 42-3 and the selection rule $\Delta n = \pm 1$ may apply only to these states. In most terrestrial applications, only the ground and first excited states are significantly populated, so the small-amplitude approximation applies.) For typical diatomic molecules, ω is on the order of 10^{14} s^{-1}, in the infrared region of the spectrum. Consequently, study of molecular vibrations involves infrared spectroscopy.

As we found in Chapter 40, the minimum energy of a quantum harmonic oscillator is the ground-state energy $E_0 = \tfrac{1}{2}\hbar\omega$, from Equation 42-3 with $n = 0$. Thus a molecule can never have zero vibrational energy, although Equation 42-2 shows that it *can* have zero rotational energy.

● **EXAMPLE 42-2**   MOLECULAR ENERGIES

An HCl molecule is in its vibrational ground state. Its classical vibration frequency is $f = 8.66 \times 10^{13}$ Hz. If its rotational and vibrational energies are nearly equal, what are its rotational quantum number and its angular momentum?

Solution
The rotational and vibrational energies are given by Equations 42-2 and 42-3, respectively, the latter with $n = 0$ to give the ground state. Equating these energies gives

$$\frac{\hbar^2}{2I} j(j + 1) = \tfrac{1}{2}\hbar\omega = \tfrac{1}{2}hf,$$

or

$$j(j + 1) = \frac{4\pi^2 If}{h},$$

where we used $\hbar = h/2\pi$. We could solve this quadratic equation for j, but from Example 42-1 we already know that transitions between rotational energy levels involve microwave photons that have much lower energy than the infrared photons associated with vibrational transitions. Therefore, we anticipate that a highly excited rotational state is necessary for the rotational energy to approximate the ground-state vibrational energy. So j should be large, and we approximate $j + 1$ by j. Then we have approximately

$$j = \sqrt{\frac{4\pi^2 If}{h}} = \sqrt{\frac{(4\pi^2)(2.65 \times 10^{-47} \text{ kg·m}^2)(8.66 \times 10^{13} \text{ Hz})}{6.63 \times 10^{-34} \text{ J·s}}}$$

$$= 12,$$

to the nearest integer. Equation 42-1 then gives the molecular angular momentum:

$$J = \sqrt{j(j+1)}\hbar \simeq \sqrt{j^2}\hbar = j\hbar$$

$$= (12)(1.06\times10^{-34} \text{ J·s}) = 1.3\times10^{-33} \text{ J·s},$$

where again we use the approximation $j(j+1) \simeq j^2$.

EXERCISE What photon energy would be needed to increase by one both the vibrational and rotational quantum numbers of the molecule in Example 42-2?

Answer: 0.39 eV

Some problems similar to Example 42-2: 11, 14, 15, 41, 42 ●

Molecular Spectra

A molecule in a state characterized by vibrational quantum number n and rotational quantum number j can undergo transitions obeying the selection rules $\Delta n = \pm 1$ and $\Delta j = \pm 1$. If molecules could not rotate, the molecular spectrum would consist of a single line at the classical vibration frequency, corresponding to transitions among adjacent vibrational states. But each vibrational level corresponds to an infinity of rotational states, with $j = 0, 1, 2, 3, \ldots$, and having different rotational energies. The resulting energy-level diagram, shown in Fig. 42-11, is therefore much more complicated. At typical temperatures, only the ground and first vibrational levels are significantly populated, but with energy distributed among many rotational levels. As a result, molecular spectra typically show a rich structure, with many lines corresponding to the different transitions of Fig. 42-11. Figure 42-12 is a spectrum of HCl, taken with a high-resolution infrared spectrometer that resolves the individual spectral lines. With a lower resolution instrument, the pattern shows up simply as a broad band, and we often speak of infrared absorption bands in describing the effect of molecules on infrared radiation. For example, absorption bands of atmospheric carbon dioxide limit the escape of infrared radiation from Earth. Carbon dioxide emissions from burning fossil fuels have increased atmospheric CO_2 levels to the point where this "trapping" of infrared radiation—the "greenhouse effect"—may result in a global warming trend, with serious climatic implications for the twenty-first century.

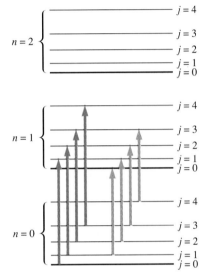

FIGURE 42-11 Energy-level diagram showing the ground state and first two vibrational excited states of a diatomic molecule. Each vibrational level (n value) is split into infinitely many levels corresponding to different rotational energies. Here four rotational states are shown for each vibrational level. Colored arrows indicate allowed photon-absorbing transitions from the vibrational ground state to the first excited state; these obey the selection rules $\Delta n = \pm 1$, $\Delta j = \pm 1$. The allowed transitions form two bands of higher and lower energy transitions, corresponding to $\Delta j = +1$ and $\Delta j = -1$. These bands form the two distinct sets of spectral lines shown in Fig. 42-12.

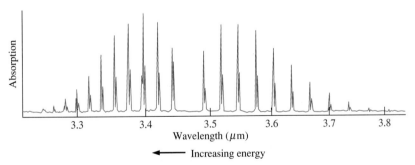

FIGURE 42-12 Absorption spectrum of HCl, taken with an infrared spectrophotometer. The lines shown result from transitions between the vibrational ground and first excited state. The left-hand group of lines are from transitions in which the rotational quantum number j increases by 1; for the right-hand group j decreases by 1. These two groups correspond, respectively, to the left-hand and right-hand sets of transitions in Fig. 42-11. Each line is split because of the presence of the two chlorine isotopes ^{35}Cl and ^{37}Cl, whose different masses lead to different rotational inertias and therefore different rotational energies. The higher peaks are from the more abundant isotope ^{35}Cl.

■ **APPLICATION** A STAR WARS LASER

High-power lasers have been proposed for the ballistic missile defense system once called "star wars." The technological requirements for a missile defense laser are staggering: the laser must deliver at least 30 MW and be able to focus its beam on a target thousands of kilometers distant while the target itself moves at several km/s.

Thirty megawatts is an awesome laser power, equivalent to some 60 billion of the typical helium-neon lasers used in introductory physics demonstrations. How can this much power be generated, especially in a space-based system with no connection to terrestrial energy sources? A possible answer is the chemical laser, whose operation relies on molecular excited states.

In the preceding chapter we saw that laser action requires a population inversion, with a majority of atoms or molecules in excited states. Stimulated emission then results in an intense beam of coherent photons. In a chemical laser, a chemical reaction produces the lasing molecules; the population inversion occurs because the molecules form in vibrational excited states. Among the reactions considered for star wars chemical lasers are those involving hydrogen fluoride (HF) and deuterium fluoride (DF).

Figure 42-13 shows an HF or DF laser. Two sets of nozzles deliver supersonic streams of fluorine and hydrogen (or deuterium). They react to produce HF or DF; in the HF reaction the majority of molecules are in the $n = 2$ excited state. The excited molecules enter a region between curved mirrors, where stimulated emission produces the intense laser beam. Still moving supersonically, the de-excited molecules exit the device. Newly formed excited molecules enter continuously, giving a continuous output beam.

Emission from chemical lasers, like other molecular vibrational transitions, lies in the infrared region of the spectrum. The HF laser's emission is in a band centered at about 2.8 μm, while the DF band is centered at 3.8 μm. Figure 42-14 shows that atmospheric transmission of infrared radiation varies drastically between these two wavelengths; in consequence the HF laser would have to be space-based, while the DF laser could operate from the ground.

The technological challenges of a 30-MW missile-defense laser are daunting. The device must be readied for firing within minutes after months of dormancy. A space-based HF laser affords little opportunity for testing and maintenance. The exhaust of spent HF gas acts as a rocket thrust, disturbing the aim of the weapon and also contaminating optical surfaces. The fundamental limit of optical diffraction requires huge mirrors for aiming the beam (see Example 37-9). Finally, present laser designs must be scaled upward by more than an order of magnitude in power (Fig. 42-15). These technical reasons, coupled with political realities of the 1990s, have led to greatly decreased emphasis on star wars laser weapons.

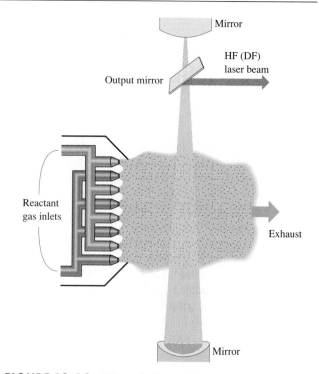

FIGURE 42-13 Schematic diagram of a hydrogen fluoride chemical laser. Fluorine and hydrogen emerge from two sets of nozzles to form HF molecules in vibrational excited states. The molecules de-excite through stimulated emission between two mirrors, producing an intense beam of coherent photons. The diagonal mirror extracts the beam.

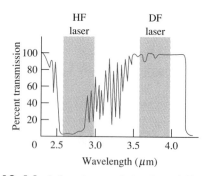

FIGURE 42-14 Infrared transmission through Earth's atmosphere, as a function of wavelength. Wiggles in the curve result from vibrational excitation of atmospheric molecules. Superimposed are the output bands of HF and DF lasers, showing that the atmosphere is opaque to HF emission but nearly transparent to DF emission. HF lasers for missile defense must therefore be based in space.

FIGURE 42-15 In a test experiment, the beam from a high-power chemical laser blows apart a stationary missile. Despite the test's apparent success, the concept of laser-based anti-missile defense has proven technologically and economically unfeasible.

42-3 SOLIDS

We have already seen that molecular bonding mechanisms can join relatively few atoms to form a molecule, or many to form a solid. In the lowest energy state, the atoms of a solid are arranged in a regular, repeating pattern; the solid is then **crystalline.** Sometimes solids form without their atoms having the opportunity to arrange themselves into a crystal structure; such solids are termed **amorphous.** Glass is a common amorphous solid. Their inherent randomness makes amorphous materials difficult to analyze theoretically, so we will concentrate here on crystalline solids.

Crystal Structure

The hallmark of a crystalline solid is the regular, repetitive arrangement of its atoms. If we examine a crystal with very fine resolution, we find that a basic arrangement of a few atoms repeats throughout the crystal (Fig. 42-16). This basic arrangement is the **unit cell.** Different crystalline materials have different unit cells; the structure of the unit cell is determined by the individual atoms and the bonding mechanism that acts between them. Sometimes the same underlying matter may assume different structures, depending on how the solid was formed; this is the case with diamond and graphite, both crystalline forms of pure carbon.

As with individual molecules, properties like atomic separation in a crystalline solid are determined by the interplay of attractive and repulsive interac-

FIGURE 42-16 (*a*) In the cesium chloride crystal, each cesium ion is surrounded by eight chlorine ions, located at the corners of a cube. The unit cell is a cube with one-eighth of a chlorine ion at each corner and a cesium ion at the center. (*b*) A cesium chloride crystal is a periodic array of unit cells. (*c*) Sodium chloride is different; in this crystal each ion is surrounded by only six nearest neighbors of the opposite type.

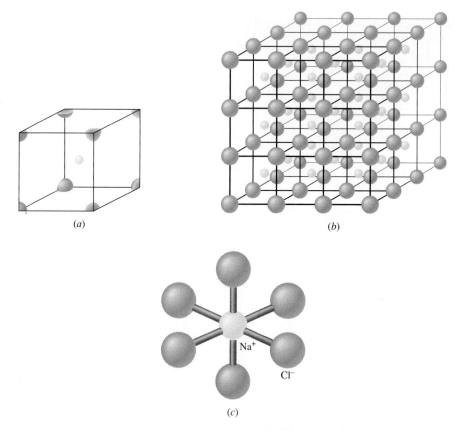

(*a*)

(*b*)

(*c*)

tions. The situation is complicated, however, because an individual atom experiences forces from many other atoms in the surrounding crystal. Crystals whose bonding is primarily ionic are most amenable to simple mathematical treatment.

For ionic crystals, we can consider the individual ions as point charges, and the attraction between them as resulting from the electrostatic force. Consider, for example, the NaCl structure shown in Fig. 42-16*c*. Each sodium ion is surrounded by six nearest chlorine ions, each some distance *r* away. The potential energy of a singly ionized positive sodium ion in the potential of each negative chlorine ion is $-ke^2/r$. So the contribution to the potential energy of the six nearest chlorine ions is $-6ke^2/r$, with the minus sign indicating an attractive interaction. But then there are 12 sodium ions a distance $\sqrt{2}r$ from the sodium in question; they give rise to a repulsive force and consequently a positive potential energy of value $+12ke^2/\sqrt{2}r$. At a distance of $\sqrt{3}r$ there are eight more chlorines, giving a potential energy $-8ke^2/\sqrt{3}r$. The result is that the electrostatic potential energy of the sodium ion can be written

$$U_1 = -\alpha \frac{ke^2}{r},$$

where $\alpha = 6 - 12/\sqrt{2} + 8/\sqrt{3} - \ldots$, and is called the **Madelung constant.** Many terms in the series above are required to compute α accurately, showing that the effect of distant ions is significant in determining the energy of an ion in the crystal. For the NaCl structure, α is approximately 1.748.

As ions are brought closer together, they experience the repulsive effect of the exclusion principle, as we discussed in Section 40-1. This repulsion is described approximately by a potential energy of the form $U_2 = A/r^n$, where A and n are constants. So the total potential energy of an ion in the crystalline solid is

$$U = U_1 + U_2 = -\alpha \frac{ke^2}{r} + \frac{A}{r^n}.$$

At equilibrium the potential energy is a minimum, corresponding to zero net force on the ion. Differentiating the potential energy with respect to r and setting dU/dr to zero to find the minimum, we have

$$0 = \frac{\alpha ke^2}{r_0^2} - \frac{nA}{r_0^{n+1}},$$

where r_0 designates the equilibrium separation. Solving for A gives $A = \alpha ke^2 r_0^{n-1}/n$, so the potential energy becomes

$$U = -\alpha \frac{ke^2}{r_0} \left[\frac{r_0}{r} - \frac{1}{n} \left(\frac{r_0}{r} \right)^n \right]. \qquad (42\text{-}4)$$

The value of U at the equilibrium separation r_0 is designated U_0 and is called the **ionic cohesive energy.** The magnitude of U_0 represents the energy needed to remove an ion entirely from the crystal. The cohesive energy is sometimes given in kcal/mol, in which case its magnitude is the energy per mole needed to break an entire crystal into its constituent ions (see Problem 17). Figure 42-17 shows a plot of the potential-energy function given by Equation 42-4.

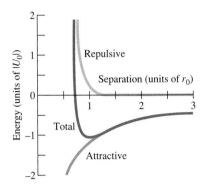

FIGURE 42-17 Potential-energy function for an ionic crystal, showing contributions of the attractive and repulsive terms.

● **EXAMPLE 42-3** POTENTIAL-ENERGY FUNCTION FOR THE NaCl CRYSTAL

For NaCl, the ionic cohesive energy is measured to be -7.84 eV. The equilibrium separation can be determined from density considerations (see Problem 16); its value is 0.282 nm. Use these values, and the Madelung constant $\alpha = 1.748$, to determine the exponent n in Equation 42-4 for NaCl.

Solution
Evaluating Equation 42-4 at $r = r_0$ gives

$$U_0 = -\alpha \frac{ke^2}{r_0} \left(1 - \frac{1}{n} \right).$$

Solving for n, we have

$$n = \left(1 + \frac{U_0 r_0}{\alpha ke^2} \right)^{-1}$$

$$= \left[1 + \frac{(-7.84 \text{ eV})(1.6 \times 10^{-19} \text{ J/eV})(0.282 \times 10^{-9} \text{ m})}{(1.748)(9.0 \times 10^9 \text{ N·m}^2/\text{C}^2)(1.6 \times 10^{-19} \text{ C})^2} \right]^{-1}$$

$$= 8.22.$$

The large value of this exponent shows how strongly resistant the NaCl crystal is to compression. In Problem 21 you can calculate the repulsive force involved in maintaining this "stiffness."

EXERCISE Find the potential energy of an ion in an NaCl crystal if the atomic separation throughout the crystal has been (a) compressed by 20% or (b) expanded by 20%.

Answers: (a) -4.36 eV; (b) -7.20 eV

Some problems similar to Example 42-3: 18–20, 43, 44

●

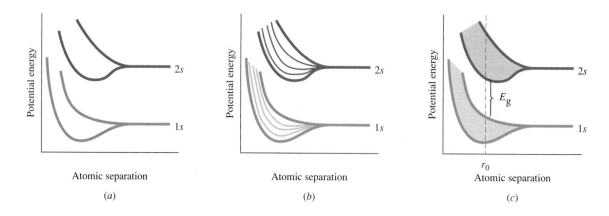

(a)

(b)

(c)

FIGURE 42-18 (a) Energy levels of the 1s and 2s states as a pair of atoms are brought close together. When the atoms are widely separated, both 1s levels and both 2s levels have the same energy. But the exclusion principle causes the levels to separate as the atoms approach. (b) With five atoms, each level splits into a group of five closely spaced levels. (c) In a crystalline solid, the large number of atoms results in essentially continuous energy bands, separated by gaps: r_0 marks the equilibrium separation, and E_g the band-gap energy.

Band Theory

Quantum-mechanical analysis of a solid containing 10^{23} atoms or so might seem a hopeless task. But the regularity of a crystalline solid makes that problem, while not an easy one, at least amenable to mathematical treatment. The physical regularity of the solid is reflected mathematically in the properties of the wave function; specifically, the wave function for a crystalline solid in equilibrium is itself periodic. This must be the case because equivalent points in two different unit cells have exactly the same physical properties.*

Here, we will not attempt to solve the Schrödinger equation for a crystal, or even to write the solutions. But we can see qualitatively what some properties of those solutions must be. Consider two identical atoms, initially widely separated, as they are brought closer together. When the atoms are far apart, they are described by identical wave functions and associated energy-level diagrams; a given electron state, for example, has exactly the same energy in each atom. But as the atoms move closer, their wave functions begin to overlap to form a single wave function that characterizes the entire composite system. Because of the exclusion principle, two electrons that were in identical states in the two widely separated atoms can no longer be in the same state. This effect manifests itself as a separation of what were originally identical energy levels. Figure 42-18a shows schematically what happens to the energies of two states as a pair of atoms are brought close together. As more and more atoms come together, initially identical energy levels split into ever more finely spaced levels (Fig. 42-18b). In a crystalline solid, there are so many atoms that each energy level splits into an essentially continuous **band** of allowed energies (Fig. 42-18c). Often, the bands arising from distinct single-atom states are separated by **band gaps,** as shown in Fig. 42-18c. An electron can have any energy between the top and bottom of a band, but electron energies in the band gaps are forbidden. The situation is like that of a single atom, where electrons are allowed only certain discrete energies, except now the discrete levels have broadened into bands.

We're usually interested in the properties of a solid at or near its equilibrium state, designated r_0 in Fig. 42-18c. There the solid is characterized by an energy-level diagram in which the energy levels are those of Fig. 42-18c at the value $r = r_0$. Figure 42-19 shows such a diagram.

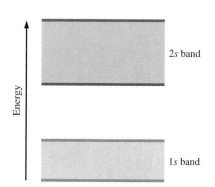

FIGURE 42-19 Energy-level diagram for the equilibrium separation of Fig. 42-18c, showing the band structure of the 1s and 2s states. The vertical axis still represents energy, but now the horizontal axis has no significance.

*Actually this periodicity holds exactly only if the crystal structure extends forever in all directions. But it is a good approximation for crystals consisting of many unit cells, as long as we do not consider points close to the surface of the material.

Conductors, Insulators, and Semiconductors

Figure 42-18 and 42-19 are highly simplified; actually many more states contribute to the band structure of a solid. Furthermore, the splitting and shifting process depicted in Fig. 42-18 can result in overlapping bands. Figure 42-20 shows the band structure of sodium, in which the $3s$ and $3p$ bands overlap. Note that the high-energy band containing electrons—here the $3s/3p$ band—is not completely full, meaning that energy levels in the upper portion of the band are not occupied by electrons.

We can determine which of the allowed energy levels of a solid will be occupied in the same way we established the electronic structure of atoms: by placing a given total number of electrons in the lowest possible levels consistent with the exclusion principle. In some materials, like the sodium of Fig. 42-20, that filling process results in the highest-energy occupied band being only partially full. But in others, like the material shown in Fig. 42-21, the highest occupied band is completely full.

In the difference between Figs. 42-20 and 42-21 lies the difference between electrical conductors and insulators. We introduced electrical conductors in Chapter 23, and elaborated on conduction mechanisms in Chapter 27. Our discussion was necessarily sketchy and qualitative, because a full description of electrical conduction in solids requires quantum mechanics. We're now ready to understand, in quantum-mechanical terms, the distinctions among insulators, conductors, and semiconductors.

We defined a conductor as a material containing charges that are free to move in response to an electric field, thereby establishing an electric current. Classically, there's no problem with this: we apply an electric field, and if an electron is "free," it will be accelerated by the field and gain energy. But quantum-mechanically, an electron can gain energy only by jumping from one allowed energy level to another. When an electric field is applied to a solid, an electron's energy can increase only if there is a higher, unoccupied energy level for it to jump into.

Consider sodium, a conductor whose band structure is shown in Fig. 42-20. If there are N atoms in a sample of sodium, then each band arises from the splitting of N distinct atomic energy levels. Each band can therefore hold N times as many electrons as the corresponding atomic level. In sodium, all the $1s$ electrons (a total of $2N$) from the individual atoms are in the $1s$ band of the solid, all the $2s$ electrons (again, $2N$ of them) are in the $2s$ band, and all the $2p$ electrons (a total of $6N$) are in the $2p$ band. But the $3s$ shell of an individual sodium atom contains only one of its two possible electrons; in the solid, this means that the corresponding $3s$ band is only half full, as suggested in Fig. 42-20. Because energy levels within a band are very closely spaced, an electron near the top of the filled portion of the $3s$ band needs to gain only a very small amount of energy to move into one of the unoccupied states within the same band. Therefore, it is very easy for an electric field to promote electrons to higher energy states, and for this reason sodium is a good conductor.

In other materials, as shown in Fig. 42-21, the uppermost occupied band is completely full. An electron in this band cannot move to another state in the band because all such states are full. For an electron to gain energy at all, it must jump the band gap to the lowest unoccupied band. Since this jump requires a relatively large energy, an applied electric field of reasonable magnitude cannot increase electron energies, so no current flows and the material is an insulator.

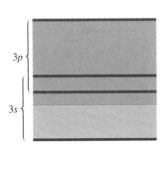
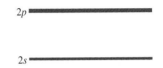
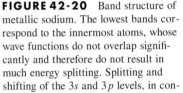

FIGURE 42-20 Band structure of metallic sodium. The lowest bands correspond to the innermost atoms, whose wave functions do not overlap significantly and therefore do not result in much energy splitting. Splitting and shifting of the $3s$ and $3p$ levels, in contrast, is so great that the bands arising from these states overlap. Blue indicates states containing electrons; note that the $3s/3p$ band is only partially full.

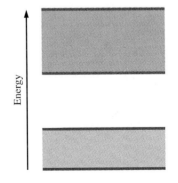

FIGURE 42-21 Band structure for an insulator. Blue indicates occupied states, tan unoccupied states. The highest occupied band is completely full, so there are no nearby unoccupied states to which electrons can jump in response to an applied electric field.

Whether a solid is an insulator or a conductor depends on whether the highest occupied band is full or only partially full. If the energy bands arising from individual atomic states always remained distinct, then we could predict the electrical properties of a solid by noting whether the outermost subshell of its individual atoms was full or not. For example, sodium has the electronic structure $1s^2 2s^2 2p^6 3s^1$; its outermost subshell—the $3s$ subshell—can hold two electrons but has only one. So we might expect the $3s$ band of solid sodium to be partially full, and the material to be a conductor—which turns out to be the case. But magnesium, one element beyond sodium in the periodic table, has the structure $1s^2 2s^2 2p^6 3s^2$; since its outermost subshell is full, we might expect the solid to be an insulator. But look again at Fig. 42-20, the band structure of sodium. Here the splitting and shifting of the energy levels has resulted in an overlap between the $3s$ and $3p$ levels. Magnesium exhibits a similar band structure, and the $3s$ electrons from its individual atoms do not fill the combined $3s/3p$ band. Electrons in the uppermost occupied states find adjacent unoccupied states into which they can move in response to an electric field, so the material is a conductor. Predicting the electrical characteristics of a material, as in the case of sodium and magnesium, is complicated by the possibility of overlapping bands. But solid-state physicists have developed sophisticated approximation schemes, based ultimately on the Schrödinger equation, that allow accurate theoretical calculation of band structures.

Metallic Conductors

In Chapter 27 we developed a semiclassical model to describe electrical conduction in metals, but we found that classical physics could not account correctly for all aspects of metallic conduction. In particular, classical predictions of the temperature dependence of electrical conductivity proved quite inconsistent with actual measurements (see Fig. 27-5 and related text).

Since the conduction electrons in a metal are not bound to particular atoms, their behavior can be approximated by assuming that the metal contains a gas of free electrons. Quantum-mechanically, the conduction electrons are confined to the metal by a three-dimensional square well potential like the one we discussed in Chapter 40. By counting the quantum states available to electrons in such a potential, it can be shown that the number of states per unit energy interval increases with energy. (You can see this trend already in Fig. 40-29, showing the first few states in a three-dimensional box.) We will not carry out this counting procedure; the result, however, is given by

$$g(E) = \left(\frac{2^{7/2} \pi m^{3/2}}{h^3} \right) E^{1/2}, \tag{42-5}$$

where E is the energy, m the electron mass, and $g(E)$ the number of states per unit volume per unit energy interval at the energy E. The quantity $g(E)$ is called the **density of states.** The calculation leading to Equation 42-5 accounts for the two possible spin states at each energy level.

At absolute zero, electrons fill the lowest available energy states according to the exclusion principle, which permits only one electron in a given state. In terms

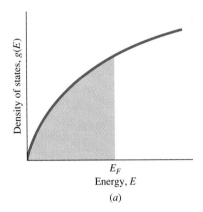

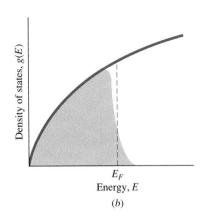

(a)   *(b)*

FIGURE 42-22 Occupation of energy states by electrons in a metal. Curve is the density of states given by Equation 42-5, while shaded region indicates occupied energy levels. (*a*) At absolute zero, all states up to the Fermi energy E_F are completely full, and those above empty. (*b*) At higher temperatures, thermal excitation promotes a few electrons above the Fermi energy.

of band theory, the result is the partially filled conduction band that we described in the preceding section. In terms of Equation 42-5, the number of electrons increases with energy up to the maximum energy needed to accommodate all the electrons. That maximum energy is called the **Fermi energy.** In a metal at absolute zero, the Fermi energy is the energy of the highest filled state in the conduction band. All energy levels below the Fermi energy are fully occupied, and all those above the Fermi energy are empty, as shown in Fig. 42-22*a*.

At temperatures above absolute zero, thermal excitation promotes some electrons to levels above the Fermi energy, leaving some levels just below that energy vacant (see Fig. 42-22*b*). But the Fermi energy in most metals is on the order of 1 to 10 eV, much higher than the thermal energy at typical temperatures. Therefore, the electron distribution is altered only slightly from its form at absolute zero. The result is that conduction in metals is provided almost entirely by electrons whose energies are near the Fermi energy, regardless of temperature. It is this situation that causes the mean electron speed to be different from the classical thermal speed, and makes the temperature dependence of the electrical conductivity in a metal quite different from the classical prediction. Problem 26 compares the classical and quantum results for electron speed in a metal.

● **EXAMPLE 42-4**   ELECTRON DENSITY AND THE FERMI ENERGY

Use Equation 42-5 to derive a relation between the Fermi energy E_F and the density n of conduction electrons in a metal. Then calculate the electron density in magnesium, whose Fermi energy is 7.13 eV.

Solution

Equation 42-5 gives $g(E)$, the number of states per unit volume per unit energy available to conduction electrons in a metal. Multiplying by some small energy interval dE then gives $g(E)\, dE$, the number of states per unit volume in the energy interval between E and $E + dE$. At absolute zero, all states up to the Fermi energy are occupied, and all those above the Fermi energy are vacant. Therefore, if we sum the number of states per unit volume in all the energy intervals from $E = 0$ to $E = E_F$, we will have the total number of occupied states per unit volume or, equivalently, the density of conduction electrons. Since $g(E)$ is a continuously varying function, that summation becomes an integration, and we have

$$n = \int_{E=0}^{E=E_F} g(E)\, dE = \int_0^{E_F} \left(\frac{2^{7/2}\, \pi m^{3/2}}{h^3} \right) E^{1/2}\, dE$$

$$= \left(\frac{2^{7/2} \pi m^{3/2}}{h^3} \right)\left(\frac{E^{3/2}}{3/2} \right) \Bigg|_0^{E_F} = \left(\frac{2^{9/2} \pi m^{3/2}}{3h^3} \right) E_F^{3/2}.$$

This is the desired relation between electron density and Fermi energy.

For magnesium, with $E_F = 7.13$ eV, we have

$$n = \left(\frac{2^{9/2}\pi m^{3/2}}{3h^3}\right)E_F^{3/2}$$

$$= \left(\frac{2^{9/2}\pi(9.11\times10^{-31}\text{ kg})^{3/2}}{(3)(6.63\times10^{-34}\text{ J·s})^3}\right)$$

$$\times\ [(7.13\text{ eV})(1.6\times10^{-19}\text{ J/eV})]^{3/2}$$

$$= 8.61\times10^{28}\text{ m}^{-3}.$$

This value is consistent with conduction electron densities we encountered in Chapter 27.

[If you try the above calculation on your calculator, you may get an error indication, showing that powers of 10 involved are outside the calculator's range. Try grouping the quantity $m^{3/2}/h^3$ as $(\sqrt{m}/h)^3$ to get some cancellation of the large negative exponents.]

EXERCISE Lithium has 4.6×10^{28} conduction electrons per cubic meter. Find the Fermi energy for lithium.

Answer: 4.71 eV

Some problems similar to Example 42-4: 22–24, 46 ●

Semiconductors

In Chapter 27 we discussed semiconductors—the materials at the heart of our modern electronic world—and gave a simple classical explanation of their properties. Here we see how band theory gives a more correct quantum-mechanical explanation of semiconductors.

Our band diagram for an insulator (Fig. 42-21) is strictly correct only at absolute zero temperature. At this temperature, the highest occupied band—called the **valence band**—is completely full and the next band—the **conduction band**—is completely empty. At any other temperature, random thermal energy may give an occasional electron enough energy to leap the gap between the valence and conduction bands. Such electrons would then be available for acceleration by an electric field, and the material would have nonzero electrical conductivity. In good insulators, though, the band gaps are relatively large (several eV or more), so the number of electrons thermally excited to the conduction band is negligible. But in some materials, notably silicon and germanium, the band gap is relatively small (on the order of 1 eV; see Table 42-1). Even at room temperature, random thermal excitation in low-band-gap materials promotes a significant number of electrons to the conduction band. As a result these materials conduct electricity, although their conductivity is much lower than in

▲ **TABLE 42-1** BAND-GAP ENERGIES FOR SELECTED SEMICONDUCTORS (AT 300 K)

SEMICONDUCTOR	BAND-GAP ENERGY (eV)
Si	1.14
Ge	0.67
InAs	0.35
InP	1.35
GaP	2.26
GaAs	1.43
GaSb	0.78
CdS	2.42
CdSe	1.74
ZnO	3.2
ZnS	3.6

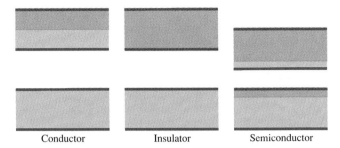

FIGURE 42-23 Band structures for a conductor, an insulator, and a semiconductor. Blue and tan show, respectively, occupied and unoccupied states. The semiconductor's small band gap allows thermal excitation of some electrons into the conduction band, giving the material a modest electrical conductivity.

metallic conductors. A material whose conductivity results from thermal excitation across a small band gap is a **semiconductor.** Figure 42-23 shows schematically the band-structure differences between conductors, insulators, and semiconductors. Note that a semiconductor really has the same band structure as an insulator and would, at absolute zero, be an insulator.

In Chapter 27 we described how the addition of small quantities of impurities—a process called doping—could radically alter the electrical conductivity of semiconductors. In terms of band theory, a dopant such as arsenic, with five valence electrons, adds so-called **impurity levels** just a few tenths of an eV below the conduction band (Fig. 42-24a). Thermal energy readily promotes electrons from these impurity levels to the conduction band, greatly increasing the conductivity of the material. In this case, as we defined it in Chapter 27, the doped material becomes an **N-type semiconductor,** N-type because its predominant charge carriers are *negative* electrons. If we add a dopant like aluminum with only three valence electrons, we create a **P-type semiconductor,** whose dominant charge carriers are *positive* **holes,** or absences of electrons in the crystal lattice. Figure 42-24b shows how the holes occur when electrons in the valence band are thermally excited into the nearby impurity levels.

The diodes, transistors, and integrated circuits we discussed in Chapter 27 are technological manifestations of our mastery of the physics of semiconductors. Theoretical understanding coupled with technological precision now enables us to fabricate materials with arbitrarily chosen band gaps. Light-emitting diodes (LEDs), commonly used as indicator lights and for digital displays, involve semiconductors whose band-gap energies correspond to visible-light photons. As electrons jump from the conduction to the valence band, the device emits visible light (Fig. 42-25). A related device is the semiconductor laser, in which a population inversion is established by pumping excess electrons into the conduction band. Stimulated emission then results in a coherent beam of light as electrons fall across the band gap. Semiconductor lasers are used to drive light signals in fiber-optic communications.

Conversely, a material with a band gap corresponding to visible-light photons can absorb light energy, in the process promoting electrons to the conduction band and creating electron-hole pairs and driving an electric current. The result is the conversion of light energy into electricity. A semiconductor device designed for such conversion is called a **photovoltaic cell** (Fig. 42-26). Photovoltaic cells have long been used to generate electricity in remote applications

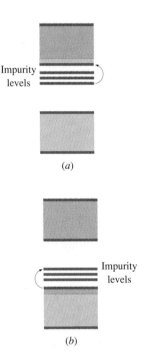

Impurity levels

(a)

Impurity levels

(b)

FIGURE 42-24 Band structure in doped semiconductors. In (a), the dopant introduces impurity levels just below the conduction band. Thermal energy excites electrons from these levels to the conduction band, making an N-type semiconductor whose charge carriers are predominantly negative electrons. In (b), dopant impurity levels lie just above the valence band, and are initially vacant. Thermal energy excites electrons to the impurity levels, leaving positive holes that make the material a P-type semiconductor.

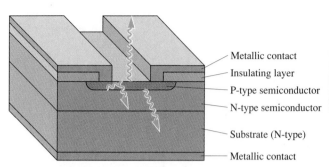

FIGURE 42-25 Construction of a light-emitting diode. Photons are emitted at the *PN* junction, where electrons drop from the conduction band to the valence band. Much of the light never escapes the device, making it relatively inefficient.

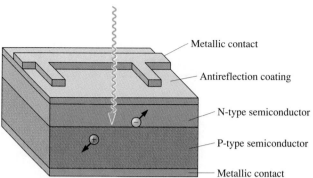

FIGURE 42-26 Structure of a photovoltaic cell. Electron-hole pairs form near the *PN* junction, driving a current through the metallic contacts at top and bottom of the cell and on to an external load. Antireflection coating maximizes light transmission into the device.

like automated weather stations, electric fences for livestock control, and spacecraft. But increasingly, they are also being used to power individual buildings and even to replace central power plants (Fig. 42-27). For years, solar cells have been too expensive to compete with more conventional means of large-scale electric power generation. Recently, though, the capital cost per watt of power generation capability for photovoltaic cells dropped close to the ever-rising cost per watt for large nuclear power plants (Fig. 42-28). With vigorous ongoing research, and with production of photovoltaic cells nearly doubling every year, photovoltaic solar energy conversion may soon make significant contributions to our electrical energy supply.

FIGURE 42-27 Photovoltaic cells are increasingly competitive with more traditional sources of electricity. This photovoltaic installation in California produces 6.4 MW of electric power.

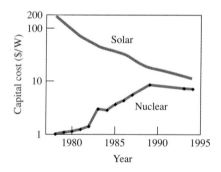

FIGURE 42-28 Capital cost per watt of generating capacity for photovoltaic and nuclear power plants. Nuclear data are based on actual power-plant costs, adjusted for the average time plants are off line; dots mark individual power plants going on line in the year indicated. Solar data are based on an average solar input of 250 W/m² and on power-plant costs twice that of the photovoltaic cells alone.

● **EXAMPLE 42-5** ENGINEERING THE BAND GAP

Optical fibers used in communication systems exhibit the least signal attenuation at wavelengths on the order of 1.5 μm. What should be the band gap in an infrared diode laser used to drive a fiber at this wavelength?

Solution
We want 1.5-μm photons to be emitted as electrons drop across the band gap, so the gap energy should be that of a 1.5 μm photon. Equating photon energy to frequency and then wavelength, we have

$$E = hf = \frac{hc}{\lambda} = \frac{(6.63 \times 10^{-34} \text{ J·s})(3.00 \times 10^8 \text{ m/s})}{1.5 \times 10^{-6} \text{ m}}$$

$$= 1.33 \times 10^{-19} \text{ J} = 0.829 \text{ eV}.$$

Of the materials listed in Table 42-1, GaSb comes closest to this band-gap energy. By carefully adjusting the composition of the semiconductor material, the desired gap energy can be achieved more precisely.

EXERCISE What is the maximum wavelength of light that can produce electron-hole pairs in gallium phosphide (GaP)?

Answer: 550 nm

Some problems similar to Example 42-5: 28–31 ●

42-4 SUPERCONDUCTIVITY

In Chapter 27 we introduced **superconductivity**—the complete loss of electrical resistance in certain materials at low temperature. First discovered in mercury by H. Kamerlingh Onnes in 1911, superconductivity was for decades limited to a few individual elements and alloys at temperatures below about 20 K. Then, in 1986, a series of rapid breakthroughs led to materials that are superconducting at temperatures on the order of 100 K (Fig. 42-29). The first of these breakthroughs—surmounting the "30 K barrier"—brought the 1987 Nobel Prize in physics to J. Georg Bednorz and K. Alex Müller of IBM's Zurich Research Laboratory (Fig. 42-30). By early 1988 the highest superconducting temperature had reached 125 K, and in 1993 researchers found hints of possible superconductivity at 250 K—the temperature of a cold winter day in the northern United States (see Fig. 27-15 for a temperature history of superconductors). Research on superconductivity is one of the most exciting areas of contemporary physics. A host of new superconducting materials have been found—including even doped versions of the remarkable compound buckminsterfullerine described earlier in this chapter. The ultimate goal of a room-temperature superconductor, once thought beyond reach, may one day be achieved.

Table 42-2 shows **transition temperatures** for selected superconductors. In general, materials that are relatively poor conventional conductors make the best superconductors; note that aluminum, the best conventional conductor in Table 42-2, has the lowest T_c. Pure elements tend to have lower transition temperatures than alloys and other compounds. Niobium has the highest elemental T_c, and niobium compounds are the most widely used of the more established low-temperature superconductors. The newer high-T_c superconductors are complex structures whose essential ingredient is a metal-oxide compound; they come in many forms with a range of T_c's (Fig. 42-31).

To date, cooling requirements of conventional superconductors have limited their use to specialized applications including large, high-field electromagnets and very sensitive magnetic-field measuring devices (Fig. 42-32). But the discovery of high-temperature superconductors, which can be cooled with inexpensive liquid nitrogen ($T = 77$ K, and a price per liter less than that of orange juice), may propel superconducting technology into everyday use. Potential uses for superconductivity include electromagnets for energy storage, for medical uses including nuclear magnetic resonance imaging, for fusion and high-energy physics research, and for separation of materials for purification and recycling; electric power generation and transmission; electric motors and magnetically levitated mass transportation vehicles; superconducting computers; magnetometers for brain-wave imaging; and even mechanical devices like low-friction

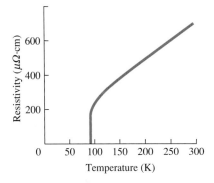

FIGURE 42-29 Resistivity versus temperature for a thin film of $YBa_2Cu_3O_x$ superconductor. The superconducting transition occurs at 93 K, well above the 77 K boiling point of liquid nitrogen.

FIGURE 42-30 1987 Nobel laureates Bednorz and Müller discovered a new class of high-temperature superconductors.

▲ **TABLE 42-2**
SUPERCONDUCTING
TRANSITION TEMPERATURES

MATERIAL	T_c, K
Al	1.20
Sn	3.72
Pb	7.19
Nb	9.26
La_3In	10.4
NbN	16.0
V_3Si	17.1
Nb_3Sn	18.1
$Rb\text{-}C_{60}$	28
La-Sr-Cu-O	40
Y-Ba-Cu-O	93
Ti-Ba-Ca-Cu-O	125

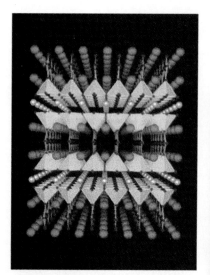

FIGURE 42-31 Structure of yttrium-barium-copper-oxide superconductor, whose transition temperature is 93 K. Conduction occurs along planes defined by the green pyramids, which represent copper oxide.

bearings. Figure 42-33 shows two applications of high-temperature superconductors.

To make use of a superconductor's ability to carry electric current without loss, it is advantageous to form the material into flexible wires, tapes, or thin films. High-temperature superconductors are ceramics that are brittle and are difficult to form into elongated structures. Considerable effort by materials scientists and others has resulted in wires and ribbon-shaped conductors of high-temperature superconductor (Fig. 42-34). High-temperature superconducting films have also been deposited successfully on silicon, paving the way for semiconductor chips whose interconnections will be superconducting. Already high-T_c thin film electronic components for microwave systems are commercially available.

Superconductivity and Magnetism

The hallmark of a superconductor is its zero electrical resistance; currents established in superconductors—called **supercurrents**—have persisted for years without diminishing. Another property is the so-called **Meissner effect,** wherein a superconductor excludes magnetic flux from its interior. Faraday's law suggests that a *changing* magnetic flux should not be possible inside a

(a)   (b)

FIGURE 42-32 Applications for conventional superconductors include (a) superconducting cables and (b) electromagnets, such as this unit under construction for the now defunct Superconducting Super Collider.

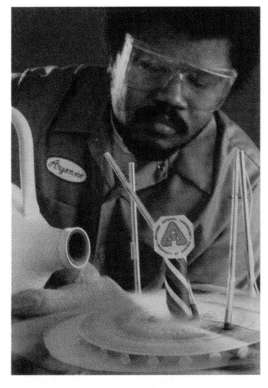

(a)

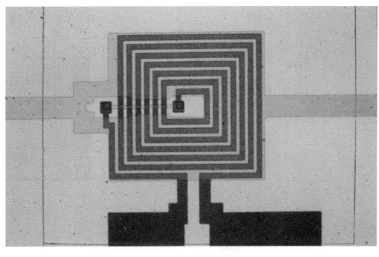

(b)

FIGURE 42-33 Applications of high-temperature superconductors. (a) Richard McDaniel of Argonne National Laboratory pours liquid nitrogen to cool an experimental motor made with high-T superconductors. (b) This SQUID (superconducting quantum interference device) uses high-temperature superconductors to measure very small magnetic fields. Manufactured by Conductus Inc., the device is among the first commercial products using high-T superconductors.

superconductor, for such a change would result in a finite electric field and therefore an infinite current in the zero-resistance superconductor. But the Meissner effect goes beyond this, disallowing even static magnetic fields from the interior of the superconductor. Figure 42-35a shows a piece of potentially superconducting material immersed in a uniform magnetic field. The material is not yet superconducting, and magnetic field lines pass through it. But as it cools below its transition temperature, the material becomes superconducting and the magnetic flux is excluded, as shown in Fig. 42-35b. Figure 42-35c shows that

FIGURE 42-34 Coil being wound from a ribbon of high-temperature superconductor at the American Superconductor Corporation.

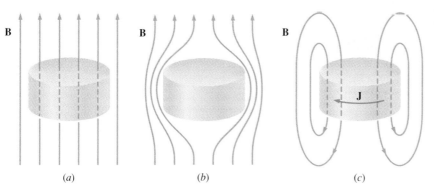

(a)   (b)   (c)

FIGURE 42-35 The Meissner effect. (a) Above the superconducting transition temperature, magnetic field lines penetrate the material. (b) In the superconducting state, magnetic flux is excluded from the material. (c) The overall field in (b) is the sum of the original uniform field and the field associated with persistent supercurrents **J** (arrow) in the superconductor.

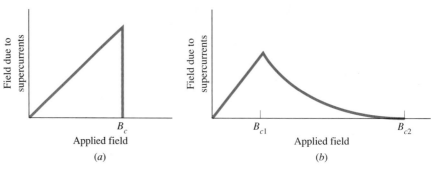

FIGURE 42-36 A small cubical magnet floats above a wafer of high-temperature superconductor. The superconductor is cooled in a bath of liquid nitrogen at 77 K.

FIGURE 42-37 Responses of (a) type I and (b) type II superconductors to applied magnetic fields. Horizontal axis is the applied field, while vertical axis is the magnitude of the fields generated by the induced supercurrents. For low fields, the induced field is equal in magnitude to the applied field but opposite in direction; the two cancel to give no net field inside the superconductor. Superconductivity ceases abruptly in a type I superconductor at the critical field B_c, while a type II superconductor gradually allows magnetic flux to penetrate while remaining superconducting above the lower critical field B_{c1}. Only at the upper critical field B_{c2} does superconductivity cease altogether.

the overall field can be considered the sum of the original applied field and a dipole-like field generated by supercurrents induced within the superconductor. The Meissner effect is dramatically illustrated by placing a magnet on a piece of potentially superconducting material. When the material is cooled below its transition temperature, the magnet levitates and floats above the superconductor (Fig. 42-36). This levitation is a result of the repulsive interaction between the magnetic moments of the permanent magnet and the superconductor, the latter resulting from persistent currents that flow in the superconductor to cancel the flux from the permanent magnet. As we indicated in Section 31-5, the complete exclusion of magnetic flux in a superconductor makes it a diamagnetic material with magnetic susceptibility -1.

As the strength of an externally applied magnetic field increases, so do the supercurrents and resulting magnetic field of the superconductor. But beyond a certain field strength, called the **critical field,** the external magnetic field begins to alter the superconducting state, and the superconductor no longer excludes magnetic flux. This alteration can happen in two ways, as shown in Fig. 42-37; a given superconducting material is classified as **type I** or **type II superconductor** depending on which response it exhibits. In a type I superconductor, magnetic fields beyond the critical field abruptly destroy the superconducting state (Fig. 42-37a), and the material becomes a normal conductor. A type II superconductor, in contrast, has a lower and an upper critical field, as shown in Fig. 42-37b. At the lower critical field, the material begins to allow flux penetration but maintains its zero-resistance state until the much higher upper critical field is reached. What is happening between the two critical fields of a type II superconductor is that distinct regions of normal conductivity develop, centered on magnetic field lines and forming a regular array (Fig. 42-38). Between the lower and upper critical fields, the fraction of the material in the superconducting state diminishes, until at the upper critical field superconductivity stops altogether. Type I superconductors tend to be pure elements like lead or mer-

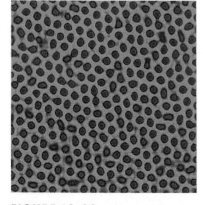

FIGURE 42-38 Structure of a type II superconductor between the upper and lower critical fields shows a regular array of concentrated magnetic flux regions (red). Each region is a zone of normal (nonsuperconducting) material. The flux regions are about 1 μm apart.

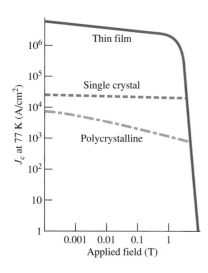

FIGURE 42-39 Critical current density J_c versus applied magnetic field in thin film, single crystal, and polycrystalline forms of $YBa_2Cu_3O_7$ superconductor at 77 K.

cury, while type II superconductors are generally alloys or more complicated mixtures. High-temperature superconductors are all of type II.

The destruction of superconductivity by magnetic fields has enormous practical significance, for a major application of superconductivity is to carry large electric currents without energy loss, especially in electromagnets. But electric current is the source of magnetic field (Chapter 30), so there is a maximum current in a given superconductor at which the magnetic field of that current destroys the superconducting state. For type I superconductors, the critical field is so low as to make these materials useless in high-current applications like electromagnets. But the upper critical fields of type II superconductors are generally much higher, so most practical applications of superconductors use type II materials. Critical fields of the high-temperature superconductors are as high as 100 T, and current densities greater than 10^6 A/cm^2 have been achieved in thin-film superconducting samples at 77 K (Fig. 42-39).

■ APPLICATION MAGLEV TRANSPORTATION

Maglev (short for magnetically levitated) vehicles using superconducting magnets offer the promise of high-speed transportation in densely populated urban corridors. Maglev systems are currently under development in several countries, especially Germany and Japan. Japanese maglev prototypes have carried passengers at speeds in excess of 400 km/h, and a 22-km German-built maglev system in Orlando, Florida, may become the world's first commercial maglev enterprise.

Engineers are pursuing a number of different maglev configurations (Fig. 42-40). In the Japanese system, on-board superconducting magnets interact with metal coils in vertical walls of a fixed guideway. The German Transrapid uses conventional magnets slung underneath a steel rail. A proposed American design called the magneplane would use magnetic repulsion between on-board superconducting magnets and coils in a curved guideway.

Maglev vehicles are propelled as well as levitated by magnetic forces. The vehicle and its guideway constitute a linear electric motor, much like a conventional motor that has been "unrolled" to produce straight-line motion. Alternating current supplied to coils in the guideway produces an alternating magnetic field that alternatively pushes and pulls on the vehicle through interaction with its on-board magnets.

Prototype maglevs have used either conventional electromagnets or low-temperature superconductors. Successful development of high-temperature superconducting magnets would spur maglev development, and the discovery of a room-temperature superconductor could make maglev systems very attractive.

(a)

(b)

FIGURE 42-40 Maglev designs. (a) The Japanese MLU002 prototype uses superconducting levitation magnets whose fields interact with coils in the vertical walls of the guideway. (b) The German Transrapid 07 uses conventional magnets slung under an iron rail, thus lifting the vehicle upward. (c) Magneplane, an American proposal shown here in cross section, would use superconducting magnets to levitate the vehicle some 15 cm above a curved guideway.

Guideway

Superconducting levitation magnet

Superconducting propulsion magnet

Propulsion magnet

(c)

Theories of Superconductivity

Superconductivity is a purely quantum-mechanical phenomenon; classical physics is totally inadequate to explain its existence. A successful theory of conventional low-T_c superconductors, called the **BCS theory** after its originators, was formulated in 1957 by John Bardeen, Leon Cooper, and John Robert Schrieffer; the trio shared the 1972 Nobel Prize in Physics.

In the BCS theory, superconductivity results from a quantum-mechanical pairing of electrons that leads to a lower-energy state in which electron pairs move through the crystal lattice with no energy loss to the ions. This absence of energy loss is what gives the superconductor its zero electrical resistance. (Contrast with the collisional origin of electrical resistance discussed in Section 27-2.) The electron pairing itself involves an interaction with the ion lattice; one

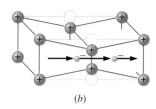
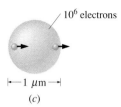

10^6 electrons

$\vdash$ 1 μm $\dashv$

(a)   (b)   (c)

FIGURE 42-41 Electron pairing in the BCS theory. (a) A normal conductor, with uncorrelated electrons. (b) In a superconductor, one electron passes through the lattice, deforming it slightly. About 10^{-12} s later a second electron passes and experiences the potential of the distorted lattice. The motions of the two electrons are therefore correlated. (c) Part (b) is not to scale; paired electrons are typically 1 μm apart, and there are about a million other electrons in their vicinity. The long-range coherent motion of all these electron pairs gives rise to superconductivity.

electron of the pair slightly deforms the lattice, and the second electron is attracted by the slight positive charge of the deformed lattice site, resulting in a new minimum-energy state (Fig. 42-41). But the paired electrons are not physically close; typically, a million other electrons, each paired with another distant electron, may lie between two electrons constituting a pair (Fig. 42-41c). The result of this long-range pairing is a coherence in the motion of the conduction electrons that extends throughout the superconductor. Like dancers in a well-choreographed dance, the electrons all move together in a way that precludes energy loss to the ion lattice.

High-temperature superconductors are not fully understood, although they almost certainly involve quantum-mechanical pairing of charge carriers—probably holes rather than electrons. The mechanism of the pairing is less clear; one promising candidate involves magnetic interactions, athough other mechanisms are under investigation. Superconductivity should provide a continuing challenge to theorists and experimentalists for some time to come.

CHAPTER SYNOPSIS

Summary

1. The bonding of atoms into molecules involves a balance between attractive and repulsive forces, producing stable structures when the potential energy is a minimum.
   a. **Ionic bonding** occurs between atoms near opposite ends of the periodic table, when one atom gives up an electron to the other. The resulting ions are strongly attracted.
   b. In **covalent bonding,** two or more atoms share their outermost electrons in common orbitals. Molecules such as O_2, N_2, and H_2O are covalently bonded.
   c. **Hydrogen bonds** are weak bonds involving the protons of hydrogen-containing molecules.
   d. **Van der Waals bonding** is very weak and results from electric forces between molecular dipoles.
   e. **Metallic bonding** results from sharing of a free electron gas among all the ions of the crystal lattice.

2. The energy-level structure of molecules is very rich. In addition to atomic-like excitation of electrons to different energy levels, internal motion also contributes to the quantization of molecular energy:
   a. The **rotational energy** of molecules is quantized, with the allowed levels given by

   $$E_{\text{rot}} = \frac{\hbar^2}{2I} j(j + 1),$$

   where I is the molecule's rotational inertia and the integer j the rotational quantum number. Spacings between rotational energy levels are measured in milli-electron-volts.
   b. The **vibrational energy** is also quantized, with quantized energy levels being approximately those of the quantum harmonic oscillator:

   $$E_{\text{vib}} = (n + \tfrac{1}{2})\hbar\omega.$$

Ground-state vibrational energies are on the order of 0.1 eV.

   c. Selection rules limit rotational and vibrational transitions to those for which $\Delta j = \pm 1$ and $\Delta n = \pm 1$.

3. **Solids** may be held together by any of the bonding mechanisms that work for molecules. The minimum-energy configuration for a solid consists of a regular array of atoms that compromise a **crystal.** The energy needed to disassemble a crystalline solid is its **cohesive energy.**

   a. In solids, individual atomic electron energy levels spread into **energy bands,** separated by **band gaps.**

   b. Electrical **conductors** have their highest bands partially filled with electrons, making it easy for electric fields to promote electrons to higher energy and thus allowing current to flow. The **Fermi energy** is the energy of the highest occupied level in a metallic conductor.

   c. **Insulators** have their highest band completely full, so excitation of electrons is impossible unless they gain enough energy to jump the substantial band gap.

   d. **Semiconductors** are like insulators, but with smaller band gaps. Random thermal energy promotes some electrons across this gap to the **conduction band,** giving these materials modest conductivity. Doping with impurities introduces additional levels that result in high concentrations of electrons (*N*-type semiconductor) or holes (*P*-type semiconductor).

4. The sudden loss of all electrical resistance below a **critical temperature** marks the onset of **superconductivity.** All superconductors exhibit the **Meissner effect,** wherein magnetic flux is excluded from the interior of the material. In a **type I superconductor** the exclusion is complete up to a **critical field** at which superconductivity ceases. In a **type II superconductor,** flux exclusion is complete up to a lower critical field, above which flux penetration gradually occurs until superconductivity ceases at an upper critical field. The **BCS theory** invokes the correlated motion of electron pairs to explain low-temperature superconductors.

Terms You Should Understand

(Pairs are closely related terms whose distinction is important; number in parentheses is chapter section where term first appears.)

ionic, covalent, hydrogen, van der Waals, and metallic bonding (42-1)

unit cell (42-2)

band, band gap (42-3)

conductor, insulator, semiconductor (42-3)

conduction band, valence band (42-3)

Fermi energy (42-3)

superconductor, type I and type II (42-4)

transition temperature (42-4)

Meissner effect (42-4)

critical field (42-4)

Symbols You Should Recognize

α (42-3) E_F (42.3)

r_0 (42-3) T_c (42.4)

Problems You Should Be Able to Solve

evaluating molecular rotational and vibrational energy levels and associated spectra (42-2)

determining molecular bond lengths from rotational spectra (42-2)

evaluating ionic energies and related quantities (42-3)

relating Fermi energy and electron density (42-3)

relating band-gap energies and light emission or absorption (42-3)

Limitations to Keep in Mind

Models developed in this chapter to describe the properties of molecules and solids are only approximations; in particular, most real molecules do not act like simple harmonic oscillators for vibrational excitations with $n > 1$.

QUESTIONS

1. Why is the exclusion principle crucial to the existence of stable molecules?

2. Why do ionically bonded materials have high melting points?

3. Ionic bonds clearly result from electrostatic attraction between ions. In what way do covalent bonds also involve electrostatic attraction?

4. Does it make sense to distinguish individual NaCl molecules in a salt crystal? What about individual H_2O molecules in an ice crystal? Explain.

5. Why is the hydrogen bond unique to hydrogen?

6. Is it useful to think of the highest-energy electrons "belonging" to individual atoms in an ionically bonded molecule? In a covalently bonded molecule?

7. What are the approximate relative magnitudes of the energies associated with electronic excitation of a molecule, with molecular vibration, and with molecular rotation?

8. You meet a scientist who uses a microwave technology to study molecules. What form of molecular energy is she most concerned with?

9. Radio astronomers have discovered many complex organic molecules in interstellar space. Why were these discoveries made with radio telescopes as opposed to optical telescopes?

10. In Fig. 21-24, why are the rotational states excited at lower temperatures than the vibrational states?

11. How can diamond be an insulator while graphite, also pure carbon, is a conductor of electricity? Suggest an explanation based on band theory.

12. Would you expect solid hydrogen to conduct electricity? Why or why not?

13. The Fermi energy in metals is much greater than the thermal energy at typical temperatures. Why does this make the mean speed of conduction electrons nearly independent of temperature?

14. Why does the size of the band gap determine whether a material is an insulator or a semiconductor?

15. How would you expect the conductivity of an undoped semiconductor to depend on temperature? Why?

16. Compare the relative contributions made by electrons and by holes to the electrical conductivity of undoped semiconductors, P-type semiconductors, and N-type semiconductors.

17. If you could heat an insulator to very high temperature without its melting or dissociating, might it exhibit the electrical properties of a semiconductor? Explain.

18. Name some technological innovations that might result from a room-temperature superconductor.

19. Suppose a room-temperature superconductor were discovered, but that it had a very low critical field. In what way would this limit its practical applicability?

20. Why can a magnet float motionless above a superconductor but not above an ordinary conductor? After all, bringing the magnet near either conductor induces eddy currents whose fields repel the magnet.

21. How do type I and type II superconductors differ?

22. What is the economic advantage of high temperature superconductors? After all, they still require cooling far below room temperature.

23. What role does the ion lattice play in the electron pairing described by the BCS theory?

PROBLEMS

Section 42-2 Molecular Energy Levels

1. Find the energies of the first four rotational states of the HCl molecule described in Example 42-1.

2. Find the wavelength of electromagnetic radiation needed to excite oxygen molecules (O_2) to their first rotational excited state. The rotational inertia of an oxygen molecule is 1.95×10^{-46} kg·m^2.

3. A molecule drops from the $j = 2$ to the $j = 1$ rotational level, emitting a 2.50-meV photon. If the molecule then drops to the rotational ground state, what energy photon will be emitted?

4. Calculate the wavelength of a photon emitted in the $j = 5$ to $j = 4$ transition of a molecule whose rotational inertia is 1.75×10^{-47} kg·m^2.

5. Photons of wavelength 1.68 cm excite transitions from the rotational ground state to the first rotational excited state in a gas. What is the rotational inertia of the gas molecules?

6. A molecule absorbs a photon and jumps to the next higher rotational state. If the photon energy is three times what would be needed for a transition from the rotational ground state to the first rotational excited state, between what two levels is the transition?

7. Find an expression for the energy of a photon required for a transition from the $(j - 1)$th level to the jth level in a molecule with rotational inertia I.

8. A molecule of rotational inertia I undergoes a transition from the jth rotational level to the $(j - 1)$th level. Show that the wavelength of the emitted photon is $\lambda = 4\pi^2 Ic/hj$.

9. The rotational spectrum of diatomic oxygen (O_2) shows spectral lines spaced 0.356 meV apart in energy. Find the atomic separation in this molecule. *Hint:* See Example 42-1, but remember that the oxygen atoms have equal mass.

10. Use the result given in Problem 52 to find the bond length in carbon monoxide (CO), using the fact that excitation of the first rotational state requires photons of wavelength 2.59 mm.

11. For the HCl molecule of Example 42-2, determine (a) the energy of the vibrational ground state and (b) the energy of a photon emitted in a transition between adjacent vibrational levels. Assume the rotational quantum number does not change.

12. The classical vibration frequency for diatomic hydrogen (H_2) is 1.32×10^{14} Hz. What is the spacing between its vibrational energy levels?

13. The energy between adjacent vibrational levels in diatomic nitrogen is 0.293 eV. What is the classical vibration frequency of this molecule?

14. Diatomic deuterium (D_2) has classical vibration frequency 9.35×10^{13} Hz and rotational inertia 9.17×10^{-48} kg·m^2. Find (a) the energy and (b) the wavelength of a photon emitted in a transition between the $n = 1, j = 1$ state and the $n = 0, j = 2$ state.

15. An oxygen molecule is in its vibrational and rotational ground states. It absorbs a photon of energy 0.19653 eV and jumps to the $n = 1, j = 1$ state. It then drops to the $n = 0, j = 2$ level, emitting a 0.19546-eV photon. Find (a) the classical vibration frequency and (b) the rotational inertia of the molecule.

Section 42-3 Solids

16. Use the 2.16 g/cm^3 density of NaCl to calculate the ionic spacing r_0 in the NaCl crystal. *Hint:* Consult Appendix D for atomic weights.

17. Express the 7.84-eV ionic cohesive energy of NaCl in kilocalories per mole of ions.

18. Lithium fluoride, LiF, has the same crystal structure as NaCl and therefore essentially the same Madelung constant α. Its ionic cohesive energy is -10.5 eV, and the value of n in Equation 42-4 is 6.25. What is the equilibrium ionic separation in LiF?

19. Determine the constant n in Equation 42-4 for potassium chloride (KCl), for which $r_0 = 0.315$ nm and $U_0 = -7.21$ eV. The crystal structure is the same as for NaCl.

20. A salt crystal contains 10^{21} sodium-chlorine pairs. How much energy would it take to compress the crystal to 90% of its normal size?

21. (a) Differentiate Equation 42-4 to obtain an expression for the force on an ion in an ionic crystal. (b) Use your result to find the force on an ion in NaCl if the crystal could be compressed to half its equilibrium spacing (see Example 42-3 for relevant parameters). Compare with the electrostatic attraction between the ions at a separation of $\frac{1}{2}r_0$. Your result shows how very "stiff" this ionic crystal is.

22. The Fermi energy in aluminum is 11.6 eV. What is the density of conduction electrons in aluminum?

23. Determine the Fermi energy for calcium, which has 4.6×10^{28} conduction electrons per cubic meter.

24. Metal A has twice the Fermi energy of metal B. How do their conduction electron densities compare?

25. Suppose the charge carriers in a material were protons, with density 10^{28} m^{-3}—comparable to that of electrons in a metal. What would be the order of magnitude of the Fermi energy?

26. The Fermi energy for copper is 7.0 eV. Compare the electron speed associated with this energy with the classical thermal speed for an electron at room temperature (300 K). Your result shows how the quantum and classical descriptions of metallic conduction differ.

27. The *Fermi temperature* is defined by equating the thermal energy kT to the Fermi energy, where k is Boltzmann's constant. Calculate the Fermi temperature for silver ($E_F = 5.48$ eV), and compare with room temperature.

28. What is the wavelength of light emitted by a gallium phosphide (GaP) light-emitting diode? *Hint:* See Table 42-1.

29. What is the shortest wavelength of light that could be produced by electrons jumping the band gap in a material from Table 42-1? What is the material?

30. Which material in Table 42-1 would provide the longest wavelength of light in a light-emitting diode? What is the wavelength?

31. A common light-emitting diode is made from a combination of gallium, arsenic, and phosphorous (GaAsP) and emits red light at 650 nm. What is its band gap?

32. Photons with energy less than a semiconductor's band gap are not readily absorbed by the material, so measurement of light absorption versus wavelength gives the band gap. An absorption spectrum for silicon shows no absorption for wavelengths greater than 1090 nm. Use this information to calculate the band gap in silicon, and verify its entry in Table 42-1.

33. The Sun radiates most strongly at about 500 nm, the peak of its Planck curve. The semiconductor zinc sulfide has a band gap of 3.6 eV. (a) What is the maximum wavelength abs\orbed by ZnS? (b) Would ZnS make a good photovoltaic cell? Why or why not?

34. Light-emitting diodes are available commercially with output wavelengths in the range from about 550 nm to 1300 nm. What is the corresponding range of band-gap energies?

35. A blue-green semiconductor laser being developed for long-playing compact discs emits at 447 nm (see Application: Long-Playing CDs, in Chapter 37). What is the band gap in this laser?

Section 42-4 Superconductivity

36. Pure aluminum, which superconducts below 1.20 K, exhibits a critical field of 9.57 mT. What is the maximum current that can be carried in a 30-gauge (0.255 mm diameter) aluminum superconducting wire without the field from that current exceeding the critical field? *Hint:* Where is the field greatest? Consult Example 30-4.

37. The critical magnetic field in niobium-titanium superconductor is 15 T. What current is required in a 5000-turn solenoid 75 cm long to produce a field of this strength?

38. A danger in superconducting magnet systems is "quenching"—a sudden loss of superconductivity associated with rapid resistive heating and loss of coolant. If the solenoid of the preceding example has a diameter of 5.0 cm, (a) how much magnetic energy will be released if it quenches? (b) What volume of liquid helium could be vaporized? The heat of vaporization is 2.6 kJ per liter of liquid helium.

Paired Problems

(Both problems in a pair involve the same principles and techniques. If you can get the first problem, you should be able to solve the second one.)

39. The atomic spacing in diatomic hydrogen (H_2) is 74 pm. Find the energy of a photon emitted in a transition from the first rotational excited state to the ground state.

40. Find the wavelength of electromagnetic radiation needed to excite a transition from the rotational ground state to the first rotational excited state in N_2, whose atomic spacing is 109 pm.

41. What wavelength of infrared radiation is needed to excite a transition between the $n = 0, j = 3$ state and the $n = 1$, $j = 2$ state in KCl, for which the rotational inertia is 2.43×10^{-45} kg·m^2 and the classical vibration frequency is 8.40×10^{12} Hz?

42. Find the wavelengths emitted in all possible allowed transitions between the first three rotational states in the $n = 1$ level to any states in the $n = 0$ level in diatomic hydrogen. The rotational inertia of the hydrogen molecule is 4.60×10^{-48} kg·m^2, and its classical vibration frequency is 3.69×10^{14} Hz.

43. Lithium chloride, LiCl, has the same structure and therefore the same Madelung constant as NaCl. The equilibrium separation in LiCl is 0.257 nm, and $n = 7$ in Equation 42-4. Find the ionic cohesive energy of the LiCl crystal.

44. Find the energy per unit cell needed to decrease the ionic separation in NaCl by 15%.

Supplementary Problems

45. What would be the Fermi energy in a one-dimensional infinite square well 10 nm wide and holding 100 electrons? Assume two electrons (with opposite spins) per energy level.

46. Find the fraction of conduction electrons in a metal at absolute zero that have energies less than half the Fermi energy.

47. Figure 21-24 shows that diatomic hydrogen acts like a monotomic gas at temperatures below about 100 K. Use the fact that the rotational inertia of H$_2$ is 4.6×10^{-48} kg·m^2 to show that this low-temperature behavior makes sense. *Hint:* Compare the thermal energy kT with the minimum energy needed to excite molecular rotation.

48. Repeat the preceding problem, now considering vibrational energy levels. The classical vibration frequency for H$_2$ is about 4.4×10^{13} Hz, and Fig. 21-24 shows that vibrational effects on the specific heat become important at about 2000 K.

49. The transition from the ground state to the first rotational excited state in diatomic oxygen (O$_2$) requires about 356 μeV. At what temperature would the thermal energy kT be sufficient to set diatomic oxygen into rotation? Would you ever find diatomic oxygen exhibiting the specific heat of a monatomic gas at normal pressure?

50. The density of rubidium iodide (RbI) is 3.55 g/cm^3, and its ionic cohesive energy is -145 kcal/mol. Determine (a) the equilibrium separation and (b) the exponent n in Equation 42-4 for RbI.

51. The HCl bond-length calculation of Example 42-1 assumed the molecule was not vibrating. But because the ground-state vibrational energy is not zero, the bond is actually stretched slightly by its vibration. To estimate this stretching (a) use the classical vibration frequency of 8.66×10^{13} Hz to find the ground-state vibrational energy. (b) Use the result of part (a) and the fact that the effective "spring constant" for HCl is about 480 N/m to estimate the stretching of the molecular bond due to ground-state vibration. Assume that half the energy is potential energy of the stretched "spring." (c) Compare your result with the bond length found in Example 42-1.

52. Consider a diatomic molecule consisting of different atoms with masses m_1 and m_2, separated by a distance R. (a) Find the molecule's center of mass. (b) Use your result to show that the rotational inertia about the center of mass is

$$ I = \left(\frac{m_1 m_2}{m_1 + m_2} \right) R^2. $$

The quantity $m_1 m_2 / (m_1 + m_2)$ is called the reduced mass. (c) Use this result to compute a more accurate value for the HCl bond length of Example 42-1.

53. The Madelung constant (Section 42-3) is notoriously difficult to calculate because it is the sum of an alternating series of nearly equal terms. But it can be calculated for a hypothetical one-dimensional crystal consisting of a line of alternating positive and negative ions, evenly spaced (Fig. 42-42). Show that the potential energy of an ion in this "crystal" can be written

$$ U = -\alpha \frac{ke^2}{r_0}, $$

where the Madelung constant α has the value 2 ln2. *Hint:* Study the series expansions listed in Appendix A.

FIGURE 42-42 Problem 53.

CHAPTER 43

NUCLEAR PHYSICS

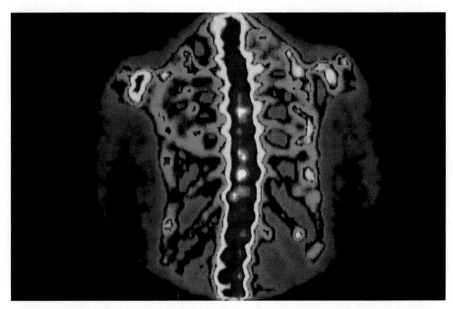

Gamma rays are high-energy photons produced in quantum transitions within the atomic nucleus. This image, taken with gamma rays, reveals the presence of cancer in human vertebrae.

The last two chapters showed how modern physics describes the structure of atoms and the way they join to form molecules and solids. Now we turn inward, to the atomic nucleus. This chapter deals with nuclear structure and the phenomenon of radioactivity; in the next chapter we consider nuclear energy and its applications. In the final chapter we go even deeper into the structure of matter, probing the innards of the neutrons and protons that make up atomic nuclei.

43-1 DISCOVERY OF THE NUCLEUS

When J. J. Thomson discovered the electron in 1897, it was not known how electrons and the equivalent positive charge were distributed in atoms. Thomson himself favored the **plum-pudding model,** with electrons embedded throughout a "pudding" of positive charge (Fig. 43-1a). In a 1909–1911 series of experiments suggested by Ernest Rutherford, Rutherford's colleague Hans Geiger and undergraduate student Ernest Marsden shot alpha particles (helium nuclei) from a radioactive source into thin gold foils. They expected the particles to pass straight through the gold or to be deflected at most very slightly. Most behaved as expected, but a few bounced back in nearly the direction they had come from (Fig. 43-2), implying collision with a much more massive object. It was, to paraphrase Rutherford, as though you shot a cannonball at a piece of tissue paper and the ball bounced back. Rutherford reasoned that nearly all the atom's mass must be concentrated in a small, massive core: the **nucleus.** A new atomic model emerged, with the tiny, positively charged nucleus at the center, surrounded by electrons in distant orbits (Fig. 43-1*b*). We now know, of course, that the distinct orbits of Rutherford's atomic model are not consistent with modern quantum mechanics. But the essence of his discovery remains: The atom consists of a tiny, massive nucleus surrounded by electrons that are, at least statistically, distant from the nucleus.

By 1920 Rutherford had proposed that nuclei other than hydrogen must contain electrically neutral as well as positively charged entities. The discovery of the neutron in 1932 confirmed this proposal. Today we know that protons and neutrons, collectively called **nucleons,** are the constituents of nuclei.

43-2 BUILDING NUCLEI: ELEMENTS, ISOTOPES, AND STABILITY

The Nuclear Force

What holds the nucleus together, given the mutual electrical repulsion of its protons? There must be another force—the **nuclear force**—acting attractively between the constituents of the nucleus. Through much of the twentieth century

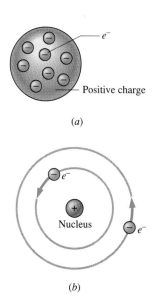

FIGURE 43-1 (*a*) Thomson's "plum-pudding" model of the atom had electrons immersed in a "pudding" of positive charge. (*b*) In Rutherford's nuclear model, electrons orbit the tiny but massive nucleus. Figure is not to scale; the nucleus is actually about 10^{-5} times the size of the atom.

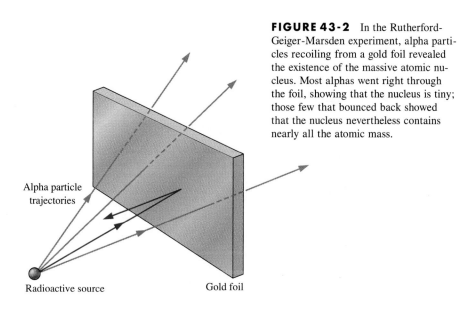

FIGURE 43-2 In the Rutherford-Geiger-Marsden experiment, alpha particles recoiling from a gold foil revealed the existence of the massive atomic nucleus. Most alphas went right through the foil, showing that the nucleus is tiny; those few that bounced back showed that the nucleus nevertheless contains nearly all the atomic mass.

the nuclear force was thought to be a fundamental force of nature, but today we recognize it as a manifestation of a more fundamental force that acts between particles called quarks that themselves make up the nucleons. We will study quarks and their interactions in Chapter 45.

The nuclear force acts attractively between pairs of nucleons—neutrons and protons, protons and protons, neutrons and neutrons—without regard for their electric charge. The force is very strong at separations less than a few fm (10^{-15} m), but it falls off exponentially with distance. The attractive nuclear force between two protons therefore dominates at short distances, but the repulsive electric force becomes dominant at separations of more than a few fm. The structure and characteristics of nuclei are determined, to a first approximation, by the interplay between the weaker but longer range electric force and the stronger but shorter range nuclear force.

Nuclear Symbols

A nucleus consists of neutrons and protons. The **atomic number,** Z, is the number of protons; it determines the electric charge (Ze) of the nucleus and, thus, the number of electrons in a neutral atom and, therefore, the atom's chemical properties. Nuclei with the same atomic number are therefore nuclei of the same **element.** The **mass number,** A, is the number of nucleons—neutrons plus protons. Specifying these two numbers is enough to identify a given nucleus. We describe nuclei with a shorthand notation, giving the element symbol used in the periodic table, preceded by a subscript for the atomic number and a superscript for the mass number. Thus, $^4_2$He is the common variety of helium, and $^{235}_{92}$U is the type of uranium nucleus used in nuclear reactors. The atomic number and element symbol are redundant since helium *means* the element with two protons in its nucleus and uranium *means* 92 protons. Sometimes, therefore, we write expressions like helium-4, He-4, or ^{4}He to mean the same thing as $^4_2$He.

Isotopes

Nuclei of a given element—determined by the atomic number Z—can have different numbers of neutrons and still make atoms with the same chemical properties. These different nuclei are called **isotopes.** Most elements have several naturally occurring isotopes. Hydrogen, for example, is most abundant as the isotope $^1_1$H, whose nucleus consists of a single proton. But about one in every 6500 hydrogen atoms is deuterium ($^2_1$H), whose nucleus contains a neutron and a proton. Although the chemical properties of isotopes are similar, their nuclear properties may differ dramatically. Uranium provides an extreme case: the more common isotope $^{238}_{92}$U cannot sustain a nuclear fission chain reaction, while the rare (0.72% of natural uranium) isotope $^{235}_{92}$U can. Uranium enriched to a high proportion of U-235 is therefore a dangerous material, suitable for making nuclear weapons—as we will see in the next chapter. Figure 43-3 shows the structure of some isotopes.

Stable Nuclei

Of the infinitely many possible combinations of neutrons and protons, only about 400 are known to be **stable** under normal conditions, meaning they do not spontaneously decay into other nuclei. Figure 43-4 shows the stable nuclei, along

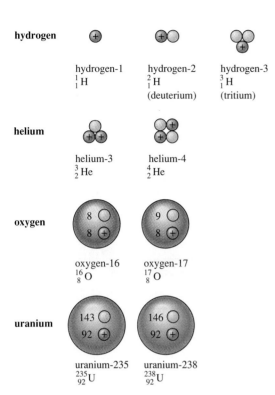

hydrogen

hydrogen-1
$^1_1$H

hydrogen-2
$^2_1$H
(deuterium)

hydrogen-3
$^3_1$H
(tritium)

FIGURE 43-3 Isotopes of a given element have the same number of protons but different numbers of neutrons. Their chemical properties are the same but their nuclear properties may differ significantly.

helium

helium-3
$^3_2$He

helium-4
$^4_2$He

oxygen

oxygen-16
$^{16}_8$O

oxygen-17
$^{17}_8$O

uranium

uranium-235
$^{235}_{92}$U

uranium-238
$^{238}_{92}$U

with many unstable ones, on a chart of atomic number Z versus neutron number $N = A - Z$. This figure shows that the lighter stable nuclei have roughly the same number of neutrons and protons, but that heavier nuclei have more neutrons than protons. Figure 43-5 suggests the reason why more neutrons are necessary in the heavier nuclei. As the nuclear size grows, the more distant protons become quite far apart compared with the range of the nuclear force.

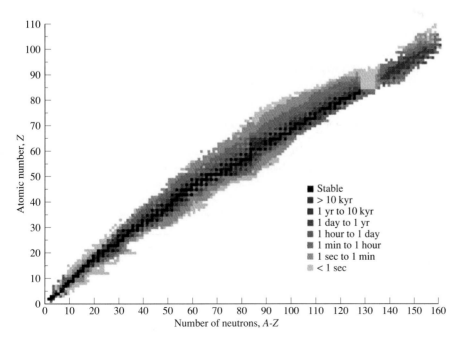

FIGURE 43-4 A chart of the nuclei, plotting all known nuclei according to the number of protons (atomic number Z) versus number of neutrons ($A - Z$). The nuclei are color coded by half-life. Note that the more massive nuclei are richer in neutrons; this is necessary to counter the electrical repulsion of the protons.

■ Stable
■ > 10 kyr
■ 1 yr to 10 kyr
■ 1 day to 1 yr
■ 1 hour to 1 day
■ 1 min to 1 hour
▥ 1 sec to 1 min
▨ < 1 sec

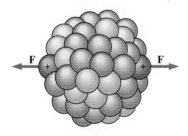

FIGURE 43-5 Two widely separated protons in a large nucleus experience significant electrical repulsion, but the attractive nuclear force between them is negligible because of its short range. More neutrons are therefore needed to keep the nucleus from flying apart.

They feel essentially only the repulsive electric force, and alone they would therefore tend to tear the nucleus apart. Extra neutrons, subject only to the nuclear force, help bind the protons into the nucleus. But even this effect has a limit, and the result is that there are no stable nuclei for $Z > 83$.

Electrical repulsion explains why neutrons are necessary and why their numbers increase with increasing atomic number. But why aren't there stable nuclei with a great many more neutrons than protons? The answer lies in the fact that the neutron is itself an unstable particle. Through an interaction involving the so-called weak force, now known to be a relative of electromagnetism, an isolated neutron decays into a proton, an electron, and an elusive particle called a neutrino. This decay also happens in nuclei with too many neutrons, but is suppressed in stable nuclei.

43-3 PROPERTIES OF THE NUCLEUS

Nuclear Masses

A nucleus of mass number A contains Z protons and $A - Z$ neutrons. Although the proton and neutron have different masses, those masses are the same to three significant figures, and thus, the mass of a nucleus is approximately A proton masses. Mass spectrometers (see Problem 29-27) give more accurate measurements of nuclear masses. It is convenient to express nuclear masses in terms of the **unified mass unit,** u, defined as one-twelfth of the mass of a neutral carbon-12 atom. The unified mass unit is very nearly 1.66057×10^{-27} kg, slightly less than the mass of the proton or neutron. High-energy physicists, cognizant of the Einstein mass-energy equivalence, often express nuclear and particle masses in MeV/c^2—a value numerically equal to the particle's rest energy in MeV. Table 43-1 lists some important masses in these various units.

Since the unified mass unit u is defined as one-twelfth the mass of the $^{12}_6C$ atom and is less than the proton or neutron mass, that means the carbon-12 atom's mass is less than that of its constituent nucleons alone—let alone the slight additional mass of its six electrons. Why is this? The answer lies in the Einstein mass-energy equivalence. The nucleus is a bound system, meaning that its total energy is *less,* by some amount ΔE, than that of its constituents when they are widely separated. Its mass is therefore less as well, by the amount $\Delta E/c^2$.

▲ **TABLE 43-1** SELECTED MASSES

	MASS, kg	MASS, u	MASS, MeV/c²
Electron	$9.109\ 39 \times 10^{-31}$	0.000 548 579	0.510 999
Proton	$1.672\ 62 \times 10^{-27}$	1.007 276	938.272
Neutron	$1.674\ 93 \times 10^{-27}$	1.008 665	939.566
^1_1H atom	$1.673\ 53 \times 10^{-27}$	1.007 825	938.783
α particle (^4_2He nucleus)	$6.644\ 66 \times 10^{-27}$	4.001 506	3727.38
$^{12}_6C$ atom	$1.992\ 65 \times 10^{-26}$	12	11 177.9
Unified mass unit (u)	$1.660\ 54 \times 10^{-27}$	1	931.494

As we emphasized in Chapter 38, there's nothing uniquely nuclear about mass-energy equivalence, and therefore, other bound systems exhibit this mass decrease as well. A water molecule, for example, has less mass than the total of two separate hydrogen atoms and one oxygen atom. The nuclear difference lies in the great strength of the nuclear force and the correspondingly large energies involved in nuclear interactions. That makes the mass difference much more obvious in nuclear situations. We will consider nuclear mass differences and energies more in the next section.

Nuclear Size

Rutherford interpreted data from the alpha-particle experiments to show that the nuclear size must be on the order of 1 fm (see Problem 1). Since Rutherford's time physicists have used high-energy electrons and other probes to learn the details of nuclear structure. The results of these studies give the **nuclear radius** R defined as the radius at which the nuclear density has fallen to half its central value. A rough, empirical formula for this radius is

$$R = R_0 A^{1/3}, \tag{43-1}$$

where $R_0 = 1.2$ fm and A is the mass number.

Equation 43-1 shows that the nuclear radius is proportional to the cube root of the mass number A. That means the nuclear volume is proportional to A itself—just what we would expect if the nucleus consisted of a tight, spherical bundle of protons and neutrons, as shown in Fig. 43-5. To the extent that the nuclear mass is the sum of the masses of its constituent particles, that means all nuclei have essentially the same ratio of mass to volume—i.e., the same density. The following example shows that nuclear density is enormously greater than the densities of ordinary materials.

● **EXAMPLE 43-1** GIBRALTAR IN A TEASPOON: NEUTRON STARS

(a) Calculate the density of nuclear matter, and determine the mass of 1 teaspoon (about 5 ml) of it. (b) The largest known agglomerations of nuclear matter are neutron stars, which sometimes form in the supernova explosions that mark the deaths of massive stars. Find the radius of a neutron star with mass equal to the Sun's.

Solution
Taking the mass of the nucleus to be approximately A proton masses, we can use Equation 43-1 to find the density of nuclear matter:

$$\rho = \frac{m}{V} = \frac{Am_p}{\frac{4}{3}\pi R^3} = \frac{Am_p}{\frac{4}{3}\pi R_0^3 A} = \frac{3m_p}{4\pi R_0^3} = \frac{(3)(1.67\times10^{-27}\text{ kg})}{(4\pi)(1.2\times10^{-15}\text{ m})^3}$$

$$= 2.3\times10^{17}\text{ kg/m}^3,$$

about 10^{14} times the density of ordinary matter. A teaspoon of nuclear matter (5 mL or 5×10^{-6} m³) then has a mass of about 10^{12} kg, about that of the Rock of Gibraltar!

Appendix E shows that the Sun's mass is about 2×10^{30} kg. The number of nucleons in this mass is then given by $A = 2\times10^{30}$ kg$/1.67\times10^{-27}$ kg/nucleon or 1.2×10^{57}. Equation 43-1 then gives the radius:

$$R = R_0 A^{1/3} = (1.2\times10^{-15}\text{ m})(1.2\times10^{57})^{1/3} = 13\text{ km},$$

about the size of a city. Another approach would have been to get the volume and then the radius from the density calculated in part (a). Our answer is approximate because Equation 43-1 does not scale exactly to gravitationally bound nuclei of stellar mass, but it is good within a factor of about two.

EXERCISE Compare the radius of a uranium-238 nucleus with that of an oxygen-16 nucleus.

Answer: $R_{\text{U-238}} = 2.46 R_{\text{O-16}}$

Some problems similar to Example 43-1: 8, 9, 11, 12 ●

Nuclear Spin

In Chapter 41 we discussed the important role of electron spin in atomic structure, and we noted that protons and neutrons, like electrons, are spin-$\frac{1}{2}$ particles—meaning they have intrinsic angular momentum of magnitude $\sqrt{s(s + 1)}\hbar$, where $s = \frac{1}{2}$. The spins of the individual nucleons, combined with any angular momentum associated with their motions within the nucleus, give the nucleus a quantized angular momentum I that obeys the same rules we've seen for other angular momenta in quantum mechanics:

$$I = \sqrt{i(i + 1)}\hbar, \tag{43-2}$$

where i, the nuclear spin quantum number, is an even or odd multiple of a half integer. The component of the nuclear angular momentum on a given axis is also quantized, just like other angular momenta, according to the rule $I_z = m_i\hbar$, where m_i ranges from $-i$ to i in steps of 1.

The values of i may be integer or half integer, depending on whether the number of nucleons is even or odd, respectively. We noted in Chapter 41 that particles with half-integer spins obey the Pauli exclusion principle, while those with integer spins do not. That can lead to profound differences in the behavior of isotopes of the same element; in particular, the spin-$\frac{1}{2}$ nucleus ^3_2He exhibits very different behavior from that of spin-zero ^4_2He, even though both are isotopes of helium (Fig. 43-6).

The angular momentum of the nucleus results in a nuclear magnetic moment, usually expressed in units of the **nuclear magneton,**

$$\mu_N = \frac{e\hbar}{2m_p} = 5.05 \times 10^{-27} \text{ J/T},$$

where m_p is the proton mass. This quantity is smaller than the Bohr magneton that we introduced in Chapter 41 by the ratio of the electron to the proton mass, showing that magnetic effects involving nuclei are much smaller. Since the proton is positive, its magnetic moment is in the same direction as the spin. There is no fully satisfactory theory to predict the exact value of this moment, but experiments show that its projection on a given axis takes the values $\pm 2.793\mu_N = \pm 1.4106 \times 10^{-26}$ J/T. Listings of fundamental constants often call this quantity "the magnetic moment of the proton," although the actual magnitude is greater by a factor of $\sqrt{3}$ because of space quantization (recall Fig. 41-11). Interestingly, the neutral neutrons as well as the charged protons contribute to the nuclear magnetic moment, giving a tantalizing hint that the neutron may itself contain charged particles whose total charge sums to zero. We'll see in Chapter 45 that this is indeed the case.

Interaction of the nuclear magnetic moment with magnetic fields—either externally applied or produced by atomic electrons—alters very slightly the energy levels of the atom (Fig. 43-7a). In hydrogen, for example, the spin-$\frac{1}{2}$ proton can have either of two orientations in the magnetic field of the electron, and the result is **hyperfine splitting** of the hydrogen ground state into two levels

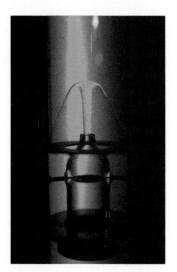

FIGURE 43-6 This fountain of liquid helium-4 occurs because the spin-0 helium-4 nuclei can all occupy the same quantum state, resulting in a "superfluid" that flows without viscosity. The phenomenon does not occur with spin-1/2 helium-3.

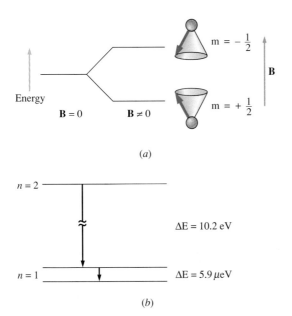

FIGURE 43-7 (*a*) A nonzero magnetic field **B** splits the energy level of the spin-$\frac{1}{2}$ proton into two levels. The upper level corresponds to magnetic quantum number $m = -\frac{1}{2}$, meaning that the spin is aligned as nearly as possible antiparallel to the field. (*b*) The two possible orientations of the proton's magnetic moment in the magnetic field of the electron split the hydrogen ground state into two levels 5.9 μeV apart.

spaced a mere 5.9 μeV apart (Fig. 43-7*b*). Transitions between these levels result in a spectral line at a radio wavelength of 21 cm. This transition is used in hydrogen masers, which serve as very accurate atomic clocks, and radio astronomers use it to detect interstellar clouds of neutral hydrogen.

● **EXAMPLE 43-2** FLIPPING PROTONS

A 2.5-T magnetic field points in the z direction. A proton has its magnetic moment oriented so the z component is opposite the field direction. It then flips into the lower energy state, where the z component of its magnetic moment is aligned with the field. Find the energy and frequency of the photon it emits in flipping between the two states.

Solution
In Chapter 29 (Equation 29-12) we found that the potential energy of a magnetic dipole of moment **μ** in a magnetic field **B** is $U = -\mathbf{μ} \cdot \mathbf{B}$. The value $μ_p = 1.41 \times 10^{-26}$ J/T for the proton magnetic moment is already the projection on an axis defined by the magnetic field direction, so we can write $U = \pm μ_p B$ for the energies of the two states. Then magnitude of the energy difference between the two states is

$$\Delta U = μ_p B - (-μ_p B) = 2μ_p B$$
$$= (2)(1.41 \times 10^{-26} \text{ J/T})(2.5 \text{ T})$$
$$= 7.05 \times 10^{-26} \text{ J} = 0.44 \ \mu\text{eV}.$$

Using $E = hf$ shows that a photon with this energy has a frequency of 106 MHz, which happens to lie in the FM broadcasting band.

EXERCISE Use the hyperfine splitting energy of 5.9 μeV in the hydrogen ground state to estimate the magnetic field experienced by the proton in the hydrogen atom.

Answer: 33.5 T

Some problems similar to Example 43-2: 13–15 ●

■ APPLICATION NUCLEAR MAGNETIC RESONANCE

Putting nuclei in an external magnetic field results in two possible energy states, as shown in Fig. 43-7a, depending on the orientation of the nuclear magnetic moments relative to the field. Irradiating nuclei in the lower energy state with electromagnetic radiation whose photon energy is equal to the energy difference between the two states will then cause nuclei to "flip" into the higher energy state. But nuclei also experience magnetic fields from the atomic electrons that surround them, and those fields are extremely sensitive to the details of the surrounding electron distribution—i.e., to the molecular structures in which the atoms and their nuclei are embedded. Measuring the exact energy needed to flip nuclear spins in a magnetic field thus provides information about molecular structure.

Figure 43-8 shows an **NMR spectrometer,** widely used for structural analysis of chemical compounds. The sample under analysis is placed in a uniform magnetic field, usually several tesla in strength and often provided by superconducting coils carrying currents that persist indefinitely. Another coil carries current at a frequency f—usually a few hundred MHz—corresponding to the photon energy hf that would flip the spin of an isolated proton in the spectrometer's magnetic field. This coil acts as an antenna, irradiating the sample with electromagnetic waves oa frequency f. If nuclei in the sample absorb the photons

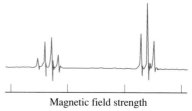

FIGURE 43-9 NMR spectrum of ethyl bromide (CH_3CH_2Br). Each group of peaks corresponds to a different magnetic field in the NMR spectrometer, reflecting the diamagnetic shielding due to different electron configurations surrounding the various nuclei, and to magnetic coupling between nuclear spins. Individual peaks within each group reflect various combinations of proton spin orientations in each part of the molecule. The horizontal axis of an NMR spectrum is usually specified as a frequency shift rather than a magnetic field value, but the two are equivalent.

in this radiation, they flip into their higher energy states and subsequently drop back, emitting radiation of frequency f in the process. A receiver coil mounted perpendicular to the transmitter coil detects this radiation.

Protons in a sample are not isolated but are surrounded by electrons that produce additional magnetic fields, and therefore, they will not generally absorb the photons of frequency f in the steady magnetic field of the spectrometer. So the magnetic field is varied slightly, using a small coil whose field superposes with the uniform background field. When this additional field exactly cancels the field at a nucleus due to the surrounding electrons, a resonance condition exists and the protons absorb photons. They flip to the upper spin state, drop back down, and produce a signal detected in the receiver coil. In all but the simplest compounds, different protons are in different electron environments, and consequently each is in resonance at a different value of the magnetic field. Scanning the field through a range of values allows detection of these different resonances (Fig. 43-9). Comparison with the resonance peaks of known molecular substructures leads to determination of the structure of the sample molecules.

Nuclear magnetic resonance is the basis of the widely used medical technique called **MRI,** for **magnetic resonance imaging.** In MRI, a person is placed inside a large solenoid. A gradient in the solenoid's magnetic field means that the nuclear magnetic resonance frequency for protons varies with position, and thus, the resonance signal can be used to provide positions of protons giving the magnetic resonance signal. A computer then uses the position information to construct an image.

Most of the MRI signal from the human body comes from fat and water. Fat gives the strongest signal because it has a higher density of protons. Thus, MRI is good for imaging soft tissues that do not show in x rays (see Fig. 43-10).

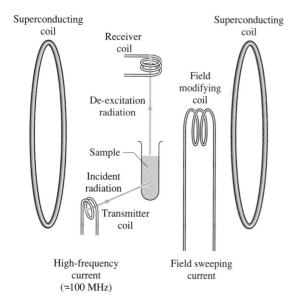

FIGURE 43-8 Schematic diagram of a nuclear magnetic resonance spectrometer. When the field-modifying coil gets the magnetic field to the value that allows nuclei to absorb photons from the transmitter coil, the nuclei flip to their higher energy states and then drop back down, emitting photons that are detected by the receiver coil. The exact fields needed for such resonance provide information about molecular structure.

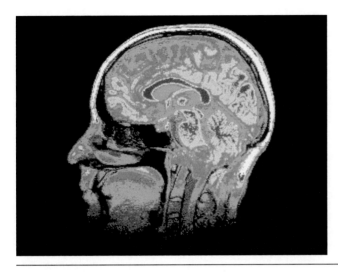

FIGURE 43-10 MRI image of a human head. Note the details of brain structure.

43-4 BINDING ENERGY

We've seen that the mass of a nucleus is less than the sum of the masses of its constituent nucleons—a reflection of the fact that the nucleus is a bound system whose energy is lower than that of its widely separated constituents. Disassembling a nucleus thus requires that we supply energy. The more tightly bound the nucleus, the more energy is needed. Quantitatively, conservation of energy and the Einstein mass-energy equivalence allow us to write

$$m_N c^2 + E_b = Z m_p c^2 + (A - Z) m_n c^2 \qquad (43\text{-}3)$$

for a nucleus of mass m_N, atomic number Z, and mass number A, where m_p and m_n are the proton and neutron masses, respectively. We can regard E_b in Equation 43-3 as the energy it would take to disassemble the nucleus; consequently, E_b is called the **binding energy.** Alternatively, E_b is the energy released when the constituent particles come together to form the nucleus.

● **EXAMPLE 43-3** POWERING THE SUN

Use information from Table 43-1 to calculate the binding energy of the helium-4 nucleus.

Solution
Table 43-1 gives the masses of the proton, neutron, and $^4_2$He nucleus. Using these values in Equation 43-3 and solving for E_b gives

$$E_b = Z m_p c^2 + (A - Z) m_n c^2 - m_N c^2$$

$$= (2)(938.272 \text{ MeV}) + (4 - 2)(939.566 \text{ MeV})$$

$$- 3727.38 \text{ MeV} = 28.3 \text{ MeV}.$$

Here we used Table 43-1's values for mass in MeV/c^2; multiplying by c^2 then gives an energy in MeV with the same numerical value.

The formation of helium through a sequence of nuclear reactions is the process that powers the Sun, and our 28.3-MeV result is very close to the actual 26.7 MeV released for each helium-4 nucleus so formed.

EXERCISE Find the binding energy of the deuterium nucleus ($^2_1$H), whose mass is 2.0136 u.

Answer: 2.19 MeV

Some problems similar to Example 43-3: 16, 19, 20–22

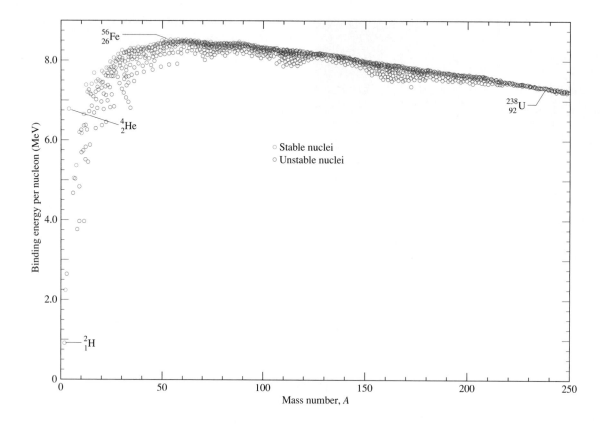

FIGURE 43-11 Binding energy per nucleon as a function of mass number. Nuclei with mass numbers near $A = 60$ are most stable. Note that the stable nuclei at a given mass number are generally those with the greatest binding energy.

In practice one often knows the atomic mass rather than the nuclear mass, but Equation 43-3 still applies because the constituent electrons then appear on both sides of the equation and so cancel—provided, as is always the case, that the binding energy of the electrons to the atom is negligible compared with the nuclear binding energy.

The binding energies of nuclei play a crucial role in determining the origin of the nuclei as well as their behavior. Figure 43-11 shows the **curve of binding energy,** which is a plot of the binding energy *per nucleon* (E_b/A), as a function of mass number. The larger the binding energy per nucleon, the more tightly bound the nucleus. Note that the curve has a broad peak in the vicinity of $A = 60$, indicating that nuclei with mass numbers near this value are most tightly bound. That means it's energetically favorable for light nuclei to join together in the process called **nuclear fusion.** But the heaviest nuclei have lower binding energies per nucleon than do the middle-weight ones, so it's energetically favorable for them to split or **fission** into lighter nuclei. We discuss these nuclear energy processes at length in the next chapter.

● **EXAMPLE 43-4** TRITIUM AND HELIUM

The mass of a helium-3 nucleus (^3_2He) is 3.01493 u, and that of a tritium nucleus (^3_1H) is 3.01550 u. Calculate the binding energy per nucleon for each.

Solution

Helium-3 consists of two protons and a neutron, so $Z = 2$ and $A = 3$; using masses from Table 43-1 in Equation 43-3 then gives

$$E_b = Zm_pc^2 + (A - Z)m_nc^2 - m_Nc^2$$

$$= (2)(938.272 \text{ MeV}) + (3 - 2)(939.566 \text{ MeV})$$

$$- 2808.39 \text{ MeV} = 7.721 \text{ MeV},$$

where we first converted the He-3 mass to MeV/c^2 by multiplying its value in u by the factor 931.494 MeV/c^2/u given in Table 43-1. For tritium, with $Z = 1$ and $A = 3$, we similarly have

$$E_b = Zm_pc^2 + (A - Z)m_nc^2 - m_Nc^2$$

$$= (1)(938.272 \text{ MeV}) + (3 - 1)(939.566 \text{ MeV})$$

$$- 2808.92 \text{ MeV} = 8.484 \text{ MeV}.$$

Since each nucleus contains three nucleons, the binding energies per nucleon for helium-3 and tritium are 2.57 MeV and 2.83 MeV, respectively.

EXERCISE The isotope iron-56 is near the peak of the binding-energy curve of Fig. 43-11, with 8.8 MeV/nucleon binding energy. Use the curve, along with data from Table 43-1, to find the nuclear mass of iron-56.

Answer: 55.92 u

Some problems similar to Example 43-4: 17, 54

Nucleosynthesis and the Origin of the Elements

Where did the chemical elements come from? The curve of binding energy holds many of the answers. Since it's energetically favorable for light nuclei to fuse together, they will do so if given enough energy to overcome the repulsive electric force. Cosmologists believe that this happened in the high-temperature conditions from about 1 minute to about 30 minutes after the big bang event that started the universe, in the process forming the helium that we now find comprising about 25% of the observed matter in the universe. Also formed were minute amounts of deuterium, lithium, beryllium, and boron—but nothing with $Z > 5$ emerged under those primordial conditions. Billions of years later stars formed, and in the interiors of the hotter stars conditions proved suitable for a process in which three helium nuclei join to make carbon. From there nuclear fusion reactions can lead to the formation of isotopes up to those near the $A = 60$ peak of the binding energy curve. In fact, the nuclei of essentially all the elements with $A < 60$ except hydrogen and helium—including most of the materials that make our bodies—were formed in the interiors of massive stars.

Once fusion builds nuclei at the peak of the binding energy curve, it then requires energy to make still more massive nuclei with their lower binding energies; therefore this process does not occur spontaneously. So where did the more massive nuclei—like silver, gold, or uranium—come from? And how did the lighter nuclei get out of the stars? Astrophysicists believe the heavier elements were formed in supernova explosions, runaway nuclear reactions in which massive stars explode, spewing their contents into interstellar space (Fig. 43-12). Energy released in the process supplied the energy needed to form the heavier elements. The expanding cloud from the supernova explosion thus distributes both lighter and heavier elements into the interstellar medium, where eons later they may find themselves incorporated into new stars, planets, and living beings.

43-5 RADIOACTIVITY

In 18v6 Henre Becqu rel of,Paris noticed that a photographic plate that had been stored near uranium compounds appeared fogged when he developed it. He reasoned that some sort of rays must have emanated from the uranium to fog the

FIGURE 43-12 An exploding supernova produces nuclei with $A > 60$, and spews them and the lighter nuclei "cooked" in the star's interior into interstellar space. Photo shows Supernova 1987A (lower right) in the Large Magellanic Cloud, a satellite galaxy of our Milky Way. This was the brightest supernova seen from Earth in several centuries.

film. Becquerel had discovered **radioactivity,** the phenomenon wherein some substances spontaneously emit either particles or photons of very high energy.

In 1898, also in Paris, the Polish student Marie Curie (Fig. 43-13) decided to do her doctoral thesis on Becquerel's rays. Working with her husband, Pierre Curie, she found that not only uranium but also thorium gave off rays, and she coined the term "radioactivity." She soon discovered the more intensely radioactive substances polonium, which she named for her native Poland, and radium. The Curies shared the 1903 Nobel Prize in Physics with Becquerel for their work on radioactivity, and Marie Curie later won the 1911 Nobel Prize in Chemistry for discovering polonium and radium. Today we know a great many radioactive isotopes.

FIGURE 43-13 Marie Curie and Pierre Curie in their Paris laboratory in 1906.

Decay Rates and Half Lives

Different radioactive isotopes emit radiation at different rates, their nuclei decaying in the process into other forms. The greater the number of radioactive nuclei present, the greater the number of decays per unit time. We can express this mathematically by writing

$$\frac{dN}{dt} = -\lambda N, \qquad (43\text{-}4)$$

where λ is a constant characteristic of a particular decay. We've seen a similar equation before, in the analysis of RC and RL circuits. We solve it the same way, by multiplying through by dt/N and integrating. Taking the integration from time 0, when there are N_0 nuclei, to an arbitrary time t, when there are N nuclei, gives

$$\int_{N_0}^{N} \frac{dN}{N} = -\lambda \int_{0}^{t} dt.$$

Evaluating the integrals, we have

$$\ln\left(\frac{N}{N_0}\right) = -\lambda t$$

or, exponentiating each side and using $e^{\ln x} = x$,

$$N = N_0 e^{-\lambda t}. \qquad (43\text{-}5a)$$

Thus, the number of nuclei drops exponentially with time. The constant λ is called the **decay constant;** its inverse $\tau = 1/\lambda$ is the **mean lifetime,** or the average time before a given nucleus decays (see Problem 71). It's often convenient to express Equation 43-5a in terms of the **half-life,** or time it takes half the nuclei in a given sample to decay. Since $2 = e^{\ln 2}$, we can write $e^{-\lambda t} = 2^{-\lambda t/\ln 2}$; then Equation 43-5a becomes

$$N = N_0 2^{-t/t_{1/2}}, \qquad (43\text{-}5b)$$

▲ **T A B L E   4 3 - 2**   SELECTED RADIOISOTOPES

ISOTOPE	HALF-LIFE	DECAY MODE	COMMENTS
Carbon-14 ($^{14}_{6}$C)	5730 years	β^-	Used in radiocarbon dating.
Iodine-131 ($^{131}_{53}$I)	8.04 days	β^-	Fission product abundant in fallout from nuclear weapons and reactor accidents; damages thyroid gland.
Oxygen-15 ($^{15}_{8}$O)	2.03 minutes	β^+	Short-lived oxygen isotope produced with cyclotrons and used for medical diagnosis in positron emission tomography (PET).
Potassium-40 ($^{40}_{19}$K)	1.25×10^9 years	β^-	Comprises 0.012% of natural potassium; dominant radiation source within the normal human body. Used in radioisotope dating.
Plutonium-239 ($^{239}_{94}$Pu)	24,110 years	α	Fissile isotope used in nuclear weapons; produced by neutron capture in $^{238}_{92}$U.
Radium-226 ($^{226}_{88}$Ra)	1600 years	α	Highly radioactive isotope discovered by Marie and Pierre Curie and which results from decay of $^{238}_{92}$U.
Radon-222 ($^{222}_{86}$Rn)	3.82 days	α	Radioactive gas formed naturally in decay of $^{226}_{88}$Ra; seeps into buildings, where it may cause serious radiation exposure.
Strontium-90 ($^{90}_{38}$Sr)	29 years	β^-	Fission product that behaves chemically like calcium; readily absorbed into bones.
Tritium ($^{3}_{1}$H)	12.3 years	β^-	Hydrogen isotope used in biological studies and to enhance yields of nuclear weapons.
Uranium-235 ($^{235}_{92}$U)	7.04×10^8 years	α	Fissile isotope comprising 0.72% of natural uranium; used as reactor fuel and in primitive nuclear weapons.
Uranium-238 ($^{238}_{92}$U)	4.46×10^9 years	α	Predominant uranium isotope; cannot sustain a chain reaction.

where

$$t_{1/2} = \frac{\ln 2}{\lambda} = \frac{0.693}{\lambda} \qquad (43\text{-}6)$$

is the half-life. (To see this, put $t = t_{1/2}$ in Equation 43-5b; then $N = \frac{1}{2}N_0$.) Table 43-2 lists some significant radioisotopes and their half-lives.

It's hard to count all the nuclei in a sample, but with suitable detectors we can easily count the number of decays in a given time interval—that is, we can measure $|dN/dt|$, which is called the **activity** of the sample. (Since the number of nuclei decreases as they decay, dN/dt is itself a negative quantity.) Equation 43-4 shows that dN/dt is proportional to N, so the activity also decays exponentially according to Equations 43-5. The SI unit of activity is the **becquerel** (Bq), equal to one decay per second. An older unit, the **curie** (Ci), is defined as 3.7×10^{10} Bq and is approximately the activity of one gram of radium-226.

● **EXAMPLE 43-5** INDOOR RADON

The radioactive gas radon-222 is a serious pollutant in many homes. The gas results from the decay of radium-226, which itself arises in the decay series of the uranium-238 that occurs naturally in rocks, soil, and building materials. It seeps through cracks in basement walls and floors, where it reaches a steady concentration with infiltration balancing radioactive decay. The mean activity of Rn-222 in American homes is about 0.05 Bq per liter of air. How many Rn-222 nuclei are there per liter?

Solution

We're given the activity $|dN/dt|$ in a 1-L sample, so we can solve Equation 43-4 for N. Table 43-2 gives 3.82 days for the half-life, so we can use Equation 43-6 to write $\lambda = 0.693/t_{1/2}$. Then we have

$$N = -\frac{1}{\lambda}\frac{dN}{dt} = \frac{t_{1/2}}{0.693}\left|\frac{dN}{dt}\right|$$

$$= \left[\frac{(3.82\text{ d})(86,400\text{ s/d})}{0.693}\right](0.05\text{ Bq})$$

$$= 2.4\times10^4\text{ Rn-222 atoms in each liter.}$$

Here we took away the minus sign and put the absolute value sign on dN/dt because the latter is a negative quantity.

Even the modest activity level calculated here presents a serious health threat: Lifetime exposure would reduce your life span by an average of 40 days, comparable to the risk of drowning and much greater than the risk from living next door to a nuclear power plant. ●

● **EXAMPLE 43-6** FALLOUT FROM CHERNOBYL

The 1986 accident at the Chernobyl nuclear power plant in the then-Soviet Ukraine spread radioactive fallout over eastern Europe and Scandinavia (Fig. 43-14). A particularly dangerous isotope released was iodine-131, which is absorbed by the thyroid gland and can cause thyroid cancer. Following the accident, I-131 levels in milk in Romania rose to 2900 Bq per liter. How long did Romanians have to wait before I-131 levels dropped below their government's 185-Bq/L safety standard?

Solution

We could solve this problem formally using Equation 43-5b. But what the equation says is that after n half-lives, the activity drops to $1/2^n$ of its initial value. So the number of half-lives needed to bring the I-131 activity from 2900 Bq/L to 185 Bq/L is given by

$$\frac{1}{2^n} = \frac{185}{2900}$$

or, inverting and taking logarithms of both sides,

$$\ln(2^n) = \ln\left(\frac{2900}{185}\right).$$

But $\ln 2^n = n \ln 2$, so

$$n = \frac{\ln(2900/185)}{\ln 2} = 3.97 \text{ half-lives}.$$

Table 43-2 gives 8.04 days for I-131's half-life, so Romanians had to wait 32 days before their milk was considered safe. Other countries, with less stringent safety standards and lower initial levels, had much shorter waits. We discuss the Chernobyl accident further in the next chapter.

FIGURE 43-14 The Chernobyl accident spread radioactive fallout over much of Europe. Here a crane works to build a concrete tomb around the damaged reactor.

TIP **Powers of Two** After n half lives, the activity of a radioactive sample is down by a factor of $1/2^n$. In estimating activity levels it's useful to note that $2^{10} = 1024$, showing that activity drops by very nearly a factor of 1000 every 10 half-lives.

EXERCISE A patient undergoing a PET scan receives a dose of radioactive oxygen-15 ($t_{1/2} = 2$ minutes). How long will it take for the activity of the dose to decline to one one-millionth of its original level?

Answer: about 40 min (20 half-lives)

Some problems similar to Examples 43-5 and 43-6: 28, 31, 35, 36, 42 ●

■ APPLICATION RADIOCARBON DATING

Archaeologists, art historians, geologists, and others use radioactive decay to date ancient materials. The radioactive isotope carbon-14 forms continually in the atmosphere through the interaction of cosmic rays with nitrogen. Carbon-14 has a 5730-year half-life. Since it acts chemically just like ordinary carbon, it gets incorporated into living organisms. While alive, an organism maintains a steady concentration of ^{14}C, established by the balance between radioactive decay and the uptake of new ^{14}C. At death, carbon uptake ceases but radioactive decay continues, so the concentration of ^{14}C decreases in relation to that of the stable isotope ^{12}C. By measuring ^{14}C activity in an ancient sample of organic material (charcoal, bone, etc.) and comparing the $^{14}C/^{12}C$ ratio with that found in living material, the age of the sample can be determined (Fig. 43-15).

Until recently, samples of 1 to 5 grams were required to count enough decays for accurate radiocarbon dating. A newer technique ionizes carbon in the sample, then uses an electrostatic accelerator to produce an ion beam. A mass spectrometer, using the fact that different mass ions move on trajectories of different radii in a magnetic field, then separates ^{14}C and ^{12}C for an accurate determination of their abundance ratio. This technique counts all the ^{14}C atoms, not just those that decay, and as a result it requires only about 1 mg of carbon. The new technique therefore provides dates for objects too valuable or too small for the earlier technique (Fig. 43-16).

The $^{14}C/^{12}C$ ratio varies slightly with time, as a result of fluctuations in cosmic ray intensity that is in turn related to solar activity. By counting growth rings in ancient trees, scientists have assigned accurate dates back 10,000 years, which provide a more accurate calibration of the radiocarbon dates. Radiocarbon dating remains quite accurate to about 20,000 years and can be used back to about 50,000 years. For longer time spans, up to the billions of years characterizing the ages of rocks, ratios of longer lived isotopes provide age information. Much knowledge of our own past, and our planet's and our solar system's, comes from radioisotope dating.

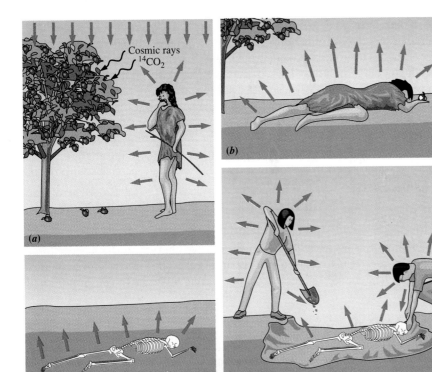

FIGURE 43-16 This piece of the Shroud of Turin, smaller than a postage stamp, was used in accelerator-based radiocarbon dating to establish the shroud's age at 700-800 years, thus refuting the belief that the shroud was the burial cloth of Jesus.

FIGURE 43-15 The principle of radiocarbon dating. (*a*) Carbon-14 formed in the atmosphere is incorporated into a living organism through the food chain. While it lives, the organism exhibits a mild, steady level of ^{14}C radioactivity (arrows). (*b*) At death, ^{14}C uptake ceases. (*c*) Much later, ^{14}C activity has decayed considerably. (*d*) Archaeologists excavate the long-dead remains. By measuring ^{14}C activity, they can infer the time since death. Note that the archaeologists, with their active ^{14}C intake, are more radioactive than their ancient ancestor.

● **EXAMPLE 43-7** ARCHAEOLOGY

Archaeologists unearth charcoal from an ancient campfire. They find that carbon-14 activity per unit mass of the sample is 7.4% that expected in living wood. Find the sample's age.

Solution

The sample contains 0.074 of its original ^{14}C radioactivity. This is the factor $1/2^n$, with n the number of half-lives elapsed. So $0.074 = 1/2^n$, or

$$2^n = \frac{1}{0.074} = 13.5.$$

Taking logarithms of both sides and noting that $\ln(2^n) = n \ln 2$, we have

$$n \ln 2 = \ln 13.5,$$

or

$$n = \frac{\ln 13.5}{\ln 2} = 3.76.$$

Since ^{14}C has a 5730-year half life, the sample's age is $(3.76)(5730 \text{ years}) = 21,500 \text{ years}$.

EXERCISE Find the age of a bone sample whose ^{14}C activity is 19% that of living bone.

Answer: 13,700 years

Some problems similar to Example 43-7: 40, 45 ●

Types of Radiation

Passing the emanations from radioactive materials through a magnetic field shows that there are three kinds of radiation, one positively charged, one negatively charged, and one neutral (Fig. 43-17). Earlier researchers gave these the names alpha, beta, and gamma radiation, respectively. Today we know that alpha radiation consists of helium-4 nuclei, that beta radiation consists of high-energy electrons (or sometimes positrons), and that gamma radiation is a stream of very-high energy photons. These radiations also differ in penetrating power: alpha particles are stopped by a sheet of paper and beta particles by several cm of matter, but some gamma rays can penetrate substantial thicknesses of concrete or lead. Radioactive nuclei—radioisotopes—are characterized not only by which type of radiation they emit, but also by the radiation energy. Uranium-238, for example, emits alpha particles with energies around 4 MeV, while carbon-14, widely used in radioactive dating, emits beta particles with maximum kinetic energy of 156 keV.

Alpha Decay

Alpha emitters are nuclei with too much positive charge, so the nuclear force cannot keep the repulsive electric force in check. They decay into a smaller nucleus with less charge and less mass by emitting a bundle of two protons and

FIGURE 43-17 There are three types of radiation, which separate in a magnetic field into a positive (α particles), negative (β particles) and neutral (γ rays).

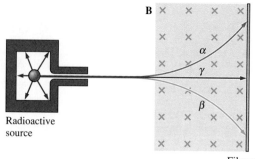

Radioactive source

Film or detector

two neutrons—an alpha particle. The process, therefore, drops the atomic and mass numbers by two and four, respectively, and symbolically we write

$$\ce{^A_Z X} \longrightarrow \ce{^{A-4}_{Z-2} Y} + \ce{^4_2 He}. \qquad (43\text{-}7)$$

Here X is the original nucleus—the **parent**—and Y the resulting **daughter** nucleus. Since an alpha particle is a helium nucleus, we write $\ce{^4_2 He}$ for the alpha particle, although this is sometimes abbreviated α. Note that the sums of the atomic numbers on both sides of this equation are equal, as are the sums of the mass numbers; this must be true in any nuclear reaction that conserves charge and the number of nucleons. Energy is released in the alpha decay, and appears as the alpha particle's kinetic energy. That the alpha's energy is less than that needed to overcome the nuclear potential barrier is one of the most direct confirmations of quantum tunneling; alpha decay is possible only because the escaping particle tunnels through the potential barrier.

Beta Decay

Beta emitters typically have too many neutrons to be stable nuclei. But they emit electrons, not neutrons. So how does that help? And where did the electron come from, if the nucleus contains only protons and neutrons? The answer lies in the **weak force,** a relative of electromagnetism, that can bring about the decay of a neutron into a proton and an electron. That's just what happens in beta decay, and the result is one less neutron and one more proton. The ejected electron carries off one unit of negative charge, so the atomic number of the remaining nucleus must increase by one to conserve charge. The total number of nucleons remains the same, however, so the mass number does not change.

In the 1920s physicists studying beta decay found a disturbing thing: the ejected electrons did not account for all the energy released in the process. Where was the rest of it? In 1930, Pauli proposed the existence of a hitherto undetected particle, which he called the **neutrino,** for "little neutral one." The neutrino is an elusive particle indeed: it probably has no mass, meaning that it travels only at the speed of light, and it interacts only through the weak force. Neutrinos emitted in nuclear reactions in the Sun's core, for example, travel right through the solid Earth with little probability of interaction. Neutrinos have nevertheless been detected through rare nuclear reactions in which they participate, and today they are used to probe conditions in the core of the Sun and in the earliest second of the universe, as we describe in Chapter 45.

The neutrino, like other particles, has an antiparticle. Ordinary beta decay, in fact, produces antineutrinos. A typical beta decay can be written

$$\ce{^A_Z X} \longrightarrow \ce{^A_{Z+1} Y} + e^- + \bar{\nu}, \qquad (43\text{-}8a)$$

where e^- is the electron and $\bar{\nu}$ is the antineutrino. Another beta-decay reaction turns a proton into a neutron, emitting a positron (anti-electron, e^+) and a neutrino:

$$\ce{^A_Z X} \longrightarrow \ce{^A_{Z-1} Y} + e^+ + \nu. \qquad (43\text{-}8b)$$

This reaction occurs in some short-lived isotopes of carbon and oxygen, and gamma rays from the subsequent annihilation of the positrons are used to produce images in the medical diagnostic procedure known as positron emission tomography (PET).

A third beta-decay process is **electron capture,** in which a nucleus captures an inner-shell atomic electron, converting a proton into a neutron and lowering the atomic number by one. A neutrino is also ejected:

$$_{Z}^{A}X + e^- \longrightarrow \: _{Z-1}^{A}Y + \nu. \tag{43-8c}$$

● EXAMPLE 43-8 NUCLEAR DECAYS

Write equations for (a) the α decay of uranium-238 and (b) the β decay of carbon-14.

Solution

Alpha decay drops the atomic number by two and the mass number by four. A glance at the periodic table shows that uranium has atomic number 92, and that the element two numbers lower is thorium. Thus the daughter nucleus is $_{90}^{234}$Th, and the reaction reads

$$_{92}^{238}U \longrightarrow \: _{90}^{234}Th + \: _{2}^{4}He.$$

Beta decay increases the atomic number by one, changing carbon ($Z = 6$) to nitrogen ($Z = 7$). Since the mass number stays the same, this reaction reads

$$_{6}^{14}C \longrightarrow \: _{7}^{14}N + e^- + \bar{\nu}.$$

Here the daughter nucleus is the abundant, stable form of nitrogen.

EXERCISE What is the daughter nucleus formed when oxygen-15 decays by positron emission?

Answer: $_{7}^{15}$N

Some problems similar to Example 43-8: 26–28, 65
●

Gamma Decay

Alpha and beta decays usually leave the daughter nucleus in an excited state. Nuclei can also be excited by collisions with high-energy particles or through photon absorption. As with atoms, nuclear excited states decay through photon emission. But—as a simple uncertainty-principle estimate (Example 39-8) shows—the energies associated with nuclear states must be on the order of MeV. The high-energy photons emitted in nuclear transitions are called gamma rays. A photon carries neither mass nor charge, so gamma emission, unlike alpha or beta, does not change the type of nucleus; we therefore write

$$_{Z}^{A}X^* \longrightarrow \: _{Z}^{A}X + \gamma, \tag{43-9}$$

where X^* designates an excited state.

Since each nucleus has a distinctive energy-level pattern, gamma-ray spectral lines can be used to identify nuclear species (Fig. 43-18). Even nonradioactive materials can be identified this way, by first exciting their nuclei through exposure to a neutron beam from a nuclear reactor and measuring the gamma-ray spectrum that results from the subsequent de-excitation.

Decay Series and Artificial Radioactivity

A few radioisotopes—like ^{40}K and ^{238}U in Table 43-2—have half-lives comparable with the age of the Earth. It's not surprising that we find these occurring

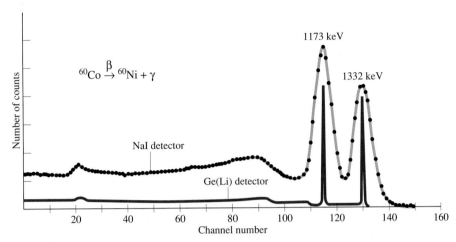

FIGURE 43-18 Gamma-ray spectra of cobalt-60. The actual decay is a two-step process, with ^{60}Co beta decaying to excited ^{60}Ni*, which then emits gamma radiation as it drops to its stable ground state. Two spectra shown are from measurements of light flashes in a sodium-iodide crystal and from a much higher resolution semiconductor detector made from germanium and lithium. Gamma-ray counts are sorted into channels corresponding to different energy ranges, so horizontal position is a measure of gamma-ray energy.

naturally. But most half-lives are much shorter. So why do we find short-lived isotopes at all? Some, like cosmic ray-produced carbon-14, result from naturally occurring nuclear reactions. Many others arise in the decay of long-lived isotopes, while some are produced artificially in nuclear reactors and particle accelerators.

Figure 43-19 shows the **decay series** for uranium-238, whose 4.46-billion-year half-life ensures that there's still plenty of it around. Even though they have short half-lives, the daughter products in this series are present wherever uranium is found. The relative abundance of each is established by a balance between its own decay and its formation in the preceding decay. Note that one of the daughters in the uranium series is the chemically inert gas radon, which readily escapes to become a serious health hazard in enclosed buildings.

In 1934 Marie Curie's daughter Irène and her husband Frédéric Joliot-Curie (Fig. 43-20) were the first to induce artificial radioactivity, by bombarding stable isotopes with alpha particles. Enrico Fermi soon demonstrated that bombardment with slow neutrons could also produce new radioactive nuclei. Today we obtain radioisotopes in a variety of ways, including bombardment with particle beams, irradiation with neutrons from nuclear reactors, or by extracting them from the by-products of nuclear fission. As one example, oxygen-15, used in positron emission tomography (PET) is produced by bombarding ordinary nitrogen-14 with cyclotron-accelerated deuterons ($^{2}_{1}$H):

$$^{14}_{7}\text{N} + {}^{2}_{1}\text{H} \longrightarrow {}^{15}_{8}\text{O} + {}^{1}_{0}n,$$

where $^{1}_{0}n$ is a free neutron (0 charge, 1 mass unit) that also results from the reaction. Because ^{15}O has a two-minute half-life, the cyclotron must be located right in the hospital doing the PET diagnosis (see Fig. 29-13).

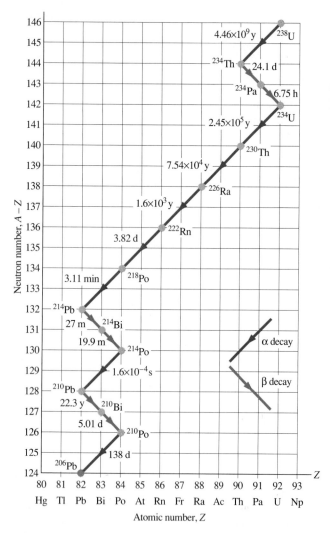

FIGURE 43-20 Irène and Frédéric Joliot-Curie, discoverers of artificial radioactivity. The couple shared the 1935 Nobel Prize in Chemistry.

FIGURE 43-19 The decay of uranium-238 results in a series of shorter lived nuclei that occur wherever uranium is present. Red arrows represent alpha decay, blue arrows beta decay. The half-life for each decay is indicated next to the arrow. The series ends with the stable isotope lead-206.

■ **APPLICATION** USES OF RADIOACTIVITY

Here we survey just a few of the myriad uses for radioactive materials in our technological civilization, some of which are shown in Fig. 43-21.

Radioactive Tracers The use of small quantities of radioactive materials allows scientists, engineers, and physicians to trace material flows in a wide range of systems. Medical techniques such as brain, thyroid, or lung scans use isotopes of materials incorporated into these organs to study organ func-

tion. The more general technique of positron-emission tomography sends short-lived positron emitters through the body to provide internal images. Biologists use radioactive tracers routinely to trace the uptake and distribution of chemical compounds in biological organisms; "labeling" carbon dioxide with ^{14}C, for example, helps scientists unravel the carbon cycle and learn, perhaps, how plants can help alleviate global warming caused by carbon emissions from fossil fuels. Engineers use radioisotope tracers to study wear in mechanical systems. In-

 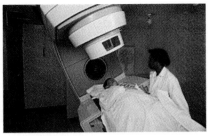 

(a)   (b)   (c)

FIGURE 43-21   Some uses of radiation. (a) Food preservation. Both rolls are two weeks old, but the one on the right was irradiated to retard spoilage. (b) Treating cancer. (c) Bomb detection. This unit at New York's Kennedy Airport uses neutrons to produce short-lived radioisotopes in airline luggage; analysis of the decay radiation permits detection of explosives.

corporating radioisotopes in an engine bearing, for example, and then measuring the radioactivity of the lubricating oil provides a measure of the rate of wear of the bearing surface.

Cancer Treatment Radiation destroys living cells, especially those that, like cancer cells, divide rapidly. By bombarding cancer tissue with radiation, the cancer cells can be selectively destroyed. Early efforts used gamma radiation from cobalt-60, but today's treatments rely on particle beams whose energy and therefore penetration can be precisely tailored to minimize damage to nearby tissue. Alternately, small amounts of radioisotopes can be embedded in the tumor region itself. Gold-98, a 65-hour beta emitter, is often used for this purpose, while injection of phosphorus-32 into the bloodstream helps fight blood cancers like leukemia.

Food Preservation High doses of radiation destroy bacteria and enzymes that cause spoilage. The food does not get radioactive, but it may undergo chemical changes that some argue could alter its nutritional value or introduce harmful compounds. Con-

siderable controversy surrounds the safety of food irradiation as more food varieties are approved for this treatment.

Insect Control Since radiation preferably damages rapidly dividing cells, high doses to the reproductive organs cause sterility. If large numbers of harmful insects are sterilized in this way, then released into the wild, they mate with normal insects but no offspring result. The population then dwindles. The medfly, a serious pest of citrus crops, has been partially controlled in this way.

Activation Analysis Bombarding materials with neutrons or other particles results in excitation of nuclei and in the production of unstable isotopes. Analyzing the resulting radiation helps identify unknown materials. This technique is used by art historians to detect art forgeries, and by archaeologists, chemists, and environmental scientists. It has the advantage of being nondestructive, in that no material need be removed from the sample. Recently, activation analysis has emerged as a leading candidate for detection of explosives in airline luggage.

Biological Effects of Radiation

In the early years of the twentieth century, radiation was believed beneficial. People actually paid for the privilege of spending time in old uranium mines, breathing the radioactive vapors. But we now know that radiation, despite its many beneficial uses in medicine, is harmful to living tissue.

Radiation works its damage because alpha, beta, and gamma particles have sufficient energy to ionize and otherwise disrupt biological molecules. Results include cell death, loss of necessary biological functions, mutations that lead to cancer, and mutations in reproductive cells that cause genetic changes. Numerous studies confirm that these effects occur. In the 1920s, watch hands were painted with radium-containing paint so they would glow in the dark. Factory workers painting the hands licked their brushes to keep the bristles together;

nearly all of them died of bone cancer or radiation-induced anemia. Uranium miners, exposed to radon gas trapped in the mines, have much higher than average lung cancer rates. Marie Curie and her daughter Irène, along with other early nuclear scientists, succumbed to leukemia and other cancers that undoubtedly resulted from radiation exposure. (Pierre Curie was spared this fate; he was killed in the street when a horse bolted from its carriage.)

The energy absorbed in a dose of radiation is a crude measure of its biological danger. The SI unit of absorbed dose is the **gray** (Gy), defined as 1 J of energy absorbed per kg of absorbing material. (An older unit, the rad, equal to 0.01 Gy, is still widely used.) A more appropriate measure of biological radiation dose is the **sievert** (Sv), which is the absorbed energy per unit mass weighted by factors that compensate for differences in biological effectiveness among different types and energies of radiation. (An older unit, the rem, is equal to 0.01 Sv.) Alpha particles, for example, cause much more damage per unit energy delivered than do gamma rays, so 1 Gy of alpha radiation is more harmful than 1 Gy of gamma radiation. But 1 Sv of alphas and 1 Sv of gammas cause essentially the same biological damage.

The biological effects of high radiation doses are well known; exposure to 4 Sv, for example, causes death in about half of its human victims. But doses in the 0.1-Sv range and below are far more controversial. There are only a few cases of exposures to populations large enough that small radiation effects can be determined with statistical accuracy. One of the most important such populations are the survivors of the Hiroshima bombing, from whom many of our radiation-exposure standards derive. But in the 1990s, half a century after the bombing, controversy continues over the doses delivered at Hiroshima. Our radiation safety standards remain, therefore, uncertain.

Even less certain are the effects of very low doses, such as the 10^{-5}-Sv average dose to people living near the 1979 Three Mile Island nuclear accident. Today these effects are estimated by linearly extrapolating from the effects of higher doses. But many radiation biologists believe that natural repair mechanisms in cells may act to diminish the effects of low doses, resulting in an overestimate of radiation dangers. On the other hand, varying susceptibility of different individuals, especially the young and unborn, may mean the danger is greater than current standards suggest.

The public has a deeply held fear of things nuclear, and of radiation in particular. Yet we're exposed to a great deal more natural radiation than we are to radiation associated with nuclear technologies. The average U.S. citizen, for example, receives about 3.6 mSv yearly; Fig. 43-22 shows the breakdown among natural and artificial sources. That number varies dramatically with geographical location—high altitudes and uranium-rich rocks lead to higher exposures—and with occupation. Airline crews, for example, often receive cosmic-ray radiation doses in excess of the doses allowed for nuclear power plant workers. And the average risk to your health from radiation—with the possible exception of indoor radon exposure in high-radon areas—is very small compared with many other risks you knowingly accept.

Detecting Radiation

Nuclear radiation is undetectable by the human senses, but a number of technological devices make detection and measurement of radiation a routine process.

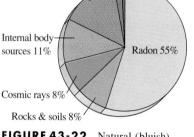

FIGURE 43-22   Natural (bluish) and artificial (reddish) sources of radiation, as percentages of the average yearly dose of 3.6 mSv. It was only in the late 1980s that the importance of radon was recognized.

Becquerel's 1896 discovery of radioactivity by the accidental fogging of photographic film showed that film can detect radiation. A more widely used detector is the **Geiger counter,** described in Fig. 43-23. Geiger counters and related detectors use radiation-induced ionization of gas atoms to produce pulses of electric current. Another class of detectors uses **scintillation**—the production of light flashes as atoms excited by incident radiation drop to lower energy states. A number of common plastics, as well as some crystalline materials and liquids, make efficient scintillators. The light flashes are usually detected by a photomultiplier (see Application: The Photomultiplier in Chapter 39) that converts them to electrical signals. Semiconductors provide another approach to radiation detection. As we discussed in the preceding chapter, electric current in semiconductors is carried by negative electrons, positive holes, or both. Radiation passing through a semiconductor can excite electrons to the conduction band, creating electron-hole pairs and producing a brief pulse of current that marks the passage of radiation. Detectors using these various mechanisms not only indicate the presence and quantity of radiation, but many can also provide accurate measures of alpha, beta, or gamma energies.

43-6 MODELS OF NUCLEAR STRUCTURE

We've seen how the interplay of nuclear and electric forces, along with the phenomenon of weak-force beta decay, determines the general mix of neutrons and protons in the stable nuclei shown in Fig. 43-4. But look more closely at that figure: There are many more stable nuclei for even values of Z, and some—those with the so-called **magic numbers** 2, 8, 20, 28, 50, 82, or 126 protons or neutrons—have many more stable nuclei than do nearby values. Why? And why do unstable nuclei decay with the particular particle energies and half-lives they exhibit? Any detailed model of the nucleus must answer these questions.

Unfortunately there is no complete nuclear theory, analogous to the atomic theory presented in Chapter 41, that explains all aspects of all nuclei. The reasons for this have to do with our still imperfect knowledge of the nuclear force, and with the fact that nucleons are so tightly packed that each interacts with many others; we therefore cannot solve for nuclear structure with a simple two-particle model like we used for the single electron and nucleus in hydrogen-like atoms. Instead, nuclear physicists resort to several models to explain different aspects of nuclear structure.

The Liquid-Drop Model

We have seen that the density of nuclear matter is essentially independent of nuclear size—just as the density of a liquid drop doesn't depend on the drop size. Treating the nucleus as a liquid drop is a reasonable aproximation for larger nuclei, whose many nucleons are like the molecules of a liquid. In this **liquid-drop model** the nuclear binding energy is analogous to the heat of vaporization, and surface-tension effects show why nucleons near the surface of a nucleus are less tightly bound. A semi-empirical formula derived from the liquid-drop model gives good agreement with observed nuclear binding energies, and the model helps explain the process of nuclear fission. A liquid-drop nucleus can rotate, vibrate, and change shape as long as its volume stays the same, and the

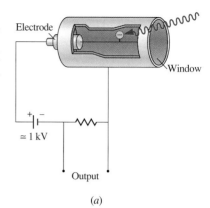

(a)

(b)

FIGURE 43-23 (a) When a high-energy particle or gamma-ray photon enters the gas-filled tube of a Geiger counter, it knocks electrons from some gas atoms. Accelerated toward the positive central electrode, the electrons ionize additional atoms to produce a measurable pulse of current in the resistor. The output can drive an amplifier and loudspeaker, a meter, a digital counter, or sophisticated electronics for determination of particle energies. (b) Worker using a Geiger counter (arrow) to check for radiation during cleanup after the Three Mile Island nuclear accident.

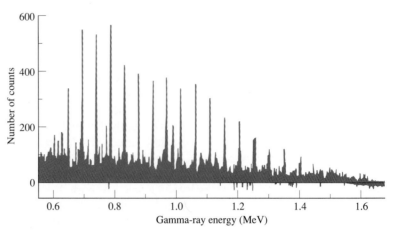

FIGURE 43-24 Regularly spaced gamma rays from dysprosium-152. The dysprosium nuclei are set spinning rapidly by collisions, and the spectral lines result as they decay through the different rotational states.

resulting quantized energy levels lead to predictions of nuclear spectral lines—analogous to those we discussed for molecules in the preceding chapter—that are in good agreement with observation (Fig. 43-24).

Because a liquid is a continuum, however, the liquid-drop model cannot explain effects resulting from small changes in nucleon number. Thus, the model sheds no light on why even atomic numbers are more stable or on why there are magic numbers with exceptional stability.

The Shell Model

FIGURE 43-25 Maria Goeppert Mayer with her daughter. Mayer and J. Hans Jensen shared the Nobel Prize for their work on the nuclear shell model.

The **nuclear shell model,** advanced in the late 1940s by physicists Maria Goeppert Mayer (Fig. 43-25) and J. Hans Jensen, treats the nucleus as having a shell structure similar to that of atoms. The shells occur because neutrons and protons obey the exclusion principle as they move in a square-well-like average potential. By including spin-orbit interaction in their model, Goeppert Mayer and Jensen explained the magic numbers as resulting from closed-shell structures similar to the electronic structures of the inert gases. The nucleons of a closed shell are tightly bound, giving the nucleus a high degree of stability. Additional nucleons are not bound into the closed-shell structure and stay primarily on the outskirts of the nucleus where they are more readily excited to higher energy levels.

Neutrons and protons in the shell model act independently, and each has its own set of quantum numbers. Closed-shell structure, and thus enhanced stability, occurs at magic numbers of either protons or neutrons, and thus, there are several stable nuclei for each magic value of atomic number Z or neutron number $A - Z$. Some nuclei, like $^{40}_{20}$Ca ($Z = 20$, $A - Z = 20$) or $^{208}_{82}$Pb ($Z = 82$, $A - Z = 126$) are "doubly magic" and show exceptional stability.

In addition to explaining the magic numbers, the shell model also predicts the angular momenta of most nuclei. It is generally most successful for lighter nuclei and for those with proton or neutron numbers near the magic values.

The Collective Model

Neither the liquid-drop nor the shell model is a complete theory of the nucleus. Each is good at explaining certain aspects of nuclear structure but neither gives the whole picture. The **collective model,** advanced in part by Niels Bohr's son Aage, combines aspects of the liquid-drop and shell models. As in the shell model, magic numbers of nucleons group into filled "cores," surrounded by "extra" nucleons orbiting much like atomic electrons. But as in the liquid-drop model, the core shape, and with it the nuclear potential, can change as the nucleons undergo collective motions that give the model its name. These motions include wave-like circulations that help account for nuclear angular momenta and magnetic moments. One remarkable prediction of the collective model is that nuclei need not be spherical. The reason is that the values of the magic numbers change if the nuclear potential changes due to distortion from spherical shape. That means nuclei with nonmagic numbers may actually be more stable if they take on nonspherical shapes. In 1975 Aage Bohr shared the Nobel Prize with James Rainwater of the United States and Ben Mottelson of Denmark for their work on nonspherical nuclei in the collective model.

CHAPTER SYNOPSIS

Summary

1. Nearly all the mass of an atom is in the **nucleons**—protons and neutrons—that make up the **atomic nucleus.** Although massive, the nucleus occupies only a tiny fraction of the atomic volume. The number of protons is the **atomic number,** Z, and determines the chemical element. The total number of nucleons is the **mass number,** A. Nuclei with the same atomic number but different mass numbers are **isotopes** of the same element.

2. Nuclei are held together by the strong but short-range **nuclear force** that acts between nucleons. The nuclear force must overcome the long-range repulsive electric force between protons, which explains why nuclei must include neutrons, why there are more neutrons than protons in the more massive nuclei, and why stable nuclei do not exist for atomic number $Z > 83$. Decay of the neutron, mediated by the **weak force,** explains why nuclei cannot have too many neutrons.

3. The mass of a nucleus differs from the sum of the masses of its constituent nucleons by E_b/c^2, where E_b is the **binding energy.** The binding energy per nucleon is greatest for nuclei with mass numbers around 60, showing that these are the most stable nuclei. All nuclei have essentially the same density, and their radii are given approximately by $R = R_0 A^{1/3}$.

4. Coupling of nucleon spins gives many nuclei a net angular momentum and thus a net magnetic moment. Flipping the magnetic moment results in slight energy differences when the nucleus is in a magnetic field, either external or arising from atomic electrons. This energy shift causes **hyperfine splitting** of the hydrogen ground-state energy and is the basis of **nuclear magnetic resonance.**

5. Unstable nuclei are **radioactive,** and decay by emitting particles. In **alpha decay** the nucleus emits a helium-4 nucleus (alpha particle), and its mass number and atomic number drop by four and two, respectively. In most **beta decays,** a neutron decays into a proton, an electron (beta particle), and an antineutrino. The electron and antineutrino escape, leaving a nucleus with its atomic number increased by one but with no change in mass number. Some beta decays emit a positron and a neutrino, converting a proton to a neutron and dropping the atomic number by one. In **gamma decay,** an excited nucleus drops to a lower energy state, emitting a high-energy photon (gamma ray) with no change in either its atomic or mass number.

6. The decay of radioactive nuclei is characterized by the **decay constant** λ or its inverse, the **mean lifetime** τ or, equivalently, the **half-life** $t_{1/2}$. A sample containing initially N_0 nuclei decays exponentially with time:

$$N = N_0 e^{-\lambda t} = N_0 e^{-t/\tau} = N_0 2^{-t/t_{1/2}}.$$

The **activity** of a radioactive sample is the number of decays per unit time. The SI activity unit is the **becquerel** (Bq), equal to one decay per second. Activities decay with the same exponential factors given in the equation above. Short-lived isotopes occur naturally as a result of **decay** series of longer lived species. Particle accelerators and nuclear reactors help create a variety of radioactive isotopes, with numerous uses in science, engineering, medicine, and industry.

7. Radiation is harmful to living tissues, in very rough proportion to the energy absorbed. The **gray** (Gy), equal to 1 J of energy absorbed per kg, is the SI measure of absorbed dose. The dose unit weighted for biological effectiveness is the **sievert** (Sv). Radiation doses of a few Sv are lethal to humans, but much controversy surrounds the effect of lower doses. A typical human receives several mSv of radiation dose yearly, most of it from natural sources.

8. There is not yet a complete theory of nuclear structure, but several models can explain different aspects of the nucleus. The **liquid-drop** model treats the nucleus as a deformable drop of constant-density liquid and is reasonably successful in predicting nuclear binding energies. The **shell model** explains the **magic numbers** of high nuclear stability in terms of closed nucleon shells analogous to the closed electron shells of inert gases. The **collective model** combines aspects of the two others and helps explain nonspherical nuclei.

Terms You Should Understand

(Pairs are closely related terms whose distinction is important; number in parentheses is chapter section where term first appears.)

nucleon (43-1)
nuclear force (43-2)
atomic number, mass number (43-2)
element, isotope (43-2)

hyperfine splitting (43-3)
nuclear magnetic resonance (NMR) (43-3)
binding energy (43-4)
radioactivity (43-5)
decay constant, mean lifetime, half-life (43-5)
alpha, beta, gamma radiation (43-5)
neutrino (43-5)
decay series (43-5)
magic numbers (43-6)
liquid-drop, shell, and collective models (43-6)

Symbols You Should Recognize

A, Z (43-2)
^4_2He and similar nuclear symbols (43-2, 43-5)
λ, τ, $t_{1/2}$ (43-5)
ν, $\bar{\nu}$ (43-5)

Problems You Should be Able to Solve

formulating nuclear symbols (43-2)
calculating nuclear radii (43-3)
evaluating energies associated with nuclear magnetic moments (43-3)
relating binding energy and nuclear mass (43-4)
using half-lives or mean lifetimes to analyze radioactive decay (43-5)
writing equations for nuclear decay reactions (43-5)

Limitations to Keep in Mind

We do not yet have a complete theory of the atomic nucleus.

QUESTIONS

1. How did alpha-particle scattering experiments reveal the existence of the nucleus?
2. Why do nuclei contain neutrons?
3. Why do different isotopes have the same chemical behavior?
4. Why are there no stable nuclei for sufficiently high atomic numbers?
5. Lighter nuclei have roughly equal numbers of protons and neutrons, but heavier nuclei have more neutrons. Why?
6. Give an order-of-magnitude comparison between the sizes of atoms and nuclei.
7. Why is the MeV/c^2 a useful unit for the masses of nuclei and elementary particles?
8. How is the existence of a peak in the curve of binding energy significant for the evolution of stars?
9. Could a nucleus have more mass than that of its separate constituents?

10. Some books mistakenly report the binding energy per nucleon as the energy needed to remove one nucleon from the nucleus. Why might this be incorrect?
11. Why might future archaeologists have problems dating samples from the second half of the twentieth century?
12. What types of radiation are affected by magnetic fields? Why?
13. Beta decay by positron emission is soon followed by a pair of 511-keV gamma rays. Why?
14. Cobalt-60 undergoes beta decay. Why, then, do gamma rays result?
15. What is the difference between a gray and a sievert?
16. How does the shell model explain magic numbers?
17. Why would it have been easier to make bombs fueled with uranium-235 a few billion years ago?
18. Why doesn't the level of carbon-14 in a living organism continue to build up until death?

19. Why are the nuclear magnetic resonance frequencies of the hydrogen nuclei in water (H_2O) and methane (CH_4) different? After all, the nuclei in both cases are just protons.
20. Why are iodine-131 and strontium-90 particularly dangerous radioisotopes?
21. List some beneficial and some harmful aspects of radiation.

22. Which model, liquid-drop or shell, would do a better job explaining (a) nuclear fission and (b) nuclear gamma-ray spectra?
23. When the Sun is very active it can lower the flux of cosmic rays entering the solar system. If this happened at some earlier time, what affect would it have on radiocarbon dating of samples from that time?

PROBLEMS

Section 43-1 Discovery of the Nucleus

1. In a head-on collision of a 9.0-MeV α particle and a nucleus in a gold foil, what is the minimum distance before electrical repulsion reverses the α particle's direction? Assume the gold nucleus remains at rest. Your answer shows how Rutherford set upper limits on the size of nuclei.
2. In a Rutherford-type experiment performed with 6.0-MeV α particles incident on iron, what is the closest approach to the center of an iron nucleus by an α particle on a head-on course?

Section 43-2 Building Nuclei: Elements, Isotopes, and Stability

3. Three of the isotopes of radon ($Z = 86$) have 125, 134, and 136 neutrons, respectively. Write symbols for each.
4. Write the symbol for the germanium isotope with 44 neutrons.
5. How do (a) the number of nucleons and (b) the nuclear charge compare in the two nuclei $^{35}_{17}Cl$ and $^{35}_{19}K$?
6. Examine Fig. 43-4 to find the element that has the most stable isotopes.
7. Use Fig. 43-4 to find atomic numbers $Z < 83$ for which there are no stable isotopes. What elements are these?

Section 43-3 Properties of the Nucleus

8. Compare the radius of the proton (the $A = 1$ nucleus) with the Bohr radius of the hydrogen atom.
9. Write the symbol for a boron nucleus with twice the radius of ordinary hydrogen (^1_1H).
10. How does the minimum distance calculated in Problem 1 compare with the radius of the gold nucleus?
11. A uranium-235 nucleus splits into two roughly equal-size pieces. What is their common radius?
12. To what size would Earth have to collapse to be at nuclear density?
13. Find the energy needed to flip the spin state of a proton in Earth's magnetic field, whose magnitude is about 30 μT.
14. An NMR spectrometer is described as a "300-MHz instrument," meaning 300 MHz is the frequency supplied to its transmitter coil to flip the spin states of bare protons. What is the strength of its unperturbed magnetic field?

15. The permanent magnet from an old NMR spectrometer has a field strength of 1.4 T. What frequency of electromagnetic radiation was used in the instrument?

Section 43-4 Binding Energy

16. What is the total binding energy of oxygen-16, given that its nuclear mass is 15.9905 u?
17. Determine the atomic mass of nickel-60, given that its binding energy is very nearly 8.8 MeV/nucleon.
18. Find the nuclear mass of plutonium-239, given its atomic mass 239.052157 u.
19. Find the total binding energy of sodium-23, given its atomic mass of 22.989767 u.
20. Iron-56 is among the most tightly bound nuclei. Its mass is 55.9206 u. Find the binding energy per nucleon, and check against Fig. 43-11.
21. The mass of a lithium-7 nucleus is 7.01435 u. Find the binding energy per nucleon.
22. Calculate the binding energy per nucleon for gold-197, given its atomic mass of 196.96654 u.
23. By what percentage is the binding energy of $^{64}_{30}Zn$ different from that of $^{64}_{29}Cu$, another isotope with the same mass number? Use atomic masses of 63.92915 for ^{64}Zn and 63.92977 for ^{64}Cu.

Section 43-5 Radioactivity

24. Use the half-life given in Table 43-2 to calculate (a) the decay constant and (b) the mean lifetime for uranium-235.
25. The decay constant for argon-46 is 0.0835 s^{-1}. What is its half-life?
26. Copper-64 can decay by any of the three beta-decay processes. Write equations for each decay.
27. Referring to Fig. 43-19, write equations describing the decays of (a) radon-222 and (b) lead-214.
28. Carbon-10 is a positron emitter with a 19.3-s half-life. (a) Write an equation to describe its decay. (b) A ^{10}C sample has initial activity 48 MBq. What will be its activity after 1 minute, in Bq?
29. A milk sample shows an iodine-131 activity level of 450 pCi/L. What is its activity in Bq/L?
30. Nuclear bomb tests of the 1950s deposited a layer of strontium-90 over Earth's surface. How long will it take

from the time of the bomb tests for (a) 99% and (b) 99.9% of this radioactive contaminant to decay?

31. In the 4.5-billion-year lifetime of the Earth, how many half-lives have passed of (a) carbon-14, (b) uranium-238, and (c) potassium-40? For each isotope, give the number of atoms remaining today from a sample of 10^6 atoms at Earth's formation.

32. You measure the activity of a radioactive sample at 2.4 MBq. Thirty minutes later, the activity level is 1.9 MBq. What is the decay constant for the radioactive material?

33. How many atoms in a 1-g sample of U-238 decay in 1 minute? *Hint:* This time is so short compared with the half-life that you can consider the activity essentially constant.

34. Plot the fraction of the initial atoms remaining in a carbon-14 sample for times from $t = 0$ to $t = 40,000$ years.

35. The Swedish standard for I-131 in milk is 2000 Bq/liter. How long would Swedes have waited following the Chernobyl accident for their milk to be considered safe, assuming the same initial contamination level as in Example 43-6?

36. If radon infiltration into a house were stopped, what fraction of the initial radon would be left after (a) 1 day, (b) 1 week, (c) 1 month?

37. Nitrogen-13 has a 10-min half life. A sample of ^{13}N contains initially 10^5 atoms. Plot the number of atoms as a function of time from $t = 0$ to $t = 1$ hour. Make your horizontal axis (time) linear, but your vertical axis logarithmic. Why is the curve a straight line? What is the significance of its slope?

38. Thorium-232 is an α emitter with a 14-billion-year half life. Radium-228 is a β^- emitter with a 5.75-year half-life. Actinium-228 is a β^- emitter with a 6.13-hour half-life. (a) What is the third daughter in the thorium-232 decay series? (b) Make a chart similar to Fig. 43-19 showing the first three decays in the thorium series.

39. How much cobalt-60 ($t_{1/2} = 5.24$ years) must be used to make a laboratory source whose activity will exceed 1 GBq for a period of 2 years?

40. Archaeologists unearth a bone (Fig. 43-26) and find its carbon-14 content is 34% of that in a living bone. How old is the archaeological find?

FIGURE 43-26 How old is the bone? (Problem 40)

41. Marie Curie and Pierre Curie won the 1903 Nobel Prize for isolating 0.1 g of radium-226 chloride (^{226}RaCl$_2$). What was the activity of their sample, in Bq and in curies?

42. Below are reported levels of iodine-131 contamination in milk for several countries affected by the 1986 Chernobyl accident, along with each country's safety guideline. Given that I-131's half-life is 8.04 days, how long did each country have to wait for I-131 levels to decline to the level it considered safe?

COUNTRY	REPORTED	SAFETY GUIDELINE
Poland	2000	1000
Sweden	2900	2000
Austria	1500	370
Germany	1184	500

LEVEL (BQ/L)

43. A mixture initially contains twice as many atoms of sodium-24 ($t_{1/2} = 15$ h) as it does of potassium-43 ($t_{1/2} = 22.3$ h). Plot the decay curves for the two isotopes, and use your graph to determine when the numbers of atoms are equal.

44. How many atoms are in a radioactive sample with activity 12 Bq and half-life 15 days?

45. Analysis of a moon rock shows that 82% of its initial K-40 has decayed to Ar-40, a process with a half-life of 1.2×10^9 years. How old is the rock?

46. Neutron activation of the stable nitrogen isotope $^{15}_7$N in a bomb-detection system turns it into unstable $^{16}_7$N, which decays by beta emission with a 7.13-s half-life. How long after activation will the N-16 activity have dropped by a factor of one million?

47. An excited state of the isotope technetium-99, designated Tc-99*, has a 6.01-hour half-life and is widely used in nuclear medicine. Tc-99* decays to Tc-99, an unstable isotope with a 2.13×10^5-year half-life. A 0.10-μg sample initially contains pure Tc-99*. Find its activity level at (a) $t = 0$, (b) $t = 1$ day, (c) $t = 1$ week, (d) $t = 1$ year, and (e) $t = 10$ years. The atomic weight of Tc-99 is 98.9 u.

48. In the preceding problem, what fraction of the activity after 1 week is due to Tc-99* decay?

49. Today, uranium-235 comprises only 0.72% of natural uranium; essentially all the rest is U-238. Use the half-lives given in Table 43-2 to determine the percentage of uranium-235 in natural uranium when Earth formed, about 4.5 billion years ago.

50. Figure 43-27 shows the activity as a function of time for a mixture of two radioisotopes. Find (a) the half-life and (b) the initial activity for each. *Hint:* Note that the activity is plotted logarithmically, and that the graph consists approximately of two straight lines connected by a curved stretch.

Section 43-6 Models of Nuclear Structure

51. For each of the magic numbers, give an isotope that has a magic number of protons and one with a magic number of neutrons.

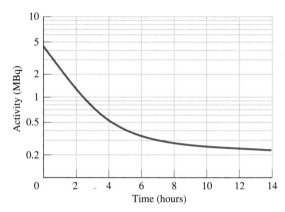

FIGURE 43-27 Problem 50.

52. List all the doubly magic stable isotopes—those with magic numbers of both protons and neutrons.

Paired Problems

(Both problems in a pair involve the same principles and techniques. If you can get the first problem, you should be able to solve the second one.)

53. The nuclear mass of $^{48}_{22}$Ti is 47.9359 u. Find the binding energy per nucleon.

54. Find the atomic mass of iridium-193, for which the binding energy is 7.94 MeV/nucleon.

55. Geologists are looking for underground sites that could store nuclear wastes securely for 250,000 years. What fraction of plutonium-239 initially in such waste would remain after that time? See Table 43-2.

56. One of the longer lived isotopes produced in nuclear explosions is cesium-137, with a 30-year half-life. Atmospheric testing by the major nuclear powers ended in 1963. What fraction of the Cs-137 in the environment at that time remains today?

57. A sample of oxygen-15 ($t_{1/2} = 2.0$ min) is produced in a hospital's cyclotron. What should be the initial activity concentration if it takes 3.5 min to get the O-15 to a patient undergoing a PET scan for which an activity of 0.5 mCi/L is necessary?

58. Technetium-99*, an excited state with a 6.01-hour half-life, is widely used in brain scans. How much Tc-99* must a manufacturer produce if it is to arrive at a hospital 3.5 hours later with 10 mg of Tc-99* remaining?

Supplementary Problems

59. If a solid sphere with the 1.9-cm radius of a ping-pong ball were made entirely of nuclear matter, what would be the acceleration due to gravity at its surface? Compare with that at Earth's surface.

60. Show that the gravitational acceleration at the surface of a sphere of nuclear matter is $g = 6.4 \times 10^7 R$, where g is in m/s^2 and R is the sphere's radius in meters. (General relativity limits this result to cases where the surface escape speed is much less than the speed of light.)

61. How cool would you have to get a material before the thermal energy kT was insufficient to excite protons in a 35-T magnetic field from their lower to upper spin state?

62. Alpha particles emitted by polonium-209 describe a circle of radius r in a magnetic field B, where the product Br has the value 0.332 T·m. What are (a) the speed and (b) the energy of the alpha particles?

63. The atomic masses of uranium-238 and thorium-234 are 238.050784 u and 234.043593 u, respectively. Find the energy released in the alpha decay of ^{238}U.

64. In the alpha decay of the preceding problem, what percentage of the original U-238 mass gets converted to energy?

65. Some human lung cancers in smokers may be caused by polonium-210, which arises in the decay series of uranium-238 that occurs naturally in fertilizers used on tobacco plants. Write equations for (a) the production and (b) the decay of ^{210}Po. (c) How might the health effects of Po-210 differ if its half-life were 1 day or 10 years instead of the actual 138 days?

66. When lithium-7 absorbs a neutron, the resulting nucleus first beta decays by electron emission, then splits into two identical pieces. Identify the pieces.

67. In the preparation of a transuranic nucleus, $^{209}_{83}$Bi and $^{54}_{24}$Cr interact to form a heavy nucleus plus a neutron. Identify the heavy nucleus.

68. It is possible but difficult to realize the alchemists' dream of synthesizing gold. One reaction involves bombarding mercury-198 with neutrons, producing, for each neutron captured, a gold-197 nucleus and another particle. Write the equation for this reaction.

69. Nickel-65 beta decays by electron emission with decay constant $\lambda = 0.275$ h^{-1}. (a) What is the daughter nucleus? (b) In a sample that is initially pure Ni-65, how long will it be before there are twice as many daughter nuclei as parent nuclei?

70. Free neutrons have a half-life of about 10 min. A nuclear fission reaction produces neutrons with kinetic energies around 1 MeV. (a) How far, on average, can such a neutron get from the reaction site before it decays? (b) Repeat for a neutron with thermal energy kT typical of room temperature.

71. The mean lifetime τ of a radioactive nucleus is the average of the times from $t = 0$ to $t = \infty$, weighted by the number dN of nuclei that decay in each time interval dt, that is,

$$\tau = \frac{\displaystyle\int_{t=0}^{\infty} t \, dN}{\displaystyle\int_{t=0}^{\infty} dN}.$$

Evaluate this quantity to show that $\tau = 1/\lambda$. *Hint:* Write dN as $|dN/dt| \, dt$.

CHAPTER 44

NUCLEAR ENERGY: FISSION AND FUSION

Nuclear energy is 10 million times more concentrated than chemical energy. One consequence is the awesome destructive power of nuclear weapons.

Nuclear energy plays a major role in the natural universe, powering the Sun and other stars. On Earth, applications of nuclear energy became a serious possibility in the 1930s, a reality in the 1940s, and a fact of life in the 1950s. Since then, the use of nuclear energy for both peaceful and military purposes has increased substantially. Proliferation of nuclear weapons throughout the world presents a serious threat to human survival. On a more positive note, nuclear fission supplies more than 15% of the world's electrical energy, and nuclear fusion raises the prospect of a cleaner, safer nuclear power source with nearly limitless fuel resources.

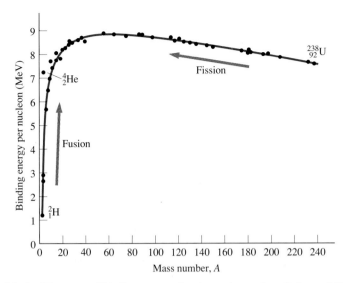

FIGURE 44-1 The curve of binding energy, showing regimes where fusion and fission can result in the release of nuclear energy.

44-1 ENERGY FROM THE NUCLEUS

In the preceding chapter we saw how the curve of binding energy leads to the possibility of nuclear energy release through **fusion**—the joining of lighter nuclei to form heavier ones nearer the peak of the binding energy curve (Fig. 44-1). Fusion in the interiors of long-dead stars created the nuclei of which we're made, while fusion in the Sun provides the energy that sustains life on Earth. Fusion energy released in our thermonuclear weapons, on the other hand, has the potential to destroy that life.

For nuclei heavier than iron, energy release can occur if a nucleus splits, or **fissions,** into two lighter nuclei. In contrast to fusion, nuclear fission seems to be of little consequence in the cosmic scheme of things. A self-sustaining fission reaction did occur naturally in what is now a uranium mine in Africa some three billion years ago, and it has provided useful information on the long-term movement of nuclear waste through the environment. Technologically, though, nuclear fission is important: it is the one major energy-releasing nuclear reaction that we can now sustain in a controlled manner to produce electricity, while uncontrolled fission reactions play an important role in nuclear weapons.

44-2 NUCLEAR FISSION

The 1932 discovery of the neutron by James Chadwick gave physicists an excellent tool for probing the atomic nucleus. Unlike protons or alpha particles, the neutron carries no electric charge and therefore can penetrate the nucleus without having to overcome the coulomb repulsion. Following earlier work by the Italian physicist Enrico Fermi, the German chemists Otto Hahn and Fritz

FIGURE 44-2 Lise Meitner and Otto Hahn. Meitner and her nephew Otto Frisch interpreted Hahn and Strassmann's experiments as evidence of neutron-induced fission of uranium. Meitner, an Austrian physicist, had fled to Sweden to escape Hitler. Earlier, as a woman in a male-dominated field, she had braved sexist policies that denied her access to laboratories when men were present. By the 1930s she had become one of the world's most respected nuclear physicists. Element 109 (meitnerium) now bears her name.

FIGURE 44-3 Painting of the first nuclear reactor, built under the stands of the University of Chicago stadium during World War II. No photographs were taken because of wartime secrecy. Note the man in the pit, manually adjusting a control rod to increase the nuclear reaction rate.

Strassmann in 1938 bombarded uranium (atomic number $Z = 92$) with neutrons. They were puzzled to find among the reaction products radioactive versions of the much lighter elements barium ($Z = 56$) and lanthanum ($Z = 57$). Physicist Lise Meitner (Fig. 44-2) and her nephew Otto Frisch soon interpreted these unusual findings, concluding that neutron bombardment had caused the uranium nuclei to fission into two parts. Word of the discovery spread throughout the world's physics community, and with it the realization that fission represented an energy source many orders of magnitude more potent than chemical reactions. It was the eve of World War II, and the military implications were obvious and ominous. The United States initiated a program to develop fission explosives, hoping to produce nuclear weapons before the Germans did. With the help of the international physics community, many of whom had fled Europe to escape Fascism, the U.S. effort succeeded. In 1945, only seven years after the discovery of fission, the world's first nuclear explosion was detonated in the New Mexico desert. A few weeks later, fission bombs devastated the Japanese cities of Hiroshima and Nagasaki, bringing World War II to an end. In the course of development work on the bomb, scientists led by Enrico Fermi had also constructed the world's first nuclear reactor. Built under the stands of the University of Chicago stadium, the reactor became operational in 1942 (Fig. 44-3).

Nuclear fission occurs when a massive nucleus splits into two lighter parts. Although the process can occur spontaneously in a variety of heavy nuclei, such **spontaneous fission** is extremely rare. More common is **induced fission,** occurring typically when a heavy nucleus absorbs a neutron. (High-energy protons and gamma rays can also induce fission.) In a common fission reaction, a ^{235}U nucleus absorbs a neutron to form a highly excited ^{236}U nucleus. The ^{236}U undergoes vigorous oscillations that deform it into a dumbbell shape (Fig. 44-4). The two ends of the dumbbell are positively charged, and they still experience the long-range electrical repulsion. But the short-range nuclear force is less effective in holding the distant ends together. If the oscillation amplitude is great enough, the nucleus reaches a point of no return at which the coulomb repulsion dominates and the two ends of the dumbbell fly apart. This fissioning occurs very rapidly; the intermediate ^{236}U nucleus lasts only about 10^{-12} s.

The two lighter nuclei formed in the fission process are **fission fragments.** The composition of the fission fragments varies, depending on exactly how a

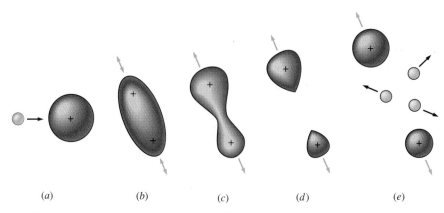

(a) (b) (c) (d) (e)

FIGURE 44-4 Neutron-induced fission of ^{235}U.

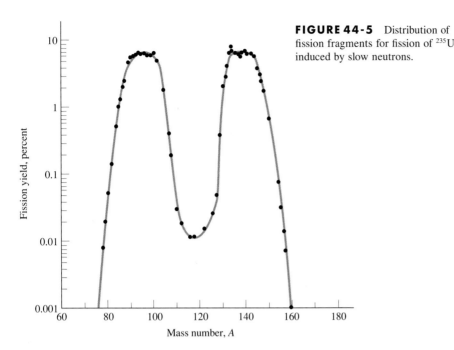

FIGURE 44-5 Distribution of fission fragments for fission of ^{235}U induced by slow neutrons.

given nucleus splits. The most likely outcome is a pair of unequal-mass fragments, with mass numbers in the ranges 90 to 100 and 135 to 145 more probable; other distributions occur with lower probability (Fig. 44-5). As we saw in the preceding chapter, the ratio of neutrons to protons for stable nuclei increases with increasing mass number. Since the fission fragments share the protons and neutrons of the parent uranium, these lighter nuclei have an excess of neutrons for their mass, and are therefore highly unstable (Fig. 44-6). This instability manifests itself in an almost immediate "boiling off" of neutrons from the fission fragments. Typically two to three neutrons are emitted for each fission event; the average number is about 2.47 neutrons per ^{235}U fission. The fragments remaining are still highly unstable and therefore very radioactive; in particular, their radioactivity greatly exceeds that of the original uranium prior to fissioning.

To summarize, induced fission of ^{235}U involves these steps:

1. A neutron strikes a ^{235}U nucleus and is absorbed, forming highly excited ^{236}U.
2. The ^{236}U undergoes violent oscillations that distort it into a dumbbell shape. In the dumbbell configuration, the electrical repulsion overcomes the strong nuclear force, and in about 10^{-12} s the ^{236}U nucleus fissions into two unequal-mass fragments.
3. The fission fragments almost immediately "boil off" several excess neutrons, leaving two nuclei that are still highly radioactive.

We can describe the fission event using the standard notation for nuclear reactions. In writing such a reaction, we generally skip the short-lived intermediate stages of ^{236}U and the fission fragments before neutron emission; the general neutron-induced fission of ^{235}U can then be written

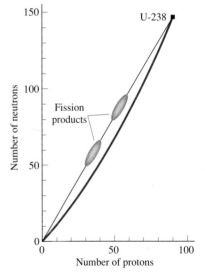

FIGURE 44-6 Chart of the stable nuclei, simplified from Fig. 43-4. Thick curve is the region of stability, extended to include the long-lived unstable isotopes through uranium. Thin straight line has the neutron-to-proton ratio of uranium and thus of fission products. The latter lie well above the stable region, and are therefore radioactive.

$$\, _0^1 n \, + \, _{92}^{235}U \, \longrightarrow \, X + Y + b \, _0^1 n, \tag{44-1}$$

where X and Y stand for the fission fragments and b is the number of neutrons emitted. The reaction conserves electric charge and the number of nucleons, so the atomic numbers Z that subscript the element symbols, and separately the mass number superscripts, must sum to the same values on both sides of the reaction. A specific example is a ^{235}U fission that produces isotopes of molybdenum (Mo) and tin (Sn):

$$\, _0^1 n \, + \, _{92}^{235}U \, \longrightarrow \, _{42}^{102}Mo + _{50}^{131}Sn + 3 \, _0^1 n$$

(This particular reaction is significant because ^{131}Sn quickly undergoes a series of beta decays leading to ^{131}I, an especially dangerous environmental contaminant that lodges in the thyroid gland; see Example 43-6.)

● EXAMPLE 44-1 A FISSION REACTION

Neutron-induced fission of ^{235}U results in the formation of ^{141}Ba, three neutrons, and a second fission fragment. What is that fragment?

Solution

From Appendix D we find that barium, Ba, has atomic number $Z = 56$. So the reaction is

$$\, _0^1 n \, + \, _{92}^{235}U \, \longrightarrow \, _{56}^{141}Ba + X + 3 \, _0^1 n,$$

where X is the unknown fragment. The atomic numbers on the left sum to 92, reflecting the presence of 92 protons. So there must be 92 protons on the right; with $Z = 56$ for barium, element X must have $Z = 92 - 56 = 36$. Consulting Appendix D, we see that element 36 is krypton, Kr.

On the left side of the equation, the neutron contributes one unit to the total mass number and the uranium contributes 235, reflecting a total of 236 nucleons participating in the reaction. With mass number $A = 141$ for the barium fragment, and with the three neutrons each contributing one mass unit, the mass number of the krypton fragment must be $236 - (141 + 3) = 92$. So the second fragment is krypton-92, or $_{36}^{92}Kr$.

EXERCISE Neutron-induced fission of ^{235}U produces rubidium-93 and two neutrons. What is the other fission product?

Answer: cesium-141

Some problems similar to Example 44-1: 4–6 ●

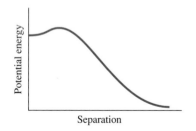

FIGURE 44-7 Potential-energy curve for ^{235}U fission. Horizontal axis is the separation of the two parts that become the fission fragments, with zero representing the original spherical nucleus. Note the barrier, associated with unbalanced forces at the nuclear surface, that must be overcome.

Something besides fission fragments and neutrons appears in the fission. That something is energy—a substantial amount of energy. Energy is released because the total binding energy of the original nucleus is greater than the sum of the binding energies of the widely separated fission fragments, as suggested in Fig. 44-7. A slight barrier, associated with unbalanced forces on the nucleons at the edge of the nucleus, keeps the unexcited uranium nucleus from fissioning. When a neutron is absorbed, the binding energy of the neutron appears as vibrational energy of the nucleus, and the fission process shown in Fig. 44-4 ensues. As the fission fragments fly apart, they gain energy from the electrical repulsion; in terms of Fig. 44-7, the configuration slides down the potential slope to the right of the barrier, converting potential to kinetic energy.

A total energy of about 200 MeV is released in a typical uranium fission event. Most of this—about 165 MeV—appears as kinetic energy of the fission fragments. Neutron kinetic energy accounts for about 5 MeV, while approximately 7 MeV are released almost instantaneously as gamma rays emitted by the excited fission fragments. After a slight delay, the fission fragments emit somewhat over 20 MeV additional energy in the form of high-speed beta particles (electrons or positrons), gamma rays, and neutrinos. In a mass of fissioning material, all but the gamma rays quickly share their energy through collisions, heating the material.

● **E X A M P L E 4 4 - 2** FISSION ENERGY

Assuming 200 MeV per fission event, how much pure ^{235}U should be fissioned to provide the same energy as the burning of 1 metric ton (1000 kg) of coal?

Solution

Consulting the "Energy Content of Fuels" table in Appendix C, we find that coal burning releases 2.9×10^7 J/kg, so 1000 kg of coal releases an energy $E = 2.9 \times 10^{10}$ J. With 1.6×10^{-19} J/eV, uranium fission releases about $(200 \times 10^6$ eV/fission$) \times (1.6 \times 10^{-19}$ J/eV$) = 3.20 \times 10^{-11}$ J/fission.

So we need a total of

$$\frac{2.9 \times 10^{10} \text{ J}}{3.2 \times 10^{-11} \text{ J/fission}} = 9.06 \times 10^{20} \text{ fission events}.$$

Each fission event involves one ^{235}U nucleus, so we need this many nuclei. The mass of a ^{235}U nucleus is approximately 235 u, so the mass of ^{235}U required is

$(9.06 \times 10^{20}$ nuclei$)(235$ u$)(1.66 \times 10^{-27}$ kg/u$) = 3.5 \times 10^{-4}$ kg.

Thus, 0.35 g of pure ^{235}U—about one one-hundredth of an ounce—can supply the same energy as a ton of coal! The ratio of coal to uranium mass is about 3×10^6, showing that uranium packs over a million times as much energy per unit mass as does coal. This ratio is typical of the difference between nuclear and chemical fuels, and shows why a nuclear power plant may be refueled once a year, while a comparable coal-burning plant needs many hundred-car trainloads of coal each week.

E X E R C I S E Estimate the explosive energy yield in joules and in megatons (see Appendix C) for a fission bomb made from 18 kg of U-235, under the unrealistic assumption that all the U-235 fissions.

Answer: 1.48×10^{15} J $= 0.35$ Mt

Some problems similar to Example 44-2: 7, 8, 41, 42 ●

Besides ^{235}U, a number of other nuclides can undergo neutron-induced fission; these are said to be **fissionable.** Fissionable nuclides that will fission with neutrons of any energy—especially relatively low thermal energy—are called **fissile.** In a fissile nucleus, the barrier in Fig. 44-7 is very low and the required neutron energy is therefore negligible. Only three fissile nuclides are known: they include the uranium isotopes ^{233}U, ^{235}U, and the plutonium isotope ^{239}Pu. By far the most important of these are ^{235}U and ^{239}Pu.

Although ^{235}U occurs naturally, it presently constitutes only about 0.72% of natural uranium (nearly all the rest—99.27%—is ^{238}U). For most uses, uranium must be enriched in ^{235}U, to several per cent for commercial power reactors and often over 90% for weapons. **Uranium enrichment** is a difficult and expensive process; since the isotopes ^{235}U and ^{238}U are chemically similar, enrichment schemes make use of the very slight mass difference between the two. Among the techniques used are centrifuging, gaseous diffusion, and selective ionization of ^{235}U by lasers (Fig. 44-8). Enrichment technology is highly sensitive because a nation possessing it can readily produce weapons-grade uranium.

Plutonium-239, with a half-life of 24,110 years, does not occur in nature. It is produced artificially by neutron bombardment of ^{238}U. This reaction first produces the highly unstable isotope ^{239}U, which undergoes beta decay with a half-life of 23.5 minutes to form ^{239}Np. The ^{239}Np again decays by beta emission, with a half-life of 2.35 days, leaving ^{239}Pu. This sequence of reactions leading to plutonium can be written

FIGURE 44-8 Uranium-enrichment technologies. (*a*) In gaseous diffusion, uranium hexafluoride gas (UF_6) passes through a membrane; the lighter, faster moving $^{235}UF_6$ (light color) is more likely to get through. After many such cells the desired concentration is reached. (*b*) United Nations worker destroying an Iraqi centrifuge following the 1991 Persian Gulf war. The high-speed centrifuge had been built to separate uranium isotopes as part of Iraq's well-advanced nuclear weapons program.

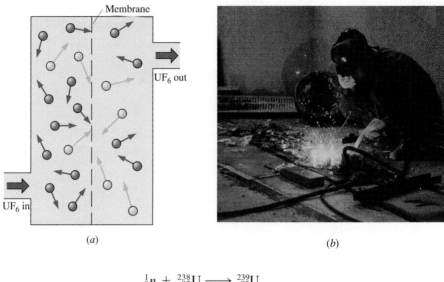

(*a*) (*b*)

$$^1_0 n + {}^{238}_{92}U \longrightarrow {}^{239}_{92}U$$

$$^{239}_{92}U \longrightarrow {}^{239}_{93}Np + e^- + \overline{\nu}$$

$$^{239}_{93}Np \longrightarrow {}^{239}_{94}Pu + e^- + \overline{\nu}.$$

Although ^{239}Pu is produced in copious amounts in nuclear reactors (see Problem 55), **reprocessing** spent reactor fuel to extract plutonium is difficult and dangerous. Contamination with other plutonium isotopes further complicates the separation of fissile plutonium. Like uranium enrichment, plutonium separation is a sensitive technology; nevertheless, all the nations except China that are known to have developed nuclear weapons chose plutonium over uranium for their first nuclear explosions. And by the early 1990s, several European countries and Japan had embarked on ambitious plutonium reprocessing programs for commercial power reactors, including intercontinental shipments of plutonium.

A number of other isotopes, most importantly ^{238}U, are fissionable with fast neutrons that bring in enough kinetic energy to overcome the potential barrier. But ^{238}U fission does not result in significant neutron emission, so, for reasons we will discuss in the next section, a self-sustaining ^{238}U fission reaction is not possible. However, fast-neutron fission of ^{238}U plays a significant role in thermonuclear weapons, as we will see later.

Chain Reactions

Neutrons induce fission, and fission itself may release neutrons—the very things needed to induce fission. This fact makes possible a **chain reaction,** in which each fission event supplies neutrons that give rise to more such events.

For a sustained chain reaction, neutrons from each fission event must, on the average, cause at least one more nucleus to fission. Otherwise the reaction will fizzle to a halt, in which case the configuration of fissile material is called a **subcritical mass.** If each fission event causes, on the average, exactly one additional fission event, then the reaction continues with a constant rate of energy release; the fissile material then comprises a **critical mass.** Recall that the average number of neutrons released in ^{235}U fission is 2.47; thus in a critical

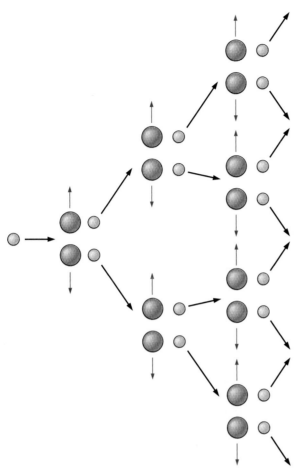

FIGURE 44-9 A supercritical chain reaction. In this case the multiplication factor $k = 2$, since two neutrons from each fission cause additional fission. In reality some neutrons are lost, and some fission events produce more than two neutrons.

reaction most of the neutrons do not cause additional fission. If, on the other hand, an average of more than one neutron from each fission causes additional fission, then the reaction rate grows exponentially with time (Fig. 44-9). In this case the configuration is **supercritical.**

Quantitatively, the criticality of a fissile mass is described by the **multiplication factor** k, which is just the average number of neutrons from a fission event that cause additional fission. A subcritical mass has $k < 1$, a critical mass has $k = 1$, and a supercritical mass has $k > 1$. The value of k is determined by several factors, including the proportion of fissile isotope, the concentration of neutron-absorbing substances, and the size and configuration of the mass. With $k = 2$, for example, a single fission event results in two more fissions; they cause a total of four more fissions, then eight, and so on. In general, the number of fissions increases by a factor of k with each successive generation of fission events; therefore, the number of fission events occurring after n such generations is k^n. The *total* number, N, of fission events that occur by the nth generation is the sum $N = 1 + k + k^2 + k^3 + \ldots + k^n$. You may recognize this sum as a geometric series; in any event, you can show by mathematical induction that it has the value

$$N = \frac{k^{n+1} - 1}{k - 1}. \tag{44-2}$$

The average time between successive generations of fission events is called the **generation time.** As the example below indicates, short generation times can lead to explosive release of nuclear energy.

● **EXAMPLE 44-3** EXPLOSION

A 14 kg sample of pure ^{235}U is assembled into a supercritical mass with multiplication factor $k = 1.8$. If the generation time between fission events is $\tau = 10$ ns, how long does it take the entire mass to fission? How much energy is released in the process?

Solution
With an approximate atomic weight of 235 u, our 14-kg mass contains a number N of nuclei given by

$$N = \frac{14 \text{ kg}}{(235 \text{ u/nucleus})(1.66 \times 10^{-27} \text{ kg/u})} = 3.59 \times 10^{25} \text{ nuclei}.$$

We need to find the number of fission generations that must occur before this number of nuclei have fissioned. Solving equation 44-2 for k^{n+1} gives

$$k^{n+1} = N(k - 1) + 1.$$

With $N \simeq 10^{25}$, the $+1$ on the right-hand side is totally negligible. Dropping it, and taking logarithms of both sides, we have

$$\ln(k^{n+1}) = \ln[N(k - 1)].$$

But $\ln(x^y) = y \ln x$, so

$$(n + 1) \ln k = \ln[N(k - 1)].$$

Solving for n then gives

$$n = \frac{\ln[N(k - 1)]}{\ln k} - 1 = \frac{\ln[(3.25 \times 10^{25})(1.8 - 1)]}{\ln(1.8)} - 1$$

$$= 99,$$

where we have rounded to the nearest whole number. Thus, the time for the entire mass to fission is 99τ—that is, 990 ns or just under one microsecond.

We can get the energy released from the "Energy Content of Fuels" table in Appendix C or by using the approximate figure of 200 MeV per fission event. Choosing the latter approach, we have

$$E = (200 \times 10^6 \text{ eV/fission})(1.6 \times 10^{-19} \text{ J/eV})$$

$$\times (3.59 \times 10^{25} \text{ fissions})$$

$$= 1.1 \times 10^{15} \text{ J}.$$

To get a feel for this number, note that it is about 3.2×10^8 kWh, or nearly 2 weeks' energy output of a large electric power plant, released instead in a microsecond! Or, comparing our 14-kg supercritical mass with the 0.35 g of ^{235}U that we found in Example 44-2 to be equivalent to a ton of coal, we find that the energy released in the 1-μs supercritical chain reaction is equivalent to burning $(14 \text{ kg})/(0.35 \times 10^{-3} \text{ kg } ^{235}U/\text{ton coal}) = 40,000$ tons of coal in one microsecond. The reaction is truly explosive, and our supercritical mass is, of course, a nuclear bomb.

EXERCISE A terrorist group designs a crude nuclear weapon made from 25 kg of ^{235}U with generation time $\tau = 2.0$ μs and multiplication factor $k = 1.08$. How long would it take this weapon to consume its fuel, assuming—as is unlikely—that it could remain together during the process.

Answer: 1.5 ms

Some problems similar to Example 44-3: 14–16, 43, 44, 48 ●

44-3 APPLICATIONS OF NUCLEAR FISSION

Rarely has a basic discovery led as quickly as nuclear fission to significant technological applications. Following the 1938 discovery of fission, the pressures of wartime brought the first nuclear reactor to criticality in 1942 and the first nuclear weapon to detonation in 1945. Since then, nuclear fission technology has built on these two very different devices. Today, nuclear reactors are used for power generation, propulsion of ships, research, and the production of plutonium for weapons. Fission weapons, too, have developed, with fission now complementing fusion in the most destructive weapons. Some milestones in fission technology include the first full-scale commercial nuclear power plant at Calder Hall in England going on line in 1956 and the launching of the first nuclear-powered submarine, the *U.S. Nautilus,* also in 1956. Other nuclear projects have fared less well; early enthusiasm for nuclear-powered rockets and aircraft was quickly dampened, although proposals for flights to Mars have revived interest in nuclear rockets.

Fission Weapons

As Example 44-3 shows, assembling a critical mass of fissile material can result in a colossal explosion. And the mass required is not large—about 15 kg for uranium-235 and less than 5 kg for plutonium-239. Weapons based on these simple facts of nuclear fission have been a frightening reality for half a century. Here and again later in this chapter we discuss the physics of nuclear weapons not out of any enthusiasm for these horrific devices, but because, for better or worse, they play a central and potentially devastating role in the modern world. The threat of war between the nuclear superpowers has waned with the end of the Cold War and the nuclear arms reduction treaties of the 1990s, but the danger that nuclear weapons will proliferate throughout the world community continues to grow. That danger arises directly from physics of the fission chain reaction and from the small quantity of fissile material needed for a critical mass.

The major technological difficulty in actually producing a fission weapon is to assemble a supercritical mass rapidly enough that the explosive chain reaction consumes most of the fissile material before it blows apart. The occurrence of spontaneous fission means that a supercritical mass, once assembled, will almost instantly explode. Fission weapons therefore start with fissile material in a subcritical configuration and, at detonation time, rapidly reconfigure the material into a supercritical arrangement. The simplest fission weapon is the gun-type device shown in Fig. 44-10. The weapon that destroyed Hiroshima was a gun-type uranium bomb whose critical mass was about the size of a large grapefruit (Fig. 44-11). The explosive yield was equivalent to about 12,500 tons (12.5 kt) of the chemical explosive trinitrotoluene (TNT). [Explosive yields of nuclear weapons are commonly expressed by giving the weight in tons of TNT that would release the same explosive energy. One kiloton (kt) is equivalent to 4.18×10^{12} J; for larger weapons the megaton (MT), equivalent to 4.18×10^{15} J, is used.] Despite its awesome destructive power, the Hiroshima weapon was quite inefficient. As the following example shows, fission actually occurred in less than 1 kg of its approximately 50 kg of enriched uranium.

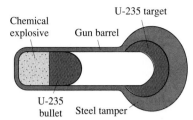

FIGURE 44-10 A gun-type fission weapon. The U-235 "bullet" and target are both subcritical masses. Igniting the chemical explosive propels the bullet into the target, forming a critical mass. The heavy steel tamper holds the fissioning mass together briefly, ensuring more complete fission.

FIGURE 44-11 The gun-type nuclear bomb that destroyed Hiroshima. The device had a mass of just over 3000 kg, of which only about 50 kg was uranium. Of that, less than 1 kg was consumed in the fission reaction—yet it yielded the explosive equivalent of 12,500 tons of chemical explosive.

● EXAMPLE 44-4   DESTRUCTION OF HIROSHIMA

What mass of ^{235}U actually fissioned to produce the 12.5-kt explosive yield of the Hiroshima bomb?

Solution

Converting kilotons to joules and then to MeV, we have for the explosive energy

$$E = \frac{(12.5 \text{ kt})(4.18 \times 10^{12} \text{ J/kt})}{(1.6 \times 10^{-19} \text{ J/eV})(10^6 \text{ eV/MeV})} = 3.27 \times 10^{26} \text{ MeV}.$$

At 200 MeV per fission event, this energy requires

$$\frac{3.27 \times 10^{26} \text{ MeV}}{200 \text{ MeV/fission}} = 1.63 \times 10^{24} \text{ fissions}.$$

Each fission event involves one ^{235}U nucleus, with an approximate mass of 235 u; therefore, the total mass M of ^{235}U undergoing fission is

$$M = (1.63 \times 10^{24} \text{ nuclei})(235 \text{ u/nucleus})(1.66 \times 10^{-27} \text{ kg/u})$$

$$= 0.64 \text{ kg}.$$

This is only about 1% of the total uranium mass in the bomb. Modern fission explosives are far more efficient, consuming about 30% of their fissile fuel.

EXERCISE   How much uranium-235 would be needed to produce 150-kt fission weapon with 30% efficiency?

Answer: 25 kg

Some problems similar to Example 44-4:   9–13   ●

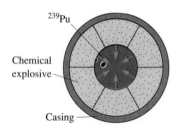

FIGURE 44-12   An implosion-type fission weapon. Chemical explosives compress a barely subcritical sphere of ^{239}Pu to critical density, initiating an explosive chain reaction.

A more sophisticated design is required for a plutonium fission weapon, where neutrons from spontaneous fission make it more likely that the weapon will "preignite" and blow apart before complete fission takes place. Consequently plutonium weapons are generally of the implosion type shown in Fig. 44-12. A barely subcritical sphere of plutonium is surrounded symmetrically by chemical explosives. When the explosives detonate, they compress the sphere to a supercritical density, and the nuclear explosion follows. The world's first nuclear explosive, detonated at the Trinity Site in New Mexico, was a 20-kt implosion-type plutonium device, as was the bomb dropped on Nagasaki.

Practical weapons have some efficiency-boosting features not shown in our simple diagrams. For example, neutron-emitting material is often embedded in the fissile mass to ensure more complete fission. For the same reason, neutron-reflecting material surrounds the uranium or plutonium to reduce neutron losses and therefore increase the multiplication factor k.

Although the construction of a fission weapon is straightforward, acquisition of weapons-grade fissile material is not. As we noted earlier, enrichment of uranium with the rare isotope ^{235}U is difficult and costly. Removal and purification of ^{239}Pu from spent reactor fuel poses similar problems. For these reasons, efforts to halt the spread of nuclear weapons have concentrated on keeping enriched uranium and plutonium under strict control, and on curbing the spread of enrichment and reprocessing technologies. Nevertheless, the spread continues. Six nations have openly tested nuclear explosives, and several more almost certainly have them. Investigation of Iraq's nuclear-weapons program following the Persian Gulf War shows that a nation seeking nuclear weapons capability can approach that goal undetected. The breakup of the Soviet Union in 1991 poses a further proliferation peril, as fissile materials and perhaps even weapons themselves become items of illicit international commerce. These developments, coupled with the immense destructive power in the hands of any nation or terrorist group possessing nuclear weapons, makes our nuclear age a dangerous time.

Nuclear Reactors

In a **nuclear reactor,** the fission chain reaction is controlled rather than explosive; that is, the reaction is held just critical, with $k = 1$ and with a constant rate of energy release. Since the maximum possible k is 2.47 (the average number of neutrons per fission event in ^{235}U), a reactor must be designed so most neutrons do not induce additional fission events; otherwise it would quickly run out of control. In commercial power reactors k is limited in part by keeping the concentration of fissile ^{235}U low—typically a few per cent—so that many neutrons are absorbed by ^{238}U instead of causing fission. Since uranium enrichment is expensive, this is also an economical approach, and in addition ensures that reactor fuel cannot readily be converted to weapons use.

Figure 44-13 shows that the cross section for neutron-induced fission is much higher for slow than for fast neutrons. (The cross section is the effective area presented by the nucleus for a given reaction, and may differ enormously from its geometric cross section.) In typical reactor fuel, the concentration of ^{235}U is low enough that fast neutrons alone could not sustain a chain reaction. Consequently most reactors include a substance called the **moderator** that slows neutrons to thermal speeds. The slow neutrons are also called **thermal neutrons,** and a reactor using thermal neutron fission is a **thermal reactor.** In a thermal reactor, neutrons collide elastically with moderator nuclei, transferring energy to them. In Chapter 11 we found that the maximum energy transfer occurs when colliding objects have equal mass; therefore, the best moderators are low-mass nuclei (see Problem 18). Water, with the protons of its hydrogen atoms serving as moderator nuclei, is the most widely used moderator in U.S. power reactors. Its main disadvantage is that it can absorb neutrons, forming the hydrogen isotope deuterium (^2H); the absorbed neutrons are then lost to the chain reaction. Canadian reactors use heavy water (deuterium oxide, ^2H_2O) as the moderator. The cross section for neutron absorption by deuterium is much lower than for ordinary hydrogen, so fewer neutrons are lost. With more neutrons available, Canadian reactors can use natural, unenriched uranium as fuel.

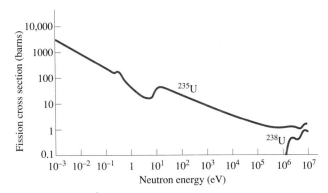

FIGURE 44-13 Cross section for neutron-induced fission of ^{235}U, as a function of neutron energy. Neutrons emitted in fission have energies on the order of 10^6 eV. The much higher cross section for thermal neutrons ($\sim 10^{-2}$ eV) shows why a moderator is important in a nuclear reactor. Curve has been smoothed to eliminate rapid fluctuations in the 1 to 1000-eV range. Also plotted is the cross section for ^{238}U fission, showing that only fast neutrons can induce fission in this isotope. Cross sections—effective areas presented by the nuclei for neutron absoption—are in barns; 1 barn = 10^{-28} m^2.

Finally, some reactors, including a few in the United States, some in Great Britain, and a substantial number in the former Soviet Union, use graphite as the moderator. Note that the moderator is *not* in place to keep the reaction under control. Quite the opposite: by slowing neutrons, the moderator greatly increases the fission rate. If the moderator were removed, the reactor would go subcritical and the chain reaction would stop.

Control of the reaction is achieved by positioning **control rods** amid the fuel. These are made of neutron-absorbing materials like cadmium or boron, and can be moved in and out of the fuel mass to alter the rate of neutron absorption and therefore adjust the multiplication factor k. By changing the number of rods inserted into different sections of the reactor core, the reaction can be brought to criticality at different power levels.

Control by mechanical movement of neutron-absorbing rods is possible only because a small fraction—about 0.65%—of the neutrons from a ^{235}U fission event are emitted not immediately but in the decay of fission fragments with half-lives from about 0.2 s up to about 1 min. In a reactor that is very nearly critical (k very close to 1), absorbing some of these **delayed neutrons** is enough to keep the chain reaction under control. In this condition, the generation time is greatly influenced by the delayed neutrons, and turns out to be about 0.1 s for a typical reactor. As the example below shows, this long generation time precludes rapid changes in reactor power, so the mechanical movement of control rods can occur fast enough to counter such changes. But if conditions changed to the extent that **prompt neutrons**—those emitted with essentially no delay—were alone sufficient to maintain the chain reaction, then the generation time in a typical reactor would be about 10^{-4} s, and it would be impossible for the relatively slow mechanical movement of a control rod to respond quickly enough to a change in neutron production. In the event of a neutron increase, the reactor could go supercritical and out of control before the control rods had moved significantly. This dangerous situation, known as **prompt criticality,** played an important role in the 1986 Chernobyl reactor accident; we will soon describe this accident in detail.

● **EXAMPLE 44-5** DELAYED NEUTRONS AND REACTOR CONTROL

Suppose a small change in operating conditions makes a reactor very slightly supercritical, with $k = 1.001$. Determine the time it would take the reactor power to double (a) if delayed neutrons establish a generation time $\tau = 0.1$ s, and (b) if prompt neutrons alone sustain the reaction, giving $\tau = 10^{-4}$ s.

Solution
Since the number of fission events increases by a factor $k = 1.001$ in each generation, so does the reactor power. The power will therefore have doubled after n generations such that $k^n = 2$. Taking logarithms of both sides of this equation, we have

$$\ln(k^n) = \ln 2.$$

But $\ln(k^n) = n \ln k$, so $n = \dfrac{\ln 2}{\ln k} = \dfrac{\ln 2}{\ln 1.001} = 693$.

With a generation time of 0.1 s, it therefore takes more than a minute for the power to double. This is ample time to sense the power increase and move control rods to absorb excess neutrons and reduce k to 1, thereby maintaining control. But with $\tau = 10^{-4}$ s, the doubling time is $(693)(10^{-4}$ s$) \simeq 0.07$ s, far too short for mechanical control. This example shows why reactor engineering and operation strive to avoid prompt criticality at all cost.

EXERCISE During the Chernobyl accident, the fission rate in the reactor increased by a factor of 4000 in 5 s. Assuming a prompt-neutron generation time of 10^{-4} s, what was the multiplication factor k responsible for this rapid increase?

Answer: 1.000166

Some problems similar to Example 44-5: 23, 24, 26–28

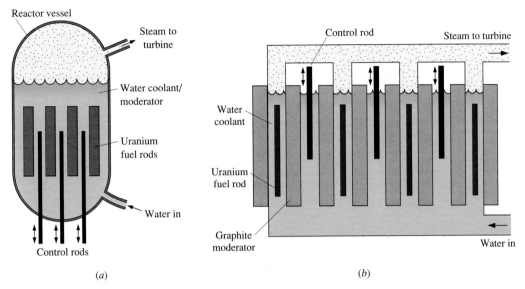

FIGURE 44-14 Reactor core designs. (a) A boiling-water reactor, one of two types commonly used in the United States, uses water for both coolant and moderator. The water boils directly in a large pressure vessel. (b) The RBMK design, used at Chernobyl and elsewhere in the former Soviet bloc, has a graphite moderator interspersed with many pipes carrying the water coolant.

In a reactor producing significant power, a third component is needed: a **coolant,** to carry off the heat generated by fission. In a power reactor, that heat is the reactor's useful product. Loss of coolant can be disastrous, as we will see later in discussing reactor accidents. Water is the most commonly used coolant, although some reactors are gas-cooled and still others use liquid sodium metal. In most water-cooled reactors, the same water serves a dual purpose, acting as both coolant and moderator. This is a safety feature, since loss of coolant also means loss of moderator, and the reaction comes to a halt.

Reactor Types

The combination of fuel, moderator, and control rods forms the **reactor core;** coolant flows through the core to remove the heat generated by fission. Reactor details vary greatly. Most U.S. power reactors are light-water reactors, using ordinary water as coolant and moderator. The core is encased in a thick steel pressure vessel, and water flows freely among the fuel rods (Fig. 44-14a). In contrast, the RBMK design, common in the former Soviet bloc, uses a solid graphite moderator; its water coolant is confined in pipes passing among the moderator blocks (Fig. 44-14b). The Canadian CANDU reactor design uses two separate loops of deuterium oxide (heavy water) as moderator and coolant. The high-temperature gas-cooled reactor (HTGR), used in England and elsewhere, has a graphite moderator and gaseous carbon dioxide or helium coolant. There is no clearly superior reactor type. Each has its own economic and safety advantages and disadvantages. U.S. light-water reactors, for example, must be shut down completely for refueling, a process that takes weeks to months. CANDU and RBMK reactors can be refueled in operation, using robot machinery. The CANDU design seems particularly safe from a proliferation standpoint

FIGURE 44-15 A pressurized-water reactor, the most common type of commercial power reactor in the United States.

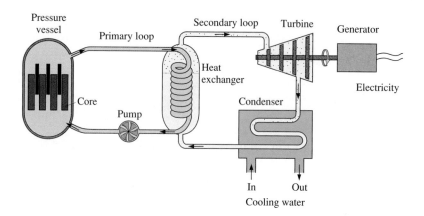

since no uranium enrichment is needed. (However, plutonium produced in Canadian-style heavy-water reactors may have been used in India's successful development of a nuclear explosive.)

Whatever the reactor details, the ultimate purpose of a nuclear power plant is to produce steam that drives a turbine that, in turn, is connected to a generator. Electromagnetic induction in the spinning generator drives electric current that is the power plant's useful product. The second law of thermodynamics (see Chapter 22, especially Fig. 22-12) limits the efficiency with which heat energy can be converted to mechanical and electrical energy; as a result, about two-thirds of the energy generated in the core of a typical power reactor is dumped to the environment as waste heat. This heat is extracted in the power plant's condenser, where spent steam from the turbine is condensed to water.

In the simplest water-cooled reactor designs, the core cooling water is itself boiled to produce steam. A U.S. light-water reactor of this type is called a **boiling-water reactor.** Although simple and economical, the BWR has the disadvantage that water circulating through the turbine and condenser is highly radioactive. A more common U.S. design is the **pressurized-water reactor** (PWR), in which the core cooling water is kept under pressure to prevent boiling. Heat is transferred from this primary coolant to a secondary loop where water boils and drives the turbine. Figure 44-15 is a diagram of a typical PWR. In CANDU and gas-cooled reactors, heat from the primary coolant is also transferred to a secondary boiling-water loop.

As fission proceeds in a reactor core, the concentration of fission fragments in the fuel increases. Before the ^{235}U is exhausted, these fission products absorb enough neutrons to interfere with the chain reaction. Fuel rods must therefore be replaced at regular intervals. In U.S. light-water reactors, about one-third of the fuel is replaced each year, so a given fuel rod remains in place for 3 years (Fig. 44-16). In Canadian CANDU and Russian RBMK reactors, coolant circulates through pipes rather than in a pressure vessel surrounding the entire reactor. In these reactors, fuel rods can be reached while the reactor is in operation, and refueling takes place on a nearly continuous basis. Another reaction that occurs in the nuclear fuel is the conversion of ^{238}U to plutonium, via the neutron-capture reaction we discussed earlier. As plutonium builds up, it begins to fission in significant quantities. Near the end of a fuel rod's 3-year residency in the reactor core, in fact, only 30% of the energy production is from ^{235}U fission. Most of the rest—54%—is from fission of ^{239}Pu, with the remain-

FIGURE 44-16 New fuel rod bundles being lowered into the core of a nuclear reactor. The blue glow is from beta radiation—high-energy electrons—interacting with the reactor's cooling water.

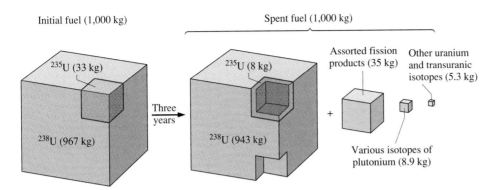

FIGURE 44-17 Evolution of 1000 kg of 3.3% enriched uranium over its 3-year stay in a reactor core. Of the 33 kg of ^{235}U initially present, 25 kg are consumed in fission. In addition, 24 kg of ^{238}U are converted to plutonium and other transuranic isotopes, some of which also undergo fission. The fuel mass remains largely ^{238}U, contaminated with 35 kg of radioactive fission products and lesser amounts of transuranic elements formed by neutron capture.

der from other plutonium isotopes. Figure 44-17 outlines the evolution of fuel in the reactor.

As Fig. 44-17 indicates, spent reactor fuel is contaminated with highly radioactive fission products. It also contains significant amounts of fissile ^{235}U and ^{239}Pu. Through the expensive, technologically complex, and dangerous technique of nuclear fuel reprocessing, these fissile isotopes can be extracted and used to make new reactor fuel or nuclear weapons. The high cost and threat of weapons proliferation in a plutonium-reprocessing economy have led some nations—notably the United States—to forgo reprocessing. Others have eagerly developed reprocessing technology, and international shipments of plutonium began in 1992. Whether or not nuclear waste is reprocessed, the spent fuel ultimately requires disposal. The presence of long-lived radioactive isotopes means that the material remains dangerous for thousands of years. To date, no entirely satisfactory method of disposal has been developed, although it is generally agreed that underground storage will prove the safest alternative (Fig. 44-18).

The thermal reactors we have been discussing use uranium fuel enriched at most slightly in ^{235}U. Although some of the ^{238}U comprising the bulk of the fuel is converted to plutonium, most remains energetically useless. In contrast, a **breeder reactor** is designed to convert large amounts of ^{238}U to plutonium, "breeding" more fissile fuel. Use of breeder reactors would greatly extend our supplies of fissile fuels, by making much of the 99.27% of natural uranium that is nonfissile ^{238}U into fissile ^{239}Pu. For this reason, breeder technology has been pursued in a number of countries, particularly in France and Japan. Breeders operate at higher temperatures and use fast instead of slow neutrons to induce fission. They have no moderator and their coolant is liquid sodium. High temperature and the use of fast neutrons make these reactors less stable and technologically more sensitive than nonbreeding designs (see Problem 23). Furthermore, a breeder-based power system inherently involves reprocessing and the movement of large amounts of plutonium—of which only a few kg suffice to make a fission weapon. Because of these technological, safety, and proliferation concerns, the future of breeder reactors is unclear.

FIGURE 44-18 Containers of radioactive waste being stored at an underground facility in France, which gets 70% of its electricity from nuclear power. The United States, which has just over 20% nuclear-generated electricity, has yet to decide on a long-term disposal site for spent fuel waste.

■ **APPLICATION** THE CHERNOBYL ACCIDENT

In April 1986, the Chernobyl nuclear power plant in the then-Soviet Ukraine exploded, spewing more than 10^{16} Bq of radioactive material into the atmosphere (Fig. 44-19). Two people were killed in the explosion, and hundreds more died of radiation sickness. Fallout affected much of eastern Europe and Scandinavia, and tens of thousands of people are expected to develop fatal cancers as a result. Details of the Chernobyl accident illustrate many aspects of basic nuclear physics and of reactor engineering and control.

Ironically, the accident occurred during a test of the power supply for the **emergency core-cooling system** (ECCS), designed to dump water on the reactor core in the event of a loss-of-coolant accident. Mismanagement of the test, serious operator errors, and reactor design all contributed to the Chernobyl disaster. Figure 44-20 outlines the sequence of events leading to the April 26 explosion.

Preparation for the test began at 1:00 A.M. on April 25 as operators slowly decreased the reactor output from its normal 3200 MW thermal power to 1600 MW, a process that took 12 hours. Then, following the test plan, they disconnected the turbine-generator and disabled the emergency core-cooling system. Although the plan called for shutting off the ECCS to prevent its coming on and disturbing the test, this move violated the reactor's operating procedures. When the operators resumed lowering the reactor power, one of them failed to set an automatic control that would have maintained a thermal power at 700 to 1000 MW. The power level plunged to a mere 30 MW.

We noted earlier that fission products act as "poisons," absorbing neutrons and thereby inhibiting fission. A particularly virulent reactor poison is xenon-135, whose cross section for neutron absorption is 4400 times that of ^{235}U. Xenon-135 forms in the 6.7-h-half-life decay of ^{135}I. In normal operation, the ^{135}Xe concentration reaches a steady level in which neutron capture destroys the isotope as quickly as ^{135}I decay creates it. But when reactor power decreases, neutron production drops and with it the destruction of ^{135}Xe. But ^{135}I continues to decay into ^{135}Xe, so xenon concentration increases. The "poisoning" effect then makes it difficult to raise the reactor power until several of the xenon's 9.2-h half-lives have passed.

At Chernobyl, an operator's error had resulted in a rapid power drop, leading to high ^{135}Xe concentration. Impatient to complete the test, operators committed another safety violation: To compensate for neutron absorption in the xenon, they withdrew too many control rods. By 1:19 A.M. on April 26, they had managed to raise the power to 200 MW, still well below the 700 MW minimum needed for the test. About the same time they turned on two additional cooling water pumps, as called for in the test procedures.

In a U.S. light-water reactor, the cooling water is also the moderator. But in a graphite-moderated reactor like Chernobyl, the dominant effect of water is to absorb neutrons. So the

FIGURE 44-19 Technicians checking for radiation inside the damaged Chernobyl nuclear power plant.

additional water required withdrawal of still more control rods. Now the reactor was in a dangerous situation: An increase in power would boil water, decreasing neutron absorption and thus increasing the fission rate. That would make the water boil even faster, increasing the power even more, and a runaway reaction could result. Worse, with so few control rods in place, the reaction might be sustained by prompt neutrons alone, resulting in a power increase too rapid to halt with mechanical control rods.

The Chernobyl operators realized they had too much water, and at 1:22 A.M. they reduced the flow. But they did not immediately reinsert control rods. Thirty seconds later a computer warned that the reactor should be shut down. Ignoring the warning, operators continued the test by diverting steam from the turbine-generator. The decreased load caused more water to boil, again reducing neutron absorption. The reactor went supercritical from prompt neutrons alone, and the power level soared by a factor of 4000 in 5 seconds. The power surge ruptured water pipes, causing a steam explosion that lifted the heavy concrete reactor cover. A second explosion followed, caused perhaps by hydrogen generated from steam reacting with the zirconium cladding on the fuel rods. The graphite moderator caught fire, and heavy smoke carried highly radioactive fission products into the atmosphere. Substantial radiation release continued for 10 days, and fallout dropped on much of Europe (see Example 43-6).

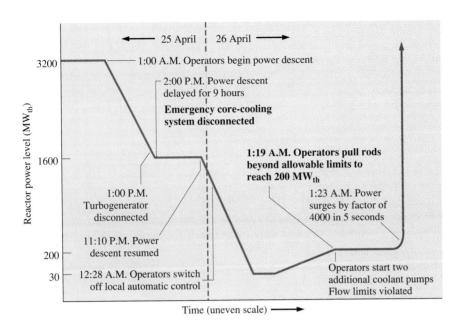

FIGURE 44-20 Sequence of events leading to the Chernobyl reactor explosion, shown as a plot of reactor power as a function of time. Time scale is not linear.

Could a Chernobyl accident happen in the United States? For commercial light-water reactors the answer is decidedly no. Loss of the water coolant/moderator in a light-water reactor immediately halts the chain reaction, making a runaway reaction impossible. But that doesn't mean light-water reactors are entirely safe. Even after the chain reaction stops, the immense heat generated by radioactive decay is enough to melt the core. A partial **meltdown** occurred in the 1979 Three Mile Island accident in Pennsylvania, but fortunately the reactor's containment structure held in nearly all the radioactivity. The threat of a hydrogen explosion during that accident—a possibility not previously considered in reactor accident scenarios—showed that we may not yet realize all the potentially dangerous situations possible in a system as complex as a nuclear power plant.

Is Nuclear Power Safe?

Opinions are strong on both sides of this question. The small but nonzero probability of a catastrophic accident makes nuclear power seem especially dangerous in the public mind. Radiation—invisible and poorly understood by most people—makes the environmental and health effects of nuclear power seem particularly insidious. And its relation to nuclear weapons makes nuclear power seem far from benign.

The important question is not whether nuclear power is safe but how it compares with other energy sources—especially its main competitor, coal. Coal mining accidents regularly kill as many people as did Chernobyl, and the latter's excess cancer fatalities are no match for the 10,000 deaths that occur *each year* in the United States alone as a result of air pollution from coal-burning power plants. And the continued use of fossil fuels, especially carbon-rich coal, has started us on the path of irreversible climatic change that could bring environmental disaster in the twenty-first century.

Statistically, the risks from nuclear power seem low compared with those of its fossil-fuel alternatives. And those risks are orders of magnitude lower than others many people willingly endure, like smoking, failing to wear seat belts, or persisting in high-fat diets. Nevertheless, unanswered questions about the long-term effects of low-level radiation, about nuclear waste disposal, and most significantly about the proliferation of nuclear technology in a politically unstable world raise doubts even for those who agree that nuclear power is statistically quite safe. Many take the compromise position that views nuclear fission as a bridge to a time when safer, more economical, and more sustainable power sources become available. In Fig. 42-28 we argued that solar photovoltaic cells may soon be one such source. Nuclear fusion, which we now discuss, may be another.

44-4 NUCLEAR FUSION

The curve of binding energy (Fig. 44-1) shows that nuclear energy can be released either by fission of heavy nuclei or by fusion of light nuclei. The binding-energy curve is steepest near its left end, showing that the most energy per nucleon is released by fusing the very lightest element—hydrogen. Indeed, the fusion reactions powering the Sun and many other stars begin with the fusion of hydrogen to form deuterium:

$$^1_1H + {}^1_1H \longrightarrow {}^2_1H + e^+ + \nu \quad (0.42 \text{ MeV}); \quad (44\text{-}3a)$$

a positron (e^+) and neutrino (ν) are also released, and the total energy liberated in the reaction is 0.42 MeV. Deuterium then fuses with hydrogen to form the helium isotope ^3_2He:

$$^1_1H + {}^2_1H \longrightarrow {}^3_2He + \gamma \quad (5.49 \text{ MeV}). \quad (44\text{-}3b)$$

Here γ represents a gamma ray, and the quantity in parentheses is the total energy released. Helium-3 nuclei from two such reactions then react to give a single ^4He nucleus and a pair of protons; this event liberates 12.86 MeV:

$$^3_2He + {}^3_2He \longrightarrow {}^4_2He + 2{}^1_1H \quad (12.86 \text{ MeV}). \quad (44\text{-}3c)$$

There is one additional energy-producing reaction associated with these events: the positron from reaction 44-3a annihilates with an electron, forming two gamma rays with a total energy of $2mc^2$ or 1.022 MeV:

$$e^+ + e^- \longrightarrow 2\gamma \quad (1.022 \text{ MeV}). \quad (44\text{-}3d)$$

The reactions 44-3a–c constitute the **proton-proton cycle.** In the full cycle, reactions 44-3a and b occur twice for each occurrence of reaction 44-3c. The net effect, including two occurrences of the annihilation reaction 44-3d, is to convert four protons and two electrons into a single helium-4 nucleus; a total of 26.7 MeV is released in the process (Fig. 44-21; see also Problem 30). In massive stars, ^4_2He is then the building block for the formation of still heavier nuclei by additional fusion reactions, as we discussed in the preceding chapter.

Although ordinary hydrogen (^1_1H) is abundant, the reaction 44-3a has a low cross section and so does not occur readily. Terrestrial fusion research has therefore focused on reactions involving the heavier hydrogen isotopes. Of most immediate interest are deuterium-deuterium (D–D) and deuterium-tritium (D–T) reactions, listed below along with the energy released in each:

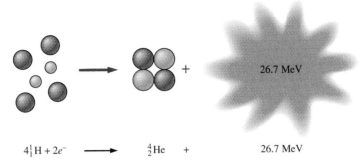

$$4\,^1_1\mathrm{H} + 2e^- \longrightarrow \,^4_2\mathrm{He} + 26.7\ \mathrm{MeV}$$

FIGURE 44-21 The net result of the proton-proton cycle in Equations 44-3a–d is the conversion of 4 protons and 2 electrons into a helium-4 nucleus, in the process releasing 26.7 MeV of energy.

$$^2_1\mathrm{H} + \,^2_1\mathrm{H} \longrightarrow \,^3_2\mathrm{He} + \,^1_0n \qquad (3.27\ \mathrm{MeV}) \qquad (44\text{-}4a)$$

$$^2_1\mathrm{H} + \,^2_1\mathrm{H} \longrightarrow \,^3_1\mathrm{H} + \,^1_1\mathrm{H} \qquad (4.03\ \mathrm{MeV}) \qquad (44\text{-}4b)$$

$$^2_1\mathrm{H} + \,^3_1\mathrm{H} \longrightarrow \,^4_2\mathrm{He} + \,^1_0n \qquad (17.6\ \mathrm{MeV}). \qquad (44\text{-}4c)$$

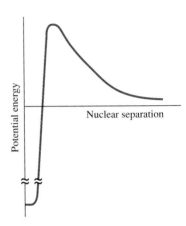

FIGURE 44-22 Potential-energy diagram for two nuclei, showing electrostatic potential barrier and deep well associated with the attractive nuclear force.

The two possible outcomes of the D–D reaction have nearly equal probability.

The electrical repulsion between nuclei makes it difficult to bring them close enough to fuse. In potential-energy terms, nuclei must overcome the potential barrier associated with the electrostatic force before they can drop into the deep potential well of the stronger but shorter-range nuclear force (Fig. 44-22). As we discussed in Chapter 40, nuclei can quantum-mechanically tunnel through the potential barrier when they lack sufficient energy to overcome it. Although the possibility of tunneling lowers the energy needed to initiate fusion, that energy still remains high. In the Sun's core, for example, fusing nuclei approach one another with energies on the order of 1 keV, corresponding to a temperature of 15 MK.

In terrestrial applications, simply achieving fusion is not sufficient for net energy production. At fusion temperatures, atoms are stripped of their electrons and the material forms a hot plasma, or ionized gas. As electrons and nuclei interact in the plasma, their speeds and directions of motion change. Since the particles are charged, those accelerations result in electromagnetic radiation that represents a loss of energy from the plasma. Radiation losses increase with increasing temperature, but power generated by fusion increases even faster (Fig. 44-23). The temperature at which fusion-generated power exceeds power loss by radiation is the **critical ignition temperature.** For the D–D reactions of Equations 44-4a and b, Fig. 44-23 shows that ignition temperature is about 6×10^8 K; for the D–T reaction of Equation 44-4c it is roughly an order of magnitude lower, or about 5×10^7 K.

The high temperatures required for fusion pose two problems: first, how to achieve those temperatures and, second, how to contain the fusing material. The stars solve both problems with their immense gravitational fields. A star is born as interstellar material—mostly hydrogen and helium—collapses under its own gravity; the resulting compression heats the material to fusion temperatures. Once fusion has begun, the star settles into an equilibrium in which the high pressure of the fusing material is balanced by the gravitational force.

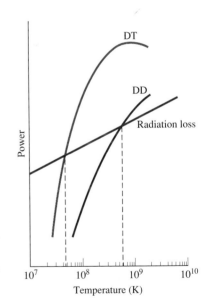

FIGURE 44-23 Power loss by radiation and power produced by D–D and D–T fusion reactions, as functions of temperature on a log-log plot. Ignition temperature (dashed lines) is the temperature at which fusion power equals radiation loss.

Gravitational confinement is not possible in terrestrial fusion applications, and we must find another means of confining the fusing material. For net fusion energy production, confinement must last long enough for the energy produced by fusion to exceed the energy needed to heat the material. The heat input required is proportional to the number of nuclei being heated, or, on a per-unit-volume basis, to the density, n. (Fusion researchers generally use the particle density, n—the number of particles per unit volume—rather than the mass density.) However, the rate of fusion energy production per unit volume is proportional to the *square* of the density. You can see this by considering half the nuclei as targets to be struck by the other half. If you double the number of target nuclei alone, you double the fusion rate. Doubling the density doubles the number of targets and the number of projectiles hitting them, and therefore quadruples the fusion rate. The total energy produced by fusion is, in turn, given by the fusion rate multiplied by the time τ during which fusion is occurring. Requiring that the fusion energy, proportional to $n^2\tau$, exceed the heating energy, proportional to n, gives a minimum value for the quantity $n\tau$—the product of density and confinement time—that must be met in an energy-producing fusion device. This condition on $n\tau$ is called the **Lawson criterion.** For the D–D and D–T reactions of Equation 44-4, the Lawson criteria are approximately

$$
\begin{aligned}
n\tau &> 10^{22} \text{ s/m}^3 \quad \text{(D–D)} \\
n\tau &> 10^{20} \text{ s/m}^3 \quad \text{(D–T).}
\end{aligned}
\tag{44-5}
$$

The factor-of-100 difference between these two Lawson criteria shows that D–T fusion is much more readily achieved.

The Lawson criterion offers a choice in the design of a fusion device: Strive for a high plasma density with a short confinement time or a lower density with a longer time. Two distinct approaches emphasize these two possibilities. **Inertial confinement** relies on the inertia of the reacting particles—that is, on their inability to be accelerated instantaneously away from the reaction site—to provide confinement; very short confinement times are required in this scheme. Inertial confinement occurs in fusion weapons and in particle-beam and laser-fusion devices. In **magnetic confinement,** the more traditional approach to controlled fusion, complicated magnetic field configurations confine the fusion plasma at lower densities but for longer times. However the Lawson criterion is met, it is of course also necessary to surpass the critical ignition temperature.

Fusion Weapons

Although peaceful fusion devices still elude scientists and engineers, fusion weapons have been with us since the 1950s. Explosives powered substantially by nuclear fusion are often called "hydrogen bombs" to distinguish them from nuclear fission explosives (incorrectly called "atomic bombs"). Fusion-based devices are more correctly called **thermonuclear weapons,** a term reflecting the high temperature needed to ignite fusion. As we will see, there is at present no such thing as a pure fusion weapon, and the distinction between fission and fusion weapons is blurred by a variety of fission-fusion hybrids.

The earliest use of fusion in nuclear weapons was to boost the yield of fission explosives. The D–D and D–T reactions of Equations 44-4a and c release

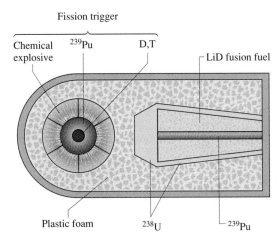

FIGURE 44-24 A fission-fusion-fission weapon. Although details remain classified, the device is known to involve a sequence in which energy from the fission trigger is focused by plastic foam onto the lithium deuteride and plutonium-239. Neutrons from plutonium fission convert lithium to tritium. Deuterium-tritium fusion ensues, releasing energy and neutrons. The neutrons induce fission in the ^{238}U surrounding the fusion fuel.

neutrons. If a small quantity of deuterium-tritium mixture is placed at the center of an implosion-type fission weapon (see Fig. 44-12), the material will be compressed to ignition temperature by the fission explosion, and the neutrons released in the subsequent fusion will ensure more complete fissioning of the uranium or plutonium. Such fusion-boosted fission weapons were first tested in 1951, and today tritium is an essential element in weapons of the U.S. nuclear arsenal.

The effort to produce fusion weapons was fraught with controversy; many scientists who had worked on fission bombs during World War II were horrified at what they had produced and felt that the world did not need the 1000-fold increase in destructive power that fusion promised. For a while it also seemed technologically impossible to produce a fusion weapon compact enough to be delivered by aircraft; early designs involved enormous refrigerators to maintain tanks of liquid deuterium and tritium. But in the early 1950s, Edward Teller and Stanislaw Ulam devised a clever configuration wherein the energy of a fission explosion is focused on a mixture of solid lithium deuteride (LiD) and plutonium-239 (Fig. 44-24). The mixture is compressed to fusion ignition temperature, while fission-produced neutrons react with lithium to produce tritium and helium:

$$^{6}_{3}\text{Li} + ^{1}_{0}n \longrightarrow ^{4}_{2}\text{He} + ^{3}_{1}\text{H}.$$

The deuterium and tritium undergo fusion that provides typically half the weapon's explosive yield. The remainder comes from fission in a layer of natural uranium (predominantly ^{238}U) surrounding the fusion fuel. Although ^{238}U cannot sustain a fission chain reaction, it will fission when bombarded by fast neutrons. In a thermonuclear weapon, the D–T fusion reaction is the source of those neutrons. The high-yield thermonuclear weapons in today's nuclear arsenals generally involve the fission-fusion-fission sequence outlined here (Fig. 44-25).

FIGURE 44-25 A 10-megaton thermonuclear test explosion in the Pacific Ocean during the 1950s.

(a)

FIGURE 44-26 Nuclear weapons have become smaller and more efficient. (a) A model of the 21-kt Nagasaki bomb beside a 335-kt MK-12A warhead. (b) The Minuteman II missile carries three MK-12A's, each capable of destroying a city. When fully implemented, the 1993 Strategic Arms Reduction Treaty (START II) will eliminate such multiple-warhead missiles based on land.

(b)

In a fission weapon, explosive yield is limited by the amount of material that can participate in the chain reaction before the weapon blows apart. But neither the fusion explosion in a thermonuclear weapon nor the fissioning of its ^{238}U requires a chain reaction. For this reason, there is no upper limit to the explosive yield of a thermonuclear weapon. Devices as large as 58 MT have been exploded, yielding almost 5000 times the energy of the fission bomb that destroyed Hiroshima. Following development of thermonuclear weapons in the 1950s, arsenals of the major nuclear powers contained many weapons in the 10- to 20-MT range. The development of ballistic missiles then led to an emphasis on smaller devices with higher yield-to-weight ratios (Fig. 44-26). The yields of strategic warheads today are typically in the 40 kT to 1 MT range, and the missiles that deliver them may carry ten or more such warheads (Fig. 44-26).

The impact of any nuclear weapon is horrifying, but thermonuclear weapons are particularly destructive. A 1-MT explosion causes third-degree burns in people as far as 11 km from the blast; shock-wave over-pressures in excess of the 5 psi needed to collapse most buildings extend out to 5 km (Fig. 44-27; see also Problem 54). Fission-fusion-fission weapons also produce a great deal of radioactive fallout from uranium fission. Depending on wind patterns, this fallout can deliver lethal radiation doses hundreds of kilometers from the blast. With tens of thousands of nuclear weapons in the world's arsenals, this application of nuclear technology is truly a threat to human civilization.

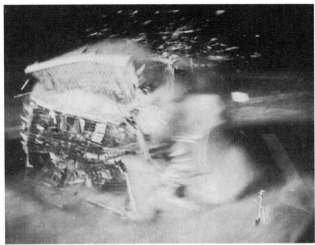

(a)

(b)

FIGURE 44-27 Destruction of a wood-frame house during a nuclear test in the 1950s. At left, house is illuminated with visible radiation from the blast. At right, 2.3 s later, shock wave arrives and the house disintegrates from the 5 psi overpressure. In a 1-Mt nuclear explosion, this level of destruction covers an area of 80 km^2.

Controlled Fusion: Inertial Confinement

In a fusion weapon, confinement occurs because the reactions are over before the explosion blows the reacting material apart. **Inertial confinement** fusion (ICF) schemes seek to apply the same strategy, with miniature fusion explosions taking place under controlled conditions (Fig. 44-28). Experimental ICF devices use laser or particle beams to compress pellets of deuterium-tritium fuel to enormous densities for confinement times on the order of 10^{-11} to 10^{-9} s. If the pellet is struck symmetrically from many directions, rapid heating of its outer layers drives an inward-propagating shock wave that compresses the material about 10,000-fold, resulting in densities higher than at the center of the Sun! The Nova laser fusion experiment at Lawrence Livermore National Laboratory in California achieves a peak power of about 10^{14} W—well above the output of all the world's electric power generating plants. The Nova device delivers its energy in 0.1 to 1 ns through 10 converging beams (Fig. 44-29). An alternative to laser fusion is the particle-beam fusion device at Sandia National Laboratory, shown in Fig. 44-30. Inertial-confinement fusion has made impressive strides in recent years, but the technological problems of delivering huge energies symmetrically to the target in a short time remain formidable.

Controlled Fusion: Magnetic Confinement

Controlled fusion research began in the 1950s with **magnetic confinement** of the fusion plasma. In Chapter 29 we showed how charged particles in a highly conducting plasma are essentially "frozen" to magnetic field lines, able to move

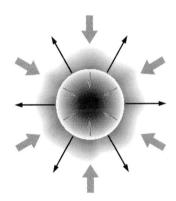

FIGURE 44-28 Implosion of a fusion pellet. Laser beams (yellow arrows) vaporize the outer surface of the pellet, which expands outward. By Newton's third law, the core of the pellet experiences an inward force that compresses it to fusion conditions.

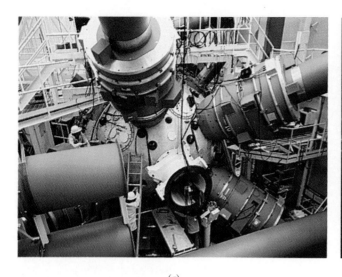

(a)

(b)

FIGURE 44-29   (a) Lasers converge on the target chamber of the NOVA laser-fusion facility at Lawrence Livermore National Laboratory. (b) For a brief instant the fusing pellet—only a few millimeters in diameter—shines like a miniature star.

FIGURE 44-30   The Particle Beam Fusion Accelerator II at Sandia National Laboratory uses intense ion beams to compress a fusion pellet. Electrical discharges result from breakdown of air at the surface of the water used as the dielectric in capacitors that shape the ion-current pulse.

easily along the field lines but not across them. There, and again in Chapter 31, we discussed some aspects of magnetic confinement in fusion reactors. Magnetic confinement is appealing because it offers the possibility of relatively long confinement times in "magnetic bottles" that keep the 100-MK fusion plasma from contact with any material surface. Because of the long confinement times—several seconds in some devices—densities needed to achieve the Lawson criterion are much lower than atmospheric density.

The first job of any magnetic confinement scheme is to create a magnetic configuration that keeps plasma away from the relatively cool walls of the device. Plasma particles can reach the walls in three general ways: (1) If magnetic field lines penetrate the walls, particles spiraling along those field lines may hit the walls (Fig. 44-31a). This mechanism is known as end loss. (2) Collisions among particles, and inhomogeneities in the magnetic field, result in particles drifting across the field lines toward the walls of the fusion device (Fig. 44-31b). (3) Plasmas are notoriously unstable. A wide variety of waves can propagate in plasma, and some of these waves can grow exponentially at the expense of particle energy, resulting in gross distortion of the plasma and field configuration that lets plasma hit the walls (Fig. 44-31c).

The most promising magnetic confinement devices eliminate end loss altogether, using a toroidal magnetic field whose field lines do not penetrate the device walls (Fig. 44-32). But bending the field lines into circles makes the field nonuniform, giving rise to cross-field drifts (Fig. 44-31b). Making the toroid larger decreases the inhomogeneities in the magnetic field and increases the time necessary for drifting particles to reach the walls. Both these trends are promising, as they suggest that appropriate confinement times should be achievable simply by building bigger devices. In practical toroidal devices, additional field components enhance confinement and suppress instabilities. A particularly promising toroidal fusion device is the **tokamak,** a Russian invention now under

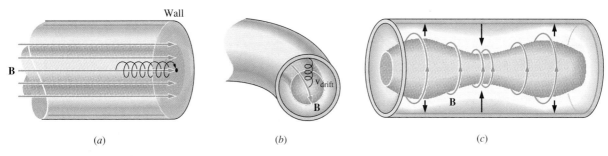

(a) (b) (c)

FIGURE 44-31 Plasma loss in magnetic confinement. (*a*) End losses occur when field lines intersect the device walls. (*b*) Curvature of the field lines results in cross-field drifts. (*c*) Large-scale instabilities distort the plasma and magnetic field. Here the so-called sausage instability causes alternate narrowing and bulging of the plasma column.

study throughout the world. We described the tokamak at the end of Section 31-4; see Fig. 31-30.

A successful fusion device must not only confine plasma but also heat it to ignition temperature. A variety of heating schemes have been tried. In tokamaks and related devices, the fusion plasma carries considerable electric current, and the I^2R joule heating increases the plasma temperature. Unfortunately, the resistance of a plasma drops with increasing temperature, so joule heating alone is unable to bring the plasma to ignition temperature. A second approach is to inject energy in the form of electromagnetic waves at the cyclotron frequency of the ions. Recall from Chapter 29 that the cyclotron frequency is the frequency at which charged particles execute circular motion about magnetic field lines; when electromagnetic waves at the same frequency are present, a resonance condition exists and the particles absorb energy from the waves. Finally, one of the most successful heating methods is the injection of high-energy beams of neutral particles into the plasma. The particles must be neutral in order to cross the magnetic field lines providing plasma confinement; once in the plasma they are ionized by collisions and thus help maintain not only the plasma temperature but also the density of charged particles. Figure 44-33 describes neutral-beam injection.

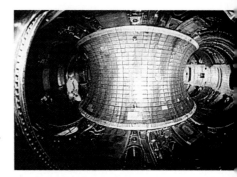

FIGURE 44-32 A toroidal fusion device has no end losses, since its magnetic field lines don't penetrate the device walls. The toroidal shape of the Tokamak Fusion Test Reactor at Princeton University shows clearly in this photo taken inside the device during its assembly.

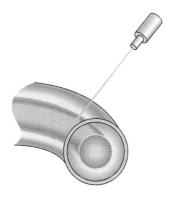

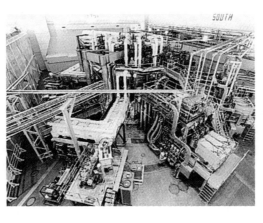

(a) (b)

FIGURE 44-33 (*a*) Neutral-beam heating of a fusion plasma. A high-energy beam of neutral atoms crosses the magnetic field that confines the ionized plasma, dumping energy that heats the plasma. (*b*) Neutral-beam injector for the Tokamak Fusion Test Reactor at Princeton.

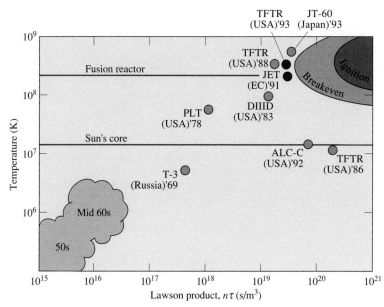

FIGURE 44-34 Progress in controlled fusion research. Breakeven occurs when fusion energy produced is equal to the energy supplied to heat the plasma; ignition is a self-sustaining fusion reaction. Individual points represent achievements of several experimental fusion devices. TFTR is the Tokamak Fusion Test Reactor at Princeton University; JET is Joint European Torus in England. The two red points represent the first experiments using deuterium-tritium plasmas.

Prospects for Fusion Energy

When work on fusion power began in the 1950s, researchers confidently predicted that limitless energy sources would be available in a few decades. But plasma confinement and heating have proved more complex and subtle than expected, and the newer inertial confinement scheme has revealed its own technical problems. Nevertheless, the promise of nearly unlimited energy—a gallon of seawater equivalent to more than 300 gallons of gasoline (see Problem 39)—remains, and progress toward controlled fusion continues. An important milestone was reached in 1991, when the Joint European Torus in England produced some 2 MW of fusion power for several seconds. Figure 44-34 summarizes progress in controlled fusion research.

Once controlled fusion proves scientifically feasible, there will remain formidable engineering challenges in the design of a practical fusion power plant. The intense neutron fluxes from D–T fusion cause severe degradation of the materials comprising the reaction chamber walls. Furthermore, neutron-capture reactions produce radioactive isotopes within the walls, greatly complicating maintenance procedures. Although fusion does not produce the plethora of radioactive materials present in fission products, handling of radioactivity—especially from D–T reactions—is still a formidable problem. Heat from D–T fusion would be extracted by a heat-transfer medium, then used to drive a conventional steam turbine and generator. Figure 44-35 shows a possible design for a D–T fusion power plant.

The first practical fusion power plants are likely to use D–T fusion because its ignition temperature and Lawson criterion are much lower than for D–D

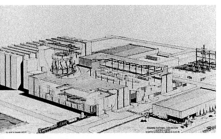

FIGURE 44-35 Possible design for a power plant using D–T fusion, showing tokamak fusion reactor in cutaway at center.

fusion. But the D–D reaction promises cleaner and more efficient power production. With D–D fusion there is no radioactive tritium fuel. And a look at the D–D reactions (Equation 44-4a and b) shows that one of the reactions produces protons instead of neutrons. Thus there is less neutron-induced radioactivity. Furthermore, high-energy protons can be extracted and passed through a magnetohydrodynamic generator, a device that uses electromagnetic induction to convert charged-particle energy directly into electricity. Use of MHD generators would bypass the conventional steam cycle and greatly increase the thermodynamic efficiency of the power plant. Even as they strive to make D–T fusion a reality, many fusion researchers have their eyes on a more distant future where D–D fusion provides much of our energy.

● **EXAMPLE 44-6** FUSION ENERGY RESOURCES

The D–D reactions of Equations 44-4a, b produce $^3_2$He and $^3_1$H, which can undergo further fusion reactions. The net result is an average release of 7.2 MeV per deuteron (see Problem 38). About 0.015% of hydrogen nuclei in seawater are actually deuterium. Use these figures to estimate how long deuterium in the world's oceans (average depth about 3 km) could supply humanity's energy needs at the current energy-consumption rate of about 10^{13} W.

Solution

Earth's surface is about three-fourths water, so the volume V of water in the oceans is approximately three-fourths of Earth's surface area multiplied by the average water depth:

$$V = \tfrac{3}{4}(4\pi R_E^2)d = (3\pi)(6.4\times10^6 \text{ m})^2(3\times10^3 \text{ m})$$

$$= 1.2\times10^{18} \text{ m}^3.$$

With water's density of 1 g/cm^3 or 1000 kg/m^3, this amounts to 1.2×10^{21} kg of water. The mass of a water molecule (H$_2$O) is 18 u, so there are about

$$\frac{1.2\times10^{21} \text{ kg}}{(18 \text{ u})(1.66\times10^{-27} \text{ kg/u})} = 4.0\times10^{46}$$

water molecules in the ocean. Each molecule contains two hydrogen nuclei, but only 0.015% of those nuclei are deuterons. So the number of deuterons in the oceans is roughly

$$(4.0\times10^{46} \text{ H}_2\text{O})(2 \text{ H/H}_2\text{O})(0.00015 \text{ D/H}) = 1.2\times10^{43} \text{ D}.$$

With an energy release of 7.2 MeV per deuteron, the energy equivalent of this oceanic deuterium is

$$(1.2\times10^{43} \text{ D})(7.2\times10^6 \text{ eV/D})(1.6\times10^{-19} \text{ J/eV}) = 1.4\times10^{31} \text{J}.$$

At 10^{13} W (= J/s), this energy source would last a time given by

$$t = \frac{1.4\times10^{31} \text{ J}}{(1\times10^{13} \text{ J/s})(3.2\times10^7 \text{ s/y})} = 4\times10^{10} \text{ y}.$$

This is 10 times as long as the Sun will continue to shine, so fusion energy resources are essentially unlimited.

EXERCISE How much water would need to be processed to power a 1000-MW D–D fusion power plant for 1 year?

Answer: 2.7×10^3 m^3 = 7.2×10^5 gal

Some problems similar to Example 44-6: 33, 34, 36 ●

Would the development of controlled fusion be a good thing? Our planet may be moving toward significant climatic change caused by the increase in atmospheric carbon dioxide from burning fossil fuels. In that context, replacement of fossil-fueled energy sources with fusion reactors might help alleviate a serious global problem. But in the longer term, the availability of virtually unlimited energy could itself lead to climatic disaster. As discussed in Chapter 19, Earth is in a state of thermal energy balance, its temperature determined by the rate of solar energy gain versus loss by radiation. Any additional source of energy can upset the balance, especially if that source becomes more than a tiny fraction of the solar input. With fusion, human energy use could easily grow to levels that would result in major global temperature increases. In fact, any energy source other than solar energy itself carries this potential danger.

CHAPTER SYNOPSIS

Summary

1. **Nuclear fission** is the splitting of heavy nuclei into two lighter **fission fragments,** with the release of energy. Fission is currently the most developed nuclear energy technology.

   a. In **fissionable** nuclei, fission may be caused by absorption of a neutron. If such **neutron-induced fission** occurs even at low neutron energies, the nucleus is termed **fissile.** A widely used fission reaction is that of the fissile isotope ^{235}U:

   $$^1_0n + {}^{235}_{92}U \rightarrow X + Y + b\,{}^1_0n,$$

   where X and Y stand for a variety of possible fission products, and b is the number of neutrons emitted. These reactions release about 200 MeV of energy and, on average, about 2.47 additional neutrons.

   b. Other fissile isotopes include ^{233}U and ^{239}Pu; the latter is produced by neutron absorption in the common uranium isotope ^{238}U.

   c. A fission **chain reaction** occurs when the neutrons from a fission event go on to induce more fission. The average number of neutrons causing additional fission is the **multiplication factor,** k. If $k < 1$ the reaction is **subcritical** and quickly comes to a halt, if $k = 1$ the reaction is **critical** and proceeds at a steady rate, and if $k > 1$ the reaction is **supercritical** and the reaction rate grows exponentially. A single fission event results, after n generations of fission have passed; in a total of N such events:

   $$N = \frac{k^{n+1} - 1}{k - 1}.$$

   The actual rate at which the reaction grows is determined by the multiplication factor k and the **generation time** between fissions.

   d. **Fission weapons** are detonated by rapidly assembling a supercritical configuration of fissile ^{235}U or ^{239}Pu. In explosive yield, several kilograms of fissile material are equivalent to tens of thousands of tons of chemical explosive.

   e. In a **nuclear reactor,** the chain reaction is controlled to maintain a steady reaction rate. In addition to the fissile fuel itself, key components of a reactor include

   - **Control rods,** made of neutron-absorbing material and moved in and out of the fuel mass to adjust the reaction rate.
   - A **moderator,** like water or graphite, that slows neutrons and thereby enhances the likelihood of fission.
   - A **coolant** that carries off heat from the nuclear reaction. In a power reactor, heat produces steam that drives a turbine connected to an electric generator.

   f. Safety of nuclear power is a matter of public and scientific debate; issues range from uranium mining to reactor accidents to disposal of radioactive waste. A related concern is the diversion of reactor fuel or reactor-produced plutonium for nuclear weapons. Statistically, however, nuclear fission may well be safer than some other means of large-scale electric power generation, especially the burning of coal.

2. **Nuclear fusion** is the joining of light nuclei to form heavier nuclei. Fusion is an important natural process, powering the stars and thereby creating elements from helium through iron. Fusion requires temperatures high enough that average particle energies are sufficient to overcome electrical repulsion. The most promising reactions for terrestrial fusion applications are the deuterium-tritium (D–T) and deuterium-deuterium (D–D) reactions:

   $$^2_1H + {}^2_1H \longrightarrow {}^3_2He + {}^1_0n + 3.27 \text{ MeV}$$

   $$^2_1H + {}^2_1H \longrightarrow {}^3_1H + {}^1_1H + 4.0 \text{ MeV}$$

   $$^2_1H + {}^3_1H \longrightarrow {}^4_2He + {}^1_0n + 17.6 \text{ MeV}.$$

   a. **Fusion weapons** actually involve a combination of fission, fusion, and more fission.

   b. Practical energy generation from **controlled fusion** has yet to be achieved. Any controlled fusion scheme must satisfy the **Lawson criterion,** stating that the product of density and confinement time must exceed a certain value.

   - **Inertial confinement** schemes use powerful lasers or particle beams to compress pellets of fusion fuel to high density and temperature. Fusion occurs before the reacting material has time to leave the reaction site.
   - **Magnetic confinement** schemes use magnetic fields to confine fusion plasma at low densities but for relatively long times.

Terms You Should Understand

(Pairs are closely related terms whose distinction is important; number in parentheses is chapter section where term first appears.)

fission, fusion (44-1)
fission fragment (44-2)
fissionable, fissile (44-2)
subcritical, critical, supercritical mass (44-2)
chain reaction (44-2)
multiplication factor (44-2)
generation time (44-2)
moderator (44-3)
control rods (44-3)

prompt, delayed neutrons (44-3)
proton-proton cycle (44-4)
D–T fusion, D–D fusion (44-4)
inertial confinement, magnetic confinement (44-4)
thermonuclear weapon (44-4)
kiloton, megaton (44-4)

Symbols You Should Recognize

τ (44-2)
k (44-2)
kt, Mt (44-4)

Problems You Should Be Able to Solve

calculating energy release in fission and fusion reactions (44-2, 44-4)
determining fuel requirements and energy yields for fission and fusion devices (44-3, 44-4)
analyzing super-and subcritical chain reactions (44-2)

Limitations to Keep in Mind

Descriptions of the nuclear technologies given here are highly simplified; much more subtle considerations of physics and engineering are required to make working devices.

QUESTIONS

1. If the nuclear binding-energy curve increased monotonically with atomic number, would energy release by nuclear fission be possible? By nuclear fusion? Explain.

2. On a per-nucleon basis, does fission or fusion release more energy? Explain by invoking Fig. 44-1.

3. On an energy-release-per-unit-mass basis, by approximately what factor do nuclear reactions exceed chemical reactions?

4. A chain reaction in natural uranium is very difficult to achieve today, because the fissile isotope ^{235}U comprises only 0.72% of natural uranium. Yet 3 billion years ago, a natural chain reaction occurred in what is now a uranium mine in Africa. Speculate on why this might have been possible, using the fact that the half-lives of ^{235}U and ^{238}U are about 700 million years and 4.5 billion years, respectively.

5. Explain and distinguish the roles of the control rods and moderator in a nuclear reactor.

6. Explain how delayed neutrons make possible mechanical control of a nuclear chain reaction.

7. Compare the multiplication factors in a thermal fission reactor and in a fission bomb.

8. Why are kilotons and megatons—actually units of weight—used as units of nuclear explosive yield?

9. Why is a water-moderated reactor intrinsically safer in a loss-of-coolant accident than a graphite-moderated reactor?

10. What is the essential distinction between a boiling-water reactor and a pressurized-water reactor?

11. Name an advantage of the Canadian CANDU and Russian RBMK reactor designs over U.S. light-water reactors.

12. What is prompt criticality? Why is it a dangerous situation in a nuclear reactor?

13. Name some advantages and disadvantages of nuclear fuel reprocessing.

14. Is ^{238}U fissonable? Is it fissile? Explain the distinction.

15. Uranium-238 cannot sustain a fission chain reaction. Is it therefore useless in nuclear power reactors? Explain.

16. Uranium-238 cannot sustain a fission chain reaction, yet ^{238}U fission provides about half the explosive yield of a thermonuclear weapon. How does this happen?

17. Why are the names "atomic bomb" and "hydrogen bomb" inappropriate?

18. Explain the role of ^{135}Xe in the Chernobyl reactor accident.

19. Why are fission fragments radioactive? Invoke Fig. 44-6 in giving your answer.

20. Why are high temperatures required for nuclear fusion?

21. Nuclear fusion powers the stars. Besides energy, what other important product results from stellar fusion?

22. The D–T reaction has the disadvantage of involving radioactive tritium fuel. Why is this reaction nevertheless the most likely to see first commercial use?

23. Why is there an upper limit to the yield of an effective fission weapon but not of a fusion weapon?

24. Why do we not talk of chain reactions and multiplication factors in discussing fusion?

25. Contrast the inertial and magnetic approaches to fusion confinement.

26. Why are many magnetic fusion reactors toroidal in shape?

27. What are some advantages of D–D over D–T fusion for a controlled fusion reactor?

PROBLEMS

Section 44-1 Energy from the Nucleus

1. The masses of the neutron, the deuterium nucleus, and the ^{3}He nucleus are 1.008665 u, 2.013553 u, and 3.014932 u, respectively. Use the Einstein mass-energy relation to verify the 3.27-MeV energy release in the D–D fusion reaction of Equation 44-4a.

2. In the fission reaction of Example 44-1, the masses are 235.043915 u, 140.9139 u, and 91.8973 u for ^{235}U, ^{141}Ba, and ^{92}Kr, respectively. Use these values, along with the neutron mass from the preceding problem, to find the energy released in this reaction.

3. Using the values 200 MeV for ^{235}U fission and 17.6 MeV for D–T fusion, calculate and compare the energy release per kilogram for each reaction.

Sections 44-2 and 44-3 Nuclear Fission and Its Applications

4. A ^{235}U nucleus undergoes neutron-induced fission, resulting in a ^{141}Cs nucleus, three neutrons, and another nucleus. What is that nucleus?

5. Neutron-induced fission of ^{235}U results in the fission fragments iodine-139 and yttrium-95. How many neutrons are released?

6. Write a complete equation for the neutron-induced fission of plutonium-239 that results in barium-143, 2 neutrons, and another nucleus.

7. Assuming 200 MeV per fission, determine the number of fission events occurring each second in a reactor whose thermal power output is 3200 MW.

8. How much ^{235}U would be needed to fuel the reactor of the preceding problem for one year? (Your answer is an overestimate because fission of ^{239}Pu also contributes to the power output.)

9. Find the explosive yield in equivalent tonnage of TNT for the chain reaction analyzed in Example 44-3.

10. Arguments over the feasibility of a nuclear test ban have been focused on the ability to distinguish nuclear explosions in the 1-kt range from chemical explosions and seismic events. How much U-235 would have to fission to produce an explosive yield of 1.0 kt?

11. How much uranium-235 would be consumed in a fission bomb with 20-kt explosive yield?

12. What is the efficiency of a nuclear weapon that uses 28 kg of ^{235}U to produce a 44-kt explosive yield?

13. The 1974 Threshold Test Ban Treaty limits underground nuclear tests to a maximum yield of 150 kt. (a) How much ^{235}U would be needed for a device with this yield, assuming 30% efficiency? (b) What would be the diameter of this mass, assembled into a sphere? The density of uranium is 18.7 g/cm^3.

14. Suppose that all of the approximately 2.5 neutrons per fission event caused additional fission in a U-235 mass with generation time $\tau = 10$ ns. (a) How long would it take to fission the first 10^{24} nuclei? (b) How long would it take for ten times this number to fission?

15. The effective multiplication factor in a typical nuclear weapon is about 1.5, and the generation time is about 10 ns. Under these conditions, (a) how many generations would it take to fission a 10-kg mass of ^{235}U? (b) How long would the process take?

16. What is the generation time in a fission reaction that consumes 2.0×10^{20} uranium-235 nuclei in its first 10 s, if the multiplication factor is 1.0012?

17. The temperature in a typical reactor core is 600 K. What is the thermal speed of a neutron at this temperature? *Hint:* See Section 20-1.

18. A neutron is emitted from a fission reaction with kinetic energy 2.0 MeV. If it loses 88.9% of its energy in a collision with a moderator nucleus (typical for head-on collisions in heavy water), how many such collisions must it undergo before it reaches 0.05 eV, the thermal energy at typical reactor temperatures?

19. A neutron collides elastically and head-on with a stationary deuteron in a reactor moderated by heavy water. How much of its kinetic energy is transferred to the deuteron? *Hint:* Consult Chapter 11.

20. A buildup of fission products "poisons" a reactor, dropping the multiplication factor to 0.992. How long will it take the reactor power to drop in half, assuming a generation time of 0.10 s?

21. If a reactor is shut down abruptly, neutron absorption by ^{135}Xe prevents rapid start-up. The xenon has a 9.2-h half-life. How long must reactor operators wait until the ^{135}Xe level drops to one-tenth of its peak value? (Your answer neglects formation of additional ^{135}Xe from the decay of ^{135}I.)

22. It takes n generations to fission the first N nuclei in a chain reaction with multiplication factor k, where $N \gg 1$. Find an expression for the number of generations that will pass before a total of $10N$ nuclei have been fissioned.

23. One reason breeder reactors are considered potentially dangerous is that the reaction in a breeder is sustained by fast neutrons alone; there is no moderator. The generation time for prompt neutrons is then 100 ns, down from 100 ms in a thermal reactor. For the conditions of Example 44-5, compute the power-doubling time in a breeder with $\tau = 100$ ns.

24. It is generally considered impossible for a slow-neutron reactor to blow up like a bomb, even in the event of a runaway reaction. To help confirm this, compare the time for the reaction rate to double in (a) an out-of-control reactor with prompt-neutron generation time $\tau = 100$ μs

and multiplication factor $k = 1.01$ (unrealistically high) and (b) in a bomb with $\tau = 10$ ns and $k = 1.5$.

25. In the dangerous situation of prompt criticality in a fission reactor, the generation time drops to 100 μs as prompt neutrons alone sustain the chain reaction. If a reactor goes prompt critical with $k = 1.001$, how long does it take for a 100-fold increase in reactor power?

26. During an accident, a reactor's fission rate increases by a factor of 100 in 20 s. If the generation time is 3.0×10^{-4} s, what is the multiplication factor k?

27. Operators seek to double the power output of a nuclear reactor over a period of 1 hour. If the generation time is 120 ms, by how much should the multiplication factor k be increased from 1 during this period?

28. Find the generation time in a reactor whose power increases by a factor of 4 in 5 minutes, given that the multiplication factor is $k = 1.00022$.

Section 44-4 Nuclear Fusion

29. Fusion researchers often express temperature in energy units, giving the value of kT rather than T. What is the temperature in kelvins of a 2-keV plasma?

30. Verify from Equations 44-3 that the proton-proton cycle yields a net energy of 26.7 MeV.

31. In a magnetic-confinement fusion device with confinement time 0.5 s, what density would be required to meet the Lawson criterion for D–T fusion?

32. An early tokamak, the T-4 device at Moscow's Kurchatov Institute, achieved a particle density of 3×10^{19} m^{-3} for 10 ms. By what factor would its confinement time need to be increased to meet the D–T Lawson criterion?

33. How much heavy water (deuterium oxide, ^{2}H$_2$O or D$_2$O) would be needed to power a 1000-MW D–D fusion power plant for 1 year?

34. A D–D fusion-powered car (Fig. 44-36) gets 30 miles per gallon of seawater. How many times could it circle the Earth on a gallon of pure heavy water (^{2}H$_2$O)?

FIGURE 44-36 Fusion-powered car from the film "Back to the Future" (Problem 34).

35. Inertial-confinement schemes generally involve confinement times on the order of 0.1 ns. What is the corresponding density needed to meet the Lawson criterion for D–T fusion?

36. (a) Estimate the D–D fusion energy content of Lake Erie, whose volume is 480 km^3. (b) How long would this resource supply humanity's energy needs at the current rate of 10^{13} W?

37. The proton-proton cycle consumes four protons while producing about 27 MeV of energy. (a) At what rate must the Sun consume protons to produce its power output of about 4×10^{26} W? (b) The present phase of the Sun's life will end when it has consumed about 10% of its original protons. Estimate how long this phase will last, assuming the Sun's 2×10^{30} kg mass was initially 71% hydrogen.

38. Consider deuterium undergoing fusion via the reactions of Equation 44-4a and 44-4b. Since reaction 44-4b also produces one tritium nucleus, reaction 44-4c also occurs. Furthermore, ^{3}He nuclei from reaction 44-4a fuse with deuterium, yielding additional energy:

$$^3_2\text{He} + {}^2_1\text{H} \longrightarrow {}^4_2\text{He} + {}^1_1\text{H} + 18.3 \text{ MeV}.$$

(a) How many deuterons are needed for one occurrence of each of these four reactions? (b) Assuming that reactions 44-4a and b occur with equal probability, and that all the $^3_2$He and $^3_1$H formed undergo the remaining two reactions, use the result of (a) and the energies listed for the various reactions to determine the average energy release per deuteron.

39. About 0.015% of hydrogen nuclei are actually deuterium. (a) How much energy would be released if all the deuterium in a gallon of water underwent fusion? (Use an average of 7.2 MeV per deuteron; see preceding problem.) (b) In terms of energy content, to how much gasoline does a gallon of water correspond? Gasoline's energy content is 36 kWh/gal, as listed in Appendix C.

40. The Princeton Large Torus (PLT) fusion device has a plasma column 90 cm in diameter, with particle density about 4×10^{19} m^{-3}. Heating by neutral-beam injection results in a plasma temperature about 6×10^7 K, and the confinement time is 25 ms. (a) What is the Lawson parameter for this device? (b) Keeping other factors the same, what confinement time would be needed to meet the D–T Lawson criterion?

Paired Problems

(Both problems in a pair involve the same principles and techniques. If you can get the first problem, you should be able to solve the second one.)

41. The total power generated in a nuclear power reactor is 1500 MW. How much ^{235}U does it consume in a year?

42. New Hampshire's Seabrook nuclear power plant produces electrical energy at the rate of 1.2 GW, and consumes 1311 kg of ^{235}U each year. Find (a) its thermal power

output and (b) its efficiency, assuming the plant operates continuously.

43. Two fission chain reactions have $\tau = 20$ ns, but reaction A has $k = 1.3$ and reaction B has $k = 1.4$. Compare the total number of nuclei that fission in the first μs of each reaction.

44. A fission chain reaction has $\tau = 10$ ns and $k = 1.35$. Find the total number of nuclei that fission (a) in the first 1.0 μs and (b) in the interval from 1.0 μs to 1.1 μs.

45. This problem and the next one explore the differences between magnetic and inertial-confinement fusion. (a) What is the mean thermal speed of a deuteron in the PLT device of Problem 40? (b) At this speed, how long would it take a deuteron to cross the plasma column? Using this value as a "confinement time" in the absence of magnetic confinement and the density given in Problem 40, calculate the Lawson parameter and show that it falls far short of the value needed for D–T fusion.

46. A laser fusion device compresses a fusion pellet to a diameter of 0.1 mm and a density of 250 g/cm^3. (a) If the temperature is comparable to that of the PLT plasma (see preceding problem), how long does it take a deuteron with thermal speed to cross the pellet? (b) Estimate the number density of deuterons in the pellet, approximating it as all deuterium. (c) Use the results of (a) and (b) to calculate the Lawson parameter for the pellet, and show that it exceeds the required value for D–T fusion.

Supplementary Problems

47. A laser-fusion fuel pellet has mass 5.0 mg and is composed of equal parts (by mass) of deuterium and tritium. (a) If half the deuterium, and an equal number of tritium nuclei, participate in D–T fusion, how much energy is released? (b) At what rate must pellets be fused in a power plant with 3000 MW thermal power output? (c) What mass of fuel would be needed to run the plant for 1 year? Compare with the 3.6×10^6 tons of coal needed to fuel a comparable coal-burning power plant.

48. A fission chain reaction has $\tau = 10$ ns. In the first microsecond of the reaction, a total of 2.4×10^{23} fission events occur. In the next 100 ns an additional 5.8×10^{25} fissions occur. Find the multiplication factor k.

49. Use a graphical or numerical method to find the multiplication factor necessary to fission 10^{22} uranium-235 nuclei in 2.0 μs, if the generation time is 10 ns.

50. The half-lives of ^{235}U and ^{238}U are 7.04×10^8 years and 4.46×10^9 years, respectively. Uranium-235 presently constitutes 0.72% of natural uranium, with essentially all the rest ^{238}U. Three billion years ago, a natural fission chain reaction occurred in what is now a uranium mine in Africa. What percentage of natural uranium was ^{235}U at that time? Compare with the 3% ^{235}U typical of fuel for today's power reactors.

51. Roughly half the yield of a thermonuclear weapon comes from D–T fusion and the rest from ^{238}U fission induced by fast neutrons from the D–T fusion. Each D–T reaction releases about 18 MeV and each fission about 200 MeV. Estimate the masses of fusion fuel and ^{238}U in a 1-Mt thermonuclear bomb, assuming both types of fuel react completely.

52. An important cooling mechanism in fusion plasmas is the loss of electromagnetic radiation by plasma electrons. Analysis of Maxwell's equations shows that the power radiated by an electron is proportional to the *square* of its acceleration. (a) Write an expression for the acceleration of an electron a distance r from a fully ionized nucleus of atomic number Z. (b) Show from your result that, for a given r, the radiated power is proportional to Z^2. Your result shows why contamination with high-Z materials, such as the metals of the reactor walls, can seriously degrade the plasma temperature.

53. The volume of the fireball produced in a nuclear explosion is roughly proportional to the weapon's explosive energy yield. For weapons exploded at ground level, how would the land area subject to a given level of damage scale with the weapon's yield? Your result shows one reason military strategists favor multiple smaller warheads.

54. The peak overpressure generated in a nuclear explosion is approximately

$$\Delta P = 22.4 \left(\frac{Y}{r^3} \right) + 15.8 \left(\frac{Y}{r^3} \right)^{1/2},$$

where ΔP is the overpressure in pounds per square inch (psi), Y is the weapon's yield in megatons, and r is the distance in miles from the explosion. An overpressure of 5 psi is sufficient to destroy a typical wood-frame house (see Fig. 44-27). Determine the radius of the 5-psi zone for (a) the 12.5-kt Hiroshima bomb, (b) a 300-kt MX missile warhead, (c) and the 9-Mt bombs that were deployed on U.S. bombers as recently as the 1980s.

55. Of the neutrons emitted in each fission event in a light-water reactor, an average of 0.6 neutrons are absorbed by ^{238}U, leading to the formation of ^{239}Pu. (a) Assuming 200 MeV per fission, how much ^{239}Pu forms each year in a 30% efficient nuclear power plant whose electric power output is 1000 MW? (b) With careful design, a fission explosive can be made from 5 kg of ^{239}Pu. How many potential bombs are produced each year in the power plant of part (a)? (Extracting the ^{239}Pu is not an easy job.)

56. In the D–D reaction of Equation 44-4b, assume the two deuterons approach head on with equal speeds. Use conservation of momentum to determine how the 4-MeV reaction energy is shared between the tritium nucleus and the proton that result. Assume the initial kinetic energy of the deuterons is negligible.

CHAPTER 45

FROM QUARKS TO THE COSMOS

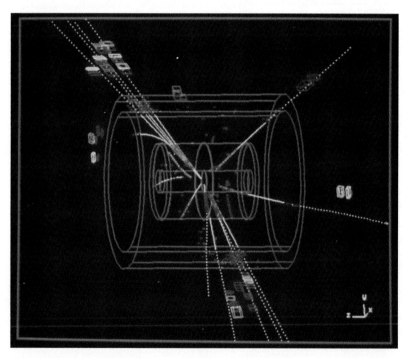

Jets of particles emerge in this image obtained with the DELPHI detector at the Large Electron-Positron Collider of the European Laboratory for Particle Physics (CERN). Production of new particles in high-energy collisions such as this one gives our best experimental evidence for the structure and organization of the fundamental particles.

The past six chapters have extended the realm of physics to the atomic and molecular scale, then down to the even smaller scale of the atomic nucleus. Here we go further still, probing the structure of nucleons themselves and trying to make sense of the host of subatomic particles nature reveals. We'll be

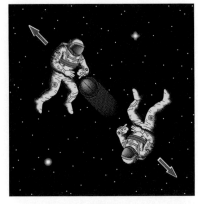

(a)

(b)

FIGURE 45-1 (a) Two astronauts toss a ball back and forth. The resulting forces cause an apparent repulsion of the two. This is analogous to the exchange of virtual photons between particles of like charge. (b) The astronauts struggle for possession of the ball, resulting in an attractive force. This is analogous to the exchange of virtual photons between particles of opposite charge.

asking questions about the ultimate nature of matter at the tiniest scales, but in the process we will find a remarkable connection with questions of the largest scale—questions about the origin and ultimate fate of the entire universe.

45-1 PARTICLES AND FORCES

By 1932 there were four "elementary" particles of matter known: the electron, the proton, the neutron, and the neutrino. In addition, there were the positron, antiparticle to the electron, and the photon of electromagnetic radiation. There were also the seemingly fundamental forces—gravity, the electromagnetic force, the nuclear or strong force, and the weak force that manifests itself in beta decay.

In Chapter 39 we saw how the photoelectric effect shows that the interaction of electromagnetic waves with matter ultimately involves individual photons—the quanta of the electromagnetic field. In the quantum-mechanical view of electromagnetism, the force between two charged particles also involves photons, now exchanged between the interacting particles. Imagine two astronauts floating in space, tossing a ball back and forth (Fig. 45-1a). Catching or throwing the ball, one astronaut gains momentum in a direction away from the other, so the exchange results in a net average repulsive force. If the two astronauts struggle for possession of the ball, each trying to wrest it from the other, then the ball mediates what appears as an attractive interaction (Fig. 45-1b). Figure 45-1 gives classical analogs for the attractive and repulsive electrical interactions involving photon exchange.

So far we've found photons emitted when a particle jumps into a lower energy state, with the photon carrying off energy equal to the energy difference between the two states. The process obviously conserves energy. But now we're saying that a single, free electron emits photons that it exchanges with another particle to produce what we call the electromagnetic force. How can that process conserve energy? The energy-time uncertainty relation (Equation 39-16) says that an energy measured in a time Δt is necessarily uncertain by an amount $\Delta E \geq \hbar/\Delta t$. The photon exchanged by two particles lasts only a short time, and therefore, its energy is uncertain. So we can't really say that energy conservation is violated. A photon created in this way and lasting only for the short time it takes to exchange with another particle is called a **virtual photon.** We never "see" the virtual photon since it's emitted by one particle and absorbed by the other.

The quantum theory of the electromagnetic interaction is called **quantum electrodynamics** (QED). Although begun by Dirac, it was finally brought to consistent form in 1948 by Richard Feynman, Sin-Itiro Tomonaga, and Julian Schwinger. The fundamental event in QED is the interaction of a photon with an electrically charged particle. Two such events joined by a common virtual photon give the quantum electrodynamical description of the electromagnetic force (Fig. 45-2). The predictions of quantum electrodynamics have been confirmed experimentally to a remarkably high precision, and today QED is our best-verified theory of physical reality.

Mesons

In 1935 the Japanese physicist Hideki Yukawa (Fig. 45-3) proposed that the nuclear force should, like the electromagnetic force, be mediated by exchange of a particle. Yukawa called his hypothetical particle a **meson.** Because the range of the strong force is limited, Yukawa argued, the meson should have finite mass. The reason for this connection between mass and range again lies in the energy-time uncertainty relation.

Consider first the electromagnetic force: It falls off as $1/r^2$ and thus has an essentially infinite range. Two particles can be very far apart and still interact electromagnetically. Since photons travel at the finite speed of light, the time Δt for a photon interaction can be arbitrarily long. The energy-time uncertainty principle $\Delta E \Delta t \geq \hbar$ thus shows that the energy uncertainty ΔE can be arbitrarily small. Thus, the possible energies for virtual photons must extend downward toward zero—and that can happen only if photons have zero rest energy.

The nuclear force, in contrast, has a finite range of about 1.5 fm. Moving at close to the speed of light, the longest a particle mediating this force would need to exist is for a time Δt given by

$$\Delta t = \frac{\Delta x}{c} = \frac{1.5 \times 10^{-15} \text{ m}}{3.0 \times 10^8 \text{ m/s}} = 5.0 \times 10^{-24} \text{ s}.$$

Then energy-time uncertainty gives

$$\Delta E \geq \frac{\hbar}{\Delta t} = \frac{1.05 \times 10^{-34} \text{ J} \cdot \text{s}}{5.0 \times 10^{-24} \text{ s}} = 2.1 \times 10^{-11} \text{ J} = 130 \text{ MeV}.$$

Since the meson energy need not go below this value, this could be its rest energy. Yukawa therefore proposed the existence of a new particle with mass 130 MeV/c^2, about 250 times that of the electron. Yukawa's prediction was eventually confirmed—but not before nature introduced yet another particle.

45-2 PARTICLES AND MORE PARTICLES

In the 1930s, the most available source of high-energy particles was cosmic radiation—streams of high-energy protons and other particles of extraterrestrial origin that, on striking Earth's atmosphere, induce nuclear and elementary-particle interactions. In 1937 the American physicist Carl Anderson (who had earlier discovered the positron) and his colleagues identified in cosmic rays a particle with a mass 207 times that of the electron. Now called the **muon,** this particle had the same charge and spin as the electron and seemed to behave much like a heavier version of the latter. Two muons were found: the negatively charged μ^- and its antiparticle μ^+. Although the muon mass was close to Yukawa's prediction, the muon interacted only weakly with nuclei and, therefore, could not be the mediator of the nuclear force. "Who ordered that?" quipped Columbia University physicist I. I. Rabi, meaning that there seemed to be no role for the new particle in the then-understood scheme of things.

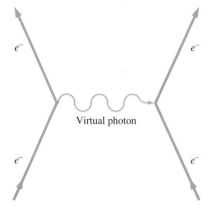

Virtual photon

FIGURE 45-2 A Feynman diagram, showing the interaction of two electrons through the exchange of a virtual photon. Time proceeds upward in the diagram, so the two electrons are initially on converging paths. After exchanging the photon their paths diverge, so the interaction represents a repulsive force.

FIGURE 45-3 Richard Feynman and Hideki Yukawa.

The real Yukawa particle was discovered 10 years later in 1947, again in cosmic rays, and turned out to have a mass about 270 times that of the electron. This time there were three related particles, now called **pions:** the positive π^+, negative π^-, and neutral π^0.

The new particles are all unstable, undergoing decays that ultimately result in the well-known stable particles. The negative pion, for example, decays with a mean lifetime of 26 ns into a negative muon and an antineutrino:

$$\pi^- \rightarrow \mu^- + \overline{\nu}.$$

The muon then decays with a 2.2-μs lifetime into an electron and a neutrino-antineutrino pair:

$$\mu^- \rightarrow e^- + \nu + \overline{\nu}.$$

■ APPLICATION   DETECTING PARTICLES

Despite their minute size, we can, remarkably, follow the trajectories of individual subatomic particles. Early particle detectors used the fact that high-energy particles create ions as they tear through matter. One of the first detectors was the **cloud chamber,** built by C. T. R. Wilson in 1910. The chamber contains a supersaturated vapor that condenses on the ions to produce a visible track. A more recent version is the **bubble chamber,** developed in 1952. Here rapid expansion brings a liquid beyond its boiling point, and bubbles form on the ions left by high-energy particles (Fig. 45-4). The **multiwire proportional chamber** works somewhat like the Geiger counter we described in Chapter 43. In the chamber, criss-crossed grids of fine wire record current pulses from electrons liberated as charged particles pass through the gas-filled chamber; analyzing the pulse distribution reveals the particle trajectory (Fig. 45-5). Development of the multiwire proportional chamber won Georges Charpak the 1992 Nobel Prize. Magnetic fields cause the trajectories to curve, allowing determination of their charge-to-mass ratios. **Scintillation detectors** give off flashes of light as particles pass through them; photomultiplier tubes detect the flashes, whose intensity gives a measure of the particle energy. **Calorimeters,** consisting of composite layers of scintillators and energy-absorbing material, analyze the showers of secondary particles produced by a single high-energy

FIGURE 45-5   A multiwire detector under construction at the Stanford Linear Accelerator Center. High-energy particles induce electrical signals in nearby wires; processing signals from the thousands of wires then allows a computer to reconstruct the particles' trajectories.

(*a*)   (*b*)

FIGURE 45-4   (*a*) Installation of the Big European Bubble Chamber (BEBC) at CERN, the European Laboratory for Particle Physics. (*b*) Subatomic particle tracks in a BEBC photo.

particle to provide accurate measurement of the original particle's energy. Modern detectors are often huge agglomerations of
several basic detector types, arranged to extract as much information as possible from particle interactions (Fig. 45-6). Detector outputs are analyzed by computer, allowing identification of
events so rare that they may occur only once in a million interaction events.

These detection schemes work for charged particles, but what
about neutrals like the neutron, the neutrino, or the π^0? Although their trajectories do not show in the detectors' particle-
trajectory images, they often lead to interactions involving
charged particles whose detection then infers the properties of
the neutral particle.

FIGURE 45-6 Technician working inside the OPAL particle
detector at CERN. OPAL consists of a variety of detector types
arranged concentrically around the point where particles collide.

Particle Classification

Soon more particles were discovered, first in cosmic rays, then with the increasingly energetic particle accelerators that were becoming available. By 1980
there were well over 100 "elementary" particles (Fig, 45-7). Physicists sought
to explain this "particle zoo," and a first step was to classify similar particles into
general categories. Early classification schemes separated particles by mass, but
a more enlightening approach is based on the fundamental forces.

Leptons are particles that do not experience the strong force. They include
the familiar electron, the muon, a more massive particle called the tau, and three
types of neutrinos, one associated with each of the charged leptons. The neutrinos are thought to be massless, although some theories suggest they should have
very small masses. Each of the leptons has an antiparticle as well. There are thus
a total of six lepton-antilepton pairs, and current theory suggests that no others
can exist. The leptons all have spin $\frac{1}{2}$, and are therefore **fermions**—a term we
introduced in Chapter 41 to describe particles that have half-integer spin and
therefore obey the exclusion principle. Leptons are believed to be true elementary point particles with zero size and no internal structure.

Hadrons are particles that do experience the strong force. They fall into two
subclasses: **mesons** have integer spin and are therefore **bosons** that do not obey
the exclusion principle. Mesons include Yukawa's pions and a host of others; all
are unstable. **Baryons** have half integer spins and are therefore fermions. They
include the familiar proton and neutron and similar but more massive particles.
The baryons are grouped into pairs, triplets, and higher multiple groupings of
closely related particles. The neutron and proton, for example, form a pair,
differing in charge and very slightly in mass. Each hadron has an antiparticle,
as do most mesons; but some neutral mesons are their own antiparticles.

The third class of particles comprises the **field particles** or **gauge bosons,**
quanta of the different force fields and "carriers" of those forces. These include
the familiar photon for the electromagnetic force, three particles called the W^+,
W^-, and Z for the weak force, a particle called the gluon, and a hypothetical
graviton that would carry the gravitational force in an as yet incomplete theory

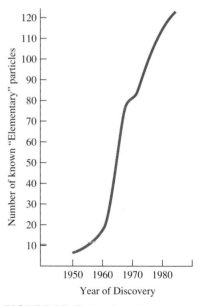

FIGURE 45-7 Total number of particles discovered, as a function of time.
The development of bubble chambers led
to the large increase during the 1960s.

▲ **TABLE 45-1** SELECTED PARTICLES

CATEGORY/ PARTICLE	SYMBOL (PARTICLE/ ANTIPARTICLE)	SPIN	MASS (MeV/c^2)	BARYON NUMBER, LEPTON NUMBERS, STRANGENESS (ANTI-PARTICLE HAS OPPOSITE SIGN)					LIFETIME(s)
				B	L_e	L_μ	L_τ	S	
Field particles									
Photon	γ, γ	1	0	0	0	0	0	0	Stable
Z^0	Z^0, Z^0	1	91177	0	0	0	0	0	$\sim 10^{-25}$
Leptons									
Electron	e^-, e^+	1/2	0.511	0	+1	0	0	0	Stable
Muon	μ^-, μ^+	1/2	105.7	0	0	+1	0	0	2.2×10^{-6}
Tau	τ^-, τ^+	1/2	1784	0	0	0	+1	0	3.0×10^{-13}
Electron neutrino	$\nu_e, \overline{\nu}_e$	1/2	0? (<13 eV)	0	+1	0	0	0	Stable
Muon neutrino	$\nu_\mu, \overline{\nu}_\mu$	1/2	0? (<270 keV)	0	0	+1	0	0	Stable
Tau neutrino	$\nu_\tau, \overline{\nu}_\tau$	1/2	0? (<35 MeV)	0	0	0	+1	0	Stable
Hadrons									
Mesons									
Pion	π^+, π^-	0	139.6	0	0	0	0	0	2.6×10^{-8}
Pion	π^0, π^0	0	135.0	0	0	0	0	0	0.83×10^{-16}
Eta	η^0, η^0	0	549	0	0	0	0	0	$\sim 5 \times 10^{-19}$
Rho	ρ^0, ρ^0	1	768	0	0	0	0	0	$\sim 4 \times 10^{-24}$
Kaon	K^+, K^-	0	494	0	0	0	0	1	1.2×10^{-8}
Kaon	$K^0, \overline{K}^0$	0	498	0	0	0	0	1	0.89×10^{-10} $5.2 \times 10^{-8*}$
Baryons									
Proton	$p, \overline{p}$	1/2	938.3	1	0	0	0	0	Stable†
Neutron	$n, \overline{n}$	1/2	939.6	1	0	0	0	0	889
Lambda	$\Lambda^0, \overline{\Lambda}^0$	1/2	1115.6	1	0	0	0	−1	2.6×10^{-10}
Sigma	$\Sigma^+, \overline{\Sigma}^+$	1/2	1189.4	1	0	0	0	−1	0.80×10^{-10}
Sigma	$\Sigma^0, \overline{\Sigma}^0$	1/2	1192.6	1	0	0	0	−1	7.4×10^{-20}
Sigma	$\Sigma^-, \overline{\Sigma}^-$	1/2	1197.4	1	0	0	0	−1	1.5×10^{-10}
Omega	Ω^-, Ω^+	3/2	2285.2	1	0	0	0	−3	0.82×10^{-10}

* The neutral kaon exists as a quantum-mechanical superposition of states with two different lifetimes.

† If unstable, proton lifetime exceeds 10^{31} years.

of quantum gravity. All the field particles are bosons, carrying spin 1 or, for the graviton, spin 2. You might think Yukawa's meson should be in this category in its role as carrier of the nuclear force. That it doesn't appear here is a hint that the nuclear force is not really fundamental; as we'll soon see, it is the gluon that plays the more fundamental role.

Table 45-1 lists just a few of the known particles.

Particle Properties and Conservation Laws

Many new particles can be characterized by known properties like mass, spin, and electric charge. Of these, spin and charge are associated with important conservation laws—the conservation of angular momentum and the conservation of electric charge. Allowed interactions among particles *must* conserve these quantities. The annihilation of an electron-positron pair, for example, is allowed because the initial particles have no net charge and neither do the resulting photons. Similarly, beta decay of the neutron produces an electron, a proton, and a neutral antineutrino and thus conserves charge:

$$n \rightarrow p + e^- + \overline{\nu}_e.$$

Here we subscript the antineutrino to indicate that it's an electron antineutrino, as opposed to the muon or tau variety.

Other particle properties appear to be conserved as well. Associated with each baryon or antibaryon is its **baryon number,** assigned the value $+1$ for a baryon and -1 for an antibaryon. All experimental evidence to date points to conservation of baryon number: In all observed particle reactions, the sums of the baryon numbers before and after the reaction are equal. An example is, again, beta decay of the neutron: The process starts with a neutron of baryon number $B = 1$ and ends up with a proton $(B = 1)$, an electron, and an antineutrino. The last two are leptons, so their baryon numbers are zero, and thus, baryon number is conserved. Some theories, which we describe shortly, suggest that baryon number is only approximately conserved. If that is so, then the proton itself is an unstable particle with a mean lifetime in excess of 10^{31} years.

Lepton number seems also to be conserved; specifically, there are three separate lepton numbers associated with the electron, the muon, the tau, and their corresponding neutrinos. Again beta decay provides an example: The neutron and proton, being baryons, have electron-lepton number zero, while the resulting electron and *anti*neutrino have electron-lepton numbers of $+1$ and -1, respectively.

In the late 1950s particles called K, Λ, Σ, and Ξ were discovered. Strange characteristics of these particles' decays could be explained by introducing a new fundamental property, called **strangeness.** Previously known particles have zero strangeness, but the new **strange particles** listed above have strangeness $+1$, -1, -1, and -2, respectively. Strangeness is conserved in strong and electromagnetic interactions, but in weak interactions its value can change. We have listed strangeness along with other particle properties in Table 45-1. We'll soon see that several other new properties are needed to characterize matter fully.

● **EXAMPLE 45-1** CONSERVATION LAWS

A pion collides with a proton, producing a neutral kaon and a lambda particle:

$$\pi^- + p \rightarrow K^0 + \Lambda^0.$$

(a) Which of electric charge, baryon number, lepton numbers, and strangeness are conserved? (b) The kaon and lambda are unstable and undergo further decays. Could the end result be an electron and a proton?

Solution
(a) Table 45-1 shows that all the particles are hadrons, so lepton numbers are zero on both sides of the equation. On the left, with a negative and a positive particle, the net charge is zero. On the right, both particles are neutral, so charge is conserved. On the left, the pion is a meson and the proton a baryon, so the total baryon number is $+1$. On the right, Table 45-1 shows that the kaon is a meson and the Λ^0 a baryon, so

baryon number, too, is conserved. Finally, neither the pion nor the proton is strange, so the total strangeness on the left is zero. Table 45-1 shows that the kaon has strangeness $+1$ and the lambda particle has strangeness -1. Thus strangeness, too, is conserved.

(b) Having an electron and a proton as the final state would conserve charge (0), strangeness (0), and baryon number (1). But it would not conserve electron-lepton number, which is zero for the original particles and would become $+1$. Adding an electron antineutrino would, however, satisfy all conservation laws.

EXERCISE Is the interaction $p + n \rightarrow p + p + n + \overline{p}$ permitted by conservation of baryon number?

Answer: Yes

Some problems similar to Example 45-1: 5–9 ●

(a)

(b)

FIGURE 45-8 *(a)* Theorists C. N. Yang (left) and T. D. Lee suggested that parity need not be conserved. *(b)* Experimentalist Chien-Shiung Wu and her collaborators verified the Lee-Yang suggestion.

Symmetries

Watch a physical process in a mirror and you would expect the image ought also to be a possible physical process. If so, then the laws of physics should exhibit **symmetry** with respect to mirror reflection. At the subatomic level, the statement that a process and its mirror image are equally likely is called **conservation of parity.** Mathematically, a system has parity $+1$ if its wave function is unchanged on reflection through the origin—that is, on a change of coordinates $x \rightarrow -x$, $y \rightarrow -y$, $z \rightarrow -z$. If the wave function changes sign then the parity is -1. Parity is conserved if its value before a system undergoes some interaction is the same, either $+1$ or -1, as it is afterwards.

In 1956 there were two particles, then called τ and θ, that seemed identical in all respects except that their decay products had different total parity. In that year theoretical physicists T. D. Lee and C. N. Yang (Fig. 45-8) noted that conservation of parity, known to hold for electromagnetic and strong interactions, had not been tested for interactions involving the weak force. They made the revolutionary suggestion that parity need not be conserved—a statement tantamount to suggesting that nature can, at a fundamental level, distinguish right- from left-handed systems that are otherwise identical.

Experimentalists quickly set out to test the Lee-Yang suggestion. A group led by Chien-Shiung Wu (Fig. 45-8) studied the beta decay of cobalt-60 by aligning ^{60}Co nuclei with a magnetic field at the very low temperature of 0.01 K. If the beta-decay electrons were emitted with equal probability along and opposite the field direction, Wu and her co-workers reasoned, then the original process and its mirror image would be equally likely (Fig. 45-9a). But what they found was a preferential beta emission opposite the field direction—a clear indication that the mirror world behaved differently (Fig. 45-9b).

Although parity might be violated, theorists held that a combination of parity reversal (P) and charge conjugation (C)—changing particles into their antiparticles—would result in physical systems whose behavior was indistinguishable. But in 1964 even this so-called CP conservation was found to be violated, albeit

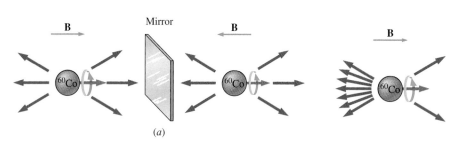

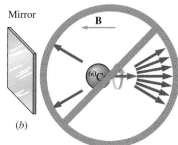

(a)   *(b)*

FIGURE 45-9 Experimental evidence for nonconservation of parity. At left of mirror, a ^{60}Co nucleus has its spin aligned with a magnetic field; in a classical picture, this would correspond to the nucleus spinning with its top part emerging from the page. Reflected in a mirror, this sense of spin does not change, so the spin vector in the mirror image still points to the right, even though the magnetic field is reversed. *(a)* In order that this mirror image be an equally likely physical situation, beta emission (blue arrows) must occur with equal probability along and against the spin direction. *(b)* Experiment shows that beta emission occurs preferentially opposite the spin direction, so the mirror-image situation at right in *(b)* does not occur.

only in the rare decay of the neutral kaon into a pion-antipion pair. The Russian physicist Andrei Sakharov suggested that this asymmetric decay into matter and antimatter might have led to a slight imbalance between the two in the early universe and thus could be responsible for the present situation in which we find large quantities of matter but essentially no antimatter.

It still appears that *CPT* conservation holds—that is, a combination of mirror reflection, charge conjugation, and reversal of the time coordinate $t \rightarrow -t$ makes a new physical process indistinguishable from the original. For most processes *C*, *P*, and *T* hold individually; for weak decays *C* and *P* are separately violated but *CP* and *T* still hold, while for the rare kaon events even time reversal does not give an equally probable physical process. There may be a deep philosophical connection here involving the direction of time, but the full implications of *CPT* symmetry and the failure of its individual components are not fully understood.

45-3 QUARKS AND THE STANDARD MODEL

The proliferation of particles distressed physicists used to finding an ultimate simplicity underlying the phenomena of nature. What kind of a universe is it that requires hundreds of "elementary" particles? Were they all really "elementary," or was there some still more fundamental level exhibiting the sought-after simplicity?

The Eightfold Way

In 1961, physicists Murray Gell-Mann and Yu val Ne'eman independently noted patterns in the groupings of the then-known particles. They plotted strangeness versus electric charge and found that the particles fell into groups that formed simple geometric patterns (Fig. 45-10). The patterns were called collectively the **Eightfold Way,** after a set of Buddhist principles for right living.

Clearly the Eightfold-Way patterns revealed some underlying organization or structure, in much the way that the periodic table of the elements hints at the building of atoms from more basic particles in an organizational scheme that produces the chemical elements. A triumph came when Gell-Mann arranged the spin-$\frac{3}{2}$ baryons into an Eightfold-Way pattern and found an empty spot where a

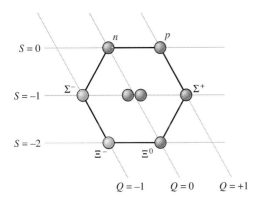

FIGURE 45-10 There are eight spin-$\frac{1}{2}$ baryons. On a plot of strangeness versus charge, they form a distinct hexagonal pattern. Similar patterns exist for other particle groups.

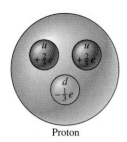

Proton

Neutron

FIGURE 45-11 Protons and neutrons consist of the quark combinations *uud* and *udd*, respectively.

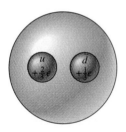

FIGURE 45-12 Mesons consist of a quark and an antiquark. The meson shown here is the π^+, made from an up quark and a down antiquark.

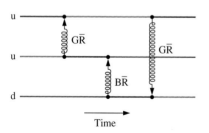

FIGURE 45-13 The color force binds quarks through the interchange of gluons. Shown here is a proton, consisting of the quark combination *uud*. The three quarks are continually exchanging gluons, changing their colors but not flavors. The composite proton remains colorless throughout.

particle with strangeness $S = -3$ should be. Experimentalists soon found the particle—the Ω^-—in bubble-chamber photographs from previous experiments.

Quarks

Success of the Eightfold Way convinced physicists that many "elementary" particles were not really elementary. In 1964 Gell-Mann and his Cal Tech colleague George Zweig independently proposed a set of three particles that could be combined to form the then-known hadrons. Gell-Mann called these particles **quarks,** a word used in James Joyce's novel *Finnegan's Wake.* Gell-Mann's quarks became known as the **up quark** (*u*), the **down quark** (*d*), and the **strange quark** (*s*); the latter was needed to give strange particles their "strangeness." For each quark there was also an antiquark.

One surprising thing about quarks is that they carry fractional electric charges. The two least massive quarks, the up and the down, carry $+\frac{2}{3}e$ and $-\frac{1}{3}e$, respectively; their antiparticles have the opposite charges. The quarks combine in pairs or triplets to make the two classes of hadrons. Baryons, like the proton and the neutron, consist of three quarks (Fig. 45-11). Mesons, in contrast, contain a quark and its antiquark (Fig. 45-12). The quarks all have spin $\frac{1}{2}$, which explains why the three-quark baryons all have odd half integer spin, and why the two-quark mesons have integer spin.

The Pauli exclusion principle precludes three particles having the same quantum numbers, and therefore, there must be an additional property distinguishing quarks. Called **color,** this property is a kind of "charge"—not to be confused with electric charge—that can take on any of three values called, whimsically, red, green, and blue. The force binding quarks of different colors is the **color force,** and the quark theory is therefore known as **quantum chromodynamics (QCD).** In QCD, particles called **gluons** play the role of photons in quantum electrodynamics, binding particles subject to the color force. Particles formed from quarks—the mesons and the baryons—are always **colorless.** This is true of mesons because they contain a quark of one color and another of its anticolor. It is true of baryons because they contain three quarks of different colors which combine to give the baryon zero net color charge. The nuclear force, once thought to be fundamental, is actually a residual manifestation of the color force, acting between the quarks in colorless particles—in much the same way that the van der Waals force between neutral gas molecules is a "residue" of the stronger electric force among the particles comprising the molecules.

Photons mediate the electromagnetic force between charged particles, but are themselves uncharged. In contrast, gluons, like the quarks they bind, carry color charge. There are eight different gluons; six carry combinations like red-antiblue ($R\overline{B}$), green-antired ($G\overline{R}$), etc.; the other two are colorless. Exchange of a colored gluon, unlike photon exchange in quantum electrodynamics, thus changes the colors of the particles involved (Fig. 45-13).

Another surprising aspect of quarks is that the color force binding them into composite particles does not decrease with separation. For that reason it appears impossible to isolate a single quark (Fig. 45-14). As a result we never see individual free particles with fractional electric charge.

Today the quark model is universally accepted. Many of the previously "elementary" particles—including protons and neutrons—are now considered

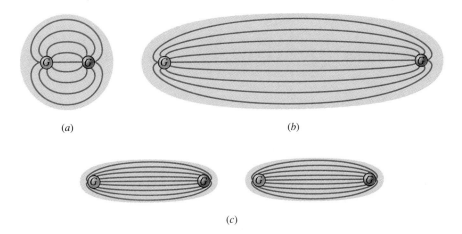

(a) (b)

(c)

FIGURE 45-14 Quark confinement in a meson consisting of a green quark and its antiquark. (*a*) Field lines represent the "color field" that joins the two. (*b*) The field remains confined as the quarks are moved apart, so the field strength stays essentially constant. Infinite energy would therefore be required to separate the quarks. (*c*) Pulling the quarks far apart builds up enough energy to create a quark-antiquark pair, with the result that the energy goes into creating new composite particles rather than isolating individual quarks.

composites of quarks. Electrons and other leptons seem, on the other hand, to be truly elementary.

The up, down, and strange quarks soon proved insufficient to account for all the observed particles. Theorist Sheldon Glashow argued for a fourth quark, which he called the **charmed quark.** Ten years later, following intensive searches, experimental teams at Brookhaven National Laboratory and the Stanford Linear Accelerator Center announced the discovery of a particle that implies the existence of the charmed quark (Fig. 45-15). The Brookhaven group called the new particle J and the Stanford group called it ψ; to this day it is called the J/ψ. The charmed and strange quarks form a related pair, similar to the up/down quark pair.

Since the J/ψ discovery, yet another pair of quarks has been proposed. A 1977 experiment confirmed the existence of the **bottom quark,** and in 1994 the quark model was fully verified in Fermilab experiments giving evidence for the **top quark.** The more exotic quarks are more massive, and therefore, through mass-energy equivalence, it takes more energy to produce particles containing them. This need for higher energy is what drives the push for ever more powerful and expensive particle accelerators (see Application: Particle Accelerators, later in this chapter.) Table 45-2 lists some properties of the six quarks.

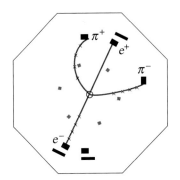

FIGURE 45-15 A computer reconstruction of one of the first ψ-particle events detected at Stanford. An electron-positron collision created the ψ particle, which decayed into four pions. Here we see the pion tracks, which happen to make a shape resembling the letter ψ.

▲ **TABLE 45-2** MATTER PARTICLES OF THE STANDARD MODEL

QUARK NAME	SYMBOL	MASS* (MeV/c²)	CHARGE	CORRESPONDING LEPTONS (SYMBOL, MASS IN MeV/c²)
Down	d	5–15	$-\frac{1}{3}e$	Electron (e, 0.511), electron neutrino (ν_e, 0†)
Up	u	2–8	$+\frac{2}{3}e$	
Strange	s	100–300	$-\frac{1}{3}e$	Muon (μ, 106), muon neutrino (ν_μ, 0†)
Charmed	c	1300–1700	$+\frac{2}{3}e$	
Bottom	b	4700–5300	$-\frac{1}{3}e$	Tau (τ, 1784), tau neutrino (ν_τ, 0†)
Top	t	1.7×10^5	$+\frac{2}{3}e$	

* Quark masses are estimates; specifying quark masses is difficult because individual quarks are not found in isolation and because the mass equivalent of hadron binding energy is very large.

† Neutrino upper mass limits are <7.3 eV, <270 keV, and <35 MeV, for ν_e, ν_μ, ν_τ, respectively.

● **EXAMPLE 45-2** QUARK COMPOSITION

Evaluate the charge and strangeness of the particle with quark composition *uds*.

Solution

From Table 45-2 we see that the net charge is $(\frac{2}{3}e) + (-\frac{1}{3}e) + (-\frac{1}{3}e) = 0$. Since the *s* quark has strangeness -1 while the up and down are not strange, the total strangeness is -1. The particle is, in fact, the Λ^0 listed in Table 45-1.

EXERCISE Find the charge and strangeness of the Ξ^-, whose quark composition is *dss*.

Answer: $-e$, -2

Some problems similar to Example 45-2: 14–17 ●

The Standard Model

We now have six **flavors** of quarks—up, down, strange, charmed, top, bottom—that seem to be truly elementary constituents of matter. As we've seen, quarks join to form all the hadrons—the baryons and mesons. But other particles, namely the leptons and the field particles, are not made from quarks. They, like the quarks, are believed to be truly indivisible, elementary particles.

In this "zoo" of elementary particles, physicists recognize three distinct "families." The up and the down quarks comprise the neutron and proton; together with the electron and its related neutrino, they account for the properties of ordinary matter. A second family consists of the strange and charmed quarks, the electron-like muon, and the muon neutrino. The quarks of this family are more massive than the up and down quarks, and the muon is more massive than the electron. More massive still are the particles of the third family, consisting of the top and bottom quarks, the electron-like tau particle, and the tau neutrino. Table 45-2 shows these three families from which all known matter is constructed.

You may be expecting that future editions of this book will tell of additional quarks, and thus of additional families of matter. Whether or not such additional families exist was an open question until 1991, when physicists at the Large Electron Positron Collider in Geneva, Switzerland, examined over half a million particle decay events to conclude that the number of different types of neutrinos that can exist is 2.99 ± 0.06. Since there is presumably a neutrino type for each family, this result seems to preclude the existence of additional families.

The theory that currently describes elementary particles and their interactions is called the **standard model.** In addition to the constituent particles of matter shown in Table 45-2, the standard model includes the photons that mediate the electromagnetic interaction, the gluons of the color force, the *W* and *Z* particles that mediate the weak force, and an as yet undetected particle called the Higgs boson, believed responsible for other particles' masses (Table 45-3). The standard model is very successful in explaining the phenomena of particle physics, but it leaves many fundamental questions unanswered. Why, for example, do the quarks and leptons have the particular masses they do? Why are there only three families of elementary particles? Why are leptons and quarks distinct? Are these particles really elementary, or are there even smaller structures at hitherto unexplored scales? Continuing theoretical work and experiments at ever higher energies may someday answer these questions.

▲ **TABLE 45-3** FIELD PARTICLES AND FORCES

PARTICLE	MASS (GeV/c^2)	ELECTRIC AND COLOR CHARGE	FORCE MEDIATED	RANGE	APPROXIMATE STRENGTH AT 1 fm (RELATIVE TO COLOR FORCE)
Graviton	0	0, 0	Gravity	Infinite	10^{-38}
$W^{\pm}$	80.2	$\pm 1, 0$	Weak	$<2.4 \times 10^{-18}$ m	10^{-13}
Z^0	91.2	0, 0			
Photon, γ	0	0, 0	Electromagnetic	Infinite	10^{-2}
Gluon, g (8 varieties)	0	0, 6 color/anticolor combinations, 2 colorless	Color	Infinite*	1
Higgs boson, H^0	40–1000?	0, 0	The Higgs boson is an as yet undetected particle needed to account for the nonzero masses of the $W^{\pm}$ and Z^0, as well as of other particles.		

*The nuclear force is the residual color force between colorless particles and has a range of about 1 fm (10^{-15} m).

45-4 UNIFICATION

We first introduced the fundamental forces of nature in Chapter 5, indicating that today physicists recognize just three such forces: gravity, the electroweak force, and the color force. It was not always that simple, though. In Chapters 23 through 34 we studied the electric and magnetic forces, first separately but then with realization that they fell under the single umbrella of electromagnetism. The unification of electricity and magnetism was a major step forward in our understanding of physical reality. Physicists continue to strive for further unification, with the ultimate hope that someday all the forces will be understood as a manifestation of a single common interaction.

Electroweak Unification

In the 1960s and early 1970s, a century after Maxwell formalized the unification of electromagnetism, physicists Steven Weinberg, Abdus Salam, and Sheldon Glashow proposed that the electromagnetic force and the weak force are really aspects of the same thing. Their theory predicted the existence of the particles W^+, Z^0, and W^-, the "carriers" of the unified electroweak force. In 1983 a huge international consortium headed by Carlo Rubbia discovered the W and Z particles, using advances in accelerator technology developed by Simon van der Meer (Fig. 45-16). That discovery confirmed the electroweak unification, and Rubbia and van der Meer joined a long list of physicists who had won the Nobel Prize for contributions to our understanding of the structure of matter (Table 45-4).

Further Unification

Electroweak unification led to the present situation in physics, which has gravity, electroweak, and color forces as the fundamental interactions of matter. A further step, the **grand unification theories** (GUTs), attempt to merge the

(a)

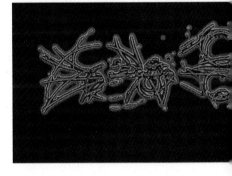

(b)

FIGURE 45-16 (a) Carlo Rubbia and Simon van der Meer at CERN, the European Laboratory for Particle Physics. (b) Computer image showing particle tracks resulting from decay of a W particle.

▲ **TABLE 45-4** NOBEL PRIZES FOR INTERPRETING
THE STRUCTURE OF MATTER

PRIZE YEAR	PRIZE RECIPIENT	RESEARCH CONTRIBUTION[+]
1903	Antoine H. Becquerel	Discovery of radioactivity (1896)
	Marie and Pierre Curie	Studies of radioactivity
1906	J. J. Thomson	Discovery of the electron and other studies (1897)
1908	Ernest Rutherford*	Disintegration of elements using alpha particles (1902)
1935	James Chadwick	Discovery of the neutron (1932)
	Irène and Frédéric Joliot-Curie	Synthesizing new radioactive elements (1934)
1936	Victor F. Hess	Discovery of cosmic rays (1911)
	Carl D. Anderson	Discovery of the positron (1932)
1938	Enrico Fermi	Production of new radioactive elements (1934–1937)
1949	Hideki Yukawa	Theoretical prediction of mesons (1934)
1951	John D. Cockcroft and Ernest T.S. Walton	Transmutation of nuclei (1932)
1957	C.N. Wang and T.D. Lee	Theoretical prediction of parity violation (1956)
1959	Emilio G. Segrè and Owen Chamberlain	Production of the antiproton (1955)
1963	Maria Goeppert Mayer and J. Hans Jensen	Nuclear shell model (1949)
	Eugene P. Wigner	Application of symmetry to elementary particle theories
1968	Luis Alvarez	Contributions to particle accelerators (1946)

[+]Number in parentheses is year discovery was made.
* Rutherford and Joliot-Curie prizes in chemistry; all others in physics.

FIGURE 45-17 The proton-decay experiment of the University of California at Irvine, the University of Michigan, and Brookhaven National Laboratory, known as IMB, contains 4000 tons of water in a salt mine in Ohio. Photomultipliers record light signals from particle reactions in the water. So far no proton decays have been detected, although the experiment did "see" neutrinos from the supernova 1987A. Here a diver makes repairs in the ultra-pure water that fills the experimental chamber.

electroweak and color forces. The simplest versions of these theories predict that the proton is not a stable particle but should decay on the very long timescale of some 10^{31} years. We can't wait that long, but we can put 10^{33} protons together (that's about 4000 tons of water) and watch for one year (Fig. 45-17). Experiments of this sort have not found the predicted decays, thus ruling out the simplest form of the GUTs. Two other GUTs predictions—that magnetic monopoles exist and that neutrinos have mass—are also under investigation. Despite the negative results from proton decay experiments, many physicists believe that grand unification will soon be achieved.

Even grand unification would still leave two forces, one of them gravity. Attempts to reconcile our current theory of gravity—Einstein's general theory of relativity—with quantum mechanics have so far made little progress. Yet such a reconciliation is a necessary prerequisite for a final unification of all known forces. In the 1980s a flurry of interest developed in so-called **superstring theories,** which picture elementary particles as fundamental vibration modes on string-like structures that may be as short as 10^{-35} m (Fig. 45-18). This model is set not in the four-dimensional spacetime to which we have become accustomed, but in a ten-dimensional spacetime with six of its dimensions

PRIZE YEAR	PRIZE RECIPIENT	RESEARCH CONTRIBUTION[+]
1969	Murray Gell-Mann	Theory classifying elementary particles (1964)
1975	Aage Bohr, Ben Mottelson, and James Rainwater	Studies of nuclear structure and dynamics
1976	Burton Richter and Samuel C. C. Ting	Discovery of the J/ψ particle, the first known particle made from charmed quarks (1973–1974)
1979	Sheldon L. Glashow, Abdus Salam, and Steven Weinberg	Unification of the electromagnetic and weak forces (1961–1972)
1980	Val L. Fitch and James W. Cronin	Experimental proof of violation of conservation of charge-parity (1964)
1984	Carlo Rubbia and Simon van der Meer	Discovery of W particles, predicted by the electroweak unification (1983); stochastic cooling for focusing particle beams (1968)
1988	Leon M. Lederman, Melvin Schwartz, and Jack Steinberger	Production of neutrino beams and discovery of the μ neutrino (1960–1962)
1990	Jerome Friedman, Henry Kendall, and Richard Taylor	Using scattering of electrons from protons and neutrons to explore the quark model
1992	Georges Charpak	Developing electronic detectors for detecting subatomic particles in accelerators

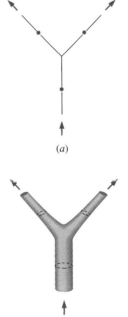

(a)

(b)

FIGURE 45-18 (a) A particle decay involving point particles in the standard model. (b) A similar decay in string theory. At a given instant in time, the particle is represented by a string-like loop. Its trajectory in spacetime traces out the cylindrical structures shown. Note that there is no single point in spacetime at which the decay takes place.

"compactified" in a way that makes them undetectable in normal interactions. To some physicists, string theories hold the promise of a "theory of everything," explaining all our observations about the behavior of the universe. To others they are another in a long list of unsuccessful attempts at a comprehensive explanation of physical phenomena. Only further research will tell.

Symmetry Breaking

Unification theories predict that phenomena that appear distinct under normal conditions will be clearly seen as one at sufficiently high energies. The observed unification represents a kind of symmetry that is "broken" as the energy level drops. A crude mechanical analog for such **symmetry breaking** is a ball rolling around the rim of a bowl. As long as the ball keeps rolling, there is no preferred angular direction. But if the ball stops, it drops into the bowl in a particular direction. The symmetry of the original situation has been broken. Analogously, at energies above 100 GeV, what we call the electromagnetic and weak forces are one and the same. But at lower energies the symmetry is broken, and we see two distinct forces. Particle accelerators now being planned will exceed the energy of electroweak symmetry breaking, allowing us to explore that interaction in its fundamental simplicity. But the energy at which symmetry breaking occurs increases to some 10^{15} GeV as we move from electroweak to grand unification, making it unlikely that we will achieve that energy in the foreseeable future. And the energy at which gravity, too, would join a single unified force is an even more remote 10^{19} GeV.

■ APPLICATION PARTICLE ACCELERATORS

Most particles are more massive than the proton and some, like the weak-force mediators $W^{\pm}$ and Z, are extremely massive. Since the more massive particles are all unstable, discovering them involves first creating them—and that requires energy of at least mc^2, with m the particle mass. That energy requirement, along with the hint of new phenomena like force unification, drives particle physicists' seemingly insatiable desire for particle accelerators of ever higher energies.

The earliest accelerators were electrostatic devices that established large potential differences between conducting electrodes and used the associated electric field to accelerate charged particles (Fig. 45-19). But such accelerators are limited to maximum energies around 20 MeV because of the difficulties of handling high voltages. We saw in Chapter 29 how this problem is cleverly circumvented in the cyclotron, an accelerator that uses a magnetic field to keep particles in circular orbits so they can gain energy on each orbit from a modest electric field. But cyclotrons work only for nonrelativistic particles, for which the cyclotron frequency is independent of particle energy. Today's high-energy experiments call for ultrarelativistic particles, whose speeds differ only minutely from the speed of light. As a result, today's accelerators are primarily variations on the **synchrotron,** a device in which the magnetic field increases with the particle energy to maintain particles in a circular orbit of fixed radius (Fig. 45-20). An alternative to the synchrotron is the **linear accelerator,** the largest of which is the Stanford Linear Accelerator shown in Fig. 38-15.

A head-on collision between two cars is much more damaging than a collision of a moving car with a stationary one since in the former case all the energy goes into damaging the cars while in the latter a great deal of it goes into accelerating the initially stationary "target" car. For the same reason head-on collisions of high-energy particles make much more energy available for creation of new particles. As a result, most of the highest energy accelerators today are colliders, with particle beams accelerated in opposite directions and brought to collide inside elaborate detectors (see Application: Detecting Particles earlier in this chapter). The largest accelerator designed so far is the Superconducting Super Collider (SSC), whose construction was halted by the U.S. Congress in 1993. With its 20-TeV

FIGURE 45-19 This van de Graaf accelerator, now at the Boston Museum of Science, develops a potential difference of 1.5 MV. It was originally used as a particle accelerator.

$(20 \times 10^{12}$ eV) colliding proton beams, the SSC should have been able to create the Higgs boson, a hypothetical particle believed responsible for symmetry breaking and for other particles' masses. It would also have created conditions that physicists believe existed after the universe was only 10^{-12} s old; more on this in the next section. Figure 45-21 compares some of the world's most energetic accelerators.

(a)

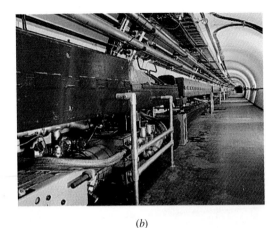

(b)

FIGURE 45-20 (a) Aerial view of Fermilab, the national accelerator laboratory outside Chicago. The main ring is 2 km in diameter and accelerates protons and antiprotons to 1 TeV. Beams of the two particle species then collide to produce scores of new particles. (b) Inside the main ring. The red structures comprise a ring of conventional electromagnets that steer the particles until their energy is large enough for injection into the superconducting magnet ring (yellow, below).

Superconducting Super Collider, U.S.
20 TeV × 20 TeV, protons-antiprotons
(cancelled 1993)

Experimental halls

Main tunnel

West campus

1 km

East campus

HEB
MEB
LEB
Linac

Injectors

Large Electron-Positron Collider, CERN, Switzerland
50 GeV × 50 GeV, electrons-positrons

Stanford Linear Collider, U.S.
50 GeV × 50 GeV, electrons-positrons

Experimental halls

UNK, Russia
3 TeV, protons

Tevatron, Fermilab, U.S.
1 TeV × 1 TeV protons-antiprotons

FIGURE 45-21 Drawings of the world's largest accelerators, shown on the same scale. The Superconducting Super Collider, had it been completed, would have measured 87 km in circumference—about the same size as the beltway surrounding Washington, DC. Note that the larger machines have substantial injector systems, consisting of one or more smaller accelerators to bring the particles to high energy before injection into the main ring. For the SSC these include a linear accelerator (linac), and low, medium, and high-energy boosters. Energies of the separate particle beams are listed for collider machines.

FIGURE 45-22   Edwin Hubble in about 1960, at the prime focus of the 5-m telescope at Palomar Observatory in California.

FIGURE 45-23   In Einstein's general theory of relativity, a finite universe would be like the 4-dimensional analog of the surface of a sphere. If the sphere expands, all the galaxies move apart at speeds proportional to their distances from each other. Yet there is no "center" to this expanding universe—i.e., to the sphere's surface.

45-5 THE EVOLVING UNIVERSE

We come at the end to the broadest possible questions about physical reality: How did the universe come to be? What is its overall structure? What will its future bring? Remarkably, these cosmological questions are closely linked with the questions of particle physics.

Expansion of the Universe

Early in the twentieth century, astronomers argued about the nature of certain fuzzy patches visible in telescope photographs. Many thought they were gas clouds scattered among the visible stars, but others made a more radical proposal: that some of these "nebulae" were gravitationally bound systems containing billions of stars and that they were almost inconceivably distant.

In the 1920s, the opening of the 2.5-m telescope at California's Mt. Wilson Observatory finally resolved the issue. There, astronomer Edwin Hubble (Fig. 45-22) proved that some nebulae were indeed distant galaxies like our own Milky Way, each containing some 10^9 to 10^{13} stars. Today cosmologists think of galaxies as individual "point particles" whose distribution traces the overall structure of the universe.

Hubble continued to study the galaxies through the 1920s, and by analyzing their spectra he made a remarkable discovery: spectral lines from distant galaxies are shifted toward the red, with the amount of shift dependent on the distance to the galaxies. The most reasonable and widely held explanation is that the redshift is caused by the Doppler effect (see Section 17-7). Then the implication of Hubble's work is that the distant galaxies are receding from us at speeds proportional to their distances. This result is known as the **Hubble law:**

$$v = H_0 d, \tag{45-1}$$

where v is the recession speed, d the distance, and H_0 the **Hubble constant.** The value of the Hubble constant is controversial, with different astronomers arguing for values that range from 15 to 30 kilometers per second per million light-years of distance. Astronomers now use the Hubble relation to find the distances to remote galaxies, measuring their redshifts and using Equation 45-1 to infer their distances.

It may sound like Hubble's law puts us right in the center of things, in grotesque violation of modern science's view that Earth and its inhabitants do not occupy a favored position in the universe. But actually the inhabitants of any other galaxy would observe the same thing: all the distant galaxies would be receding from them at speeds proportional to their distances. As long as the universe is infinite in extent, none can claim to be at the center. And if it's not infinite, then Einstein's general theory of relativity gives it a closed-curve shape which still has no center (Fig. 45-23).

The Hubble relation implies that the universe is expanding, as all the galaxies move apart. Extrapolating the expansion backwards in time suggests there was a time when all the galaxies were together. Thus Hubble's discovery is evidence that the universe may have had a definite beginning, in the form of a colossal explosion that flung the matter of the universe into an expansion that continues today. Based on additional evidence that we will describe shortly, most scientists now accept the theory that the universe began with such an explosive **big bang.** Hubble's pioneering work on the universe's origin is honored in the naming of the Hubble Space Telescope.

EXAMPLE 45-3   HOW OLD IS THE UNIVERSE?

Using the value $H_0 = 15$ km/s/Mly and assuming the expansion rate has been constant, find how long the universe has been expanding.

Solution
Today, a galaxy a distance d from our Milky Way is moving with speed $v = H_0 d$ (Equation 45-1). If the galaxy has been moving at constant speed, then $d = vt$, and the time since the expansion started must be

$$t = \frac{d}{v} = \frac{d}{H_0 d} = \frac{1}{H_0}.$$

This quantity is called the **Hubble time.** To find its value, we need to convert from the mixed units used for the Hubble constant:

$$t = \frac{1}{H_0} = \frac{1}{(15\ \text{km/s/Mly})/[3.0 \times 10^5\ (\text{km/s})/(\text{ly/y})]} = 20\ \text{Gy},$$

where we used the speed of light—3.0×10^5 km/s—to convert km/s to ly/y, thus giving the answer in years. So the universe is about 20 billion years old under the assumptions of this example. (A more accurate calculation would account for the gravitationally caused decrease in the expansion rate.)

EXERCISE What is the distance to a galaxy whose redshift shows it to be moving away from Earth at 1% of the speed of light?

Answer: 200 Mly

Some problems similar to Example 45-3:   25–29 ●

The Cosmic Background Radiation

In 1965 Arno Penzias and Robert Wilson at Bell Laboratories found a faint "noise" of microwave radiation in a satellite communications antenna they were testing. They could not eliminate the noise, which seemed to come from all directions. Theorists at Princeton pointed out that Penzias and Wilson had detected radiation dating to a much earlier era in the universe. This **cosmic microwave background radiation** is the strongest evidence for the big bang theory.

The big bang theory suggests that the universe started very hot and then cooled as it expanded, doing work against its own gravitation. At first it was so hot that any atoms that formed would be dashed apart by collisions at the high thermal energy prevailing. Thus, in its early times the universe was populated by individual charged particles. These interacted readily with electromagnetic radiation, making the universe opaque. But by about 10^5 years the temperature had dropped to several thousand kelvins, and at that point atoms of hydrogen and helium could form (there were almost no heavier nuclei present; see Chapter 43). Since neutral atoms interact much more weakly with electromagnetic radiation, the universe became transparent, and photons emitted as the atoms formed could travel throughout the universe with little chance of being subsequently absorbed. Those photons became the cosmic background radiation, permeating the entire universe.

Expansion of the universe would cool not only matter but also radiation; the big bang scenario suggests that radiation from the time the atoms formed should now be cooled to that of a blackbody at 2.7 K—a radiation distribution that peaks in the microwave region of the electromagnetic spectrum. Analysis of the microwave background shows that it has indeed precisely that blackbody char-

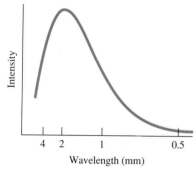

FIGURE 45-24 Spectrum of the cosmic microwave background, taken with the Cosmic Background Explorer satellite (COBE). The observational plot so perfectly matches that of a blackbody at 2.726 K that the discrepancies are less than the width of the curve drawn here. COBE was built with the specific goal of making detailed studies of the cosmic background radiation.

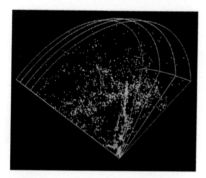

FIGURE 45-25 The positions of 1065 galaxies in a wedge-shaped region extending more than one-third of the way around the sky. The radial distance from Earth extends outward from the apex to a maximum of about 1 billion light-years. The galaxies are distributed on the edges of large bubble-like voids.

acter and temperature (Fig. 45-24). Thus, the observed cosmic microwave background is a direct reflection of conditions 10^5 years after the universe began.

An outstanding problem in cosmology is the origin of large-scale structure in the universe. Why should an initially homogeneous "soup" of individual particles eventually organize itself into large "clumps" like galaxies? Increasingly, astronomical observations show that the galaxies themselves are organized into much larger structures, concentrated on the edges of giant bubble-like voids hundreds of millions of light-years across (Fig. 45-25). How could this structure arise from an initially structureless big bang? A partial answer came in 1992, with the discovery that the cosmic microwave background itself contains very faint "ripples" that may be the "seeds" of larger structures that developed later as the universe expanded.

The Earliest Times

With the microwave background, we can "see" the universe as it was 10^5 years after its beginning. We also have a direct tie to an earlier time, from about 1 second to tens of minutes, when the lightest nuclei were forming. The first composite nuclei were the simplest: deuterium, consisting of a proton and a neutron. The rate of deuterium formation is critically sensitive to the expansion rate of the universe. If the expansion was rapid, then the density dropped fast enough that a deuteron, once formed, was unlikely to be disrupted in a subsequent collision. But a slower expansion rate would result in less deuterium, as deuterons and protons interacted to form tritium and helium-3. Very sensitive measurements of deuterium abundance, based on spectral lines from interstellar deuterium, therefore provide direct evidence for conditions early in the big bang.

We may soon be able to look further back, by studying a "neutrino background," believed to date from about 1 second after the big bang began, a time when the density became low enough that neutrinos ceased to have much interaction with other matter. Evidence for still earlier times comes indirectly from applying elementary particle physics to the behavior of matter at extreme temperatures and seeing if it predicts an evolution to conditions we do observe. Today's accelerators achieve conditions that existed only 10^{-9} s from the beginning, and the Superconducting Super Collider would have pushed that back to 10^{-12} s. Thus, these early times provide a test of elementary particle physics, and that physics in turn helps us to understand how the universe began and how it evolved to its present state. Figure 45-26 summarizes our understanding of the evolution of the universe, showing that the phenomena of particle physics and unification are inextricably tied with cosmic questions.

The Inflationary Universe

The original big bang theory has problems with several observed features of the universe. Why, for example, do we find only matter but not antimatter? Why does the whole universe seem essentially homogeneous and in thermodynamic equilibrium on the largest scales, when the most distant regions are so far apart that light could not have traveled between them in the time since the beginning?

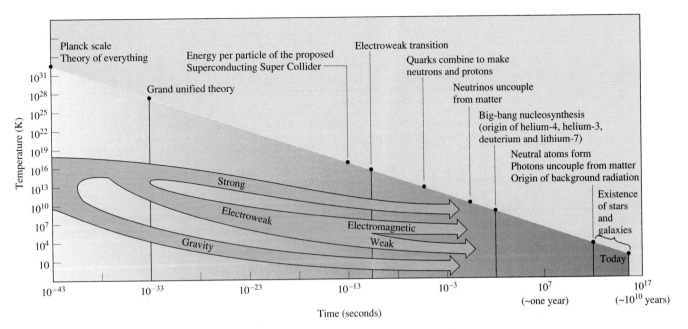

FIGURE 45-26 The evolution of the universe from the earliest times to the present. Note the highly logarithmic scale. As the universe has expanded, its average temperature and energy have fallen, creating epochs in which different physical processes dominated. At first the fundamental forces were all one, but symmetry breaking caused them to separate as the energy dropped. Present-day physics cannot address times before 10^{-43} s, which requires a merging of quantum theory and general relativity.

And why—as we elaborate in the next section—do the kinetic and potential energy of the universe seem exactly equal?

The current solution to these conundrums is the **inflationary universe** theory, first advanced by Alan Guth of MIT. This theory incorporates the ideas of grand unification and thus links the evolution of the universe directly with particle physics. At the earliest times, the temperature and thermal energy were so high that all the forces were unified. The inflationary theory proposes that, at 10^{-35} s, the universe expanded at a phenomenal rate, growing 10^{50} times more than in the original big bang theory (Fig. 45-27). This rapid expansion lasted a mere 10^{-32} s, but that short period was sufficient to establish many of the characteristics we now observe.

Because of the tremendous expansion, now-distant locations would have once been much, much closer—so close that they could have reached the thermodynamic equilibrium now evident at large scales. As the universe cooled, it acted like a very pure liquid that can cool well below its normal freezing point without freezing. Analogously, the thermal energy of the universe dropped below the value at which the forces all appear unified, but unification still held. In this supercooled state, the universe had excess energy, and this energy led to the inflation in a way that can be predicted from theory. When the symmetry was finally broken and the individual forces appeared distinct, the excess energy was transformed into the matter and energy we observe today. Thus, in this theory

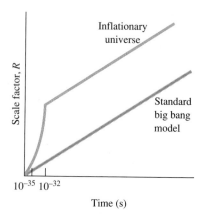

FIGURE 45-27 Expansion in the standard and inflationary big-bang theories. The scale factor R measures the amount of expansion.

we ourselves and all that we deal with are consequences of the delay in this "phase change" of the universe.

What of the Future?

Will the universe expand forever, or will the expansion eventually stop and then reverse? That question is like asking whether a spacecraft launched from Earth will eventually return or not, and the answer is the same as the one we found in Chapter 9: If the kinetic energy exceeds the magnitude of the (negative) potential energy, then the expansion will proceed forever; otherwise a collapse will eventually ensue. The ratio of potential to kinetic energy is designated Ω:

$$\Omega = \frac{|\text{ potential energy of the universe }|}{\text{kinetic energy of expansion}}.$$

This single parameter determines the universe's ultimate fate; $\Omega > 1$ implies eventual collapse, while $\Omega < 1$ means the expansion will continue forever. An alternative way of expressing the same dichotomy is in terms of the average density, as the example below illustrates.

● EXAMPLE 45-4 CRITICAL DENSITY

Assuming a spherically symmetric universe obeying Newton's laws, find the **critical density** above which the universe is destined ultimately to collapse.

Solution

Let R be a typical cosmological distance for which the Hubble relation applies, and m the mass of a test particle (i.e., a galaxy). Since the universe is homogeneous on the large scale, it suffices to determine the potential to kinetic energy ratio for this single test particle. If M is the total mass within a sphere of radius R, then

$$\Omega = \frac{GMm/R}{\frac{1}{2}mv^2} = \frac{GM/R}{\frac{1}{2}v^2}.$$

We can substitute for the expansion speed v from Hubble's law to get

$$\Omega = \frac{GM/R}{\frac{1}{2}H_0^2R^2} = \frac{2GM}{H_0^2R^3}.$$

But the density $\rho = M/\frac{4}{3}\pi R^3$, so we can solve for M/R^3 and substitute to get

$$\Omega = \frac{8\pi G\rho}{3H_0^2}.$$

The critical density ρ_c occurs when $\Omega = 1$, so

$$\rho_c = \frac{3H_0^2}{8\pi G}.$$

Although we derived this result on the assumption of a homogeneous, spherically symmetric, Newtonian universe, it is in fact correct even for a universe governed by Einstein's general relativity.

EXERCISE Find the approximate value of ρ_c, using the Hubble constant from Example 45-3.

Answer: 5×10^{-27} kg/m³

Some problems similar to Example 45-4: 30, 31 ●

Observationally, cosmologists have several ways to determine the density ρ of the universe or, equivalently, the expansion parameter Ω. By studying the most distant objects, we look back into the past to a time when the expansion rate was faster. Determining just how much faster gives a measure of how rapidly the expansion is slowing, and thus a measure of Ω. Unfortunately, our knowledge of how galaxies evolved in the early universe is not sufficient to draw firm conclusions from the limited data available at present. A more direct approach is to measure the amount of matter and then calculate the density. Using the density of visible matter alone gives considerably less than the critical density. But the motions of galaxies in large galaxy clusters imply the presence of much more matter than we can see—enough to bring the universe close to the critical density. We have no idea whether that **dark matter** is ordinary matter, perhaps in burned-out stars, or consists of neutrinos whose mass may not be zero after all, or is something entirely new. It is a sobering thought that most of the matter in the universe may be in a form about which we know essentially nothing.

One of the most remarkable features of the universe is that its density appears even close to the critical density, since it seems the density could have *any* value. The inflationary theory provides an explanation, suggesting that the entire universe we now observe was, before inflation, smaller than a proton. The tremendous expansion flattened out the curves of spacetime which, in general relativity, are the manifestation of gravity. The result is a universe right on the borderline between expansion and gravitational collapse, a universe that will expand forever but just barely so.

Understanding the Universe

With this chapter's brief survey, we have reached the limits of present understanding of the universe. In the process we have seen that particle physics and cosmology are inextricably linked. To understand our universe, we need to understand all its aspects from the largest to the smallest—and that means understanding all the forces, from gravity to the weak force; all the physical laws, from Newton's and Maxwell's on to the laws of quantum mechanics; and the nature of the particles that comprise all matter. We have touched, in many cases only briefly, on all these topics. We hope this text has given you a foundation for further understanding and appreciation of the richness and diversity of the physical universe.

CHAPTER SYNOPSIS

Summary

1. What appear in classical physics as forces are actually interactions mediated by the exchanges of particles called **field particles** or **gauge bosons;** an example is the photon of the electromagnetic force. The mass of a field particle is zero for forces with infinite range, but nonzero for short-range forces.

2. There are a large number of subnuclear particles. Major groups include the **leptons,** which do not experience the nuclear or strong force, and the **hadrons,** which do experience the strong force. Leptons include the familiar electron and two similar but more massive and unstable particles, the **muon** and the **tau,** along with three types of neutrinos. Each also has an antiparticle. **Hadrons** include the integer-spin **mesons** and the half-integer-spin **baryons,** of which the proton and neutron are familiar examples. A third group are the field particles that mediate the fundamental forces.

3. Particles exhibit a number of properties like electric charge and spin. Some, like electric charge, angular momentum, and baryon number, are conserved in all interactions. Others are conserved in only some interactions; an example is **strangeness,** which is conserved in strong and electromagnetic interactions but which may change in weak interactions. Particle systems may also exhibit **symmetries,** including invariance under parity inversion, charge conjugation, and time reversal. However, each of these symmetries is violated individually in some interactions.

4. Leptons and field particles appear to be truly elementary, but hadrons are composed of **quarks.** A quark and an antiquark together make a meson, while three quarks make a baryon. Quarks come in the six **flavors** up, down, strange, charmed, top, and bottom, and in three **colors.** Particles made from quarks are **colorless,** and the nuclear force between such particles is actually a residue of the **color force** between their quark constituents. **Gluons** are the field particles that mediate the color force. The color force does not decrease with increasing quark separation, making it impossible to isolate individual quarks.

5. The **standard model** incorporates the quarks and leptons into three families of particles, with ordinary matter involving only the first family: the up and down quark and the electron and its associated neutrino.

6. **Unification** of the fundamental forces is a goal of physics. In the 1970s, physicists Weinberg, Salam, and Glashow showed that the weak force and electromagnetism are manifestations of the same underlying force. Physicists today are striving to unify this electroweak force with the color force, to produce a **grand unification theory.** Unification with gravity is a more remote goal. At high enough energies, forces now distinct should seem as one.

The appearance of distinct forces as energy drops is called **symmetry breaking.**

7. Questions of **cosmology** are closely linked with those of particle physics. The **Hubble law** shows that the universe is expanding, providing evidence that the universe had a definite beginning in a cosmic explosion called the **big bang.** Further evidence for the big bang comes from the **cosmic microwave background radiation,** left over from the time the atoms formed, about 10^5 years after the beginning. Evidence of earlier times comes from primordial nucleosynthesis, especially of deuterium, and from applying particle physics to conditions believed to exist within the first second. The **inflationary universe** theory helps explain several features of the present universe, including its density being very nearly the **critical density** that separates a forever-expanding universe from one that will eventually collapse under its own gravity.

Terms You Should Understand

(Pairs are closely related terms whose distinction is important; number in parentheses is chapter section where term first appears.)

virtual photon (45-1)	flavor, color (45-3)
muon, pion (45-2)	standard model (45-3)
lepton, hadron, field	unification (45-4)
particle (45-2)	Hubble's law (45-5)
meson, baryon (45-2)	cosmic microwave back-
fermion, boson (45-2)	ground radiation (45-5)
strangeness (45-2)	big bang (45-5)
parity (45-2)	inflationary universe (45-5)
quark (45-3)	critical density (45-5)

Symbols You Should Recognize

uud, $u\bar{d}$, and similar quark symbols (45-3)
H_0 (45-5)
Ω, ρ_c (45-5)

Problems You Should Be Able to Solve

evaluating conserved quantities in particle interactions (45-2)
determining quark composition of baryons and mesons (45-3)
applying Hubble's law (45-5)

Limitations to Keep in Mind

This chapter deals with the frontiers of physics, and some of the material presented is likely to undergo change.

QUESTIONS

1. Why did Yukawa conclude that the particle mediating the strong force should have nonzero mass?
2. Researchers looking for Yukawa's meson found instead the muon, a particle with approximately the predicted mass. How did they know they had not found the right particle?
3. How can we detect tracks of individual particles?
4. Which interactions conserve parity? Which conserve *CPT*?
5. How are baryons fundamentally different from leptons?
6. Why was discovery of the J/ψ particle significant?
7. What coordinates are changed under the inversion processes *P* and *T*?
8. Why are we unlikely to observe an isolated quark?
9. Describe the relation between the color force and the nuclear force.
10. What is the role of gluons?
11. Classify (a) mesons and (b) baryons as fermions or bosons, and relate your classification to the particles' quark compositions.
12. A hadron decays into leptons. Is the hadron a baryon or a meson? Explain.
13. What is the difference between color and flavor?
14. Why are most particle accelerators circular?
15. Photons and neutrinos are both massless particles. So how are they different?
16. Why is the Fermilab accelerator called a "tevatron?"
17. Name the fundamental force involved in (a) binding of a proton and neutron to make a deuterium nucleus, (b) decay of a neutron into a proton, an electron, and a neutrino, (c) binding of an electron and proton to make a hydrogen atom.
18. What forces are unified in the electroweak theory?
19. What forces would be unified by GUTs?
20. Why do we need higher-energy particle accelerators to explore fully the standard model?
21. How can Hubble's law hold without the universe having a center?
22. Is it possible for a charged particle to be its own antiparticle?
23. Is it possible for a neutral particle *not* to be its own antiparticle? Explain.
24. Describe the origin of the cosmic microwave background.
25. Why is deuterium an especially good probe of conditions in the first few minutes after the big bang?
26. A graph of the Hubble relation for not-too-distant galaxies is a straight line. But the relation may curve upward for the most distant galaxies, as shown in Fig. 45-28. Why?
27. Why are particle physics and cosmology linked?
28. The Superconducting Super Collider would have created conditions like those that existed 10^{-12} s after the big bang. Explain.

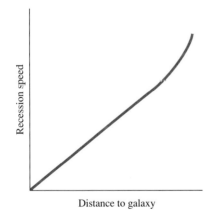

FIGURE 45-28 Why might the Hubble relation curve upward for the most distant objects (Question 26)?

PROBLEMS

Section 45-1 Particles and Forces

1. How long could a virtual photon of 633-nm red laser light exist without violating conservation of energy?
2. Some scientists have speculated on a possible "fifth force," with a range of about 100 m. Following Yukawa's reasoning, what would be the mass of the field particle mediating such a force?
3. The mass of the photon is assumed to be zero, but experiments put only an upper limit of 5×10^{-63} kg for the photon mass. What would be the range of the electromagnetic force if the photon mass were actually at this upper limit?

Section 45-2 Particles and More Particles

4. Write the equation for the decay of the positive pion into a muon and neutrino, being sure to label the type of neutrino. *Hint:* The positive muon is an antiparticle.
5. Use Table 45-1 to find the total strangeness before and after the decay $\Lambda^0 \rightarrow \pi^- + p$, and use your answer to determine the force involved in this reaction.
6. The η^0 particle is a neutral nonstrange meson that can decay into a positive, a negative, and a neutral pion. Write the reaction for this decay, and verify that it conserves charge, baryon number, and strangeness.
7. Are either or both of the following decay schemes possible for the tau particle? (a) $\tau^- \rightarrow e^- + \overline{\nu}_e + \nu_\tau$, (b) $\tau^- \rightarrow \pi^- + \pi^0 + \nu_\tau$
8. Is the interaction $p + p \rightarrow p + \pi^+$ allowed? If not, what conservation law does it violate?
9. Which of the following reactions (a) $\Lambda^0 \rightarrow \pi^+ + \pi^-$ and (b) $K^0 \rightarrow \pi^+ + \pi^-$ is not possible, and why?
10. Both the neutral kaon and the neutral ρ meson can decay into a pion-antipion pair. Which of these decays is mediated by the weak force, and why?
11. Grand unification theories suggest that the decay $p \rightarrow \pi^0 + e^+$ may be possible, in which case all matter may eventually become radiation. Are (a) baryon number and (b) electric charge conserved in this hypothetical proton decay?
12. Consider systems described by wave functions that are proportional to the terms (a) xy^2z, (b) x^2yz, and (c) xyz, where x, y, and z are the spatial coordinates. Which pairs of these systems could be transformed into each other under an interaction obeying conservation of parity?
13. What happens to the sign of the wave functions proportional to the terms (a) xy^2t and (b) xy^2t^2 under the operation PT?

Section 45-3 Quarks and the Standard Model

14. What is the quark composition of the π^-?
15. The Eightfold Way led Gell-Mann to predict a baryon with strangeness -3. What must be its quark composition?

16. The Σ^+ and Σ^- have quark compositions *uus* and *dds*, respectively. Are the Σ^+ and Σ^- each other's antiparticles? If not, give the quark compositions of their antiparticles.
17. The J/ψ particle is an uncharmed meson that nevertheless includes charmed quarks. What is its quark composition?
18. List all the possible quark triplets formed from any combination of the up, down, and charmed quarks, along with the charge of each.

Section 45-4 Unification

19. Estimate the volume of the 4000 tons of water used in the proton-decay experiment of Fig. 45-17.
20. Estimate the temperature in a gas of particles such that the thermal energy kT is enough to make electromagnetism and the weak force appear as a single phenomenon.
21. Repeat the preceding problem for the 10^{15} GeV energy of grand unification.
22. The Tevatron at Fermilab accelerates protons to an energy of 1 TeV. (a) How much is this in joules? (b) How far would a 1-g mass have to fall in Earth's gravitational field to gain this much energy?
23. (a) What would be the relativistic factor γ for a proton in the 20-TeV Superconducting Super Collider? (b) Find an accurate value for the proton's speed.
24. (a) How long would it take a 20-TeV proton to complete one circuit around the 87-km circumference of the Superconducting Super Collider, in the rest frame of the SSC? (b) By what factor is the time longer for a 55-km/h car trip around the Washington Beltway, roughly the size of the SSC?

Section 45-5 Evolution of the Universe

Note: In working these problems, take $H_0 = 15$ km/s/Mly unless otherwise indicated.

25. Express the Hubble constant in SI units.
26. What is the distance to a galaxy receding from us at 2×10^4 km/s?
27. What is the recession speed of a galaxy 300 Mly from Earth?
28. What would be the age of the universe, assuming constant expansion rate, if the Hubble constant were 25 km/s/Mly?
29. A widely used value for H_0 is 17 km/s/Mly. What age does this value imply for the universe, under the simple assumptions of Example 45-3?
30. Estimate the diameter to which the Sun would have to be expanded for its average density to be the critical density of Example 45-4.
31. Find the critical density if the Hubble constant is 34 km/s/Mly.

Supplementary Problems

32. A baryon called Λ^0 has mass 1116 MeV/c^2. Find the minimum speed necessary for the particles in a proton-antiproton collider to produce lambda-antilambda pairs.

33. Muonium is a hydrogen-like atom consisting of a proton and a muon. What would be (a) the size and (b) the ground-state energy of muonium? The muon's mass is 207 times that of the electron.

34. (a) By what factor must the magnetic field in a proton synchrotron be increased as the proton energy increases by a factor of 10? Assume the protons are highly relativistic, so $\gamma \gg 1$. (b) By what factor must the diameter of the accelerator be increased to raise the energy by a factor of 10 without changing the magnetic field?

35. How long would it take the proton of Problem 24 to circle the SSC, as measured in the proton's reference frame? Assume you can apply special relativity even though the path is curved.

36. A galaxy's hydrogen-β spectral line, normally at 486.1 nm, appears at 495.4 nm. (a) Use the Doppler shift of Chapter 17 to find the galaxy's recession speed, and (b) infer the distance to the galaxy. Are you justified in using Chapter 17's nonrelativistic Doppler formulas? Use $H_0 = 15$ km/s/Mly.

37. Use Wien's law (Equation 39-2) to find the wavelength at which the 2.7-K cosmic microwave background has its peak intensity.

38. One of the earliest measurements of the cosmic microwave background was made at a wavelength of 7 cm. By what factor would the intensity have been greater had the measurement been made at the wavelength of peak intensity? See preceding problem.

39. Many particles are far too short-lived for their lifetimes to be measured directly. Instead, tables of particle properties often list "width," measured in energy units, and indicating the width of distribution of measured rest energies for the particle. For example, the Z^0 has a mass of 91.18 GeV and a width of 2.5 GeV. Use the energy-time uncertainty relation to estimate the corresponding lifetime.

40. A mix of particles starts with equal numbers of the three types of sigma particles listed in Table 45-1. What is the relative portion of each after (a) 5×10^{-20} s and (b) 5×10^{-10} s? Give your answer in a reference frame in which the particles are at rest.

APPENDIX A

MATHEMATICS

A-1 ALGEBRA AND TRIGONOMETRY

Quadratic Formula

If $ax^2 + bx + c = 0$ then $x = \dfrac{-b \pm \sqrt{b^2 - 4ac}}{2a}$

Circumference, Area, Volume

Where $\pi \simeq 3.14159\ldots$

circumference of circle	$2\pi r$
area of circle	πr^2
surface area of sphere	$4\pi r^2$
volume of sphere	$\frac{4}{3}\pi r^3$
area of triangle	$\frac{1}{2}bh$
volume of cylinder	$\pi r^2 \ell$

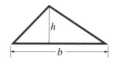

Trigonometry

definition of angle (in radians): $\theta = \dfrac{s}{r}$

2π radians in complete circle
1 radian $\simeq 57.3°$

Trigonometric Functions

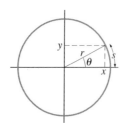

$$\sin \theta = \frac{y}{r}$$

$$\cos \theta = \frac{x}{r}$$

$$\tan \theta = \frac{\sin \theta}{\cos \theta} = \frac{y}{x}$$

Values at Selected Angles

$\theta \rightarrow$	0	$\dfrac{\pi}{6}$ (30°)	$\dfrac{\pi}{4}$ (45°)	$\dfrac{\pi}{3}$ (60°)	$\dfrac{\pi}{2}$ (90°)
$\sin\theta$	0	$\dfrac{1}{2}$	$\dfrac{\sqrt{2}}{2}$	$\dfrac{\sqrt{3}}{2}$	1
$\cos\theta$	1	$\dfrac{\sqrt{3}}{2}$	$\dfrac{\sqrt{2}}{2}$	$\dfrac{1}{2}$	0
$\tan\theta$	0	$\dfrac{\sqrt{3}}{3}$	1	$\sqrt{3}$	∞

Graphs of Trigonometric Functions

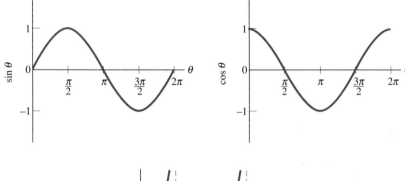

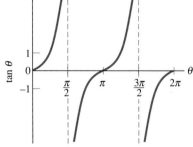

Trigonometric Identities

$\sin(-\theta) = -\sin\theta$

$\cos(-\theta) = \cos\theta$

$\sin\left(\theta \pm \dfrac{\pi}{2}\right) = \pm\cos\theta$

$\cos\left(\theta \pm \dfrac{\pi}{2}\right) = \mp\sin\theta$

$\sin^2\theta + \cos^2\theta = 1$

$\sin 2\theta = 2\sin\theta\cos\theta$

$$\cos 2\theta = \cos^2 \theta - \sin^2 \theta = 1 - 2\sin^2 \theta = 2\cos^2 \theta - 1$$

$$\sin(\alpha \pm \beta) = \sin\alpha\cos\beta \pm \cos\alpha\sin\beta$$

$$\cos(\alpha \pm \beta) = \cos\alpha\cos\beta \mp \sin\alpha\sin\beta$$

$$\sin\alpha \pm \sin\beta = 2\sin\left[\frac{1}{2}(\alpha \pm \beta)\right]\cos\left[\frac{1}{2}(\alpha \mp \beta)\right]$$

$$\cos\alpha + \cos\beta = 2\cos\left[\frac{1}{2}(\alpha + \beta)\right]\cos\left[\frac{1}{2}(\alpha - \beta)\right]$$

$$\cos\alpha - \cos\beta = -2\sin\left[\frac{1}{2}(\alpha + \beta)\right]\sin\left[\frac{1}{2}(\alpha - \beta)\right]$$

Laws of Cosines and Sines   Where A, B, C are the sides of an arbitrary triangle and α, β, γ the angles opposite those sides:

Law of cosines

$$C^2 = A^2 + B^2 - 2AB\cos\gamma$$

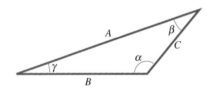

Law of sines

$$\frac{\sin\alpha}{A} = \frac{\sin\beta}{B} = \frac{\sin\gamma}{C}$$

Exponentials and Logarithms

Graphs

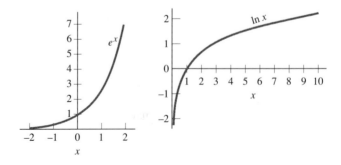

Exponential and Natural Logarithms Are Inverse Functions

$$e^{\ln x} = x, \quad \ln e^x = x \quad e = 2.71828\ldots.$$

Exponential and Logarithmic Identities

$a^x = e^{x \ln a}$ $\ln(xy) = \ln x + \ln y$

$a^x a^y = a^{x+y}$ $\ln\left(\dfrac{x}{y}\right) = \ln x - \ln y$

$(a^x)^y = a^{xy}$ $\ln\left(\dfrac{1}{x}\right) = -\ln x$

$\log x \equiv \log_{10} x = \ln(10)\ln x \approx 2.3\ln x$

Expansions and Approximations

Series Expansions of Functions
Note: $n! = n(n-1)(n-2)(n-3)\cdots(3)(2)(1)$

$e^x = 1 + x + \dfrac{x^2}{2!} + \dfrac{x^3}{3!} + \cdots$ (exponential)

$\sin x = x - \dfrac{x^3}{3!} + \dfrac{x^5}{5!} - \cdots$ (sine)

$\cos x = 1 - \dfrac{x^2}{2!} + \dfrac{x^4}{4!} - \cdots$ (cosine)

$\left. \right\}$ (x in radians)

$\ln(1 + x) = x - \dfrac{x^2}{2} + \dfrac{x^3}{3} - \cdots$ (natural logarithm)

$(1 + x)^p = 1 + px + \dfrac{p(p-1)}{2!}x^2 + \dfrac{p(p-1)(p-2)}{3!}x^3 + \cdots$

(binomial, valid for $|x| < 1$)

Approximations For $|x| \ll 1$, the first few terms in the series provide a good approximation; that is,

$e^x \simeq 1 + x$

$\sin x \simeq x$

$\cos x \simeq 1 - \tfrac{1}{2}x^2$ for $|x| \ll 1$

$\ln(1 + x) \simeq x$

$(1 + x)^p \simeq 1 + px$

Expressions that do not have the forms shown may often be put in the appropriate form. For example:

$$\frac{1}{\sqrt{a^2 + y^2}} = \frac{1}{a\sqrt{1 + \dfrac{y^2}{a^2}}} = \frac{1}{a}\left(1 + \frac{y^2}{a^2}\right)^{-1/2}.$$

For $y^2 \ll a^2$, this may be approximated using the binomial expansion $(1 + x)^p \simeq 1 + px$, with $p = -\frac{1}{2}$ and $x = y^2/a^2$:

$$\frac{1}{a}\left(1 + \frac{y^2}{a^2}\right)^{-1/2} \simeq \frac{1}{a}\left(1 - \frac{1}{2}\frac{y^2}{a^2}\right).$$

Vector Algebra

Vector Products

$$\mathbf{A} \cdot \mathbf{B} = AB\cos\theta$$

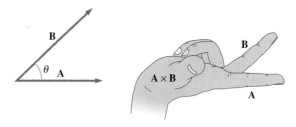

$|\mathbf{A} \times \mathbf{B}| = AB\sin\theta$, with direction of $\mathbf{A} \times \mathbf{B}$ given by right-hand rule:

Unit Vector Notation   An arbitrary vector $\mathbf{A}$ may be written in terms of its components A_x, A_y, A_z and the unit vectors $\hat{\mathbf{i}}, \hat{\mathbf{j}}, \mathbf{k}$ that have length 1 and lie along the x, y, z axes:

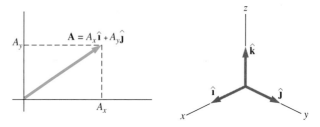

In unit vector notation, vector products become

$$\mathbf{A} \cdot \mathbf{B} = A_x B_x + A_y B_y + A_z B_z$$

$$\mathbf{A} \times \mathbf{B} = (A_y B_z - A_z B_y)\hat{\mathbf{i}} + (A_z B_x - A_x B_z)\hat{\mathbf{j}} + (A_x B_y - A_y B_x)\hat{\mathbf{k}}$$

Vector Identities

$$\mathbf{A} \cdot \mathbf{B} = \mathbf{B} \cdot \mathbf{A}$$

$$\mathbf{A} \times \mathbf{B} = -\mathbf{B} \times \mathbf{A}$$

$$\mathbf{A} \cdot (\mathbf{B} \times \mathbf{C}) = \mathbf{B} \cdot (\mathbf{C} \times \mathbf{A}) = \mathbf{C} \cdot (\mathbf{A} \times \mathbf{B})$$

$$\mathbf{A} \times (\mathbf{B} \times \mathbf{C}) = (\mathbf{A} \cdot \mathbf{C})\mathbf{B} - (\mathbf{A} \cdot \mathbf{B})\mathbf{C}$$

A-2 CALCULUS

Derivatives

Definition of the Derivative If y is a function of x [$y = f(x)$], then the **derivative of y with respect to x** is the ratio of the change Δy in y to the corresponding change Δx in x, in the limit of arbitrarily small Δx:

$$\frac{dy}{dx} = \lim_{\Delta x \to 0} \frac{\Delta y}{\Delta x}.$$

Algebraically, the derivative is the rate of change of y with respect to x; geometrically, it is the slope of the y versus x graph—that is, of the tangent line to the graph at a given point:

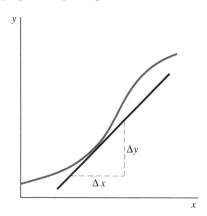

Derivatives of Common Functions Although the derivative of a function can be evaluated directly using the limiting process that defines the derivative, standard formulas are available for common functions:

$$\frac{da}{dx} = 0 \quad (a \text{ is a constant})$$

$$\frac{dx^n}{dx} = nx^{n-1} \quad (n \text{ need not be an integer})$$

$$\frac{d}{dx}\sin x = \cos x$$

$$\frac{d}{dx}\cos x = -\sin x$$

$$\frac{d}{dx}\tan x = \frac{1}{\cos^2 x}$$

$$\frac{de^x}{dx} = e^x$$

$$\frac{d}{dx}\ln x = \frac{1}{x}$$

Derivatives of Sums, Products, and Functions of Functions

1. Derivative of a constant times a function

$$\frac{d}{dx}[af(x)] = a\frac{df}{dx} \quad (a \text{ is a constant})$$

2. Derivative of a sum

$$\frac{d}{dx}[f(x) + g(x)] = \frac{df}{dx} + \frac{dg}{dx}$$

3. Derivative of a product

$$\frac{d}{dx}[f(x)g(x)] = g\frac{df}{dx} + f\frac{dg}{dx}$$

Examples

$$\frac{d}{dx}(x^2\cos x) = \cos x\frac{dx^2}{dx} + x^2\frac{d}{dx}\cos x = 2x\cos x - x^2\sin x$$

$$\frac{d}{dx}(x\ln x) = \ln x\frac{dx}{dx} + x\frac{d}{dx}\ln x = (\ln x)(1) + x\left(\frac{1}{x}\right) = \ln x + 1$$

4. Derivative of a quotient

$$\frac{d}{dx}\left[\frac{f(x)}{g(x)}\right] = \frac{1}{g^2}\left(g\frac{df}{dx} - f\frac{dg}{dx}\right)$$

Example

$$\frac{d}{dx}\left(\frac{\sin x}{x^2}\right) = \frac{1}{x^4}\left(x^2\frac{d}{dx}\sin x - \sin x\frac{dx^2}{dx}\right) = \frac{\cos x}{x^2} - \frac{2\sin x}{x^3}$$

5. Chain rule for derivatives
   If f is a function of u and u is a function of x, then

$$\frac{df}{dx} = \frac{df}{du}\frac{du}{dx}.$$

Examples

a. Evaluate $\dfrac{d}{dx}\sin(x^2)$. Here $u = x^2$ and $f(u) = \sin u$, so

$$\frac{d}{dx}\sin(x^2) = \frac{d}{du}\sin u\frac{du}{dx} = (\cos u)\frac{dx^2}{dx} = 2x\cos(x^2).$$

b. $\dfrac{d}{dt}\sin\omega t = \dfrac{d}{d\omega t}\sin\omega t\dfrac{d}{dt}\omega t = \omega\cos\omega t.$ (ω a constant)

c. Evaluate $\dfrac{d}{dx} \sin^2 5x$. Here $u = \sin 5x$ and $f(u) = u^2$, so

$$\frac{d}{dx} \sin^2 5x = \frac{d}{du} u^2 \frac{du}{dx} = 2u \frac{du}{dx} = 2 \sin 5x \frac{d}{dx} \sin 5x$$

$$= (2)(\sin 5x)(5)(\cos 5x) = 10 \sin 5x \cos 5x = 5 \sin 2x.$$

Second Derivative The second derivative of y with respect to x is defined as the derivative of the derivative:

$$\frac{d^2 y}{dx^2} = \frac{d}{dx}\left(\frac{dy}{dx}\right).$$

Example

If $y = ax^3$, then $dy/dx = 3ax^2$, so

$$\frac{d^2 y}{dx^2} = \frac{d}{dx} 3ax^2 = 6ax.$$

Partial Derivatives When a function depends on more than one variable, then the partial derivatives of that function are the derivatives with respect to each variable, taken with all other variables held constant. If f is a function of x and y, then the partial derivatives are written

$$\frac{\partial f}{\partial x} \quad \text{and} \quad \frac{\partial f}{\partial y}.$$

Example

If $f(x, y) = x^3 \sin y$, then

$$\frac{\partial f}{\partial x} = 3x^2 \sin y \quad \text{and} \quad \frac{\partial f}{\partial y} = x^3 \cos y.$$

Integrals

Indefinite Integrals Integration is the inverse of differentiation. The **indefinite integral,** $\int f(x)\,dx$, is defined as a function whose derivative is $f(x)$:

$$\frac{d}{dx}\left[\int f(x)\,dx\right] = f(x).$$

If $A(x)$ is an indefinite integral of $f(x)$, then because the derivative of a constant is zero, the function $A(x) + C$ is also an indefinite integral of $f(x)$, where C is any constant. Inverting the derivatives of common functions listed in the preceding section gives some common integrals (a more extensive table appears at the end of this appendix).

$$\int a\,dx = ax + C \qquad\qquad \int \cos x\,dx = \sin x + C$$

$$\int x^n\,dx = \frac{x^{n+1}}{n+1} + C, \quad n \neq -1 \qquad \int e^x\,dx = e^x + C$$

$$\int \sin x\,dx = -\cos x + C \qquad \int x^{-1}\,dx = \ln x + C$$

Definite Integrals   In physics we are most often interested in the **definite integral,** defined as the sum of a large number of very small quantities, in the limit as the number of quantities grows arbitrarily large and the size of each arbitrarily small:

$$\int_{x_1}^{x_2} f(x)\,dx \equiv \lim_{\substack{\Delta x \to 0 \\ N \to \infty}} \sum_{i=1}^{N} f(x_i)\,\Delta x,$$

where the terms in the sum are evaluated at values x_i between the limits of integration x_1 and x_2; in the limit $\Delta x \to 0$, the sum is over all values of x in the interval.

The definite integral is used whenever we need to sum over a quantity that is changing—for example, to calculate the work done by a variable force (Chapter 7), the entropy change in a system whose temperature varies (Chapter 22), or the flux of an electric field that varies with position (Chapter 24).

The key to evaluating the definite integral is provided by the **fundamental theorem of calculus.** The theorem states that, if $A(x)$ is an *indefinite* integral of $f(x)$, then the *definite integral* is given by

$$\int_{x_1}^{x_2} f(x)\,dx = A(x_2) - A(x_1) \equiv A(x)\Big|_{x_1}^{x_2}.$$

Geometrically, the definite integral is the area under the graph of $f(x)$ between the limits x_1 and x_2:

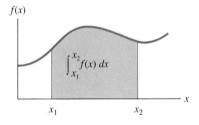

Evaluating Integrals   The first step in evaluating an integral is to express all varying quantities within the integral in terms of a single variable. For example, in evaluating $\int E\,dr$ to calculate an electric potential (Chapter 25), it is necessary first to express E as a function of r. This procedure is illustrated in many examples throughout this text; Example 23-7 provides a typical case.

Once an integral is written in terms of a single variable, it is necessary to manipulate the integrand—the function being integrated—into a form whose integral you know or can look up in tables of integrals. Two common techniques are especially useful:

1. **Change of variables**
   An unfamiliar integral can often be put into familiar form by defining a new variable. For example, it is not obvious how to integrate the expression

$$\int \frac{x\,dx}{\sqrt{a^2 + x^2}}.$$

where a is a constant. But let $z = a^2 + x^2$. Then

$$\frac{dz}{dx} = \frac{da^2}{dx} + \frac{dx^2}{dx} = 0 + 2x = 2x,$$

so $dz = 2x\,dx$. Then the quantity $x\,dx$ in our unfamiliar integral is just $\frac{1}{2}dz$, while the quantity $\sqrt{a^2 + x^2}$ is just $z^{1/2}$. So the integral becomes

$$\int \tfrac{1}{2} z^{-1/2}\,dz = \frac{\tfrac{1}{2}z^{1/2}}{(1/2)} = \sqrt{z},$$

where we have used the standard form for the integral of a power of the independent variable. Substituting back $z = a^2 + x^2$ gives

$$\int \frac{x\,dx}{\sqrt{a^2 + x^2}} = \sqrt{a^2 + x^2}.$$

2. **Integration by parts**
   The quantity $\int u\,dv$ is the area under the curve of u as a function of v between specified limits. In the figure below, that area can also be expressed as the area of the rectangle shown minus the area under the curve of v as a function of u. Mathematically, this relation among areas may be expressed as a relation among integrals:

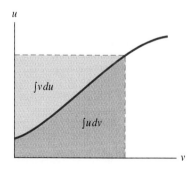

$$\int u\,dv = uv - \int v\,du. \qquad \text{(integration by parts)}$$

This expression may often be used to transform complicated integrals into simpler ones.

Example

Evaluate $\int x \cos x \, dx$. Here let $u = x$, so $du = dx$. Then $dv = \cos x \, dx$, so $v = \int dv = \int \cos x \, dx = \sin x$. Integrating by parts then gives

$$\int x \cos x \, dx = (x)(\sin x) - \int \sin x \, dx = x \sin x + \cos x,$$

where the $+$ sign arises because $\int \sin x \, dx = -\cos x$.

Table of Integrals [More extensive tables are available in many mathematical and scientific handbooks; see, for example, **Handbook of Chemistry and Physics** (Chemical Rubber Co.) or Dwight, **Tables of Integrals and Other Mathematical Data** (Macmillan).] Increasingly, sophisticated computer software is used instead of tables for the symbolic evaluation of integrals; among the most widely used are *Maple*, *Mathematica*, and *Derive*.

In the expressions below, a and b are constants. An arbitrary constant of integration may be added to the right-hand side.

$$\int e^{ax} \, dx = \frac{e^{ax}}{a}$$

$$\int \sin ax \, dx = -\frac{\cos ax}{a}$$

$$\int \cos ax \, dx = \frac{\sin ax}{a}$$

$$\int \tan ax \, dx = -\frac{1}{a} \ln(\cos ax)$$

$$\int \sin^2 ax \, dx = \frac{x}{2} - \frac{\sin 2ax}{4a}$$

$$\int \cos^2 ax \, dx = \frac{x}{2} + \frac{\sin 2ax}{4a}$$

$$\int x \sin ax \, dx = \frac{1}{a^2} \sin ax - \frac{1}{a} x \cos ax$$

$$\int x \cos ax \, dx = \frac{1}{a^2} \cos ax + \frac{1}{a} x \sin ax$$

$$\int \frac{dx}{\sqrt{a^2 - x^2}} = \sin^{-1}\left(\frac{x}{a}\right)$$

$$\int \frac{dx}{\sqrt{x^2 \pm a^2}} = \ln(x + \sqrt{x^2 \pm a^2})$$

$$\int \frac{dx}{x^2 + a^2} = \frac{1}{a} \tan^{-1}\left(\frac{x}{a}\right)$$

$$\int \frac{x \, dx}{\sqrt{a^2 - x^2}} = -\sqrt{a^2 - x^2}$$

$$\int \frac{x \, dx}{\sqrt{x^2 \pm a^2}} = \sqrt{x^2 \pm a^2}$$

$$\int \frac{dx}{(x^2 \pm a^2)^{3/2}} = \frac{\pm x}{a^2 \sqrt{x^2 \pm a^2}}$$

$$\int x e^{ax} \, dx = \frac{e^{ax}}{a^2}(ax - 1)$$

$$\int x^2 e^{ax} \, dx = \frac{x^2 e^{ax}}{a} - \frac{2}{a}\left[\frac{e^{ax}}{a^2}(ax - 1)\right]$$

$$\int \frac{dx}{a + bx} = \frac{1}{b} \ln(a + bx)$$

$$\int \frac{dx}{(a + bx)^2} = -\frac{1}{b(a + bx)}$$

$$\int \ln ax \, dx = x \ln ax - x$$

THE INTERNATIONAL SYSTEM OF UNITS (SI)

This material is from the United States edition of the English translation of the sixth edition of "Le Système International d'Unités (SI)," the definitive publication in the French language issued in 1991 by the International Bureau of Weights and Measures (BIPM). The year the definition was adopted is given in parentheses.

unit of length (meter): The meter is the length of the path traveled by light in vacuum during a time interval of 1/299 792 458 of a second. (1983)

unit of mass (kilogram): The kilogram is the unit of mass; it is equal to the mass of the international prototype of the kilogram. (1889)

unit of time (second): The second is the duration of 9 192 631 770 periods of the radiation corresponding to the transition between the two hyperfine levels of the ground state of the cesium-133 atom. (1967)

unit of electric current (ampere): The ampere is that constant current which, if maintained in two straight parallel conductors of infinite length, of negligible circular cross section, and placed 1 meter apart in vacuum, would produce between these conductors a force equal to 2×10^{-7} newton per meter of length. (1948)

unit of thermodynamic temperature (kelvin): The kelvin, unit of thermodynamic temperature, is the fraction 1/273.16 of the thermodynamic temperature of the triple point of water. (1957) Also, the unit kelvin and its symbol K should be used to express an interval or a difference of temperature.

unit of amount of substance (mole): (1) The mole is the amount of substance of a system that contains as many elementary entities as there are atoms in 0.012 kilogram of carbon 12. (1971) (2) When the mole is used, the elementary entities must be specified and may be atoms, molecules, ions, electrons, other particles, or specified groups of such particles.

unit of luminous intensity (candela): The candela is the luminous intensity, in a given direction, of a source that emits monochromatic radiation of frequency 540×10^{12} hertz and that has a radiant intensity in that direction of (1/683) watt per steradian. (1979)

▲ SI BASE AND SUPPLEMENTARY UNITS

	SI UNIT	
QUANTITY	**NAME**	**SYMBOL**
Base Unit		
Length	meter	m
Mass	kilogram	kg
Time	second	s
Electric current	ampere	A
Thermodynamic temperature	kelvin	mol
Amount of substance	mole	mol
Luminous intensity	candela	cd
Supplementary Units		
Plane angle	radian	rad
Solid angle	steradian	sr

▲ SI PREFIXES

FACTOR	**PREFIX**	**SYMBOL**
10^{24}	yetta	Y
10^{21}	zetta	Z
10^{18}	exa	E
10^{15}	peta	P
10^{12}	tera	T
10^{9}	giga	G
10^{6}	mega	M
10^{3}	kilo	k
10^{2}	hecto	h
10^{1}	deka	da
10^{0}	—	—
10^{-1}	deci	d
10^{-2}	centi	c
10^{-3}	milli	m
10^{-6}	micro	μ
10^{-9}	nano	n
10^{-12}	pico	p
10^{-15}	femto	f
10^{-18}	atto	a
10^{-21}	zepto	z
10^{-24}	yocto	y

▲ SOME SI DERIVED UNITS WITH SPECIAL NAMES

			SI UNIT	
QUANTITY	**NAME**	**SYMBOL**	**EXPRESSION IN TERMS OF OTHER UNITS**	**EXPRESSION IN TERMS OF SI BASE UNITS**
Frequency	hertz	Hz		s^{-1}
Force	newton	N		$m \cdot kg \cdot s^{-2}$
Pressure, stress	pascal	Pa	N/m^2	$m^{-1} \cdot kg \cdot s^{-2}$
Energy, work, heat	joule	J	$N \cdot m$	$m^2 \cdot kg \cdot s^{-2}$
Power	watt	W	J/s	$m^2 \cdot kg \cdot s^{-3}$
Electric charge	coulomb	C		$s \cdot A$
Electric potential, potential difference, electromotive force	volt	V	J/C	$m^2 \cdot kg \cdot s^{-3} \cdot A^{-1}$
Capacitance	farad	F	C/V	$m^{-2} \cdot kg^{-1} \cdot s^4 \cdot A^2$
Electric resistance	ohm	Ω	V/A	$m^2 \cdot kg \cdot s^{-3} \cdot A^{-2}$
Magnetic flux	weber	Wb	$V \cdot s$	$m^2 \cdot kg \cdot s^{-2} \cdot A^{-1}$
Magnetic field	tesla	T	Wb/m^2	$kg \cdot s^{-2} \cdot A^{-1}$
Inductance	henry	H	Wb/A	$m^2 \cdot kg \cdot s^{-2} \cdot A^{-2}$
Radioactivity	becquerel	Bq	1 decay/s	s^{-1}
Absorbed radiation dose	gray	Gy	J/kg, 100 rad	$m^2 \cdot s^{-2}$
Radiation dose equivalent	sievert	Sv	J/kg, 100 rem	$m^2 \cdot s^{-2}$

APPENDIX C

CONVERSION FACTORS

The listings below give the SI equivalents of non-SI units. To convert from the units shown to SI, multiply by the factor given; to convert the other way, divide. For conversions within the SI system see table of SI prefixes in Appendix B, Chapter 1, or inside front cover. Conversions that are not exact by definition are given to, at most, 4 significant figures.

Length

1 inch (in.) = 0.0254 m
1 foot (ft) = 0.3048 m
1 yard (yd) = 0.9144 m
1 mile (mi) = 1609 m
1 nautical mile = 1852 m

1 angstrom (Å) = 10^{-10} m
1 light-year (ly) = 9.46×10^{15} m
1 astronomical unit (AU) = 1.5×10^{11} m
1 parsec = 3.09×10^{16} m
1 fermi = 10^{-15} m = 1 fm

Mass

1 slug = 14.59 kg
1 metric ton (tonne; T) = 1000 kg

1 unified mass unit (u) = 1.660×10^{-27} kg

Force units in the English system are sometimes used (incorrectly) for mass. The units given below are actually equal to the number of kilograms multiplied by g, the acceleration of gravity.

1 pound (lb) = weight of 0.454 kg
1 ton = 2000 lb = weight of 908 kg

1 ounce (oz) = weight of 0.02835 kg

Time

1 minute (min) = 60 s
1 hour (h) = 60 min = 3600 s

1 day (d) = 24 h = 86 400 s
1 year (y) = 365.2422 d* = 3.156×10^7 s

*The length of the year changes very slowly with changes in Earth's orbital period.

Area

1 hectare (ha) = 10^4 m^2 1 acre = 4047 m^2
1 square inch (in.^{2}) = 6.452×10^{-4} m^2 1 barn = 10^{-28} m^2
1 square foot (ft^2) = 9.290×10^{-2} m^2 1 shed = 10^{-30} m^2

Volume

1 liter (L) = 1000 cm^3 = 10^{-3} m^3 1 gallon (U.S.; gal) = 3.785×10^{-3} m^3
1 cubic foot (ft^3) = 2.832×10^{-2} m^3 1 gallon (British) = 4.546×10^{-3} m^3
1 cubic inch (in.^{3}) = 1.639×10^{-5} m^3
1 fluid ounce = 1/128 gal = 2.957×10^{-5} m^3
1 barrel = 42 gal = 0.1590 m^3

Angle, Phase

1 degree (°) = $\pi/180$ rad = 1.745×10^{-2} rad
1 revolution (rev) = 360° = 2π rad
1 cycle = 360° = 2π rad

Speed, Velocity

1 km/h = (1/3.6) m/s = 0.2778 m/s   1 ft/s = 0.3048 m/s
1 mi/h (mph) = 0.4470 m/s   1 ly/y = 3.00×10^8 m/s

Angular Speed, Angular Velocity, Frequency, and Angular Frequency

1 rev/s = 2π rad/s = 6.283 rad/s (s^{-1})   1 rev/min (rpm) = 0.1047 rad/s (s^{-1})
1 Hz = 1 cycle/s = 2πs^{-1}

Force

1 dyne = 10^{-5} N   1 pound (lb) = 4.448 N

Pressure

1 dyne/cm^2 = 0.10 Pa   1 lb/in.^{2} (psi) = 6.895×10^3 Pa
1 atmosphere (atm) = 1.013×10^5 Pa   1 in. H$_2$O (60°F) = 248.8 Pa
1 torr = 1 mm Hg at 0°C = 133.3 Pa   1 in. Hg (60°F) = 3.377×10^3 Pa

Energy, Work, Heat

1 erg = 10^{-7} J   1 Btu* = 1.054×10^3 J
1 calorie* (cal) = 4.184 J   1 kWh = 3.6×10^6 J
1 electron-volt (eV) = 1.602×10^{-19}J   1 megaton (explosive yield; Mt)
1 foot-pound (ft·lb) = 1.356 J   = 4.18×10^{15} J

* Values based on the thermochemical calorie; other definitions vary slightly.

Power

1 erg/s $= 10^{-7}$ W

1 horsepower (hp) $= 746$ W

1 Btu/h (Btuh) $= 0.293$ W

1 ft·lb/s $= 1.356$ W

Magnetic Field

1 gauss (G) $= 10^{-4}$ T

1 gamma (γ) $= 10^{-9}$ T

Radiation

1 curie (ci) $= 3.7 \times 10^{10}$ Bq

1 rad $= 10^{-2}$ Gy

1 rem $= 10^{-2}$ Sv

▲ ENERGY CONTENT OF FUELS

ENERGY SOURCE	ENERGY CONTENT
Coal	2.9×10^7 J/kg $= 7300$ kWh/ton $= 25 \times 10^6$ Btu/ton
Oil	43×10^6 J/kg $= 39$ kWh/gal $= 1.3 \times 10^5$ Btu/gal
Gasoline	44×10^6 J/kg $= 36$ kWh/gal $= 1.2 \times 10^5$ Btu/gal
Natural gas	55×10^6 J/kg $= 30$ kWh/100 ft^3 $= 1000$ Btu/ft^3
Uranium (fission)	
Normal abundance	5.8×10^{11} J/kg $= 1.6 \times 10^5$ kWh/kg
Pure U-235	8.2×10^{13} J/kg $= 2.3 \times 10^7$ kWh/kg
Hydrogen (fusion)	
Normal abundance	7×10^{11} J/kg $= 3.0 \times 10^4$ kWh/kg
Pure deuterium	3.3×10^{14} J/kg $= 9.2 \times 10^7$ kWh/kg
Water	1.2×10^{10} J/kg $= 1.3 \times 10^4$ kWh/gal $= 340$ gal gasoline/gal
H_2O	
100% conversion, matter to energy	9.0×10^{16} J/kg $= 931$ MeV/u $= 2.5 \times 10^{10}$ kWh/kg

THE ELEMENTS

The atomic weights of stable elements reflect the abundances of different isotopes; values given here apply to elements as they exist naturally on Earth. For stable elements, parentheses express uncertainties in the last decimal place given. For elements with no stable isotopes (indicated in bold-face), sets of most important isotopes are given. (Exceptions are the unstable elements thorium, protactinium, and uranium, for which atomic weights reflect natural abundances of long-lived isotopes.) See also periodic table inside back cover.

ATOMIC NUMBER	NAMES	SYMBOL	ATOMIC WEIGHT
1	Hydrogen	H	1.00794 (7)
2	Helium	He	4.002602 (2)
3	Lithium	Li	6.941 (2)
4	Beryllium	Be	9.012182 (3)
5	Boron	B	10.811 (5)
6	Carbon	C	12.011 (1)
7	Nitrogen	N	14.00674 (7)
8	Oxygen	O	15.9994 (3)
9	Fluorine	F	18.9984032 (9)
10	Neon	Ne	20.1797 (6)
11	Sodium (Natrium)	Na	22.989768 (6)
12	Magnesium	Mg	24.3050 (6)
13	Aluminum	Al	26.981539 (5)
14	Silicon	Si	28.0855 (3)
15	Phosphorus	P	30.973762 (4)
16	Sulfur	S	32.066 (6)
17	Chlorine	Cl	35.4527 (9)
18	Argon	Ar	39.948 (1)
19	Potassium (Kalium)	K	39.0983 (1)
20	Calcium	Ca	40.078 (4)
21	Scandium	Sc	44.955910 (9)
22	Titanium	Ti	47.88 (3)
23	Vanadium	V	50.9415 (1)
24	Chromium	Cr	51.9961 (6)
25	Manganese	Mn	54.93805 (1)
26	Iron	Fe	55.847 (3)
27	Cobalt	Co	58.93320 (1)

ATOMIC NUMBER	NAMES	SYMBOL	ATOMIC WEIGHT
28	Nickel	Ni	58.69 (1)
29	Copper	Cu	63.546 (3)
30	Zinc	Zn	65.39 (2)
31	Gallium	Ga	69.723 (1)
32	Germanium	Ge	72.61 (2)
33	Arsenic	As	74.92159 (2)
34	Selenium	Se	78.96 (3)
35	Bromine	Br	79.904 (1)
36	Krypton	Kr	83.80 (1)
37	Rubidium	Rb	85.4678 (3)
38	Strontium	Sr	87.62 (1)
39	Yttrium	Y	88.90585 (2)
40	Zirconium	Zr	91.224 (2)
41	Niobium	Nb	92.90638 (2)
42	Molybdenum	Mo	95.94 (1)
43	**Technetium**	**Tc**	**97, 98, 99**
44	Ruthenium	Ru	101.07 (2)
45	Rhodium	Rh	102.90550 (3)
46	Palladium	Pd	106.42 (1)
47	Silver	Ag	107.8682 (2)
48	Cadmium	Cd	112.411 (8)
49	Indium	In	114.82 (1)
50	Tin	Sn	118.710 (7)
51	Antimony (Stibium)	Sb	121.75 (3)
52	Tellurium	Te	127.60 (3)
53	Iodine	I	126.90447 (3)
54	Xenon	Xe	131.29 (2)
55	Cesium	Cs	132.90543 (5)
56	Barium	Ba	137.327 (7)
57	Lanthanum	La	138.9055 (2)
58	Cerium	Ce	140.115 (4)
59	Praseodymium	Pr	140.90765 (3)
60	Neodymium	Nd	144.24 (3)
61	**Promethium**	**Pm**	**145, 147**
62	Samarium	Sm	150.36 (3)
63	Europium	Eu	151.965 (9)
64	Gadolinium	Gd	157.25 (3)
65	Terbium	Tb	158.92534 (3)
66	Dysprosium	Dy	162.50 (3)
67	Holmium	Ho	164.93032 (3)
68	Erbium	Er	167.26 (3)
69	Thulium	Tm	168.93421 (3)
70	Ytterbium	Yb	173.04 (3)
71	Lutetium	Lu	174.967 (1)
72	Hafnium	Hf	178.49 (2)
73	Tantalum	Ta	180.9479 (1)
74	Tungsten (Wolfram)	W	183.85 (3)
75	Rhenium	Re	186.207 (1)
76	Osmium	Os	190.2 (1)
77	Iridium	Ir	192.22 (3)
78	Platinum	Pt	195.08 (3)
79	Gold	Au	196.96654 (3)
80	Mercury	Hg	200.59 (3)
81	Thallium	Tl	204.3833 (2)
82	Lead	Pb	207.2 (1)
83	Bismuth	Bi	208.98037 (3)

ATOMIC NUMBER	NAMES	SYMBOL	ATOMIC WEIGHT
84	Polonium	Po	209, 210
85	Astatine	At	210, 211
86	Radon	Rn	211, 220, 222
87	Francium	Fr	223
88	Radium	Ra	223, 224, 226, 228
89	Actinium	Ac	227
90	Thorium	Th	232.0381 (1)
91	Protactinium	Pa	231.03588 (2)
92	Uranium	U	238.0289 (1)
93	Neptunium	Np	237, 239
94	Plutonium	Pu	238, 239, 240, 241, 242, 244
95	Americium	Am	241, 243
96	Curium	Cm	243, 244, 245, 246, 247, 248
97	Berkelium	Bk	247, 249
98	Californium	Cf	249, 250, 251, 252
99	Einsteinium	Es	252
100	Fermium	Fm	257
101	Mendelevium	Md	255, 256, 258, 260
102	Nobelium	No	253, 254, 255, 259
103	Lawrencium	Lr	256, 258, 259, 261
104	Rutherfordium	Rf	257, 259, 260, 261
105	Hahnium	Ha	260, 261, 262
106	Seaborgium	Sg	259, 260, 261, 263
107	Nielsbohrium	Ns	261, 262
108	Hassium	Hs	264, 265
109	Meitnerium	Mt	266

ASTROPHYSICAL DATA

SUN, PLANETS, PRINCIPAL SATELLITES

BODY	MASS (10^{24} kg)	MEAN RADIUS (10^6 m EXCEPT AS NOTED)	SURFACE GRAVITY (m/s²)	ESCAPE SPEED (km/s)	SIDEREAL ROTATION PERIOD* (days)	MEAN DISTANCE FROM CENTRAL BODY† (10^6 km)	ORBITAL PERIOD	ORBITAL SPEED (km/s)
Sun	1.99×10^6	696	274	618	36 at poles 27 at equator	2.6×10^{11}	200 My	250
Mercury	0.330	2.44	3.70	4.25	58.6	57.6	88.0 d	48
Venus	4.87	6.05	8.87	10.4	−243	108	225 d	35
Earth	5.97	6.37	9.81	11.2	0.997	150	365.3 d	30
Moon	0.0735	1.74	1.62	2.38	27.3	0.385	27.3 d	1.0
Mars	0.642	3.38	3.74	5.03	1.03	228	1.88 y	24.1
Phobos	9.6×10^{-9}	9-13 km	0.001	0.008	0.32	9.4×10^{-3}	0.32 d	2.1
Deimos	2×10^{-9}	5-8 km	0.001	0.005	1.3	23×10^{-3}	1.3 d	1.3
Jupiter	1.90×10^3	69.1	26.5	60.6	0.414	778	11.9 y	13.0
Io	0.0889	1.82	1.8	2.6	1.77	0.422	1.77 d	17
Europa	0.478	1.57	1.3	2.0	3.55	0.671	3.55 d	14
Ganymede	0.148	2.63	1.4	2.7	7.15	1.07	7.15 d	11
Callisto	0.107	2.40	1.2	2.4	16.7	1.88	16.7 d	8.2
and 13 smaller satellites								
Saturn	569	56.8	11.8	36.6	0.438	1.43×10^3	29.5 y	9.65
Tethys	0.0007	0.53	0.2	0.4	1.89	0.294	1.89 d	11.3
Dione	0.00015	0.56	0.3	0.6	2.74	0.377	2.74 d	10.0
Rhea	0.0025	0.77	0.3	0.5	4.52	0.527	4.52 d	8.5
Titan	0.135	2.58	1.4	2.6	15.9	1.22	15.9 d	5.6
Iapetus	0.0019	0.73	0.2	0.6	79.3	3.56	79.3 d	3.3
and 12 smaller satellites								
Uranus	86.6	25.0	9.23	21.5	−0.65	2.87×10^3	84.1 y	6.79
Ariel	0.0013	0.58	0.3	0.4	2.52	0.19	2.52 d	5.5
Umbriel	0.0013	0.59	0.3	0.4	4.14	0.27	4.14 d	4.7
Titania	0.0018	0.81	0.2	0.5	8.70	0.44	8.70 d	3.7
Oberon	0.0017	0.78	0.2	0.5	13.5	0.58	13.5 d	3.1
and 11 smaller satellites								
Neptune	103	24.0	11.9	23.9	0.768	4.50×10^3	165 y	5.43
Triton	0.134	1.9	2.5	3.1	5.88	0.354	5.88 d	4.4
and 7 smaller satellites								
Pluto	0.015	1.2	0.4	1.2	−6.39	5.91×10^3	249 y	4.7
Charon	0.001	0.6			−6.39	0.02	6.39 d	0.2

*Negative rotation period indicates retrograde motion, in opposite sense from orbital motion. Periods are sidereal, meaning the time for the body to return to the same orientation relative to the distant stars rather than the Sun.

†Central body is galactic center for Sun, Sun for planets, and planet for satellites.

ANSWERS TO ODD-NUMBERED PROBLEMS

CHAPTER 1

3. 100,000 times bigger
5. 0.108 782 775 7 ns
7. 10^8
9. 0.79 rad
11. 28 g
13. 10^6
15. 7%
17. Yes, by 7 mi/h
19. 30 AU
21. (a) 0.0032 mpg; (b) 0.0014 km/L
23. Approximately 0.0175 rad
25. L/T^2
27. (c) Actually, the speed is given by $v = \sqrt{\lambda g/2\pi}$
29. 2.5×10^6 m
31. 7.4×10^6 m/s^2
33. (a) 2.5×10^{-4} mm^2; (b) 1.6×10^{-2} mm
35. 280 K
37. 41 m
39. 7
43. $\sim 1.3 \times 10^6$
45. 3600 (this assumes a 12-hour watch)
47. (a) $\sim 1.4 \times 10^{18}$ m^3; (b) $\sim 1.3 \times 10^{12}$
49. ~ 10 times as much in stars
51. 1
53. $M \cdot L/T^2$
55. $\sim 10^4$
57. $d_{Sun}/d_{moon} = 380$; $R_{Sun} \sim 7 \times 10^5$ km
(a) 4 μm; (b) 7500
59. (a) $\sim 10^{28}$; (b) $\sim 10^{14}$
63. $\sqrt{F_0/\mu}$; this is in fact the equation for the wave speed

CHAPTER 2

1. 10.16 m/s
3. 5.431 m/s

5. (a) 24 km north; (b) 9.6 km/h; (c) 16 km/h; (d) 0; (e) 0
7. 26.7 km/h = 7.42 m/s
9. 48 mi/h
11. 1 m/s = 2.24 mi/h
13. 51 ft/s = 35 mi/h
15. (a) 2 d, 17 h; (b) 70 km/h
17. 2.6 h later; 1800 km from New York
21. (a) 2.0 m/s; (b) 0; (c) -5.0 m/s; (d) 1.2 m/s; (e) 0.17 m/s
23. (a) $b - 2ct$; (b) at 6.9 s after launch
25. (a) $t = 0$ s, 0.13 s, 2.5 s; (b) $v = 3bt^2 - 2ct + d$;
(c) $v_0 = 1.0$ m/s; (d) $t = 0.065$ s, 1.7 s
27. 0
29. 11.9 m/s^2
31. 31 s
33. (a) 126 m/s; (b) 0.46 m/s^2
35. $a = 6bt - 2c$
37. 100 m by both methods
41. (a) 46 m/s^2; (b) 61 s
43. (a) 2.0 m/s^2; (b) 150 m
45. 27 ft/s^2
47. (a) $t = 2v_0/a$; (b) $v = v_0$
49. 22 m/s
51. (a) 0.42 m/s^2; (b) toward Chicago; (c) 1.1 km
53. $a = 125g = 1200$ m/s^2
55. Yes, $a = 370$ m/s^2
57. No collision; 10 m apart
59. 4.6×10^{-3} m/s^2
61. Collide at 12 km/h
63. 11.3 m/s
65. (a) 27 m; (b) 4.7 s
67. Venus
69. 273 m
71. 2.0 m/s
73. 2.4 s

75. (a) 25 km/h; (b) 13 km/h

77. 3.6 s; 8.3 m/s

79. 0.196 s

81. 5.0 s; 17 km/h

83. (a) 7.0 m/s; (b) in 2.3 s

85. (a) 10.45 m/s; (b) 8.98 m/s; (c) 8.88 m/s; (d) 1.83 m/s

87. (a) $\frac{1}{2}(v_1 + v_2)$; (b) $\dfrac{2v_1 v_2}{v_1 + v_2}$

89. (b) 3.8 s; (c) 19 m; (d) 100 m

91. (a) $v = \omega x_0 \cos \omega t$, $a = -\omega^2 x_0 \sin \omega t$; (b) $v_{max} = \omega x_0$, $a_{max} = \omega^2 x_0$

93. (a) 33 m; (b) 8%

95. 1.19 m

CHAPTER 3

1. 260 m, 7.9° N of W

3. 702 km, 21.3° east of north

5. 18.9 units long, 7.1° E of S

7. (a) 47.8 cm; (b) 30.4 cm

9. 409 km, 79.6° west of north

11. 130°

13. 1.24A at 234° to the horizontal

15. −1.5, 2.5

17. $\mathbf{A} + \mathbf{B} = 4.8\hat{\mathbf{i}} + 0.82\hat{\mathbf{j}}$; $\mathbf{A} - \mathbf{B} = 12\hat{\mathbf{i}} + 11\hat{\mathbf{j}}$;
    $\mathbf{A} + \mathbf{C} = 4.8\hat{\mathbf{i}} + 13\hat{\mathbf{j}}$; $\mathbf{A} + \mathbf{B} + \mathbf{C} = 1.4\hat{\mathbf{i}} + 8.1\hat{\mathbf{j}}$

19. (a) $A_x = 5.9$, $A_y = 8.1$; $A'_x = 5.4$, $A'_y = 8.4$ units

21. (a) $A_x = 8.7$, $A_y = 5.0$; (b) $A_x = 9.7$, $A_y = -2.6$;
    (c) $A_x = 10$, $A_y = 0$

23. $\mathbf{C} = -15\hat{\mathbf{i}} + 9\hat{\mathbf{j}} - 18\hat{\mathbf{k}}$

25. (a) in x-y system $\Delta \mathbf{r} = 5.0\hat{\mathbf{i}} + 3.5\hat{\mathbf{j}}$ km; in x'-y' system
    $\Delta \mathbf{r} = 6.1\hat{\mathbf{i}} + 0.50\hat{\mathbf{j}}$ km; (b) $\Delta r = 6.1$ km

27. (a) 5.4 mi at 32° E of N; (b) 15 mi/h at 32° E of N

29. (a) 264 km/h, 29° west of north; (b) $-128\hat{\mathbf{i}} + 231\hat{\mathbf{j}}$ km/h

31. $19\hat{\mathbf{i}} + 4.5\hat{\mathbf{j}} + 0.26\hat{\mathbf{k}}$ km/h

33. 5.12 m/s², 41° south of west

35. (a) 3.8×10^{-3} m/s²; (b) 5.4×10^{-3} km/s²; (c) 45°

37. $-4.9\hat{\mathbf{i}} - 2.8\hat{\mathbf{j}}$ m/s²

39. $\mathbf{v} = (3bt^2 + c)\hat{\mathbf{i}} + 2dt\hat{\mathbf{j}} + e\hat{\mathbf{k}}$
    $\mathbf{a} = 6bt\hat{\mathbf{i}} + 2d\hat{\mathbf{j}}$

41. (a) 18 km/h, 14 km/h, 10 km/h, 14 km/h;
    (b) 21 km/h, 17 km/h, 13 km/h, 17 km/h

43. $30\hat{\mathbf{i}} + 64\hat{\mathbf{j}}$ km/s

45. (a) 1.6 m, 2.8 m/s, both vertically downward; (b) 2.03 m, 3.57 m/s, both at 38.1° to the vertical; (c) 9.8 m/s², vertically downward in both frames of reference

47. 16.2 m

49. (a) $\mathbf{A} = 9.06\hat{\mathbf{i}} + 4.23\hat{\mathbf{j}}$ m; $\mathbf{B} = 10.0\hat{\mathbf{j}}$ m;
    $\mathbf{C} = -7.55\hat{\mathbf{i}} + 6.56\hat{\mathbf{j}}$ m; (b) $\mathbf{D} = -1.52\hat{\mathbf{i}} - 20.8\hat{\mathbf{j}}$ m;
    (c) $D = 20.9$ m

51. (a) 1.8 km/h/s = 0.50 m/s²; (b) 142°

53. (a) 0.32 cm/s; (b) 0.034 cm/s²; (c) 90°

55. 25° upstream

57. (a) 1.30×10^6 m; (b) 7.78 km/s; (c) 9.19 m/s², almost g

59. $-b\hat{\mathbf{i}} + a\hat{\mathbf{j}}$, $b\hat{\mathbf{i}} - a\hat{\mathbf{j}}$

61. $\frac{\sqrt{2}}{2}\hat{\mathbf{i}} + \frac{\sqrt{2}}{2}\hat{\mathbf{j}}$

65. 53.8°, 13.9 km/h

CHAPTER 4

1. 0°

3. $1.3\hat{\mathbf{i}} + 2.3\hat{\mathbf{j}}$ m/s

5. 4.49 m/s² at 58° below the x axis

7. (a) 23°; (b) 5.4×10^3 km

9. 1.09 m

11. (a) $t = 18$ s; (b) 300 m; (c) 22 m/s, at 120° to x axis

13. (a) 2.6×10^{17} cm/s², upward; (b) parabolic

15. (a) 1.4 s; (b) 10 m

17. 5.7 m/s

19. 1.27 m/s

21. 34 nm

23. 8.3 m/s at 61°

25. Yes

27. −14.6 m/s

29. (a) 6.64 km/s; (b) 16.0 min; (c) 8.28 km/s

31. 1.1 s

33. 1090 m

37. (a) 8.8 m; (b) 0.53 m

39. 11.2 m/s

43. 31.2° or 65.7°

45. 2.8×10^{-3} m/s²

47. 54 min

49. 20 cm

51. 0.344 ns

53. 14 s

55. (a) 148 m/s; (b) 0.974 m/s²

57. $t = \sqrt{r/a_t}$

59. 89 m/s

61. 32 m

63. 19 m

65. 300 m, 119 m

67. $\mathbf{v}_0 = 6.36\hat{\mathbf{i}} + 10.3\hat{\mathbf{j}}$ m/s, or 12.1 m/s at 58.3°

69. 83°

75. 7.2 m/s at 77° to horizontal

77. (a) $\tan \theta_0 = \dfrac{v_0^2}{gx} \pm \sqrt{\dfrac{v_0^4}{g^2 x^2} - 1}$;
    (b) taking $x \simeq 220$ m from the graph gives angles of 29.8° and 60.2°

79. 38°

81. v_x^2 / g

83. 892 m/s²

87. (a) $\tan^{-1}\left(\tan \theta_0 - \dfrac{gt}{v_0 \cos \theta_0} \right)$;
    (b) $\tan^{-1}\left(\tan \theta_0 - \dfrac{gx}{v_0^2 \cos^2 \theta_0} \right)$

CHAPTER 5

1. 3.8 MN

3. 1.5×10^3 kN

5. 10^{-4} N

7. 9.0×10^{22} m/s^2

9. (a) $m_B = 3.0m_A$; (b) 2.0 m/s^2

11. Quadruples

13. (a) 11 m; (b) 24 m; (c) 43 m; (d) 53 m

15. $F_{driver} = 5.7$ kN; $F_{passenger} = 125$ kN, 22 times greater

17. 5.77 N at 72.2° to the x axis

19. Venus

21. (a) 3.3 N; (b) 12 oz

23. 9.1×10^3 kg

25. 8.3×10^{12} m/s^2

27. 6 m/s^2

29. 385 N

31. 2.9 m/s^2, downward

33. (a) 3.1×10^7 N; (b) 9.4×10^2 N

35. $a > g$

37. 0.53 s

39. 55 kN

41. 2.0 N

43. 1.3×10^{-21} cm

45. (a) 5.26 kN; (b) 1.08 kN; (c) 494 N; (d) 589 N

47. 130 N

49. 33 cm

51. 1.9 m/s^2

53. 830 g

55. (a) 132 cm; (b) 127 cm; (c) 120 cm; (d) 40 m/s^2

57. 30.7 kN

59. At least 2.9 cm

61. Apparent weight is 55% of actual weight

63. 240 N

65. 4.3 cm

67. 14 N

69. 7.2 m

71. (a) 0.40mg; (b) 2.40mg; (c) 1.40mg

73. (a) 16 kN; (b) 1.5 kN

75. (a) $a = \dfrac{m_f - m_s}{m_s} g$;

(b) $y = \dfrac{m_f a_s h}{(m_f - m_s)(g + a_s)}$

77. $\ell + nm(a + g)/k$, where n is the spring number measured from the bottom

81. 900 N

CHAPTER 6

1. $2.7\hat{\imath} + 5.5\hat{\jmath}$ N, or 6.1 N at 64° to the x axis

3. (a) 2.0 kN; (b) 1.4 kN

7. 43 cm

9. 530 N; 3.6 times the weight

11. 98 N in horizontal string; $98\sqrt{2}$ N in vertical string

13. 230 N in short rope; 84 N in long rope

15. (a) 6.3 m/s^2; (b) 0.44 s

17. Right-hand mass 2.5 times left-hand mass

19. (a) 7.1 kg; (b) 3.9 kg

21. Left to right, 56.9 N, 34.4 N, 89.2 N

23. 8.18×10^{-8} N

27. (a) 13 m/s; (b) 4.4°

29. 132 m

31. (a) 310 N; (b) 0; (c) nothing

33. (a) 35°; (b) 22°

35. 37°

37. 17 m/s

39. 490 N

41. 45 kg

43. 0.18

45. 340 N; 1.6 times the weight

47. (a) 8.0°; (b) 0.50 m/s^2

49. 0.38 s

51. 4.1 m/s^2, accident deceleration is 21 times greater

53. 12 cm

55. (a) 1.6 m/s^2; (b) 3.3 N

57. 4.2 m/s^2

59. 95 km/h

61. 0.12

65. 0.72

67. 6.9 m/s^2

69. 3.45 rev/min

71. $\mu_s \geq 0.24$

73. (a) 10.0 N; (b) 10.4 N; (c) 10.4 N; (d) 90°, 73.5°, 90°

75. 0.70 m/s^2, upward

77. 7.75 m/s

79. 1.40 s

81. (a) 0.12; (b) toward the inside of the turn

83. 6.3 m/s

87. $\mu_k = \dfrac{v_0^2}{2gx_1} - \dfrac{x_2^2}{4x_1 h}$

93. 28 cm

CHAPTER 7

1. 1.3 kJ

3. 59 kJ

5. 9.6 MJ

7. (a) 400 J; (b) 31 kg

9. 5.9 MN

11. (a) 370 J; (b) 0.26

15. (a) $\hat{\imath} \cdot \hat{\imath} = \hat{\jmath} \cdot \hat{\jmath} = \hat{k} \cdot \hat{k} = 1$;
    (b) $\hat{\imath} \cdot \hat{\jmath} = \hat{\jmath} \cdot \hat{k} = \hat{k} \cdot \hat{\imath} = 0$

17. (a) 14; (b) −12; (c) −16

21. (a) 45°; (b) 111°; (c) 66°

23. (a) 630 J; (b) 0

25. 25°

27. (a) 60 kJ; (b) 20 kJ; (c) 80 kJ

29. (a) 360 J; (b) 350 J; (c) 357.5 J; (d) 359.375 J

31. $k_B = 8k_A$

33. 190 J

35. (a) 30 J; (b) 56 J; (c) 72 J

37. $F_0\left(x + \dfrac{x^2}{2\ell_0} + \dfrac{\ell_0^2}{\ell_0 + x} - \ell_0\right)$

39. (a) 0; (b) $2FR$; (c) πFR

43. 90 J
45. (a) 14 GJ; (b) 3.3 MJ; (c) 28 J
47. 2.46×10^7 m/s $= 0.082c$
49. 2.3 kJ
51. 42 cm
53. (a) 24 J (b) 18 m/s
55. 4.1 m
57. (a) None
59. (a) 60 kW; (b) 1 kW; (c) 40 W
61. 9.4 MJ
63. (a) 36 MW; (b) 1.1 MW
67. 2.7 h
69. 300 million gal/day
71. 2.1 MJ
73. $7.7°$
75. 1.6 m
77. $\frac{1}{2}F_0 x_0$
79. $70.5°$
81. (a) 28 kJ; (b) 18 kJ
83. (a) 450 W; (b) 8.0 kJ
85. (a) $W = \frac{1}{2}bt^2$; (b) $a = \sqrt{b/m}$
87. (a) 71 kW·y; (b) 93 kW·y
89. (a) $P = mgv \cos(vt/R)$
91. (a) 33 J; (b) 167 J
93. 6.3 J

CHAPTER 8

1. (a) $-2\mu mg\ell$; (b) $-\sqrt{2}\mu mg\ell$
3. (a) 0; (b) $F_0 a$
5. (a) 170 MJ; (b) -7.6 MJ
9. (a) 1.07 J; (b) 1.12 J
11. 7.5 J
13. 54 cm
15. 0.53 J
17. $U = -\frac{1}{3}ax^3 - bx$
19. $U = F_0\left(x + \dfrac{x^2}{2\ell} + \dfrac{\ell^?}{\ell + x} - \ell\right)$
21. 50 m/s (180 km/h)
23. 96 m
27. 6.5 m/s
29. 26 MN/m
31. 15 km/h
35. 2.6 m/s
39. $x = \pm 69$ cm
43. (a) $U = -\dfrac{a}{2}x^2 + \dfrac{b}{4}x^4$; (b) $x = 0.66$ m, $x = 2.1$ m
45. (a) -6.7 N; (b) 0; (c) 4.5 N
49. (a) 30 cm; (b) 10.4 N/m
51. 44%
53. 19%
55. 0.36
57. 2.6 m/s
59. 0.036

61. 62 cm from left end of frictional zone
63. 1000 MW, twice that of the coal plant
65. $x = 2(h_1 - h_2)$
69. (a) $v = \sqrt{\sqrt{2}g\ell}$; (b) $T = \dfrac{3}{\sqrt{2}}mg$
71. $x = \pm 24$ cm
73. 14 m
75. (a) 2.53×10^5 m/s; (b) 2.91×10^5 m/s;
    (c) 2.93×10^5 m/s
77. 54.6 mJ
79. 75 cm
81. (a) 1.74 cm; (b) 0.78 cm; (c) 7.4×10^7 m/s
83. (a) $v = \left[\dfrac{x^2}{m}\left(a - \frac{1}{2}bx^2\right)\right]^{1/2}$; (c) $v_{\max} = a/\sqrt{2mb}$
85. $v = [2ax^{|b+1|}/m\,|b + 1|]^{1/2}$
87. (a) 11 m/s; (b) $x = \pm 1.4$ m

CHAPTER 9

1. $R_E/\sqrt{2}$
3. 58%
5. 8.6 kg
7. 46 nN
9. 440 m
11. 1.2×10^{-7}
13. 3.1 km/s
15. 1.8 days
17. 1.0 hour
19. 2.6×10^{41} kg, about 10^{11} solar masses
21. 6.3×10^{10} m
23. 2.64×10^{10} m
25. 2.47 times Earth's orbital radius
27. No
29. 3.2×10^9 J
31. 530 km
33. 58 MJ
35. $R_E/99 = 64$ km; underestimate
37. 3%
39. $\sqrt{2}$
41. 5.8×10^6 m
43. (a) 11.2 km/s; (b) 9.74 km/s; (c) no
47. $v = \sqrt{2GM\left(\dfrac{3}{R} + \dfrac{1}{r}\right)}$
49. 8.1×10^{11} m, just beyond Jupiter
53. 109 min
55. 15 km/s, 23 km/s
57. 7.95 km/s
59. 8.85×10^5 m
61. (a) 9.0×10^{10} m; (b) 5.3×10^{11} J; (c) 38 km/s
63. 4.60×10^{10} m
65. (b) 8.8 mm; (c) 2.9 km
67. 1400 km lower

CHAPTER 10

1. 0.75 m from the center
3. $X = 50$ cm; $Y = 69$ cm; with origin at lower left
5. $\ell/2\sqrt{3}$ along the perpendicular bisector of any side
7. $X = 44$ cm, $Y = 55$ cm, with origin at lower left
9. $0.115a$ above the vertex of the missing triangle
11. 6.5 pm from the oxygen
13. $\mathbf{R} = \left(t^2 + \dfrac{10}{3}t + \dfrac{7}{3}\right)\hat{\mathbf{i}} + \left(\dfrac{2}{3}t + \dfrac{8}{3}\right)\hat{\mathbf{j}};$

$\mathbf{V} = \left(2t + \dfrac{10}{3}\right)\hat{\mathbf{i}} + \dfrac{2}{3}\hat{\mathbf{j}}; \mathbf{A} = 2\hat{\mathbf{i}}$
15. $\sim 10^{-10}$ m, the diameter of a hydrogen atom
17. $m_{\text{mouse}} = \frac{1}{4}m_{\text{bowl}}$
19. 3.0 m/s in the negative x direction
21. 21 g, $-x$ direction
23. 460 m/s
25. 8.4 km/h
27. (a) 0.14 N/m²; (b) 0.014 mm
29. (a) 0.99 m; (b) 3.9 m/s
31. 3.9 km/h
33. $26\hat{\mathbf{i}} + 16\hat{\mathbf{j}}$ m/s
35. 1100 kg
37. (a) 3.3×10^6 N (b) 3.4×10^5 kg
39. 0.22
41. $K_{\text{cm}} = 1.1\times10^{-14}$ J before and after;
$K_{\text{int}} = 0$ before, $K_{\text{int}} = 1.3\times10^{-14}$ J after
43. before: $K_{\text{cm}} = 1.6$ MJ, $K_{\text{int}} = 21$ kJ; after: $K_{\text{cm}} = 1.6$ MJ, $K_{\text{int}} = 0$
45. (a) 1.7 cm above the bottom; (b) 2.7 cm above the bottom
47. 20 m
49. $-47\hat{\mathbf{i}} - 68\hat{\mathbf{j}}$ m/s
51. (a) 3.5 m; (b) 1.3 m/s; (c) 0
53. (a) 0.096 m/s²; (b) 6.2 m/s
55. 9.3 m/s
57. (a) $m = \dfrac{\pi\rho h^2}{2a}$; (b) $Z = \dfrac{2}{3}h$
59. (a) 2×10^4 kg·m/s; (b) 2×10^8 J; (c) yes
61. (a) 37.7°; (b) 0.657 m/s
63. 5.8 s after explosion
65. (a) thrust $= m\dfrac{dv}{dt} = [(1 + f)V_{\text{ex}} - V]\dfrac{dM_{\text{in}}}{dt}$; (b) 1504 lb
67. $v_1 = \left(\dfrac{m_2 kx^2}{m_1^2 + m_1 m_2}\right)^{1/2}$; $v_2 = \left(\dfrac{m_1 kx^2}{m_2^2 + m_1 m_2}\right)^{1/2}$

CHAPTER 11

1. 95 N·s
3. 4.3×10^3 N, $7.1mg$
5. (a) $-2.6\hat{\mathbf{i}} + 0.74\hat{\mathbf{j}}$ N·s; (b) $-51\hat{\mathbf{i}} + 14\hat{\mathbf{j}}$ N
7. (a) 150 N·s, upward; (b) 3.0 kN, about 5 times your weight

9. (a) 7.3 MN·s; (b) 5.6 MN
11. (a) 6.8×10^{-3} N·s; (b) 2.3 N
13. $\Delta P/P = 2\%$
15. 12 ms
17. (a) 6.2 mi/h; (b) 12%
19. 19 kg
23. 10^{21} kg
25. $4.0\hat{\mathbf{i}} + 21.5\hat{\mathbf{j}}$ Mm/s
29. 1.3 μJ
31. 120°
33. 46 m/s
35. $v_{1f} = -11$ Mm/s; $v_{2f} = 6.9$ Mm/s
37. $3 + 2\sqrt{2} \approx 5.8$; it doesn't matter which is more massive
39. $v_A = -\frac{1}{3}v$; $v_B = \frac{2}{9}v$; $v_C = \frac{8}{9}v$
45. 22°
47. $(v_{1i} = 0.833$ m/s, $v_{2i} = 1.22$ m/s, $\theta_{2i} = 28.3°)$;
$(v_{1i} = 1.20$ m/s, $v_{2i} = 1.12$ m/s, $\theta_{2i} = 31.2°)$
49. 13 m/s at 27° to horizontal
51. $v_A = -\frac{1}{5}v_0\hat{\mathbf{i}}$; $v_B = \frac{3}{5}v_0\hat{\mathbf{i}} + \frac{1}{5}\sqrt{3}v_0\hat{\mathbf{j}}$ $v_C = \frac{3}{5}v_0\hat{\mathbf{i}} - \frac{1}{5}\sqrt{3}v_0\hat{\mathbf{j}}$
53. 350 N
55. 52 km/h at 33° north of east
57. $m_{\text{truck}} = 7.6m_{\text{car}}$
59. (a) $m_1 = 3m_2$; (b) $v_{2f} = 2v$
61. $v_{1f} = 1.66$ m/s; $v_{2f} = 0.703$ m/s; $\theta_{2f} = 67°$ clockwise from initial velocity of the first ball
63. (a) 12.0 m; (b) 15.4 m/s
65. 0.88
69. 44
71. $v_{1\,\text{kg}} = 4.0$ m/s; $v_{4\,\text{kg}} = 1.0$ m/s at 50° clockwise from the x axis
73. $v_{1200} = 2.2$ km/h, $v_{1800} = 18$ km/h
75. (a) $v_1 = 0.28v_0$, $v_2 = 0.48v_0$; (b) 3, $0.26v_0$, $0.31v_0$

CHAPTER 12

1. (a) 7.27×10^{-5} rad/s; (b) 1.75×10^{-3} rad/s;
(c) 1.45×10^{-4} rad/s; (d) 31.4 rad/s
3. (a) $v = (\pi/30)\omega r$; (b) $v = 2\pi\omega r$;
(c) $v = (\pi/180)\omega r$
5. (a) 66 rpm; (b) 3.7 s
7. (a) 21.7 rad/s, 207 rpm; (b) 34.7 rad/s, 331 rpm
9. (a) 0.068 rpm/s; (b) 7.1×10^{-3} rad/s²
11. (a) 12 min; (b) 2.2×10^4
13. 1.3 rad/s²
15. (a) 2.0 s; (b) 1.0 rev
17. 1.2 m
19. 0.079 N·m
21. 0.15 N·m
23. (a) 0.70 N·m, counterclockwise; (b) $\mathbf{F}_1$ and $\mathbf{F}_5$
25. 22 cm
27. (a) $\frac{2}{3}m\ell^2$; (b) $\frac{2}{3}m\ell^2$; (c) $\frac{4}{3}m\ell^2$
29. 45 kg·m²
33. (a) 9.7×10^{37} kg·m²; (b) 2.6×10^{19} N·m

35. (a) 1.29×10^{38} kg·m²; (b) 6.45×10^{33} N·m
39. (a) 430 min; (b) 1900 rev
41. 1900 N·m
43. 170 rpm
45. $m_{\text{pulley}} = 0.49$ kg; $m_1 = 0.41$ kg; $m_2 = 0.58$ kg
47. (a) 450 J; (b) 140 W
49. 0.089%
51. 12.2 rad/s
53. 7.0 m/s
55. $v = \sqrt{\dfrac{2gh}{\alpha + 1}}$
57. 17%
59. hollow
61. (a) 0.156; (b) 0.070 rad/s
63. $\dfrac{253}{512} MR^2 = 0.494 MR^2$
65. (a) 310 N; (b) 165 kg
67. (a) 3.5 m/s; (b) 24%
69. $\dfrac{27}{10} R$
71. $\omega = \sqrt{2A/I}$
73. $I = \frac{1}{2} Mb^2$
75. $2\sqrt{2g/3R}$
77. $\frac{1}{2} Mg\ell \cos \phi$

CHAPTER 13

1. 63 rad/s, west
3. 0.52 rad/s², $-37°$
5. 16.6 rad/s
7. (a) $-z$; (b) z; (c) in the x-y plane, 45° clockwise from the x axis
9. (a) $-12\hat{\mathbf{k}}$ N·m; (b) $36\hat{\mathbf{k}}$ N·m; (c) $12\hat{\mathbf{i}} - 36\hat{\mathbf{j}}$ N·m
11. $-17\hat{\mathbf{k}}$ N·m
13. Parallel to the x axis or 120° clockwise from x axis
17. $F_x = 1.33 F_y - 3.13$ N
19. 414 kg·m²/s
21. 7.6 rad/s
23. 0.017 kg·m²/s
25. 37 kg·m²/s
27. 2.7×10^5 kg·m²/s, out of page in Fig. 13-30
29. (a) $-4.2\hat{\mathbf{k}}$ kg·m²/s; (b) 0; (c) $6.4\hat{\mathbf{k}}$ kg·m²/s
31. (a) 0.17 rev/s; (b) 386 J
33. (a) 142 rpm; (b) 21%
35. 2.5 days
37. (a) 23.7 rpm; (b) 3.49 mJ
39. (a) $\dfrac{2M\omega_0}{2M + 3m}$; (b) $\dfrac{M\omega_0}{M + 6m}$; (c) same as (b)
41. (a) 0.537 rad/s; (b) 6.44 m/s; (c) 207 N
43. 6.0 cm
45. $I = \dfrac{mgd}{2\omega\Omega}$
47. 1.05×10^{-34} kg·m²/s $= \hbar$
49. $\tan^{-1}(\frac{1}{2}) = 26.6°$

51. 0.37 rev/s
53. 22 g
55. $\dfrac{I}{I + mR^2}$
57. (a) 1.61; (b) 2.22
61. Sun's rotation 2.8%; Jupiter's orbital motion 60%
63. $v = \left[\dfrac{8(m + M)g\ell}{m^2} \left(\dfrac{1}{4}m + \dfrac{1}{3}M \right) \right]^{1/2}$
65. Both wheels have stopped rotating about their axes, but the whole contraption is now rotating about the center-line between the two.

CHAPTER 14

1. (a) $\frac{1}{3}\ell F_3 - \frac{2}{3}\ell F_2 + \ell F_1 = 0$;
   (b) $\frac{1}{2}\ell F_1 - \frac{1}{6}\ell F_2 - \frac{1}{6}\ell F_3 + \frac{1}{2}\ell F_4 - \frac{1}{2}\ell F_5 = 0$
3. (b) $\boldsymbol{\tau}_{\text{origin}} = -7\hat{\mathbf{k}}$ N·m
5. (a) A vector of length $\sqrt{2}\, F$, oriented 45° clockwise from the negative y axis, applied at the point $x = 0$ m, $y = +1$ m (or anywhere on the line $y = (1 + x)$ m. (b) Not possible; the first two vectors sum to zero but produce a nonzero torque, so any other vector applied to balance torques will upset force balance.
7. Both sets have $-F_1 + F_2 \sin\phi + F_3 = 0$, $-F_2 \cos\phi + F_4 = 0$; torque equations are (a) $\ell_2 F_4 \cos\phi - \ell_2 F_3 \sin\phi - \ell_1 F_2 = 0$; (b) $(\ell_2 - \ell_1)F_2 - \ell_2 F_1 \sin\phi = 0$.
9. (a) $\tau_A = \frac{1}{2}\ell mg$; (b) $\tau_B = 0$; (c) $\tau_C = \frac{1}{2}\ell mg$
11. $m_2 = 0.384 m_1$
13. (a) 61 cm from left end; (b) 1.4 m from left end
15. 120 N
17. 11.7 kN
19. (a) 40 N·m; (b) 1300 N
21. Vertical forces both 73.5 N, downward, horizontal forces both 33.6 N, away from door jamb at top, toward jamb at bottom
23. 5.0 kN; tension
25. 0.87
27. 500 N
29. 50 kN
31. 6.05 kN
33. $\frac{1}{8}$
35. Maximum height of CM is at sphere center; lower for clown (b)
37. Two equilibria for $|a| > 2\sqrt{3}$; one metastable, other unstable
39. 1.2 m
41. 170 N
43. 74 kg
45. $\frac{1}{2}(\sqrt{2} - 1)mgs$
47. $\mu = \dfrac{\sin 2\theta}{3 + \cos 2\theta}$
49. $mg/2k$
51. 28°

53. Left scale 16.3 N; right scale 22.9 N
55. Tip
57. Slide
59. (a) $0.44mg$, at 12° to Earth's polar axis;
    (b) $0.036mgR_E$, out of the plane of Fig. 14-55

PART 1   CUMULATIVE PROBLEMS

1. 16.5 m from the post
3. $a = \dfrac{2g[(m_1 + m_2)\sin\theta - \mu m_1 \cos\theta]}{2m_1 + 3m_2}$
5. (a) $v = \frac{2}{7}\omega R$; (b) $\Delta x = \dfrac{2\omega^2 R^2}{49\mu g}$

CHAPTER 15

1. $T = 0.780$ s; $f = 1.28$ Hz
3. 11.5 fs
5. $A = 20$ cm, $\omega = \pi/2$ s^{-1}, $\phi = 0$;
   $A = 30$ cm, $\omega = 2.0$ s^{-1}, $\phi = -\pi/2$;
   $A = 40$ cm, $\omega = \pi/2$ s^{-1}, $\phi = \pi/4$
7. 63.3 kg
9. 1.7 kN/m
11. 0.69 s
13. (a) $\pi\sqrt{m/k}$; (b) $v_0\sqrt{m/k}$
17. (a) 1.0 cm; (b) 6.2 m; (c) 3.6 km
19. 0.11 N·m/rad
21. (a) $2\pi\sqrt{\ell/g}$; (b) $2\pi\sqrt{2\ell/3g}$; (c) $2\pi\sqrt{2\ell/g}$; (d) infinite
23. 0.34 s
25. $R = \sqrt{2\kappa/k}$
27. Within 1 μm
33. 5.0 g
35. $\omega^2 = \dfrac{k_1 k_2}{m(k_1 + k_2)}$
37. a and b are the amplitudes in the x and y directions, respectively
39. 400 J, or 1.4×10^{-3} of the total KE
41. $t = (0.14 + n)$ s, $t = (0.53 + n)$ s, n an integer;
    $x = \pm37$ cm
45. $\omega = \sqrt{2k/3M}$
47. 34
49. 77% at 0.90ω; 66% at 1.1ω
53. (a) 19 s^{-1}; (b) 0.33 s; (c) 92 m/s^2
55. (a) 6.5 cm; (b) 0.51 s
57. 1.64 s
59. 300 g
61. (a) $E_2 = \frac{1}{4}E_1$; (b) $a_{2\,max} = \frac{1}{4}a_{1\,max}$
63. 2.1 m/s^2
65. 0.44
67. $2\pi\sqrt{10ga/7}$
69. $f = 0.54$ Hz; $A = 22$ cm; $\phi = -0.11$ rad
71. $T = 2\pi\sqrt{R/g}$
73. $T = 2\pi\sqrt{mL/2F_0}$

1. 3.4 s
3. 3.38 m
5. 1.81×10^8 m/s $= 0.604c$
7. (a) 400 nm; (b) 0.3 mm
9. 11 m
11. (a) 0.58 m^{-1}; (b) 1.53 s^{-1}
13. (a) 13.7 s^{-1}; (b) 0.393 cm^{-1};
    (c) $y = 2.5\cos(0.393x + 13.7t)$
15. (a) 25 cm; (b) 0.37 Hz; (c) 12 m; (d) 4.4 m/s
17. $y = \dfrac{2}{(x - 3t)^4 + 1}$
19. (a) 3.0 m; (b) 1.5 s; (c) 2.0 m/s; (d) $+x$
21. 250 m/s
23. (a) 7.6 N; (b) 1.7 m/s
25. 364 m/s
27. 94 N
29. 7.64 g/cm^3
31. 585 m/s
33. 9.9 W
35. 35 cm
37. $4\pi^2 A^2 F/\lambda$
39. 12 mW/m^2
41. (a) 9.1 kW/m^2; (b) 0.88 W/m^2
43. (a) 6.4 kW/m^2; (b) 4.9 W/m^2
45. 5.1 m
47. (a) 2 cm; (b) pulse 1 at $x = 0$, direction $+x$, pulse 2 at
    $x = 5$, direction $-x$; (c) $t = 2.5$ s
49. Every 6 s
51. 5.34 m
53. $\sqrt{gh}$
55. (a) 1.5 cm; (b) 63 cm; (c) 11 ms; (d) 56 m/s;
    (e) 18 W
57. $v = \sqrt{k\ell(\ell - \ell_0)/m}$
59. 10 m
61. Every 30 s
63. $u < 0.063v$
67. 5.2 km
69. 67 m

CHAPTER 17

3. $\lambda = 34$ cm; $T = 1.0$ ms; $\omega = 6.3\times10^3$ s^{-1};
   $k = 0.18$ cm^{-1}
5. 0.29 s
7. 0.14 kg/m^3
9. monatomic
11. 739 m/s
13. 190 m/s
15. 4.4 nm
17. (a) 3.8 mW/m^2; (b) 96 dB; (c) 1.8 N/m^2; (d) 1.6 μm
19. 1 kHz to 6.5 kHz
21. (a) 3.2 μW/m^2, 0.051 N/m^2; (b) 3.2×10^{-13} W/m^2;
    1.6×10^{-5} N/m^2

23. (a) 20 dB; (b) approximately 250 Hz
27. 6.3 m
29. 1
31. 3.2 km/s
33. 0.75 mm
35. 39 μs
37. (a) $\frac{2}{3}A$; (b) $\frac{1}{3}A$
39. (a) 0.12 s; (b) 1.1 cm; (c) 0.86 cm
41. 7
43. (a) 280 Hz; (b) 70 Hz; (c) 210 Hz
49. (a) 16.6 cm; (b) 424–457 Hz
51. 0.33 Hz
53. 91 Hz
55. 253 m/s
57. 43 m/s = 154 km/h
59. (a) 2800 Hz; (b) 933 Hz
61. 25 m/s
63. $u/v = 1/\sin 45° = 1.4$
65. (a) 5.5×10^{-4} W/m²; (b) 87 dB
67. (a) 112 m/s; (b) 4, 5
69. 960 m/s
71. 0.445 s
73. (a) $\lambda = 5.0$ m, $f = 0.56$ Hz; (b) $\lambda = 2.5$ m; $f = 0.79$ Hz
75. 1.36
79. 16 kHz
81. $\dfrac{4\sqrt{\mu_1\mu_2}}{(\sqrt{\mu_1} + \sqrt{\mu_2})^2}$

CHAPTER 18

1. 1.2 kg
3. 10^{-14}
5. (a) 81 N; (b) 65 N
7. 1 in. H$_2$O = 249 Pa
9. 2×10^7 N, or 2000 tons
11. 21 N
13. 0.25 m²
15. No; $F = 2.3 \times 10^4$ N, or 2.5 tons
17. 1700 kg/m³
19. ~90 m
21. 890 Pa gauge
23. 8.1×10^{10} N
25. 3.6 mm
27. 93 cm higher in the eye
29. 8.11×10^3 kg
31. 44 kg
33. 0.75 %
35. 59 g
37. 27 m
39. (a) 49 kg; (b) 2500 kg
43. (a) 1.8×10^4 m³/s; (b) 1.5 m/s
45. 1.75 m/s
47. (a) $h_2 = h_1$; (b) $h_2 = h_1 - \dfrac{3v^2}{2g}$
49. 14.3 m

51. 7.2 cm³/s
53. (a) no; (b) yes
55. 13
57. 70%
59. $A\sqrt{2gh}$
61. (a) 14 m/s; (b) 2.2 m
65. (a) 25 L/s; (b) 55 m/s; (c) 1.8 kPa
67. $t = \dfrac{A_0}{A_1}\sqrt{\dfrac{2h}{g}}$
69. $P = P_a + \rho gh_0 + \dfrac{1}{2}\rho\omega^2 r^2$; (b) $h = h_0 + \dfrac{\omega^2 r^2}{2g}$
71. (a) $\rho(h) = \dfrac{P_0}{h_0 g}e^{-h/h_0}$; (b) 5.7 km

PART 2 CUMULATIVE PROBLEMS

1. (a) $\ell = \dfrac{4M}{\pi d^2 \rho}$; (b) $T = 4\sqrt{\dfrac{\pi M}{d^2 \rho g}}$
5. 17.2 cm

CHAPTER 19

1. 720
3. 20°C
5. −40
7. −196°C, −321°F
9. (a) 138 kPa; (b) 33.4 kPa; (c) 233 kPa
11. 1.37 L
13. 586 mm
15. 240 kcal
17. 0.36 kg
19. 24 days
21. (a) 23 kJ; (b) 337 kJ; (c) 65 kJ
23. 7.5 kW
25. 2.4 kg
27. (a) 560 g; (b) 0.27 K/s
29. 1.8 kg
31. 0.70 K
33. 1.6 K/s
35. 197 g
37. 56.2°C
39. 3.7 kW
41. 0.293 W
43. 25 ft²·°F·h/Btu
45. 200 W
47. (a) 12.3 ft²·°F·h/Btu; (b) 715 Btu/h
49. Will save 10 gallons/month
51. 23°C
53. 80%
55. −25°C
57. Drop by 5.9 K
59. 24°C
61. 480 W
63. 1151 K

65. −2.5°C
67. (a) $87; (b) $10
69. 4.65%
71. $\Delta T_{copper} = 0.16$ K; $\Delta T_{iron} = 2.1$ K
73. 2.9 J
77. 10 hours

CHAPTER 20

1. 2.6 m³
3. 1.8 MPa
5. (a) 27 L; (b) 330 K
7. 2.7×10^7
9. 11 L
11. 515 kPa
13. (a) 1.27 atm; (b) 0.0268 mol; (c) 0.786 atm
15. 2.88×10^3 K
17. 10^{10} K; gas molecules would dissociate first
19. 268 K, compared with ideal gas 292 K
21. 1.76 MPa
23. (a) 9.1×10^{20}; (b) 2.0×10^{20}
25. 22 kJ
27. 3.9 kg
29. 5.96 MJ
31. 1.3×10^{10} kg
33. 564 W
35. 44 minutes
37. 48 min
39. (a) 117 s
41. 3.55 MJ
43. 64°C
45. 177 g
47. 135 g solid in 865 g liquid, at 234 K
49. 5.0 kg
51. 4.9°C
53. 1.00021 cm
55. 3.9 km
57. 43.6 mm³
59. 307 K
61. (a) Drop by 0.0115 km³; (b) increase by 0.048 km³
63. 120 mol/m³
65. 19 kW
67. 79 g
69. 1.2 kg ice, 0.80 kg water, all at 0°C
71. 50 min
73. (a) 61 h; (b) 52 h
79. 34 km

CHAPTER 21

1. 29 kJ
3. Increases by 250 J
5. 140 kW
7. 0.02°C
9. $2P_1V_1$
11. 1.2 kJ

13. 4.3 kJ
15. 190 K
17. 1.99 kJ
19. (a) 399 J; (b) 264 kPa
21. (a) 571 kPa; (b) 438 J
23. 440°C
25. $V = 0.18V_0$
27. (b) Gas does 13 J of work; (c) 22 J heat lost from gas
29. (a) 300 K, 1.5 kJ; (b) 336 K, 0 J; (c) 326 K, 429 J
31. (a) 39.9 kPa; (b) 83.3 kPa; (c) 80.2 kJ
33. 928 J
35. (a) 211 J; (b) 12.9 L
37. 75°C
39. 128 J
41. (a) $\frac{9}{2}R$; (b) $\frac{11}{9}$
43. 57.7%
45. 79%
47. 20 mol
49. Drops 23.1 K
51. 343 K
53. 28 kPa
55. (a) 598 J; (b) 2500 J flows in
57. 25 m
59. $\frac{4}{3}P_1V_1$
63. (a) 2.5 kJ; (b) 447 K
69. (a) $M = \dfrac{P_0A}{g}\left[\dfrac{h_1}{h_2} - 1\right]$; (b) $M = \dfrac{P_0A}{g}\left[\left(\dfrac{h_1}{h_2}\right)^\gamma - 1\right]$

CHAPTER 22

1. (a) 12!/6!, or 6.7×10^5 states; (b) about 1 in 1000
3. 5×10^{24} J, assuming oceans cover 75% of Earth to an average depth of 3 km; this is about 20,000 times annual use
5. (a) 27%; (b) 7.0%; (c) 77%
7. 0.95 K
9. 52% winter, 48% summer
11. (a) 1.75 GW; (b) 43%; (c) 505 K
13. 2×10^7 kg/s, slightly more than the Mississippi's flow
15. (a) 39%; (b) 550 J; (c) 190°C
17. 53.3 kJ
19. (a) 4.3; (b) maximum COP = 11
23. (a) $COP_{summer} = 13$, $COP_{winter} = 3.5$; (b) 0.076 J; (c) 0.22 J
25. (a) 561.7 J; (b) 464.1 J; (c) 97.66 J; (d) 17.4%;
    (e) $T_c = 403$ K, $T_h = 487$ K
29. 718 K
31. 1.22 kJ/K
33. 8.9°C
35. 1.36×10^8 J/K
39. (a) 53 J/K; (b) 74 J/K; (c) 0
41. (a) −109 J/K; (b) 122 J/K; (c) 13 J/K
43. $\Delta S_{AB} = 68.5$ J/K, $\Delta S_{BC} = 45.7$ J/K, $\Delta S_{CA} = -114.2$ J/K
45. 470 kPa
47. (a) 69%; (b) 967 K
49. Decrease in $T_{minimum}$
51. $58

53. (a) $W = Q = 345$ J; (b) $e = 24\%$
55. 598 J/K
57. About 30 km for 200 kg of water and a 60 kg bather
59. 166 MW
61. (a) $1 - 5^{1-\gamma}$; (b) $3T_{min}(5^{\gamma-1})$; (c) $e_{Carnot} = 1 - \frac{1}{3}(5^{1-\gamma})$
63. (a) 7.94; (b) 5.26; (c) $P_n = 2.96P_s$
65. (a) $T_h = T_{h0}e^{-P_0t/mc(T_{h0}-T_c)}$;

 (b) $P = 0$ at $t = \dfrac{mc(T_{h0} - T_c)}{P_0} \ln\left(\dfrac{T_{h0}}{T_c}\right)$

67.

	P	V	T	$U - U_A$	$S - S_A$
A					
B			$3.4\,T_0$	$6.0\,P_0 V_0$	$3.1\,P_0 V_0/T_0$
C	$1.5\,P_0$	$2.2\,V_0$	$3.4\,T_0$	$6.0\,P_0 V_0$	$3.8\,P_0 V_0/T_0$
D			$3.0\,T_0$	$5.0\,P_0 V_0$	$3.8\,P_0 V_0/T_0$

PART 3 CUMULATIVE PROBLEMS

1. $e = 1 - r^{1-\gamma}\left[\dfrac{r_c^\gamma - 1}{\gamma(r_c - 1)}\right]$

3. $W = an^2\left(\dfrac{1}{V_2} - \dfrac{1}{V_1}\right) + nRT \ln\left(\dfrac{V_2 - bn}{V_1 - bn}\right)$

5. (a) $t_1 = \dfrac{L_f M T_h}{P_h T_0}$; (b) $P = P_h\left[1 - \dfrac{T_0}{T_h}e^{P_h(t-t_1)/McT_h}\right]$,

 (c) $t_2 = t_1 + \dfrac{McT_h}{P_h} \ln\dfrac{T_h}{T_0}$, with L_f the heat of fusion of ice and c the specific heat of water.

CHAPTER 23

1. Several coulombs
3. (a) uud; (b) udd
5. About 10^9 N; about 10^6 times typical human weight
7. 8.2×10^{-8} N
9. $21.8\ \mu$C
11. $-3.3\ \mu$C
13. $14\hat{\imath} - 7.4\hat{\jmath}$ N
17. $15\ \mu$C
19. $1.6\hat{\imath} - 0.33\hat{\jmath}$ N
21. $\dfrac{kq^2}{a^2}\left(\sqrt{2} + \dfrac{1}{2}\right)$
23. $q_2 = 143\ \mu$C; $q_3 = 116\ \mu$C
25. 3.8×10^9 N/C
27. (a) 2.2 MN/C; (b) 77 N
29. 5.15×10^{11} N/C
31. (a) 2.0 MN/C, upward; (b) 0.82 MN/C, downward; (c) 58 MN/C, downward
33. $-4e$
35. (a) $\mathbf{E} = \dfrac{2kqy}{(a^2 + y^2)^{3/2}}\hat{\jmath}$; (b) $y = \pm a/\sqrt{2}$
37. (a) $8.0\hat{\jmath}$ GN/C; (b) $190\hat{\jmath}$ MN/C; (c) $216\hat{\jmath}$ kN/C
39. 39 pm

43. 2.1 MN/C
45. 339 kN/C, upward
47. $-\dfrac{2k\lambda_0}{\pi\ell}\hat{\imath}$
49. -137 nC
53. 1.1 kN/C
55. (a) $2.5\ \mu$C/m; (b) 3.0×10^5 N/C; (c) 1.8 N/C
57. 3.3×10^{-12} kg
59. (a) 1.35 cm; (b) reverses direction, accelerates and exits field region at 3.8×10^5 m/s
61. $\ell\sqrt{eE/md}$
63. 2.8 Mm/s
65. $-14\ \mu$C/m
67. (a) 3.0 mN·m; (b) 11 mJ
69. (b) Attractive
71. $x = -8.09$ nm
73. $\dfrac{k\lambda_0}{\ell}\hat{\imath}$
75. $2\sqrt{2}kQ/\pi a^2$
77. 1.4 cm
79. $-4q$, a distance $3a$ to the right of $-q$
81. 50.7 kN/C, downward
83. 7.0 cm; 0.54 μC
85. (a) 5.3×10^{-12} N, to right; (b) 5.3×10^{-12} N, to left

CHAPTER 24

1. $+3\ \mu$C
5. (a) 1.7 kN·m²/C; (b) 1.2 kN·m²/C; (c) 0
7. 490 N·m²/C
9. $\pi R^2 E$
11. (a) $-q/\varepsilon_0$; (b) $-2q/\varepsilon_0$; (c) 0; (d) 0
13. 4.9×10^4 N·m²/C
15. (a) 0.69 MN·m²/C (b) -0.69 MN·m²/C (c) 0
17. 10 kN/C
19. 1.8×10^{12} N/C
21. (a) 2.2×10^5 N/C, outward; (b) 2.5×10^4 N/C, outward; (c) 4.0×10^3 N/C, inward
23. (a) $8kQ/R^2$, inward; (b) $kQ/4R^2$, inward; (c) (a) would not change, (b) would become 0
25. (a) 3.6 MN/C; (b) 3.8 MN/C; (c) (a) would not change, (b) would nearly double
27. as $1/r$
29. 6.3 μC/m³
33. 3.6 mC/m³
35. 58 nC/m²
37. $E_1 = \sigma/2\varepsilon_0$, left; $E_2 = \sigma/2\varepsilon_0$, right; $E_3 = 3\sigma/2\varepsilon_0$, right; $E_4 = \sigma/2\varepsilon_0$, right
39. 18 N/C
41. (a) $x < 1.83$ cm; (b) $x > 54.5$ cm
43. 1.6×10^5 N/C
45. (a) $\rho = 0$; (b) $\sigma = 4.0$ mC/m²; (c) other charges would destroy the symmetry, making σ nonuniform
47. (a) 0.50 μC/m²; (b) 56 kN/C
49. (b) $-Q$

51. 1.8 MN/C

53. (a) 0; (b) 180 kN/C; (c) 0; (d) 20 kN/C

57. (a) $4kq/R^2$; (b) $3kq/4R^2$

59. (a) 0; (b) 1.3 MN/C; (c) 0

61. (a) 1.9×10^{11} N/C; (b) 3.6×10^{10} N/C

63. $\dfrac{\rho r}{3\varepsilon_0} - \dfrac{\rho a^3}{3\varepsilon_0 r^2}$

65. $\frac{1}{3}E_0 a^2$

67. 0.39 μs

69. $E_{in} = \rho_0 x^2/2\varepsilon_0 d$; $E_{out} = \rho_0 d/8\varepsilon_0$

73. (a) 0; (b) $(ac/\varepsilon_0 r^2)(e^{-1} - e^{-r/a})$;
    (c) $(ac/\varepsilon_0 r^2)(e^{-1} - e^{-b/a})$

77. $+10.6$ μC/m² on both outer faces, ± 36.9 μC/m² on inner faces

CHAPTER 25

1. 600 μJ

3. 3.0 kV

5. 910 V

7. 5.6 kV/m

9. Proton and He⁺ both gain 100 eV $= 1.6 \times 10^{-17}$ J; α gains 200 eV $= 3.2 \times 10^{-17}$ J

11. 4.5 V

13. 0.23 MC

17. 6.1 μC

19. 27.2 V

21. $Q = 5.4$ nC, $r = 17$ cm

23. (a) 442 kV; (b) 9.2 Mm/s

25. kQ/R.

27. $V(x) = -\frac{1}{2}ax^2$

29. 52 nC/m

31. $x = -a/2$, $x = a/4$

33. (a) 2.6 kV; (b) 1.8 kV; (c) 0

35. 12 μm

37. $2kQ/R$

39. $2\pi k\sigma(\sqrt{x^2 + b^2} - \sqrt{x^2 + a^2})$

41. (a) $V(x, y) = -E_0(x + y) = -150(x + y)$ V/m;
    (b) 150V

45. (a) $\mathbf{E} = -ay\hat{\mathbf{i}} - ax\hat{\mathbf{j}}$

47. $\mathbf{E} = 10\hat{\mathbf{i}} + 5.8\,\hat{\mathbf{j}}$ V/m

49. (a) 4 V; (b) $E_x = 1$ V/m, $E_y = -12$ V/m, $E_z = 3$ V/m

51. $\mathbf{E} = \dfrac{kQx}{(x^2 + a^2)^{3/2}}\hat{\mathbf{i}}$

53. $E = V_0/R$, radially outward

55. 3 kV

57. (a) 34 kV, -9.0 kV; (b) 12.6 kV on each; (c) 24 nC

59. (a) 43 kV; (b) 1.7 MV/m; (c) 540 V; (d) 0

61. 1.55 keV $= 2.47 \times 10^{-16}$ J

63. (a) $\dfrac{2kqx}{x^2 - a^2}$; (b) $\dfrac{2kq}{x}$

65. (a) 27 kV; (b) no change

67. -7.5 V

69. (a) $x = -3$ m, 0 m, 1 m; (b) $\mathbf{E} = (3x^2 + 4x - 3)\hat{\mathbf{i}}$;
    (c) $x = -1.87$ m, 0.535 m

71. (a) 7.2 kV; (b) 14.4 kV

73. 14 cm, 1.7 nC

75. 23 nC/m

77. (a) $V(x) = \dfrac{k\lambda_0 x}{\ell^2}\left[x \ln\left(\dfrac{2x + \ell}{2x - \ell}\right) - \ell\right]$;

    (b) $\frac{1}{12}\lambda_0\ell$; (c) $\dfrac{k\lambda_0\ell}{12x}$

79. $y^2 + (x - \frac{5}{3}a)^2 = (\frac{4}{3}a)^2$; i.e., a circle

81. $\dfrac{kq}{2a}\ln\left(\dfrac{\sqrt{2} + 1}{\sqrt{2} - 1}\right) \approx \dfrac{0.881\,kq}{a}$

CHAPTER 26

1. $3kq^2/\ell$

3. 4.88 kJ

5. $v = 2q\sqrt{\dfrac{k}{m\ell}}$

7. $6kq^2/a$

9. (a) 2.0 MV/m; (b) 9.9 kV; (c) 5.5 mJ

11. (a) 0.74 μC; (b) 40 kV

13. $\frac{1}{2}kQ^2\left(\dfrac{1}{a} - \dfrac{1}{b}\right)$

15. (b) $dW = 2kq\,dq/a$; (c) $W = kQ^2/a$

17. 1 km³

19. 9×10^{30} J/m³

21. 24 μJ

25. $(kQ^2/R)(2^{2/3} - 1) = 0.60$ mJ

27. $U/\ell = \pi a^4 \rho^2/16\varepsilon_0$

29. ± 14 C

31. 6.5 mF

33. 0.74 nF

35. 55 pF

39. 70 nF

41. 1 μF stores 15 times as much energy

43. (a) 30 μF; (b) 0.1 μF; (c) 0.01 μF

45. (a) 5.0 kW; (b) 250 μF; (c) 0.50 W

47. (a) Increases by factor of 2.5; (b) drops to 40% of its original value

51. Equal

53. (a) 6.0 μF; (b) 0.55 μF; (c) 0.83 μF, 1.3 μF, 1.5 μF, 2.2 μF, 2.8 μF, 3.7 μF

55. (a) 2 series pairs in parallel or two parallel pairs in series; (b) 4 in series

57. 0.86 μF

61. $\pm 1\%$

63. (a) 64 V; (b) drops from 16 mJ to 14 mJ

65. (a) 3.5 mm; (b) 87 kV

67. 126 pF

69. (a) 50 nC, 170 nC; (b) 23 μJ, 77 μJ

71. 1.4 mm

73. (a) 0.90 J; (b) 1.8 J; energy comes from work done by agent separating the plates

75. (a) 1.2 μF; (b) 24 μC

77. 13 min

77. 13 min

79. $\dfrac{kq^2}{a}\left(2\sqrt{3}-\dfrac{15}{2}\right) \approx -4.04\dfrac{kq^2}{a}$

81. (a) $C = \frac{1}{2}(\kappa + 1)C_0$; (b) $U = \dfrac{C_0 V_0^2}{\kappa + 1}$;

   (c) $F = \dfrac{2C_0 V_0^2(\kappa - 1)}{L(\kappa + 1)^2}$, into capacitor

83. (a) 75 mF; (b) 1.4×10^{12} J

85. (a) $Q^2/2\varepsilon_0 a$; (b) $Q^2/\varepsilon_0 a$; (a) is right since (b) includes the fields of *both* plates, and a plate doesn't experience a force from its own field

87. $31d^2kp^2/1280\ell^5$

89. 7.2 pF/m

CHAPTER 27

1. 9.4×10^{18}
3. 2.9×10^5 C
5. (a) 480 m; (b) 50 mA; (c) 14 μA
7. 0.17 mm/s
9. 1.23 cm/s
11. 0.31 mA
13. (a) 37 A/m^2; (b) 86%
15. (a) $2t^2 - t^3$; (b) $t = 2$ s
17. 0.17 V/m
19. 2.2×10^{-6} $\Omega\cdot$m
21. (a) 5.95×10^7 $(\Omega\cdot$m$)^{-1}$; (b) 4.5 $(\Omega\cdot$m$)^{-1}$
23. 6.4×10^{-15} s
25. 25 Ω
27. 2.34 mA
29. 25 mA
31. 1.5 kA
33. (a) 17 mΩ; (b) 86 A; (c) 18 MA/m^2; (d) 1.7 V/m
35. $d_{Al} = 1.26\, d_{Cu}$
37. (a) 2.07 cm; (b) 2.60 cm; (c) aluminum
39. 34 mΩ
41. 1.38 kW
43. 160 μA
45. 240 Ω
47. 48 W
49. 960 W·h = 3.5 MJ
51. Resistor with more power has $\sqrt{2}$ times greater diameter
53. (a) 150 A; (b) 3.4 km
55. 0.54 mA
57. 0.94 Ω
59. (a) 8.7 kA; (b) 15%
61. 2.9 A
63. 2.5 A
65. 203 A
67. 17 J/K

CHAPTER 28

5. 1.4 hours
7. 6.0 V
9. 229 kΩ

11. 1.5 A
13. 0.02 Ω
15. 45 Ω
17. 30 A
19. R_1 for each
21. 24
23. (a) 162 Ω; (b) 125 mW
25. (a) $\dfrac{R_1 \mathscr{E}}{R + 2R_1}$; (c) $\frac{1}{2}\mathscr{E}$
27. 2.45 W
29. $(\mathscr{E}_2 R_1 - \mathscr{E}_1 R_2)/(R_1 + R_2)$
31. $\frac{7}{5}R$
33. $I_1 = 2.79$ A, $I_2 = 2.36$ A, $I_3 = 0.429$ A
35. $\mathscr{E}_2 > 5.49$ V
37. 1.6% low
39. 1.2 kW41.24.99 kΩ
41. 24.99 kΩ
43. (a) 20 V; (b) 3.0 mA
49. (a) 9.0 V; (b) 1.5 ms; (c) 0.32 μF
51. (a) 0.35 s; (b) 0.17 s
53. 3.4 μF
55. (a) $I_1 = 25$ mA, $I_2 = 0$, $V_C = 0$; (b) $I_1 = I_2 = 10$ mA, $V_C = 60$ V; (c) $I_1 = 0$, $I_2 = 10$ mA, $V_C = 60$ V; (d) $I_1 = I_2 = 0$, $V_C = 0$
57. (a) $3\mathscr{E}/4R$; (b) $2\mathscr{E}/3R$
59. 3.4 kΩ
61. (a) 4.51 V; (b) 35.2 Ω
63. 1.07 A, left to right
65. (a) 13.0 V; (b) 2.23 mA
67. 15.9 ms
69. 83 μs
71. (a) 0; (b) 1.0 A; (c) 0.75 A; (d) 0; (e) 1.0 A; (f) 3.0 A; (g) 3.0 A; (h) 1.0 A
73. (b) 22 hours
75. $I_1 = \dfrac{\mathscr{E}_1 R_2 + \mathscr{E}_1 R_3 - \mathscr{E}_2 R_3}{R_1 R_2 + R_1 R_3 + R_2 R_3}$;

   $I_2 = \dfrac{\mathscr{E}_1 R_3 - \mathscr{E}_2 R_1 - \mathscr{E}_2 R_3}{R_1 R_2 + R_1 R_3 + R_2 R_3}$;

   $I_3 = \dfrac{\mathscr{E}_1 R_2 + \mathscr{E}_2 R_1}{R_1 R_2 + R_1 R_3 + R_2 R_3}$;

CHAPTER 29

1. (a) 1.6 mT; (b) 2.3 mT
3. (a) 2.0×10^{-14} N; (b) 1.0×10^{-14} N; (c) 0
5. (a) $-1.1\hat{\imath} + 1.5\hat{\jmath} + 1.7\hat{k}$ mN
7. $\mathbf{B} = 0.13\hat{k}$ T; $\mathbf{v}_2 = -14\hat{\imath}$ km/s
9. 7.86×10^{-14} N
11. 40.1° or 140°
13. (a) $1.2\hat{\imath} + 0.45\hat{\jmath}$ fN; (b) $-1.2\hat{\imath} + 35\hat{\jmath} - 15\hat{k}$ fN
15. 3.9 mm
17. 1.5 mT
19. $r_{proton} = 43\, r_{electron}$
21. 30 mT; yes
23. 1.3 μs

25. (a) 15 MHz; (b) 19 MeV; (c) 6467
29. 0.43%
31. 1.1 mm, 2.6 mm
33. 0.38 N
35. (a) 49 mT; (b) 0.73 N/m
37. 43 kN
39. 0.12 T
41. 21 mN, diagonally toward the upper right
45. 76 mT
47. (a) 1.1 mA·m^2; (b) 1.0 mN·m
49. 0.15 A·m^2
51. (a) 0.35 A·m^2; (b) 4.2×10^{-2} N·m
53. 1.97×10^{-25} J = 1.23 μe V
55. 42$\hat{\imath}$ + 88$\hat{\jmath}$ − 25$\hat{k}$ fN
57. 30 km
59. 0.27 N, to right
61. 0.010 T
63. 6.8 mm
65. (a) 12.8 N/m; (b) 24.9 cm
67. $\tan^{-1}(mg/2IaB)$
69. 77 mT
73. $\mu = 9.25 \times 10^{-24}$ A·m^2

CHAPTER 30

1. 12 cm
3. 1.23 mT
5. 2.8×10^9 A
7. 0.875 cm^2
9. 732
11. Between the wires, 2.0 cm from center of 5.0-A wire
13. 23° west of magnetic north
15. $\mu_0 I/4a$, into page
17. $\dfrac{\mu_0 I}{4ab}(b - a)$, out of the page
19. 5 μN
21. 3.8 mm
23. 13.2 μN/m, at 71.6° below right-pointing horizontal
25. 23.5 mN
27. 7.0 A
29. 24 A
31. 123 mA
33. (a) 0; (b) 0.36 mT; (c) 1.9 mT; (d) 0.50 mT
35. (a) $\mu_0 J_0 r^2/3R$; (b) $\mu_0 J_0 R^2/3r$
37. (a) 5.3 mm; (b) maximum; (c) 130 A
39. 0.83 mT
41. (a) $\sqrt{2}\mu_0 J_s/2$ both inside and outside (but inside and outside fields are mutually perpendicular)
43. (a) 0; (b) $\dfrac{\mu_0 I(r^2 - a^2)}{2\pi r(b^2 - a^2)}$; (c) $\dfrac{\mu_0 I}{2\pi r}$
45. 17 T
47. (a) 38 mT; (b) 5.9 μT
49. 3.3 m
51. $1/2\pi nR$
53. $\chi_M = -1.6 \times 10^{-3}$; diamagnetic

55. 7.2×10^3
57. $I_{outer} = 2I_{inner}$, in opposite direction
59. Out of page
61. 1 km
63. (a) 8.0 μT; (b) 4.0 μT; (c) 0
65. (a) 0.40 mT; (b) 1.0 nT
67. (a) $\dfrac{2\sqrt{2}\mu_0 Ia^2}{\pi(a^2 + 4x^2)\sqrt{a^2 + 2x^2}}$; (b) $\dfrac{\mu_0 Ia^2}{2\pi x^3} = \dfrac{\mu_0 \mu}{2\pi x^3}$
69. 55A

CHAPTER 31

3. 1.4×10^{-4} T·m^2
5. 160 T/s
7. (a) 0.30 A; (b) 0.20 A
9. $B^2\ell^3 v/R$
11. 39 V
13. 0.32 V
15. 9.0 μA, counterclockwise in Fig. 31-5
17. (a) $I = -25$ mA for $0 < t < 2$ s; $I = 0$ for 2 s $< t < 3$ s; $I = +25$ mA for 3 s $< t < 5$ s; (b) $P = 3.1$ mW for $0 < t < 2$ s and 3s $< t < 5$ s; otherwise 0
19. (a) Left to right; (b) 140 μA; (c) 28 μA
21. (a) $I_R = -I_{peak}\cos\omega t$, where (b) $I_{peak} = 110$ mA; (c) 0
23. 0.16 s
25. 7.1 mV
27. (a) Downward; (b) $(B\ell v)^2/R$
29. $v_{final} = \mathcal{E}/B\ell$; R affects time to reach final speed
31. (a), (b) Both 6.7 mA, counterclockwise; (c) 0.44 mW in both cases
33. (a) 25 mA; (b) 1.25 mN; (c) 2.5 mW; (d) 2.5 mW
35. (a) 77 mV; (b) 94 mV
37. 1.08 T/ms
39. (a) 530 V/m; (b) counterclockwise (c) 4.0 keV
41. (a) 32.1 V/m; (b) 1.6 ms, 5.7 ms
43. $E = \frac{1}{2}bh$, to the left above the field region and to the right below
45. $\frac{16}{3}B_0 x_0^2$
47. 42 mA, clockwise
49. $mgR/B^2\ell^2$
51. $\frac{1}{2}BR^2\omega$
53. 58 T/ms
55. $I = at^2$, with $a = 0.81$ A/s^2
59. (a) 2.4 km; (b) 2.89 kW; (c) $6.94; (d) their generators get harder to turn, so they use more fuel
61. (a) $(NAB\omega\sin\omega t)^2/R$; (b) $(NAB\sin\omega t)^2\omega/R$
65. $V = \frac{1}{2}B\ell\theta_0^2\sqrt{g\ell}\sin(2t\sqrt{g/\ell})$

CHAPTER 32

1. 120 V
3. 12 V
5. (a) $-2\pi fMI_p\cos 2\pi ft$; (b) 133 mH
7. $\mu_0 n^2 \pi R^2 \ell$
9. $\dfrac{\mu_0\ell}{2\pi}\ln\left(\dfrac{a + w}{a}\right)$

11. 3.2 mH

13. 40 kV

15. 1350

17. (a) 2.1 A; (b) −1.5 A (minus indicates direction reversal)

19. 26 A/ms

21. 11 A

25. 130 Ω

27. 15 s

29. (a) 0.11 H; (b) 20 mA

31. (a) 76 mA; (b) 4.4 V (c) 7.6 V; (d) 2.2 A/s;
    (e) 0.58 W

33. 400 Ω

35. 50 Ω

37. (a) 2.0 A; (b) 6.0 A; (c) 60 V

39. 100 mA

41. (a) 52 J; (b) 1.8 s

43. (a) 2.5 kW; (b) no

45. 15 W

47. (a) 5.7 MJ; (b) 31 mΩ; (c) 39 s

49. 9.9×10^8 J/m^3

51. 10^{11} times that of gasoline; 2600 times that of U-235

53. Smaller by factor $1/4n^2R^2$

55. 10^{18} J

57. $|\mathcal{E}| = 2Mbt$

59. 48 H

61. (a) 4.4 A; (b) 0; (c) 70 V

63. $(\mu_0 I^2 \ln 100)/4\pi$

65. (a) $L = \dfrac{\mu_0 N^2 \ell}{2\pi} \ln\left(\dfrac{R + \ell}{R}\right)$

67. (a) $P = I_0^2 R e^{-2Rt/L}$

69. $E/B = 1/\sqrt{\mu_0\varepsilon_0} = 3.0 \times 10^8$ m/s $= c$

CHAPTER 33

1. $V = 325 \sin(314t)$, with V in volts and t in seconds.

3. 9.9 V

5. (a) 350 mA; (b) 1.50 kHz

9. $V_{\text{rms}} = V_p/\sqrt{3}$

11. $I_1 = I_p \cos\phi$, $I_2 = I_p \sin\phi$

13. 45 mA

15. (a) 804 Ω; (b) 48.2 Ω; (c) 2.41 Ω

17. 1.47 μF

19. 15.9 kHz

21. (a) 10 mA; (b) 14 V; (c) 7.7 mA

23. 2.7 mC

25. 0.32 A

27. 8.23 kHz or 5.17×10^4 s^{-1}

29. 4.9 pF to 42 pF

31. 3.65 F

33. (a) 1.63 A; (b) 4.58 μs

35. (a) 63 V; (b) 89 V; (c) 764 mA; (d) 99 μJ

37. (a) $1/\sqrt{2}$; (b) 1/2; (c) $-1/\sqrt{2}$; (d) 1/2

39. Close B, wait 0.70 s, close A and simultaneously open B, wait 0.35 s, open A

41. 0.22 μF

43. 16.2 V

45. (a) 0.43 μH; (b) $R > 16$ Ω

47. (a) $V_{Rp} = 23.86$ V, $V_{Cp} = 28.19$ V, $V_{Lp} = -28.05$ V;
    (b) $V_{R\,\text{rms}} = 16.9$ V, $V_{C\,\text{rms}} = 256$ V, $V_{L\,\text{rms}} = 255$ V

49. (a) above; (b) current lags by approximately 38°

51. (a) I leads V by 63.6°; (b) I lags V by 81.7°

53. 500 W

55. (a) 80 Ω; (b) 97 Hz

57. (a) 5.5%; (b) 9.1%; (c) large

59. 5.0 mA

61. (a) 8.91 V; (b) 1.38 kΩ

63. $4f_1$

65. 13.2 V

67. 0.64 Ω

69. 8.25 V

71. 36.2 nF, 11.5 nF

73. $C = \dfrac{L[\ln(U_0/U_1)]^2}{4\pi^2 N^2 R^2}$

77. $R = 400$ Ω, $L = 67$ mH, $C = 0.094$ μF

79. 1.3 kN/m

CHAPTER 34

1. 1.3 nA

5. (a) 0; (b) 8.33×10^{-13} T; (c) 9.26×10^{-13} T

7. $-x$

9. (a) $\sqrt{2}E$; (b) $(\hat{\mathbf{j}} - \hat{\mathbf{i}})/\sqrt{2}$

13. 0.24 s

15. 1 ns

17. 600 m

19. 2.32

21. 5000 km

23. 27.3 MHz

25. 15 kV/m

27. 1.1 pT

29. 63°

31. 30°

33. $0.34S_0$

35. $0.304S_0$

37. $S = S_0 \cos^2 \omega t$, with $\omega = 20\pi$ s^{-1}

39. 12 GW/m^2

41. 2.7×10^{-10} W/m^2

43. (a) 157 W/m^2; (b) 344 V/m; (c) 1.15 μT

45. 1.1%

47. 3.9×10^{26} W

49. (a) 8.3×10^{-26} W/m^2; (b) 7.9×10^{-12} V/m

51. (a) 4.6 kW; (b) 52.5 mV/m

53. (a) as $1/r$; (b) as $1/r^2$

55. 6.0 mPa

57. 5.0 μN

59. 6.18×10^3 years

61. $E_p = 40$ V/m, $B_p = 0.13$ μT

63. 1/4

65. 4.2 μN

67. 17.7 μPa
71. (a) 1.0 MV/m; (b) 4.3 mm; (c) 64 mJ;
    (d) 2.1×10^{-10} kg·m/s; (e) 64 W
73. Long, with cylindrical symmetry
75. 0.3 μm

PART 4 CUMULATIVE PROBLEMS

1. 6.0 MV
3. 15.7 mA, top to bottom
5. (b) $c/\sqrt{\kappa}$

CHAPTER 35

1. 15°
3. ±0.5°
5. (a) 2; (b) 210°
9. 126 nm
11. 1.57
13. Ethyl alcohol
15. 1.83
17. 5.1 m
19. (a) 3.20×10^{14} Hz; (b) 937 nm
21. (a) 49.8°; (b) 42.2°; (c) 22.4°
23. (a) 61.3°; (b) 80.9°; (c) there is none
27. $\sqrt{2}$
29. 2.62×10^8 m/s
31. 53.5°
35. 36.5° to 38.1°
37. About 3.4 cm
39. 98.2%
41. 1.76
43. 2.3
45. 1.522
47. 1.88
49. 1.96×10^8 m/s
51. 2.32 m
53. 28.8°
55. (a) base; (b) 63.2°
57. 35.2°
59. $n = 1.17$
65. (b) 42.1°

CHAPTER 36

1. Mirror surface 55° to horizontal
3. 5 m
5. (a) Image height is 1/4 object height; (b) inverted
7. (a) 12.4 cm in front of mirror; (b) 3.22 times
9. (a) 24.3 cm behind the mirror; (b) 29 mm; (c) virtual
11. 1.1 m
13. 75 cm, 15 cm
15. 40 cm
17. Each 1.0 m
19. 12 cm
21. 47.7 mm

23. 2.3 mm
25. (a) Real, inverted, 7.7 cm high; (b) virtual, upright, 7.7 cm high
27. ($\ell = 30$ cm, $M = -2$); ($\ell = 60$ cm, $M = -\frac{1}{2}$)
29. 40 cm
33. $\ell' = -67.9$ cm, closer
35. 2
37. $R_{\text{plano-convex}} = \frac{1}{2} R_{\text{double convex}}$
39. Virtual image, 81 cm from lens on same side as object
41. (a) 40 cm; (b) 160 cm; (c) −170 cm (the lens becomes a diverging lens)
43. (a) 15.3 cm; (b) 1.63
45. Real image, magnified 2.74 times
47. (a) 5.0 m; (b) 37 cm
49. (a) Enlarged 2.9 times at 110 mm; (b) 11
53. 1.85 cm
55. −17.7 cm
57. (a) 2.0 m; (b) 3.0 m in front of the mirror
59. 25 cm
61. 96 cm
63. ($\ell = 85.4$ cm, $m = -1.81$); ($\ell = 155$ cm, $m = -0.552$)
65. Real image, 109 cm on other side of lens
67. $h' = -f\alpha$
69. No, one is virtual, the other real
73. (a) 1.1 m from primary; (b) elliptical

CHAPTER 37

1. 103 nm
3. 424 nm, 594 nm
5. (a) 75.2 nm; (b) violet
7. 368 nm
9. 5
11. 236
13. 375 nm (i.e., $\frac{3}{4}\lambda$)
15. 545 nm
17. 1022
19. 83.8 cm
21. (a) 95 cm; (b) 3.8 mm
23. 1.25 mm
25. 17.7 cm, 44.7 cm
27. 0.034°
29. (a) 17.1 cm; (b) 20.0 cm
31. 4
33. (a) 4.8°, 9.7°; (b) 2.9°, 6.8°
35. (a) 8.97°; (b) 51.3°
37. (a) 2nd; (b) 1st
39. 6th order
41. 3200
43. 415 nm, 581 nm
45. Echelle grating has 60% greater resolving power
47. 2000
49. $a/\lambda = 1$
51. 26.6°

53. 0.0162
55. 1.3 m
57. 0.025 mm
59. 6.9 km
61. For diameters greater than 14 cm
63. 77.2 nm
65. 7645
67. 0.01 nm
69. 46 m
71. 484 nm
75. 5542
77. Every 2.8 s
79. 6.0 km/s
81. 3.0 μm
83. $m_0 = (n - 1)d \sin \alpha / \lambda_0$

PART 5 CUMULATIVE PROBLEMS

1. $m_0 = (n - 1)d (\sin\alpha)/\lambda_0$
3. 3.2 min
5. (b) $x = \dfrac{\sqrt{3}}{2a} \ln\left[2(1 - ay - \sqrt{\tfrac{1}{4} - 2ay + a^2y^2})\right]$

CHAPTER 38

1. (a) 4.50 h; (b) 4.56 h; (c) 4.62 h
3. (a) $v/c = \tan\theta$, with θ the amplitude of the sine curve. Here $\theta = 20''$ (20/3600 degree), giving $v = 29$ km/s; (b) the orbit is nearly circular
5. 41 cm
7. (a) 8.26 h; (b) 6.28 h
9. (a) 0.857c; (b) 9.69 min
11. 0.14c
13. $c/\sqrt{2}$
15. Twin A: 83 years; twin B: 40 years
17. 0.96c
19. 0.996c
21. (a) B first, by 5.74 min; (b) A first, by 13.1 min; (c) essentially simultaneous
23. A, by 6.4 My
27. (a) 21.7 m; (b) 12.4 m
29. 0.918c
31. 3.4
33. (a) 2.1 MeV; (b) 1.6 MeV
35. 0.36c
37. 5.3×10^{-8} kg·m/s, about the same as the insect
39. 5.9×10^{-30} kg
41. (a) 0.020c; (b) 0.55c; (c) 0.94c; (d) essentially c
47. (a) -0.75×10^{10} ly^2; (b) 99×10^{10} ly^2; events can be causally related only if $(\Delta s)^2 > 0$
49. 0.274c
51. (a) 16 ns; (b) 4.8 m
53. (a) 4.2 ly; (b) -2.4 years (i.e., B occurs earlier)
55. 0.395c
57. 0.9999953c

59. 598 m
61. 0.95c
65. (a) 1 cm; (b) 3 km; (c) 0.03 ly

CHAPTER 39

1. 16
3. 10 μm
5. 1.4 eV
7. 500 nm
9. 4.1 kK–7.2 kK
11. $R_{200\text{ nm}} = 0.056 R_{500\text{ nm}}$
15. 191 nm; ultraviolet
17. $P_{\text{blue}} = 1.44 P_{\text{red}}$
19. (a) 1.7×10^{28} s^{-1}; (b) 3.2×10^{15} s^{-1}; (c) 1.3×10^{18} s^{-1}
21. 223 nm
23. (a) 1.12×10^{15} Hz; (b) 2.80 eV
25. Copper, nickel, silicon
27. 17.9%
29. (a) 243 nm; (b) 2.3 eV
31. 295 eV
33. 255 keV
35. β $(n_1 = 5)$
37. 91.2 nm; ultraviolet
39. 253 nm
41. $n = 7$
43. 2.95 eV
45. (a) 26 pm; (b) 40.8 eV
47. (a) 3.7×10^{-63} m; (b) 73 nm
49. $v_p = 5.46\times10^{-4} v_e$
51. 151 V
53. 110 eV
55. 6.3×10^7 m/s
57. 126 nm
59. 23 keV
61. 21.1 pm
63. 1 ps
65. (a) 8.2 eV; (b) 6.1×10^{-8}
67. (a) 1.32 μm; (b) $R_{690\text{ nm}} = 0.27 R_{\text{max}}$
69. No
71. 9.72 pm
73. (a) 10.2 eV; (b) $n = 2, n = 1$
75. The proton
77. 6×10^5 m/s
79. 18 pm
85. $\tfrac{1}{2}m_e c^2[(\gamma - 1) + \sqrt{(\gamma - 1)(\gamma + 3)}]$

CHAPTER 40

1. m$^{-1/2}$
3. $1/\sqrt{a}$
5. $A = \sqrt{\dfrac{\sqrt{2}}{a\sqrt{\pi}}}$
7. $U = \infty$ for $|x| > b$; $U = E - \hbar^2/m(b^2 - x^2)$ for $-b \leq x \leq b$

9. 1.43 pm
11. Electron
13. (a) 3.48×10^{-19} J = 2.18 eV; (b) 571 nm
15. 21.3 μm
17. E_1 = 200 MeV
19. 2.38 nm
21. At $x = \frac{1}{4}L$ or $\frac{3}{4}L$
23. 0.196
25. 33.0 eV
27. 493 N/m
29. $n > 99$
33. (a) 0.168; (b) $10^{-7.9 \times 10^{36}}$
35. 9.87×10^{-11} J = 616 MeV
39. (a) 1×10^{10} K; (b) 2×10^{13} K
41. 0.23 MeV
43. 0.987
45. 2.66×10^{14} Hz
47. 0.93 nm
49. $n = 3$
51. (a) 4; (b) 2
57. (a) No bound states; (b) one bound state, at $\varepsilon = 6.44$;
    (c) two bound states, at $\varepsilon = 7.53$, $\varepsilon = 29.3$

CHAPTER 41

3. (a) $1/\pi a_0^3 e = 7.91 \times 10^{29}$ m^{-3};
   (b) $1/a_0 e = 6.95 \times 10^9$ m^{-1}
5. $\sqrt{42} \hbar$
7. -1.51 eV
9. $n = 4$, $\ell = 3$
11. 1.7×10^{40}
13. $3d$
15. (a) Possible with $n = 5$, $\ell = 3$; (d) possible with $n = 2$,
    $\ell = 1$
17. (a) 3.66×10^{-34} J·s; (b) 3.17×10^{-34} J·s
23. (a) $\dfrac{\sqrt{15}}{2} \hbar$; (b) 4
25. $\frac{3}{2}, \frac{5}{2}$
29. (a) 2 at $n = 1$, 2 at $n = 2$, 1 at $n = 3$; (b) $19h^2/8mL^2$
31. (a) 5; (b) $\frac{5}{2}\hbar\omega$
33. $1s^2 2s^2 2p^6 3s^2 3p^6 3d^1 4s^2$
35. $1s^2 2s^2 2p^6 3s^2 3p^6 3d^{10} 4s^1$
37. $n_2 = 3$
39. 2.19 eV
41. 3.193 eV
43. 0.34 nm
45. 980 V
47. (a) 18.7 eV; (b) 9.5%
49. (a) 645 nm ($4 \rightarrow 3$), 502 nm ($5 \rightarrow 4$),
    410 nm ($6 \rightarrow 5$); (b) 2
51. 0.11
53. $\sqrt{2}\hbar$
55. (a) $3D_{5/2}$; (b) 6
57. 1.09 keV
63. (b) He: 54.4 eV; O: 870 eV; Fe: 9.19 keV; Pb: 91.4 keV;
    U: 115 keV

CHAPTER 42

1. 0, 2.63 meV, 7.89 meV, 15.8 meV
3. 1.25 meV
5. 9.40×10^{-46} kg·m^2
7. $\hbar^2 j/I$
9. 0.121 nm
11. (a) 0.179 eV; (b) 0.359 eV
13. 7.07×10^{13} Hz
15. (a) 4.73×10^{13} Hz; (b) 1.95×10^{-46} kg·m^2
17. 181 kcal/mol
19. 10.2
21. (a) $F = \dfrac{\alpha k e^2}{r_0^2}\left[\left(\dfrac{r_0}{r}\right)^{n+1} - \left(\dfrac{r_0}{r}\right)^2\right]$;   (b) 3.0×10^{-6} N,
    about 260 times the electrostatic attraction
23. 4.7 eV
25. 1 meV
27. 6.35×10^4 K, over 200 times room temperature
29. 345 nm; ZnS
31. 1.91 eV
33. (a) 345 nm; (b) no; not much of the solar energy will be
    absorbed
35. 2.78 eV
37. 1.79 kA
39. 15.2 meV
41. 35.8 μm
43. -8.40 eV
45. 9.41 eV
47. E_{min} = 15 meV; kT = 8.6 meV
49. 4.1 K; no, this is below the boiling point of O_2
51. (a) 2.87×10^{-20} J = 179 meV; (b) 8 pm; (c) about 6%
    of bond length

CHAPTER 43

1. 25 fm
3. $^{211}_{86}$Ra, $^{220}_{86}$Ra, $^{222}_{86}$Ra
5. (a) same; (b) $q_K = q_{Cl} + 2e$
7. $Z = 43$, technetium and $Z = 61$, promethium
9. $^8_5$B
11. 5.9 fm
13. 5.3×10^{-12} eV
15. 60 MHz
17. 59.93 u
19. 187 MeV
21. 5.61 MeV/nucleon
23. 0.037%
25. 8.3 s
27. (a) $^{222}_{86}$Rn $\rightarrow$ $^{218}_{84}$Po + $^4_2$He; (b) $^{214}_{82}$Pb $\rightarrow$ $^{214}_{83}$Bi + e^- + $\bar{\nu}$
29. 16.7 Bq/L
31. (a) 7.85×10^5, essentially 0; (b) 1.009, 5.0×10^5;
    (c) 3.6, 8.2×10^4
33. 7.5×10^5
35. 4.3 days
37. Slope $= -\log_{10} e/\tau$, with τ the mean lifetime
39. 3.096×10^{17} atoms = 31 μg

41. 2.78 GBq = 75.2 mCi
43. At t = 45.8 h
45. 3.0×10^9 y
47. (a) 19.5 GBq; (b) 1.22 GBq; (c) 138 Bq; (d) 62.8 Bq; (e) 62.8 Bq
49. 23%
51. 2: $^3_2$He, $^4_2$He; 8: $^{16}_8$O, $^{15}_7$N; 20: $^{40}_{20}$Ca, $^{39}_{19}$K; 28: $^{58}_{28}$Ni, $^{52}_{24}$Cr; 50: $^{119}_{50}$Sn, $^{89}_{39}$Y; 82: $^{208}_{82}$Pb, $^{138}_{56}$Ba; 126: no magic Z, $^{208}_{82}$Pb
53. 8.7 MeV/nucleon
55. 0.076%
57. 1.68 mCi/L
59. 1.2×10^6 m/s^2, more than 10^5 times g_{Earth}
61. 0.072 K
63. 4.3 MeV
65. (a) $^{210}_{83}$Bi $\rightarrow$ $^{210}_{84}$Po + e^- + $\bar{\nu}$; (b) $^{210}_{84}$Po $\rightarrow$ $^{206}_{82}$Pb + $^4_2$He
67. $^{262}_{107}$Ns
69. (a) copper-65; (b) 4.0 h

CHAPTER 44

3. 8.2×10^{13} J/kg for fission; 3.4×10^{14} J/kg for fusion; fusion is about 4 times greater
5. 2
7. 1.0×10^{20} fissions/s
9. 263 kt
11. 1.02 kg
13. (a) 25.6 kg; (b) 13.8 cm
15. (a) 142; (b) 1.42 μs
17. 3.86 km/s
19. 88.9%
21. 31 h
23. 69 μs
25. 0.46 s
27. $\Delta k = 2.3 \times 10^{-5}$
29. 23 MK

31. 2×10^{20} m^{-3}
33. 454 kg
35. 10^{30} m^{-3}
37. (a) 3.7×10^{38} proton/s; (b) 7.3×10^9 years
39. 4.4×10^{10} J; (b) 340 gal
41. 576 kg
43. $N_B = 33 N_A$
45. (a) 8.6×10^5 m/s; (b) 4×10^{13} s/m^3, less than 1 millionth of the Lawson criterion
47. (a) 1.1 GJ; (b) 2.8 pellets/s; (c) 450 kg
49. 1.28
51. 6.0 kg fusion fuel; 26 kg ^{238}U
53. Area $\propto$ yield$^{2/3}$
55. (a) 783 kg; (b) 157

CHAPTER 45

1. 3×10^{-16} s
3. 7×10^{19} m
5. $S = -1$ for Λ^0; $S = 0$ for decay products; weak force
7. Both
9. (a), violates conservation of baryon number
11. (a) No; (b) yes
13. (a) Stays same; (b) changes sign
15. sss
17. $c\bar{c}$
19. 3.6×10^3 m^3, about 1 million gallons
21. 10^{28} K
23. (a) 2.13×10^4; (b) 0.999 999 998 9c
25. 1.6×10^{-18} s^{-1}
27. 4.5×10^3 km/s
29. 18 Gy
31. 2×10^{-26} kg/m^3
33. (a) 2.6×10^{-13} m; (b) -2.8 keV
35. 13.7 ns
37. 1.07 mm
39. 2.6×10^{-25} s

PHOTO CREDITS

1	J. Nettis/H. Armstrong Roberts
2	R. du Buisson/The Stock Market
3T	NASA
3B	UPI/Bettmann
5T	B. Paulson/H. Armstrong Roberts
5B	National Inst. of Standard & Technology
8	G. Anderson/The Stock Market
9	Sidney Harris
10	Anglo-Australian Telescope Board
11T	Courtesy, Digital Instrument
11B	Mark Wagner/Tony Stone Images
14	NASA
15	Courtesy, Texas Instruments
21	The Globus Brothers/The Stock Market
22	Gunter Ziesler/Peter Arnold, Inc.
32	Courtesy Boeing Corporation
34	Loren Winters
35	Jay M. Pasachoff
36	Soames Summerhays/Photo Researchers
45	Mark D. Phillips/Photo Researchers
46	Philip Bailey/The Stock Market
66	Kalt/Aefa/H. Armstrong Roberts
69	Loren Winters
71L	Terry Vine/Tony Stone Images
71R	Jean-Francois Causse/Tony Stone Images
77	U.S. Department of Defense
87	Bob Martin/ALLSPORT USA
90	David Madison/Duomo Photography Inc.
91	Steve Lissau/Rainbow
93	National Portrait Gallery, London
95	Larry Mulvehill/Photo Researchers
97	Jay M. Pasachoff
100	NASA
102	NASA
103	NASA
108	1986/UNIVERSAL PRESS SYNDICATE. Reprinted with permission. All Rights Reserved.
110	Tom Pantages
114	Eric Neurath/Stock Boston
115	Martha Swope
119	Camerique/H. Armstrong Roberts
127	David Parker/SPL/Photo Researchers
130	Jon Porter/Great America
131	Jay M. Pasachoff
137	Creators Syndicate
139	Douglas T. Mesney/The Stock Market
151	Vince Streano/The Stock Market
152L	Gerard Vandystadt/Photo Researchers
152R	Benelux Press/West Light
159	W. Hill/Camerique/H. Armstrong Roberts
167	Ted Horowitz/The Stock Market
169	Dennis O'Clair/Tony Stone Images
177	© The Harold E. Edgerton 1992 TRUST/Palm Press
178	Nathan Bilow/Tony Stone Images
180	Exroy/Explorer/Photo Researchers
183	Courtesy Northeast Utilities
184	David Stoecklein/The Stock Market
193	Fermi National Accelerator Laboratory
195ALL	Erik Borg
204	NASA
205	The Granger Collection, New York

726	Fermi National Accelerator Laboratory	815L	Jay M. Pasachoff
728TL	Richard Megna/Fundamental Photographs	823	NASA
728TR	Richard Megna/Fundamental Photographs	833	"Copyright 1994 Tektronix, Inc., All Rights Reserved"
730T	Paul J. Sutton/Duomo Photography Inc.	845	Erik Borg
730B	Ivan Massar	853BR	Camerique/H. Armstrong Roberts
731	Peter Arnold, Inc.	853TL	Chris Rogers/The Stock Market
732	Jay M. Pasachoff	853CL	H. Abernathy/H. Armstrong Roberts
733C	Lionel F. Stevenson/Photo Researchers	853BL	G. Fritz/H. Armstrong Roberts
733T	Princeton Plasma Physics Laboratory	853TC	Coco McCoy/Rainbow
733B	Dr. L. A. Frank, Univ. of Iowa	853TR	Coco McCoy/Rainbow
736T	Japan National Railways	853CR	Michael Mathers/Peter Arnold, Inc.
736C	Mitsuhiro Wada/Gamma-Liaison	862	Dale Boyer/Photo Researchers
736B	Mitsuhiro Wada/Gamma-Liaison	875	P. Degginger/H. Armstrong Roberts
737	Walker Scientific Company	876T	J. Amos/H. Armstrong Roberts
742L	Jay M. Pasachoff	876B	National Radio Astronomy Observatory
742R	Jay M. Pasachoff	877R	NASA
744	Lawrence Berkeley Laboratory/SPL/Photo Researchers	877L	JPL/NASA
751T	Ray Pfortner/Peter Arnold, Inc.	878	Diane Schiumo/Fundamental Photographs
751B	American Association of Physics Teachers	879R	Sepp Geitz/Woodfin Camp & Associates
754	Richard Megna/Fundamental Photographs	886L	Courtesy AT&T, Bell Laboratories
756L	National Optical Astronomy Observatories	886R	Hughes Research Laboratories
756R	National Optical Astronomy Observatories	891R	David R. Frazier Photolibrary
757	Richard Megna/Fundamental Photographs	891L	NASA
758	Boebinger, AT&T Bell Labs	895	Pete Saloutos/The Stock Market
760	National Center for Atmospheric Research	896	Schott Corporation
767	Richard Megna/Fundamental Photographs	898	Richard Megna/Fundamental Photographs
768L	Courtesy of Automatic Switch Company	899	NASA
768R	Dan McCoy/Rainbow	901	Richard Megna/Fundamental Photographs
770	Michael Seul, AT&T Bell Labs	904	John W. Dunay/Fundamental Photographs
771	Richard Megna/Fundamental Photographs	905	Richard Megna/Fundamental Photographs
782	Daniel Zirinsky/Photo Researchers	907R	Foto Forum
791L	Milt & Joan Mann/Cameramann International, Ltd.	907L	Luettge/ZEFA/H. Armstrong Roberts
791R	T. J. Florian/Rainbow	908B	R. Giovaneill & H. R. Gillett/ CSIRO National Measurement Laboratory, Australia
792T	David R. Frazier Photolibrary	908T	Alfred Pasieka/Peter Arnold, Inc.
792BL	Erik Borg	908C	Alfred Leitner/Rensselaer Polytechnic Institute
792BR	Erik Borg	909	Simon Fraser/SPL/Photo Researchers
793	Robert Mathena/Fundamental Photographs	912	Bruno J. Zehnder/Peter Arnold, Inc.
794	Erik Borg	913L	Leonard Lessin/Peter Arnold, Inc.
797	NASA	913R	Len Lessin/Peter Arnold, Inc.
800	Courtesy, GE Appliances	915T	Library of Congress
803	Courtesy of IBM	915ALL	James E. Kettler/Ohio University, Belmont
805	Jay M. Pasachoff	920	Dan McCoy/Rainbow
813	Milt & Joan Mann/Cameramann International, Ltd.	923B	Jay M. Pasachoff
		923T	The Exploratorium

INDEX

Note: Page numbers in *italics* refer to illustrations; page numbers followed by t refer to tables.

GEOPHYSICAL AND ASTROPHYSICAL DATA

EARTH

Mass	5.97×10^{24} kg
Mean radius	6.37×10^6 m
Orbital period	3.16×10^7 s (365.3 days)
Mean distance from Sun	1.50×10^{11} m
Mean density	5.5×10^3 kg/m^3
Surface gravity	9.81 m/s^2
Surface pressure	1.013×10^5 Pa
Magnetic moment	8.0×10^{22} A·m^2

SUN

Mass	1.99×10^{30} kg
Mean radius	6.96×10^8 m
Orbital period (about galactic center)	6×10^{15} s (200 My)
Mean distance from galactic center	2.6×10^{20} m
Power output (luminosity)	3.85×10^{26} W
Mean density	1.4×10^3 kg/m^3
Surface gravity	274 m/s^2
Surface temperature	5.8×10^3 K

MOON

Mass	7.35×10^{22} kg
Mean radius	1.74×10^6 m
Orbital period	2.36×10^6 s (27.3 days)
Mean distance from Earth	3.85×10^8 m
Mean density	3.3×10^3 kg/m^3
Surface gravity	1.62 m/s^2

COMPARATIVE ENERGIES

Supernova explosion	10^{46} J	World per capita energy use, per second	2×10^3 J
Sun's output, per second	4×10^{26} J	Food energy used by human body,	
Solar energy incident on Earth, per second	2×10^{17} J	per second	10^2 J
Explosive yield of MX missile		Energy to lift this book 1 foot	8 J
(10 350-kiloton nuclear warheads)	1×10^{16} J	Particle energy in Superconducting	3.2×10^{-6} J
Human energy use, per second	10^{13} J	Super Collider	(20 TeV)
1 ounce uranium-235	2.3×10^{12} J	Energy release in U-235 fission (200 MeV)	3.2×10^{-11} J
1 gallon seawater		Energy release in deuterium-tritium fusion	2.8×10^{-12} J
(with deuterium content as fusion fuel)	4.7×10^{10} J		(17.6 MeV)
1 ton of coal	2.6×10^{10} J	Energy release in forming hydrogen atom	2.2×10^{-18} J
Electrical output of large power plant,			(13.6 eV)
per second	10^9 J	Thermal energy kT at room temperature	4×10^{-21} J
1 gallon of gasoline	1.3×10^8 J		(0.025 eV)
1500-kg car going 60 mph	6.3×10^5 J	Thermal energy kT associated with cosmic	4×10^{-23} J
U.S. per capita energy use, per second	10^4 J	background radiation	(0.23 meV)